list of the elements with their symbols and atomic weights

	symbol	atomic number	atomic weight		symbol	atomic number	atomic weight		symbol	atomic number	atomic weight
Actinium	Ac	89	(227)[a]	Hafnium	Hf	72	178.49	Praseodymium	Pr	59	140.907
Aluminum	Al	13	26.9815	Hahnium	Ha	105	(260)[b]	Promethium	Pm	61	(147)
Americium	Am	95	(243)	Helium	He	2	4.0026	Protactinium	Pa	91	(231)
Antimony	Sb	51	121.75	Holmium	Ho	67	164.930	Radium	Ra	88	(226)
Argon	Ar	18	39.948	Hydrogen	H	1	1.00797	Radon	Rn	86	(222)
Arsenic	As	33	74.9216	Indium	In	49	114.82	Rhenium	Re	75	186.2
Astatine	At	85	(210)	Iodine	I	53	126.9044	Rhodium	Rh	45	102.905
Barium	Ba	56	137.33	Iridium	Ir	77	192.2	Rubidium	Rb	37	85.47
Berkelium	Bk	97	(247)	Iron	Fe	26	55.847	Ruthenium	Ru	44	101.07
Beryllium	Be	4	9.0122	Krypton	Kr	36	83.80	Rutherfordium	Rf	104	(257)[b]
Bismuth	Bi	83	208.980	Lanthanum	La	57	138.91	Samarium	Sm	62	150.35
Boron	B	5	10.811	Lawrencium	Lw	103	(257)	Scandium	Sc	21	44.956
Bromine	Br	35	79.909	Lead	Pb	82	207.19	Selenium	Se	34	78.96
Cadmium	Cd	48	112.41	Lithium	Li	3	6.939	Silicon	Si	14	28.086
Calcium	Ca	20	40.08	Lutetium	Lu	71	174.97	Silver	Ag	47	107.870
Californium	Cf	98	(249)	Magnesium	Mg	12	24.312	Sodium	Na	11	22.9898
Carbon	C	6	12.01115	Manganese	Mn	25	54.9380	Strontium	Sr	38	87.62
Cerium	Ce	58	140.12	Mendelevium	Md	101	(256)	Sulfur	S	16	32.064
Cesium	Cs	55	132.905	Mercury	Hg	80	200.59	Tantalum	Ta	73	180.948
Chlorine	Cl	17	35.453	Molybdenum	Mo	42	95.94	Technetium	Tc	43	(99)
Chromium	Cr	24	51.996	Neodymium	Nd	60	144.24	Tellurium	Te	52	127.60
Cobalt	Co	27	58.9332	Neon	Ne	10	20.183	Terbium	Tb	65	158.924
Copper	Cu	29	63.54	Neptunium	Np	93	(237)	Thallium	Tl	81	204.37
Curium	Cm	96	(247)	Nickel	Ni	28	58.71	Thorium	Th	90	232.038
Dysprosium	Dy	66	162.50	Niobium	Nb	41	92.906	Thulium	Tm	69	168.934
Einsteinium	Es	99	(254)	Nitrogen	N	7	14.0067	Tin	Sn	50	118.69
Erbium	Er	68	167.26	Nobelium	No	102	(253)	Titanium	Ti	22	47.90
Europium	Eu	63	151.96	Osmium	Os	76	190.2	Tungsten	W	74	183.85
Fermium	Fm	100	(253)	Oxygen	O	8	15.9994	Uranium	U	92	238.03
Fluorine	F	9	18.9984	Palladium	Pd	46	106.4	Vanadium	V	23	50.942
Francium	Fr	87	(223)	Phosphorus	P	15	30.9738	Wolfram	W	74	183.85
Gadolinium	Gd	64	157.25	Platinum	Pt	78	195.09	Xenon	Xe	54	131.30
Gallium	Ga	31	69.72	Plutonium	Pu	94	(242)	Ytterbium	Yb	70	173.04
Germanium	Ge	32	72.59	Polonium	Po	84	(210)	Yttrium	Y	39	88.905
Gold	Au	79	196.967	Potassium	K	19	39.098	Zinc	Zn	30	65.37
								Zirconium	Zr	40	91.22

[a] Approximate values for radioactive elements are listed in parentheses.
[b] Name and symbol not officially approved.

chemistry
the central science

The cover photo shows the surface of a high-purity crystal of arsenic (magnified about 1100 times) after it has been subjected to a vacuum at 282°C for 4 hours. The pyramidlike objects are actually pits in the surface caused by the evaporation of arsenic molecules. This photo was taken in connection with a study on the molecular processes by which molecules evaporate from crystalline solids. Photos such as this show that the most rapid evaporation occurs where imperfections are found in the crystal. (*Photo courtesy of Gerd M. Rosenblatt, Carl A. Hultman, and Michael B. Dowell, Department of Chemistry, The Pennsylvania State University*)

chemistry

PRENTICE-HALL, INC., ENGLEWOOD CLIFFS, NEW JERSEY 07632

the central science

THEODORE L. BROWN
UNIVERSITY OF ILLINOIS

H. EUGENE LeMAY, JR.
UNIVERSITY OF NEVADA

Library of Congress Cataloging in Publication Data

Brown, Theodore L
   Chemistry: the central science.

   Includes index.
    1. Chemistry. I. LeMay, Harold Eugene, 1940–
joint author. II. Title.
QD31.2.B78   540   76-22159
ISBN 0-13-128769-9

to those from whom we have learned,
and to our wives and children

chemistry: the central science
THEODORE L. BROWN | H. EUGENE LeMAY, JR.

Printed in the United States of America

Technical illustrations by Vantage Art, Inc.

PRENTICE-HALL INTERNATIONAL, INC., *London*
PRENTICE-HALL OF AUSTRALIA PTY. LIMITED, *Sydney*
PRENTICE-HALL OF CANADA, LTD., *Toronto*
PRENTICE-HALL OF INDIA PRIVATE LIMITED, *New Delhi*
PRENTICE-HALL OF JAPAN, INC., *Tokyo*
PRENTICE-HALL OF SOUTHEAST ASIA PTE. LTD., *Singapore*
WHITEHALL BOOKS LIMITED, *Wellington, New Zealand*

contents

8 chemical bonding *204*

9 geometries of molecules; molecular orbitals *239*

10 chemistry of the atmosphere *278*

11 liquids, solids, and intermolecular forces *308*

12 solutions *343*

preface

The preface is almost always the last part of a textbook to be written. Only when the book has assumed its final form can an author declare to the reader what he thinks he has accomplished. Thus, what is for students the beginning is, in one sense, for authors the culmination of a long, sometimes difficult journey. We have been at work on this book for more than 3 years. We have written, revised, and then revised again; we have put in, taken out, moved around; we have talked with our colleagues, with our students, and with our editors; we have studied industrial literature; we have spent time in libraries and in our laboratories. And while all this activity was taking place, we asked ourselves over and over again the same question: What is it that you as students need and should receive from the text you use in your study of chemistry? We believe that we have found some valuable new answers to that question, and in this preface we want to tell you (and your instructor) about these answers and about the ideas and concepts that have informed our writing of the text. We also want to tell you how we think you can best make use of this book.

Most of you are studying chemistry because it has been declared an essential part of the curriculum in which you are enrolled. That curriculum may be agriculture, dental hygiene, electrical engineering, geology, microbiology, metallurgy, paleontology, or one of many other related areas of study. It is fair to ask why it is that so many diverse areas of study should all relate in an essential way to chemistry. The answer is that chemistry is, by its nature, the *central* science. In any area of human activity that deals with some aspect of the material world, there must inevitably be a concern for the fundamental character of the materials involved—their endurance, their interactions with other materials, their changes under a given set of conditions. This is true whether the materials involved are the rocks beneath our feet, the colors used by a Renaissance painter, or the blood cells of a child born with sickle-cell anemia. It is very likely that chemistry plays an important role in the profession to which you now aspire, or may decide later to pursue. You will be a better professional, a more creative and knowledgeable person, if you understand the chemical concepts applicable to your work and are able to apply these concepts as needed.

The relationship of chemistry to professional goals is important, and this factor provides reason enough for you to study chemistry. There is, however, an even more important reason. Because chemistry is so central and so intimately involved in almost every aspect of our contact with the material world, this science is an integral part of our culture. The involvement of chemistry in our lives goes much deeper than the well-known advertising slogan, "Better things for better living through chemistry." In addition to all the obvious ways in which we use the products of chemical research and production—plastic bags, children's toys, counter tops, weed and insect killers, photographic films—we indirectly use thousands of chemical products via the foods we eat, the cars we drive, the medical care we receive, and so forth. During the past several years, we have become increasingly aware that our use of chemicals has had a profound and frightening effect on our environment. Indeed, many scientists are convinced that we have so intensely polluted this planet and so unthinkingly sowed the seeds of future pollution that the fate of civilization is all but sealed. Whether this is so remains to be seen; however, if you are to be a responsible citizen, you will surely need to be informed on many complex issues involving chemistry and the use of chemicals. Because vested interests have a powerful stake in public policy, the public often is presented with conflicting information and claims. You can more fully appreciate and analyze the complex issues put before you if you understand the fundamental principles involved and keep them in mind during your reading and study.

With all of these considerations in mind, you should now be impatient and eager to begin your study of chemistry. Now that you are ready to go, we should say something about how this book can best help you. You might first take a few minutes to glance through the table of contents. The particular sequence of chapters that we have chosen is one that we feel promotes a natural unfolding of the science of chemistry. However, the order in which the chapters of the book are covered in the classroom will be determined by your instructor. You should not be disturbed if the order is not the same as the order in the book. The book has been written so as to make allowance for alternative chapter orders and, in some instances, for the complete omission of certain chapters. Notice that several chapters interspersed throughout the book deal with the chemical aspects of the world in which we live; the air, the earth, and the waters on the earth's surface. In these chapters we have attempted to connect the chemical facts and principles introduced in other, usually earlier, chapters to the familiar (and sometimes not so familiar) aspects of our surroundings on earth. Your instructor, the person who will guide you through this book, may not feel that there is sufficient time to cover some or all of the materials in these chapters. We suggest that you read them anyway; they will help you appreciate the many ways in which chemical concepts and observations are related to contemporary life.

If you should at some point encounter a term or concept you are expected to know but can't remember, use the index at the back of the book. A good index is a rarity; we have worked hard to make your index in this book as complete and accurate as possible. Use it often. (Remember the index also when you later use the book as a reference, after having finished the course. It can help you find what you want more quickly than any other means.)

The difficulties that many chemistry students experience often can be traced to faulty exposition and confusing explanations in their text. This book has been worked on very thoroughly by many people to ensure that it is as clear, concise, and free of confusion as possible. However, you may find that a single reading of a chapter will not suffice if you are to use the book effectively as a learning tool. We suggest that you read every assigned chapter as early as possible, preferably before the material is covered in lecture. This will make you aware of important concepts and terms even before they are treated by the lecturer. Later, you will need to go through the assigned sections of the book much more carefully, making sure that you understand the new terms and problems put before you. We have inserted a great many *sample exercises* into the text, so that you might have clearly worked out examples of problem solving of various types. You should study these exercises carefully, noting every aspect of them, especially if numerical problem solving is involved.

The review section at the end of each chapter is an integrated package designed to help you determine whether you have in fact learned all the material assigned you in each chapter. The *summary* points out the highlights of the chapter; sometimes we say things a little differently in the summary in order to add an extra element of understanding to what you have gotten from the chapter itself. The *key terms* that you should know are also collected for your convenience. The *learning goals* are placed at the end of the chapter to enable you to test yourself. You should make sure that you can meet each learning goal. This can best be done if you state a definition and then check it, write a formula and then check it, or solve a problem and then check it. It may happen, of course, that your instructor will not have covered part of the material in a chapter. You can then skip over the learning goals for this material, but you should still read the complete summary and learn all of the key terms. By learning even nonrequired terms and concepts you can expand your chemical vocabulary with little effort.

The *exercises* at the back of each chapter are designed to test your understanding of the materials covered in the chapter. They are grouped according to topic, except for a number of general exercises. The purpose of the general exercises is to test your ability to solve a problem when it is not clearly identified as to topic. Also, some of the questions in this category require the application of material from more than one topic area. Problems marked with brackets are, in general, a little more difficult to solve than the others. We have prepared a solutions manual that contains detailed answers to all the end-of-chapter exercises; you should consult this manual only after working out problems on your own.

Finally, you should note that there are several appendices following Chapter 25. These are designed to aid you in various ways. You should get acquainted with what is there by glancing through them before the course gets under way. In particular, note that answers are provided to many of the end-of-chapter exercises. Color question numbers in the text indicate that the answer to the question is in Appendix F.

Your instructor may have elected to have you purchase the *Student's Guide* designed for use with the text. This guide, written by Professor James C. Hill, of California State University, Sacramento, is a nicely organized and well-written supplement to the text. You will find it filled with helpful ideas, problem-solving techniques, and fresh insights into the materials presented in the text. We are very happy that Jim has agreed to write the study guide; we feel that it is valuable learning aid for use with the text.

Most general chemistry courses involve laboratory as well as classroom work. There is a very good reason for this. Chemistry is an experimental science; the entire theoretical structure of chemistry is based on the results of laboratory experiments. As you study chemistry, you should try to relate what you learn in the classroom and from the text to operations and observations made in the course of your laboratory work. A very fine laboratory manual for use with this text has been written by Professors John H. Nelson and Kenneth C. Kemp of the Department of Chemistry, University of Nevada, Reno. We believe that it is also an important learning tool in your study of chemistry.

During the many years that we have been practicing chemists, we have found chemistry to be an exciting intellectual challenge and an extraordinarily rich and varied part of our human cultural heritage. We hope that all the hassles you must face regarding course grades will not keep you from sharing with us some of that enthusiasm and appreciation. We have, in effect, been engaged by your instructor to help you learn chemistry. We are confident that we've done that job well. In any case, we would appreciate your writing us, either to tell us of the book's shortcomings, so that we might do better, or of its virtues, so that we'll know where we have helped you most.

Theodore L. Brown
School of Chemical Sciences
University of Illinois
Urbana 61801

H. Eugene LeMay, Jr.
Department of Chemistry
University of Nevada
Reno 89557

acknowledgments

This book owes its final shape and form to the assistance and hard work of many people. Several colleagues reviewed the manuscript and helped us immensely by sharing their insights and criticizing our initial writing efforts. We would like especially to thank Professors David Adams, North Shore Community College; Robert S. Boikess, Douglass College of Rutgers University; John Burmeister, University of Delaware; Wade Freeman, University of Illinois, Chicago Circle; Carol J. Grimes, University of California, Irvine; Charles G. Haas, Pennsylvania State University; James C. Hill, California State University, Sacramento; Kenneth C. Kemp, University of Nevada, Reno; J. J. Lagowski, University of Texas, Austin; Robert H. Marshall, Memphis State University; Robert J. Munn, University of Maryland; John H. Nelson, University of Nevada, Reno; Fred Redmore, Highland Community College; Spencer Seager, Weber State College; Bassam Z. Shakhashiri, University of Wisconsin, Madison; Stephen P. Tanner, University of West Florida; and Lawrence E. Wilkins, Santa Monica College.

We deeply appreciate the assistance of the following members of Prentice-Hall's College Division: David R. Esner, Director, Product Development Department, William L. Gibson, Chemistry Editor, Harry A. McQuillen, former Chemistry Editor, and Marvin Warshaw, College Art Director, who provided guidance and counsel throughout the project; Mary Helen Fitzgerald, Editorial Assistant, who assisted us with photo research; Lorraine Mullaney, Book Designer, who is responsible for the book's attractive and stylish format; and Raymond Mullaney, Development Editor, who worked with us during the final revision of the manuscript and then transformed it into a finished book.

Finally, special thanks are due to Evelyn Abolt and Marguerite Meyer, who so ably typed manuscript copy from our rough drafts.

introduction: some basic concepts

1

Perhaps the only thing permanent about our world is change. All around us are numerous examples of change in ourselves and our environment. Trees change color in autumn, iron rusts, snow melts, paint peels, seeds become flowers, and logs burn. We grow up, we grow old. Living plants and animals undergo continual change, and even dead plants and animals continue to change as they decay. Such changes have long fascinated people and have prompted them to look more closely at nature's working in hopes of better understanding themselves and their environment.

Understanding change is closely tied to understanding the nature and composition of matter. Matter is the physical material of the universe; it is anything that occupies space and has mass. Chemistry is the science that is primarily concerned with matter and the changes that it undergoes. Therefore, as we begin our study of chemistry, our primary focus will be on matter. First, however, let's sketch a somewhat broader picture of chemistry.

Chemistry is a changing science. Therefore, the questions that chemists seek to answer are constantly changing also. Because of this, we might define chemistry as what chemists do. In many regards this definition is unsatisfactory. Nevertheless, it does suggest that chemistry itself changes as chemists absorb new information from other fields, tackle new problems, or reexamine old ones in new ways. One of the important activities of chemists is the synthesis of new materials or the improvement in the ways of making old ones. This aspect of chemistry has had great impact on our lives; chemists have synthesized new fibers, medicines, fertilizers, pesticides, and structural materials. Many new chemicals never find any commercial use but are nevertheless important to chemists in answering subtle questions about matter and its changes. In designing ways to synthesize new materials, it is useful to know the factors that determine how fast and to what extent the required changes proceed. Such knowledge allows chemists to improve, avoid, or control many changes of matter. This knowledge is necessary, for example, in devising ways to clean up automobile exhaust or to make fertilizers at lower cost. Chemists are also interested in determining the identity and concentration of substances. Such analysis may involve determining the quality of a soap in a manufacturing operation, the concentration of a pollutant in the air, the amount of gold in a potential ore, the amount of mercury in a lake, the identity of the substances in some physiologically active mixture, or the chemicals resulting from the utilization of a drug in the body. Chemists are interested not only in determining what things are made of, but also in discovering the ways their composition and structure are related to properties. For example, what makes a particular substance poisonous or sweet or hard or explosive?

The intent of this text is to introduce you to basic chemical facts and theories, not as ends in themselves, but as means to help you understand the material world and to recognize the constraints and opportunities it provides. We hope that this text will provide not only a firm foundation for further scientific studies, whether they be in chemistry or some other field, but also a background to enable you to evaluate scientific information found in news media and semiscientific periodicals. In the remainder of this chapter we shall consider some background material useful to your studies—the metric system, uncertainty in measurement, and problem solving in chemistry. We shall also briefly explore the historical and philosophical background of chemistry and the scientific approach to problems.

**1.1
the emergence
of chemistry:
a historical
perspective**

Chemistry has two roots. First, it is rooted in the craft traditions such as metallurgy, brewing, tanning, and dyeing, which provided a practical understanding of how matter behaves. Second, it can also be traced to the philosophers of ancient Greece who concerned themselves with questions of the basic nature of matter. Through the years the growth of chemistry has reflected both man's desire to solve problems and his innate curiosity and desire to understand his surroundings without regard to the practical application of that understanding.

Metallurgy, the science and art of obtaining and working with metals, exemplifies the development of chemical knowledge through the craft traditions. This craft developed for many years and achieved considerable sophistication without any theoretical framework that would explain metallurgical operations and guide their development. Developments were made largely through trial and error and through accidental discoveries. For the most part, the early pattern of discovery of metals followed their ease of recovery from ores, the earthy mixtures that are mined as sources of metals. Gold was one of the first metals used because it is found in nature in an uncombined, metallic state, for example, as gold nuggets. Copper, which is more abundant, was not used until about 3500 B.C. At that time processes for obtaining copper metal from its ores were discovered. This discovery was undoubtedly accidental. It could have occurred when some copper ore was dropped or thrown into the coals of a fire. Copper metal can be obtained by heating copper ore with charcoal. Methods for obtaining iron, which is much more abundant than copper, did not develop until about 1500 B.C.

People probably attempted to understand changes in matter even before they began to use these changes to their advantage. We know that many early explanations presumed the existence of supernatural powers. In contrast to this approach, the early Greeks sought to understand matter and its changes purely on the basis of logic. However, they were not especially concerned with using these ideas as guides to improve their crafts. Modern science differs from the approach of the Greeks; it depends not only on logic but also on the systematic gathering of facts and on careful observations. Furthermore, the ideas of science are widely used to guide the development of our technologies.

The Greeks, principally Aristotle (384–322 B.C.) and Plato (427–347 B.C.), had proposed that all of nature was composed of four elements: fire, earth, air, and water. This idea provides a logical explanation for many observations. For example, a green log can burn, producing smoke (air) and flame (fire), leaving behind ashes (earth), and perhaps even giving a fleeting glimpse of sap (water). The idea of four elements was extended by emphasizing the basic properties of the elements. These were coldness, hotness, dryness, and wetness. Each element, in its ideal form, had two associated properties as shown schematically in Figure 1.1. For example, water was the wet, cold element. The Greeks believed that an element could be changed into any other element by altering its properties. For example, water could be changed to earth by replacing the property of wetness with the property of dryness.

This concept and its associated logic persisted for over 1000 years, influencing thought through the Middle Ages. However, by the time of the Middle Ages the list of elements and their associated properties had grown. Operating in this framework of logic the alchemists sought to convert common or ordinary metals into gold (Figure 1.2). Their attempts were based on the idea that addition of the essential quality or property of "nobility" to these metals would cause them to "grow" into gold. In the course of their futile attempts to bring about such changes the alchemists discovered new chemicals and developed new ways for working with them.

The Greek ideas of elements, as further developed during the Middle Ages, spawned the idea of a combustible property associated with flammable materials. The element associated with this property was known as *phlogiston* after the Greek word for fire. At this point, the idea of basic or elementary properties of matter had become quite confused with that of elements. The concept of phlogiston was used to explain many observations. It was believed that objects could burn only so long as they still contained phlogiston. After phlogiston had escaped from an object it was no longer flammable.

Early scientists noted that when metals are heated in air they lose

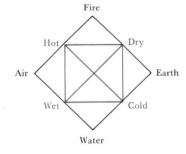

FIGURE 1.1 A schematic representation of the four elements of the Greeks and the four properties associated with these elements.

their metallic properties such as their luster and are converted to powdery materials that were known as calxes.* Under appropriate conditions, iron can be converted to rust by heating it in air. Rust is a calx of iron. According to phlogiston theory, a metal is converted to its calx when it loses phlogiston. In the eighteenth century an English clergyman and self-trained amateur scientist named Joseph Priestley (Figure 1.3) heated the red calx of mercury and thereby generated mercury metal together with a gas (oxygen). His experiment is shown in Figure 1.4. Priestley found that objects burn more vigorously in this gas than in air. He called the gas "dephlogisticated air" because its properties suggested that the gas had a particularly large capacity for phlogiston.

THE BIRTH OF MODERN CHEMISTRY

In 1772 a wealthy French nobleman named Antoine Lavoisier (Figure 1.5) began experimenting with combustion, the act of burning. By weighing objects before and then after combustion, Lavoisier observed that burning objects gain weight. This effect is easily observed when metals are converted to their calxes; Lavoisier showed that it is also true when nonmetals like sulfur are burned. Furthermore, he observed that when combustion is carried out in a closed container, there is no change in weight; the weight of all of the substances in the container is the same before and after combustion even though they change form. There was

*The word "calx" is only of historical interest and is no longer used in chemistry.

FIGURE 1.2 *The Alchemist,* a seventeenth-century painting by the Flemish painter David Teniers (1582–1649). *(Bettmann Archive)*

FIGURE 1.3 Joseph Priestley (1733–1804); Priestley became interested in chemistry at the age of 39, perhaps through his personal acquaintance with Benjamin Franklin. Because he lived next door to a brewery where he could obtain carbon dioxide, his initial studies involved this gas and were later extended to other gases. Because he was suspected of sympathizing with the American and French Revolutions, his church, home, and laboratory in Birmingham were burned by a mob in 1791. Priestley had to flee in disguise. He eventually emigrated to the United States in 1794 where he lived his remaining years in relative seclusion in Pennsylvania. Although his discovery of "dephlogisticated air" (oxygen) eventually led to the downfall of the phlogiston theory, Priestley stubbornly continued to support this theory even after strong evidence had brought it into serious question. Priestley was a scientific conservative, although he was very liberal in his religious and political views. (*Library of Congress*)

no evidence for the loss of phlogiston. Instead, the experiments indicated that when a substance burns it gains something from the air. The weight gained by the burning sample was the same as the weight lost by the air. This is why there is no net change in weight when combustion is carried out in a closed container. These quantitative measurements could not be explained using the phlogiston idea. Lavoisier also found that combustion in dephlogisticated air, which he renamed oxygen, gives the same products as combustion in air. It was concluded that air contains oxygen.

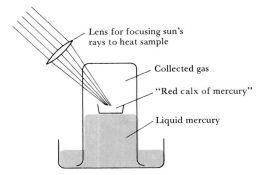

Lens for focusing sun's rays to heat sample

Collected gas

"Red calx of mercury"

Liquid mercury

FIGURE 1.4 Priestley's experiment showing how he prepared "dephlogisticated air" (oxygen).

On the basis of these observations, Lavoisier rejected the phlogiston theory. In its place he proposed that when an object burns, oxygen is removed from air and becomes incorporated into the burning object. In this view, the combustion of coal, which contains carbon, involves a reaction between oxygen and carbon. The product of this reaction is a gaseous substance called carbon dioxide. The reaction can be summarized as

$$\text{Carbon} + \text{oxygen} \longrightarrow \text{carbon dioxide}$$

where the arrow may be read as "produces" or "yields." Because combustion could be understood better without reference to phlogiston, that idea slowly died. Today we define combustion as a rapid reaction accompanied by heat and light; we find that most common combustion reactions do indeed involve oxygen as Lavoisier suggested.

Air is only 20 percent oxygen; consequently, combustion is much more vigorous in pure oxygen than in air. The fire that broke out on the Apollo 1 rocket in January 1967 and resulted in the deaths of astronauts Grissom, White, and Chaffee burned so very fast because the space capsule was filled with pure oxygen. Because of that incident, a mixture of 60 percent oxygen and 40 percent nitrogen has been used in space capsules ever since. A fire may be extinguished by placing a blanket over it because the blanket excludes the oxygen. People in a closed room can be suffocated by a fire because it removes oxygen from air.

FIGURE 1.5 Antoine Lavoisier, after a painting by Louis David. Lavoisier (1743–1794) conducted studies that led to the downfall of the phlogiston theory and the birth of modern chemistry. Unfortunately, Lavoisier's career was cut short by the French Revolution. He was not only a member of the French nobility but also a tax collector. He was guillotined in 1794 during the final months of the Reign of Terror. He is now generally considered to be the father of modern chemistry because of his reliance on carefully controlled experiments and his use of quantitative measurements. The tribunal that sentenced him to death, however, declared that France had no need for scientists. The great mathematician Lagrange, then living in Paris, remarked: "It took but a moment to cut off that head, though a hundred years perhaps will be required to produce another like it." (*Courtesy Burndy Library*)

Lavoisier is generally considered the father of modern chemistry because of his reliance on carefully controlled experiments and his use of quantitative measurements and not merely qualitative observations. In his studies Lavoisier also abandoned the ancient idea that if a material contains a particular element it must have the properties of that element. He adopted an idea proposed in 1661 by Robert Boyle. Boyle proposed that elements are the basic substances out of which all other substances can be made and into which all other substances can be decomposed or resolved. We will pursue the idea of elements more fully in Chapter 2.

1.2 the scientific approach

The fundamental activity of science is making careful observations. These may be of both a qualitative and quantitative nature and often involve controlled experiments. Scientists seek general relations that will unify their observations. A concise verbal statement or a mathematical equation that summarizes a broad variety of observation and experience is known as a scientific **law.** A familiar example is the law of gravity; it summarizes our experience that what goes up must come down. We also seek to understand our laws. A tentative explanation is called a **hypothesis.** A hypothesis is useful only if it can be used to make predictions that can be tested by further experiments and thereby verified or refuted. A hypothesis that continually withstands such tests is called a **theory.** A theory may serve to unify a broad area and may provide a basis for explaining many laws. Such is the case with the atomic theory of matter, which we will begin to examine in Chapter 2.

There is no fail-proof, step-by-step scientific method that scientists use. The approaches of various scientists depend on their temperament, circumstances, and training. Rarely will two scientists approach the same problem in the same way. The scientific approach involves doing one's utmost with one's mind to understand the workings of nature. Just because we can spell out the results of science so concisely or neatly in textbooks does not mean that scientific progress is smooth, certain, and predictable. The path of any scientific study is likely to be irregular and uncertain; progress is often slow and many promising leads turn out to be dead ends. Through the course of our studies we will see that serendipity (fortunate accidental discovery) has played an important role in the development of science. What we will often miss discussing are the doubts, conflicts, clashes of personalities, and revolutions of perception that have led to our present ideas. We should also remember that our theories are not chiseled into stone; they are tentative.

SELECTION OF THEORIES

No hypothesis or theory can ever be exposed to all the possible tests necessary for absolute verification. However, a hypothesis or theory can be disproved by obtaining experiment results inconsistent with it. Scientific advance depends on such disproofs to eliminate faulty hypotheses and theories. In the absence of such disproofs, scientists often choose between hypotheses or theories by comparing the ability of each to explain the evidence at hand; the one that explains the facts better is chosen.

It has been suggested that scientific progress is most rapid when scientists are open and imaginative enough to formulate several alternative hypotheses to explain their observations. Experiments can then be designed to test these alternatives, thereby excluding some of them. This approach involves a continual search for alternative hypotheses. When you hear of a new scientific hypothesis or theory, you should ask yourself what experiment could disprove that hypothesis. On hearing a scientific experiment described, you should ask what hypothesis it disproves. A person who works with only a single hypothesis can become strongly attached to it. Research can then become a strenuous and devoted attempt to force nature into the conceptual boxes supplied by that hypothesis.

In the end it is the collective judgment of the scientific community that effectively decides between theories. One of the most important activities of a scientist is therefore public disclosure of scientific results through publication. The authority of science does not rest ultimately on the individual who has done the work, but rather on whether others can repeat the work and obtain the same results or extend the work in a self-consistent fashion.

THE DEVELOPMENT OF SCIENCE

In the traditional view science progresses by the gradual accumulation of factual knowledge and of ever more encompassing and useful laws and theories to unify it. Indeed, this may serve as an explanation of what is meant by scientific progress. The term "useful" can be taken to mean better problem-solving ability. It might also suggest that progress involves movement toward greater technological benefits. If this is true, we must ask whether the benefits involve only an increase in the quantity of "things" or whether they reflect a consideration for the quality of life. Similarly, progress is often thought to involve our increased ability to manipulate nature. Hopefully we are beginning to realize that this ability also means learning how to live in harmony with nature. The idea that our goal is to "conquer nature" suggests exploitation and ignores the fact that we are part of nature. Finally, scientific progress is sometimes viewed as a movement toward an increasingly true picture of nature. Indeed, science has been defined as the inductive search for truth about nature. Many people believe that the more encompassing and useful our theories, the truer their representation of nature. Critics of this view say science merely seeks self-consistency in the hope that it is truth.

There is, however, another view that contrasts with that of the evolutionary or cumulative development of science. The philosopher and historian Thomas S. Kuhn has suggested that there are discontinuities in the development of science that involve revolutionary changes in the ways scientists perceive and approach nature. Kuhn suggests that at any given time scientists operate under a set of ideas about what nature is like that is rather universally accepted as general truth. This intellectual matrix of ideas and beliefs, which have been called **paradigms,** gives direction to experimental efforts and influences what scientists are able to perceive. "Facts" may be distorted by our expectations or totally overlooked because they were not anticipated. In designing experiments, the types of questions asked are conditioned by the types of answers expected. All people must make judgments, and scientists are not immune to the frailties and possibilities of prejudice that beset all persons. A scientific revolution occurs when a paradigm such as the phlogiston theory undergoes change.

Persons involved in such revolutions often refer to a flash of insight or a new perception, indicating a new way of seeing things. These revolutions often involve young scientists or persons who are new to a particular field of science and therefore are not deeply committed to its prevalent paradigms. These revolutions are often triggered by observations that fail to yield to explanations in terms of the theories of the day. Kuhn suggests the following:

> Sometimes a normal problem, one that ought to be solvable by known rules and procedures, resists the reiterated onslaught of the ablest members of the group within whose competence it falls. On other occasions a piece of equipment designed and constructed for the purpose of normal research fails to perform in the anticipated manner, revealing an anomaly that cannot, despite repeated effort, be aligned with professional expectation. In these and other ways besides, normal science repeatedly goes astray. And when it does—when, that is, the professional can no longer evade the anomalies that subvert the existing tradition of scientific practice—then begin the extraordinary investigations that lead the practice of science. . . . They are the tradition-shattering complements to the tradition-bound activity of normal science.*

Scientific revolutions, whether they be discontinuities in the development of science or part of its evolutionary development, are always traumatic, or disturbing. This is because scientists resist changes of concepts that they have relied on as the basis of their scientific thinking. Priestley's resistance to Lavoisier's new chemistry and his adherence to the phlogiston theory represents a common reaction to paradigm change. We can gain some empathy for this trauma by trying to identify some of our current paradigms. If it is difficult to do so, it is in part because it is traumatic to imagine that certain of our fundamental ideas may not be representative of the way nature will be viewed in the future. We believe, for example, that chemistry can be understood completely by reference to only three fundamental particles—the proton, neutron, and electron. Furthermore, we believe that we can understand complex systems by understanding their parts (that it is possible and desirable to fragment nature for purposes of investigation).

1.3 measurement and the metric system

Lavoisier's studies of combustion (Section 1.1) should impress upon us the importance of quantitative measurements. This idea is simply common sense to us now, although it has not always been so. Consider a person who is ill. Using only sense perception we may conclude that this person is running a fever, but we are not certain, and another person may disagree with our conclusion. To tell accurately, we use a thermometer,

*Thomas S. Kuhn, *The Structure of Scientific Revolutions* (Chicago: The University of Chicago Press, 1962), p. 5.

FIGURE 1.6 A road sign along an interstate highway in Ohio, showing distance in both metric and English-system units. (*Ohio Department of Transportation*)

and much can depend upon the result of our measurement—for example, whether the person's temperature is 98.6°F or 102°F. This example suggests three points about measurement. First, our five senses are the most important tools we have, but they are limited. We therefore resort to various instruments to extend and quantify our sense perceptions. Second, an advantage of quantitative data is that they allow different people to obtain the same results, thus avoiding many arguments based on opinions. Third, measurements depend on a standard of reference. For instance, there is a considerable difference between 102°F and 102°C. (The meaning of °F and °C will be discussed shortly.)

The standards used in science are those of the **metric system.** This is the system of weights and measures used throughout most of the world; the United States is also moving toward adopting it in many facets of society. The weights of most canned products in the grocery store are now given in grams as well as in ounces; there are even a few highway signs that show distance in both miles and kilometers (Figure 1.6).

According to international agreement reached in 1960, certain basic metric units and units derived from them are to be preferred in scientific use. The preferred units are known as International System units (**SI** units). The basic units of the SI system are given in Table 1.1. Non-SI units are to be progressively discouraged and with time phased out. Adoption of SI

TABLE 1.1

basic SI units

PHYSICAL QUANTITY	NAME OF UNIT	ABBREVIATION
Mass	Kilogram	kg
Length	Meter	m
Time	Second	s or sec
Electric current	Ampere	A
Temperature	Kelvin	K
Luminous intensity	Candela	cd
Amount of substance	Mole	mol

TABLE 1.2

selected prefixes used in the metric system

PREFIX	ABBREVIATION	MEANING	EXAMPLE
Mega-	M	10^6	1 megameter (Mm) $= 1 \times 10^6$ m
Kilo-	k	10^3	1 kilometer (km) $= 1 \times 10^3$ m
Deci-	d	10^{-1}	1 decimeter (dm) $= 0.1$ m
Centi-	c	10^{-2}	1 centimeter (cm) $= 0.01$ m
Milli-	m	10^{-3}	1 millimeter (mm) $= 0.001$ m
Micro-	μ[a]	10^{-6}	1 micrometer (μm) $= 1 \times 10^{-6}$ m
Nano-	n	10^{-9}	1 nanometer (nm) $= 1 \times 10^{-9}$ m
Pico-	p	10^{-12}	1 picometer (pm) $= 1 \times 10^{-12}$ m

[a]This is the Greek letter μ (pronounced "mew").

units is an attempt to further systematize the metric system. However, until SI units are fully adopted by practicing scientists, it is necessary to be aware of both SI units and the non-SI units that are still in use. Whenever we first encounter a non-SI unit in the text, the proper SI unit will also be given.

The primary standards used for the basic SI units are selected on the basis of their being reproducible, unchanging, and capable of use for precise measurement. Basically, however, they are otherwise arbitrarily selected. For example, the kilogram is defined as the mass of a standard platinum-iridium cylinder that is stored at the International Bureau of Weights and Measures at Sèvres, France.

The metric system employs a series of prefixes to indicate decimal fractions of the various basic measurements. The most common of these are given in Table 1.2.* In using the metric system and in working problems throughout this text, it is important to have a comfortable familiarity with exponential notation. If you are unfamiliar with exponential notation or want to review it, refer to Appendix A.1.

LENGTH

The basic SI unit of length is the meter (m). From the comparisons between metric and English-system measurements given in Table 1.3 we can see that the meter is only slightly longer than a yard. A diagram-

*It is interesting that the monetary system of the United States is decimal: 0.01 of a dollar is a cent (centidollar) while 0.001 of a dollar (a tenth of a cent) is a mill (millidollar). Extending this usage, we could theoretically refer to $1000 as a kilodollar or kilobuck.

11

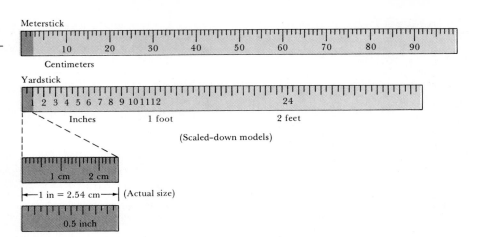

FIGURE 1.7 A comparison of common measures of length.

matic comparison between metric and English-system measures of length is made in Figure 1.7. We shall consider interconversion of English and metric system measures more closely in Section 1.5. For the moment it is more important that we clearly understand the use of the prefixes that are given in Table 1.2.

MASS

The basic SI unit of **mass** * is the kilogram (kg). As shown in Table 1.3 and Figure 1.8, a kilogram is equal to 2.2 pounds. The unit of mass used most frequently in chemistry is the gram (g), which is 1/1000 of a kilogram. The mass of an object is determined by balancing it against a set of known masses using a device known as a balance. Several types of common laboratory balances are shown in Figure 1.9.

VOLUME

In measuring volume, units such as the cubic meter (m^3), which is the volume of a cube 1 m on each edge, or related units such as cubic centimeters (cm^3 or cc) are used. Another common measure of volume is

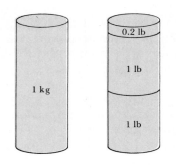

FIGURE 1.8 A comparison of common measures of mass.

*Mass and weight are often incorrectly thought to be the same. Mass is a measure of the amount of material in an object; the weight of that object, however, depends not only on its mass, but also on the attractive force of gravity. In outer space, where gravitational forces are very weak, an astronaut may be weightless, but he is not massless. In fact, he has the *same* mass as he has on earth. Nevertheless, it is common practice to use the terms mass and weight interchangeably.

TABLE 1.3

metric-English system equivalents

LENGTH	MASS	VOLUME
1 meter = 1.094 yards	1 kilogram = 2.205 pounds	1 liter = 1.06 quarts
2.54 centimeters = 1 inch	453.6 grams = 1 pound	1 cubic foot = 28.32 liters

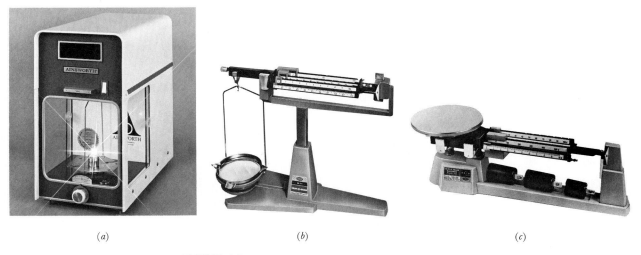

(a) (b) (c)

FIGURE 1.9 Three common types of laboratory balances: (*a*) analytical balance; (*b*) triple-beam balance of the stirrup type; (*c*) triple-beam platform balance. (*a, Denver Instrument Company, Ainsworth Division; b and c, Ohaus Scale Corp.*)

the liter (l.), a volume roughly the size of a quart (refer to Table 1.3 and to Figure 1.10). A liter is the volume occupied by 1 cubic decimeter. There are 1000 ml in a liter, and each milliliter is the same volume as a cubic centimeter. Thus milliliter and cubic centimeter are commonly used interchangeably in expressing volume. The liter is the first metric unit that we have encountered that is not an SI unit.

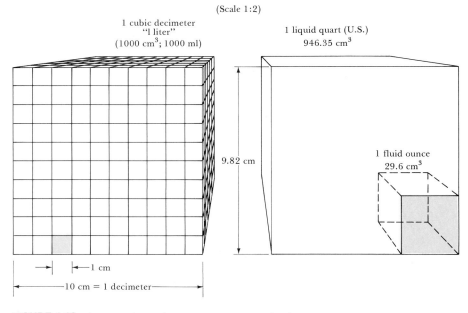

FIGURE 1.10 A comparison of common measures of volume.

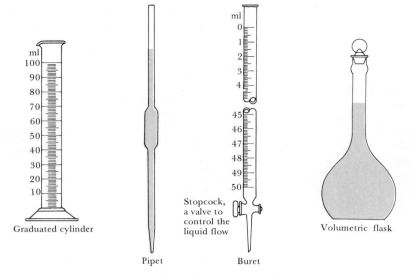

FIGURE 1.11 Common devices used in chemistry laboratories to measure volume.

The devices most frequently used in chemistry to measure volume are illustrated in Figure 1.11. Pipets and burets allow delivery of liquids with more accurately known volumes than do graduated cylinders. Volumetric flasks are used to prepare accurately a designated volume of solution.

DENSITY

Density is a quantity widely employed by chemists to identify substances. It is defined as the amount of mass in a unit volume of the substance:

$$\text{Density} = \text{mass/volume}$$

[1.1]

Density is commonly expressed in units of grams per cubic centimeter (g/cm^3 or $g\,cm^{-3}$). The densities of some common substances are recorded in Table 1.4.

TABLE 1.4

densities of some selected substances

SUBSTANCE	DENSITY (g/cm^3)
Air	0.002
Balsa wood	0.16
Water	1.00
Table salt	2.16
Iron	7.9
Gold	19.32

SAMPLE EXERCISE 1.1

Calculate the density of mercury if 100 g occupies a volume of 7.36 cm³.

Solution:

$$\text{Density} = \frac{\text{mass}}{\text{volume}} = \frac{100\ g}{7.36\ cm^3} = 13.6\ g/cm^3$$

Density is an **intensive property,** meaning that it is a property that does not depend on the amount of material chosen. By contrast, both volume and mass are **extensive properties,** because they depend on the amount of sample. Anyone who suggests that iron weighs more than air is referring to this intensive property, density; a pound of air weighs the same as a pound of iron.

SAMPLE EXERCISE 1.2

Consider the following description, labeling each property or characteristic as intensive or extensive: "The yellow solid, which weighed 5 g, had an unusual shape."

Solution: Shape and mass are extensive properties. Color is an intensive property.

TEMPERATURE

The temperature scales commonly employed in scientific studies are the Celsius (or centigrade) and Kelvin scales; as previously noted, the Kelvin is the SI unit of temperature. The Celsius scale is based on assignment of 0°C to the freezing point of water and 100°C to its boiling point at sea level. The corresponding temperatures in the Fahrenheit scale are 32°F and 212°F. There are 100° between the freezing point and boiling point of water in the Celsius scale, whereas there are 180° between these points on the Fahrenheit scale. Consequently, the Celsius and Fahrenheit scales are related as follows:*

$$°C = \frac{100}{180}(°F - 32°) = \frac{5}{9}(°F - 32°) \qquad [1.2]$$

The Kelvin scale is based on the properties of gases, and its origins will be considered more fully in Chapter 5. Zero on this scale corresponds to $-273.15°C$, and the size of a Kelvin is the same as a degree Celsius. The Kelvin and Celsius scales are therefore related by Equation [1.3].†

$$K = °C + 273.15° \qquad [1.3]$$

Further comparisons between the Celsius, Kelvin, and Fahrenheit scales are made in Figure 1.12 and Table 1.5.

We sense temperature as a measure of the hotness or coldness of an object. Indeed, temperature determines the direction of heat flow; heat always flows spontaneously from a substance at high temperature to one at low temperature. Thus we feel the influx of energy when we touch a hot stove, and we know that the stove is at a higher temperature than our hand. The temperature of an object is an intensive property, whereas its

---

*Using Equation [1.2] it can be show that $-40°C = -40°F$. This fact permits another simple method of converting between Celsius and Fahrenheit scales. To convert a temperature from Fahrenheit to Celsius add 40°, multiply the result by $\frac{5}{9}$, and then subtract 40°. To convert from Celsius to Fahrenheit add 40°, multiply the result by $\frac{9}{5}$, and then subtract 40°.

†According to SI convention a degree sign (°) is not used with the Kelvin scale. Thus we write 273 K and not 273°K.

TABLE 1.5

some comparisons of Fahrenheit, Celsius, and Kelvin temperatures

Absolute zero	−460°F	−273°C	0 K
Freezing point of water	32°F	0°C	273 K
Average room temperature	70°F	21°C	294 K
Normal body temperature	98.6°F	37°C	310 K
Boiling point of water	212°F	100°C	373 K

heat content is an extensive property. This difference can be seen by comparing a match and a bonfire: although both may be burning at the same temperature, the bonfire produces more heat because of its larger size.

SAMPLE EXERCISE 1.3

If a weatherman predicts that the temperature for the day will reach 30°C, what is the predicted temperature in °F? Should you wear a coat?

Solution:

$$°C = \frac{5}{9}(°F − 32°)$$

$$30° = \frac{5}{9}(°F − 32°)$$

$$\left(\frac{9}{5}\right)(30°) = °F − 32°$$

$$54° + 32° = °F$$

$$86° = °F$$

If the weather report is correct, you obviously have no need for a coat.

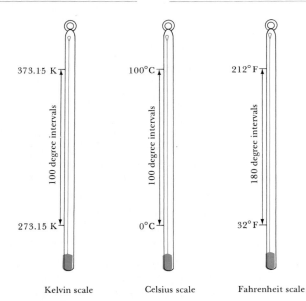

FIGURE 1.12 A comparison of the Fahrenheit, Celsius, and Kelvin temperature scales.

All measurements have some degree of uncertainty; how great the uncertainty is depends on both the accuracy of the measuring device and the skill of its operator. For example, the magnitude of the uncertainty associated with weighing an object will depend on the type of balance employed. On a triple-beam platform balance (Figure 1.9), the mass of a sample substance can be measured to the nearest 0.1 g; mass differences less than this cannot be detected on this balance. We might therefore indicate the mass of a dime measured on this balance as 2.2 ± 0.1 g; the ±0.1 (read plus or minus 0.1) is a measure of the accuracy of the measurement. It is important to have some indication of how accurately any measurement is made; the ± notation is one way to accomplish this. We could measure the mass of the dime more accurately on an analytical balance, to the nearest 0.0001 g if we are careful. We might therefore report the mass as 2.2405 ± 0.0001 g. It is common to drop the ± notation with the understanding that *there is uncertainty of at least one unit in the last digit of the measured quantity;* that is, measured quantities are reported in such a way that only the last digit is uncertain. All of the digits, including the uncertain one, are called significant digits or, more commonly, **significant figures**. The number 2.2 has two significant figures, while the number 2.2405 has five significant figures.

SAMPLE EXERCISE 1.4

What is the difference between 4.0 g and 4.00 g?

Solution: Many people would say there is no difference, but a scientist would note the difference in the number of significant figures between the two measurements. The number 4.0 has two significant figures while 4.00 has three. This implies that the second measurement has been made more accurately. A mass of 4.0 g indicates that the mass of the sample must be between 3.95 g and 4.05 g, closer to 4.0 g than to 3.9 g or 4.1 g. A mass of 4.00 g means that the sample must have a mass between 3.995 and 4.005 g, closer to 4.00 g than to 3.99 g or 4.01 g.

The following rules apply to determining the number of significant figures in a measured quantity:

1. All nonzero digits are significant—457 cm (three significant figures); 0.25 g (two significant figures).
2. Zeros between nonzero digits are significant—1005 kg (four significant figures); 1.03 cm (three significant figures).
3. Zeros to the left of the first nonzero digits in a number are not significant; they merely indicate the position of the decimal point—0.02 g (one significant figure); 0.0026 cm (two significant figures).
4. When a number ends in zeros that are to the right of the decimal point, they are significant—0.0200 g (three significant figures); 3.0 cm (two significant figures).
5. When a number ends in zeros that are not to the right of a decimal point, the zeros are not necessarily significant—130 cm (two or three

significant figures); 10,300 g (three, four, or five significant figures). The way to remove this ambiguity is described below.

Use of standard exponential notation avoids the potential ambiguity of whether the zeros at the end of a number are significant (rule 5). For example, a mass of 10,300 g can be written in exponential notation showing three, four, or five significant figures:

1.03×10^4 g   (three significant figures)

1.030×10^4 g   (four significant figures)

1.0300×10^4 g   (five significant figures)

In these numbers all the zeros to the right of the decimal point are significant (rules 2 and 4).

The rules we have stated apply to nonintegral measured quantities; for these cases the number of significant figures is indicative of the associated uncertainty of the measurement. It is important to distinguish these nonintegral numbers from exact, integral ones. For example, there are exactly 3 ft in a yard, exactly four people in my immediate family, exactly 1000 g in a kilogram, and exactly 12 eggs in a dozen eggs. These are examples of exact numbers, numbers with no associated uncertainty. They can be considered to have an infinite number of significant figures. The number 1 in the metric-English system equivalents given in Table 1.3 is an exact number.

In carrying measured quantities through calculations the rule used is that the accuracy of the result is limited by the least accurate measurement. *In multiplication and division the result must be reported as having no more significant figures than the measurement with the fewest significant figures.* This limitation is illustrated in the following example and in Sample Exercises 1.5 and 1.6.*

$$(6.221)(5.2) = 32.3492 \longrightarrow \text{round off to } 32$$

Fewest significant
figures (2)

Result has 2
significant figures

The rule used in multiplication and division cannot be used *for addition and subtraction.* For these operations *the result should be reported to the same number of decimal places as that of the term with the least number of decimal places.* The following is an example:

$$\begin{array}{r} 20.4 \\ 1.322 \\ 83 \\ \hline 104.722 \end{array} \longrightarrow \text{round off to } 105$$

This number limits
the number of significant
figures in the result

---

*If the number to be dropped in rounding off is less than 5, it is simply dropped—2.23 rounds off to 2.2. If it is more than 5, the preceding number is increased by 1—2.17 rounds off to 2.2. If the number to be dropped is 5, by convention the preceding number is not changed if the preceding number is even—2.25 rounds off to 2.2; if the preceding number is odd, it is increased by 1—2.15 rounds off to 2.2.

In future calculations, answers will be given to the proper number of significant figures; the round-off process will not be shown.

SAMPLE EXERCISE 1.5

How many significant figures are there in each of the following numbers? (a) 4.003; (b) 6.023×10^{23}; (c) 5000; (d) the sum $15.3 + 0.2334$; and (e) the product $(16)(5.7793)$.

Solution: (a) Four; the zeros are significant figures. (b) Four; the exponential term does not add to the number of significant figures. (c) One; the zeros are not significant unless otherwise indicated. In standard exponential notation, the number is 5×10^3. (d) Three; the sum, expressed to the proper 3 significant figures, is 15.5. (e) Two; the product, expressed to the proper 2 significant figures, is 92.

SAMPLE EXERCISE 1.6

If a certain United States nickel weighs 4.9556 g and has a density of 8.8 g/cm³, what is its volume?

Solution:

$$\text{Density} = \frac{\text{mass}}{\text{volume}}$$

Therefore,

$$\text{Volume} = \frac{\text{mass}}{\text{density}} = \frac{4.9556 \text{ g}}{8.8 \text{ g/cm}^3} = 0.56 \text{ cm}^3$$

Notice that the result should have only two significant figures.

1.5 dimensional analysis— an approach to problem solving

Before we go on, perhaps a word of caution is in order. Sometimes students have little difficulty reading their chemistry text or following the lecture and yet have difficulty on exams. In some instances the problem is lack of familiarity with terms. Often the problem is that the students have a passive but not an active understanding of the material. They can see how someone else has worked a problem, but they are unable to work any on their own. An active understanding involves being able to use the material in new situations, including especially working problems that are not identical to those used as sample exercises in the text. It is important to use the problems at the end of each chapter to test yourself to determine how well you are able to use the material in the chapter. Colored numbers indicate problems whose answers can be found in Appendix F. Bracketed numbers indicate problems of above-average difficulty. Wherever possible we have used an approach that can be referred to as **dimensional analysis** in solving problems. If you develop a facility with this approach, which is illustrated in the following discussion, your work will be much easier. If you need a review of basic mathematics, refer to Appendix A.

A number reported for a measured quantity is meaningless unless its units are specified. If units are treated as algebraic quantities, they can be carried through all calculations and will indicate whether the calculation has been performed correctly. This approach is illustrated in the following examples and in the sample exercises that follow.

Consider the conversion of mass from pounds to kilograms. If a man weighs 175 lb, what is his mass in kilograms? From Table 1.3 we have the following relationship: 1 kg = 2.205 lb. We can therefore write the following equalities:

$$1 = \frac{1 \text{ kg}}{2.205 \text{ lb}}; \quad 1 = \frac{2.205 \text{ lb}}{1 \text{ kg}}$$

These equalities, which can be read as 1 kg per 2.205 lb and 2.205 lb per kilogram, are referred to as **unit conversion factors**. Multiplication of a quantity by these factors changes the units in which the quantity is expressed but not its value. To convert pounds to kilograms we choose the unit conversion factor that cancels pounds:

$$? \text{ kg} = (175 \text{ lb}) \left(\frac{1 \text{ kg}}{2.205 \text{ lb}} \right) = 79.4 \text{ kg}$$

Now consider a more complex conversion of units, the calculation of the number of inches in 3.00 km. We can begin by writing the equality we are working toward:

$$? \text{ in.} = 3.00 \text{ km}$$

From the relations shown in Tables 1.2 and 1.3 and from our basic knowledge of the English system we can write the following equalities:

$$1 \text{ km} = 1000 \text{ m}; \quad 1 \text{ m} = 1.094 \text{ yd}; \quad 1 \text{ yd} = 36 \text{ in.}$$

Therefore,

$$\frac{1000 \text{ m}}{1 \text{ km}} = 1; \quad \frac{1.094 \text{ yd}}{1 \text{ m}} = 1; \quad \frac{36 \text{ in.}}{1 \text{ yd}} = 1$$

If we multiply 3.00 km by these factors we have

$$? \text{ in.} = (3.00 \text{ km}) \left(\frac{1000 \text{ m}}{1 \text{ km}} \right) \left(\frac{1.094 \text{ yd}}{1 \text{ m}} \right) \left(\frac{36 \text{ in.}}{1 \text{ yd}} \right)$$
$$= 1.18 \times 10^5 \text{ in.}$$

Each conversion factor is applied so as to cancel the units of the preceding factor. This converts kilometers successively from kilometers to meters to yards to inches. Because we are left with the proper units, we know that the problem has been correctly set up. (There are exactly 1000 m in a kilometer and exactly 36 in. in a yard; the result therefore has three significant figures.)

Because 1 m = 1.094 yd, it is also true that 1 m/1.094 yd = 1. If this factor is applied instead of its reciprocal as above, the result is

$$? \text{ in.} = (3.00 \text{ km}) \left(\frac{1000 \text{ m}}{1 \text{ km}} \right) \left(\frac{1 \text{ m}}{1.094 \text{ yd}} \right) \left(\frac{36 \text{ in.}}{1 \text{ yd}} \right)$$
$$= 9.88 \times 10^4 \frac{\text{m}^2 \text{ in.}}{\text{yd}^2}.$$

Clearly, the units do not cancel to give the desired units, inches, so we know that the result is incorrect.

SAMPLE EXERCISE 1.7

You have to pour 2.0 cubic yards (yd^3) of concrete for a patio. What is this volume in cubic meters (m^3)?

Solution:

$$? \; m^3 = (2.0 \; yd^3) \left(\frac{1 \; m}{1.094 \; yd} \right) \left(\frac{1 \; m}{1.094 \; yd} \right) \left(\frac{1 \; m}{1.094 \; yd} \right)$$

$$= (2.0 \; yd^3) \left(\frac{1 \; m}{1.094 \; yd} \right)^3$$

$$= 1.5 \; m^3$$

SAMPLE EXERCISE 1.8

You are approaching a city and see a sign indicating a speed limit of 40 km/hr. What is the corresponding speed in miles per hour?

Solution:

$$? \; \frac{mi}{hr} = \left(40 \; \frac{km}{hr} \right) \left(\frac{1000 \; m}{1 \; km} \right) \left(\frac{1.094 \; yd}{1 \; m} \right) \left(\frac{1 \; mi}{1760 \; yd} \right)$$

$$= 25 \; \frac{mi}{hr}$$

SAMPLE EXERCISE 1.9

The acid in an automobile battery (a solution of sulfuric acid) has a density of 1.2 g/cm^3. What is the mass (in grams) of 200 ml (2.00×10^2 ml) of this acid?

Solution:

$$? \; g = (2.00 \times 10^2 \; ml) \left(\frac{1 \; cm^3}{1 \; ml} \right) \left(1.2 \; \frac{g}{cm^3} \right)$$

$$= 2.4 \times 10^2 \; g = 240 \; g$$

Notice that density can be thought of as a unit conversion factor for converting volume to mass or *vice versa*.

SAMPLE EXERCISE 1.10

What is the mass (in grams) of an aluminum block whose dimensions are 2.0 in. $\times$ 3.0 in. $\times$ 4.0 in. and whose density is 2.7 g/cm^3?

Solution:

$$? \; g = (2.0 \; in.)(3.0 \; in.)(4.0 \; in.)$$

$$\times \left(\frac{2.54 \; cm}{1 \; in.} \right) \left(\frac{2.54 \; cm}{1 \; in.} \right) \left(\frac{2.54 \; cm}{1 \; in.} \right) \left(\frac{2.70 \; g}{1 \; cm^3} \right)$$

$$= 1.1 \times 10^3 \; g$$

Dimensional analysis cannot be used on all problems that we shall work, and you should feel free to abandon it and work problems stepwise if you wish. However, you should *always* carry units throughout all of your calculations, making sure that they cancel properly. Whenever you finish a calculation, look at both the units and magnitude of your answer and ask yourself whether your answer makes any sense. This will help you to avoid making some embarrassingly simple errors.

FOR REVIEW

summary

We have defined chemistry as the study of the properties, composition, and changes of **matter**. Chemistry has two origins: (1) the craft traditions such as metallurgy, and (2) the more philosophical search for basic understanding of matter. Modern chemistry rests upon certain scientific **laws** arrived at through both qualitative observations and quantitative measurements. **Hypotheses** are devised to provide a tentative explanation for the laws. If the hypotheses are successful they become **theories.**

Measurements are made using the **metric system,** which is based on the decimal system. Uncertainties associated with measurements can be expressed by use of **significant figures.** We have seen that when measured quantities are carried through calculations, it is important to keep track of units.

learning goals

Having read and studied this chapter, you should be able to:

1. Use the metric system and list the basic metric units and the common prefixes.
2. Interconvert metric and English-system measurements using dimensional analysis.
3. Perform calculations involving density.
4. Determine the number of significant figures in a derived quantity.

key terms

Among the more important terms and expressions used for the first time in this chapter are the following:

Combustion (Section 1.1) is a process that proceeds so rapidly that flames or light are produced.

Density (Section 1.3) is mass per unit volume.

A **hypothesis** (Section 1.2) is a trial idea or explanation; a tentative theory.

An **intensive property** (Section 1.3) is one that is independent of the amount of material under consideration; an **extensive property** (Section 1.3) depends on the amount of material under consideration.

A scientific **law** (Section 1.2) is a concise verbal or mathematical statement of a relationship between phenomena.

Mass (Section 1.3) is a measure of the amount of "stuff" in an object. It measures the resistance of a stationary object to be moved. In SI units, mass is measured in kilograms.

Matter (introduction) is the physical material (the "stuff") of the universe; it is anything that occupies space and has mass.

A **paradigm** (Section 1.2) is a set of basic ideas and beliefs that is rather universally believed.

Significant figures (Section 1.4) are the digits that indicate the precision with which a measurement has been made—those digits of a measured number that have uncertainty only in the last digit.

A **theory** (Section 1.2) is an explanation of a set of related observations.

EXERCISES

general exercises

1.1* Which of the following are matter? (a) steak; (b) air; (c) dust; (d) music; (e) heat; (f) love; (g) water; (h) salt

1.2 Distinguish between (a) law and theory; (b) mass and weight; (c) m^3 and liter; (d) heat and temperature; (e) intensive and extensive properties.

[1.3]† What is meant by scientific progress?

[1.4] List a concept about life or the world that you somehow vaguely know is true. Would it be appropriate to call this one of your paradigms?

1.5 How would the following observations be explained by phlogiston theory? (a) When charcoal burns, very little residue remains. (b) Burning is more rapid in pure oxygen than in air.

[1.6] Lavoisier made the following statement: "Chemists have made a vague principle of phlogiston which is not strictly defined, and which in consequence accommodates itself to every explanation into which it is pressed." What is the danger associated with very general theories?

1.7 List each of the following as an intensive or extensive property: (a) volume; (b) mass; (c) temperature; (d) color; (e) density; (f) shape; (g) melting point; (h) resistance to corrosion.

metric system

1.8 Indicate whether the following units measure mass, volume, or length: (a) ml; (b) mm^3; (c) km; (d) μg.

1.9 What are the basic SI units for length, mass, volume, and time?

1.10 What prefixes indicate the following multipliers? (a) $\frac{1}{100}$; (b) 1×10^{-3}; (c) 1×10^3

1.11 What multipliers of a meter are indicated by the following prefixes? (a) milli-; (b) micro-; (c) centi-; and (d) kilo-

1.12 Express 10.2 g in kilograms and in milligrams.

1.13 Express 8.77 m in kilometers and in millimeters.

significant figures

1.14 How many significant figures are there in each of the following numbers? (a) 6.2×10^{22}; (b) 5.001; (c) the product $(3.26)(5.0 \times 10^{23})$; (d) the

result of $(3.57)(7.577)/(2.844)$; and (e) the sum $2.1 + 5.778$

1.15 How many significant figures are there in each of the following numbers? (a) 0.0027; (b) 3.750×10^2; (c) the quotient $3.07/5.2$; (d) the difference $57.23 - 5.2$; and (e) the result of $(6.22)^2(5.7892)/(3.2)$

1.16 Criticize the following comments. (a) A fossil was studied 18 years ago and found to be 2,500,000 years old. It must now be 2,500,018 years old. (b) The population of a city is 106,000. You know of a family of five that recently moved from the city. The population must now be 105,995.

1.17 Express the number 35,000 in exponential notation to show (a) two significant figures; (b) four significant figures.

metric-English conversions and dimensional analysis

1.18 If a man is 6 ft tall, what is his height in kilometers?

1.19 What is the distance of the 220-yd dash expressed in meters?

1.20 How many yards are there in 1000 km?

1.21 A cut of meat sells for $2.19 per pound. What is its price per kilogram?

1.22 If gasoline costs 60¢ per gallon, what is its cost per liter?

1.23 An automobile engine has a displacement of 305 cubic inches. What is its displacement in liters?

1.24 A metric ton is 1000 kg. What is this mass in pounds?

1.25 What is the speed in kilometers per hour of an auto traveling at 55 mph?

1.26 A fifth of an alcoholic beverage like wine or whiskey is $\frac{1}{5}$ of a gallon. How many liters is this?

1.27 A grain (gr) is a unit of mass used in pharmaceutical work; 15 gr = 1 g. How many grams are there in a 5 gr tablet of aspirin?

1.28 A block has dimensions of 2.0 cm × 1.0 cm × 3.5 mm. What is its volume in cubic centimeters? In liters? In cubic inches?

1.29 A block has dimensions of 1.0 in. × 316 in. × 2.2 ft. What is its volume in cubic centimeters? In liters? In cubic meters?

*Color question numbers indicate problems whose answers may be found in Appendix F.

†Brackets indicate problems of above-average difficulty.

1.30 If a person has a temperature of 104°F, what is his body temperature in °C? In K?

1.31 Convert each of the following temperatures into the indicated scales. (a) 72°F to °C; (b) 0°F to °C; (c) −10°C to °F; (d) 25°C to K; (e) 57°F to K

1.32 Ethylene glycol, used in antifreeze, freezes at −17.4°C. What is its freezing point in °F and in K?

density

1.33 A gold nugget weighing 153 g was found to have a volume of 7.92 ml. What is its density?

1.34 According to legend, Archimedes was asked to authenticate a crown suspected of being gold-plated rather than solid gold. If the crown weighed 2800 g and had a volume of 150 ml, was it pure gold? (See Table 1.4.)

1.35 What is the mass of 4.00 l. of mercury (density = 13.6 g/cm^3)?

1.36 Iron has a density of 7.20 g/cm^3. Calculate the mass of 1.5×10^2 cm^3 of iron.

1.37 Air has a density of 1.29 g/l. at 0°C. What is the volume of 5.0 g of air?

[1.38] Calculate the density of a block of metal 0.25 m × 10 cm × 5 mm that weighs 0.9 kg.

[1.39] Calculate the density of a block of wood 0.5 in. × 0.25 in. × 0.5 ft that weighs 9 g.

1.40 A graduated cylinder weighs 101.22 g. When 25.3 ml of liquid are added, its mass plus that of the liquid is 116.57 g. Calculate the density of the liquid.

[1.41] A bottle that can hold 200 g of water can hold only 71.8 g of ether. What is the density of ether?

[1.42] How many grams of NaOH are contained in 50 ml of a solution if the density is 1.53 g/ml and the solution is 50.5 percent NaOH by weight?

our chemical world: atoms, molecules, and ions

2

Our present understanding of the changes we see around us—such as the melting of ice and the burning of wood—is intimately tied to our understanding of the nature and composition of matter. For example, before we can hope to understand what is happening when ice melts we must know what ice is—what it is composed of. It is possible to resolve or separate matter into a great variety of different **pure substances**. These are materials or portions of matter whose composition and intrinsic properties are uniform throughout. For example, seawater can be separated into several different pure substances, the most abundant being water and ordinary table salt (sodium chloride).

In this chapter we shall examine the composition of matter. We shall attempt to answer many fundamental questions: What types of pure substances are there? How can matter be separated into pure substances? Can pure substances be broken into simpler components? How do substances differ at the microscopic or atomic level? How do we represent the compositions of substances and how do we name them?

2.1
states of matter

Before we get too carried away in answering the questions we have posed, it is useful to note that matter exists in three states: gas (also known as vapor), liquid, or solid. A **gas** has neither a shape of its own nor a fixed volume. It takes the shape and volume of any container into which it is introduced. It can be compressed to fit a small container; it will expand to occupy a large one. Air is a gas.

A **liquid** has no specific shape; it assumes the shape of the portion of any container that it occupies. It does not expand to fill the entire container; it has a specific volume. Furthermore, a liquid is only slightly compressible. Water and gasoline are common liquids.

A **solid** has a firmness that is not associated with either gases or liquids. It has a fixed volume and shape. Like liquids, solids are not compressible to any appreciable degree. Numerous objects around us are solids—nails, coins, salt, and sugar to name a few.

The state of a substance depends on temperature. Above 100°C, water exists as a gas, known as water vapor. Between 0°C and 100°C, it exists as a liquid. Below 0°C, it exists as a solid—ice.

Changes of state, such as the change of ice to liquid water, are examples of **physical changes**. Physical changes are ones that do not involve creation of new substances; they involve no change in the composition of the specimen of matter under consideration. **Chemical changes,** also called **chemical reactions,** involve conversion of one substance into another. Every pure substance has a unique set of **properties** or characteristics that allows us to recognize it and distinguish it from other substances. **Chemical properties** are those properties that refer to the way a substance is able to change into other substances (its reactivity, how it "reacts"). The **physical properties** of a substance are those that do not involve a change in the chemical identity of the substance.

SAMPLE EXERCISE 2.1

Chlorine is a greenish-yellow gas with a density of 3.21 g/l. It can be changed to a liquid by cooling to −34.6°C; it reacts explosively with sodium to form sodium chloride (table salt). Which of these properties are physical properties and which are chemical?

Solution: The color and density of chlorine and the temperature at which it changes state, from a gas to a liquid, are all physical properties. They do not involve a change of chlorine into any other substance. The ability of chlorine to react explosively with sodium is a chemical property. In reacting with sodium, chlorine is changed into a different substance, sodium chloride.

2.2
the
classification
of matter

All specimens of matter can be classified either as pure substances or as **mixtures** of two or more substances. Most matter around us consists of mixtures. Mixtures are characterized by variable composition and by the fact that they can be separated by physical means. That is, mixtures can be separated by taking advantage of differences in physical properties such as boiling points. For example, we can recognize that blood is a mixture because its composition may vary in many ways, such as in its iron content. Furthermore, blood can be separated into two components, packed cells and plasma, by centrifugation, a physical method of separation. Some common methods of separating mixtures are discussed in Section 2.3.

The term *homogeneous* is used to describe portions of matter that are uniform throughout. Homogeneous mixtures are known as **solutions**. Mixtures that are not homogeneous are said to be *heterogeneous*. When salt and water are mixed, a homogeneous mixture, or solution, forms. The

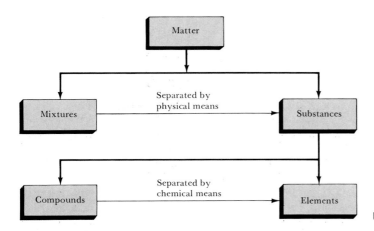

FIGURE 2.1 The classification of matter.

salt is said to dissolve in the water. When clay and water are mixed, no solution forms. The resultant mixture is heterogeneous.

A pure substance is a homogeneous material with a constant, invariable composition and a distinct set of intrinsic properties. Substances are of two types: compound substances and elementary substances. We normally refer to these as *compounds* and *elements*, respectively. Most substances are **compounds,** substances composed of two or more elements united chemically in definite proportions by mass. Compounds can be decomposed by chemical reaction into elements. **Elements** are the simplest substances; they cannot be decomposed or resolved by chemical change into simpler substances.* The classification of matter into mixtures, substances, compounds, and elements is summarized in Figure 2.1. Before we examine compounds and elements more closely, let's consider some of the ways a mixture can be separated into its component parts.

2.3 separation of mixtures

Chemists often need to separate mixtures into their component substances. For example, separation procedures are used both to determine the composition of mixtures and to purify substances. A large number of separation procedures have been developed. Three of the most common of these are filtration, distillation, and chromatography.

FILTRATION

Solids are readily separated from liquids by passing the mixture through a filter, a barrier with many small openings. This method is called **filtration.** This is a simple method for separating sediment from water in the course of the treatment of water for drinking purposes, because the particles of sand and clay making up the sediment do not pass through the filter.

Filtration is often used in conjunction with procedures that take advantage of different solubilities of substances (that is, differences in the abilities of substances to mix to form solutions). For example, a mixture of 10 g of table salt and 10 g of baking soda can be partially separated by

---

*We shall consider a more sophisticated and probably more satisfactory criterion for defining an element later in this chapter.

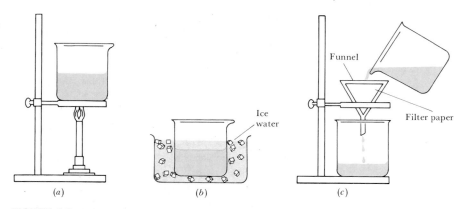

FIGURE 2.2 Schematic representation of a laboratory recrystallization and filtration. (*a*) The solids are dissolved in a minimum amount of water. (*b*) The mixture is cooled. (*c*) The mixture is filtered.

dissolving it in 100 ml of water at about 70°C and then lowering the temperature of the solution to 0°C. The table salt exhibits little change in solubility as its temperature is lowered, and it remains in solution. The baking soda, however, is much less soluble at 0°C than at 70°C, and about 9 g of this substance will separate from solution at the lower temperature. The solution is filtered to remove the baking soda. The procedure is summarized in Figure 2.2.

DISTILLATION

In **distillation,** differences in the volatilities of substances (that is, differences in the ease with which substances form gases) are utilized. Imagine that we wish to remove salt from seawater so that the water can be used

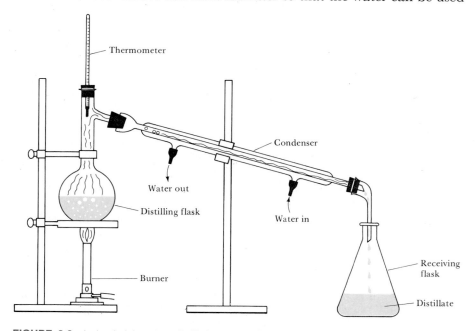

FIGURE 2.3 A simple laboratory distillation setup. Cool water circulating through the jacket of the condenser causes the liquid to condense.

for drinking purposes. The seawater could be heated in an apparatus like that shown in Figure 2.3. Water vaporizes at a much lower temperature than salt, and so the water boils off, leaving a residue of salt in the distilling flask. The water is condensed by cooling elsewhere in the system and collected. The liquid obtained by condensation of vapor in a distillation is known as the **distillate.**

If a mixture of several volatile substances is distilled, the vapor will be richer in the more volatile component. For example, the distillate obtained from the distillation of wine has a higher alcohol content than the wine, because alcohol is more volatile than water. Small percentages of minor components that impart flavor and aroma also are found in the distillate, which is known as brandy. However, a single simple distillation does not effect complete separation of the components of wine. If several components of a mixture have similar volatilities, repeated distillations may be necessary for complete separation.

A fractionating column effects in a single operation what may require several simple distillations. This procedure is called **fractional distillation.** The column, shown in Figure 2.4, has a packing such as glass beads that provides cooling space where part of the vapor condenses as it moves upward from the distilling flask. The condensed liquid is richer in the least volatile component. As the condensed liquid trickles down the beads toward the distillation flask, it comes in contact with fresh vapor moving upward from the flask. Because the vapor is hotter than the liquid on the beads, heat interchange occurs. As a result, the more volatile part of the liquid vaporizes and the less volatile part of the vapor condenses.

Thus the vapor becomes further enriched in the more volatile component. Because many such heat interchanges occur along the column, only the most volatile component or components of the mixture are finally able to reach the condenser and escape.

Fractional distillation is used to separate crude oil into fractions. Commercially this is done using equipment of very sophisticated design. The fractions referred to as gasoline, kerosene, and lubricating oil differ in volatilities. The more volatile component, gasoline, boils in the approximate range of 60–150°C, whereas kerosene and lubricating oil boil in the 150–250°C and 250–350°C ranges, respectively.

CHROMATOGRAPHY

In **chromatography,** separation is achieved by utilization of differences in the degree to which various substances are adsorbed onto the surface of an inert material. (An inert material is one that does not undergo a chemical change.) The difference between adsorption and absorption should be noted; as illustrated in Figure 2.5, adsorption is a surface

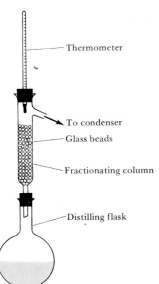

FIGURE 2.4 A distilling flask with a fractionating column attached.

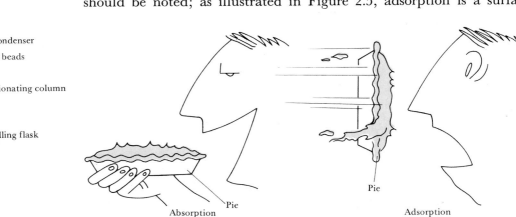

FIGURE 2.5 Adsorption is a surface phenomenon.

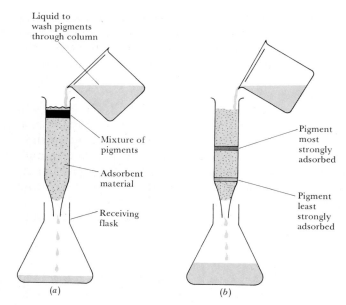

FIGURE 2.6 Separation by column chromatography: (a) initial stage; (b) after some time has elapsed.

phenomenon. The name chromatography arose from its use in separating pigments; it means literally "graphing of colors." For example, a solution containing the colored pigments of a leaf may be washed through a column packed with alumina (aluminum oxide). The various components will move through the column at different speeds due to differences in the degree to which they are adsorbed. This type of separation, illustrated in Figure 2.6, is known as **column chromatography**. By using various instruments instead of visual inspection to determine the location of the components, it is not necessary to restrict the technique to colored materials.

When the adsorbent material is paper and the solution containing the mixture moves upward through the paper, the technique is called **paper chromatography**. As the liquid moves upward on the paper, it carries the mixture along. The components that are adsorbed most strongly to the paper move most slowly. The technique is illustrated in Figure 2.7.

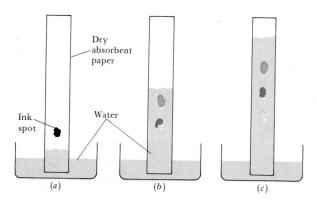

FIGURE 2.7 Separation of ink into components by paper chromatography. (a) Water begins to move up paper. (b) Water moves past ink spot, lifting different components of the ink at different rates. (c) Water has separated ink into three different components.

When the mixture to be separated is a gas swept through a column by an inert gas such as helium, the technique is called **gas chromatography**. The column is coated with a nonvolatile liquid that absorbs the gas. Consequently the technique is also known as gas-liquid chromatography and abbreviated *glc*.

**2.4
pure
substances**

Elements are the basic substances out of which all matter is composed. In light of the seemingly endless variety in our world, it is perhaps surprising that there are only 105 known elements. Not all of these are of equal importance or abundance. Ninety percent, by weight, of the portion of the earth to which we have access for raw materials is composed of only five elements: oxygen, silicon, aluminum, iron, and calcium. Over 90 percent of the human body is composed of just three elements: oxygen, carbon, and hydrogen. At the other extreme, about 20 elements are either found in nature in only minute traces or are man-made and are therefore available in only very small quantities.

Some of the more familiar elements are listed in Table 2.1, together with the chemical symbols used to denote them. All of the known elements are listed on the inside front cover of this text. It can be seen that the abbreviation or symbol for an element consists of one or two letters with the first letter capitalized. These symbols are usually derived from the English name (first and second columns of Table 2.1), but sometimes they are derived instead from a foreign name (third column). You will need to know these symbols and to learn others as we encounter them in the text.

COMPOUNDS

As was mentioned earlier, compounds are substances composed of two or more elements united chemically in definite proportions by mass. We can gain clearer insight into these substances by examining a common example, water. With the discovery of methods of generating electricity, it was found that water could be decomposed into the elements hydrogen and oxygen, as shown in Figure 2.8. Pure water consists of 89 percent oxygen and 11 percent hydrogen by mass, irrespective of its source. Furthermore, the properties of water are clearly unique and much different from those of its constituent elements as seen in Table 2.2. In forming compounds the elements lose their characteristic properties.

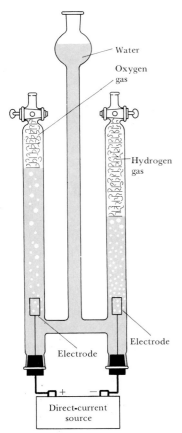

FIGURE 2.8 Decomposition of the compound water into the elements hydrogen and oxygen by passing a direct electric current through it (electrolysis).

TABLE 2.1
some common elements and their symbols

Carbon (C)	Aluminum (Al)	Copper (Cu, from *cuprum*)
Fluorine (F)	Barium (Ba)	Iron (Fe, from *ferrum*)
Hydrogen (H)	Calcium (Ca)	Lead (Pb, from *plumbum*)
Iodine (I)	Chlorine (Cl)	Mercury (Hg, from *hydrargyrum*)
Nitrogen (N)	Helium (He)	Potassium (K, from *kalium*)
Oxygen (O)	Magnesium (Mg)	Silver (Ag, from *argentum*)
Phosphorus (P)	Platinum (Pt)	Sodium (Na, from *natrium*)
Sulfur (S)	Silicon (Si)	Tin (Sn, from *stannum*)

The observation that the elemental composition of a pure compound is always the same is known both as the **law of constant composition** and the **law of definite proportions.** Although this law has been known for over 150 years, the general belief persists among some people that there is a fundamental difference between compounds prepared in the laboratory and the corresponding compounds found in nature. However, a pure compound has the same composition and properties regardless of source. Both man and nature must use the same elements and operate under the same natural laws. Differences in composition and properties between substances indicate that the compounds are not the same or that at least one is impure. Harmful chemicals such as strychnine are made by nature as well as by people. Nature is capable of making compounds such as proteins that man is not yet able to make. Man, on the other hand, has succeeded in preparing many compounds, such as certain pesticides, not found in nature.

TABLE 2.2

comparison of water, hydrogen, and oxygen

	WATER	HYDROGEN	OXYGEN
Physical state[a]	Liquid	Gas	Gas
Normal boiling point	100°C	−253°C	−183°C
Density[a]	1.00 g/ml	0.090 g/l.	1.43 g/l.
Combustible?	No	Yes	No

[a]At room temperature and atmospheric pressure.

SAMPLE EXERCISE 2.2

Identify the following as element, compound, or mixture: milk, gold, table salt, ink.

Solution: Gold (Au) is an element (refer to the table of elements on the back inside cover of the text). Table salt is a compound that we have now mentioned several times. It is composed of the elements sodium (Na) and chlorine (Cl). Both milk and ink are mixtures and are recognized as such by their variable compositions.

2.5
the atomic
theory

The classification of matter as mixtures and substances, compounds and elements, is central to modern chemistry. It helps us to systematize many chemical facts. However, it also raises a number of questions. Why is one element different from another? Why is a compound different from a mixture? Why do elements combine to form compounds? These questions are connected with the facts that we have already discussed. We need a theory to help us explain these facts. We therefore shift our attention now to a discussion of our present theory of matter, the atomic theory. We shall find that this theory helps us answer the questions we have raised and provides us with a mental picture of matter.

The seeds of the atomic theory go back at least to the time of the ancient Greeks. The Greeks pondered a seemingly abstract question: Can matter be divided endlessly into smaller and smaller pieces, or is it composed of some ultimate particle that cannot be further divided? The main line of Greek thought, following the views of Aristotle and Plato, was that matter was continuous. However, some Greek philosophers, notably Democritus, disagreed with this view and argued that matter was composed of small indivisible particles that Democritus called *atomos,* meaning indivisible. This atomic concept was also central to the natural philosophy of the Roman poet and philosopher Lucretius, who lived in the first century B.C. He wrote a famous poem, *De Rerum Natura* (On the Nature of Things), in which he elaborated on the atomic view of matter.

Even if it were granted that matter is atomic in nature, the question arises how the atoms of different substances differ from one another. Lucretius suggested that the atoms of substances that have a bitter taste have barbs on their surfaces that scrape the tongue, whereas the atoms of substances with a bland taste must have a smooth surface. Not much improvement in the atomic view of matter occurred in the 18 centuries following Lucretius. The philosophical ideas of Plato and Aristotle, neither of whom accepted the atomistic view of matter, held sway in European thought for many centuries. Even though the atomic idea was occasionally revived, as in the "corpuscular" philosophy of the Renaissance, early proponents of the particulate theory of matter relied largely on intuition to support their views. During this long period, however, there was a thin, intermittent stream of experimental work. Much of it was prompted by erroneous notions, such as the alchemical belief that common metals such as lead might be transformed into precious metals. Nevertheless, experience of how chemical substances react with one another accumulated, and more quantitative methods of studying chemical reactions were developed. The way was prepared for a new and more meaningful statement of an atomic theory. It came, in the early years of the nineteenth century, from John Dalton, an English schoolteacher (Figure 2.9). Dalton's atomic theory, published in the period

FIGURE 2.9 John Dalton (1766–1844) was the son of a poor English weaver. Because he was a Quaker, Dalton's quiet life-style stands in contrast to the life-styles of Priestley and Lavoisier. Dalton began teaching at the age of 12; he spent most of his years in Manchester, where he taught both grammar school and college. His lifelong interest in meteorology led him to study gases and hence to chemistry and eventually to the atomic theory. It was perhaps because Dalton's training was not in chemistry that he was able to approach problems with a viewpoint different from that of chemists of his time. (*Library of Congress*)

1803–1807, was strongly tied to experimental observation. His efforts were so successful that his theory has dominated our thinking since his time and has had to undergo little revision.

The basic postulates of Dalton's theory were as follows:

1. Each element is composed of extremely small particles called atoms.
2. All atoms of a given element are identical.
3. Atoms of different elements have different properties (including different masses).
4. Atoms of an element are not changed into different types of atoms by chemical reactions; atoms are neither created nor destroyed in chemical reactions.
5. Compounds are formed when atoms of more than one element combine.
6. In a given compound, the relative number and kind of atoms are constant.

This theory provides us with a mental picture of matter. As represented schematically in Figure 2.10, we visualize an element as being composed of tiny particles called atoms. **Atoms** are the basic building blocks of matter; they are the smallest units of an element that can combine with other elements. Compounds involve atoms of two or more elements combined in definite arrangements. Mixtures do not involve the intimate interactions between atoms that are found in compounds.

Dalton's theory embodies several simple laws of chemical combination that were known at the time. Because atoms are neither created nor destroyed in the course of chemical reactions (postulate 4), it is readily evident that matter is neither created nor destroyed in such reactions. Thus we have the **law of conservation of matter**, which was discovered by Lavoisier. This law is one of the principal topics of Chapter 3. The law of constant composition, which was cited in Section 2.4, is explained by postulate 6: In a given compound the relative number and kind of atoms is constant. A third law discovered by Dalton and consistent with his theory is the **law of multiple proportions**: When two elements

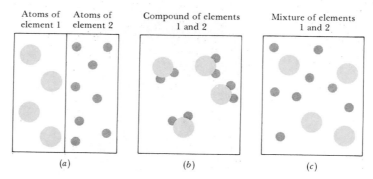

Atoms of   Atoms of   Compound of elements   Mixture of elements
element 1   element 2   1 and 2   1 and 2

(a)   (b)   (c)

FIGURE 2.10 Difference between elements, compounds, and mixtures as visualized through atomic theory. (a) Elements are composed of small particles called atoms. All atoms of a given element are identical. Atoms of different elements are different. (b) Compounds involve atoms of two or more elements combined in definite arrangements. (c) Mixtures have variable compositions. There is no restriction on the relative numbers of atoms of elements 1 and 2.

combine to form more than one compound, the masses of one element that combine with a given mass of the other element are in the ratio of small whole numbers. For example, the substances water and hydrogen peroxide both consist of the elements hydrogen and oxygen. In water there are 8.0 g of oxygen for each gram of hydrogen, whereas in hydrogen peroxide there are 16.0 g of oxygen for each gram of hydrogen. The ratio of the masses of oxygen that combine with a gram of hydrogen in these compounds is in the ratio of the small whole number two: Hydrogen peroxide has twice as much oxygen per unit mass of hydrogen as water does. Using the atomic theory, we understand this to mean that hydrogen peroxide contains twice as many oxygen atoms per hydrogen atom as does water. We now know that water contains one oxygen atom for each two hydrogen atoms, whereas hydrogen peroxide contains two oxygen atoms for each two hydrogen atoms.

Thus we see that the atomic theory ties together many observations and helps us explain them. To our earlier question of what makes one element different from another we can now answer that they have different types of atoms. Unfortunately, this explanation begs a further question: How are the atoms of different elements different from each other? We need to consider the structure of the atom to answer this question. This topic is taken up in the next two sections. We shall see that as we begin to understand the structure of the atom, we shall begin to understand many more aspects of matter.

2.6
the structure of the atom

Dalton and his contemporaries viewed the atom as an indivisible object. However, data slowly accumulated to indicate that the atom had a substructure of smaller particles. Only three of these subatomic particles are of interest to our current understanding of chemistry: the **proton, neutron,** and **electron.**

Protons have a positive charge; neutrons are uncharged; electrons have a negative charge. The charges on the proton and electron are equal in magnitude. Because atoms have no net electrical charge, there are equal numbers of electrons and protons in an atom. The protons and neutrons reside together in a very small volume within the atom known as the **nucleus.** Most of the rest of the atom is empty space in which the electrons move. The electrons are attracted to the nucleus and kept from flying off completely free in space by the attraction that exists between particles of unlike electrical charge (coulombic or electrostatic attraction). * Atoms have diameters on the order of 1 Å (that is, 10^{-10} m)†; the diameters of their nuclei are much smaller, on the order of 10^{-4} Å. If the

*Two bodies of the same charge repel each other. A positively charged body and a negatively charged one will be attracted to each other. The force of the interaction (F) between two charged bodies is given by Coulomb's law: $F = Q_1Q_2/d^2$ where Q_1 and Q_2 are the magnitudes of the charges on bodies 1 and 2, and d is the distance between them. The formula indicates that doubling the magnitude of one of the charges will double the force of attraction or repulsion. Doubling the distance of separation between the charges will reduce the force by one-fourth.

†The angstrom (Å), which is 10^{-10} m, is widely used to describe atomic and molecular dimensions because it permits these dimensions to be expressed without exponents. For example, the diameter of the chlorine atom is about 1.8 Å (1.8×10^{-10} m). However, the angstrom is not an SI unit of measure. In SI units, atomic dimensions are probably most conveniently expressed in picometers or nanometers. In these units, the diameter of the chlorine atom is 180 pm or 0.18 nm.

atom were scaled upward in size so that the nucleus were 2 cm in diameter (about the diameter of a 25-cent piece), the atom would have a diameter of 200 m (about twice the length of a football field). The proton and neutron have approximately equal mass, but the electron weighs only about 1/1835 as much as either of these. Therefore the tiny nucleus carries most of the mass of the atom. Indeed, the density of the nucleus is on the order of 10^{13}–10^{14} g/cm^3. A matchbox full of material of such density would weigh over $2\frac{1}{2}$ billion tons. Astrophysicists have suggested that the matter in the interior of a collapsed star may reach approximately this density before it explodes as a supernova.

Several of the details of the atom and its subatomic particles, which we have just stated, are represented pictorially in Figure 2.11 and are summarized in Table 2.3. Because the electrons are at the periphery of the atom, they play the major role in chemical reactions. In later chapters we shall examine electrons and their behavior more closely.

TABLE 2.3

comparison of the proton, neutron, and electron

PARTICLE	CHARGE	MASS
Proton	Positive (1+)	1.67×10^{-24} g
Neutron	None (neutral)	1.67×10^{-24} g
Electron	Negative (1−)	9.11×10^{-28} g

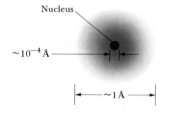

Nucleus

$\sim 10^{-4}$ Å

~ 1 Å

FIGURE 2.11 Schematic representation of the structure of the atom. The nucleus, which contains positive protons and neutral neutrons, is the location of virtually all of the mass of the atom. The rest of the atom is mainly empty space in which the light, negatively charged electrons move.

The identity of an element depends on the number of protons in the nucleus of an atom of that element. In fact we may define an element as a substance whose atoms all have the same number of protons. Thus all atoms of the element carbon have six protons and six electrons. Most also have six neutrons although some have more. Atoms of a given element that differ in number of neutrons, and consequently in mass, are called **isotopes.** The symbol $^{12}_{6}$C or simply ^{12}C (read "carbon twelve," carbon-12) is used to represent the carbon atom with six protons and six neutrons. The number of protons, which is called the **atomic number,** is shown by the subscript. Since all atoms of a given element have the same atomic number, this subscript is redundant and hence often omitted. The superscript is called the **mass number** and is the total number of protons plus neutrons in the atom. Some carbon atoms contain six protons and eight neutrons and are consequently represented as ^{14}C (read carbon-14). Normally, subscripts and superscripts are used with the symbol for an element only when reference is made to a particular isotope of that element. Three isotopes of oxygen and their chemical symbols are shown schematically in Figure 2.12. The term **nuclide** is applied in a general way to a nucleus with a specified number of protons and neutrons. For example, the nucleus of $^{16}_{8}$O is referred to as the $^{16}_{8}$O nuclide.

SAMPLE EXERCISE 2.3

How many protons, neutrons, and electrons are there in ^{197}Au?

Solution: According to the list of elements given in the inside cover of this text, gold has an atomic number of 79. Consequently ^{197}Au has 79 protons, 79 electrons and $197 - 79 = 118$ neutrons.

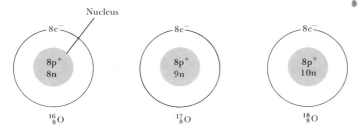

FIGURE 2.12 Some isotopes of oxygen (p^+ = proton, n = neutron, e^- = electron).

SAMPLE EXERCISE 2.4

Write the nuclear isotope symbols for the three isotopes of hydrogen, with mass numbers of 1, 2, and 3.

Solution: Because all three of the hydrogen isotopes must have the same number of protons, 1, the three symbols are: ^1_1H; ^2_1H; ^3_1H.

On the atomic level gold, oxygen, and carbon differ in terms of the number of protons, neutrons, and electrons their respective atoms contain. These subatomic particles, however, are common to all substances. We can therefore state that an atom is the smallest representative sample of an element, because breaking the atom into subatomic particles destroys its identity.

In order to change a base or common metal like lead, atomic number 82, to gold, atomic number 79, requires removal of 3 protons from the nucleus of the lead atom. Because the nucleus is extremely small and buried in the heart of the atom, and because of the very strong binding forces between particles in the nucleus, this removal is exceedingly difficult. The energies required to cause changes in the nucleus are enormously greater than the energies associated with even the most vigorous chemical reactions. Thus, we still agree with Dalton that atoms of an element are not changed into different types of atoms by chemical reactions. Therein lies the futility of the alchemists' attempts to change base metals to gold.

2.7
the structure of the atom: a historical perspective

The concept of an atom and more especially the model of an atom built of protons, neutrons, and electrons that we have been discussing is not self-evident to even the most careful observer of nature. The model is the result of a wide variety of experiments. For reasons of economy of space and time, we shall examine only three of these: the studies of cathode rays, of radioactivity, and of the scattering of alpha particles by thin metal foils.

CATHODE RAYS AND ELECTRONS

In the mid-1800s a number of investigators began to study electrical discharge through evacuated tubes. Radiation is produced within such tubes when voltages become high enough to permit current flow (about 1000 volts). This radiation became known as **cathode rays** because it emanated from the negative electrode, or cathode. This and a number of additional facts suggested that the radiation consisted of a stream of

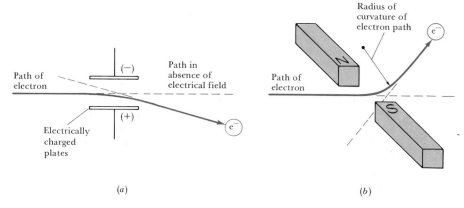

Path of
electron

(−)

Path in
absence of
electrical field

Electrically
charged
plates

(+)

e⁻

(a)

Radius of
curvature of
electron path

N

Path of
electron

e⁻

S

(b)

FIGURE 2.13 The behavior of a negatively charged particle such as an electron moving through an electric field (a) and a magnetic field (b).

negatively charged particles that were named "electrons." For example, the rays travel in straight lines in the absence of magnetic or electric fields. However, they are deflected by magnetic and electric fields in a manner expected for negatively charged particles. The behavior of a negatively charged particle in a magnetic field and in an electric field is shown in Figure 2.13. The movement of the rays can be determined because of their ability to cause certain materials, including glass, to give off light, or fluoresce. In fact, a television picture tube is a cathode-ray tube; the television picture results from fluorescence from the television screen. Because the rays (electrons) were found to be independent of the nature of the cathode material, it was deduced that they are a basic component of all matter.

In 1897 the British physicist J. J. Thomson (Figure 2.14) was able to

FIGURE 2.14 J. J. Thomson (1856–1940). (*Copyright Cavendish Laboratory, University of Cambridge*)

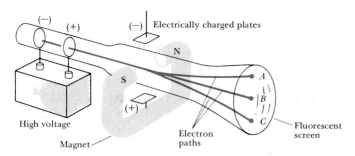

(−) (+) (−) | Electrically charged plates

N

S

(+)

High voltage

Magnet

Electron paths

A
B
C

Fluorescent screen

FIGURE 2.15 Cathode-ray tube with perpendicular magnetic and electric fields.

measure the ratio of the electrical charge to the mass of the electron using a cathode-ray tube such as that shown schematically in Figure 2.15. When only the magnetic field is turned on, the electron strikes point A of the tube. When the magnetic field is off and the electric field is on, the electron strikes point C. When both the magnetic and electric fields are off or when they are balanced so as to cancel each other's effects, the electron strikes point B. By carefully and quantitatively determining the effects of magnetic and electric fields on the motion of the cathode rays, Thomson was able to determine the charge-to-mass ratio of 1.76×10^8 coul/g.*

The external force exerted on an electron moving in a magnetic field is given by Hev, where H is the strength of the magnetic field, e is the charge on the electron, and v is its velocity. The motion of the electron is determined by the balance of this force and the tendency of the electron to continue its straight-line motion (the centrifugal force on the electron). This latter force is given by the relation mv^2/r, where m is the mass of the electron, v is its velocity, and r is the radius of the curved path taken by the electron as it moves through the magnetic field (refer to Figure 2.13). Thus the path taken by the electron must be consistent with the equation

$$Hev = \frac{mv^2}{r}$$

By rearranging this equation, we see that the radius of the path of the electron moving through the magnetic field is given by $r = mv^2/Hev = mv/He$. The larger the radius, the smaller the deflection of the particle from the straight-line path that it would take in the absence of the magnetic field. Thus, the more massive the particle, the greater its velocity, the

smaller the magnetic field, and/or the smaller charge on the particle, the greater the tendency of the particle to continue its straight-line motion.

The relation given above can also be arranged into the form $e/m = v/rH$. The velocity of the electron can be determined by balancing the effects of the magnetic field against a perpendicular electric field so that the electron moves in a straight line. The magnitude of the force exerted by the electric field on the electron is given by Ee where E is the strength of the electric field. For the electron moving in a straight line through the balanced magnetic and electric fields

$$Hev = Ee$$

so that

$$v = \frac{E}{H}$$

Thus, from the first relation, $e/m = v/rH = E/H^2r$, where all of the quantities on the right side of the equation can be determined experimentally. Thus, Thomson's experiment permitted determination of the ratio e/m for the electron.

In 1909, Robert Millikan of the University of Chicago determined the charge on the electron by measuring the effect of an electric field on the rate at which charged oil droplets fall under the influence of gravity. The apparatus that he used is shown schematically in Figure 2.16. The

*The coulomb (coul or C) is the SI unit for electrical charge.

rate at which the droplets fall in air is determined by their size and mass. By watching a particular droplet, Millikan could measure its rate of fall and calculate from this its mass. The experiment was arranged so that a source of radioactivity was near the droplets. The radioactive source caused charges to form. The charged particles floating around in the air would often become attached to an oil droplet. When an electrical charge was applied to the plates, charged oil droplets in the region between the plate would be acted upon by the electric field. Their fall would be either accelerated or retarded, depending on the charge on the droplet and the polarity of the voltage applied to the plates. By carefully measuring the effects of the electrical field on the movements of many droplets, Millikan found that the charge on the oil drops was always an integral multiple of 1.60×10^{-19} coul, which he deduced was the charge of the electron. The mass of the electron, 9.11×10^{-28} g, was then calculated by combining Millikan's value of the charge with Thomson's charge-to-mass ratio:

$$\text{Mass} = \frac{1.60 \times 10^{-19} \text{ coul}}{1.76 \times 10^{8} \text{ coul/g}} = 9.11 \times 10^{-28} \text{ g}$$

RADIOACTIVITY

The discovery of radioactivity by the French scientist Henri Becquerel in 1896 provided additional evidence for the complexity of the atom. Becquerel's imagination had been captured by W. C. Roentgen's discovery of X rays, which had been reported in January 1896. Roentgen had been quick to grasp the practical importance of his discovery, and within a short time X rays had been used in medicine. Members of the international scientific community also sensed that this was something big. Becquerel was well aware that certain substances, upon exposure to sunlight, become luminous, a phenomenon referred to as fluorescence. He sought to determine whether such fluorescent substances gave off X rays. In his initial experiments, Becquerel chose to work with a fluorescent uranium mineral. He placed this in the sunlight over a photographic plate that had been carefully wrapped to protect it from the direct radiation of the sun. When the plate was developed, he found the image of the mineral on the plate. Toward the end of February 1896, Becquerel incorrectly reported that penetrating rays, presumably X rays,

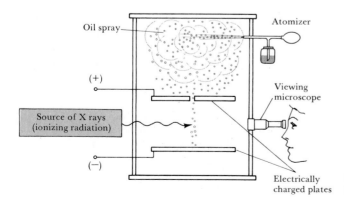

FIGURE 2.16 Schematic representation of Millikan's apparatus for studying the rate of fall of oil droplets.

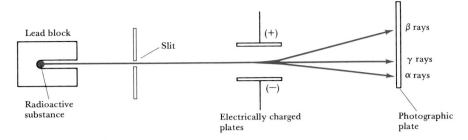

FIGURE 2.17 Behavior of alpha (α), beta (β), and gamma (γ) rays in an electric field.

could be induced by sunlight and emitted as part of fluorescence. However, the weather turned bad, and Becquerel had to postpone further studies. While the sun stayed behind the clouds, Becquerel kept the mineral and the wrapped photographic plate in a desk drawer. On March 1, 1896, he decided to develop the plate, not expecting to find any images. He was surprised to find very intense silhouettes. Becquerel concluded correctly this time that the mineral was producing a spontaneous radiation and referred to this phenomenon as **radioactivity.**

Further study of the nature of this radiation, principally by the British scientist Ernest Rutherford, revealed three types of radiation—alpha (α), beta (β), and gamma (γ) radiation. Each type differed in electrical behavior and penetrating ability.* The behavior of these three types of radiation in an electric field is shown in Figure 2.17. The α rays are streams of helium nuclei (known as α particles) and as such each particle bears a 2+ charge; they are stopped by paper. The β radiation consists of streams of high-speed electrons (known as β particles) and hence each particle has a 1− charge; they have 100 times the penetrating ability of α particles. The γ rays are high-energy radiation like X rays; they do not consist of particles. The γ rays have 1000 times the penetrating ability of α particles. These comparisons are summarized in Table 2.4.

RUTHERFORD AND THE NUCLEAR ATOM

By 1909 Rutherford had firmly established that α rays consisted of helium nuclei with a 2+ charge. Once he had unraveled the nature of α rays, he began to use them to study the structure of the atom. By this time it was well accepted that the atom was electrical in nature and contained electrons. The prevalent model of the atom, as developed by J. J. Thomson, pictured the atom as a cloud of positive charge in which negatively charged electrons were embedded like seeds in a watermelon.

In 1910 Rutherford and his co-workers performed an experiment

*While Rutherford was extending Becquerel's discovery by investigating the nature of the radiation, Marie Slodowska Curie, a Polish student working in Paris, took Becquerel's suggestion and began to search for radioactivity in other substances. When Mme. Curie presented her doctoral thesis, it was described as the greatest single contribution of any doctoral thesis in the history of science. Among other things, two new elements, polonium and radium, had been discovered. In 1903, Becquerel, Mme. Curie, and her husband, Pierre, were jointly awarded the Nobel Prize in physics. In 1908, Rutherford received the prize, and in 1911 Mme. Curie won a second Nobel Prize, this time in chemistry.

TABLE 2.4

summary of the properties of alpha, beta, and gamma rays

| | TYPE OF RADIATION | | |
	α	β	γ
Charge	2+	1−	0
Mass	6.64×10^{-24} g	9.11×10^{-28} g	0
Relative penetrating power	1	100	1000
Identity	$_{2}^{4}$He nuclei	Electrons	High-energy radiation

that led to the downfall of Thomson's model. Rutherford was studying the manner of scattering of a narrow beam of α particles as they passed through a thin gold foil. He had found slight scattering, on the order of 1 degree, which was consistent with Thomson's model. One day Hans Geiger, an associate of Rutherford's, suggested that Ernest Marsden, a 20-year-old undergraduate working in their laboratory, get some experience in conducting such experiments. Rutherford suggested that Marsden see if α particles were scattered through large angles. In his own words:

> I may tell you in confidence that I did not believe they would be since we knew that the α particle was a very massive particle with a great deal of energy. . . . Then I remember two or three days later Geiger coming to me in great excitement and saying, "We have been able to get some α particles coming backwards." . . . It was quite the most incredible event that has ever happened to me in my life. It was almost as if you fired a 15-inch shell into a piece of tissue paper and it came back and hit you.

What Rutherford and his co-workers had observed was that the vast majority of α particles passed directly through the foil without deflection. Only a few underwent deflection, some even bouncing back in the direction from which they had come, as shown in Figure 2.18.

By 1911 Rutherford was able to explain these observations as follows: Most α particles pass directly through the foil because most of the atom is

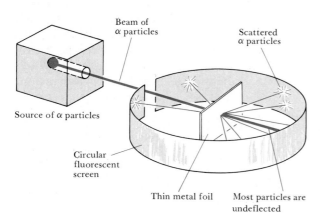

Beam of α particles

Scattered α particles

Source of α particles

Circular fluorescent screen

Thin metal foil

Most particles are undeflected

FIGURE 2.18 Rutherford's experiment on the scattering of α particles.

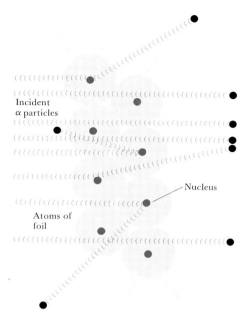

FIGURE 2.19 Rutherford's model explaining his experiment with α particles (the metal foil is actually several thousand atoms thick).

Incident
α particles

Nucleus

Atoms of
foil

empty space. The few α particles that undergo deflection do so because they have come close to the small nucleus of the gold atom in which the positive charge of the atom resides. Because the gold atom has 79 protons, the repulsion between the nucleus with a 79+ charge and the α particle coming directly at it is strong enough to deflect the α particle backward in space. This is represented schematically in Figure 2.19. It was in this way that the concept of the nuclear atom was born.

**2.8
the periodic
table:
a preview**

Many elements show very strong similarities to each other. For example, lithium (Li), sodium (Na), and potassium (K) are all soft, very reactive metals. The elements helium (He), neon (Ne), and argon (Ar) are very nonreactive gases. If the elements are arranged in order of increasing atomic number, their chemical and physical properties are found to show a repeating or periodic pattern. For example, each of the soft, reactive metals, lithium, sodium, and potassium, comes immediately after one of the nonreactive gases, helium, neon, and argon, as shown in Figure 2.20. The arrangement of elements in order of increasing atomic number, with elements having similar properties placed in vertical columns, is known

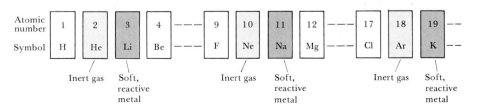

FIGURE 2.20 Arrangement of elements by atomic number illustrating the periodic or repeating pattern in properties that is the basis of the periodic table.

as the **periodic table**. The periodic table is shown in Figure 2.21 and is also given on the front inside cover of your text for easy reference. In most classrooms where chemistry is taught, large periodic tables are hung on the walls—a testimony to their usefulness.

The elements in a column of the periodic table are known as a **family** or **group**. They are identified as group 1A, 2A, and so forth, as shown at the top of the periodic table. For example, three familiar elements that have similar properties are copper (Cu), silver (Ag), and gold (Au), which occur together in group 1B. Some groups are also described by a family name. The members of group 1A—lithium, sodium, potassium, rubidium (Rb), cesium (Cs), and francium (Fr)—are known as the **alkali metals.** The members of group 2A—berylium (Be), magnesium (Mg), calcium (Ca), strontium (Sr), barium (Ba), and radium (Ra)—are known as the **alkaline earth metals.** The members of group 7A—fluorine (F), chlorine (Cl), bromine (Br), iodine (I), and astatine (At)—are known as the **halogens.** The members of group 8A—helium (He), neon (Ne), argon (Ar), krypton (Kr), xenon (Xe), and radon (Rn)—are known as the **noble gases, inert gases** or **rare gases.**

The elements in a family of the periodic table have similar properties because they have the same type of arrangement of electrons at the periphery of their atoms. We shall postpone a discussion of electronic arrangements until Chapters 6 and 7. We shall take a much closer look at the periodic table in Chapter 7. Meanwhile, however, we can use it to correlate the behaviors of elements and to help us remember many facts. You will find the table to be of great importance as you study the remainder of this chapter.

SAMPLE EXERCISE 2.5

Which of the following elements would you expect to show the greatest similarity in chemical and physical properties: Li, Be, F, S, Cl?

Solution: The elements F and Cl should be most alike because they are in the same family (group 7A, the halogen family).

One pattern that is evident when elements are arranged in the periodic table is the grouping together of the **metallic elements.** These elements, which are grouped together on the left side of the periodic table, share many characteristic properties, such as luster and high electrical and heat conductivity. The metallic elements are separated from the **nonmetallic elements** by the diagonal line that runs across the right side of the periodic table from boron (B) to astatine. Note that the majority of the elements are metallic. The nonmetals, which occupy the upper right side of the periodic table, are gases, liquids, or crystalline solids. They lack those physical characteristics that distinguish the metallic elements. Many of the elements that lie along the line that separates metals from nonmetals, such as antimony (Sb), possess properties intermediate between those of metals and nonmetals. These elements are often referred to as **semimetals** or **metalloids.**

FIGURE 2.21 The periodic table of elements.

Active metals

Transition metals

Nonmetals

1A	2A	3B	4B	5B	6B	7B		8B		1B	2B	3A	4A	5A	6A	7A	8A
1 H 1.00797																	2 He 4.0026
3 Li 6.939	4 Be 9.0122											5 B 10.811	6 C 12.01115	7 N 14.0067	8 O 15.9994	9 F 18.9984	10 Ne 20.183
11 Na 22.9898	12 Mg 24.312											13 Al 26.9815	14 Si 28.086	15 P 30.9738	16 S 32.064	17 Cl 35.453	18 Ar 39.948
19 K 39.098	20 Ca 40.08	21 Sc 44.956	22 Ti 47.90	23 V 50.942	24 Cr 51.996	25 Mn 54.9380	26 Fe 55.847	27 Co 58.9332	28 Ni 58.71	29 Cu 63.54	30 Zn 65.37	31 Ga 69.72	32 Ge 72.59	33 As 74.9216	34 Se 78.96	35 Br 79.909	36 Kr 83.80
37 Rb 85.47	38 Sr 87.62	39 Y 88.905	40 Zr 91.22	41 Nb 92.906	42 Mo 95.94	43 Tc (99)	44 Ru 101.07	45 Rh 102.905	46 Pd 106.4	47 Ag 107.870	48 Cd 112.41	49 In 114.82	50 Sn 118.69	51 Sb 121.75	52 Te 127.60	53 I 126.9044	54 Xe 131.30
55 Cs 132.905	56 Ba 137.33	57 *La 138.91	72 Hf 178.49	73 Ta 180.948	74 W 183.85	75 Re 186.2	76 Os 190.2	77 Ir 192.2	78 Pt 195.09	79 Au 196.967	80 Hg 200.59	81 Tl 204.37	82 Pb 207.19	83 Bi 208.980	84 Po (210)	85 At (210)	86 Rn (222)
87 Fr (223)	88 Ra (226)	89 †Ac (227)	104 Rf (257)	105 Ha (260)													

*Lanthanide series

58 Ce 140.12	59 Pr 140.907	60 Nd 144.24	61 Pm (147)	62 Sm 150.35	63 Eu 151.96	64 Gd 157.25	65 Tb 158.924	66 Dy 162.50	67 Ho 164.930	68 Er 167.26	69 Tm 168.934	70 Yb 173.04	71 Lu 174.97

†Actinide series

90 Th 232.038	91 Pa (231)	92 U 238.03	93 Np (237)	94 Pu (242)	95 Am (243)	96 Cm (247)	97 Bk (247)	98 Cf (249)	99 Es (254)	100 Fm (253)	101 Md (256)	102 No (253)	103 Lw (257)

45

2.9

molecules and ions

We have seen that the atom is the smallest representative sample of an element. However, elements do not necessarily exist as single atoms. In fact, only the noble gas elements are normally found as isolated atoms. Most matter is composed of molecules or ions, which are formed from atoms.

MOLECULES

Molecules are composed of combinations of tightly bound atoms. The resultant assembly or package of atoms behaves in many ways as a single object, just as a television set composed of many parts can be recognized as a single object. The nature of the forces (bonds) that bind the atoms together will be examined in Chapter 8.

When elements exist in molecular form, they contain only one type of atom. They are said to be **homoatomic.** For instance, the element oxygen, as it is normally found in air, consists of molecules composed of pairs of oxygen atoms. This molecular form of oxygen is represented by the **chemical formula** O_2 (read O-two). The subscript in this formula indicates that two oxygen atoms are present in each molecule. The molecule is said to be **diatomic.** Oxygen also exists in a form known as ozone, which consists of three bound oxygen atoms. Correspondingly, the chemical formula for ozone is O_3. Ozone and "normal" oxygen exhibit quite different chemical properties.

The elements that normally occur as diatomic molecules are hydrogen, oxygen, nitrogen, and the halogens. Their location in the periodic table is shown in Figure 2.22. When we speak of the substance hydrogen, we mean H_2 unless we indicate explicitly otherwise. Likewise, when we speak of oxygen, nitrogen, or any of the halogens, we are referring to O_2, N_2, F_2, Cl_2, Br_2, or I_2. Thus the properties of oxygen and hydrogen listed earlier in Table 2.2 are those of O_2 and H_2. In other forms, these elements behave much differently.

Molecules of compounds contain more than one type of atom; they are said to be **heteroatomic.** For example, a molecule of water consists of two hydrogen atoms and one oxygen atom. It is therefore represented by the chemical formula H_2O (read H-two-O). Another compound com-

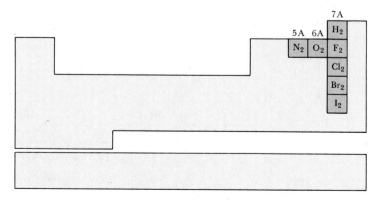

FIGURE 2.22 Elements that exist as diatomic molecules.

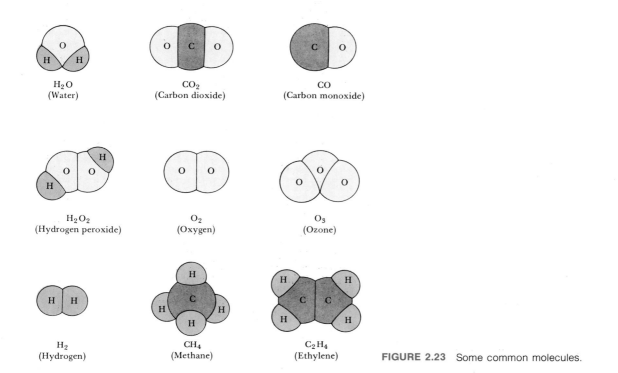

H_2O
(Water)

CO_2
(Carbon dioxide)

CO
(Carbon monoxide)

H_2O_2
(Hydrogen peroxide)

O_2
(Oxygen)

O_3
(Ozone)

H_2
(Hydrogen)

CH_4
(Methane)

C_2H_4
(Ethylene)

FIGURE 2.23 Some common molecules.

posed of these same elements is hydrogen peroxide, H_2O_2. These two compounds have quite different properties. A number of common molecules are shown in Figure 2.23. Pay close attention to how the chemical formula of each molecule reflects its composition.

SAMPLE EXERCISE 2.6

Which of the molecules shown in Figure 2.23 are compounds and which are elements?

Solution: The compounds are H_2O, CO_2, CO, H_2O_2, CH_4, and C_2H_4. The elements are O_2, O_3, and H_2 (one type of atom).

Chemical formulas such as those shown in Figure 2.23, which indicate the actual numbers and types of atoms within a molecule, are known as **molecular formulas.** It is easier to determine the relative number of each type of atom in a molecule than the actual number of atoms. A chemical formula that gives only the relative number of atoms of each type in a molecule is known as an **empirical** or **simplest formula.** The subscripts in such a formula are smallest whole-number ratios. For example, hydrogen peroxide, whose molecular formula is H_2O_2, has an empirical formula of HO; ethylene, whose molecular formula is C_2H_4, has an empirical formula of CH_2.

Glucose, a substance also known as blood sugar and as dextrose, has the chemical formula $C_6H_{12}O_6$. Is this the empirical formula for glucose?

Solution: $C_6H_{12}O_6$ is not the empirical formula, because the empirical formula has subscripts that are smallest whole-number ratios. The smallest ratios are obtained by dividing each subscript by the largest common factor, in this case 6. The empirical formula for glucose is CH_2O. Some formulas, such as H_2O and CO_2, are both empirical and molecular formulas.

In experimentally determining the molecular formula for a substance, its empirical formula is determined first. The total mass of the molecule is then determined. From the empirical formula and the mass of the molecule, its molecular formula can be determined. But now we're getting ahead of ourselves; we'll consider how empirical and molecular formulas are determined when we get to Section 3.9.

Often the chemical formulas of molecules are written so as to show the relative arrangements of the atoms. For example, the formulas for water and hydrogen peroxide can be written as follows:

Such formulas are known as **structural formulas.** The lines between the symbols for the elements represent the bonds that hold the respective atoms together.

IONS

Whereas the nucleus of an atom is unchanged by ordinary chemical processes, atoms readily gain or lose electrons. These transactions take place at the periphery of the atom. If electrons are removed or added to a neutral atom, a charged particle called an **ion** is formed. For example, the sodium atom, which has 11 protons and 11 electrons, can lose an electron. The resulting ion has 11 protons and 10 electrons; its net charge is consequently $1+$. This ion is symbolically represented as Na^+. The net charge on an ion is represented by a superscript; $+$, $2+$, and $3+$ mean a net charge resulting from loss of one, two, or three electrons, respectively. The superscripts $-$, $2-$, and $3-$ represent net charges resulting from gain of one, two, or three electrons, respectively. The formation of the Na^+ ion from a Na atom is shown schematically below:

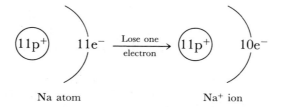

Chlorine, with 17 protons and 17 electrons, can gain an electron in chemical reactions, producing the Cl^- ion:

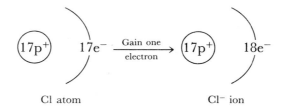

Cl atom　　　　　　　　Cl⁻ ion

In general, metal atoms lose electrons most readily, whereas nonmetal atoms tend to gain electrons.

Groups of atoms joined together as in a molecule but having a net positive or negative charge are known as **polyatomic ions.** An example is NO_3^-, known as the nitrate ion. We shall consider further examples in the next section.

The chemical properties of ions are greatly different from those of the atoms from which they are derived. The change of an atom or molecule to an ion is like that from Dr. Jekyll to Mr. Hyde: Although the body may be essentially the same (plus or minus a few electrons), the personality is much different.

Many atoms gain or lose electrons so as to end up with the same number of electrons as the closest noble gas. The members of the noble gas family are chemically very nonreactive and form very few compounds. We might deduce that this is because they have very stable electron arrangements. Nearby elements can obtain these same stable arrangements by losing or gaining electrons. For example, loss of one electron from an atom of sodium leaves it with the same number of electrons as the neutral neon atom (atomic number 10). Similarly, when chlorine gains an electron it ends up with 18, the same as argon (atomic number 18). We shall content ourselves with this simple observation in explaining the formation of ions until later chapters in which we consider chemical bonding.

SAMPLE EXERCISE 2.7

Predict the charges expected for the most stable ions of barium and oxygen.

Solution: Barium has atomic number 56. The nearest noble gas is xenon, atomic number 54. Barium can obtain the stable arrangement of 54 electrons by losing two of its electrons, thereby forming the Ba^{2+} ion.

Oxygen has atomic number 8. The nearest noble gas is neon, atomic number 10. Oxygen can obtain this stable electron arrangement by gaining two electrons, thereby forming an ion of 2− charge, O^{2-}.

Some compounds are composed of positively and negatively charged ions instead of heteroatomic molecules. For example, NaCl, which is ordinary table salt, consists of equal numbers of Na^+ and Cl^- ions. These ions are arranged in a three-dimensional array as shown in Figure 2.24. Because there is one Na^+ ion for each Cl^- ion, the overall structure has no net charge; the compound is neutral. Because there is no discrete molecule of NaCl, we are able to write only an empirical formula. This same situation is found with other compounds that are composed of ions.

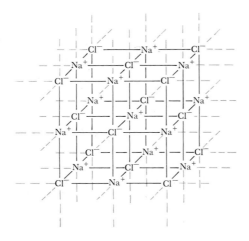

FIGURE 2.24 The arrangement of ions in sodium chloride (NaCl).

2.10

naming of inorganic compounds

The number of distinct, known compounds is in the millions. It is therefore important that there be some systematic way of naming them. The situation would be hopelessly complicated if each compound had a special name independent of all others. Therefore, a series of relatively simple rules has been developed to name compounds. Nevertheless, many simple compounds such as H_2O (water) and NH_3 (ammonia) are still identified by their traditional names.

In this section we shall consider the basic rules for naming inorganic compounds, the compounds that early chemists associated with the nonliving portion of our world.* You will have frequent occasion to refer back to this section as you encounter new substances and as your knowledge of chemistry grows.

The names of compounds that are composed of ions are derived from the names of the ions. The common ions are listed in Table 2.5, together with their names. You may find it necessary to memorize this list. However, there are some general rules that will make your task easier. Consider first the names of the positive ions, which we call **cations**.

CATIONS

Monatomic cations take the name of the element:

    Na^+ sodium ion Zn^{2+} zinc ion

If an element can form more than one positive ion, the positive charge of the ion is indicated by a Roman numeral in parentheses following the name of the metal:

    Fe^{2+} iron(II) ion Cu^+ copper(I) ion
    Fe^{3+} iron(III) ion Cu^{2+} copper(II) ion

*The classification of compounds as either organic or inorganic is one of the oldest classification schemes in chemistry. Organic compounds were first associated totally with plants and animals; however, a great number of organic compounds have now been prepared that do not occur in nature. All organic compounds contain carbon, usually in combination with hydrogen, oxygen, nitrogen, or sulfur. We shall discuss the chemistry of organic compounds in Chapter 24, although we shall have many occasions to illustrate chemical principles using these compounds as examples prior to that time.

TABLE 2.5

common ions

POSITIVE IONS (CATIONS)	NEGATIVE IONS (ANIONS)
Ammonium (NH_4^+)	Acetate ($C_2H_3O_2^-$)
Copper(I) or cuprous (Cu^+)	Bromide (Br^-)
Hydrogen (H^+)	Chloride (Cl^-)
Silver (Ag^+)	Chlorate (ClO_3^-)
Sodium (Na^+)	Cyanide (CN^-)
Potassium (K^+)	Fluoride (F^-)
Barium (Ba^{2+})	Hydrogen carbonate (HCO_3^-)
Calcium (Ca^{2+})	or bicarbonate
Cobalt(II) or cobaltous (Co^{2+})	Hydrogen sulfate (HSO_4^-)
Copper(II) or cupric (Cu^{2+})	or bisulfate
Iron(II) or ferrous (Fe^{2+})	Hydroxide (OH^-)
Lead(II) or plumbous (Pb^{2+})	Iodide (I^-)
Magnesium (Mg^{2+})	Nitrate (NO_3^-)
Manganese(II) or manganous (Mn^{2+})	Perchlorate (ClO_4^-)
Mercury(II) or mercuric (Hg^{2+})	Permanganate (MnO_4^-)
Tin(II) or stannous (Sn^{2+})	Carbonate (CO_3^{2-})
Zinc (Zn^{2+})	Chromate (CrO_4^{2-})
Aluminum (Al^{3+})	Oxide (O^{2-})
Chromium(III) or chromic (Cr^{3+})	Phosphate (PO_4^{3-})
Iron(III) or ferric (Fe^{3+})	Sulfate (SO_4^{2-})
	Sulfide (S^{2-})
	Sulfite (SO_3^{2-})

An older method still widely used for distinguishing between two differently charged ions of a metal is to use the endings *-ous* or *-ic;* these endings represent the lower and higher charged ions, respectively. They are used together with the root of the Latin name of the element:

Fe^{2+}	ferrous ion	Cu^+	cuprous ion
Fe^{3+}	ferric ion	Cu^{2+}	cupric ion

The only common polyatomic cations are those given below:

NH_4^+ ammonium ion Hg_2^{2+} mercury(I) or mercurous ion

The name mercury(I) ion is given to Hg_2^{2+} because it can be considered to consist of two Hg^+ ions. Mercury also occurs as the monatomic Hg^{2+} ion, which is known as the mercury(II) or mercuric ion.

ANIONS

Negative ions are called **anions.** Monatomic anions are named by dropping the ending of the name of the element and adding the ending *-ide:*

H^-	hydride ion	O^{2-}	oxide ion	N^{3-}	nitride ion
F^-	fluoride ion	S^{2-}	sulfide ion	P^{3-}	phosphide ion

Only a few common polyatomic ions end in *-ide:*

OH^- hydroxide ion CN^- cyanide ion O_2^{2-} peroxide ion

The endings *-ite* and *-ate* are used with anions containing oxygen, which are known as **oxyanions.** When an element has two oxyanions, the name of the one that contains the most oxygen ends in *-ate;* the name of the one with less oxygen ends in *-ite:*

NO_2^- nitrite ion $\qquad$ SO_3^{2-} sulfite ion

NO_3^- nitrate ion $\qquad$ SO_4^{2-} sulfate ion

When the series of anions of a given element extends to three or four members, as with the oxyanions of the halogens, prefixes are also employed. The prefix *hypo-* indicates less oxygen, whereas the prefix *per-* indicates more oxygen:

ClO^- $\quad$ hypochlorite ion (less oxygen than chlorite)

ClO_2^- $\quad$ chlorite ion

ClO_3^- $\quad$ chlorate ion

ClO_4^- $\quad$ perchlorate ion (more oxygen than chlorate)

Many polyatomic anions that have high charges readily add one or more hydrogen ions to form anions of lower charge. These ions are named by prefixing the word hydrogen or dihydrogen, as appropriate, to the name of the hydrogen-free anion. An older method, which is still used, is to use the prefix *bi-:*

HCO_3^- $\quad$ hydrogen carbonate (or bicarbonate) ion

HSO_4^- $\quad$ hydrogen sulfate (or bisulfate) ion

$H_2PO_4^-$ $\quad$ dihydrogen phosphate ion

IONIC COMPOUNDS

Compounds composed of ions are named by giving the name of the positive ion first:

NaCl $\qquad$ sodium chloride

$Cu(NO_3)_2$ copper(II) nitrate

Notice how parentheses are used in writing the formula for the second compound, indicating the presence of two NO_3^- ions. Rewriting the formula to show the charges on the ions, we have $Cu^{2+}(NO_3^-)_2$. This shows the 2+ charge of the copper(II) ion, which balances the 2− charge of the two NO_3^- ions. The total positive and negative charges of the ions are equal (2+ and 2−), making the compound neutral, as it must be.

SAMPLE EXERCISE 2.8

Name the following compounds: (a) K_2SO_4; (b) $Ba(OH)_2$.

Solution: (a) Because K^+ is the potassium ion and SO_4^{2-} is the sulfate ion, the name of this compound is potassium sulfate. (b) Because Ba^{2+} is the barium ion and OH^- is the hydroxide ion, the name of this compound is barium hydroxide.

SAMPLE EXERCISE 2.9

Many common toothpastes contain stannous fluoride, and foods can be enriched in iron by addition of iron(II) carbonate (ferrous carbonate). Write the chemical formulas for these two compounds.

Solution: The stannous ion, also known as the tin(II) ion, is Sn^{2+}. The fluoride ion is F^-. Two F^- ions are needed to balance the positive charge of Sn^{2+}, forming a neutral compound. The formula is therefore SnF_2.

The iron(II) ion is Fe^{2+}, while the carbonate ion is $CO_3{}^{2-}$. The compound is $FeCO_3$, one Fe^{2+} ion balancing the charge of one $CO_3{}^{2-}$ ion.

These examples should indicate the importance of remembering the charges of the common ions.

ACIDS

There is an important class of compounds known as acids which are named a special way. These compounds will be discussed further in Section 3.3 and then extensively considered in Chapter 16. These compounds all contain hydrogen. Their names are related to those of the ions that we have already discussed. Anions whose names end in *-ide* have associated acids whose names have a *hydro-* prefix and an *-ic* ending:

Cl^- chloride ion—HCl hydrochloric acid
S^{2-} sulfide ion—H_2S hydrosulfuric acid

Anions whose names end in *-ate* have associated acids whose names end in *-ic;* anions whose names end in *-ite* have associated acids whose names end in *-ous:*

ClO^- hypochlorite ion—$HClO$ hypochlorous acid
$ClO_2{}^-$ chlorite ion—$HClO_2$ chlorous acid
$ClO_3{}^-$ chlorate ion—$HClO_3$ chloric acid
$ClO_4{}^-$ perchlorate ion—$HClO_4$ perchloric acid

SAMPLE EXERCISE 2.10

Name the following acids: (a) HNO_3; (b) H_2SO_4; (c) H_2SO_3.

Solution: (a) Because $NO_3{}^-$ is the nitrate ion, HNO_3 is nitric acid. (b) Because $SO_4{}^{2-}$ is the sulfate ion, H_2SO_4 is sulfuric acid. (c) Because $SO_3{}^{2-}$ is the sulfite ion, H_2SO_3 is sulfurous acid.

OTHER COMPOUNDS

We shall consider most other rules for naming compounds as the need arises. We shall also occasionally encounter common names that are still widely used for certain compounds, especially in applied science. When we do, we shall also give their systematic names. However, before we end this discussion, we shall consider the systematic names of compounds

TABLE 2.6

prefixes used in naming multiple compounds formed between nonmetals

PREFIX	MEANING
Mono-	1
Di-	2
Tri-	3
Tetra-	4
Penta-	5
Hexa-	6

formed between two nonmetal elements. These compounds are named largely as if the compounds were ionic:

NF_3 nitrogen fluoride

When several compounds form between the same two nonmetals, the prefixes listed in Table 2.6 are used to describe the varying number of atoms of one of the elements:

CO	carbon monoxide	SO_2	sulfur dioxide
CO_2	carbon dioxide	SO_3	sulfur trioxide

FOR REVIEW

summary

Matter exists in three states: **gas, liquid,** and **solid.** Most matter consists of a **mixture** of substances. Mixtures can be either **homogeneous** or **heterogeneous;** homogeneous mixtures are called **solutions.** Mixtures can be resolved into two types of **pure substances: compounds** and **elements.** A variety of techniques can be used to separate mixtures into their components. We considered three of these: **filtration, distillation,** and **chromatography.** Each substance has a unique set of **chemical** and **physical properties** that can be used to identify it.

Atoms are the basic building blocks of matter; they are the smallest units of an element that can combine with other elements. Atoms are composed of a **nucleus** (containing **protons** and **neutrons**) and **electrons** that move around the nucleus. We considered some of the historically significant experiments that led to this model of the atom: Thomson's experiments on the behavior of cathode rays (a stream of electrons) in magnetic and electric fields; Millikan's oil-drop experiment; Becquerel's and Rutherford's studies of radioactivity; and Rutherford's studies of the scattering of α particles by thin metal foils.

Elements can be classified by **atomic number,** the number of protons in the nucleus of an atom. All atoms of a given element have the same atomic number. The **mass number** of an atom is the sum of the number of protons and neutrons. Atoms of the same element that differ in mass number are known as **isotopes.**

The **periodic table** is an arrangement of elements in order of increasing atomic number with elements with similar properties placed in vertical columns. The elements in a vertical column are known as a **periodic family** or **group.** The **metallic elements,** which comprise the majority of the elements, are on the left side of the table; the **nonmetallic elements** are located on the right side.

Atoms can combine to form **molecules.** Atoms can also either gain or lose electrons, thereby forming charged particles called **ions.**

Each element has a one- or two-letter symbol. These symbols are combined to represent compounds, as in the formula H_2O for water. Three types of formulas were considered: **simplest** (or empirical) **formulas, molecular formulas,** and **structural formulas.** We also discussed the basic rules for naming inorganic compounds.

learning goals

Having read and studied this chapter, you should be able to:

1. Differentiate between the three states of matter.
2. Distinguish between elements, compounds, and mixtures.

3. Give the chemical symbols for the elements discussed in this chapter.
4. Distinguish between the physical and chemical properties of a substance.
5. Describe the separation of mixtures by filtration, distillation, and chromatography.
6. Describe the composition of an atom in terms of protons, neutrons, and electrons.
7. Give the approximate size, approximate mass, and the charge of an atom, proton, neutron, and electron.
8. Write the chemical symbol for an element, having been given its mass number and atomic number, and perform the reverse operation.
9. Write the symbol and charge for an atom or ion, having been given the number of protons, neutrons, and electrons, and perform the reverse operation.
10. Cite the evidence for the existence of subatomic particles and for the nuclear structure of the atom.
11. Use the periodic table to predict the charges of monatomic ions.
12. Write the simplest formula for a compound, having been given the charges of the ions from which it is made.
13. Write the name of a simple inorganic compound, having been given its chemical formula, and perform the reverse operation.

key terms

Among the more important terms and expressions used for the first time in this chapter are the following:

Adsorption (Section 2.3) is the adhesion of a thin layer of atoms, molecules, or ions to the surface of a substance.

Alpha (α) particles (Section 2.7) are helium nuclei (consisting of two protons and two neutrons) emitted from the nucleus of certain radioactive atoms.

The **atomic number** (Section 2.6) of an element is the number of protons in the nucleus of one of its atoms.

Beta (β) particles (Section 2.7) are electrons emitted from the nuclei of certain radioactive atoms.

A **chemical change** (Section 2.1) is a change in which new substances form.

Chemical properties (Section 2.1) are those properties of a substance that describe its composition and reactivity—how a substance "reacts" or changes into other substances.

A **compound** (Section 2.2) is a substance composed of two or more elements united chemically in definite proportions by mass. A compound can consist of heteroatomic molecules or of ions.

An **electron** (Section 2.6) is a negatively charged subatomic particle found outside the atomic nucleus. It forms a part of all atoms. It has a mass about $\frac{1}{1835}$ times that of a proton.

An **element** (Section 2.2) is a substance whose atoms all have the same atomic number; there are 105 known elements. Elements cannot be separated into simpler substances by ordinary chemical means.

An **empirical formula** or **simplest formula** (Section 2.9) is a chemical formula that shows the kinds of atoms and their relative numbers in a substance.

Gamma (γ) radiation (Section 2.7) (or **gamma rays**) is a form of radiation similar to X rays that is emitted by radioactive atoms.

Ions (Section 2.9) are electrically charged atoms or groups of atoms (polyatomic ions). Ions can be positively or negatively charged, depending on whether electrons are lost (positive) or gained (negative) from the atoms.

Isotopes (Section 2.6) are atoms of the same element containing different numbers of neutrons and therefore having different masses.

The **mass number** (Section 2.6) is the sum of the number of protons and neutrons in the nucleus of a particular atom.

A **molecular formula** (Section 2.9) is a chemical formula that indicates the actual number of atoms of each element in one molecule of a substance.

A **molecule** (Section 2.9) is a chemical combination of two or more atoms.

A **neutron** (Section 2.6) is a neutral particle found in the nucleus of an atom; it has approximately the same mass as a proton.

A **physical change** (Section 2.1) is a change (such as a phase change) that occurs with no change in chemical properties.

Physical properties (Section 2.1) are those properties (such as color and freezing point) that can be measured without changing the composition of a substance.

A **proton** (Section 2.6) is a positively charged subatomic particle found in the nucleus of any atom.

Radioactivity (Section 2.7) is the spontaneous disintegration of an unstable atomic nucleus with accompanying emission of alpha, beta, or gamma radiation.

A **solution** (Section 2.2) is a homogeneous mixture.

A **structural formula** (Section 2.9) is a formula that shows not only the number and kind of atoms in a molecule, but also the arrangement of the atoms.

EXERCISES

elements, compounds, and mixtures and states of matter

2.1 Write the chemical symbol for a single atom of each of the following: (a) oxygen; (b) hydrogen; (c) nitrogen; (d) gold; (e) sulfur; (f) helium; (g) carbon; (h) sodium.

2.2 What chemical elements are represented by the following symbols? (a) He; (b) Pb; (c) Sn; (d) Cl; (e) O; (f) H; (g) C; (h) N; (i) S

2.3 Classify each of the following as element, compound or mixture: (a) iron; (b) milk; (c) air; (d) salt; (e) sugar; (f) water; (g) wood; (h) blood; (i) wine.

2.4 Which of the following are elements and which are compounds? (a) I_2; (b) CO_2; (c) S_8; (d) NH_3; (e) H_2O_2; (f) O_3

2.5 How do molecules of elements and compounds differ?

2.6 Identify each of the following substances as gases, liquids, or solids under ordinary conditions: (a) mercury; (b) iron; (c) aluminum; (d) oxygen; (e) alcohol; (f) water; (g) hydrogen; (h) helium; (i) chlorine.

2.7 List as many properties as possible that would allow you to distinguish between a solid, a liquid, and a gas.

2.8 Indicate whether the following are physical or chemical properties: (a) color; (b) flammability; (c) behavior toward acids; (d) density.

2.9 Classify the following as a chemical or as a physical change: (a) condensation of steam; (b) burning of wood; (c) souring of milk; (d) melting of iron; (e) bending of an aluminum rod; (f) electrolysis of water.

2.10 List as many properties as possible that would allow you to distinguish between (a) silver and aluminum; (b) water and alcohol.

2.11 Which of the following terms is associated with chemical properties and which with physical properties? (a) particle size; (b) melts; (c) tarnishes; (d) corrodes; (e) burns; (f) mass

2.12 How would you separate the following mixtures? (a) sand from water; (b) salt from water; (c) alcohol from water; (d) the color pigments in a leaf; (e) components in air; (f) components in gasoline

atoms, molecules, and ions

2.13 Identify each of the following elements if a neutral atom of each (a) contains 3 protons; (b) has an atomic number of 13; (c) contains 10 electrons.

2.14 Fill in the gaps in the following table.

Symbol	$^{12}_{6}C$	$^{16}_{8}O^{2-}$			
Protons	6	8	12		9
Neutrons	6	8	13	12	10
Electrons	6	10		10	10
Net charge	0	2−	0	1+	

2.15 Fill in the gaps in the following table.

Symbol	$^{19}F^{-}$	^{40}Ar	
Protons			13
Neutrons			14
Electrons			
Net charge			0

2.16 Fill in the gaps in the following table.

Symbol	$^{9}_{4}Be^{2+}$	
Atomic number		35
Mass number		80
Net charge		1−

2.17 Show diagrammatically (as done in Figure 2.12) the two isotopes of chlorine having mass numbers of 35 and 37. Give the chemical symbol for each.

2.18 Which of the following are pairs of isotopes? (a) $^{14}_{6}C$; (b) $^{15}_{7}N$; (c) $^{16}_{8}O$; (d) $^{14}_{7}N$; (e) $^{12}_{6}C$.

2.19 (a) Explain why ^{18}O and ^{16}O have essentially identical chemical properties. (b) Explain why the symbols for an alpha particle and a beta particle could be given as $^{4}_{2}\alpha$ and $^{0}_{-1}\beta$.

2.20 Explain why beta particles are deflected more than are alpha particles when they move through the same electric field.

2.21 How does the path of a positively charged particle through a magnetic field differ if the following changes are made? (a) The magnetic field strength is doubled. (b) The mass of the particle is doubled. (c) The charge of the particle is doubled.

2.22 Describe the contributions to atomic theory made by the following persons: (a) Dalton; (b) Rutherford; (c) Thomson; (d) Millikan; (e) Becquerel.

[2.23] In what ways would the results of Rutherford's experiments have been different if he had used aluminum foil instead of gold foil?

periodic table, ions, and chemical formulas

2.24 Which of the following elements would you expect to show the greatest similarity in chemical and physical properties: Ca, Li, Al, Ba, Cl?

2.25 Categorize the following elements as alkali metals, alkaline earth metals, halogens, or noble gases: Na, F, K, Ne, Ca, Ar.

2.26 Which of the following elements are metals and which are nonmetals: Li, Zn, Se, Ce, Br, and Xe?

2.27 Using the periodic table, predict the charges on ions of K, Mg, Al, F, and O.

2.28 Predict the formula of the compound formed by potassium and oxygen by using the periodic table to determine the probable charges on the ions.

2.29 Write the chemical formula for each of the following: (a) water; (b) a diatomic molecule of oxygen; (c) the compound formed between Ca^{2+} and I^{-} ions.

2.30 The formula of the compound formed by aluminum and oxygen is Al_2O_3. Predict the formula of the compound formed between (a) gallium and oxygen; (b) aluminum and sulfur; (c) gallium and sulfur.

2.31 Which of the following are simplest formulas and which are molecular formulas? (a) H_2O_2; (b) $C_{12}H_{22}O_{11}$; (c) H_2O; (d) NaCl; (e) NaOH; (f) O_2

2.32 Which of the following are simplest formulas and which are molecular formulas? (a) O_3; (b) N_2O_4; (c) CO_2; (d) $C_6H_{12}O_6$; (e) $NaHCO_3$; (f) CH_4

naming compounds

2.33 Write the simplest formula for each of the following compounds: (a) calcium carbonate; (b) sodium hydroxide; (c) potassium sulfate; (d) iron(III) chloride; (e) nitrous acid; (f) hydrobromic acid; (g) ammonium bromide; (h) phosphorous acid; (i) sodium sulfite.

2.34 Write the simplest formula for each of the following compounds: (a) mercurous chloride; (b) copper(II) sulfate; (c) bromic acid; (d) sulfurous acid; (e) calcium chloride; (f) perchloric acid; (g) hydrocyanic acid; (h) cobalt(II) nitrate; (i) potassium cyanide.

2.35 Assume that you encounter the following phrases in your reading. What is the chemical formula for each compound mentioned? (a) Potassium chlorate is used as a laboratory source of oxygen. (b) Sodium hypochlorite is used as a household bleach. (c) Ammonia is important in the synthesis of fertilizers such as ammonium nitrate. (d) Hydrofluoric acid is used to etch glass. (e) The smell of rotten eggs is due to hydrogen sulfide. (f) When hydrochloric acid is added to sodium bicarbonate (baking powder), carbon dioxide gas forms.

2.36 Give the chemical name for each of the following compounds: (a) $Sr(NO_3)_2$; (b) $SnBr_4$; (c) $AgNO_3$; (d) KCN; (e) $Al(OH)_3$; (f) $NaHSO_4$; (g) $(NH_4)_2SO_4$; (h) HIO_3; (i) $CuBr$.

2.37 Give the chemical name for each of the following compounds: (a) $Ca(HCO_3)_2$; (b) Na_2CO_3; (c) $CoCl_2$; (d) $Ba(OH)_2$; (e) Na_2O_2; (f) Na_2O; (g) K_2S; (h) HI; (i) H_2SeO_4.

general exercises

2.38 Distinguish between the members of the following pairs: (a) homogeneous and heterogeneous; (b) gas and liquid; (c) atomic number and mass number; (d) chemical and physical properties; (e) Ca and Ca^{2+}; (f) hydrochloric acid and chloric acid; (g) iron(II) and iron(III); (h) sodium carbonate and sodium bicarbonate; (i) H_2O and H_2O_2; (j) metal and nonmetal; (k) H and He; (l) chloride and chlorate.

2.39 Distinguish between the members of the following pairs: (a) alpha and beta particles; (b) proton and neutron; (c) ion and atom; (d) molecular formula and empirical formula; (e) chemical change and physical change; (f) solid and liquid; (g) O and O^{2-}; (h) C and Ca; (i) sodium sulfate and sodium sulfite; (j) sodium hydrogen phosphate and sodium dihydrogen phosphate; (k) cuprous chloride and cupric chloride; (l) O_2 and O_3.

conservation of mass; stoichiometry

3

The laws of nature are the basic rules under which nature operates. However, it is not so much that objects obey the laws of nature, but rather that the laws of nature describe the behavior of objects. In this chapter, we shall consider the law of conservation of mass, which is one of the fundamental laws of chemistry. This law is the basis for understanding many of the quantitative relationships that exist between substances undergoing chemical changes. The study of these quantitative relationships is known as **stoichiometry** (pronounced stoy-key-AHM-uh-tree), a word derived from the Greek words *stoicheion* ("element") and *metron* ("measure").

3.1
law of conservation of mass

Studies of countless chemical reactions have shown that the total mass of all substances present after a chemical reaction is the same as the total mass before the reaction. This observation is embodied in the **law of conservation of mass:** There are no *detectable* changes in mass in any chemical reaction.* More precisely, *atoms are neither created nor destroyed during a chemical reaction;* instead, they merely exchange partners or become otherwise rearranged. The simplicity with which this law can be

---

*In Chapter 20, we shall discuss the relationship between mass and energy summarized by the equation $E = mc^2$ (E is energy, m is mass, and c is the speed of light). We shall find that whenever an object loses energy it loses mass, and whenever it gains energy it gains mass. These changes in mass are too small to detect in chemical reactions. However, for nuclear reactions, such as those involved in the atomic and hydrogen bombs, the energy changes are enormously larger; in these reactions there are detectable changes in mass.

stated should not mask its significance. As with many other scientific laws, this law has implications far beyond those evident when the law was first formulated.

An immediate qualitative application of the law of conservation of mass relates to our attempts to conserve our natural resources and minimize pollution. We have generally gone about our business acting as if this law did not exist. For example, we have too often adopted a flush-it-and-forget-it attitude toward waste disposal. We have removed our wastes from our immediate sight by discharging them into lakes and streams and into the air. However, the law of conservation of mass reminds us that for all practical purposes we really haven't been able to throw anything away. If we discharge a substance into a lake to get rid of it, it will be diluted and spread throughout the lake. It may even change into another substance as a result of a chemical reaction, but it is never gotten rid of in an absolute sense. The atoms may rearrange themselves into new compounds, but they are still there. Thus the law of conservation of mass indicates that we are really not consumers, only converters. We have an essentially unchanging storehouse of atoms that can be converted into a variety of compounds. (See Figure 3.1.)

When we speak of consuming a resource, what we are really doing is speaking of converting it into less useful, less available forms that are either too dilute to be economically feasible sources or that require too much energy to reclaim. For example, coal, which contains carbon, is obtained from concentrated deposits and is combusted as a fuel. The combustion produces principally carbon dioxide gas (CO_2) that is spread through the atmosphere. The carbon atoms are thereby both diluted and incorporated into the very stable CO_2 molecule. It would require the expenditure of a great deal of energy to collect these molecules from the

FIGURE 3.1 Untreated domestic sewage entering streams and lakes is symptomatic of the way society ignores the law of conservation of mass. Here, raw sewage is shown entering the Mississippi River in northern Minnesota. (*USDA-SCS photo by C. H. Aubol*)

atmosphere and to reclaim the carbon atoms from the CO_2. Indeed, other scientific laws indicate that this requires more energy than we would get from subsequent reuse of the carbon as a fuel.

3.2
chemical
equations

Scientists usually represent reactions in a concise fashion using chemical symbols. For example, the combustion of carbon in coal involves a reaction with oxygen (O_2) in the air to form gaseous carbon dioxide (CO_2). This reaction is represented as

$$C + O_2 \longrightarrow CO_2 \qquad\qquad [3.1]$$

We read the + sign to mean "reacts with" and the arrow as "produces." Carbon and oxygen are referred to as **reactants** and carbon dioxide as the **product** of the reaction. Such a statement of a reaction in terms of chemical symbols and formulas is known as a **chemical equation**. Because there are equal numbers of carbon and oxygen atoms in the reactants and products in Equation [3.1], the equation is consistent with the law of conservation of mass and is said to be *balanced*. A balanced equation contains equal numbers of each type of atom on each side of the equation.

A slightly more complicated situation is encountered when methane (CH_4), the principal component of natural gas, burns producing carbon dioxide (CO_2) and water (H_2O). The combustion is "supported by" oxygen (O_2), meaning that oxygen is involved as a reactant. The unbalanced equation is

$$CH_4 + O_2 \longrightarrow CO_2 + H_2O \qquad\qquad [3.2]$$

The reactants are shown to the left of the arrow, the products to the right. Notice that the reactants and products both contain one carbon atom. However, the reactants contain more hydrogen atoms (four) than the products (two). If we place a coefficient 2 in front of H_2O, indicating formation of two molecules of water, there will be four hydrogens on each side of the equation:

$$CH_4 + O_2 \longrightarrow CO_2 + 2H_2O \qquad\qquad [3.3]$$

Before we continue to balance this equation, let's make sure that we clearly understand the distinction between a coefficient in front of a formula and a subscript in a formula. Refer to Figure 3.2. Notice that changing a subscript in a formula, such as from H_2O to H_2O_2, changes the identity of the chemical involved. The substance H_2O_2, hydrogen peroxide, is quite different from water. The subscripts in the chemical formulas should never be changed in balancing an equation. On the other hand, placing a coefficient in front of a formula merely changes the amount and not the identity of the substance; $2H_2O$ means two molecules of water, $3H_2O$ means three molecules of water, and so forth. Now let's continue balancing Equation [3.3]. There are equal numbers of carbon and hydrogen atoms on both sides of this equation; however,

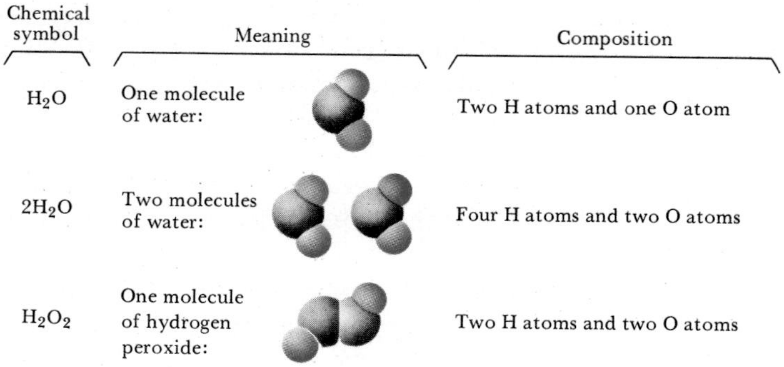

Chemical symbol	Meaning		Composition
H_2O	One molecule of water:		Two H atoms and one O atom
$2H_2O$	Two molecules of water:		Four H atoms and two O atoms
H_2O_2	One molecule of hydrogen peroxide:		Two H atoms and two O atoms

FIGURE 3.2 Illustration of the difference in meaning between a subscript in a chemical formula and a coefficient in front of the formula.

there are more oxygen atoms among the products (four) than among the reactants (two). If we place a coefficient 2 in front of O_2 there will be equal numbers of oxygen atoms on both sides of the equation:

$$CH_4 + 2O_2 \longrightarrow CO_2 + 2H_2O \qquad [3.4]$$

The equation is now balanced. There are four oxygen atoms, four hydrogen atoms, and one carbon atom on each side of the equation. The balanced equation is shown schematically in Figure 3.3.

Now, let's look at a slightly more complicated example, analyzing stepwise what we are doing as we balance the equation. Combustion of octane (C_8H_{18}), a component of gasoline, produces CO_2 and H_2O. The balanced chemical equation for this reaction can be determined by using the following four steps.

First, the reactants and products are written in the unbalanced equation:

$$C_8H_{18} + O_2 \longrightarrow CO_2 + H_2O \qquad [3.5]$$

Before a chemical equation can be written the identities of the reactants and products must be determined. We shall make some additional comments about this in Section 3.3. In the present example this information was given to us in the verbal description of the reaction.

Second, the number of atoms of each type on each side of the

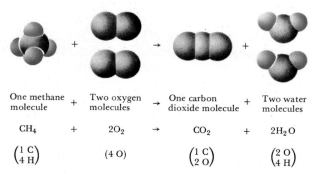

One methane molecule	+	Two oxygen molecules	→	One carbon dioxide molecule	+	Two water molecules
CH_4	+	$2O_2$	→	CO_2	+	$2H_2O$
$\begin{pmatrix} 1\ C \\ 4\ H \end{pmatrix}$		$(4\ O)$		$\begin{pmatrix} 1\ C \\ 2\ O \end{pmatrix}$		$\begin{pmatrix} 2\ O \\ 4\ H \end{pmatrix}$

FIGURE 3.3 The balanced chemical equation for the combustion of CH_4. The drawings of the molecules involved calls attention to the conservation of atoms through the reaction.

equation is determined. In the reaction above there are 8C, 18H, and 2O among the reactants, and 1C, 2H, and 3O among the products; clearly, the equation is not balanced, because the number of atoms of each type differ from one side of the equation to the other.

Third, to balance the equation, coefficients are placed in front of the chemical formulas to indicate different quantities of reactants and products, so that the same number of atoms of each type appears on both sides of the equation. To decide what coefficients to try first, it is often convenient to focus attention on the molecule with the most atoms, in this case C_8H_{18}. This molecule contains 8C, all of which must end up in CO_2 molecules. Therefore, we place a coefficient 8 in front of CO_2. Similarly, the 18H end up as $9H_2O$. At this stage the equation reads

$$C_8H_{18} + O_2 \longrightarrow 8CO_2 + 9H_2O \qquad [3.6]$$

Although the C and H atoms are now balanced, the O atoms are not; there are 25O atoms among the products but only 2 among the reactants. It takes $12.5O_2$ to produce 25O atoms among the reactants:

$$C_8H_{18} + 12.5O_2 \longrightarrow 8CO_2 + 9H_2O \qquad [3.7]$$

However, this equation is not in its most conventional form, because it indicates a half of an O_2 molecule among the reactants, and half an O_2 molecule does not exist as such. Therefore, we must go on to the next step.

Fourth, for most purposes a balanced equation should contain the smallest possible whole-number coefficients. Therefore, we multiply each side of the equation above by two, removing the fraction and achieving the following balanced equation:

$$2C_8H_{18} + 25O_2 \longrightarrow 16CO_2 + 18H_2O \qquad [3.8]$$

16C, 36H, 50O		16C, 36H, 50O
Reactants	Δ	Products

The atoms are inventoried below the equation to show graphically that the equation is indeed balanced. You might note that although atoms are conserved, molecules are not—the reactants contain 27 molecules while the products contain 34. All in all, this approach to balancing equations is largely trial and error. It is much easier to verify that an equation is balanced than to actually balance one, so practice in balancing equations is essential.

It should also be noted that the physical state of each chemical in a chemical equation is often indicated parenthetically using the symbols (*g*), (*l*), (*s*), and (*aq*) to indicate gas, liquid, solid, and aqueous (water) solution, respectively. Thus the balanced equation above can be written:

$$2C_8H_{18}(l) + 25O_2(g) \longrightarrow 16CO_2(g) + 18H_2O(l) \qquad [3.9]$$

Sometimes an upward arrow (↑) is employed to indicate the escape of a gaseous product, whereas a downward arrow (↓) identifies a precipitating solid (that is, a solid that separates from solution during the reaction). Often the conditions under which the reaction proceeds are indicated above the arrow between the two sides of the equation. For example, the temperature or pressure at which the reaction occurs could be so indicated. The symbol Δ above the arrow, as in questions 3.31 and 3.32 at the end of the chapter, is often used to indicate the addition of heat.

SAMPLE EXERCISE 3.1

Balance the following equation:

$$Na(s) + H_2O(l) \longrightarrow NaOH(aq) + H_2(g)$$

Solution: A quick inventory of atoms reveals that there are equal numbers of Na and O atoms on both sides of the equation, but that there are two H atoms among reactants and three H atoms among products. To increase the number of H atoms among reactants, we might place a coefficient 2 in front of H_2O:

$$Na(s) + 2H_2O(l) \longrightarrow NaOH(aq) + H_2(g)$$

Now we have four H atoms among reactants but only three H atoms among the products. The H atoms can be balanced with a coefficient 2 in front of NaOH:

$$Na(s) + 2H_2O(l) \longrightarrow 2NaOH(aq) + H_2(g)$$

If we again inventory the atoms on each side of the equation, we find that the H atoms and O atoms are balanced but not the Na atoms. However, a coefficient 2 in front of Na gives two Na atoms on each side of the equation:

$$2Na(s) + 2H_2O(l) \longrightarrow 2NaOH(aq) + H_2(g)$$

If the atoms are inventoried once more we find two Na atoms, four H atoms, and two O atoms on each side of the equation. The equation is therefore balanced.

3.3
chemical
reactions

Our discussion in Section 3.2 focused on how to balance chemical equations given the reactants and products for the reactions. You were not asked to predict the products for a reaction. Students sometimes ask how the products are determined. For example, how do we know that sodium metal (Na) reacts with water (H_2O) to form H_2 and NaOH as shown in Sample Exercise 3.1? These products are identified by experiment. As the reaction proceeds, there is a fizzing or bubbling where the sodium is in contact with the water (if too much sodium is used the reaction is quite violent, and so small quantities would be used in our experiment). If the gas is captured, it can be identified as H_2 from its chemical and physical properties. After the reaction is complete, a clear solution remains. If this is evaporated to dryness, a white solid will remain. From its properties this solid can be identified as NaOH. However, it is not necessary to perform an experiment every time we wish to write a reaction. We can predict what will happen if we have seen the reaction or a similar one before. So far we have seen too little chemistry to predict the products for many reactions. Nevertheless, even now you should be able to make some predictions. For example, what would you expect to happen when potassium metal is added to water? We have just discussed the reaction of sodium metal with water, for which the balanced chemical equation is

$$2Na(s) + 2H_2O(l) \longrightarrow 2NaOH(aq) + H_2(g) \qquad [3.10]$$

Because sodium and potassium are in the same family of the periodic table (the alkali metal family, family 1A), we would expect them to behave similarly, producing the same types of products. Indeed, this prediction is correct, and the reaction of potassium metal with water is

$$2K(s) + 2H_2O(l) \longrightarrow 2KOH(aq) + H_2(g) \qquad [3.11]$$

Now let's consider a totally different example, the combustion of propane gas (C_3H_8). Again we should be able to predict what the reaction will be and then write a balanced chemical equation. We need to remember some similar reaction to make our prediction. In Section 3.2 we considered three combustion reactions, the combustion of carbon, Equation [3.1], the combustion of CH_4, Equation [3.4], and the combustion of C_8H_{18}, Equation [3.8]. Remember that combustion is a rapid reaction that produces a flame and that most of these involve O_2 as a reactant. Notice that in each of the three combustion reactions given in the previous section, carbon atoms end up in CO_2 molecules among the products and hydrogen atoms end up in H_2O molecules. We would expect that combustion of C_3H_8 would be analogous. The prediction is correct and the balanced equation is

$$C_3H_8(g) + 5O_2(g) \longrightarrow 3CO_2(g) + 4H_2O(l) \qquad [3.12]$$

If we looked at further examples, we would find that combustion of compounds containing oxygen atoms as well as carbon and hydrogen also produces CO_2 and H_2O.

ACID-BASE NEUTRALIZATION REACTIONS

Before we go very far in any study of chemistry, we begin to encounter the reactions of hydrochloric acid (HCl), nitric acid (HNO_3), sulfuric acid (H_2SO_4), and sodium hydroxide (NaOH). These substances are very common, especially in the chemistry laboratory. Perhaps you've already encountered them. At any rate, let's take this opportunity to examine briefly these compounds and one of their important reactions. As their names indicate, HCl, HNO_3, and H_2SO_4 belong to a group of compounds known as acids. It turns out that NaOH is a member of a different class of compounds known as bases. For the moment we will define an **acid** as a substance that can lose a H^+ ion in water solution, whereas we define a **base** as a substance that can lose an OH^- (hydroxide) ion in water solution. These definitions are somewhat restrictive, and we shall consider broader definitions when we take a closer look at acids and bases in Chapter 16. In agreement with our definition of an acid, HCl, HNO_3, and H_2SO_4 ionize, forming $H^+(aq)$ when added to water: *

$$HCl(aq) \longrightarrow H^+(aq) + Cl^-(aq) \qquad [3.13]$$

$$HNO_3(aq) \longrightarrow H^+(aq) + NO_3^-(aq) \qquad [3.14]$$

$$H_2SO_4(aq) \longrightarrow 2H^+(aq) + SO_4^{2-}(aq) \qquad [3.15]$$

*Sulfuric acid, H_2SO_4, ionizes in a stepwise fashion:

$H_2SO_4(aq) \longrightarrow H^+(aq) + HSO_4^-(aq)$
$HSO_4^-(aq) \longrightarrow H^+(aq) + SO_4^{2-}(aq)$

Sodium hydroxide and other bases such as $Ca(OH)_2$ ionize in water forming $OH^-(aq)$:

$$NaOH(aq) \longrightarrow Na^+(aq) + OH^-(aq) \qquad [3.16]$$

$$Ca(OH)_2(aq) \longrightarrow Ca^{2+}(aq) + 2OH^-(aq) \qquad [3.17]$$

Aqueous solutions of ammonia (NH_3) also act as a base. However, we could not have predicted this behavior from its formula. Nevertheless, it is observed that ammonia reacts partly with water to form the hydroxide ion:

$$NH_3(aq) + H_2O \longrightarrow NH_4^+(aq) + OH^-(aq) \qquad [3.18]$$

The aqueous solution of ammonia is therefore often labeled "ammonium hydroxide" (NH_4OH), although NH_4OH cannot be isolated as a pure compound from the solution. We mention these facts because aqueous ammonia is also a common laboratory base.

The reaction between an acid and a base is known as a **neutralization** reaction, because the properties of the acid and those of the base are destroyed, or neutralized. Acids have characteristic properties. They have a sour taste;* they can change the colors of certain dyes in a specific way (for example, the dye known as litmus is changed from blue to red by acids); they react with certain metals such as zinc to form H_2 gas. These properties are destroyed by reaction with a base. Likewise, bases have characteristic properties that are destroyed by reaction with an acid. Bases have a bitter taste; they have a slippery or soapy feel; they are able to affect the colors of certain dyes in a different way than acids (for example, they are able to change litmus from red to blue).

In a neutralization reaction, acids and bases react to form water plus a second compound known as a salt. The following example illustrates the reaction:

$$HCl(aq) + NaOH(aq) \longrightarrow H_2O + NaCl(aq) \qquad [3.19]$$
$$Acid + Base \longrightarrow Water + Salt$$

Each available H^+ from an acid is able to react with one OH^- from a base to form water:

$$H^+(aq) + OH^-(aq) \longrightarrow H_2O \qquad [3.20]$$

$[H_3O^+] + [OH^-] \longrightarrow 2H_2O$

If two OH^- are available from the base, it will take two H^+ to completely neutralize them:

$$2HCl(aq) + Ca(OH)_2(aq) \longrightarrow 2H_2O + CaCl_2(aq) \qquad [3.21]$$

In this example, the salt incorporates the Ca^{2+} ion from $Ca(OH)_2$ and the two Cl^- ions from $2HCl$. In like manner, if an acid can supply two

---

*There are a number of common acids that you have undoubtedly tasted—for example, ascorbic acid (vitamin C), acetylsalicylic acid (aspirin), and citric acid (in citrus fruits). These exhibit the characteristic sour taste of acids.

H^+, it will take two OH^- to completely neutralize them:

$$H_2SO_4(aq) + 2NaOH(aq) \longrightarrow 2H_2O + Na_2SO_4(aq) \quad [3.22]$$

In this example, the two Na^+ ions from $2NaOH$ and the SO_4^{2-} ion from H_2SO_4 are incorporated into the neutral Na_2SO_4 salt.

SAMPLE EXERCISE 3.2

Write a balanced chemical equation for the following reactions: (a) the combustion of ethyl alcohol, C_2H_5OH; (b) the reaction between silane, Si_2H_6, and O_2; (c) the reaction between phosphoric acid, H_3PO_4, and $NaOH$.

Solution: In this exercise we must predict what reaction will occur before a balanced chemical equation can be written.

(a) In a combustion of a compound containing C, H, and O, CO_2 and H_2O are formed. The unbalanced equation is

$$C_2H_5OH + O_2 \longrightarrow CO_2 + H_2O$$

In order that there be two C atoms and six H atoms on each side of the equation (the number in C_2H_5OH), we write

$$C_2H_5OH + O_2 \longrightarrow 2CO_2 + 3H_2O$$

There are now seven O atoms among the products. The oxygens are balanced if the reactants involve three O_2:

$$C_2H_5OH + 3O_2 \longrightarrow 2CO_2 + 3H_2O$$

(b) The key to this reaction is noticing that Si and C are in the same periodic family. We would therefore predict SiO_2 as a product. The balanced equation is

$$2Si_2H_6 + 7O_2 \longrightarrow 4SiO_2 + 6H_2O$$

(c) The key to predicting the products of this reaction is recognizing it as an acid-base neutralization. The balanced equation is

$$H_3PO_4 + 3NaOH \longrightarrow 3H_2O + Na_3PO_4$$

3.4 interpretation of balanced equations: an introduction

A balanced equation implies a quantitative relation between the reactants and the products involved in a chemical reaction. Thus complete combustion of 2 molecules of C_8H_{18} requires exactly 25 molecules of O_2, no more and no less, as shown in Equation [3.9]. Although it is not possible to count directly the number of molecules of each type in any reaction, this count can be made indirectly if the mass of each molecule is known. This calculation is illustrated in Sample Exercise 3.3.

SAMPLE EXERCISE 3.3

If each O_2 molecule weighs 5.3×10^{-23} g, how many O_2 molecules are there in 1.0 g of O_2?

Solution:

$$\text{Molecules } O_2 = (1.0 \text{ g } O_2)\left(\frac{1 \text{ molecule } O_2}{5.3 \times 10^{-23} \text{ g } O_2}\right)$$

$$= 1.9 \times 10^{22} \text{ molecules } O_2$$

Indeed, this indirect approach is the one taken to obtain quantitative information about the amounts of substances involved in any chemical transformation. Therefore, before we can pursue the quantitative aspects of chemical reactions further, we must explore the concept of atomic and molecular weights.

Because different types of atoms contain different numbers of subatomic particles, they differ in mass. Atomic weights* were originally assigned on a relative scale, with hydrogen given a value of 1. Relative weights were assigned because it was impossible for scientists in the nineteenth century to determine the actual weights of atoms. Modern atomic weights are based on the assignment of a mass of exactly 12 for the ^{12}C isotope of carbon (Section 2.6). Although no units were originally attached to the weights of atoms and molecules, it is now common to refer to these weights as **atomic mass units** (amu). Thus ^{12}C has an atomic weight of exactly 12 amu. The atomic weights of elements are reported as average values, reflecting the relative abundances of each isotope of each element. Naturally occurring chlorine is 75.53 percent ^{35}Cl, which has a mass of 34.969 amu, and 24.47 percent ^{37}Cl, which has a mass of 36.966 amu. The average atomic weight for chlorine can be calculated to four significant figures from this information:

$$\text{Av AW}\dagger = (75.53\%)(34.969 \text{ amu}) + (24.47\%)(36.966 \text{ amu})$$
$$= 26.41 \text{ amu} + 9.05 \text{ amu}$$
$$= 35.46 \text{ amu}$$

The last digit in this calculation is uncertain; the accepted value for the atomic weight of chlorine, to five significant figures, is 35.453 amu. Two methods of determining atomic weights are considered in the next two sections of this chapter. The atomic weights of the elements are listed below the symbol for the element on the periodic table found on the front inside cover of this text. They are also listed in the table of elements on the front inside cover.

The **molecular weight** of a molecule is merely the sum of the weights of each atom it contains. For example, H_2SO_4, sulfuric acid, has a molecular weight of 98.0 amu:

$$\text{MW} = 2(\text{AW of H}) + \text{AW of S} + 4(\text{AW of O})$$
$$= 2(1.0 \text{ amu}) + 32.0 \text{ amu} + 4(16.0 \text{ amu})$$
$$= 98.0 \text{ amu}$$

Here we have rounded off the atomic weights so that our result has three significant figures. We shall round off the atomic weights in this way for most problems.

As we noted in the last chapter, not all compounds exist as discrete molecules and not all chemical formulas that we encounter are molecular formulas. Rather than talking about molecular weights we can talk about formula weights. The **formula weight** is defined as the sum of the atomic weights of the atoms in the formula. The formula weight of $C_6H_{12}O_6$ is $6(12.0 \text{ amu}) + 12(1.0 \text{ amu}) + 6(16.0 \text{ amu}) = 180 \text{ amu}$. This is also the molecular weight of this substance since $C_6H_{12}O_6$ is a molecular formula.

---

*Of course, what is really meant is *atomic masses,* but the term *atomic weights* is most commonly used.

†The abbreviation AW is used for atomic weight, MW for molecular weight, and FW for formula weight. The SI abbreviation for amu is simply u, as in 35.46 u.

The formula weight of NaCl is 23.0 amu + 35.5 amu = 58.5 amu. It is common practice to use the term molecular weight even with substances like NaCl, which do not exist in molecular form. Thus, although we might say that the molecular weight of NaCl is 58.5 amu, we would be more correct to say that this is its formula weight.

SAMPLE EXERCISE 3.4

(a) Calculate the formula weight of sucrose, $C_{12}H_{22}O_{11}$ (table sugar); (b) calculate the percentage of carbon in this compound.

Solution: (a) The formula weight of $C_{12}H_{22}O_{11}$ is calculated as follows:

FW = 12(12.0 amu) + 22(1.0 amu) + 11(16.0 amu)

= 342 amu

(b) The total mass of carbon in $C_{12}H_{22}O_{11}$ is 12(12.0 amu).

$$\% \text{ C} = \frac{\text{Total mass of C in } C_{12}H_{22}O_{11}}{\text{Mass of } C_{12}H_{22}O_{11}} \times 100$$

$$= \frac{12(12.0 \text{ amu})}{342 \text{ amu}} \times 100 = 42.1\%$$

3.6
atomic and molecular weights— a historical perspective

Historically, the problem of establishing a relative scale of weights for atoms was filled with difficulties and uncertainties. In attempting to determine atomic weights, Dalton incorrectly assumed a law of simplicity: If two elements, A and B, form a single compound, the combining ratio will be one atom of A for each atom of B. Thus, he incorrectly guessed the formula of water to be HO. Dalton's wrong assumptions about formulas led to wrong atomic weights, which led back again to wrong formulas for new compounds. The situation became so confusing that some chemists suggested abandoning the concept of atoms.

The key to the solution of this problem, which plagued scientists through much of the nineteenth century, was found in studies of gases. In 1808, the French chemist Joseph Gay-Lussac began a series of studies on the combining volumes of gases involved in chemical reactions with each other. He found that the combining volumes and product volumes of gases are in small, whole-number ratios when temperature and pressure are constant. For example, when hydrogen and oxygen gases combine to form water vapor, two volumes of hydrogen combine completely with one volume of oxygen to form two volumes of water vapor. Likewise, one volume of hydrogen combines with one volume of chlorine to form two volumes of hydrogen chloride. However, Gay-Lussac did not understand the reason for these relationships.

In 1811, the Italian physicist Amedeo Avogadro suggested that Gay-Lussac's law of combining volumes for gases implied that equal volumes of all gases, measured at the same pressure and temperature, contain the same number of molecules. The way this proposal explains Gay-Lussac's observations is illustrated in Figure 3.4, where the combustion of hydrogen is used as an example. Notice that the relative volumes of combining gases correspond to the coefficients of the balanced chemical equations. Notice also that the explanation assumes that hydrogen and oxygen are diatomic.

atomic and molecular weights—
a historical perspective

conservation of mass;
stoichiometry

Observation: Two volumes hydrogen + One volume oxygen → Two volumes water vapor

Explanation:

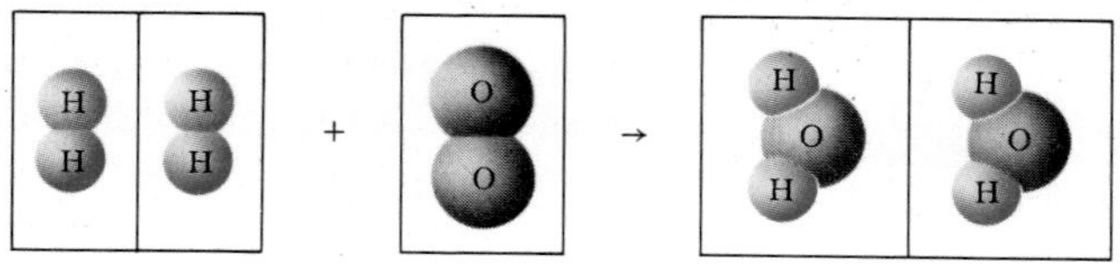

$$2H_2\,(g) \quad + \quad O_2\,(g) \quad \rightarrow \quad 2H_2O(g)$$

Equation:

FIGURE 3.4 Gay-Lussac's experimental observation of combining volumes shown together with Avogadro's explanation of this phenomenon.

SAMPLE EXERCISE 3.5

How many liters of NH_3 will be produced when 2 l. of N_2 combine with 6 l. of H_2 (all volumes measured at the same temperature and pressure)?

Solution: The balanced equation is

$$N_2(g) + 3H_2(g) \longrightarrow 2NH_3(g)$$

From the coefficients it can be deduced that each liter of N_2 combines with 3 l. H_2 to produce 2 l. NH_3. Thus 4 l. NH_3 will form from 2 l. N_2 and 6 l. H_2.

Unfortunately, Avogadro's idea had little impact until nearly 50 years later when a fellow countryman, Stanislao Cannizzaro, clearly pointed out how it could be used to determine atomic and molecular weights. The general approach involves comparisons of the densities of gases to find molecular weights. These molecular weights can then be used together with elemental composition percentages to find atomic weights. We shall illustrate this approach by using the experimental data in Table 3.1 to calculate the molecular weights of oxygen and water and to deduce the atomic weight of oxygen. We shall then see how the chemical formula for water can be determined.

According to Avogadro's hypothesis 1 l. of any gas (measured at the same temperature and pressure) should contain the same number of molecules. Thus there are as many molecules in 0.0658 g of hydrogen as in 1.046 g of oxygen because these are the weights of 1 l. of each (Table 3.1). This result implies that each oxygen molecule weighs 1.046/0.0658 times as much as a hydrogen molecule. We arbitrarily assume that the

TABLE 3.1

densities and elemental compositions of hydrogen, oxygen, and water vapor

GAS	DENSITY AT $100°c$, 1 atm[a]	ELEMENTAL COMPOSITION
Hydrogen	0.0658 g/l.	100% H
Oxygen	1.046 g/l.	100% O
Water vapor	0.598 g/l.	11.1% H, 88.9% O

[a] 1 atm means a pressure of one atmosphere, the average atmospheric pressure at sea level.

mass of a hydrogen atom is 1 amu. In interpreting Gay-Lussac's law of combining volumes, Avogadro had suggested that hydrogen is diatomic, H_2 (Figure 3.4). If we accept this result, diatomic hydrogen must have a molecular weight of 2 amu. The mass of any oxygen molecule is then

$$\left(\frac{1.046}{0.0658}\right)(2 \text{ amu}) = 32 \text{ amu}$$

Avogadro had also suggested that oxygen is diatomic, O_2. Its atomic weight must therefore be half of its molecular weight: 16 amu. Reasoning similarly, and using the density of water vapor given in Table 3.1, we see that each water molecule must weigh

$$\left(\frac{0.598}{0.0658}\right)(2 \text{ amu}) = 18 \text{ amu}$$

Because water has a molecular weight of 18 amu and contains 88.9 percent oxygen, one water molecule contains $(0.889)(18 \text{ amu}) = 16 \text{ amu}$ of oxygen, which is the mass of one oxygen atom. Likewise, one water molecule contains $(0.111)(18 \text{ amu}) = 2 \text{ amu}$ of hydrogen, which is the mass of two hydrogen atoms. The correct formula for water can thereby be deduced to be H_2O.

The conclusion that oxygen has an atomic weight of 16 amu was checked by early chemists who examined a large number of oxygen-containing substances in the manner outlined above. When many substances, such as those shown in Table 3.2, are examined, it is found that 16 amu is the smallest mass of oxygen found in these compounds. Furthermore, larger masses are multiples of 16 amu. This result further verifies our original conclusion that the mass of a single atom of oxygen is 16 amu.

TABLE 3.2

weight of oxygen per molecular unit in a series of compounds

SUBSTANCE	WEIGHT OF OXYGEN PER MOLECULAR UNIT (amu)	MODERN FORMULA
Oxygen gas	32.0	O_2
Water	16.0	H_2O
Hydrogen peroxide	32.0	H_2O_2
Nitric oxide	16.0	NO
Nitrous oxide	16.0	N_2O
Sulfur trioxide	48.0	SO_3

3.7
the mass spectrometer

The most accurate means presently available for determining atomic weights is provided by the **mass spectrometer**. This instrument, shown schematically in Figure 3.5, is similar to that used by Thomson to measure the e/m ratio of the electron (Section 2.7). In the mass spectrometer, a beam of gaseous atoms or molecules is allowed to flow into a vacuum chamber. This beam is intercepted at one point by a stream of energetic electrons that can cause ionization. The positive ions so formed

conservation of mass;
stoichiometry

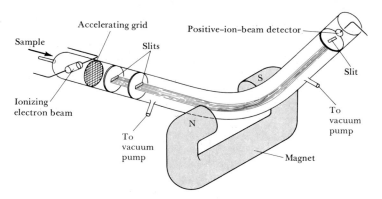

FIGURE 3.5   A schematic diagram of a modern mass spectrometer.

are accelerated down the tube by a high voltage and pass through two slits, so that there is only a narrow beam of the ions. The system is pumped to a high vacuum to remove most of the gas molecules in the spectrometer. Consequently, the ions do not undergo a significant number of disturbing collisions with other particles in their passage down the tube. When the ions reach the region of the magnet, they are deflected by the magnetic field. The extent to which they are deflected depends on the charge-to-mass ratio, e/m, for the ions. Assuming that only singly charged ions are involved (that is, ions from which only one electron has been removed), the degree to which the ions have been deflected then depends on the mass of the ion. By continuously changing the strength of the magnetic field, or by changing the accelerating voltage, the ions of varying e/m can be caused to enter the slit at the end of the instrument and thereby reach the detector. Thus it is possible to scan over a wide range of e/m. The e/m scale is then easily converted into a mass scale, because the electrical charge of the ion (e) is always the same. A graph of the intensity of signal from the detector versus the atomic weight is called a **mass spectrum.**

Historically, the mass spectrometer provided the first unambiguous evidence for the existence of isotopes. Suppose we allow a stream of mercury vapor to enter the mass spectrometer. If all the atoms of mercury

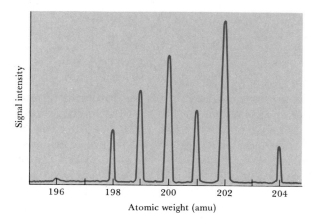

FIGURE 3.6   The mass spectrum of Hg^+ ions in mercury vapor.

TABLE 3.3

mass spectrum of mercury

MASS NUMBER	ATOMIC WEIGHT (amu)	FRACTIONAL ABUNDANCE
196	195.965	0.0014
198	197.967	0.10039
199	198.967	0.1683
200	199.968	0.2312
201	200.970	0.1323
202	201.970	0.2979
204	203.973	0.0685

were in fact identical, all of them upon ionization would possess the same e/m ratio and would therefore appear at the same place in the mass spectrum. But the mass spectrum actually observed for mercury is as shown in Figure 3.6. It is clear that there are several kinds of atoms in mercury, differing slightly in their masses.

Using the mass spectrometer, it is possible to measure relative values of the ratio e/m for ions with great accuracy. It is also possible to measure with high accuracy the relative numbers of the different isotopes of an element. As we have previously noted, the atomic weight is the weighted average of the masses of all the isotopes of that element. For example, from the mass spectrum of mercury, Figure 3.6, we obtain the data shown in Table 3.3. The atomic weight of mercury, 200.59 amu, is the sum of the masses of the various isotopes, each multiplied by its fractional abundance:

$$\begin{aligned} AW &= (0.0014)(195.965 \text{ amu}) + (0.10039)(197.967 \text{ amu}) + \cdots \\ &\quad + (0.0685)(203.973) \\ &= 200.59 \text{ amu} \end{aligned}$$

It is remarkable that samples of elements taken from various sources show the same isotopic abundances.

3.8
the chemical mole

In Section 3.4 we indicated that the concept of atomic weights is important because it provides a means of counting atoms indirectly. It now remains for us to discuss how numbers of atoms are related to atomic weights. The concept that brings together these ideas is the mole concept. A **mole*** of any element is defined as the amount of that element that contains the same number of atoms as exactly 12 g of ^{12}C. It has been determined experimentally that the number of atoms in this quantity of ^{12}C is 6.022×10^{23}. This number is given a special name, **Avogadro's number,** in honor of Amedeo Avogadro, who proposed that equal volumes of gases contain equal numbers of molecules (Section 3.6). For most purposes, we will use 6.02×10^{23} for Avogadro's number throughout the text, and this number should be committed to memory.

*The term mole comes from the Latin word *moles* meaning "a mass." The term *molecule* is the diminutive form of this word and means "a small mass."

A mole of ions, molecules, or anything else contains Avogadro's number of these objects:

$$1 \text{ mole of } {}^{12}C \text{ atoms} = 6.02 \times 10^{23} \; {}^{12}C \text{ atoms}$$

$$1 \text{ mole of } H_2O \text{ molecules} = 6.02 \times 10^{23} \; H_2O \text{ molecules}$$

$$1 \text{ mole of } NO_3^- \text{ ions} = 6.02 \times 10^{23} \; NO_3^- \text{ ions}$$

The concept of a mole as being 6.02×10^{23} of something is analogous to the concept of a dozen as 12 of something or a gross as 144 of something.

Because a ${}^{24}Mg$ atom weighs twice as much as a ${}^{12}C$ atom, a mole of ${}^{24}Mg$ must weigh twice as much as a mole of ${}^{12}C$. A ${}^{12}C$ atom weighs exactly 12 amu, whereas a ${}^{24}Mg$ atom weighs 24.0 amu. Because a mole of ${}^{12}C$ atoms weighs 12 g (by definition), a mole of ${}^{24}Mg$ atoms must weigh 24.0 g. In fact, a mole of atoms of any element has a mass in grams numerically equal to the atomic weight of a single atom:

One ${}^{12}C$ atom weighs 12 amu; 1 mole of ${}^{12}C$ weighs 12 g.

One ${}^{24}Mg$ atom weighs 24.0 amu; 1 mole of ${}^{24}Mg$ weighs 24.0 g.

One Au atom weighs 197 amu; 1 mole of Au weighs 197 g.

We can generalize this idea to include molecules and ions: The mass of a mole of formula units of any substance (that is, 6.02×10^{23} of them) is always equal to the formula weight expressed in grams:

One H_2O molecule weighs 18.0 amu; 1 mole of H_2O weighs 18.0 g.

One NO_3^- ion weighs 62.0 amu; 1 mole of NO_3^- weighs 62.0 g.

One NaCl unit weighs 58.5 amu; 1 mole of NaCl weighs 58.5 g.

Further examples of mole relationships are shown in Table 3.4.

The first entries in Table 3.4, those for N and N_2, point out the importance of stating the chemical form of a substance exactly when we

TABLE 3.4
mole relationships

NAME	FORMULA	FORMULA WEIGHT (amu)	WEIGHT OF 1 MOLE OF FORMULA UNITS (g)	NUMBER AND KIND OF PARTICLES IN 1 MOLE
Atomic nitrogen	N	14.0	14.0	6.02×10^{23} N atoms
Molecular nitrogen	N_2	28.0	28.0	$\left\{ \begin{array}{l} 6.02 \times 10^{23} \; N_2 \text{ molecules} \\ 2(6.02 \times 10^{23}) \; N \text{ atoms} \end{array} \right.$
Silver	Ag	108	108	6.02×10^{23} Ag atoms
Silver ions	Ag^+	108[a]	108	$6.02 \times 10^{23} \; Ag^+$ ions
Barium chloride	$BaCl_2$	208	208	$\left\{ \begin{array}{l} 6.02 \times 10^{23} \; BaCl_2 \text{ units} \\ 6.02 \times 10^{23} \; Ba^{2+} \text{ ions} \\ 2(6.02 \times 10^{23}) \; Cl^- \text{ ions} \end{array} \right.$

[a] Recall that the electron has negligible mass; thus ions and atoms have essentially the same mass.

use the mole concept. Suppose you read that 1 mole of nitrogen is produced in a particular reaction. You might interpret this statement to mean 1 mole of nitrogen atoms (14.0 g). Unless otherwise stated, what was probably meant is a mole of N_2 (28.0 g), because N_2 is the usual chemical form of the element. However, to avoid ambiguity it is always best to state explicitly the chemical form being discussed.

SAMPLE EXERCISE 3.6

What is the weight of one mole of glucose, $C_6H_{12}O_6$?

Solution: By adding the weights of the atoms in glucose we find it to have a formula weight of 180 amu:

$$
\begin{aligned}
6C \text{ atoms} = 6(12.0 \text{ amu}) = &\quad 72.0 \text{ amu} \\
12H \text{ atoms} = 12(1.0 \text{ amu}) = &\quad 12.0 \text{ amu} \\
6O \text{ atoms} = 6(16.0 \text{ amu}) = &\quad \underline{96.0 \text{ amu}} \\
&\quad 180.0 \text{ amu}
\end{aligned}
$$

Hence one mole of $C_6H_{12}O_6$ weighs 180 g.

SAMPLE EXERCISE 3.7

How many C atoms are there in 1 mole of $C_6H_{12}O_6$?

Solution: There are 6.02×10^{23} $C_6H_{12}O_6$ molecules in 1 mole. Each molecule contains 6C atoms; hence there are $6(6.02 \times 10^{23})$C atoms:

$$
\begin{aligned}
C \text{ atoms} = &(1 \text{ mole } C_6H_{12}O_6) \\
&\times \left(\frac{6.02 \times 10^{23} \text{ molecules}}{1 \text{ mole}}\right)\left(\frac{6C \text{ atoms}}{1 \text{ molecule}}\right) \\
= &\ 3.61 \times 10^{24} \text{ C atoms}
\end{aligned}
$$

To illustrate how the mole concept and Avogadro's number allow us to interconvert masses and number of particles, let's calculate the number of copper atoms in a penny. A penny weighs 3 g, and we'll assume that it is 100 percent copper:

$$
\begin{aligned}
Cu \text{ atoms} = &(3 \text{ g Cu}) \left(\frac{1 \text{ mole Cu}}{63.5 \text{ g Cu}}\right)\left(\frac{6.02 \times 10^{23} \text{ Cu atoms}}{1 \text{ mole Cu}}\right) \\
= &\ 3 \times 10^{22} \text{ Cu atoms}
\end{aligned}
$$

Notice how we were able to use dimensional analysis (Section 1.5) in a straightforward manner to go from grams to numbers of atoms; the conversion sequence is g $\longrightarrow$ mole $\longrightarrow$ number.

We might reflect momentarily on the number of copper atoms in a penny, 3×10^{22}. This is a tremendously large number. It becomes more impressive when we realize that the entire United States could be covered to a depth of 3 mi with this number of ice cubes, each 1 in. on an edge. We should remember that Avogadro's number is even larger.

SAMPLE EXERCISE 3.8

How many moles of glucose, $C_6H_{12}O_6$, are there in (a) 538 g and (b) 1.00 g of this substance?

Solution: (a) One mole of $C_6H_{12}O_6$ weighs 180 g (Sample Exercise 3.6). Therefore there must be more than 1 mole in 538 g.

Moles $C_6H_{12}O_6$ =

$$(538 \text{ g } C_6H_{12}O_6) \left(\frac{1 \text{ mole } C_6H_{12}O_6}{180 \text{ g } C_6H_{12}O_6} \right)$$

$$= 2.99 \text{ moles}$$

(b) In this case there must be less than 1 mole.

Moles $C_6H_{12}O_6$ =

$$(1.00 \text{ g } C_6H_{12}O_6) \left(\frac{1 \text{ mole } C_6H_{12}O_6}{180 \text{ g } C_6H_{12}O_6} \right)$$

$$= 5.56 \times 10^{-3} \text{ mole}$$

The conversion of mass to moles and of moles to mass are frequently encountered in calculations using the mole concept. Notice that the number of moles is always the mass divided by the mass of one mole (the formula weight expressed in grams).

SAMPLE EXERCISE 3.9

What is the mass, in grams, of 0.433 mole of $C_6H_{12}O_6$?

Solution: Because this is less than 1 mole, the mass will be less than 180 g, the mass of 1 mole.

Grams $C_6H_{12}O_6$ =

$$(0.433 \text{ mole } C_6H_{12}O_6) \left(\frac{180 \text{ g } C_6H_{12}O_6}{1 \text{ mole } C_6H_{12}O_6} \right)$$

$$= 77.9 \text{ g}$$

Notice that the mass of a certain number of moles of a substance is always the number of moles times the mass of one mole.

SAMPLE EXERCISE 3.10

How many glucose molecules are there in 5.23 g of $C_6H_{12}O_6$?

Solution: Because this is less than a mole, there should be fewer than 6×10^{23} molecules.

Molecules $C_6H_{12}O_6$ =

$$(5.23 \text{ g } C_6H_{12}O_6) \left(\frac{1 \text{ mole } C_6H_{12}O_6}{180 \text{ g } C_6H_{12}O_6} \right)$$

$$\times \left(\frac{6.02 \times 10^{23} \text{ } C_6H_{12}O_6 \text{ molecules}}{1 \text{ mole } C_6H_{12}O_6} \right)$$

$$= 1.75 \times 10^{22} \text{ molecules}$$

SAMPLE EXERCISE 3.11

What is the mass, in grams, of 1.00×10^{23} molecules of $C_6H_{12}O_6$?

Solution:

Grams $C_6H_{12}O_6$ = $(1.00 \times 10^{23} \text{ molecules})$

$$\times \left(\frac{1 \text{ mole}}{6.02 \times 10^{23} \text{ molecules}} \right) \left(\frac{180 \text{ g}}{1 \text{ mole}} \right)$$

$$= 29.9 \text{ g}$$

3.9
simplest
formulas from
analyses

Before we use the mole concept to determine the masses of substances involved in chemical reactions, let's see how it is used in deducing the formulas of chemical substances. For example, consider ascorbic acid (vitamin C). The simplest formula of this compound can be determined by combustion of a weighed amount in an apparatus such as that shown in Figure 3.7. The amount of CO_2 produced can be measured by the mass increase in the CO_2 absorber. Combustion of 1.000 g of ascorbic acid produces 1.500 g of CO_2. Because CO_2 is $(12.0/44.0)100 = 27.3$ percent carbon, the original sample must have contained $(0.273)(1.500 \text{ g}) = 0.410$ g of carbon. Similarly, it can be determined that the compound contains 0.045 g of hydrogen. If it is established that ascorbic acid contains only C, H, and O, the amount of oxygen in the compound must be $1.000 \text{ g} - (0.410 \text{ g} + 0.045 \text{ g}) = 0.545$ g.

The simplest formula indicates the relative number of atoms of each type in the compound. Because the number of atoms is proportional to the number of moles, the simplest formula also indicates the relative number of moles of each element in the compound. The number of moles of each element in 1.000 g of ascorbic acid is

$$\text{Moles C} = (0.410 \text{ g C})\left(\frac{1 \text{ mole C}}{12.0 \text{ g C}}\right) = 0.0342 \text{ mole C}$$

$$\text{Moles H} = (0.045 \text{ g H})\left(\frac{1 \text{ mole H}}{1.01 \text{ g H}}\right) = 0.045 \text{ mole H}$$

$$\text{Moles O} = (0.545 \text{ g O})\left(\frac{1 \text{ mole O}}{16.0 \text{ g O}}\right) = 0.0341 \text{ mole O}$$

The relative number of moles of each element can be found by dividing by the smallest number, 0.0341. The ratio of $C:H:O$ is thus $1:1.32:1$, which is the same as $3:4:3$. The simplest formula is thus $C_3H_4O_3$. In order to determine the molecular formula an experiment must be performed to determine molecular weight. The molecular weight of ascorbic acid is 176 amu. The formula weight for $C_3H_4O_3$ is $3(12.0 \text{ amu}) + 4(1.0 \text{ amu}) + 3(16.0 \text{ amu}) = 88.0$ amu. Thus there are two of these formula units in the molecule, and the molecular formula is consequently $C_6H_8O_6$.

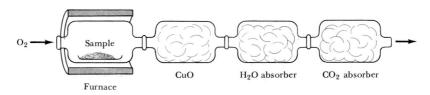

FIGURE 3.7 Apparatus for determining the percentage of carbon and hydrogen in a compound. Copper oxide serves to oxidize traces of carbon and carbon monoxide to carbon dioxide, and hydrogen to water. Magnesium perchlorate, $Mg(ClO_4)_2$, is used to absorb water, whereas sodium hydroxide, NaOH, absorbs carbon dioxide.

SAMPLE EXERCISE 3.12

Phosgene, a poison gas used during World War I, contains 12.1 percent C, 16.2 percent O, and 71.7 percent Cl. What is the empirical formula of phosgene?

Solution: For simplicity, we may assume that we have 100 g of material. The number of moles of each element is then

$$\text{Moles C} = (12.1 \text{ g})\left(\frac{1 \text{ mole C}}{12.0 \text{ g}}\right) = 1.01 \text{ moles C}$$

$$\text{Moles O} = (16.2 \text{ g})\left(\frac{1 \text{ mole O}}{16.0 \text{ g}}\right) = 1.01 \text{ moles O}$$

$$\text{Moles Cl} = (71.7 \text{ g})\left(\frac{1 \text{ mole Cl}}{35.5 \text{ g}}\right) = 2.02 \text{ moles Cl}$$

The simplest ratio, found by dividing each number by the smallest, 1.01, is $C:O:Cl = 1:1:2$ and the empirical formula is $COCl_2$. Because other experiments show that the molecular weight of the phosgene molecule is 99 amu, $COCl_2$ is also the molecular formula.

3.10
quantitative information from balanced equations

The mole concept provides a key to placing the quantitative information available in a balanced chemical equation on a practical, macroscopic level. Consider the following balanced equation:

$$2H_2(g) + O_2(g) \longrightarrow 2H_2O(l) \qquad [3.23]$$

The coefficients tell us that two molecules of H_2 react with each molecule of O_2 to form two molecules of H_2O. Therefore $2(6 \times 10^{23})$ molecules of H_2 will react with 6×10^{23} molecules of O_2 to form $2(6 \times 10^{23})$ molecules of H_2O. This is the same as saying that 2 moles of H_2 react with 1 mole of O_2 to form 2 moles of H_2O. The point is that the coefficients in a balanced equation can be interpreted *both* as the *relative numbers of molecules* (or formula units) involved in a reaction *and* as the *relative number of moles*. These interpretations are summarized in Table 3.5. Notice that 4.04 g of H_2 will react with each 32.00 g of O_2 to form 36.04 g of H_2O, because these are the masses of 2 moles of H_2, 1 mole of O_2 and 2 moles of H_2O, respectively. Notice also that the sum of the masses of the reactants equals the mass of the product as it must in any chemical reaction according to the law of conservation of mass. The quantities 2 moles H_2, 1 mole O_2, and 2 moles H_2O which are related by Equation [3.23] are called stoichiometrically equivalent quantities. We can represent this as

Mole ratio: 2 moles H_2 ≃ 1 mole O_2 ≃ 2 moles H_2O

Weight ratio: 4.04 g H_2 ≃ 32.0 g O_2 ≃ 36.0 g H_2O

where the symbol ≃ is taken to mean stoichiometrically equivalent to. These stoichiometric relations can be used to give conversion factors to relate quantities of reactants and products in a chemical reaction. For example, the number of moles of H_2O produced from 1.57 moles of O_2 can be calculated as follows:

$$\text{Moles H}_2\text{O} = (1.57 \text{ moles O}_2)\left(\frac{2 \text{ moles H}_2\text{O}}{1 \text{ mole O}_2}\right)$$

$$= 3.14 \text{ moles H}_2\text{O}$$

TABLE 3.5

interpretations of equations

	$2H_2$	+	O_2	$\longrightarrow$	$2H_2O$
Molecular ratio: 2 molecules		React with	1 molecule	To form	2 molecules
	$2(6 \times 10^{23})$ molecules	React with	6×10^{23} molecules	To form	$2(6 \times 10^{23})$ molecules
Mole ratio: 2 moles		React with	1 mole	To form	2 moles
Weight ratio: $2(2.02)$ g = 4.04 g		React with	32.00 g	To form	$2(18.02)$ g = 36.04 g

As a different example, consider the following reaction:

$$2CuFeS_2 + 5O_2 \longrightarrow 2Cu + 2FeO + 4SO_2 \qquad [3.24]$$

This equation describes a process in the smelting of copper using chalcopyrite ($CuFeS_2$) as the mineral source of the copper. Using the mole concept we can calculate the mass of Cu that can be produced from 1.00 g of chalcopyrite. From Equation [3.24] we can write the following stoichiometric relationships:

Mole ratio: 2 moles $CuFeS_2 \simeq$ 2 moles Cu

Weight ratio: $2(183$ g$)$ $CuFeS_2 \simeq 2(63.5$ g$)$ Cu

Using dimensional analysis we have

$$\text{Grams Cu} = (1.00 \text{ g } CuFeS_2)\left(\frac{127 \text{ g Cu}}{366 \text{ g } CuFeS_2}\right) = 0.346 \text{ g Cu}$$

We can similarly calculate the amount of SO_2 produced in the production of this quantity of copper using the following stoichiometric relationships:

Mole ratio: 2 moles $CuFeS_2 \simeq$ 4 moles SO_2

Weight ratio: $2(183$ g$)$ $CuFeS_2 \simeq 4(64.1$ g$)$ SO_2

Using dimensional analysis we have

$$\text{Grams } SO_2 = (1.00 \text{ g } CuFeS_2)\left(\frac{256 \text{ g } SO_2}{366 \text{ g } CuFeS_2}\right) = 0.700 \text{ g } SO_2$$

It is interesting to note that the mass of SO_2 produced in this reaction is approximately twice the mass of the copper. Consequently, considerable air pollution from sulfur dioxide is often generated in the vicinity of copper smelters (Figure 3.8).

FIGURE 3.8 Extended exposure to high concentrations of SO_2 and other pollutants can cause extensive damage to plants, animals, and structural materials. This photograph, taken in the Chest Creek Watershed, Clearfield County, Pennsylvania, shows the effects of air and water pollution. Timber in the center background has been killed by fumes from the burning mine-refuse pile at upper center. Stream pollution by mine acid and coal sedimentation is shown at center, whereas soil erosion is evident in the foreground. (*USDA-SCS*)

SAMPLE EXERCISE 3.13

How much water is produced in the combustion of 1.00 g of glucose, $C_6H_{12}O_6$: $C_6H_{12}O_6(s) + 6O_2(g) \longrightarrow 6CO_2(g) + 6H_2O(l)$?

Solution: For this reaction we have

$$1 \text{ mole } C_6H_{12}O_6 \simeq 6 \text{ moles } H_2O$$
$$180 \text{ g } C_6H_{12}O_6 \simeq 6(18.0 \text{ g})H_2O$$

Therefore,

$$\text{Grams } H_2O = (1.00 \text{ g } C_6H_{12}O_6)\left(\frac{108 \text{ g } H_2O}{180 \text{ g } C_6H_{12}O_6}\right)$$
$$= 0.600 \text{ g } H_2O$$

This type of problem can also be solved in a single step by stepwise conversion of grams $C_6H_{12}O_6$ to moles $C_6H_{12}O_6$ to moles H_2O to grams H_2O:

$$\text{Grams } H_2O = (1.00 \text{ g } C_6H_{12}O_6)\left(\frac{1 \text{ mole } C_6H_{12}O_6}{180 \text{ g } C_6H_{12}O_6}\right)$$
$$\times \left(\frac{6 \text{ moles } H_2O}{1 \text{ mole } C_6H_{12}O_6}\right)\left(\frac{18.0 \text{ g } H_2O}{1 \text{ mole } H_2O}\right)$$
$$= 0.600 \text{ g } H_2O$$

We may note that an average man ingests 2 l. of water daily and eliminates 2.4 l. The difference is produced in metabolism of foodstuffs as above. (Metabolism is a general term used to describe all the processes of a living animal or plant.) The desert rat (kangaroo rat) is able to take great advantage of its metabolic water to help it survive in the dry desert. In fact, it apparently never drinks water.

LIMITING REAGENT

In many situations an excess of one or more substances is available for chemical reaction. Some will therefore be left over when the reaction is complete. For example, consider again the combustion of hydrogen:

$$2H_2(g) + O_2(g) \longrightarrow 2H_2O(l)$$

Suppose that 2 moles of H_2 and 2 moles of O_2 are available for reaction. The balanced equation tells us that only 1 mole of O_2 is required to completely consume 2 moles of H_2, thereby forming 2 moles of H_2O; 1 mole of O_2 will therefore be left over at the end of the reaction. The amount of the substance that is completely consumed in any reaction will determine how much product is formed. This reagent is called the **limiting reagent.** In this example, H_2 is the limiting reagent.

SAMPLE EXERCISE 3.14

Part of the SO_2 that is introduced into the atmosphere ends up being converted to sulfuric acid, H_2SO_4. The net reaction is

$$2SO_2(g) + O_2(g) + 2H_2O(l) \longrightarrow 2H_2SO_4(l)$$

How much H_2SO_4 can be prepared from 5.0 moles of SO_2, 1.0 mole of O_2, and an unlimited quantity of H_2O?

Solution: The number of moles of O_2 needed for complete consumption of 5.0 moles of SO_2 is

$$\text{Moles } O_2 = (5.0 \text{ moles } SO_2)\left(\frac{1 \text{ mole } O_2}{2 \text{ moles } SO_2}\right)$$

$$= 2.5 \text{ moles } O_2$$

This quantity of O_2 is not available; therefore, all of the SO_2 cannot be consumed; O_2 must be the limiting reagent. We use the quantity of the limiting reagent, O_2, to calculate the quantity of H_2SO_4 prepared:

$$\text{Moles } H_2SO_4 = (1.0 \text{ mole } O_2)\left(\frac{2 \text{ moles } H_2SO_4}{1 \text{ mole } O_2}\right)$$

$$= 2.0 \text{ moles } H_2SO_4$$

We might note that in forming 2.0 moles of H_2SO_4, 2.0 moles of SO_2 are required. Therefore 3.0 moles of SO_2 are left over.

SAMPLE EXERCISE 3.15

When solutions of lead nitrate, $Pb(NO_3)_2$, and sodium chloride, $NaCl$, are mixed, the following reaction occurs:

$$Pb(NO_3)_2(aq) + 2NaCl(aq) \longrightarrow$$
$$PbCl_2(s) + 2NaNO_3(aq)$$

How many grams of $PbCl_2$ can be prepared from 1.00 g of $Pb(NO_3)_2$ and 1.00 g of $NaCl$?

Solution: The number of moles of available reactants is as follows:

$$\text{Moles } Pb(NO_3)_2 =$$

$$(1.00 \text{ g } Pb(NO_3)_2)\left(\frac{1 \text{ mole } Pb(NO_3)_2}{331 \text{ g } Pb(NO_3)_2}\right)$$

$$= 0.00302 \text{ mole } Pb(NO_3)_2$$

$$\text{Moles } NaCl = (1.00 \text{ g } NaCl)\left(\frac{1 \text{ mole } NaCl}{58.5 \text{ g } NaCl}\right)$$

$$= 0.0171 \text{ mole } NaCl$$

The number of moles of $NaCl$ needed for complete reaction of $Pb(NO_3)_2$ is

$$\text{Moles } NaCl =$$

$$(0.00302 \text{ mole } Pb(NO_3)_2)\left(\frac{2 \text{ moles } NaCl}{1 \text{ mole } Pb(NO_3)_2}\right)$$

$$= 0.00604 \text{ mole } NaCl$$

Thus there is enough $NaCl$ for complete reaction of $Pb(NO_3)_2$ with some $NaCl$ left over; $Pb(NO_3)_2$ is therefore the limiting reagent. We use the quantity of the limiting reagent, $Pb(NO_3)_2$, to calculate the quantity of $PbCl_2$ formed:

$$1 \text{ mole } Pb(NO_3)_2 \simeq 1 \text{ mole } PbCl_2$$
$$331 \text{ g } Pb(NO_3)_2 \simeq 278 \text{ g } PbCl_2$$

Therefore,

$$\text{Grams } PbCl_2 =$$

$$(1.00 \text{ g } Pb(NO_3)_2)\left(\frac{278 \text{ g } PbCl_2}{331 \text{ g } Pb(NO_3)_2}\right)$$

$$= 0.840 \text{ g } PbCl_2$$

3.11

molarity and solution stoichiometry

Having taken the concept of moles this far, let's take it a step further. Let's consider its use in dealing with solutions. As our experience with chemistry grows, we shall encounter an ever-increasing number of reactions that occur in solution, especially ones in which water is a component. Liquids such as water are important media for many types of chemical reactions. For example, the reaction between NaCl and $Pb(NO_3)_2$, cited in Sample Exercise 3.15, takes place in aqueous (water) solution. In discussing such solutions it is often convenient to call one component the **solvent** and the others **solutes**. The component of a solution whose physical state is preserved when the solution is formed is known as the solvent. For example, when sodium chloride (a solid) is mixed with water, the resultant solution is a liquid. Consequently, water is referred to as the solvent and sodium chloride as the solute. If all components of a solution are in the same state, the one present in greatest amount is called the solvent.

The term **concentration** is used to denote the amount of solute dissolved in a given quantity of solvent or solution. The method for expressing concentration that is most useful for discussing solution stoichiometry is **molarity**. The molarity of a solution is defined as the number of moles of solute in a liter of solution (soln):

$$\text{Molarity} = \frac{\text{moles solute}}{\text{volume of soln in liters}} \qquad [3.25]$$

A 1.50 molar solution (written 1.50 M) contains 1.50 moles of solute in every liter of solution. To make a liter of 0.150 M sucrose, $C_{12}H_{22}O_{11}$, in water requires 0.150 mole of $C_{12}H_{22}O_{11}$. This is diluted to a total volume of 1 l. A volumetric flask, which is a flask calibrated to contain a precise volume of liquid, is used for this purpose. The operation is shown in Figure 3.9.

SAMPLE EXERCISE 3.16

Calculate the molarity of a solution made by dissolving 23.4 g of sodium sulfate (Na_2SO_4) in enough water to form 125 ml of solution.

Solution:

$$\text{Molarity} = \frac{\text{moles } Na_2SO_4}{\text{liters soln}}$$

$$\text{Moles } Na_2SO_4 =$$

$$(23.4 \text{ g } Na_2SO_4)\left(\frac{1 \text{ mole } Na_2SO_4}{142 \text{ g } Na_2SO_4}\right)$$

$$= 0.165 \text{ mole } Na_2SO_4$$

$$\text{Molarity} =$$

$$\left(\frac{0.165 \text{ mole } Na_2SO_4}{125 \text{ ml soln}}\right)\left(\frac{1000 \text{ ml soln}}{1 \text{ liter soln}}\right)$$

$$= 1.32 \frac{\text{moles } Na_2SO_4}{\text{liter soln}}$$

$$= 1.32 \, M$$

One advantage to the expression of concentration in molarity is that it allows us to measure out a solution volume of known concentration and readily calculate the number of moles of solute dispensed. Molarity can be used to interconvert volume and moles just as density can be used to interconvert mass and volume (Sample Exercise 1.9). Calculation of the

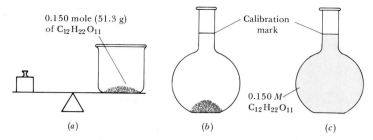

FIGURE 3.9 Procedure for preparation of 1-l. of a 0.150 M solution of $C_{12}H_{22}O_{11}$. (a) Weigh out 0.150 mole (51.3 g) of $C_{12}H_{22}O_{11}$ (MW = 342 amu). (b) Add the $C_{12}H_{22}O_{11}$ to a 1-l. volumetric flask. (c) Add water until the solution reaches the calibration mark.

number of moles of HNO_3 in 2.0 l. of 0.200 M HNO_3 solution illustrates conversion of volume to moles:

$$\text{Moles } HNO_3 = (2.0 \text{ l. soln})\left(0.200 \frac{\text{mole } HNO_3}{\text{l. soln}}\right)$$

$$= 0.40 \text{ mole } HNO_3$$

Notice how dimensional analysis can be used in this conversion if we express molarity as moles HNO_3/l. soln. Notice also that to obtain moles we multiplied liters and molarity: moles = liters $\times$ M. This same expression for moles can be obtained directly by algebraic rearrangement of Equation [3.25]. The use of molarity to convert moles to volume can be illustrated by calculating the volume of 0.30 M HNO_3 solution required to supply 2.0 moles of HNO_3:

$$\text{Liters soln} = (2.0 \text{ moles } HNO_3)\left(\frac{1 \text{ l. soln}}{0.30 \text{ mole } HNO_3}\right)$$

$$= 6.7 \text{ l. soln}$$

In this case we needed to apply the reciprocal of molarity to convert moles to volume: liters = moles $\times$ $1/M$.

SAMPLE EXERCISE 3.17

How many grams of Na_2SO_4 are required to make 350 ml of 0.50 M Na_2SO_4?

Solution: Because

$$M \, Na_2SO_4 = \frac{\text{moles } Na_2SO_4}{\text{l. soln}}$$

$$\text{Moles } Na_2SO_4 = (0.350 \text{ l. soln})\left(0.50 \frac{\text{mole } Na_2SO_4}{\text{l. soln}}\right)$$

$$= 0.175 \text{ mole } Na_2SO_4$$

Because each mole of Na_2SO_4 weighs 142 g, the required number of grams of Na_2SO_4 is

$$(0.175 \text{ mole } Na_2SO_4)\left(\frac{142 \text{ g } Na_2SO_4}{1 \text{ mole } Na_2SO_4}\right)$$

$$= 24.8 \text{ g } Na_2SO_4$$

It is also possible to work this problem by direct conversion of milliliters to liters to moles to grams. In doing so, we use molarity as a conversion factor between volume and moles:

$$\text{Grams } Na_2SO_4 = (350 \text{ ml soln})\left(\frac{1 \text{ l. soln}}{1000 \text{ ml soln}}\right)$$

$$\times \left(0.50 \frac{\text{mole } Na_2SO_4}{\text{l. soln}}\right)\left(\frac{142 \text{ g } Na_2SO_4}{1 \text{ mole } Na_2SO_4}\right)$$

$$= 24.8 \text{ g } Na_2SO_4$$

DILUTION

It is often convenient to make a solution of a certain concentration from a more concentrated solution. For example, suppose you need to prepare a liter of 0.10 M HNO_3 solution from a solution of 1.0 M HNO_3. The desired solution will contain 0.10 mole of HNO_3. Therefore, you need to remove 0.10 mole of HNO_3 from the 1.0 M solution. There is 0.10 mole of HNO_3 in 100 ml of the 1.0 M solution. Thus, 100 ml is withdrawn from the 1.0 M HNO_3 solution using a pipet and added to a 1 l. volumetric flask. It is then diluted to 1 l. as shown in Figure 3.10.*

When more solvent is added to a solution, thereby diluting it, the number of moles of solute remains unchanged:

$$\text{Moles solute before dilution} = \text{moles solute after dilution} \quad [3.26]$$

Because moles = $M \times$ liters, we can write:

$$\text{(Initial molarity)(initial volume)} = \text{(final molarity)(final volume)}$$
$$M_{initial}V_{initial} = M_{final}V_{final} \quad [3.27]$$

SAMPLE EXERCISE 3.18

How much 3.0 M H_2SO_4 would be required to make 500 ml of 0.10 M H_2SO_4?

Solution: Using Equation [3.27], $M_{initial}V_{initial} = M_{final}V_{final}$, we can write

$$V_{initial} = \frac{M_{final}V_{final}}{M_{initial}}$$
$$= \frac{(0.10 \ M)(500 \ ml)}{(3.0 \ M)}$$
$$= 17 \ ml$$

We see that if we start with 17 ml of 3.0 M H_2SO_4 and dilute it to a total volume of 500 ml, the desired 0.10 M solution will be obtained.

TITRATION

The procedure by which a solution of known concentration is added to another solution until the chemical reaction between the two solutes is complete is known as a **titration**. Titrations are widely used in chemistry to analyze the compositions of mixtures. The solution whose concentration is known is called the **standard solution**. In titrations the standard solution is slowly added from a buret to a solution that contains a known volume or known mass of solute. The latter solution is commonly referred to as the **unknown**. The point at which stoichiometrically equivalent quantities of substances have been brought together is known as the **equivalence point** of the titration.

*In diluting an acid or base, the acid or base should be added to water, then further diluted by addition of more water. Adding water directly to an acid or base can cause spattering because of the intense heat generated.

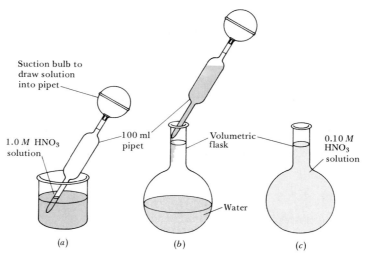

FIGURE 3.10 Procedure for preparation of 1-l. of 0.10 M HNO₃ by dilution of 1.0 M HNO₃. (*a*) Draw 100 ml of the 1.0 M solution into a pipet. (*b*) Add this amount of 1.0 M HNO₃ to a small amount of water in a 1-l. volumetric flask. (*c*) Add additional water to dilute the solution to a total volume of 1-l. The result is a 0.10 M HNO₃ solution.

In order to titrate an unknown with a standard solution, there must be some way to determine when the equivalence point of the titration has been reached. In acid-base titrations, organic dyes known as acid-base **indicators** are used for this purpose. For example, the dye known as phenolphthalein is colorless in acidic solution but is red in basic solution. If phenolphthalein is added to an unknown solution of acid, the solution will be colorless. Standard base can then be added from the buret until the solution barely turns from colorless to red. This indicates that the acid has been neutralized, and the drop of base that caused the solution to become colored has no acid to react with. The solution therefore becomes basic, and the dye turns red. The experimental procedure for the titration of NaOH and H₂SO₄ is summarized in Figure 3.11.

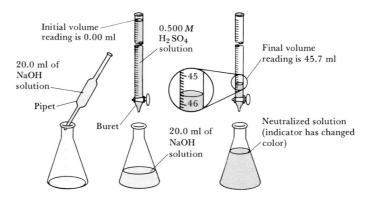

FIGURE 3.11 Procedure for titration of an unknown base against a standardized solution of H₂SO₄. (*a*) A known quantity of base is added to a flask. (*b*) An acid-base indicator is added, and standardized H₂SO₄ is added from a buret. (*c*) Equivalence point is signaled by a color change of the indicator. The concentrations and volumes shown correspond to the example discussed in Sample Exercise 3.20.

SAMPLE EXERCISE 3.19

What volume of 0.500 M NaCl is required to react completely with 0.200 moles of $Pb(NO_3)_2$? The chemical equation for this reaction is

$$2NaCl(aq) + Pb(NO_3)_2(aq) \longrightarrow$$
$$2NaNO_3(aq) + PbCl_2(s)$$

Solution: According to the reaction equation

$$2 \text{ moles NaCl} \simeq 1 \text{ mole } Pb(NO_3)_2$$

Therefore,

Moles NaCl =

$$(0.200 \text{ mole } Pb(NO_3)_2)\left(\frac{2 \text{ moles NaCl}}{1 \text{ mole } Pb(NO_3)_2}\right)$$

$$= 0.400 \text{ mole NaCl}$$

Because

$$\text{Liters NaCl soln} = \frac{\text{moles NaCl}}{M \text{ NaCl soln}}$$

$$\text{Liters NaCl soln} = \frac{0.400 \text{ mole NaCl}}{0.500 \text{ mole NaCl/l. soln}}$$

$$= 0.800 \text{ l.}$$

Therefore, 800 ml of the 0.500 M NaCl solution could be measured out and added to the 0.200 mole of $Pb(NO_3)_2$. This problem can also be solved by direct conversion of moles $Pb(NO_3)_2$ to moles NaCl to volume of NaCl solution:

$$\text{Liters NaCl soln} = (0.200 \text{ mole } Pb(NO_3)_2)$$

$$\times \left(\frac{2 \text{ moles NaCl}}{1 \text{ mole } Pb(NO_3)_2}\right)\left(\frac{1 \text{ l. NaCl soln}}{0.500 \text{ mole NaCl}}\right)$$

$$= 0.800 \text{ l.}$$

SAMPLE EXERCISE 3.20

One method used commercially to peel potatoes is to soak them in a solution of NaOH for a short time, and then spray off the peel after the potatoes are removed from the NaOH. The concentration of NaOH is normally in the range of 3 to 6 M. The NaOH is analyzed periodically to determine its ability to peel potatoes rapidly. In one such analysis, 45.7 ml of 0.500 M H_2SO_4 are required to react completely with a 20.0 ml sample of NaOH solution:

$$H_2SO_4(aq) + 2NaOH(aq) \longrightarrow$$
$$2H_2O(l) + Na_2SO_4(aq)$$

What is the concentration of the NaOH solution?

Solution:

Moles H_2SO_4 =

$$(45.7 \text{ ml soln})\left(\frac{1 \text{ l. soln}}{1000 \text{ ml soln}}\right)\left(0.500 \frac{\text{mole } H_2SO_4}{\text{l. soln}}\right)$$

$$= 2.28 \times 10^{-2} \text{ mole } H_2SO_4$$

According to the balanced equation, 1 mole $H_2SO_4 \simeq 2$ moles NaOH. Therefore,

Moles NaOH =

$$(2.28 \times 10^{-2} \text{ mole } H_2SO_4)\left(\frac{2 \text{ moles NaOH}}{1 \text{ mole } H_2SO_4}\right)$$

$$= 4.56 \times 10^{-2} \text{ mole NaOH}$$

Knowing the number of moles of NaOH present in 20.0 ml of solution allows us to calculate the concentration of this solution:

$$M_{\text{NaOH}} = \frac{\text{moles NaOH}}{\text{l. soln}} =$$

$$\left(\frac{4.56 \times 10^{-2} \text{ mole NaOH}}{20.0 \text{ ml soln}}\right)\left(\frac{1000 \text{ ml soln}}{1 \text{ l. soln}}\right)$$

$$= 2.28 \frac{\text{moles NaOH}}{\text{l. soln}} = 2.28 \text{ } M$$

SAMPLE EXERCISE 3.21

The quantity of Cl^- in a water supply is determined by titrating the sample against $AgNO_3$:

$$AgNO_3(aq) + Cl^-(aq) \longrightarrow AgCl(s) + NO_3{}^-(aq)$$

What mass of chloride ion is present in a 10.0 g sample of the water if 20.2 ml of 0.100 M $AgNO_3$ is

required to react with all of the chloride in the sample?

Solution:

Moles $AgNO_3$ =

$$(20.2 \text{ ml soln}) \left(\frac{1 \text{ l. soln}}{1000 \text{ ml soln}} \right) \left(0.100 \frac{\text{mole AgNO}_3}{\text{l. soln}} \right)$$
$$= 2.02 \times 10^{-3} \text{ mole AgNO}_3$$

From the balanced equation we see that 1 mole $AgNO_3 \simeq 1$ mole Cl^-. Therefore the sample must contain 2.02×10^{-3} mole Cl^-:

Moles $Cl^- =$
$$(2.02 \times 10^{-3} \text{ mole AgNO}_3) \left(\frac{1 \text{ mole Cl}^-}{1 \text{ mole AgNO}_3} \right)$$
$$= 2.02 \times 10^{-3} \text{ mole Cl}^-$$

The number of moles of Cl^- can then be converted to grams:

Grams $Cl^- =$
$$(2.02 \times 10^{-3} \text{ mole Cl}^-) \left(\frac{35.5 \text{ g Cl}^-}{1 \text{ mole Cl}^-} \right)$$
$$= 7.17 \times 10^{-2} \text{ g Cl}^-$$

We might note that the percentage of Cl^- in the water is

$$\% \text{ Cl}^- = \frac{7.7 \times 10^{-2} \text{ g Cl}^-}{10.0 \text{ g soln}} \times 100 = 0.717\%$$

Chloride ion is one of the major ions in water and sewage. Ocean water contains 1.92 percent Cl^-. Whether or not water containing Cl^- exhibits a salty taste depends on the other ions present. If the accompanying ions are Na^+ ions, a salty taste may be detected with as little as 0.03 percent Cl^-.

FOR REVIEW

summary

The **law of conservation of mass** states that there are no detectable changes in mass in any chemical reaction. This indicates that there are the same number of atoms of each type present after a chemical reaction as there were before the reaction. A **balanced equation** shows equal numbers of each type of atom on each side of the equation and is thereby consistent with the law of conservation of mass. We discussed how equations are balanced by placing coefficients in front of the chemical formulas for the substances involved in the reaction. We have also seen how it is possible to predict the products of simple reactions by analogy to known reactions and by use of the periodic table.

The coefficients in a balanced equation can be interpreted as either the relative number of **formula units** involved in the reaction or the relative number of moles. A **mole** of any substance is **Avogadro's number** (6.022×10^{23}) of formula units of that substance. The mass of a mole of atoms, molecules, or ions is the formula weight expressed in grams. For example, a single molecule of H_2O weighs 18 amu; a mole of H_2O weighs 18 g. We have seen how the mole concept can be used to determine the simplest formula of a compound and to calculate the quantities involved in chemical reactions. The concept of **molarity** is useful in dealing with the quantities involved in reactions that take place in solution. We have seen that molarity serves as a conversion factor for interconverting volume and moles.

learning goals

Having read and studied this chapter you should be able to:

1. Balance chemical equations.
2. Predict the products of a chemical reaction, having seen a suitable analogy.
3. Interconvert number of moles, mass in grams, and number of atoms, ions, or molecules.
4. Calculate the simplest formula of a compound, having been given appropriate analytical data such as elemental percentages or the quantity of CO_2 and H_2O produced by combustion.

5. Calculate the molecular formula, having been given the empirical formula and molecular weight.

6. Calculate the mass of a particular substance produced or used in a chemical reaction (mass-mass problems).

7. Determine the limiting reagent in a reaction.

8. Define molarity.

9. Calculate volume, moles, or molarity of a solution given two of these quantities and use these ideas in dilution and titration problems.

key terms

Some of the more important terms and expressions used for the first time in this chapter are the following:

Atomic weight (Section 3.5) or atomic mass is the mass of an atom relative to a mass of exactly 12 assigned to the ^{12}C isotope of carbon.

Atomic mass unit (amu) (Section 3.5) is the unit in which the masses of individual atoms and molecules are usually expressed. For example, the mass of a sulfur atom is 32.1 amu; 6.022×10^{23} amu $= 1$ g.

A **balanced equation** (Section 3.4) is a chemical equation in which all atoms appear in equal numbers on both sides of the equation.

Formula weight (Section 3.5) is the weight of the collection of atoms represented by a chemical formula. For example, the formula weight of NO_2 (46.0 amu) is the sum of the weights of a nitrogen and two oxygen atoms. If the formula is the molecular formula of the substance, the formula weight is the **molecular weight** of the substance.

A **mole** (Section 3.8) is a collection of **Avogadro's number** (6.022×10^{23}) of objects; for example, a mole of H_2O is 6.022×10^{23} H_2O molecules.

Molarity (Section 3.11) is the concentration of a solution expressed as moles of solute per liter of solution; abbreviated M.

A **titration** (Section 3.11) is the process of reacting a solution of unknown concentration with one of known concentration (a **standardized solution**). The procedure is commonly used to determine the concentrations of solutions of unknown concentration.

EXERCISES

balancing equations

3.1   What is the difference between O_2 and $2O$?

3.2   Balance the following equations:

(a) $Ba(OH)_2(aq) + HCl(aq) \longrightarrow BaCl_2(aq) + H_2O(l)$

(b) $CaCl_2(aq) + Na_3PO_4(aq) \longrightarrow Ca_3(PO_4)_2(s) + NaCl(aq)$

(c) $CH_3OH(l) + O_2(g) \longrightarrow CO_2(g) + H_2O(l)$

(d) $NH_4NO_3(aq) + NaOH(aq) \longrightarrow NH_3(aq) + H_2O(l) + NaNO_3(aq)$

(e) $PCl_5(s) + H_2O(l) \longrightarrow H_3PO_4(aq) + HCl(aq)$

3.3   Balance the following equations:

(a) $H_2SO_4(aq) + NaOH(aq) \longrightarrow Na_2SO_4(aq) + H_2O(l)$

(b) $Al_4C_3(s) + H_2O(l) \longrightarrow Al(OH)_3(s) + CH_4(g)$

(c) $C_6H_6(l) + O_2(g) \longrightarrow CO_2(g) + H_2O(l)$

(d) $HI(g) + Cl_2(g) \longrightarrow HCl(g) + I_2(s)$

(e) $HCl(g) + NH_3(g) \longrightarrow NH_4Cl(s)$

3.4   Write balanced chemical equations for each of the following reactions: (a) Ammonia, NH_3, is combusted, thereby reacting with O_2 from the air to form N_2 and H_2O. (b) Methanol, CH_3OH, reacts with oxygen gas to form CO_2 and H_2O. (c) Nitroglycerin, $C_3H_5(NO_3)_3$, decomposes to form N_2, O_2, CO_2, and H_2O.

3.5   The objectionable taste of chlorine in highly chlorinated water can be removed by passing the water through a bed of charcoal. The charcoal, which is primarily carbon, reacts with the chlorine as follows:

$$C(s) + Cl_2(g) + H_2O(l) \longrightarrow CO_2(g) + HCl(aq)$$

Balance this equation.

predicting products

3.6 Predict the products and write the balanced chemical equation for each of the following reactions: (a) the combustion of ethane, C_2H_6; (b) the combustion of propanol, C_3H_7OH; (c) the reaction between nitric acid and calcium hydroxide, $Ca(OH)_2$; (d) the reaction between lithium metal and water; (e) the reaction between sulfuric acid and potassium hydroxide, KOH.

3.7 Given the following reactions:

$$2HCl + CaCO_3 \longrightarrow CaCl_2 + CO_2 + H_2O$$
$$Zn + 2HCl \longrightarrow H_2 + ZnCl_2$$
$$Na_2O + H_2O \longrightarrow 2NaOH$$

Predict what will happen in the following cases and write a balanced chemical equation for each reaction: (a) hydrochloric acid is added to $BaCO_3$; (b) nitric acid is added to $CaCO_3$; (c) potassium oxide, K_2O, is added to water; (d) zinc metal is added to hydrobromic acid, HBr; (e) nitric acid is added to zinc metal.

atomic and molecular weights

3.8 Calculate the molecular weight of (a) H_3PO_4; (b) C_2H_5OH; (c) O_2; (d) H_2O_2.

3.9 Calculate the formula weight of (a) $Ba(NO_3)_2$; (b) $C_6H_{12}O_6$; (c) H_2SO_4; (d) $Ca_3(PO_4)_2$.

3.10 What is the difference between molecular weight and formula weight?

3.11 A series of molecules containing the hypothetical element X are analyzed as shown below:

WT. OF X PER MOLECULE

Substance 1	10 amu
Substance 2	20 amu
Substance 3	15 amu

What are the possible atomic weights of element X?

3.12 The mass spectrum of neon indicates that it consists of three isotopes with masses 20.0, 21.0, and 22.0 amu with relative abundances of 90.9 percent, 0.3 percent, and 8.8 percent, respectively. Calculate the average atomic weight of neon.

3.13 The mass spectrum of carbon indicates that it contains 98.89 percent ^{12}C and 1.11 percent ^{13}C. Calculate the average atomic weight of carbon.

quantities of mass, moles, atoms, and molecules

3.14 Calculate the mass of 1 mole of (a) H_2O_2; (b) H_2SO_4; (c) $C_{12}H_{22}O_{11}$; (d) $Ba(OH)_2$.

3.15 Calculate the percentage of hydrogen, sulfur, and oxygen in H_2SO_4.

3.16 Calculate the percentage of sodium, oxygen, and hydrogen in $NaOH$.

3.17 Calculate the number of moles in (a) 10.0 g of H_2SO_4; (b) 5.0 g of O_2; (c) 3.0 g of Ne; (d) 1.32×10^{20} atoms of neon; (e) 6.02×10^{24} molecules of H_2SO_4; (f) 3.01×10^{23} molecules of O_2.

3.18 Calculate the number of moles in (a) 10.0 g of $Ca(NO_3)_2$; (b) 8.6 g of Ar; (c) 0.30 g of N_2; (d) 6.02×10^{24} molecules of N_2; (e) 1.66×10^{23} atoms of Ar; (f) 1.66×10^{23} formula units of $Ca(NO_3)_2$.

3.19 Calculate the number of oxygen atoms in (a) 5.0 g of O_2; (b) 5.0 g of $KClO_3$.

3.20 What is the mass (in grams) of (a) 2.0×10^{23} atoms of oxygen; (b) 6.022×10^{23} molecules of $C_6H_{12}O_6$; (c) 0.67 mole of $C_6H_{12}O_6$; (d) 1.55 mole of O_2?

3.21 What is the difference between a mole and a molecule?

3.22 Epinephrine (adrenaline) is a hormone secreted in the bloodstream in times of danger or stress. Its chemical formula is $C_9H_{13}O_3N$. (a) What is the molecular weight of epinephrine? (b) What is the mass of one mole of epinephrine? (c) What is the percentage of oxygen in epinephrine? (d) How many moles are there in 1.00 g of this substance? (e) How many molecules are there in 1.00 g of this substance? (f) How many carbon atoms are there in 1.00 g of this substance? (g) What is the mass of 1.00×10^{23} molecules?

3.23 If 0.45 mole of a substance has a mass of 55 g, what is its molecular weight?

3.24 Hemoglobin is a complex protein molecule that carries oxygen from the lungs to the tissues where it is released. The molecular weight of hemoglobin is 68,000 amu. Each molecule contains four iron atoms that are the sites where the oxygen is bound. (a) What is the percentage iron in hemoglobin? (b) How much hemoglobin would contain 1.00 g of iron?

chemical formulas

3.25 Nicotine has a simplest formula of C_5H_7N and a molecular weight of 162.2 amu. What is its molecular formula?

3.26 Ethylene glycol, the substance used as the primary component of most antifreeze solutions, has a simplest formula of CH_3O and a molecular weight of 62.1 amu. What is its molecular formula?

3.27 Determine the simplest formulas for compounds with the following compositions: (a) 15.8 percent carbon and 84.2 percent sulfur; (b) 40.0 percent carbon, 6.7 percent hydrogen, and 53.3 percent oxygen; (c) 36.5 percent sodium, 0.8 percent hydrogen, 24.6 percent phosphorus, and 38.1 percent oxygen.

3.28 Determine the simplest formulas for the following substances: (a) limestone, which consists of 40.0 percent Ca, 12.0 percent C, and 48.0 percent O; (b) ethanol, which consists of 52.2 percent C, 13.0 percent H, and 34.8 percent O.

3.29 Cyclopropane, a substance used with oxygen as a general anesthetic, contains only two elements, carbon and hydrogen. When 1.00 g of this substance is completely combusted, 3.14 g of CO_2 and 1.29 g of H_2O are produced. What is the simplest formula of the cyclopropane?

[3.30] Butane has a simplest formula of C_2H_5 and a density of 2.67 g/l. at $0°C$ and 1 atm pressure. Under these same conditions, the density of hydrogen is 0.0899 g/l. Calculate the molecular weight of butane. What is its molecular formula?

mass relations in chemical reactions

3.31 What mass of PbO is obtained by heating 100 g of $PbCO_3$ according to the following equation:

$$PbCO_3(s) \xrightarrow{\Delta} PbO(s) + CO_2(g)?$$

3.32 Calculate the mass of $KClO_3$ required to produce 10.0 g of O_2 via the following reaction:

$$2KClO_3(s) \xrightarrow{\Delta} 2KCl(s) + 3O_2(g)$$

3.33 What mass of carbon dioxide will form when 10.0 lb of carbon is completely combusted?

3.34 How much zinc is needed to produce 10.0 g of H_2 according to the following reaction:

$$Zn(s) + 2HCl(aq) \longrightarrow ZnCl_2(aq) + H_2(g)$$

3.35 A mixture of 4.00 g of H_2 and 6.78 g of O_2 are caused to react to form H_2O. How much H_2, O_2, and H_2O remain after the reaction is complete?

3.36 The reaction

$$N_2(g) + 3H_2(g) \longrightarrow 2NH_3(g)$$

is the basis of the Haber process for the synthesis of ammonia. (a) How many moles of ammonia are produced by reaction of 1.00 mole of N_2 and 3.00 moles of H_2? (b) What is the maximum amount of ammonia that can be prepared from 1.00 mole of N_2 and 1.00 mole of H_2? (c) What is the maximum amount of ammonia that can be prepared from 2.50 moles of N_2 and 10.0 moles of H_2? How much N_2 or H_2 is left over? (d) If 1.00 l. of N_2 and 1.00 l. of H_2, measured at the same pressure and temperature, react, how many liters of NH_3 will be produced?

[3.37] Many recipes call for the addition of vinegar or sour milk to a dough containing baking soda (sodium bicarbonate, $NaHCO_3$). Vinegar and sour milk contain acids that react with the baking soda and thereby liberate $CO_2(g)$, which leavens the dough. How much vinegar, which is 5 percent acetic acid ($HC_2H_3O_2$), is required to consume completely 50.0 g of baking soda according to the reaction

$$NaHCO_3(s) + HC_2H_3O_2(aq) \longrightarrow$$
$$NaC_2H_3O_2(aq) + CO_2(g) + H_2O(l)$$

Baking powder consists of baking soda together with a solid acid which reacts when moistened.

3.38 In developing black-and-white film, unexposed AgBr is removed from the film with sodium thiosulfate ($Na_2S_2O_3$). The reaction is as follows:

$$AgBr(s) + 2Na_2S_2O_3(aq) \longrightarrow$$
$$Na_3[Ag(S_2O_3)_2](aq) + NaBr(aq)$$

How much $Na_2S_2O_3$ is required to remove 1.0 g of AgBr?

[3.39] If a company emits 4.5 tons of SO_2 per day and is able to construct a plant capable of converting 70 percent of this SO_2 to sulfuric acid, H_2SO_4, using the net reaction

$$2SO_2(g) + O_2(g) + 2H_2O(l) \longrightarrow$$
$$2H_2SO_4(aq)$$

how much H_2SO_4 can be produced per day?

solution stoichiometry

3.40 Calculate the molarity of each of the following solutions: (a) 50 g of H_2SO_4 in 350 ml of solution; (b) 5.0 ml of 3.0 M H_2SO_4 solution diluted to a total volume of 250 ml; (c) 0.30 mole of HNO_3 in 450 ml of solution; (d) a solution formed by mixing 50.0 ml of 0.20 M NaCl and 100.0 ml of 0.10 M NaCl (assume 150.0 ml total volume).

3.41 Chlorox laundry bleach contains approximately 50 g of NaOCl (sodium hypochlorite) per liter of solution. What is the molarity of this solution?

3.42 Fluoridated water supplies contain a maximum of 1 mg of fluoride ion per liter of solution. Calculate the molarity of the solution.

3.43 Describe how you would prepare each of the following solutions: (a) 200 ml of 0.200 M $AgNO_3$ from pure $AgNO_3$; (b) 500 ml of 0.200 M HCl from 1.00 M HCl.

3.44 Describe how you would prepare 100 ml of 0.100 M glucose solution (a) given solid glucose, $C_6H_{12}O_6$; (b) given a liter of 2.00 M glucose solution.

3.45 Calculate the number of moles of solute present in each of the following solutions: (a) 50.0 ml of 0.25 M H_2SO_4; (b) a solution made by dissolving 27.5 g of NaCl in enough water to form a liter of solution; (c) in 50.0 ml of the solution described in (b).

3.46 Calculate the volume of solution needed to provide (a) 0.200 mole of NaCl given a 1.00 M solution of this substance; (b) 3.2 moles of $C_6H_{12}O_6$ given a 0.50 M solution of substance.

3.47 How many moles of NaOH would be required to react with all of the H^+ from (a) 0.10 mole H_2SO_4; (b) 50.0 ml of 0.10 M HNO_3; (c) 50.0 ml of 0.10 M H_3PO_4.

3.48 What volume of 0.10 M HCl is required to react completely with 5.0 g of zinc according to the reaction

$$Zn(s) + 2HCl(aq) \longrightarrow H_2(g) + ZnCl_2(aq)?$$

[3.49] Calculate the molarity of OH^-, Cl^-, and Ca^{2+} present in a solution made by mixing equal volumes of 0.20 M $Ca(OH)_2$ and 0.20 M HCl.

3.50 If it takes 38.7 ml of 1.90 M NaOH to neutralize 10.0 ml of the sulfuric acid, H_2SO_4, in a battery, what is the molarity of the sulfuric acid solution?

3.51 The approximate concentration of hydrochloric acid, HCl, in the stomach (stomach acid) is 0.17 M. Calculate the mass of the following antacids required to neutralize 50.0 ml of this acid: (a) bicarbonate of soda, $NaHCO_3$; (b) aluminum hydroxide, $Al(OH)_3$.

[3.52] Vinegar contains acetic acid, $HC_2H_3O_2$. Titration of 5.00 g of vinegar with 0.100 M NaOH requires 33.0 ml to reach the equivalence point. (a) What is the weight percentage of $HC_2H_3O_2$ in vinegar? (b) If the vinegar has a density of 1.005 g/ml, what is the molarity of $HC_2H_3O_2$ in vinegar?

general exercises

3.53 Aspirin, $C_9H_8O_4$, is produced from salicylic acid, $C_7H_6O_3$, and acetic anhydride, $C_4H_6O_3$:

$$C_7H_6O_3 + C_4H_6O_3 \longrightarrow C_9H_8O_4 + HC_2H_3O_2$$

(a) How much salicylic acid is required to produce 1000 kg of aspirin, assuming that all of the salicylic acid is converted to aspirin? (b) How much salicylic acid would be required if only 70 percent of the salicylic acid is converted to aspirin?

3.54 Fill in the gaps in the following table involving solutions of sulfuric acid, H_2SO_4.

MOLARITY	VOLUME	MOLES
3.00 M	200 ml	
	750 ml	2.00
0.200 M		0.100

3.55 (a) Write a balanced chemical equation for the combustion of pentane, C_5H_{12}. (b) Calculate the number of grams of H_2O produced by combustion of 10.0 g of pentane.

[3.56] Calculate the mass of CO_2 produced and the mass of O_2 consumed when a car travels 100 mi given the following facts and assumptions: The car gets a gas mileage of 18 mi per gallon. Assume that the fuel is 100 percent octane, C_8H_{18}, whose density is 0.70 g/ml. Furthermore, assume that the fuel is completely combusted to CO_2 and H_2O. (1 gal = 3.79 l.)

3.57 How many milliliters of 0.200 M NaOH are required to react completely with 30.0 ml of 0.150 M H_3PO_4 solution?

3.58 Caffeine has an elemental composition of 49.5 percent carbon, 5.15 percent hydrogen, 28.9 percent nitrogen, and 16.5 percent oxygen. What is the empirical formula for this compound?

3.59 A student adds 50.0 ml of 0.10 M barium chloride solution to a solution containing 5.0 g of sodium sulfate. The reaction produces a precipitate of $BaSO_4$:

$$Na_2SO_4(aq) + BaCl_2(aq) \longrightarrow$$
$$BaSO_4(s) + 2NaCl(aq)$$

How many grams of $BaSO_4$ can be produced?

energy relationships in chemical systems

4

Our focus in the past two chapters has been on matter. We have considered the classification of matter into elements and compounds, the structure of matter in terms of atoms, ions, and molecules, and the changes of matter in chemical reactions. An important feature of chemical reactions is that they involve changes in energy. Chemistry is therefore concerned with both matter and energy. Our focus in this chapter is on energy and its involvement in chemical processes. Most energy produced in modern society comes from chemical reactions, especially combustion of coal, petroleum products, and natural gas.

For a long period in our history, people relied solely on the energy provided by their own bodies to perform work. Slowly they learned how to utilize animals, wind, water, and fire to enable them to accomplish more work. The largest jump in energy consumption came with the industrial revolution, a revolution that was really an energy revolution; the steam engine was the primary invention. Machines were developed that derived their energy largely from coal, then from oil and natural gas. Energy use changed transportation from travel by foot and animal to travel by autos and rockets; it changed communication from face-to-face talking and from handwriting to mass printing and telecommunications. Utilization of energy is the underpinning of our civilization.

4.1
the nature of energy

What is energy? We can't see, touch, or smell it as we can matter; it is a more abstract concept to us. Therefore, it is best that we carefully define energy and consider the concept from a general point of view before we get too deeply into our discussions. **Energy** is the capacity to do work or to transfer heat. Work (w) is defined and measured by the product of the net force (f) and the distance (d) through which that force moves:

$$w = f \times d \qquad [4.1]$$

That is, work is the movement of an object against some force, and energy is the "something" that is required to perform work. For example, energy is required to perform the work associated with lifting a book against the force of gravity. Work is also required to separate a positively charged ion from a negatively charged one because of the attractive force that exists between them. Matter is the substance of the universe; energy is the mover of the substance.

FORMS OF ENERGY

Energy is found in a variety of forms such as heat (thermal energy), light (radiant energy), chemical energy, mechanical energy, and electrical energy. The various forms or types of energy can also be classified as either kinetic or potential energy. **Kinetic energy** is the energy of motion. The magnitude of the kinetic energy of an object depends on its mass (m) and its velocity (v):

$$E_k = \tfrac{1}{2}mv^2 \qquad [4.2]$$

This equation tells us what we may have realized intuitively: The larger the mass of a moving object and the greater its velocity or speed, the more work it can do. The SI unit of energy is the **joule** (J). A joule is $1 \text{ kg m}^2/\text{sec}^2$. This is the kinetic energy possessed by a mass of 2 kg moving at a velocity of 1 m per second:

$$E_k = \tfrac{1}{2}(2 \text{ kg})(1 \text{ m/sec})^2 = 1 \text{ kg m}^2/\text{sec}^2 = 1 \text{ J}$$

Potential energy is the energy stored in an object by virtue of its position or composition. For example, the ram of a pile driver can be lifted against the force of gravity as shown in Figure 4.1. Once lifted it

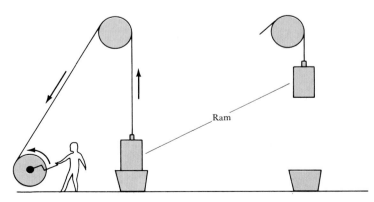

FIGURE 4.1 The ram of the pile driver has potential energy because of its position. This energy is supplied by the person who lifts the ram.

has the potential to do work; the work is accomplished by allowing the ram to fall. The ram therefore has potential energy. In a similar way two unlike-charged particles that are separated against the force of their electrostatic attraction for each other have potential energy.

Let's return to the pile driver for a few more moments. Work is required to lift the ram of the pile driver. The energy to perform this work could be supplied by the brute force of a person. In our society, machines have replaced people in many such operations, and a simple engine such as that shown in Figure 4.2 could be used to lift the ram. In this case, the energy stored in the coal used to fuel the engine is converted to heat that is transferred to the cylinder. As the gas within the cylinder is heated it expands, thereby moving the piston. This moves the wheel that lifts the ram. Energy is thereby transferred from the coal to the ram.

Although energy can be transferred from one object to another and its forms interconverted, no energy is created or destroyed in these processes. This statement is a summary of our experience with energy and is known as the **law of conservation of energy** or the **first law of thermodynamics.**

If energy is conserved in any chemical or physical change of matter, you may well wonder why there is any concern over energy resources and why the term "energy crisis" appears in the popular press. The situation with energy is very much like that with matter. In discussing the law of conservation of mass, we spoke of resources being "depleted" in the sense of being converted into forms that are less and less useful to people. The resources become too dilute or can be extracted only through the expenditure of a great deal of energy. Basically the same is true of energy; in any energy transformation some energy is always converted to heat that gets spread throughout the environment and consequently cannot be used to do work. For instance, a common automobile is only about 10 percent efficient in converting the chemical energy of gasoline into mechanical energy used to propel the car; the remainder goes into the environment as heat. Indeed, most of the energy associated with the coal

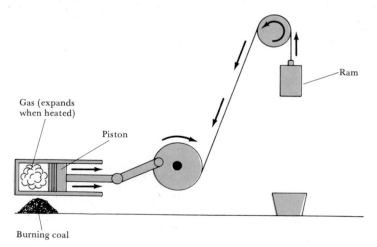

FIGURE 4.2 A simple engine that uses the expansion of a gas upon heating to lift the ram of the pile driver. The potential energy of the coal is transferred to the ram.

and petroleum we burn is lost as heat. The heat lost to our environment through our various energy conversions is known as **thermal pollution.**

The combustion of coal mentioned in the introduction of this chapter is only one of many familiar chemical reactions that produce energy. Consider the combustion of a piece of magnesium ribbon:

$$2Mg(s) + O_2(g) \longrightarrow 2MgO(s) + \text{heat and light} \qquad [4.3]$$

Most of us have seen this reaction whether we realize it or not. Flash bulbs are filled with magnesium ribbon and oxygen gas. Passage of an electric current through the magnesium causes it to ignite, producing heat and light.

   We can label the magnesium and oxygen atoms in this reaction as our system. Everything around this system, including the container in which the reaction takes place, is called the surroundings. Labeling the portion of the universe that is under examination as the **system** and the rest of the universe as the **surroundings** allows us to clarify what goes on in any energy transfer. Because energy is neither created nor destroyed, any energy lost by the system must be gained by the surroundings, and any energy gained by the system must be supplied by the surroundings. In our example, energy is lost by the system to the surroundings as the atoms of the system rearrange from magnesium metal and oxygen gas into magnesium oxide, MgO. Chemical and physical changes that give off heat to their surroundings are described as being **exothermic.** Those that absorb heat are labeled as **endothermic.** We can represent exothermic reactions as shown schematically in Figure 4.3.

   An example of an endothermic reaction is the decomposition of water into its elements:

$$\text{Energy} + 2H_2O(l) \longrightarrow 2H_2(g) + O_2(g) \qquad [4.4]$$

As we noted in Chapter 2, this reaction can be carried out by supplying electrical energy to the water. This endothermic process is represented schematically in Figure 4.4.

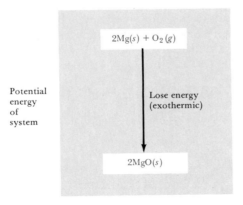

Potential energy of system

2Mg(s) + O_2 (g)

Lose energy (exothermic)

2MgO(s)

FIGURE 4.3 A potential-energy diagram for an exothermic reaction. Heat is given off by the system to its surroundings, thereby decreasing the energy content of the system.

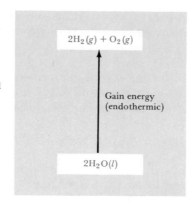

Potential
energy
of
system

FIGURE 4.4 A potential-energy diagram for an endothermic reaction. Heat is absorbed by the system from the surroundings, thereby increasing the energy content of the system.

Traditionally, the energy changes accompanying chemical reactions have been measured in calories (cal). A calorie is the amount of energy required to raise the temperature of 1 g of water by 1°C, from 14.5 to 15.5°C.* A kilocalorie, 1000 cal, is the same as the Calorie (capitalized) used in nutrition. One calorie is the same amount of energy as 4.184 joules:

$$1 \text{ cal} = 4.184 \text{ J} \tag{4.5}$$

In most places in this text the joule (J) or kilojoule (kJ) is used as the unit of energy.

4.3
enthalpy

Most chemical reactions, and especially those in living organisms, take place under the essentially constant pressure of the atmosphere. The heat of any reaction run under constant pressure is called the heat content change or the **enthalpy change** of the system. Enthalpy is represented by H, and the enthalpy change is given the symbol ΔH (read delta H).† By convention the enthalpy change for a reaction (ΔH_{rxn}) is taken to be the enthalpy or heat content of the products minus that of the reactants:

$$\Delta H_{rxn} = H(\text{products}) - H(\text{reactants}) \tag{4.6}$$

If the enthalpy of the products is less than that of the reactants, ΔH will be negative in sign; furthermore, because this heat is lost by the system to the surroundings, the process will be exothermic. Thus a negative value of ΔH represents an exothermic reaction, whereas a positive value represents an endothermic one. For the combustion of methane, the principal component of natural gas, we can write:

$$CH_4(g) + 2O_2(g) \longrightarrow CO_2(g) + 2H_2O(l) \qquad \Delta H = -890 \text{ kJ} \tag{4.7}$$

---

*The temperature interval 14.5 to 15.5°C is specified because the energy required to raise the temperature of water by 1°C is slightly different at different temperatures. However, it is constant to three significant figures, 1.00 cal, over the entire liquid range of water, from 0 to 100°C.

†The symbol Δ is commonly used to denote *change*. For example, a change of volume can be represented by ΔV.

The coefficients in the balanced equation are taken to represent the number of moles of reactants giving the associated enthalpy change.

As a consequence of the law of conservation of energy, the amount of heat associated with a reaction is directly proportional to the amount of substances involved. Thus combustion of 1 mole of CH_4 produces 890 kJ of heat, whereas combustion of two moles produces 1780 kJ.

SAMPLE EXERCISE 4.1

How much heat is produced when 4.50 g of methane gas are burned in a constant-pressure system?

Solution: According to Equation [4.7], 890 kJ are produced when a mole of CH_4 is burned. A mole of CH_4 has a mass of 16.0 g.

Heat produced =

$$(4.50 \text{ g } CH_4)\left(\frac{1 \text{ mole } CH_4}{16.0 \text{ g } CH_4}\right)\left(\frac{890 \text{ kJ}}{\text{mole } CH_4}\right)$$
$$= 250 \text{ kJ}$$

In a related fashion, the enthalpy change for a reaction is equal in magnitude but opposite in sign to ΔH for the reverse reaction. For example:

$$CO_2(g) + 2H_2O(l) \longrightarrow CH_4(g) + 2O_2(g) \qquad \Delta H = +890 \text{ kJ}$$
$$[4.8]$$

If more energy were produced by combustion of CH_4 than required for the reverse reaction, it would be possible to use these processes to create an unlimited supply of energy. Some CH_4 could be combusted, and that portion of the energy necessary to reform CH_4 could be saved. The rest could be used to do work of some type. After CH_4 is reformed, it could again be combusted, and so forth, continually supplying energy. This, of course, is clearly contrary to our experience—such behavior does not obey the law of conservation of energy. This observed situation is shown in Figure 4.5.

The enthalpy change of a reaction also depends on the state of the

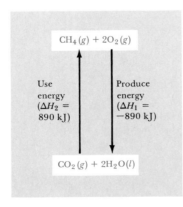

FIGURE 4.5 If a substance is combusted and then reformed from its combustion products, the total energy change must be zero ($\Delta H_1 + \Delta H_2 = 0$). In general, the net energy change for a series of reactions that eventually regenerate the original substances in their original conditions is zero.

reactants and products. If the product in the combustion of methane, Equation [4.7], were gaseous H_2O instead of liquid H_2O, ΔH would be -802 kJ instead of -890 kJ. Less heat is available for transfer to the surroundings because 88 kJ are used to vaporize 2 moles of water:

$$2H_2O(l) \longrightarrow 2H_2O(g) \qquad \Delta H = 88 \text{ kJ} \qquad [4.9]$$

Therefore, the states of the reactants and products must be specified. Furthermore, it is assumed that the reactants and products are at the same temperature, 25°C, unless otherwise indicated.

4.4 Hess's law

One specialized statement of the first law of thermodynamics that is especially useful to chemistry is known as Hess's law of constant heat summation. According to **Hess's law**, if a reaction can be carried out in a series of steps, ΔH for the reaction will be equal to the sum of the enthalpy changes for each step; the enthalpy changes are additive. For example, the enthalpy change for the combustion of methane to form carbon dioxide and gaseous water can be calculated from ΔH for the vaporization of water and ΔH for the combustion of methane giving liquid water:

$$CH_4(g) + 2O_2(g) \longrightarrow CO_2(g) + 2H_2O(l)$$
$$\Delta H = -890 \text{ kJ}$$

(Add) $\qquad\qquad 2H_2O(l) \longrightarrow 2H_2O(g) \qquad \Delta H = 88 \text{ kJ}$

$$CH_4(g) + 2O_2(g) + 2H_2O(l) \longrightarrow CO_2(g) + 2H_2O(l) + 2H_2O(g)$$
$$\Delta H = -802 \text{ kJ}$$

Net equation:
$$CH_4(g) + 2O_2(g) \longrightarrow CO_2(g) + 2H_2O(g)$$
$$\Delta H = -802 \text{ kJ}$$

To obtain the net equation, the sum of the reactants of the two equations are placed on one side of the arrow, and the sum of the products, on the other. Because $2H_2O(l)$ occurs on both sides of the arrow, it can be canceled like an algebraic quantity that is on both sides of an equal sign.

Hess's law provides a useful means of calculating energy changes that are difficult to measure directly. For instance, it is not possible to measure directly the heat of combustion of carbon to form carbon monoxide. Combustion of 1 mole of carbon with $\frac{1}{2}$ mole O_2 produces not only CO, but also CO_2, leaving some carbon unreacted. However, the heat of the reaction forming CO can be calculated as shown in Sample Exercise 4.2.

SAMPLE EXERCISE 4.2

The heat of combustion of C to CO_2 is -393.7 kJ/mole CO_2, whereas that for combustion of CO to CO_2 is -283.3 kJ/mole CO_2. Calculate the heat of combustion of C to CO.

Solution:

$$2C(s) + 2O_2(g) \longrightarrow 2CO_2(g) \qquad \Delta H = -787.4 \text{ kJ}$$
$$2CO_2(g) \longrightarrow 2CO(g) + O_2(g)$$
$$\Delta H = 566.6 \text{ kJ}$$
$$2C(s) + O_2(g) \longrightarrow 2CO(g) \qquad \Delta H = -220.8 \text{ kJ}$$

The enthalpy change for formation of 1 mole of CO is $-110.4\,\text{kJ}$. Notice that we had to turn the combustion reaction for CO around in order to obtain CO as a product when we added equations.

SAMPLE EXERCISE 4.3

Given the following reactions and their respective enthalpy changes,

$$C_2H_2(g) + \frac{5}{2}O_2(g) \longrightarrow 2CO_2(g) + H_2O(l)$$

$$\Delta H = -1300.0\ \text{kJ/mole}\ C_2H_2$$

$$C(s) + O_2(g) \longrightarrow CO_2(g)$$

$$\Delta H = -393.5\ \text{kJ/mole}\ C$$

$$H_2(g) + \frac{1}{2}O_2(g) \longrightarrow H_2O(l)$$

$$\Delta H = -285.9\ \text{kJ/mole}\ H_2$$

calculate ΔH for the reaction

$$2C(s) + H_2(g) \longrightarrow C_2H_2(g)$$

Solution: Because we want to end up with C_2H_2 we turn the first equation around; the sign of ΔH is therefore changed. Because we need to start with $2C(s)$, we multiply the second equation and its ΔH by two. We then add the resultant equations and their enthalpy changes in accordance with Hess's law:

$$2CO_2(g) + H_2O(l) \longrightarrow C_2H_2(g) + \frac{5}{2}O_2(g)$$

$$\Delta H = +1300.0\ \text{kJ}$$

$$2C(s) + 2O_2(g) \longrightarrow 2CO_2(g)$$

$$\Delta H = -787.0\ \text{kJ}$$

$$H_2(g) + \frac{1}{2}O_2(g) \longrightarrow H_2O(l)$$

$$\Delta H = -285.9\ \text{kJ}$$

$$\overline{2C(s) + H_2(g) \longrightarrow C_2H_2(g)}$$

$$\Delta H = 227.1\ \text{kJ}$$

When the three equations are added, there are $2CO_2$, $\frac{5}{2}O_2$ and H_2O on both sides of the arrow. These are canceled in writing the net equation.

4.5
state functions

In a more general way, the first law of thermodynamics tells us that enthalpy change depends only on the initial and final conditions or states of the system and not on how the change is accomplished. That is, ΔH is independent of the number and kind of steps by which the reaction is carried out. Functions, like ΔH, for which this condition is true are referred to as **state functions**. Potential energy is a state function. The potential energy of a body in a gravitational field depends on its mass (m), its height above its final resting place (h), and the pull of gravity (g):

$$E = mgh \qquad\qquad [4.10]$$

The magnitude of a body's potential energy does not depend on how the body gets to its height or how it gets down. This relationship is illustrated in Figure 4.6. State functions are also described as path-independent quantities, because they do not depend on how a change from initial to final states occurs, but only on the initial and final states themselves. The distance traveled between two points depends on the path and is therefore not a state function.

energy relationships
in chemical systems

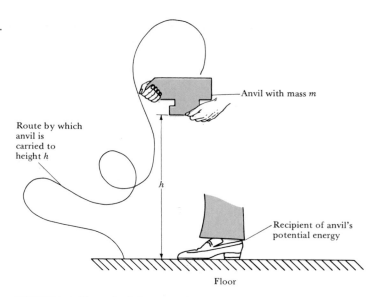

FIGURE 4.6 The potential energy of an anvil above a floor is a state function. The magnitude of the potential energy of the anvil does not depend on how it proceeded from the floor to its elevated position. The potential energy change depends only on its initial and final positions.

For the reaction of methane, CH_4, and oxygen, O_2, to form CO_2 and H_2O, we may envision the reaction to occur either directly or with the initial formation of CO, which is subsequently combusted. This set of choices is illustrated in Figure 4.7; because ΔH is a state function, either path produces the same energy, $\Delta H_1 = \Delta H_2 + \Delta H_3$. Again we may note that if this were not so it would be possible to create energy continually, a procedure in conflict with the first law of thermodynamics. Thus the first law of thermodynamics teaches us that we can never expect to obtain more (or less) energy from a chemical reaction by changing the method of carrying out the reaction (although we may affect the portion of the energy that is converted to work compared to that lost to the environment as heat).

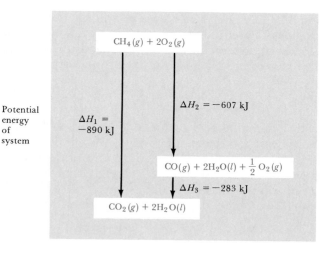

FIGURE 4.7 The quantity of heat generated by combustion of CH_4 is independent of whether the reaction takes place in one or more steps ($\Delta H_1 = \Delta H_2 + \Delta H_3$).

The first law of thermodynamics permits calculation of the enthalpy changes for many reactions from a few tabulated values. It is convenient to summarize much of this data in terms of **standard heats of formation**.* The heat of formation of a compound, ΔH_f, is the enthalpy change involved in forming a mole of that compound from its constituent elements. A standard enthalpy change, $\Delta H°$, is one that takes place with all reactants and products in their **standard states**. That is, all substances are in the forms most stable at 25°C and at standard atmospheric pressure (see Chapter 5). For example, the standard heat of formation, $\Delta H_f°$, for ethanol (C_2H_5OH) is the enthalpy change for the following reaction:

$$2C(\text{graphite}) + 3H_2(g) + \frac{1}{2}O_2(g) \longrightarrow C_2H_5OH(l) \qquad [4.11]$$

The elemental source of oxygen is O_2 and not O or O_3, because O_2 is the stable form of oxygen at 25°C and standard atmospheric pressure. Likewise, the elemental source of carbon is graphite and not diamond, because the former is the stable (lowest energy) form at 25°C and standard atmospheric pressure. The conversion of graphite to diamond requires the addition of energy as shown in Equation [4.12]:

$$C(\text{graphite}) \longrightarrow C(\text{diamond}) \qquad \Delta H° = 1.88 \text{ kJ} \qquad [4.12]$$

A number of standard heats of formation are given in Table 4.1, and additional values are given in Appendix E. By convention the standard heat of formation of the stable form of any element is zero.

The standard enthalpy change for any reaction can be found from the relationship

$$\Delta H°_{\text{rxn}} = \Sigma n \, \Delta H_f° \text{ (products)} - \Sigma m \, \Delta H_f° \text{ (reactants)} \qquad [4.13]$$

where Σ (sigma) means "the sum of" and n and m are the stoichiometric coefficients of the chemical equation. For example, $\Delta H°$ for the combustion of glucose, Equation [4.14], is given by Equation [4.15]:

$$C_6H_{12}O_6(s) + 6O_2(g) \longrightarrow 6CO_2(g) + 6H_2O(l) \qquad [4.14]$$

$$\begin{aligned} \Delta H°_{\text{rxn}} = &[6 \, \Delta H_f°(CO_2) + 6 \, \Delta H_f°(H_2O)] \\ &- [\Delta H_f°(C_6H_{12}O_6) + 6 \, \Delta H_f°(O_2)] \end{aligned} \qquad [4.15]$$

Using the heats of formation recorded in Table 4.1, this process gives:

---

*It would be more precise to refer to heats of formation as enthalpy changes of formation. However, the usual practice is to use the more general term heat of formation. Recall that the enthalpy change is the heat measured at constant pressure.

TABLE 4.1

standard heats of formation, ΔH_f°, at 25°C

SUBSTANCE	FORMULA	ΔH_f° (kJ/mole)
Acetylene	$C_2H_2(g)$	226.7
Ammonia	$NH_3(g)$	−46.19
Benzene	$C_6H_6(l)$	49.04
Calcium carbonate	$CaCO_3(s)$	−1207.1
Calcium oxide	$CaO(s)$	−635.5
Carbon dioxide	$CO_2(g)$	−393.5
Carbon monoxide	$CO(g)$	−110.5
Diamond	$C(s)$	1.88
Ethane	$C_2H_6(g)$	−84.68
Ethanol	$C_2H_5OH(l)$	−277.7
Ethylene	$C_2H_4(g)$	52.30
Hydrogen bromide	$HBr(g)$	−36.23
Hydrogen chloride	$HCl(g)$	−92.30
Hydrogen fluoride	$HF(g)$	−268.6
Hydrogen iodide	$HI(g)$	25.9
Glucose	$C_6H_{12}O_6(s)$	−1260
Methane	$CH_4(g)$	−74.85
Methanol	$CH_3OH(l)$	−238.6
Silver chloride	$AgCl(s)$	−127.0
Sodium bicarbonate	$NaHCO_3(s)$	−947.7
Sodium carbonate	$Na_2CO_3(s)$	−1130.9
Sodium chloride	$NaCl(s)$	−411.0
Sucrose	$C_{12}H_{22}O_{11}(s)$	−2221
Water	$H_2O(l)$	−285.9
Water vapor	$H_2O(g)$	−241.8

$$\Delta H_{rxn}^\circ = \left[(6 \text{ moles } CO_2) \left(-393.5 \frac{kJ}{\text{mole } CO_2} \right) \right.$$

$$\left. + (6 \text{ moles } H_2O) \left(-285.9 \frac{kJ}{\text{mole } H_2O} \right) \right]$$

$$- \left[(1 \text{ mole } C_6H_{12}O_6) \left(-1260 \frac{kJ}{\text{mole } C_6H_{12}O_6} \right) \right.$$

$$\left. + (6 \text{ moles } O_2) \left(0 \frac{kJ}{\text{mole } O_2} \right) \right]$$

$$= -2816 \text{ kJ}$$

The general relationship given by Equation [4.13] follows directly from the fact that ΔH is a state function. This fact allows calculation of the enthalpy change for any reaction from the energies required to convert the initial reactants into elements and then combine the elements into the desired products. This reaction pathway is shown in Figure 4.8 for the combustion of glucose.

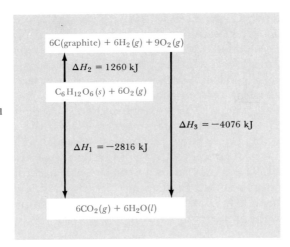

FIGURE 4.8 Because ΔH is a state function, $\Delta H_1 = \Delta H_2 + \Delta H_3$. Note that ΔH_2 is $-\Delta H_f^\circ(C_6H_{12}O_6)$ while ΔH_3 is $6\,\Delta H_f^\circ(H_2O) + 6\,\Delta H_f^\circ(CO_2)$. This is the same result as given in Equation [4.15].

SAMPLE EXERCISE 4.4

Calculate the standard enthalpy change for the reaction

$$CaCO_3(s) \longrightarrow CaO(s) + CO_2(g)$$

Solution: Using heats of formation given in Table 4.1:

$$\begin{aligned}
\Delta H_{rxn}^\circ &= \Delta H_f^\circ(CaO) + \Delta H_f^\circ(CO_2) - \Delta H_f^\circ(CaCO_3) \\
&= -635.5\ kJ - 393.5\ kJ + 1207.1\ kJ \\
&= 178.1\ kJ
\end{aligned}$$

SAMPLE EXERCISE 4.5

Compare the quantity of heat produced by combustion of 1.00 g of glucose ($C_6H_{12}O_6$) with that produced by 1.00 g of sucrose ($C_{12}H_{22}O_{11}$).

Solution: The example worked in the text gave $\Delta H^\circ = -2816\ kJ$ for the combustion of a mole of glucose. The molecular weight of glucose is 180 amu. Therefore, the heat produced per gram is

$$\left(-2816\frac{kJ}{mole}\right)\left(\frac{1\ mole}{180\ g}\right) = -15.6\frac{kJ}{g}$$

For sucrose

$$\begin{aligned}
C_{12}H_{22}O_{11}(s) + 12O_2(g) \longrightarrow \\
12CO_2(g) + 11H_2O(l)
\end{aligned}$$

$$\begin{aligned}
\Delta H_{rxn}^\circ &= [12\,\Delta H_f^\circ(CO_2) + 11\,\Delta H_f^\circ(H_2O)] \\
&\quad - [\Delta H_f^\circ(C_{12}H_{22}O_{11}) + 12\,\Delta H_f^\circ(O_2)] \\
&= 12(-393.5\ kJ) + 11(-285.9\ kJ) \\
&\quad + 2221\ kJ \\
&= (-4722 - 3145 + 2221)\ kJ \\
&= -5646\ kJ/mole\ C_{12}H_{22}O_{11}
\end{aligned}$$

The molecular weight of sucrose is 342 amu. Therefore, the heat produced per gram of sucrose is

$$\left(-5646\frac{kJ}{mole}\right)\left(\frac{mole}{342\ g}\right) = -16.5\frac{kJ}{g}$$

Both sucrose and glucose are carbohydrates. As a rule of thumb the energy obtained from the combustion of a gram of carbohydrate is 17 kJ (4 kcal or 4 Cal).

4.7

measurement of energy changes: calorimetry

Although we have been discussing the heat effects accompanying chemical or physical changes, we have not yet indicated how any of these can be measured. Measurement of heat effects is known as **calorimetry;** the techniques and equipment employed in calorimetry depend on the nature of the process being studied. Combustion reactions are generally studied by means of a **bomb calorimeter,** a device shown schematically in Figure 4.9. The sample to be examined, for instance a fuel, is placed in a cup within a sealed vessel called a bomb. The bomb, which is designed to withstand high pressures, has an inlet valve for adding oxygen under pressure and also has electrical contacts to initiate the combustion reaction. After the sample has been placed in the bomb, the bomb is sealed and filled with oxygen. It is then placed in a large insulated container and covered with an accurately measured quantity of water. The combustion reaction is initiated by passing an electrical current through a fine wire that is in contact with the sample; the sample ignites when the wire gets hot. The temperature of the water is carefully measured before and after combustion. In keeping with the law of conservation of energy, the heat lost by the burning sample is gained by its surroundings, namely the water and calorimeter parts:

$$\text{Heat lost by sample} = \text{heat gained by water} \\ + \text{heat gained by calorimeter} \quad [4.16]$$

It takes 1.00 cal (4.184 J) to raise the temperature of 1 g of water by 1°C. Therefore, if the temperature of 500 g of water increased 3.00°C, it must have absorbed 1500 cal:

$$\text{Calories} = (500 \text{ g})(3.00°C)\left(\frac{1.00 \text{ cal}}{1 \text{ g-°C}}\right) = 1500 \text{ cal}$$

$$(1500 \text{ cal})\left(\frac{4.184 \text{ J}}{1 \text{ cal}}\right) = 6280 \text{ J}$$

The heat absorbed by the calorimeter is determined experimentally by combusting a sample that gives off a known quantity of heat. For

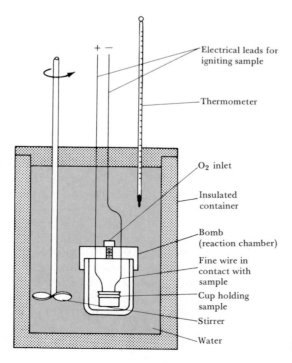

Electrical leads for
igniting sample

Thermometer

O_2 inlet

Insulated
container

Bomb
(reaction chamber)

Fine wire in
contact with
sample

Cup holding
sample

Stirrer

Water

FIGURE 4.9 A bomb calorimeter.

example, if it is known that combustion of 1.00 g of a sample produces 8.00 kJ of heat, and if it is experimentally determined that this heat raises the temperature of 500 g of water by 3.00°C, the heat gained by the calorimeter can be calculated:

$$\begin{aligned} \text{Heat gained by calorimeter} &= \text{heat lost by sample} \\ &\quad - \text{heat gained by water} \\ &= 8000\text{ J} - (500\text{ g})(3.00°\text{C})\left(\frac{4.184\text{ J}}{\text{g-}°\text{C}}\right) \\ &= 1720\text{ J} \end{aligned}$$

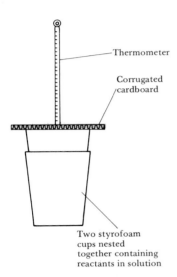

Thermometer

Corrugated cardboard

Two styrofoam cups nested together containing reactants in solution

FIGURE 4.10 A "coffee-cup" calorimeter.

The heat absorbed by the calorimeter is conveniently expressed as its "water equivalent," meaning the mass of water that would absorb the same quantity of heat when its temperature is increased the same quantity as that of the calorimeter. In the example, the quantity of water that would absorb 1720 J in increasing 3.00°C in temperature is 137 g:

$$\begin{aligned} \text{Water equivalent} &= \left(\frac{1720\text{ J}}{3.00°\text{C}}\right)\left(\frac{1\text{ g-}°\text{C}}{4.184\text{ J}}\right) \\ &= 137\text{ g} \end{aligned}$$

Sample Exercise 4.6 illustrates the calculation of a heat of reaction.

SAMPLE EXERCISE 4.6

The temperature of 3.50 kg of water is raised by 1.17°C when 1.00 g of hydrazine, N_2H_4, is burned in a bomb calorimeter. If the calorimeter has a water equivalent of 450 g, what is the quantity of heat given off by this sample?

Solution:

$$\begin{aligned} \text{Joules} &= (3500\text{ g} + 450\text{ g})(1.17°\text{C})\left(\frac{4.184\text{ J}}{1\text{ g-}°\text{C}}\right) \\ &= 19{,}300\text{ J or } 19.3\text{ kJ} \end{aligned}$$

Because a bomb calorimeter is a constant-volume device, but not necessarily a constant-pressure device, the heat changes it measures are not ΔH values. However, the relationship between ΔH and heat changes at constant volume is known, so the results can be changed to ΔH values. The difference between the heat changes at constant pressure and those at constant volume are usually small compared to the heat changes themselves. For many reactions, such as those occurring in solution, it is a simple matter to control pressure so that ΔH can be measured directly. One constant-pressure calorimeter is shown in Figure 4.10. Such simple, "coffee-cup" calorimeters are often used in freshman chemistry labs to illustrate the principles of calorimetry. Because the calorimeter is not sealed, the reaction occurs under the essentially constant pressure of the atmosphere. The heat change of the reaction is determined from the temperature increase of a known quantity of solution in the calorimeter, as shown in Sample Exercise 4.7.

105

SAMPLE EXERCISE 4.7

When 50 g of 1.0 M HCl (containing 0.050 moles HCl) and 50 g of 1.0 M NaOH (containing 0.050 moles NaOH) are mixed in a "coffee-cup" calorimeter, the temperature of the resultant solution increases from $21.0°C$ to $27.5°C$. The water equivalent of the calorimeter is 4 g. Assuming that it takes 4.184 J to increase the temperature of 1.00 g of the solution by $1.00°C$, calculate the heat of the reaction

$$HCl(aq) + NaOH(aq) \longrightarrow H_2O(l) + NaCl(aq)$$

Solution:

$$Joules = (100 \text{ g} + 4 \text{ g})(6.5°C)\left(\frac{4.184 \text{ J}}{1.00 \text{ g-}°C}\right)$$
$$= 2800 \text{ J}$$

Thus the reaction gives off 2800 J of heat. To put this on a molar basis we need to know the heat produced per mole of HCl and NaOH:

$$J/mole = \frac{-2800 \text{ J}}{0.050 \text{ mole}} = -56,000 \frac{\text{J}}{\text{mole}}$$

$$\text{or } -56 \frac{\text{kJ}}{\text{mole}}$$

4.8

fuel values of fuels and foods

Most common chemical reactions used to produce heat are combustion reactions. The energy released when a fuel or food is combusted is known as its **fuel value**. Because all heats of combustion are exothermic, it is common to report fuel values without their associated negative sign. Furthermore, because fuels and foods are usually mixtures, fuel values are reported on a gram rather than a mole basis. For example, the fuel value of octane, C_8H_{18}, a component of gasoline, is the heat produced by combustion of 1 g of this substance as shown in Equation [4.17]:

$$2C_8H_{18}(l) + 25O_2(g) \longrightarrow 16CO_2(g) + 18H_2O(g) \qquad [4.17]$$

The enthalpy change for this reaction is $\Delta H = -10,920$ kJ. Because each mole of C_8H_{18} weighs 114 g, the fuel value of octane is 47.9 kJ/g:

$$\left(\frac{10,920 \text{ kJ}}{2 \text{ moles } C_8H_{18}}\right)\left(\frac{1 \text{ mole } C_8H_{18}}{114 \text{ g } C_8H_{18}}\right) = 47.9 \frac{\text{kJ}}{\text{g } C_8H_{18}}$$

In accordance with the first law of thermodynamics, the fuel value of a substance is the same no matter how or where it reacts as long as the products of the reaction are the same. Thus a bomb calorimeter can often be used to measure the fuel values of fuels or foods. Unquestionably, this procedure is much easier than having to measure the heat given off in an engine or in our bodies.

FOODS

Most of the energy our bodies need comes from carbohydrates and fats. Carbohydrates are decomposed in the stomach into glucose, $C_6H_{12}O_6$. Glucose is soluble in blood and is known as blood sugar. It is transported by the blood to cells where it reacts with O_2 in a series of steps producing CO_2, H_2O, and energy:

$$C_6H_{12}O_6(s) + 6O_2(g) \longrightarrow$$
$$6CO_2(g) + 6H_2O(l) \qquad \Delta H° = -2816 \text{ kJ}$$

The breakdown of carbohydrates is rapid, so that their energy is quickly supplied to the body. However, the body stores a very small amount of carbohydrates. The average fuel value of carbohydrates is 17 kJ/g (4 kcal/g).

Like carbohydrates, fats produce CO_2 and H_2O in both their metabolism and their combustion in a bomb calorimeter. The reaction of stearin, $C_{57}H_{110}O_6$, a typical fat, is as follows:

$$2C_{57}H_{110}O_6(s) + 163O_2(g) \longrightarrow$$
$$114CO_2(g) + 110H_2O(l) \qquad \Delta H^\circ = -75,520 \text{ kJ}$$

The chemical energy from foods that is not used either to maintain body temperature or for muscular activity or for the organization of the atoms in food into body parts is stored in the form of fats. Fats are well suited to serve as the body's energy reserve for at least two reasons: (1) They are insoluble in water, which permits their storage in the body; (2) they produce more energy per gram than either proteins or carbohydrates, which makes them efficient energy sources on a weight basis. The average fuel value of fats is 38 kJ/g (9 kcal/g).

In the case of proteins, metabolism produces less energy than combustion, because the products are different; N_2 is released in the bomb calorimeter, whereas in the body the nitrogen ends up mainly as urea, CH_4N_2O. Proteins are used by the body mainly as building materials for construction of organ walls, skin, hair, and so forth. On the average, the metabolism of protein produces 17 kJ/g (4 kcal/g). The fuel values for a variety of common foods are shown in Table 4.2.

The amount of energy the body requires varies considerably depending on such factors as body weight, age, and muscular activity. The average adult requires about 6300 kJ (1500 kcal) a day while at rest in a warm room. When a person is doing average work, his energy requirement is increased to about 10,000–13,000 kJ (2500–3000 kcal). This is about the same amount of energy as that consumed by a 100-watt light bulb operating for a 24-hr period.

TABLE 4.2
the fuel values and compositions of some common foods

	APPROXIMATE COMPOSITION			FUEL VALUE	
	protein	fat	carbohydrate	kJ/g	kcal/g
Apples (raw)	0.4%	0.5%	13%	2.7	0.64
Beer	1	0	4	20	4.8
Bread (white, enriched)	9	3	52	12	2.8
Cheese (cheddar)	28	37	4	20	4.7
Eggs	13	10	0.7	7	1.6
Fudge	2	11	81	17	4.1
Green beans (frozen)	1	—	5.5	9	2.2
Hamburger	22	30	—	15	3.6
Milk	3.3	4.0	5.0	3.0	0.74
Peanuts	26	39	22	24	5.6

FUELS

Several common fuels are compared in Table 4.3. Notice that an increase in the percentage of carbon or hydrogen in the fuel increases the fuel value. For example, the fuel value of bituminous coal is greater than that of wood because of its greater carbon content.

Coal, oil, and natural gas, which are presently our major sources of energy, are known as **fossil fuels.** All are thought to have formed over millions of years from the decomposition of plants and animals. All are presently being depleted far more rapidly than they are formed. **Natural gas** consists of gaseous hydrocarbons, compounds of hydrogen and carbon. It varies in composition but contains primarily methane (CH_4), with small amounts of ethane (C_2H_6), propane (C_3H_8), and butane (C_4H_{10}). **Oil,** which is also known as petroleum, is a liquid composed of hundreds of compounds. Most of these compounds are hydrocarbons, with the remainder being mainly organic compounds containing sulfur, nitrogen, or oxygen. **Coal,** which is solid, contains hydrocarbons of high molecular weight as well as compounds containing sulfur, oxygen, and nitrogen. The sulfur in oil and coal is important from the standpoint of air pollution, as we shall discuss in Chapter 10.

Hydrogen (H_2) is a very attractive fuel because of its high fuel value and because its combustion produces water, a "clean" chemical that produces no negative environmental effects. However, hydrogen cannot be used as a primary energy source because there is so little H_2 in nature. Most hydrogen is produced by the decomposition of water or hydrocarbons. This decomposition requires energy; in fact, because of heat losses, more energy must be used to generate hydrogen than can be reclaimed when the hydrogen is subsequently used as a fuel. However, should large, cheap sources of energy become available because of technical advances, perhaps in areas such as nuclear or solar power generation, a portion of this energy could be used to generate hydrogen. The hydrogen could then serve as a convenient energy carrier. It would be cheaper to transport hydrogen using existing gas pipelines than to transport electrical energy; hydrogen is both portable and storable. Because present industrial technology is based on combustible fuels, hydrogen could replace oil and natural gas as these fuels become scarcer and more expensive.

TABLE 4.3

fuel values and compositions of some common fuels

| | APPROXIMATE ELEMENTAL COMPOSITION | | | FUEL VALUE |
	C	H	O	(kJ/g)
Wood (pine)	50%	6%	44%	18
Anthracite coal (Pennsylvania)	82	1	2	31
Bituminous coal (Pennsylvania)	77	5	7	32
Charcoal	100	0	0	34
Crude oil (Texas)	85	12	0	45
Gasoline	85	15	0	48
Natural gas	70	23	0	49
Hydrogen	0	100	0	142

The average energy consumption per person in the United States each day amounts to about 1.3×10^6 kJ. This amount of energy is about 100 times our average food-energy requirement. The use of energy has been increasing each year as shown in Figure 4.11. Presently about 30 percent of the world's annual production of energy is consumed in the United States.

The shifting importance of different sources of energy is shown in Figure 4.12. Until 1850, wood supplied about 90 percent of the energy used in the United States. Coal became increasingly important until it supplied about 75 percent of our energy by 1910. Presently natural gas supplies 31 percent of our energy, whereas 46 percent is supplied by oil, 19 percent by coal, about 2 percent by hydroelectric power, and slightly more than 2 percent by nuclear power.

The sources and uses of energy in the United States are shown in Figure 4.13. About 26 percent of our energy sources are used to generate electricity, whereas the remaining 74 percent is consumed directly as fuel. About 40 percent of the energy is used by industry to produce consumer goods. The largest fraction of this energy is used to produce and refine metals such as aluminum and iron. About 25 percent of the energy is used for transportation. The remainder is used by commercial establishments or in our homes. Much of this energy is used for heating and air conditioning.

Our increasing appetite for energy is one of the causes of our current energy crisis. There are, however, at least two other general factors involved: (1) concern over the development of "clean" energy sources and (2) the depletion of traditional nonrenewable fuel sources, especially natural gas and petroleum. Our increasing appetite for energy has

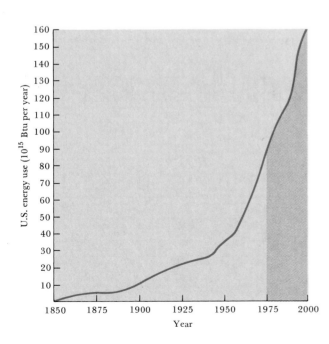

FIGURE 4.11 Energy consumption in the United States by year (1 Btu = 1.06×10^3 J). (*After S. F. Singer, "Human Energy Production as a Process in the Biosphere." Copyright 1970 by Scientific American, Inc. All rights reserved.*)

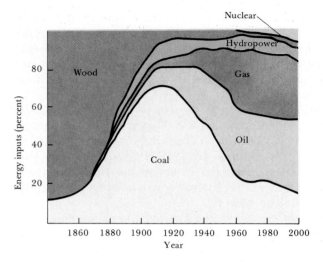

FIGURE 4.12 Pattern of energy consumption in the United States by year. (*After S. F. Singer, "Human Energy Production as a Process in the Biosphere." Copyright 1970 by Scientific American, Inc. All rights reserved.*)

required increasing dependence on imports, especially of petroleum. However, the ultimate problem related to the use of fossil fuels is that we must eventually run out of them. Meanwhile, we shall be forced to rely on increasingly expensive sources of these fuels.

FUTURE ENERGY SOURCES

According to some projections, we could run out of oil and natural gas by the end of the twentieth century unless alternate energy sources are found and per capita energy consumption levels off. There is, therefore, considerable interest in developing alternate energy sources. Much research is presently focusing on nuclear and solar energy and on the development of ways to use coal more efficiently. There is also interest in increased use of geothermal energy (the heat within the earth) and in use of winds and

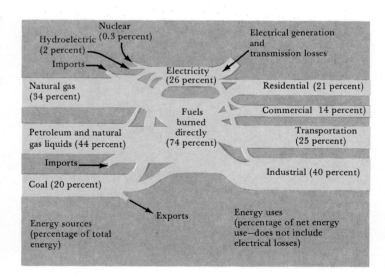

FIGURE 4.13 Pattern of energy utilization in the United States showing both sources and uses. (*After T. R. Dickson,* Introduction to Chemistry, *2nd ed. Wiley, 1975. Based on graphical scheme by Earl Cook, Texas A & M University.*)

tides to generate usable power. Experts predict that these three energy sources can each make a small but important contribution to the overall energy picture, but they will have little overall impact for the foreseeable future. We shall not discuss them further. In the following paragraphs we shall examine coal and solar energy very briefly. We shall postpone discussion of nuclear energy until Chapter 20.

Coal is the most abundant fossil fuel; it constitutes 80 percent of the fossil fuel reserves of the United States and 90 percent of those of the world. However, use of coal presents a number of problems. Coal produces more air pollution than does other fuels. It is often expensive and dangerous to mine. Furthermore, most of the remaining rich deposits of coal are in the western United States, whereas energy use is greatest along the East Coast; shipping coal great distances adds significantly to its cost. Some experts feel that coal could be used most effectively if it were converted into a gaseous form, called synthetic natural gas. Such conversion is a good way to remove sulfur from coal, thereby decreasing air pollution. Furthermore, synthetic natural gas would be easily transported in pipelines and could supplement our diminishing supplies of natural gas. Gasification of coal requires addition of hydrogen to coal. Typically, the coal is pulverized and treated with superheated steam. The product contains a mixture of CO, H_2, and CH_4, all of which can be used as fuels. However, conditions are maintained to maximize production of CH_4. A simplified schematic showing some of the reactions that occur is given in Figure 4.14.

Solar energy is the world's largest energy source. The solar energy

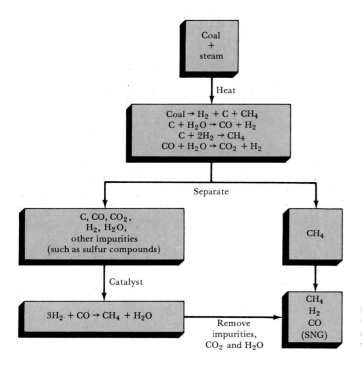

FIGURE 4.14 The basic processes involved in the gasification of coal to form synthetic natural gas (SNG). A catalyst is a substance that is able to increase the speed of a reaction without being consumed in the reaction.

incident on only 0.1 percent of the land area of the United States is equivalent to all of the energy that we currently use. The problem with the use of solar energy is that it is very dilute and fluctuates with time and weather conditions. Devices to convert solar energy to electrical energy are not presently very efficient. One possible way of using solar energy is through "energy plantations." These involve rapid and high-density growth of a mixture of plants. The plants could then be burned to produce energy. Solar energy can presently be used to great advantage to augment traditional means of heating homes; solar water heaters and careful placement of walls and windows of houses can be used to enhance the heat that is captured by a dwelling.

Another long-range development that could alleviate current problems is development of ways to remove oil from oil shale. Oil shale is a fine-grained rock that contains hydrocarbons of high molecular weight. When heated to about 900°C an oil can be obtained from the rock. There are large deposits of shale oil in the southwestern United States. However, it is presently too costly to obtain the oil using existing technologies. Furthermore, vast amounts of rock would have to be processed, probably requiring strip mining.

Before we leave this discussion, we should note that we have emphasized meeting the energy crisis by increasing energy supplies. However, we could also attack the problem by cutting back per capita energy use and by reducing energy waste. Such energy conservation is probably our most important option.

FOR REVIEW

summary

Energy is the ability to do work or transfer heat. Energy can be transferred from one object to another and its forms interconverted, but it is neither created nor destroyed in these processes. This is a statement of the **law of conservation of energy**, also known as the **first law of thermodynamics**. Energy changes are measured in **calories** or **joules** (1 cal = 4.184 J). The heat of any chemical or physical change measured under constant pressure is known as the **enthalpy change** for that process and is represented by the symbol ΔH. For **exothermic reactions** (reactions in which the **system** loses energy to its **surroundings**), ΔH is negative; for **endothermic reactions**, ΔH is positive. Enthalpy change is a **state function**, meaning that it depends on the initial and final conditions of the system and not on how the change from initial to final conditions is carried out. Because of this, enthalpy changes for reactions can be calculated from the **heats of formation** of reactants and products. This is an application of **Hess's law**. The measurement of the heat changes accompanying chemical reactions is known as **calorimetry**. The heat produced by the combustion of a material is known as its **fuel value**. We briefly considered the fuel values of some common foods and fuels including **fossil fuels**. We also discussed patterns and trends in energy use.

learning goals

Having read and studied this chapter, you should be able to:

1. Distinguish between kinetic energy and potential energy.

2. Calculate the enthalpy change, ΔH, for any reaction, having been given enthalpies of formation or other suitable ΔH values.

3. Draw a potential energy diagram such as Figure 4.8, having been given suitable enthalpy changes.

4. Calculate the heat of combustion of a substance, having been given the water equivalent of a calorimeter and other necessary information (mass of water and temperature increase).

5. Convert heats of reaction from a mole to a gram basis.

6. Describe how different food sources differ in fuel value and perform the interconversions between different heat units used to describe fuel values.

key terms

Among the more important terms and expressions used for the first time in this chapter are the following:

An **endothermic reaction** (Section 4.2) is a reaction that absorbs heat from its surroundings.

Energy (Section 4.1) is the ability to do work or transfer heat.

Enthalpy (Section 4.3) is the heat change accompanying a reaction that occurs at constant pressure.

An **exothermic reaction** (Section 4.2) is a reaction that releases heat to its surroundings.

The **fuel value** (Section 4.8) of a substance or mixture is the quantity of heat produced by combustion of 1 g of this material.

The **heat of combustion** (Section 4.8) of a substance is the enthalpy change associated with combustion of a mole of that substance.

The **heat of formation** (Section 4.6) of a substance is the enthalpy change associated with the formation of a mole of that substance from the most stable forms of its elements.

Kinetic energy (Section 4.1) is the energy of motion.

Potential energy (Section 4.1) is the energy of position or composition.

The **standard state** (Section 4.6) of a substance is the state in which the pure substance is most stable at 1 atm pressure and 25°C; it is the reference state used in tabulating standard heats of reaction.

The **surroundings** (Section 4.2) are everything around the system.

The **system** (Section 4.2) in a study of energy changes is the portion of our universe under investigation.

EXERCISES

general concept of energy

4.1 Which possesses more kinetic energy, a 1000 kg car traveling at 80 km/hr or a 2000 kg car traveling at 40 km/hr. How much kinetic energy does each possess in joules; in calories?

4.2 What is the potential energy of a 150 kg mass at a height of 10 m above the ground (the gravitational constant, $g = 9.81$ m/sec^2)? Give your answer in both joules and calories.

4.3 Give an example of the conversion of chemical energy into some other form of energy.

4.4 Classify each of the following as possessing potential or kinetic energy or both: (a) a moving train; (b) a block of wood at rest on the ground; (c) an electron moving through a magnetic field. Explain your answers.

heats of reaction

4.5 Calculate the quantity of heat given off by combustion of 10.0 g of (a) methane (CH_4); (b) carbon; (c) hydrogen (H_2).

4.6 Which of the following reactions are associated with the heat of formation of a substance and which are associated with the heat of combustion?

(a) $C_2H_6(g) + \dfrac{7}{2} O_2(g) \longrightarrow$
$\qquad 2CO_2(g) + 3H_2O(l)$

(b) $C_2H_6(g) + Cl_2(g) \longrightarrow C_2H_5Cl(g) +$
$\qquad HCl(g)$

(c) $2C(graphite) + 3H_2(g) \longrightarrow C_2H_6(g)$

(d) $2C(g) + 6H(g) \longrightarrow C_2H_6(g)$

4.7 Calculate the enthalpy changes for the following reactions. Indicate which are exothermic and which are endothermic.

(a) $3H_2(g) + N_2(g) \longrightarrow 2NH_3(g)$

(b) $2Fe(s) + \dfrac{3}{2} O_2(g) \longrightarrow Fe_2O_3(s)$

(c) $2HCl(g) + I_2(s) \longrightarrow 2HI(g) + Cl_2(g)$

(d) $NaHCO_3(s) + HCl(g) \longrightarrow$
$\qquad NaCl(s) + CO_2(g) + H_2O(l)$

4.8 The heat of formation of TNT, trinitrotoluene, $C_7H_5(NO_2)_3$, is -35.4 kJ/mole. What is the heat associated with the decomposition of TNT according to the reaction

$2C_7H_5(NO_2)_3(s) \longrightarrow$
$\qquad 7C(s) + 7CO(g) + 3N_2(g) + 5H_2O(g)$

4.9 Nitroglycerin, $C_3H_5(NO_3)_3(l)$, decomposes forming $N_2(g)$, $O_2(g)$, $CO_2(g)$, and $H_2O(l)$. (a) Write a balanced chemical equation for this reaction. (b) If ΔH_f° for nitroglycerin is -363.8 kJ/mole, calculate the enthalpy change for this decomposition reaction.

[4.10] Given

$4NH_3(g) + 5O_2(g) \longrightarrow$
$\qquad 4NO(g) + 6H_2O(l) \qquad \Delta H = -1169$ kJ

calculate the heat of formation of $NO(g)$.

4.11 The heat of vaporization of methanol, $CH_3OH(l)$, is 37.4 kJ/mole. Calculate the heat of formation of $CH_3OH(g)$.

4.12 Given the following reactions and their associated enthalpy changes:

$\dfrac{1}{2} H_2(g) + \dfrac{1}{2} F_2(g) \longrightarrow HF(g) \quad \Delta H = -259$ kJ

$\dfrac{1}{2} H_2(g) \longrightarrow H(g) \qquad\qquad\quad \Delta H = 218$ kJ

$\dfrac{1}{2} F_2(g) \longrightarrow F(g) \qquad\qquad\quad \Delta H = 79$ kJ

calculate ΔH for the reaction

$H(g) + F(g) \longrightarrow HF(g)$.

4.13 Given the following reactions and their associated enthalpy changes:

$CH_4(g) \longrightarrow C(g) + 4H(g)$	$\Delta H =$	1660 kJ
$O_2(g) \longrightarrow 2O(g)$	$\Delta H =$	490 kJ
$2H(g) + O(g) \longrightarrow H_2O(g)$	$\Delta H =$	-930 kJ
$C(g) + 2O(g) \longrightarrow CO_2(g)$	$\Delta H =$	-1610 kJ

calculate ΔH for the combustion of CH_4 to form $H_2O(g)$ and $CO_2(g)$.

4.14 Diagram the reactions shown in Question 4.13 in a fashion similar to Figure 4.8. Show the relationship between reactants $(CH_4 + 2O_2)$ and products $(CO_2 + 2H_2O)$ and the reaction path through intermediate atoms (C, 4H, 4O).

4.15 Write the balanced equation for the combustion of C_2H_5OH. Calculate ΔH° for this combustion.

calorimetry

4.16 How much energy is required to increase the temperature of 150 g of H_2O by 2.0°C?

4.17 What temperature increase can be expected if 5.00 J of energy is added to 10.0 g of water?

4.18 How much heat is produced by combustion of 1.00 g of a substance if this combustion causes the temperature of a calorimeter containing 5.00×10^3 g of H_2O and having a water equivalent of 200 g to increase by 2.03°C?

4.19 If 3.0×10^3 J raises the temperature of a calorimeter containing 120 g of H_2O from 20.0° to 25.0°C, what is the water equivalent of the calorimeter?

4.20 What is the heat of combustion of benzene (C_6H_6) on a mole basis if 1.00 g of benzene, combusted in a calorimeter containing 3500 g H_2O and with a water equivalent of 335 g, causes a temperature increase of 2.61°C?

4.21 The heat of combustion of glutaric acid, $H_2C_5H_6O_4(s)$, is 2154 kJ/mole. What is the temperature increase of the calorimeter and its contents if 1.50 g of glutaric acid is combusted in a bomb calorimeter having a water equivalent of 430 g and containing 1.500 kg of water?

fuel values

4.22 If an average adult needs 3000 kcal of energy a day to subsist on, what mass of peanuts does he need to eat to meet his energy needs for the day? Use Table 4.2.

[4.23] Sodium cyclamate, $C_6H_{12}NSO_3Na$, was once a popular nonsugar sweetener in soft drinks before it was removed from the market by the FDA. What is the Calorie savings in using 1 mg of cyclamate (heat of combustion $= 3.226 \times 10^3$ kJ/mole) in place of 30 mg of table sugar (sucrose, $C_{12}H_{22}O_{11}$, for which the heat of combustion is 5.646×10^3 kJ/mole). Cyclamate is about 30 times sweeter than sugar.

[4.24] In areas where coal is cheap, it is sometimes converted to a gaseous mixture of CO and H_2, known as "water gas," by the following reaction

$$C(s) + H_2O(g) \xrightarrow{600°C} H_2(g) + CO(g)$$

Compare the fuel value of 1 mole of carbon with that of a mixture of 1 mole H_2 and 1 mole CO. What advantages and what disadvantages are there to using water gas as a fuel instead of coal?

4.25 Compare the fuel value of the "bottled gas" propane, C_3H_8, with that of natural gas, CH_4. The heat of formation of propane is 103.8 kJ/mole.

4.26 Calculate the fuel value of cheddar cheese (Table 4.2), using the average energy values of carbohydrates, fats, and proteins given in the text and assuming that these types of substances are the only ones providing energy.

general exercises

4.27 Distinguish between the following pairs of terms: (a) exothermic and endothermic; (b) heat of formation and heat of combustion; (c) kinetic energy and potential energy; (d) calorie and Calorie.

4.28 How much energy is produced by the complete combustion of 1.00 g of diamonds?

4.29 Describe the common types of fossil fuels.

4.30 When 5.00 g of ammonium nitrate is added to 100 g of water, the temperature decreases from 25.0° to 17.3°C. If the calorimeter has a water equivalent of 5 g of water, calculate the enthalpy change for this solution process.

5

the properties of gases

Gases have played an important role in the development of science, particularly of chemistry. The properties of gases were not systematically studied until the seventeenth century. Even then there were gross misconceptions about the nature of gases. For example, Robert Boyle, a pioneer in the study of gases, had this to say about gas pressure in 1660:* "Imagine the air to be such a heap of little bodies, lying upon one another, as may be resembled to a fleece of wool. For this . . . consists of many slender and flexible hairs, each of which may indeed, like a little spring, be still endeavoring to stretch itself out again." It is evident from this that Boyle thought of the air as a mass of little particles with spring. When the air was compressed, he imagined the gas particles were squeezed together, and each of the little springs was compressed.

Once the concept of atoms and molecules became established, the study of gases played an important role in further developments in chemistry. During the period 1805–1815, Gay-Lussac studied the manner in which volumes of certain gases combined chemically to produce volumes of other gases as products. His results were correctly interpreted by Amedeo Avogadro, in the form of Avogadro's hypothesis, which we discussed briefly in Chapter 3. Some time later, in about 1858, Cannizzaro used Avogadro's hypothesis to establish the first generally useful method for determining atomic weights. Later still, the random motions

*Robert Boyle was born in Munster, Ireland, in 1627. In 1661 he published a book called *The Sceptical Chymist,* from which the quotation is taken. Boyle's book had a great influence in the development of new scientific methods and ways of analyzing experimental observations.

of a dust particle suspended in air, called Brownian motion, formed important evidence in support of the kinetic molecular theory of matter, which we shall also be discussing in this chapter.

5.1
characteristics
of gases

Under appropriate conditions, most substances can exist in any of the three states of matter. The substance H_2O, for example, is familiar to us as liquid water, ice, or water vapor. Frequently a substance exists in all three separate states of matter, or phases, at the same time. A Thermos flask containing a mixture of ice and water at $0°C$ has a certain pressure of water vapor in the gas phase over the liquid and solid phases.

Normally the three states of matter differ very obviously from one another. Gases differ dramatically from solids and liquids in several respects. A gas expands to fill its container, whereas the volumes of solids and liquids are not determined by the container. The corollary of this is that gases are highly compressible. When pressure is applied to a gas, its volume readily contracts. Liquids and solids, on the other hand, are not very compressible at all. Great pressures must be applied·to cause the volume of a liquid or solid to diminish by even as much as 5 percent.

Two or more gases form homogeneous mixtures in all proportions, regardless of how different the gases may be. Liquids, on the other hand, often do not form homogeneous mixtures. For example, when water and gasoline are poured into a bottle, the water vapor and gasoline vapors above the liquids form a homogeneous gas mixture. The two liquids, by contrast, remain largely separate; each dissolves in the other to only a slight extent.

The characteristic properties of gases arise because the individual molecules of a gas are relatively far apart. In a liquid, the individual molecules are close together and take up perhaps 70 percent of the total space. By comparison, in the air we breathe, the molecules take up only about 0.1 percent of the volume. In liquids, the molecules are constantly in contact with neighbors. They experience attractive forces for one another; this is what keeps the liquid together. However, when a pair of molecules come close together, repulsive forces prevent any closer approach. These attractive and repulsive forces differ from one substance to another. The result is that different liquids behave differently. By contrast, the molecules of a gas are well separated and are not much influenced by one another. As we shall see in more detail later, gas molecules are in constant motion, and they frequently collide. On the average, though, they remain fairly far apart. For example, in air, the average distance between molecules is about ten times as great as the sizes of the molecules themselves. Each molecule thus tends to behave as though the others weren't there. The relative degree of isolation of the molecules causes different gases to behave similarly, even though they are made up of different molecules.

To define properly the state, or condition, of a gas, it is necessary to assign values to certain variable quantities—namely, temperature, volume, quantity of gas, and pressure. We have already discussed temperature scales (Section 1.3). The absolute temperature scale, K, is the appropriate one to employ in working problems involving gases. Volume is measured in an appropriate volume unit, usually liters. The quantity of gas is measured in terms of the number of moles. Now let's consider pressure, what it is, and how it is measured.

5.2 pressure

In general terms, **pressure** carries with it the idea of a force, something that tends to move something else in a given direction. Thus, a "high-pressure" salesman might try to move a customer to buy something. In scientific terms pressure is a force that acts on a given area. One of the common units of force in our still largely nonmetric society is the pound. Air pressure in tires is measured in pounds per square inch.*

In the International System of units (SI) adopted by a worldwide conference in 1960, pressure should be expressed in units of pascals (Pa). It is convenient, however, to define a *standard atmospheric pressure* and relate other pressures to that. This can be done using a mercury **barometer,** such as shown in Figure 5.1. A barometer is formed by filling a long glass tube, which is closed at one end, with mercury and inverting it in a dish of mercury. Care must be taken that no air gets into the tube. When the tube is inverted in this manner, some of the mercury runs out, but a column remains.

The mercury surface outside the tube experiences the full force of the earth's atmosphere over each unit area, but the surface of the mercury within the tube does not. The height of the column of mercury in the tube depends on the pressure of the atmosphere. Suppose that we look more closely at the base of the tube (Figure 5.2) and imagine a surface inside the tube that is level with the surface of the mercury outside (the dashed line in Figure 5.2). The force which the atmosphere exerts on the outside surface is transferred through the liquid to the surface within the tube. There is then "1 atmosphere" of pressure per unit area pushing mercury upward across the boundary. But at the same time, the mercury above the boundary in the tube is being forced downward by the same gravitational force which is pulling the atmosphere toward earth. When the downward force of the liquid mercury equals the upward force of mercury being pushed by the atmosphere there will be no net flow of mercury. **Standard atmospheric pressure** corresponds to a pressure sufficient to support a column of mercury 760 mm in height. This is the average pressure exerted by the atmosphere at sea level. Gas pressures are thus commonly given in units of mm Hg or in units of atmospheres. One atmosphere (1 atm) corresponds to 760 mm Hg. A mm Hg pressure is also referred to as a **torr,** after the Italian scientist Evangelista Torricelli, who invented the barometer.

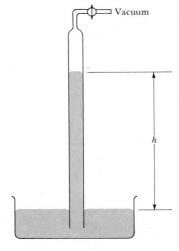

FIGURE 5.1 A mercury barometer.

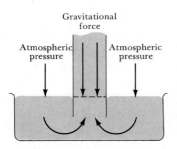

FIGURE 5.2 Balancing of atmospheric pressure and gravitational force on mercury in a mercury barometer.

*It is really not correct to say, "I carry 32 pounds of pressure in my rear tires." (The pressure should be stated as pounds *per square inch*.) Because the earth's atmosphere is pushing on the tire from the outside, the thing we call tire pressure is the *excess* pressure on the inside over that exerted from outside.

SAMPLE EXERCISE 5.1

(a) Convert 0.605 atm to torr. (b) Convert 3.5×10^{-4} mm Hg to atmospheres.

Solution: (a) Because 1 atm = 760 torr, we can employ the conversion factor 1 atm/760 torr = 1.

Conversion of atm to torr is made by multiplying the number of atm by the factor 760 torr/1 atm:

$$(0.605 \text{ atm})\left(\frac{760 \text{ torr}}{1 \text{ atm}}\right) = 460 \text{ torr}$$

Pressures are sometimes expressed in units other than atm or mm Hg. Remember that pressure is a force per unit area. In SI units (Section 1.3), the unit of force is the **Newton** (N) defined as 1 kg-m/sec². In SI units, pressure is expressed in N/m² or pascals (Pa);

SAMPLE EXERCISE 5.2

Convert atmospheric pressure from units of N/m^2 to dynes/cm².

Solution: The two conversion units we need are

$$1 \text{ dyne} = 1 \times 10^{-5} \text{ N or } 1 = \frac{1 \text{ dyne}}{1 \times 10^{-5} \text{ N}}$$

Because

$$1 \text{ m} = 100 \text{ cm}, \quad 1 \text{ m}^2 = 1 \times 10^4 \text{ cm}^2,$$

$$\text{or } 1 = \frac{1 \text{ m}^2}{1 \times 10^4 \text{ cm}^2}$$

These, then, are other and equally valid ways of expressing the atmospheric pressure. The **bar** is yet another unit of pressure; it is defined as 1.0×10^6 dynes/cm²; almost, but not quite exactly, an atmosphere. In work dealing with high pressures, kilobars are often used. A pressure of 250 kilobars, for example, corresponds to approximately 250,000

Notice that the units cancel in the required manner.

(b) To convert from torr to atm, we must multiply by the conversion factor 1 atm/760 torr:

$$(3.5 \times 10^{-4} \text{ torr})\left(\frac{1 \text{ atm}}{760 \text{ torr}}\right) = 4.6 \times 10^{-7} \text{ atm.}$$

1 atm $= 1.013 \times 10^5$ Pa. It is more common to express pressure in units of dynes per cm². One dyne equals 10^{-5} N. Thus we can carry out the conversion from N/m² to dynes/cm² by two simple unit conversions.

Then,

$$\left(1.013 \times 10^5 \frac{N}{m^2}\right)\left(\frac{1 \text{ dyne}}{1 \times 10^{-5} \text{ N}}\right)\left(\frac{1 \text{ m}^2}{1 \times 10^4 \text{ cm}^2}\right)$$

$$= 1.013 \times 10^6 \frac{\text{dyne}}{\text{cm}^2}$$

Notice again that units cancel properly to give the desired answer.

times atmospheric pressure. Pressures even greater than this exist deep within the earth. The pressure at the center of the earth is estimated to be about 3.5 million bars. At the other extreme of the pressure scale, units of millibars (1000 millibars = 1 bar) are frequently used by meteorologists and others concerned with atmospheric pressures.

5.3
Boyle's law

In contrast with liquids, gases are highly compressible. If we begin with a given volume of air and apply pressure to that volume while keeping its temperature constant, the volume contracts in roughly direct proportion to the pressure as shown in Figure 5.3. Thus if we have 1 l. of gas at a pressure of 1 atm and then increase the pressure to 2 atm, the volume reduces to about $\frac{1}{2}$ l. In general, the volume of a given quantity of gas at constant temperature is inversely proportional to the pressure:

$$V \propto \frac{1}{P}$$

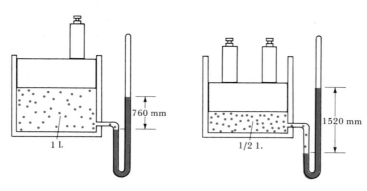

FIGURE 5.3   Gas pressure versus volume.

(The symbol $\propto$ means "varies as," or "is proportional to.") The relationship between pressure and volume at constant temperature was first made clear in experiments performed by Robert Boyle, and is known as **Boyle's law.**

Another way of expressing the relationship found by Boyle is to say that, *at constant temperature, the product of the volume and pressure of a given amount of gas is a constant:*

$$P \times V = \text{constant} \tag{5.1}$$

The value of the constant depends on how much gas is in the cylinder.

SAMPLE EXERCISE 5.3

The pressure of nitrogen gas in a 12-l. tank at 27°C is 2300 lb/in.² What volume would the gas in this tank have at 1 atm pressure (14.7 lb/in.²)?

Solution: Qualitatively we recognize that if the pressure decreases, volume must increase. Assuming that there is an inverse proportionality between pressure and volume, we look for a pressure factor that will increase the volume when the pressure decrease occurs. This pressure factor should be the ratio of the higher to the lower pressure:

$$V = (12 \text{ l.})(\text{pressure factor})$$

$$= (12 \text{ l.})\left(\frac{2300 \text{ lb/in.}^2}{14.7 \text{ lb/in.}^2}\right)$$

$$= 1880 \text{ l.}$$

From the viewpoint of Equation [5.1], we can reason that if the quantity of gas and temperature do not change, the constant in the equation must remain constant, even though P and V individually change. Thus if we have two different sets of conditions for the same quantity of gas at constant temperature, we can write

$$P_1 V_1 = P_2 V_2$$

In our example, P_1 is 2300 lb/in.², P_2 is 14.7 lb/in.², V_1 is 12 l., and V_2 is unknown. Inserting all the known quantities and solving for V_2 we obtain:

$$(2300 \text{ lb/in.}^2)(12 \text{ l.}) = (14.7 \text{ lb/in.}^2)(V_2)$$

$$V_2 = \left(\frac{2300 \text{ lb}}{1 \text{ in.}^2}\right)(12 \text{ l.})\left(\frac{1 \text{ in.}^2}{14.7 \text{ lb}}\right)$$

$$= 1880 \text{ l.}$$

Boyle's law expresses the very important fact that a gas is compressible; the more it is pressed upon, the denser it gets. As explained in

Chapter 1, density is a measure of the mass contained within a unit volume. Let us go back to the experiment shown in Figure 5.3. We have a certain quantity of gas, let us say dry air, in the cylinder. The gas, consisting primarily of O_2 and N_2 molecules, has a certain mass. Let us say that the gas is at about room temperature, 25°C, and a pressure of 1 atm, and that the volume of the cylinder is 1 l. One liter of dry air under these conditions has a mass of 1.185 g. Its density is therefore 1.185 g/liter (or, as we usually write it, 1.185 g/l.). Now, let us compress the gas to half the volume. There is still 1.185 g of air in the cylinder, because the act of compressing the gas does not change the number of gas molecules. But the density is now 1.185 g/0.5 l., or 2.370 g/l. Thus, the density has doubled in the course of halving the volume in which the gas is contained. If we squeezed the gas still further, the density would continue to increase. Thus, at constant temperature, *gas density is directly proportional to pressure.*

5.4
temperature scales; Charles's law

The volume of a gas changes not only with changes in pressure, but also with changes in temperature. Gases that are heated expand if their volume is not restricted. For example, hot air balloons are "lighter than air" because they are filled with hot gases that have expanded and thus have a lower density than the atmosphere. The quantitative aspects of this behavior could be studied with an apparatus such as that shown in Figure 5.4. A certain quantity of gas is contained in the cylinder and is under a constant pressure determined by the mass of the piston and weight. The volume of the gas changes as the temperature is altered. Suppose we measure the temperature in degrees K. A graph of volume versus temperature for a particular gas sample is shown in Figure 5.5 as

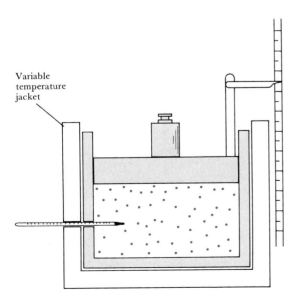

Variable temperature jacket

FIGURE 5.4 Apparatus for observing the volume of a quantity of gas versus temperature at constant pressure.

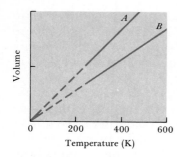

FIGURE 5.5 Volume versus temperature for a fixed quantity of gas at constant pressure.

line A. Now we change the gas sample by altering the quantity of gas, but maintaining the same pressure. A new set of volume-versus-temperature data is obtained, shown in Figure 5.5 as line B. In each case the volume is directly proportional to absolute temperature. For example, if the temperature is doubled, the volume of the gas is doubled. The proportionality between volume and temperature under these conditions can be expressed by the equation

$$V = qT \quad \text{(pressure constant)} \qquad [5.2]$$

where q is a proportionality constant. Another way of expressing this relationship is

$$\frac{V}{T} = q$$

That is, the ratio of the volume to the absolute temperature for a gas sample at constant pressure is a constant, q. The value of q varies for each sample, depending on the quantity of gas and the pressure. For example, it is larger for the sample that produced line A in Figure 5.5 than for the sample that produced line B. Equation [5.2] is one form of **Charles's law,** which states that *at constant pressure, the volume of a given quantity of gas is proportional to absolute temperature.*

SAMPLE EXERCISE 5.4

A large natural gas storage tank is arranged so that the pressure is maintained at 2.2 atm. On a cold day in December, when the temperature is $-15°C\,(4°F)$, the volume of gas in the tank is 28,500 ft³. What is the volume of the same quantity of gas on a warm July day when the temperature is $31°C\,(88°F)$?

Solution: We recognize intuitively that when the gas is heated at constant pressure it will expand. Thus we will need to multiply the gas volume at the lower temperature by a factor greater than one. In gas-law problems, temperatures must always be expressed in terms of an absolute temperature scale, that is, in degrees K. After converting temperatures to degrees K, we have the following values for initial and final conditions:

	TEMPERATURE (K)	PRESSURE (atm)	VOLUME (ft³)
Initial	258	2.2	28,500
Final	304	2.2	?

Because pressure and quantity of gas are constant, we can use Equation [5.2]. Let us call the initial volume and temperature V_1 and T_1, and the final values V_2 and T_2; we have

$$V_1 = qT_1 \qquad V_2 = qT_2$$
$$\frac{V_1}{T_1} = q \qquad \frac{V_2}{T_2} = q$$
$$\frac{V_1}{T_1} = \frac{V_2}{T_2}$$

Substituting from the data above, we have

$$\frac{28,500 \text{ ft}^3}{258 \text{ K}} = \frac{V_2}{304 \text{ K}}$$
$$V_2 = \left(\frac{304 \text{ K}}{258 \text{ K}}\right)(28,500 \text{ ft}^3)$$
$$= 33,600 \text{ ft}^3$$

$$V_1 T_2 = V_2 T_1$$

We know that when a confined gas is heated at constant volume its pressure increases. For example, a popcorn kernel bursts open under the

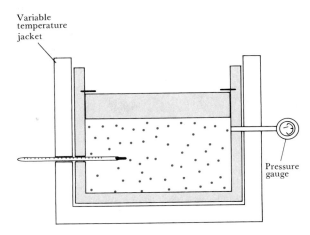

Variable
temperature
jacket

Pressure
gauge

FIGURE 5.6 Apparatus for observing the pressure of a gas versus temperature at constant volume.

pressure of steam that is formed when the kernel is heated in oil. The experimental apparatus shown in Figure 5.6 can be used to observe the effect of changing temperature on the pressure of a gas maintained at constant volume. The pins in the wall of the cylinder restrain the piston to a fixed position. The pressure of the gas changes as the temperature is changed, in the manner shown in Figure 5.7. Each of the lines shown applies for a particular sample of gas and a particular volume. The equation that describes the changing pressure with absolute temperature in each case is of the form

$$P = jT, \text{ or } \frac{P}{T} = j \qquad [5.3]$$

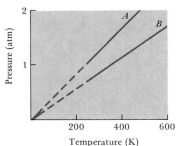

FIGURE 5.7 Pressure versus temperature for a fixed quantity of gas at constant volume.

where j is a constant that depends on the particular sample of gas and its volume. Equation [5.3] represents another form of Charles's law: *At constant volume, the pressure of a given quantity of gas is proportional to absolute temperature.*

SAMPLE EXERCISE 5.5

The gas pressure in an aerosol can is 1.5 atm at 25°C. Assuming that the gas inside obeys Charles's law, what would the pressure be if the can were heated to 450°C?

Solution: Let us proceed, as in Sample Exercise 5.4, by writing down the initial and final conditions of temperature, pressure, and volume that the problem gives us. (Remember that we must convert temperatures to degrees K.)

	VOLUME	PRESSURE (atm)	TEMPERATURE (K)
Initial	V_1	1.5	298
Final	V_1	P_2	723

Since volume and quantity of gas are constant in this problem, we can use equation (5.3). This gives us

$$\frac{P_1}{T_1} = j \qquad \frac{P_2}{T_2} = j$$

$$\frac{P_1}{T_1} = \frac{P_2}{T_2}$$

Substituting into the equation with the quantities given above, we have

$$\frac{1.5 \text{ atm}}{298 \text{ K}} = \frac{P_2 \text{ atm}}{723 \text{ K}}$$

$$P_2 \text{ atm} = (1.5 \text{ atm})\left(\frac{723 \text{ K}}{298 \text{ K}}\right) = 3.6 \text{ atm}$$

$$P_1 T_2 = P_2 T_1$$

We see that the pressure has increased. This is the answer we intuitively expect, because the pressure of a gas heated under constant volume increases. It is evident from this example why aerosol cans carry the warning not to incinerate.

5.5

Dalton's law of partial pressures

Suppose the gas with which we are concerned is not a single kind of gas particle, but is rather a mixture of two or more different substances. We might suppose that the total pressure exerted by the gas mixture is the sum of pressures due to the individual components. Each of the individual components, if present alone under the same temperature and volume conditions as the mixture, would exert a pressure that we term the **partial pressure**. John Dalton was the first to observe that *the total pressure of a mixture of gases is just the sum of the pressures that each gas would exert if it were present alone:*

$$P_t = P_1 + P_2 + P_3 + \ldots$$ [5.4]

This is known as Dalton's law of partial pressures.

AVOGADRO'S HYPOTHESIS

Amedeo Avogadro, another of the great nineteenth-century scientists, made a very important hypothesis about gases, which turned out to be correct. He suggested that *equal volumes of gases at the same temperature and pressure contain equal numbers of molecules.* This was a most important hypothesis, one that was not at all obvious at the time. Suppose, for example, that we have three 1-l. bulbs containing H_2, N_2, and Ar, respectively (Figure 5.8), and that each gas is at the same pressure and temperature. According to Avogadro's hypothesis, those bulbs contain *equal numbers of gaseous particles.* Note that one of the gases is made up of atoms, and the other two of diatomic molecules; note also that the masses of the substances in the bulbs differ greatly. As far as pressure, temperature, and volume relationships are concerned, however, what counts is

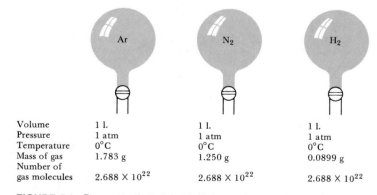

Volume	1 l.	1 l.	1 l.
Pressure	1 atm	1 atm	1 atm
Temperature	0°C	0°C	0°C
Mass of gas	1.783 g	1.250 g	0.0899 g
Number of gas molecules	2.688×10^{22}	2.688×10^{22}	2.688×10^{22}

FIGURE 5.8 Demonstration of Avogadro's hypothesis. Note that argon gas consists of argon atoms; we can regard these as one-atom molecules.

that there are equal numbers of gaseous particles in the bulbs, at the same temperature. Avogadro's hypothesis holds true for so-called ideal gases, that is, those that obey both Boyle's law and Charles's law (see Section 5.7). The three gases shown in Figure 5.8 behave essentially as ideal gases under the conditions of temperature and pressure given there.

From the data presented in Figure 5.8, we can calculate the molar volume of a gas at standard temperature and pressure, abbreviated STP, and defined as 0°C and 1 atm pressure.

SAMPLE EXERCISE 5.6

From the data presented in Figure 5.8, calculate the volume of 1 mole of an ideal gas at 1 atm pressure and 0°C (STP).

Solution: The gases in the flasks shown in Figure 5.8 are under standard conditions. In each 1-l. flask there are 2.688×10^{22} molecules. A mole of gas would consist of 6.022×10^{23} molecules (Section 3.8). Thus we have

$$\left(\frac{6.022 \times 10^{23} \text{ molecules}}{1 \text{ mole}} \right) \left(\frac{1 \text{ l.}}{2.688 \times 10^{22} \text{ molecules}} \right)$$
$$= 22.40 \text{ l./mole}$$

5.6
earth's atmosphere

One very good reason for learning about the properties of gases is that we live in a gaseous environment, the earth's atmosphere. The atmosphere is in many ways like the liquid sea of water that covers three-fourths of the earth's surface. Creatures that live in the depths of the oceans experience a vastly different environment than those that live near the surface. In a similar way, the environment that we experience at the bottom of the atmospheric sea is very different from that which we would experience if we were a few hundred kilometers up, above most of the earth's atmosphere. Because most of us have never been very far from the earth's surface, we tend to take for granted the many ways in which the atmosphere determines the environment in which we live. In this section, we shall examine some of the important physical characteristics of our planet's atmosphere in light of what we have just learned of the properties of gases.

The temperature of the atmosphere varies greatly with altitude. The reasons for the temperature variations are complex, and we shall not attempt to relate them here. Nevertheless, it is important to have an idea of what the temperature variation is, because the boundaries between the various regions of the atmosphere are determined by maxima and minima in the temperature profile, as shown in Figure 5.9.

The temperature of the atmosphere just above earth's surface normally decreases with increasing altitude, and reaches a minimum value at an elevation of about 12 km. In this region, called the **tropopause,** the temperature is about 215 K ($-60°C$). Above this elevation the temperature increases to about 275 K in the region of 50 km and then begins again to decrease. The altitude at which the temperature reaches

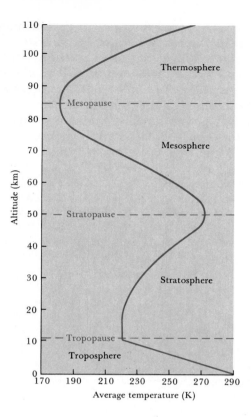

FIGURE 5.9 Temperature variations in the atmosphere at altitudes below 110 km. (*Adapted from "U.S. Standard Atmosphere, 1962." Washington, D.C.: Government Printing Office*)

a maximum is called the **stratopause.** Above the stratopause the temperature drops to an even lower value than at the tropopause. The region of this second temperature minimum is called the **mesopause.** Above the mesopause, the temperature rises rapidly, in the region called the **thermosphere.** Note that the regions of temperature minima and maxima are denoted by the suffix -*pause*. The regions between these are denoted by the suffix -*sphere*. The boundaries between different regions are important because mixing of the atmosphere across the boundaries is relatively slow. Thus, for example, pollutant gases generated in the troposphere find their way into the stratosphere only very slowly.

The **troposphere** is the region of the atmosphere in which nearly all of us live out our entire lives. When Shakespeare has Hamlet speak of "this most excellent canopy, the air, look you, this brave o'erhanging firmament, this majestical roof fretted with golden fire," he is speaking of the troposphere. Howling winds and soft breezes, rain, sunny skies, all that we normally think of as weather occurs in this region. Even when we fly in a modern jet aircraft between distant cities, we are still in the troposphere, though we may be near the tropopause.

In contrast to the temperature changes that occur in the atmosphere, the pressure of the atmosphere decreases in a quite regular way with increasing elevation, as shown in Figure 5.10. A unit volume of the atmosphere at the surface experiences a pressure that results from the entire column of atmosphere above it. All of this gas is attracted to the earth by gravitational force and thus exerts pressure on whatever is below it. At the surface this accumulated pressure amounts, as we have seen, to 760 torr. At a higher elevation, however, the pressure experienced by a unit volume of gas is less, because there is a smaller total mass of gas in the column above it. Because this unit volume of gas experiences a lower pressure, it has a lower density. Thus, with an increase in elevation above the earth's surface, the atmosphere is less and less compressed, and therefore the density is lower and lower.

If we were to measure the pressure in one of the earth's oceans, we would find that it varies nearly linearly with depth. That is, the increase in pressure in going from say 100 to 200 m depth would be about the same as the increase in pressure in going from 1100 to 1200 m depth. This is true because water is not a very compressible fluid. When pressure is applied to it, it does not contract by very much. Thus, the density of a cubic meter of water at the surface of the ocean is not much less than the density of a cubic meter of water at the bottom. The tremendous weight of all the water in the column above it changes the volume per unit weight of water at the bottom only a little.

We see from Figure 5.10, however, that atmospheric pressure drops off much more rapidly at lower elevations than at higher. The explanation for this characteristic of the atmosphere lies in its compressibility, expressed quantitatively in Boyle's law. Gases are very different from liquids in this regard. As a result of the atmosphere's compressibility, the pressure decreases from an average value of 760 torr at sea level to 2.3×10^{-3} torr at 100 km, to only 1.0×10^{-6} torr at 200 km.

COMPOSITION OF THE ATMOSPHERE

The atmosphere is an extremely complex system. Its temperature and pressure change over a wide range with altitude, as we have already seen. The atmosphere is subjected to bombardment by radiation and energetic particles from the sun and by cosmic radiation from outer space. This

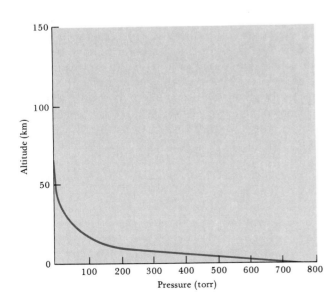

FIGURE 5.10 Variation in atmospheric pressure with altitude. At 50 km altitude the pressure has declined to about 1 torr. At still higher altitudes the pressure continues to decline, although this cannot be shown on the scale of the figure. For example, at 100 km the pressure has declined to 2.3×10^{-3} torr.

barrage of energy has profound chemical effects, especially on the outer reaches of the atmosphere. In addition, the earth's gravitational field causes some separation of atoms and molecules according to mass. In other words, lighter atoms and molecules tend to rise to the top. As a result of all these factors, the composition of the atmosphere is not constant. However, it is useful to know the composition of the atmosphere in the region near the earth's surface. Table 5.1 shows the composition of dry air near sea level. We note that although traces of a great many substances are present, only a few dominate. The two diatomic molecules N_2 and O_2 make up about 99% of the entire atmosphere. Essentially all the remainder except for carbon dioxide is made up of the monatomic rare gases.

TABLE 5.1

composition of dry air near sea level

COMPONENT[a]	CONTENT (MOLE FRACTION)	MOLECULAR WEIGHT
Nitrogen	0.78084	28.013
Oxygen	0.20948	31.998
Argon	0.00934	29.948
Carbon dioxide	0.000314	44.0099
Neon	0.00001818	20.183
Helium	0.00000524	4.003
Methane	0.000002	16.043
Krypton	0.00000114	83.80
Hydrogen	0.0000005	2.0159
Nitrous oxide	0.0000005	44.0128
Xenon	0.000000087	131.30

[a]Ozone, sulfur dioxide, nitrogen dioxide, ammonia, and carbon monoxide are present as trace gases in variable amounts.

Note that the contribution of each component of the atmosphere listed in Table 5.1 is given in terms of its mole fraction. This is simply the total number of moles of a particular component in a given sample of air, divided by the total number of moles of all the components in that sample. The partial pressure of a given component in the atmosphere is given by the total atmospheric pressure times the mole fraction of that component. This statement comes from Dalton's law of partial pressures and Avogadro's hypothesis. By using it we can write the total pressure P_t for a gas mixture consisting of gases with mole fractions X_1, X_2, and so on, as:

$$P_t = X_1 P_t + X_2 P_t + X_3 P_t + \ldots$$
$$= P_t(X_1 + X_2 + X_3 + \ldots) \qquad [5.5]$$

Because the X's must all add up to one, this equation is clearly correct.

SAMPLE EXERCISE 5.7

What is the partial pressure of CO_2 in dry air when the total dry air pressure is 735 torr?

Solution: Referring to Table 5.1, we see that the mole fraction of CO_2 is 3.14×10^{-4}. This means that the fractional contribution of CO_2 to a unit total pressure of 1 would be 3.14×10^{-4}:

$$\text{Pressure } CO_2 = (735 \text{ torr } P_t)\left(\frac{3.14 \times 10^{-4} \text{ torr } CO_2}{1 \text{ torr } P_t}\right)$$

$$= 0.231 \text{ torr } CO_2$$

A proper understanding of earth's atmosphere requires an understanding not only of the earth's physical properties, but also of its chemical properties as well. When we discuss in Chapter 10 the chemistry that occurs in the atmosphere, we shall see that the chemistry that occurs is to some extent dependent on the physical properties that we have learned about in this chapter. Thus, for example, the chemical reactions occurring in the troposphere are vastly different from those observed in the mesosphere.

5.7
the ideal-gas law

In the preceding sections, we have met several "laws" that relate the variables temperature, pressure, volume, and n, the number of moles of gas, which define the state of a gas. We can summarize each law by expressing it in a form that shows the proportionality between volume and one of the other variables, with the remaining variables being held constant.

Boyle's law: $\quad V \propto \dfrac{1}{P} \qquad T, n$ constant

Charles's law: $V \propto T \qquad P, n$ constant
$\qquad\qquad\quad P \propto T \qquad V, n$ constant

Avogadro's
hypothesis: $\quad V \propto n \qquad P, T$ constant

These three proportionality relations can be combined in a single expression:

$$V \propto \frac{nT}{P} \qquad\qquad [5.6]$$

To change the proportionality to an equation we insert a proportionality constant, R, called the molar gas constant:

$$PV = nRT \qquad\qquad [5.7]$$

The molar gas constant shows up in many different contexts and in various units. Table 5.2 shows values for R in a few of the more commonly employed units. Any gas for which the pressure, volume, and temperature are related in accordance with Equation [5.7] is called an **ideal gas.**

The **ideal-gas equation,** Equation [5.7], expresses a quantitative relationship between all four of the quantities that could possibly be varied

129

in describing the state of a gas. An equation of this type is referred to as an **equation of state,** because when we have specified all the variables we have defined the state of the system. Although real gases don't exactly obey the ideal-gas equation, the difference between ideal and real behavior is often so small that it can be ignored. Thus, we may use the ideal-gas equation to solve problems of practical importance involving gases.

TABLE 5.2

values for the molar gas constant in various units

UNITS OF R	NUMERICAL VALUE
Liter-atm/K-mole	0.08206
Calories/K-mole	1.987
Ergs/K-mole	8.317×10^7
Joules/K-mole	8.317
Btu/K-mole	0.00788

We are often faced with the need to calculate the effect on the other variables of changing one or more of the variables of a gas system. Problems of this sort may be solved by direct application of Equation [5.7].

SAMPLE EXERCISE 5.8

A 0.823-l. sample of nitrogen gas is contained in a cylinder at 27°C and at a pressure of 766 torr. What is the pressure of the gas if the volume is changed to 0.456 l. at the same temperature?

Solution: Because the change in state of the nitrogen gas occurs at constant temperature, the right-hand side of Equation [5.7] is a constant. (Note that the ideal-gas equation thus gives us Boyle's law, which states that the product of pressure and volume at constant temperature is a constant.) If we use subscripts 1 and 2 to denote the pressure and volume before and after the changes, we must have

$$P_1 V_1 = nRT = \text{constant}$$
$$P_2 V_2 = nRT = \text{same constant}$$
$$P_1 V_1 = P_2 V_2$$

Inserting the known numerical values,

$$(766 \text{ torr})(0.823 \text{ l.}) = P_2 (0.456 \text{ l.})$$
$$P_2 = (766 \text{ torr}) \left(\frac{0.823 \text{ l.}}{0.456 \text{ l.}} \right)$$
$$= 1380 \text{ torr}$$

Charles's law also follows directly from the ideal-gas equation.

Using Equation [5.7] it is easy to solve problems in which temperature, pressure, and volume are all subject to change. We can always solve such problems by using Boyle's law and Charles's law, applying each in succession. It is convenient, however, to have the use of Equation [5.7], in which all the variables are combined in a single expression.

SAMPLE EXERCISE 5.9

A quantity of helium gas occupies a volume of 16.5 l. at 78°C and 45.6 atm pressure. What is its volume at STP?

Solution: We can rearrange the ideal-gas Equation [5.7] to the form

$$\frac{PV}{T} = nR$$

As long as the total quantity of gas, n, is constant, PV/T is a constant. If we represent the initial and final conditions of pressure, temperature, and volume by subscripts 1 and 2, we can write

$$\frac{P_1 V_1}{T_1} = nR = \text{constant}$$

$$\frac{P_2 V_2}{T_2} = nR = \text{constant}$$

$$\frac{P_1 V_1}{T_1} = \frac{P_2 V_2}{T_2}$$

Let us next write down all we know of the initial and final values of pressure, temperature, and volume (remember always to convert temperatures to absolute temperature).

	PRESSURE (atm)	VOLUME (l.)	TEMPERATURE (K)
Initial	45.6	16.5	351
Final	1	V_2	273

Putting in all the quantities we know, we obtain

$$\frac{(45.6 \text{ atm})(16.5 \text{ l.})}{351 \text{ K}} = \frac{(1 \text{ atm})(V_2)}{273 \text{ K}}$$

$$V_2 = \left(\frac{45.6 \text{ atm}}{1 \text{ atm}}\right)\left(\frac{273 \text{ K}}{351 \text{ K}}\right)(16.5 \text{ l.}) = 585 \text{ l.}$$

To solve such a problem by successive applications of Charles's law and Boyle's law, we might have reasoned as follows: The change in temperature from 78°C (351 K) to 0°C (273 K) will cause the volume to decrease by the factor $\frac{273}{351}$. The change in pressure from 45.6 atm to 1 atm will cause the volume to increase by the factor $\frac{45.6}{1}$. The new volume is then

$$\text{New volume} = 16.5 \text{ l.} \left(\frac{273 \text{ K}}{351 \text{ K}}\right)\left(\frac{45.6 \text{ atm}}{1 \text{ atm}}\right)$$

$$= 585 \text{ l.}$$

Some of the most useful calculations using the ideal-gas equation involve measurements and calculation of the gas density. Density has the units of mass per unit volume. We can arrange the gas equation to obtain:

$$\frac{n}{V} = \frac{P}{RT} \qquad [5.8]$$

Now n/V has the units of moles per liter. Suppose we multiply the numerator in this equation by molecular weight (MW), which is the number of grams in 1 mole of a substance:

$$\frac{n(\text{MW})}{V} = \frac{P(\text{MW})}{RT} \qquad [5.9]$$

But the product of the two quantities on the left is density, since the units multiply as follows:

$$\frac{\text{Moles}}{\text{Liter}} \times \frac{\text{grams}}{\text{mole}} = \frac{\text{grams}}{\text{liter}}$$

Thus the density of the gas is given by the expression on the right in Equation [5.9]:

$$d = \frac{P(\text{MW})}{RT} \qquad [5.10]$$

Sample Exercises 5.10 and 5.11 illustrate the use of this relation.

SAMPLE EXERCISE 5.10

What is the density of carbon dioxide gas at 745 torr and 65°C?

Solution: The molecular weight of carbon dioxide, CO_2, is $12.0 + (2)(16.0) = 44.0$ g/mole. If we are to use 0.08206 l.-atm/K-mole for R we must convert pressure to atmospheres. We have then

$$d = \frac{\left(\frac{745}{760}\ \text{atm}\right)(44.0\ \text{g/mole})}{(0.08206\ \text{l.-atm/K-mole})(338\ \text{K})}$$

$$= 1.56\ \text{g/l.}$$

The problem can be turned around a bit to determine the molecular weight of a gas from its density as shown in Sample Exercise 5.11.

SAMPLE EXERCISE 5.11

A large flask fitted with a stopcock is evacuated and weighed; its mass is found to be 134.567 g. It is then filled to a pressure of 735 torr at 31°C with a gas of unknown molecular weight, and then reweighed; its mass is 137.456 g. The flask is then filled with water and again weighed; its mass is now 1067.9 g. Assuming that the ideal-gas law applies, what is the molecular weight of the unknown gas?

Solution: First we must determine the volume of the flask. This is given by difference in weights of the empty flask and the flask filled with water, divided by the density of water at 31°C (essentially 1):

$$V = \frac{1067.9\ \text{g} - 134.6\ \text{g}}{1\ \text{g/cm}^3} = 933.3\ \text{cm}^3$$

The next task is to calculate the number of moles of gas that will occupy this volume at the temperature and pressure indicated.

$$n = \frac{PV}{RT}$$

We must be careful to insert all quantities in the appropriate units. When this is done we obtain:

$$n = \frac{\left(\frac{735}{760}\ \text{atm}\right)(0.933\ \text{l.})}{(0.08206\ \text{l.-atm/K-mole})(304\ \text{K})}$$

$$= 0.0362\ \text{moles}$$

We now have the number of moles of gas. We know also that this number of moles weighs 137.456 g − 134.567 g = 2.889 g.

The molecular weight is then simply the mass divided by the number of moles which that mass represents:

$$\frac{2.889\ \text{g}}{0.0362\ \text{moles}} = 79.81\ \text{g/mole}$$

One of the experiments that occasionally comes up in the course of laboratory work is the determination of the number of moles of gas collected from a chemical reaction. Sometimes this gas is collected over water. For example, potassium chlorate $(KClO_3)$ may be decomposed by heating solid $KClO_3$ in a test tube with an arrangement as shown in Figure 5.11. The balanced equation for the reaction is

$$2KClO_3 \longrightarrow 2KCl + 3O_2$$

The oxygen gas is collected in the bottle initially filled with water and inverted in the water pan.

The volume of gas collected is measured by raising or lowering the bottle as necessary until the water levels inside and outside the bottle are the same. When this condition is met, the pressure inside the bottle is equal to the atmospheric pressure outside. But the total pressure inside is the sum of the pressure of gas collected and the water vapor in equilibrium with liquid water. To compute the number of moles of gas collected, a correction must be applied for the partial pressure of water

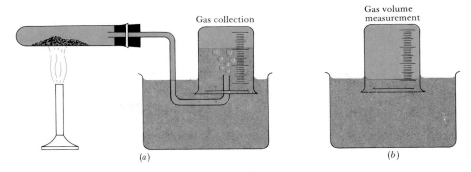

FIGURE 5.11 (a) Collection of a gas over water. (b) When the gas has been collected the bottle is raised or lowered to equalize pressures inside and outside before measuring volume of gas collected.

vapor in the collection bottle. We shall go through a complete analysis of such an experiment in Sample Exercise 5.12 to show how the stoichiometric relationships come together.

SAMPLE EXERCISE 5.12

A 2.55 g sample of ammonium nitrite (NH_4NO_2) is heated in a test tube connected as shown in Figure 5.11. The ammonium nitrite is expected to decompose according to the equation

$$NH_4NO_2 \longrightarrow N_2 + 2H_2O$$

If it does decompose in this way, will a 1-l. flask be sufficiently large to contain the N_2? The water and gas temperature are 26°C, and the barometric pressure is 745 torr.

Solution: We begin by calculating the number of moles of N_2 gas formed.

$$2.55 \text{ g } NH_4NO_2 \left(\frac{1 \text{ mole } NH_4NO_2}{64.0 \text{ g } NH_4NO_2} \right)$$
$$\times \left(\frac{1 \text{ mole } N_2}{1 \text{ mole } NH_4NO_2} \right) = 0.0398 \text{ mole } N_2$$

To predict the volume of N_2 gas that is to be collected, we might be tempted to just calculate the volume that would be occupied by 0.0398 moles N_2 at 745 torr. But we must also take account of the partial pressure of water vapor at 26°C, because the

gas within the bottle is saturated with water vapor. From a table of water vapor pressure versus temperature in the lab manual or in a chemistry handbook (see also Appendix D), we can determine that the vapor pressure of water at 26°C is 25 torr. The pressure of nitrogen gas in the flask when the water levels inside and out have been equalized is thus $745 - 25 = 720$ torr. We must calculate the predicted volume of gas using this pressure ($\frac{720}{760} = 0.947$ atm). Rearranging Equation [5.7] we obtain:

$$V = \frac{nRT}{P}$$

Inserting all known quantities in correct units we obtain:

$$V = \frac{(0.0398 \text{ moles})(0.08206 \text{ l.-atm/K-mole})(299 \text{ K})}{0.947 \text{ atm}}$$
$$= 1.03 \text{ l.}$$

The answer to our problem is that a 1-l. bottle will *not* be large enough to contain the gas we wish to collect.

5.8 kinetic-molecular theory of gases

The work of Boyle, Charles, Avogadro, Dalton, and others led to empirical relationships which are embodied in the ideal-gas law. But an empirical law is not the same as a model. To *understand* gases we must have a theoretical model, capable of being expressed in mathematical language, that yields a relationship between the variables that define the state of the gas. That is, we need a model that eventually yields the ideal-gas equation.

It had been suggested very early by Bernoulli (1738) and later by others that the properties of gases as expressed in Boyle's law and Charles's law could be accounted for if it were assumed that the particles of a gas are in ceaseless motion but have no forces whatever between them. In this model, the temperature of the gas was related to the average energy of motion of the gas particles. These ideas culminated in a new theory, called the **kinetic-molecular theory of gases,** which was published in its most complete and satisfactory form by Rudolf Clausius in 1857.

In the kinetic theory, several assumptions are made about the detailed nature of gases:

1. Gases consist of large numbers of molecules in ceaseless motion. A molecule will be taken to mean a particle consisting of one or more atoms that move together as a unit. When we say that the particles are in ceaseless motion, we mean that the average energy of all the particles does not change with time, so long as the temperature of the gas remains constant.

2. The volume of all the molecules of the gas is very small in comparison to the total volume in which the gas is contained. In other words, most of the space occupied by the gas is empty space.

3. The time during which a collision between two molecules occurs is negligibly short in comparison with the time between collisions. This means that at any one instant, if we could "freeze" the action in a gas, a very small fraction of the molecules would be undergoing collision with another molecule.

4. No attractive or repulsive forces operate between the molecules up to the point at which they collide. In other words, the gas molecules behave like tiny billiard balls.

5. The average kinetic energy of the molecules is proportional to absolute temperature.

From these hypotheses, and by applying a few basic laws of physics, it is possible to derive the ideal-gas law. In addition, a few important qualitative deductions about gas properties can be made from the postulates.

First of all, postulate 1 leads to the deduction that the pressure that the gas exerts is due to the impacts of gas molecules on the wall of the container. The more impacts, the higher the pressure. When a gas at constant temperature is compressed into a smaller volume, the observed increase in pressure is due to an increased number of collisions per unit time with a unit area of container wall. Postulate 2 comes in at this point, because we must assume that even after compression, the volume of the gas molecules is still negligible in comparison with the total container volume.

Secondly, postulate 1 leads us to conclude that the average velocity of gas molecules is constant if the temperature is constant. This follows because the average velocity is directly related to average energy. Kinetic energy is expressed as one-half the product of mass times the square of the velocity or speed. The *average* kinetic energy (ϵ) is thus given by

$$\epsilon = \tfrac{1}{2}mu^2 \qquad\qquad [5.11]$$

where u represents the *average* speed of the molecules.*

A change in temperature of a gas represents a change in the average kinetic energy (that is, energy of motion) of the gas molecules. The molecules of a gas move about at random in their container, colliding with one another and with the walls of the enclosure. At any one instant some of them are moving rapidly, others slowly. It is possible to make measurements of the distribution of molecular speeds within a gas. For nitrogen at 0°C the distribution curve looks like the black line in Figure 5.12. For nitrogen at 100°C the distribution is like that shown by the color line in the figure. At higher temperatures, the distribution is shifted toward higher speeds. If a gas is heated in a fixed volume, the pressure increases because of the increased number of collisions of gas molecules with the wall of the container, and from the increased energy of the average collision. When a gas is cooled, the pressure decreases because the collisions are less frequent and of lower average energy. We can imagine that there is a temperature at which the average energy of motion of the gas molecules would be zero. This corresponds to absolute zero. Thus the pressure-temperature behavior of this ideal gas would define an absolute temperature scale. For many years a gas thermometer provided the most accurate definition of absolute zero.

*The quantity u in Equation [5.11] is not exactly the average speed, but it is a close enough approximation for our purposes to assume that it is.

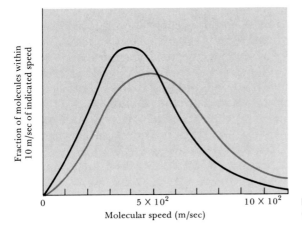

FIGURE 5.12 Distribution of molecular speeds for nitrogen at 0°C (black line) and 100°C (color line).

A sample of O_2 gas initially at STP is transferred from a 2-l. container to a 1-l. container at constant temperature. What effect does this change have on (a) the average speed of the O_2 molecules, (b) the average kinetic energies of O_2 molecules, (c) the total number of collisions of O_2 molecules with the container walls in a unit time, and (d) the number of collisions of O_2 molecules with a unit area of container wall in a unit time?

Solution: (a) The average speed of molecules is determined only by the absolute temperature. Therefore the average speed is not changed by the compression of O_2 from 2 to 1 l. at constant temperature. (b) The average kinetic energy of O_2 molecules does not change either, because this is also dependent only on temperature. (c) The total number of collisions with the container walls in a unit time must increase, because the molecules are moving within a smaller volume, but with the same average speed as before. Under these conditions they must encounter a wall more frequently. (d) The number of collisions with a unit area of wall increases, because the total number of collisions with the walls is higher, and the area of wall is smaller than before.

5.9

molecular diffusion; Graham's law

We have already made reference to the fact that the molecules of a gas do not all move at the same speed. Instead, the molecules are distributed over a range of speeds, as shown for nitrogen at two different temperatures in Figure 5.12. The distribution of molecular speeds depends on the mass of the gas molecules. Because the average kinetic energy of the molecules in any gas is determined only by temperature, it follows that the quantity $\frac{1}{2}mu^2$, which is kinetic energy, must have the same value for two gases at the same temperature, though their masses may differ. This in turn means that molecules of larger mass must have smaller average speeds. From the kinetic molecular theory it can be shown that the average speed has the value

$$u = 1.59 \sqrt{\frac{RT}{MW}} \qquad [5.12]$$

This relationship tells us that average speed at a given temperature is inversely proportional to the square root of the molecular weight. In other words, lighter molecules have higher average speeds. The distribution of molecular speeds is also skewed to higher speeds for gases of lower molecular weights, as shown for several gases in Figure 5.13.

The dependence of molecular speeds on mass has several interesting consequences. Thomas Graham discovered in about 1830 that the effusion rates of gases are inversely related to the square roots of their molecular weights. (*Effusion* refers to the escape of a gas through a tiny hole.)* Assume that we have two gases at the same initial pressure contained in identical containers, each with an identical pinhole in one wall. Let the rate of effusion be called r. Graham's law states that

---

*Effusion is related to, but is not quite the same as, *diffusion*. The latter term refers to the spread of one substance throughout a space, or throughout a second substance. For example, the molecules of a perfume diffuse through a room.

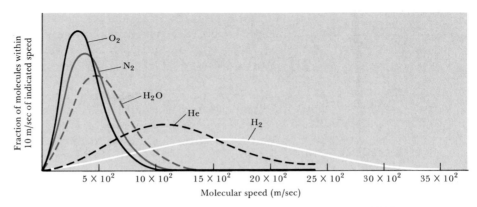

FIGURE 5.13 The distribution of molecular speeds for different gases at 25°C.

$$\frac{r_1}{r_2} = \sqrt{\frac{MW_2}{MW_1}} \qquad [5.13]$$

This equation follows from our previous discussion if we assume that the rate of effusion is proportional to the average speed of the molecules. Because R and T are constant, we have from Equation [5.13]:

$$\frac{r_1}{r_2} = \frac{u_1}{u_2} = \sqrt{\frac{MW_2}{MW_1}} \qquad [5.14]$$

We see many applications of these ideas in the properties of gases. One popular lecture demonstration, the hydrogen fountain, is illustrated in Figure 5.14. It makes use of the fact that the diffusion of hydrogen through the walls of the porous cup is faster than diffusion of atmospheric gases out of the cup through the wall. Thus an excess pressure builds up in the enclosure. Another example is seen in the behavior of toy balloons that have been filled with helium. Helium diffuses outward through the balloon surface more rapidly than the atmospheric gases diffuse in, and so the balloon collapses much more rapidly than it would if filled with air.

In the course of the effort during World War II to develop the atomic bomb, it was necessary to separate the relatively low-abundance uranium isotope ^{235}U (0.7 percent) from the much more abundant ^{238}U (99.3 percent). This was done by converting the uranium into a volatile compound, UF_6, which boils at 56°C. The gaseous UF_6 was allowed to diffuse from one chamber into a second through a porous barrier. Because of the slight difference in molecular weights, the relative rates of passage through the barrier for $^{235}UF_6$ and $^{238}UF_6$ are not exactly the same. The ratio of diffusion rates is given by the square root of the ratio of molecular weights, Equation [5.14]:

$$\frac{r_{235}}{r_{238}} = \sqrt{\frac{352}{349}} = 1.0043$$

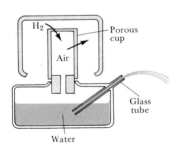

FIGURE 5.14 A hydrogen-fountain demonstration of the greater rate of diffusion of hydrogen as compared with air. A large container filled with H_2 gas is placed over the porous cup containing air. Hydrogen diffuses into the cup more rapidly than the molecules of air diffuse outward. As a result, the pressure inside the vessel increases, and water is pushed out the glass tube, which is open to the outside.

Thus the gas initially appearing on the opposite side of the barrier would be very slightly enriched in the lighter molecule. The diffusion process was repeated thousands of times, leading to a nearly complete separation of the two nuclides of uranium.

5.10
molecular
collisions;
mean free paths

We can see from the horizontal scale of Figure 5.13 that the speeds of molecules are really quite high. Translated into more familiar units, the average speed of N_2 at room temperature, 515 m/sec, corresponds to 1850 km/hr, or 1150 mi/hr. Yet we know that molecules don't travel through the atmosphere from one place to another at these speeds. For example, if a vial of a substance possessing a strong odor is opened at one end of a room, it is some time, perhaps a few minutes, before the odor is noted by a person at the other end. Therefore, the movement of gas molecules in the atmosphere is much slower than the molecular speeds would suggest. We must ascribe these slower diffusion rates to collisions between molecules.

According to the assumptions made in deriving the ideal-gas laws, the actual volumes of the gas molecules are negligible in comparison with the volume of the space enclosing the gas. At the same time, however, we know that gas molecules can't have *zero* volume. If they did, there could be no collisions between molecules, and there would be nothing to impede the flow of gas molecules from one place to another at rates commensurate with their speeds. But because molecules do have finite volumes, they collide with one another and thus suffer interruptions in their paths. The collisions are in fact very frequent for a gas at atmospheric pressure—about 10^{10} times per second for each molecule. The paths of the gas molecules are therefore interrupted very often. Diffusion of gas molecules, depicted in Figure 5.15, thus consists of a random motion, first in this direction, then that, at one instant at high speed, the next at low speed, with no overall direction to the motion of the gas as a whole except when there is a pressure difference. The average distance traveled by a molecule between collisions is referred to as the **mean free path**. This distance depends on the effective radius of the molecules, because larger molecules are more likely to undergo a collision. It depends also on the number of molecules in a unit volume—the larger the number of molecules per unit volume, the more likely is collision. For molecules of the atmosphere at atmospheric pressure, the mean free path is only about 10^{-5} cm. On the other hand, at an elevation of 100 km, just above the mesopause (Figure 5.9), the mean free path is on the order of 10 cm.

FIGURE 5.15 Schematic illustration of the diffusion of a gas molecule. For the sake of clarity all the other gas molecules in the container are not shown. The path of the molecule of interest begins at the dot. Each short segment of line represents travel between collisions. The path traveled by the molecule is often described as a "random walk."

5.11
intermolecular
forces

Although the ideal-gas law is a very useful description of gases, all real gases fail to obey the relationship to greater or lesser degree. The extent to which a real gas departs from ideal behavior may be seen by rearranging Equation [5.7] slightly:

$$\frac{PV}{RT} = n \qquad\qquad [5.15]$$

For a mole of gas, $n = 1$, the quantity PV/RT should therefore equal 1 if the gas is ideal. Figure 5.16 shows the quantity PV/RT plotted as a function of pressure for a few gaseous substances, as compared with the expected behavior of an ideal gas. It is clear from the figure that real gases are simply not ideal. However, the pressures shown are very high; at more ordinary pressures, in the range from 1 to 10 atm, the deviations from ideal behavior are not so large, and the ideal-gas equation can be used without serious error.

In the kinetic-molecular theory, it is assumed that molecules experience no forces between them until they come into direct contact. After collision they fly apart again like colliding billiard balls. However, real molecules actually do possess attractive forces between them. These forces are quite short range, which means that they do not come into play until the distances separating the molecules are very short.

The degree to which a gas departs from the ideal behavior depends on the relative importance of the attractive forces as compared with the average kinetic energy of the molecules. When the attractive forces are small, they do not noticeably influence the behavior of the molecules; the gas behaves more or less like an ideal gas. However, when the attractive forces are relatively large, the quantity PV/RT for a mole of the gas is lower than that predicted by the ideal gas equation. As an example, Figure 5.16 shows that at the same temperature, PV/RT for CO_2 is much lower at intermediate pressures than for either H_2 or N_2, suggesting the presence of greater attractive forces between CO_2 molecules.

The relative importance of intermolecular attractive forces depends also on the pressure. The higher the pressure, the smaller the average distance between molecules. Because the attractive forces are operative only at small distances between molecules, their importance increases rapidly with increasing pressure, assuming temperature remains constant.

If it were not for the intermolecular attractive forces between gas molecules, it would be impossible to liquefy gases. As it is, however, all gases liquefy at some temperature above absolute zero. Liquefaction

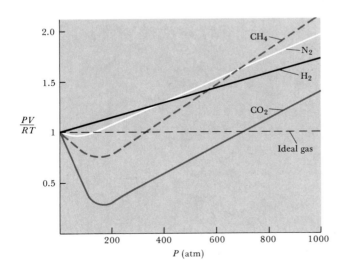

FIGURE 5.16 PV/RT versus pressure for several gases at 300 K.

occurs when the attractive forces dominate over the kinetic energies of the gas molecules. From data on the boiling points of various low-boiling liquids,* Table 5.3, we can draw certain generalizations. For the rare gas elements, a series of monatomic gases, the boiling point increases with increasing atomic weight. This suggests that the interatomic attractive forces increase with increasing mass of the atoms. We note also that the boiling points of molecules are higher than for monatomic species of about the same weight. For example, argon boils at a lower temperature than does oxygen, although its molecular mass is higher; similarly, krypton boils at a lower temperature than does chlorine. The origin of the attractive forces is the subject of later discussion (Chapter 11). We need only note here that such forces do exist, and that they vary considerably from one substance to another.

5.12
critical
temperature
and pressure

As the temperature of a gas is raised, the kinetic energies of the molecules increase in relation to intermolecular attractions, and the gas becomes more difficult to liquefy. Consequently, the pressure required to condense a gas to a liquid increases. Finally, a temperature is reached at which no amount of pressure, however great, causes the gas to pass from gas to liquid state. The highest temperature at which a gas liquefies is called its **critical temperature.** The **critical pressure** is the pressure required to bring about liquefaction at this critical temperature. The critical temperatures and pressures of gases are of considerable practical importance to engineers and others working with gases. The relative values of these quantities among different gases also provide some measure of the relative importance of intermolecular forces. Table 5.3 shows the critical temperatures and pressures for the same gases whose boiling points were just discussed. The critical temperature increases in the same way as does the boiling point of the liquid.

*Equilibrium between the gaseous and liquid states is discussed in more detail in Chapter 11. Boiling and liquefaction are simply opposite sides of the same coin. The boiling point of a substance is the temperature at which liquefaction of the substance occurs when the gas pressure is 1 atmosphere.

TABLE 5.3

boiling points and critical temperatures and pressures of some substances having low molecular weight

SUBSTANCE	MOLECULAR WEIGHT	BOILING POINT (K)	CRITICAL TEMPERATURE (K)	CRITICAL PRESSURE (atm)
He	4	4.2	5.2	2.26
Ne	20	27	44.4	25.9
Ar	40	87	151	48
Kr	84	121	210	54
Xe	131	164	290	58
Rn	222	211	377	62
H_2	2	20	33.2	12.8
N_2	28	77	126	33.5
O_2	32	90	154	49.7
Cl_2	71	239	417	76.1

We have seen that real gases show several interesting kinds of behavior that would not be expected for an ideal gas. These real-gas characteristics were shown to arise from the fact that molecules possess finite volume and experience attractive forces with other molecules. But in what way do these real-gas properties alter the equation of state that relates the three variables—temperature, pressure, and volume?

In the equation of state for an ideal gas, $PV = nRT$, the quantity V is supposed to represent the volume of the container. But this assumes that the molecules themselves occupy none of that volume. If in fact the molecules occupy space, the *free* volume, that is, the volume left in which the molecules may move about, is less than V.

It is the finite volume of the molecules that causes the PV/RT term shown in Figure 5.16 to be greater than 1 at higher pressures. We can modify the ideal-gas equation to correct for this by subtracting a term from V:

$$P(V - nb) = nRT \qquad [5.16]$$

In this equation, the quantity nb represents the volume effectively occupied by the molecules. It is a product of the number of moles of gas, and a constant b, different for each gas, which measures the effective volume per mole of gas molecules. From a study of the pressure-volume-temperature properties of gases it is possible to determine the values for the constant b for gases.

We might suppose that b is simply Avogadro's number (N) times the volume of each molecule, that is,

$$b = N\left(\frac{4}{3}\pi r^3\right)$$

where r is the radius of a gas molecule. Actually, however, b must be larger than this because of an effect called *excluded volume*. Suppose that each molecule has a radius r. We then ask, how much volume does one molecule of radius r keep any other molecule of radius r from occupying? It is this excluded volume that the constant b must represent. Figure 5.17 shows how large this excluded volume is.

Because the molecules act like billiard balls, the closest the center of one molecule can get to the center of another is with their surfaces touching. But if each molecule has a radius r, this means that a second molecule is excluded from a spherical volume of radius $2r$. The volume of this sphere, as shown, is

$$\frac{4}{3}\pi(2r)^3 = 8\left(\frac{4}{3}\pi r^3\right)$$

If we divide this excluded volume between the two molecules, we have

$$b = 4N\left(\frac{4}{3}\pi r^3\right)$$

From a study of the pressure-volume-temperature properties of gases it is possible to determine the values for the constant b for gases.

Correction of the gas-law expression for the intermolecular attractions is not quite so simple. We expect that attractive forces between the molecules will result in a lower pressure for the gas than we would have for an equal quantity of ideal gas under the same conditions. Thus the P term in Equation [5.16] must be replaced by the pressure of the real gas plus a correction term. The correction term is positive, because the real pressure plus the correction must equal the ideal pressure:

$$(P + \text{correction})(V - nb) = nRT \qquad [5.17]$$

The interactions between molecules that give rise to the nonideal behavior take effect when two molecules come close together. The number

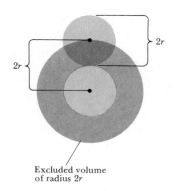

2r

2r

Excluded volume
of radius 2r

FIGURE 5.17 An illustration of the manner in which one sphere of radius *r* excludes space from another.

of such interactions increases as the square of the number of molecules per unit volume. The number of molecules per unit volume is proportional to n/V. Thus our correction to the pressure has the form of a constant, a, times $(n/V)^2$. The corrected gas equation of state thus is as follows:

$$\left(P + \frac{an^2}{V^2}\right)(V - nb) = nRT \qquad [5.18]$$

This gas equation is called the **van der Waals equation**, after Johannes van der Waals, who presented it in 1873. The values for the constants a and b must be determined by measuring the quantity PV/RT for the constant temperature over a wide range of pressure. The extent to which the gas behavior deviates from the ideal, as illustrated in Figure 5.16, can be used to choose values for a and b. These constants are chosen so that the equation predicts as closely as possible the observed pressure-volume-temperature behavior. Table 5.4 lists values for the van der

TABLE 5.4

van der Waals constants for gas molecules

SUBSTANCE	a (l.²-atm/mole²)	b (l./mole)
He	0.0341	0.02370
Ne	0.211	0.0171
Ar	1.34	0.0322
Kr	2.32	0.0398
Xe	4.19	0.0510
H_2	0.244	0.0266
N_2	1.39	0.0391
O_2	1.36	0.0318
Cl_2	6.49	0.0562
CO_2	3.59	0.0427
CH_4	2.25	0.0428
CCl_4	20.4	0.1383

Waals constants a and b for several substances. Note that both constants increase as the mass of the molecules and the complexity of their structure increases. To get some feeling for the magnitudes of the departures from ideal behavior, let's calculate these departures for CO_2 at STP.

SAMPLE EXERCISE 5.14

Calculate the correction terms to pressure and volume for CO_2 at STP, using the data in Table 5.4 and compare with the ideal gas values for P and V.

Solution: From the Equation [5.18] we see that the volume correction term is given by nb. Since $n = 1$, nb equals 0.043 liter, which is to be compared with 22.4 liter, the ideal-gas volume (Sample Exercise 5.6). The correction to volume is thus

$$\frac{0.043}{22.4} \times 100 = 0.2\%$$

The correction to pressure is given by an^2/V^2. Inserting the value of a from Table 5.4, $n = 1$ and $V = 22.4$ l.

$$\frac{an^2}{V^2} = \frac{\left(\dfrac{3.59 \text{ l.}^2\text{-atm}}{\text{mole}^2}\right) 1 \text{ mole}^2}{(22.4 \text{ l.})^2} = 0.007 \text{ atm}$$

The correction to pressure is thus $0.007 \times 100 = 0.7\%$. We conclude that the ideal-gas law is obeyed by CO_2 at STP conditions to within 1 percent.

From Equation [5.18] we can deduce the conditions under which departures from ideal behavior will become important. When the gas is at high temperature and low pressure, so that the volume per mole is large, V will be large in comparison with n. Then the correction to V in the second term is small in comparison to V itself. Also, the presence of V in the denominator in the first correction term means that when V is large this correction will also be unimportant.

FOR REVIEW

summary

In this chapter we have described and analyzed many physical properties of gases. We have related these properties to the characteristics of the earth's atmosphere. The **ideal-gas equation** represents an equation of state relating the four variables—pressure, temperature, volume, and number of moles of gas—needed to completely define the state of a gaseous system. It is sufficiently accurate to serve well in describing the properties of the earth's atmosphere. It serves well also in solving problems related to gas properties that might come up in the course of chemical experiments. For more extreme conditions of pressure and temperature, the ideal-gas equation must be modified to take into account the finite volume of gas molecules and the attractive forces operating between them.

The molecules of a gas possess a distribution of molecular speeds, depending on temperature and molecular mass. In addition, because of their finite volume, the molecules undergo collisions with one another. These characteristics are of importance in determining gas diffusion and effusion rates. We shall find later that these characteristics are of importance also in determining the rates at which chemical reactions occur in the gas phase.

The fact that real gases condense to form liquids at low temperatures is evidence that attractive forces do exist between gas molecules. These attractive forces, and the finite volumes of gas molecules, are responsible for the deviations from ideal-gas behavior seen in real gases. The **van der Waals** equation is an equation of state for gases that takes account of these two forms of departure from ideal behavior.

learning goals

Having read and studied this chapter, you should be able to:

1. Describe the general characteristics of gases as compared with other states of matter and list the ways in which gases are distinctly different.
2. List the variables that are required to define the state of a gas.
3. Define 1 atm pressure and other units in which pressure is expressed. You should also understand the principle of operation of a barometer.
4. Explain the way in which pressure, volume, and temperature are related by Boyle's law, Charles's law, and the ideal-gas equation. Using these relationships, you should be able to solve problems involving changes in state for a gas.
5. Explain the concept of gas density and describe how it relates to temperature, pressure, and molecular weight.
6. Express Dalton's law of partial pressures.
7. State Avogadro's hypothesis and explain its significance.
8. Sketch the manner in which atmospheric temperature varies with altitude

and list the names of the various regions of the atmosphere and the boundaries between them.

9. Sketch the manner in which atmospheric pressure decreases with elevation and explain in general terms the reason for the decrease.

10. Describe the composition of the atmosphere with respect to the four most abundant components.

11. List and explain the assumptions on which the kinetic theory of gases is based.

12. Describe graphically how gas molecules are distributed over a range of speeds, and how that distribution changes with temperature.

13. Describe how the relative rates of diffusion or effusion of two gases depends on their relative molecular weights (Graham's law).

14. Explain the concept of mean free path and how it relates to the rates of diffusion of molecules in the gas state.

15. List the major factors responsible for deviations of gases from ideal behavior.

16. Relate the magnitudes and kinds of intermolecular forces to the properties of substances, such as boiling points and critical temperature and pressure.

17. Explain the origins of the correction terms to P and V that appear in the van der Waals equation of state for a gas.

key terms

Among the more important terms and expressions used for the first time in this chapter are the following:

According to **Avogadro's hypothesis** (Section 5.5), equal volumes of gases at the same temperature and pressure contain equal numbers of molecules.

A **barometer** (Section 5.2) is a device for measuring atmospheric pressure in terms of the height of a liquid column sustained by that pressure.

According to **Boyle's law** (Section 5.3), at constant temperature, the product of the volume and pressure of a given amount of gas is a constant.

According to **Charles's law** (Section 5.4), (1) at constant pressure, the volume of a given quantity of gas is proportional to absolute temperature, and (2) at constant volume, the pressure of a given quantity of gas is proportional to absolute temperature.

Dalton's law of partial pressures (Section 5.5) states that the total pressure of a mixture of gases is just the sum of the pressures that each gas would exert if it were present alone.

Diffusion (Section 5.9) refers to the rate at which a substance spreads into and throughout a space. Thus, a gas might diffuse throughout a room, or atmospheric oxygen might diffuse through the waters of a lake. **Effusion** refers to the rate at which a gas escapes through an orifice or hole.

The **ideal-gas equation** (Section 5.7) is an equation of state for gases that embodies Boyle's law, Charles's law, and Avogadro's hypothesis in the form $PV = nRT$.

The **kinetic-molecular theory of gases** (Section 5.8) consists of a set of assumptions about the nature of gases. These assumptions, when translated into mathematical form, yield the ideal-gas equation.

The **mean free path** (Section 5.10) in a gas sample is the average distance traveled by a gas molecule between collisions.

Pressure (Section 5.2) is a measure of the force exerted on a unit area. A gas pressure of 1 *atmosphere* (1 atm) is that required to sustain a column of mercury 760 mm in height.

The torr (Section 5.2) is a unit of pressure (1 torr = 1 mm Hg).

The van der Waals equation (Section 5.13) is an equation of state for real gases, containing terms that correct for the existence of attractive forces between molecules and for their finite volumes.

EXERCISES

characteristics of gases; pressure

5.1 Suppose Robert Boyle were to be reincarnated among us today, and still showed an interest in the properties of gases. What evidence would you cite in demonstrating to him that his conception of the nature of gases, as described in the introduction to this chapter, is incorrect? List, and provide an interpretation of, as many different experimental demonstrations as you can think of.

5.2 The density of liquid gallium is 6.09 g/cm^3 at 35°C. If this element is employed in a barometer instead of mercury, what is the height of a column of gallium sustained in the barometer at 1 atm pressure?

5.3 The bar is a unit of pressure defined as 1.00×10^6 dynes/cm^2, or 1.00×10^5 N/m^2. What is the equivalent of the bar in units of mm Hg?

5.4 Suppose that a mercury barometer such as that shown in Figure 5.1 were constructed using mercury with a few small droplets of water trapped in it, and that these rose to the top of the mercury in the tube. Would the barometer read correct atmospheric pressure? Explain.

[5.5] The density of liquid nitrogen at -196°C is 0.808 g/cm^3. The density of nitrogen gas at STP is 1.25 g/l. From these data calculate the volume per molecule in liquid nitrogen and in gaseous nitrogen. Knowing that the volume of a sphere is related to the cube of its radius, what can you say about the relative distances between nitrogen molecules in the two cases?

Boyle's law

5.6 Suppose you were responsible for filling a 10.5-l. weather balloon with 1 atm pressure of helium each day and releasing it. Assuming no waste, how many 12-l. tanks of helium at a pressure of 2250 lb/in.^{2} would you require in a year (all temperatures about 27°C)?

5.7 A McLeod gauge is a device used to measure the pressure of gases at low pressures in a so-called vacuum line. A large volume of gas from the vacuum line system is compressed to a much smaller volume at higher pressures that can be directly measured. In this manner a 0.55-l. sample of gas at unknown pressure is compressed to a volume of 0.066 cm^3, at which time it is under a pressure of 500 torr. What is the pressure in the vacuum line system?

5.8 The compression ratio for an internal combustion engine is the ratio of the pressure of gas in the cylinder when the piston is all the way in as compared with when the gas is introduced. If the compression ratio is 8.7, and the volume of the cylinder at the time the gas is admitted is 0.52 l., what is the volume at the time of maximum compression (assume temperature constant)?

5.9 The density of a particular gas in a cylinder with a volume of 3.5 l. is found to be 1.80 g/l. The gas is compressed at constant temperature until the volume is 0.65 l. What is the density of the gas under the new conditions?

Charles's law

5.10 A gas is placed in a storage tank at a pressure of 50 atm at 23°C. There is a small metal safety plug in the tank made of a metal alloy that melts at an elevated temperature. At this temperature, the gas pressure has reached 75 atm. At what temperature is the metal plug designed to melt?

5.11 A set of automobile tires are charged with 32 lb/in.^{2} of air pressure on a day when the tires are at a temperature of 24°C. Later the car is driven at high speed on a hot day, and the temperature of the tires climbs to 44°C. Assuming no increase in volume, what is the pressure in the hot tires?

5.12 A large tank contains a gas at a pressure of 2 atm and a temperature of 45°C. The tank is heated because of a fire, and its temperature rises to 430°C. A relief value on the tank is set to open when the pressure reaches 3.0 atm. Will the valve open during the fire?

5.13 What is the partial pressure in torr of argon in dry air near sea level? (See Table 5.1.)

5.14 Neon gas is used in neon signs. In a typical sign, the gas is present at 2 torr pressure at 25°C. The total volume of the tubing in a large neon sign is on the order of 2.5 l. Neon is obtained by carefully separating the components of liquid air. Assuming 90 percent efficiency in recovery of pure neon, how many liters of gaseous air at STP must be processed to make enough neon for the sign?

5.15 The density of seawater at a depth of 3000 m is typically about 1.028 g/cm³. The density at the surface is typically about 1.025 g/cm³. The pressure at the surface is 1 atm. Neglecting the small change in density with depth, calculate the pressure at a depth of 3000 m. Suppose a balloon containing 1 l. of gas at the surface were taken to this depth. What would be the volume of the gas in the balloon at 3000 m, assuming no temperature change? How much denser would the gas be at 3000 m depth in the ocean than at the surface? Why is the density change for the gas different than for liquid water?

5.16 Above 500 km elevation, where the earth's atmosphere is very thin, the most abundant gases are atomic hydrogen and helium. The reasons for hydrogen presence here are both chemical and physical, but why should helium be present in such relatively high abundance at this elevation?

the ideal-gas law

5.17 The density of dry, carbon-dioxide free air at STP is 1.2927 g/l. What is the average molecular weight of air?

5.18 0.338 g of an organic liquid was heated to vaporize it at reduced pressure. A total of 109.8 cm³ of vapor was formed at a pressure of 463 torr and at a temperature of 100°C. What is the molecular weight of the vapor?

5.19 One of the earliest reliable methods for obtaining atomic weights was the method of gas densities. From the following values for gas densities in g/l. at STP, calculate the molecular weights of each of the elements involved, and compare with the list of atomic weights given in the inside rear cover of the text. (Be sure to carry through the appropriate number of significant figures!)

ELEMENT	DENSITY (g/l.)
Hydrogen	0.08987
Nitrogen	1.25046
Argon	1.7836
Oxygen	1.4290

[5.20] On the planet Trafalmadore, the standard temperature is 288 K and standard pressure is such that it supports a column of mercury 600 mm in height. However, the force of gravity on Trafalmadore is 1.68 times greater than on earth. What is the volume of 6.023×10^{23} argon atoms at Trafalmadorian STP?

5.21 The concentration of dissolved O_2 in ocean water is expressed in units of cm³ of O_2 gas at STP per liter of water. In these units, the concentration of O_2 in sea water at a depth of about 2000 m is found to be 5 cm³/l. What is the concentration of O_2 expressed in moles O_2 per liter of water?

5.22 In a laboratory experiment, 0.340 g of Mg were reacted with dilute hydrochloric acid solution, generating hydrogen. The hydrogen gas was collected over water at 26°C and at a barometric pressure of 750 torr. Write a balanced chemical equation for the reaction and calculate the volume of hydrogen collected.

5.23 At the instant of firing, the air in the cylinder of an air hammer is at a temperature of 280°C under 50 atm pressure and has a volume of 0.3 l. What volume of air at 1 atm and 30°C is required to make up this air?

5.24 A compound containing carbon and hydrogen is found to give the following percentages of the two elements: C, 85.7 percent; H, 14.3 percent. A 0.785 g sample of the substance is vaporized at 100°C, and found to occupy a volume of 0.286 l. at 1 atm pressure. What is the molecular formula for the substance?

5.25 When a 2.45 g sample of $KClO_3$ is thermally decomposed and the O_2 gas given off collected over water, the total volume of the gas sample in the bottle is found to be 0.770 l. at 26°C. The atmospheric pressure is 740 torr. Is the amount of O_2 evolved consistent with the chemical equation given in the text? Explain.

kinetic molecular theory

5.26 How does the fact that a gas obeys Boyle's law support the idea that the volumes of gas molecules are negligible in comparison with the volume of the space in which the gas is contained?

5.27 How does the average speed of the molecules of a gas vary with absolute temperature?

5.28 One of the important assumptions of the kinetic molecular theory is that the molecules do not experience significant attractive or repulsive forces between one another. Describe an experiment that shows that this is approximately correct for gases at ordinary temperatures and pressures.

molecular speeds; diffusion and effusion

5.29 A gas of unknown molecular weight was allowed to effuse through a small opening under constant pressure conditions. It required 60 sec for 1 l. of the gas to diffuse. Under identical experimental conditions, it required 97 sec for 1 l. of oxygen gas to diffuse. Calculate the molecular weight of the unknown gas.

5.30 For hydrogen gas, the mean or average molecular speed at $0°C$ is 1693 m/sec. What is the mean molecular speed of chlorine molecules at $0°C$?

5.31 The mean molecular speed of nitrogen molecules at $0°C$ is 455 m/sec. The mean molecular speed of carbon dioxide molecules at $0°$ is 285 m/sec. Suppose we have two 1-l. flasks, one containing N_2 at STP, the other CO_2 at STP. From Avogadro's hypothesis we know that the two flasks contain equal numbers of molecules, neglecting small nonideality corrections. How do these systems differ with respect to (a) the average kinetic energies of the molecules, (b) the total number of collisions occurring per second with the container walls, and (c) the shapes of the curves showing the distribution of molecular speeds.

5.32 Suppose it has been suggested that the relatively rare isotope of carbon, ^{13}C, could be separated from the more abundant ^{12}C by using a diffusion process similar to that described in the text for UF_6, but using either CO or CO_2. Calculate the relative rates of diffusion for each pair of isotopically related compounds. Which substance would give the greater degree of separation?

5.33 The atmosphere on Trafalmadore is 79 percent krypton and 20 percent oxygen with traces of other gases. When the first Trafalmadorians visited Earth they complained of feeling chilly even though the average temperatures are the same on the two planets. The reason for their complaints is connected with the differing average molecular weights of the two atmospheres. Can you explain the effect?

[5.34] In a gas chromatograph, a sample of a mixture of substances is put into a stream of helium at the front end of a long tube packed with some suitable material (see Figure 5.18). The differing components of the mixture are delayed to differing degrees by the column material, and thus appear at different times at the outlet. One form of detector involves a little wire that is kept hotter than the gas stream. The signal coming out of the detector depends on the temperature of the wire. When the gas passing the detector contains a component of the sample the detector signal changes, and a blip results on the chart, as shown.

Why do the different components of the sample vapor cause the signal to change? Why is helium a good carrier gas in such an apparatus?

intermolecular forces; nonideality

5.35 For each of the following pairs of gases, choose the one that you would expect to deviate most from the PV/RT relationship for an ideal gas: (a) Cl_2

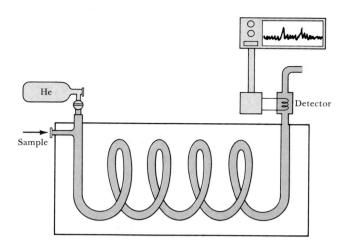

FIGURE 5.18

and I_2; (b) CH_4 and SnH_4; (c) CO and CO_2.

5.36 Suppose that the PV/RT curve shown in Figure 5.16 for CH_4 is for a temperature of 300 K. How would you expect the comparable curve for CH_4 at 500 K to look? Explain.

5.37 The standard carbon-dioxide fire extinguisher contains liquid carbon dioxide under pressure. When the valve is released the gas expands against atmospheric pressure, the temperature drops, and solid CO_2 is formed. The critical temperature for CO_2 is $31°C$. Suppose a carbon-dioxide fire extinguisher is allowed to sit in the sunlight, so that its temperature exceeds $31°C$. What occurs inside the extinguisher?

5.38 Calculate the van der Waals correction terms to the pressure and volume for argon at STP. Using these data, calculate the density of argon at STP and compare with the density calculated using the ideal-gas equation.

general exercises

5.39 Consider the experiment shown in Figure 5.19. The gas bulb contains an equimolar mixture of helium and argon at 1 atm pressure. The tube is open to the air. The stopcock connecting the flask to the tube is opened, and at various times later a tiny sample of gas is withdrawn at point A, and its density determined. Make a rough drawing of how the density of gas at point A might vary as a function of

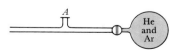

FIGURE 5.19

time, relative to the density of air. Explain the reasoning behind your answer.

5.40 Using the van der Waals b constant for carbon tetrachloride, and taking account of the excluded volume, calculate the volume of a CCl_4 molecule.

[5.41] On the planet Trafalmadore the temperature scale is called Ulp (U). A sample of helium gas at low pressure has a volume of 3.4556 l. at 12°U. The same gas sample at the same low pressure has a volume of 4.0544 l. at 135°U. From these data, assuming helium to be an ideal gas, calculate the value of absolute zero on the Ulp temperature scale. How does 1°U compare with 1°C?

5.42 Suppose a nozzle system is being designed for a rocket in which the thrust will come from reaction of H_2 with F_2, a highly exothermic reaction. The two reacting substances will be effused into the reaction zone through a series of tiny holes, all the same size. What should be the ratio of the number of holes for F_2 to those for H_2 to maintain a 1:1 stoichiometry in the reaction zone?

5.43 A large meteorological balloon weighing 32 g has a volume of 124 l. It is filled with helium at 1 atm pressure and 25°C. The lifting capacity of the balloon is equal to the difference in mass between the helium in the balloon and an equivalent volume of outside air, less the mass of the balloon itself. Calculate the lifting capacity of the balloon. Assuming that a mass less than the lifting capacity is attached to the balloon, it will rise when released. Would the balloon be expected to rise to a higher or lower maximum altitude if it is made of rubber and therefore capable of expansion or contraction, as compared with its behavior if it has a fixed volume?

5.44 Gay-Lussac was the first to carry out quantitative experiments with gases. He found that one volume of hydrogen reacted with one volume of chlorine to form *two* volumes of hydrogen chloride, HCl. This result bothered Dalton, who thought that hydrogen and chlorine were monatomic gases. What conclusion must we reach about hydrogen and chlorine if we assume Avogadro's hypothesis to be correct? Explain.

5.45 A typical ranch house has a volume of 12,000 ft^3. If the relative humidity in such a house is to be maintained at about 30 percent during winter, that is, at about 8 torr water-vapor pressure, what is the total mass of water vapor in the air of the house?

[5.46] 100 ml of water is added to a 1-l. metal can, and the can is allowed to stand at 25°C for a time, until the air inside is saturated with water vapor. Then the metal cap is screwed on tightly, and the can is heated to 100°C, the boiling point of water. What is the approximate total pressure in the can?

5.47 In most homes in winter, the temperature is higher near the ceiling than at floor level. Why is this?

5.48 The composition of the upper atmosphere differs dramatically from that near the surface. At 300 km elevation, the mean molecular weight of the atmosphere is only 19 g/mole, the temperature is about 1430 K, and the pressure is 2×10^{-7} torr. From these data, calculate the number of molecules per cm^3 at 300 km and the density of the atmosphere.

5.49 Water-well pumps, such as the familiar pitcher pump, which depend on creating a vacuum to lift water to the surface, cannot lift water from a depth greater than about 32 ft. Why is this so?

5.50 Suppose that the volume of an engine cylinder at the time gas is admitted is 0.75 l., and that at the time of firing the volume is 0.15 l. Suppose also that the temperature of the incoming gas is 35°C, and that at the time of firing the gas is at a temperature of 375°C. What is the ratio of the pressure of the gas at time of firing to that at time of admission?

5.51 Red blood cells contain hemoglobin, which is responsible for transport of oxygen from the lungs to the body's cells. Residents of the Andes mountains in Peru (elevation 15,000 ft) are found to have a larger number of red blood cells than their fellow countrymen who live in Lima (elevation 500 ft). Suggest a reason for the difference.

the electronic structures of atoms

6

One of the goals of the chemist is an understanding of how and why molecules are formed from atoms. However, before we can properly understand how atoms come together to form the great variety of known compounds, we must know something of the architecture of atoms themselves. As we take up this subject, we shall find that there is a close relationship between the properties of light, or radiant energy, and the behaviors of subatomic particles. We begin our study of atomic structure, then, by considering the characteristics of radiant energy.

6.1
radiant energy

Different kinds of radiant energy—such as the warmth from a glowing fireplace, the light reflected off snow in the mountains, and the X rays used by a dentist—*seem* very different from one another. Yet they share certain fundamental characteristics. All types of radiant energy, also called **electromagnetic radiation,** move through a vacuum at a speed of 2.9979250×10^8 m/sec, the "speed of light." This is one of the most accurately known physical constants. For our purposes, however, it will be sufficient to use 3.00×10^8 m/sec.

All radiant energy has wavelike characteristics, analogous to those of waves that move through water. A cork bobbing on water as waves pass by is not swept along with the wave. Rather, it moves up and down with the wave motion. Water waves are the result of energy imparted to the water, perhaps by the dropping of a stone, the movement of a boat, or the force of wind on the water surface. This energy is expressed as the up and down movement of the water.

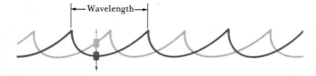

FIGURE 6.1 Characteristics of water waves. The distance between corresponding points on each wave is called the wavelength. The number of times per second that the cork bobs up and down is called the frequency.

If we look at a cross section of a water wave (Figure 6.1), we see that it is periodic in character. That is, the wave form repeats itself at regular intervals. The distance between identical points on successive waves is called the **wavelength.** As the wave passes, the cork floating on the surface moves up and down. The number of times per second that the cork moves through a complete cycle of upward and then downward motion is called the **frequency** of the wave.

In a similar way, radiant energy has a characteristic frequency and wavelength associated with it. These are illustrated schematically in Figure 6.2. The frequency of the radiation is the number of cycles that occur in 1 second as the radiation flows past a given point. Short wavelengths correspond to high frequency, and long wavelengths to low frequency, as shown in Figure 6.2. The wavelength and frequency are thus related. The product of frequency, ν (nu), and wavelength, λ (lambda), equals the speed of light, c:

$$\nu\lambda = c \qquad\qquad [6.1]$$

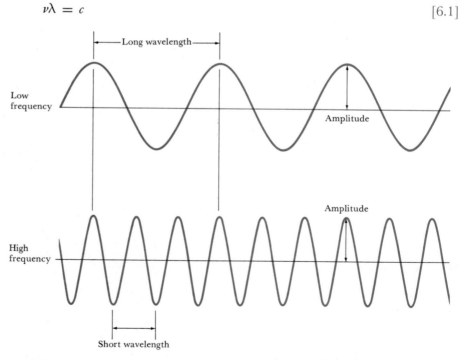

FIGURE 6.2 Characteristics of radiant energy. The wave moves through space with the speed of light. The frequency of the high frequency radiation in this illustration is three times that of the low frequency radiation. In the time that the low frequency, long wavelength radiation goes through one cycle, the high frequency, short wavelength radiation goes through three cycles. The *amplitude* of the wave refers to its intensity. Here both waves have the same amplitude.

TABLE 6.1

wavelength units for electromagnetic radiation

UNIT	SYMBOL	LENGTH (m)	TYPE OF RADIATION
Ångstrom	Å	10^{-10}	X ray, ultraviolet, visible
Nanometer (or millimicron)[a]	nm ($m\mu$)	10^{-9}	Ultraviolet, visible
Micron	μ	10^{-6}	Infrared
Millimeter	mm	10^{-3}	Infrared
Centimeter	cm	10^{-2}	Microwaves
Meters	m	1	TV, radio

[a]The millimicron is the same size unit as the nanometer; use of the latter is preferred.

Wavelength is expressed in units of length per cycle; the unit of length chosen depends on the type of radiation as shown in Table 6.1. Figure 6.3 shows the ranges of wavelength that characterize the various types of radiant energy. Frequency is given in units of hertz (Hz), which is cycles per second. For example, the frequency of an AM radio station might be written as 810 kilohertz (kHz), or 810,000 cycles/sec. Because it is understood that cycles are involved, the units are normally given simply as $\sec^{-1}$.

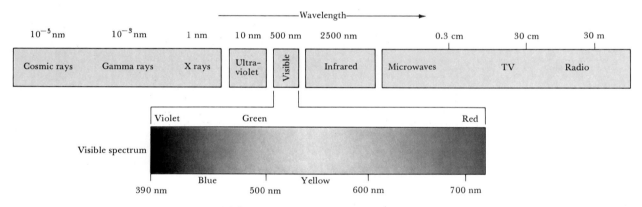

FIGURE 6.3 Wavelengths of electromagnetic radiation characteristic of various regions of the spectrum.

SAMPLE EXERCISE 6.1

The yellow light given off by a sodium lamp has a wavelength of 589.2 nm. What is the frequency of this radiation?

Solution: We can rearrange Equation [6.1] to give $\nu = c/\lambda$. We insert the value for c and λ, and then convert nm to m. This gives us

$$\nu = \left(\frac{3.00 \times 10^8 \text{ m/sec}}{589.2 \text{ nm}}\right)\left(\frac{10^9 \text{ nm}}{1 \text{ m}}\right)$$

$$= 5.09 \times 10^{14}/\text{sec}$$

Note that frequency has units of $\sec^{-1}$.

A particular source of radiant energy may emit a single wavelength, as in the light from a laser (Figure 6.4), or may contain many different wavelengths, as in the radiations from a lightbulb or a star. Radiation composed of a single wavelength is termed **monochromatic.** When the radiation from a source such as a star is separated into its monochromatic components, a **spectrum** such as shown in Figure 6.5 results. Radiations of differing wavelength are separated by dispersing them in a prism.

Radiations of different wavelengths affect matter differently. For example, overexposure of a part of your body to infrared radiation may cause a "heat burn," overexposure to visible and near-ultraviolet light causes sunburn and suntan, and overexposure to X radiation causes tissue damage, possibly even cancer. These diverse effects are due to differences in the energy of the radiations. Radiations of high frequency and short wavelength are more energetic than radiations of lower frequency and longer wavelength. The quantitative relation between frequency and energy was developed at the turn of the century in the far-reaching and revolutionary *quantum theory* of Max Planck (Figure 6.6).

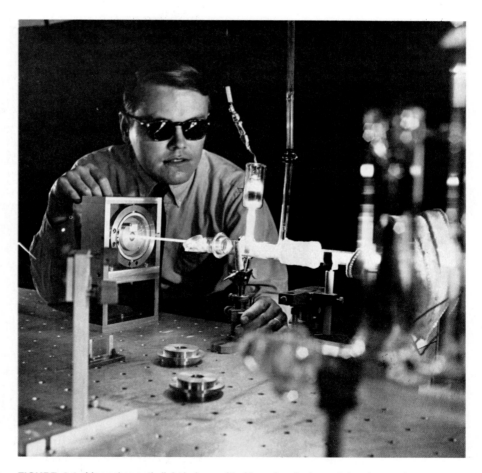

FIGURE 6.4 Monochromatic light being emitted from the discharge tube of a metal-vapor laser. (*Bell Telephone Laboratories*)

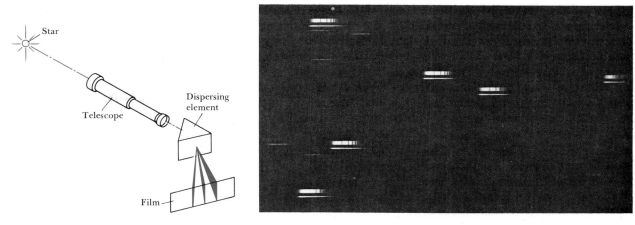

FIGURE 6.5 The drawing shows the spectrum of wavelengths present in radiation from a hypothetical star. The dispersing element used to separate radiations of different wavelengths in this illustration is a prism. The photograph shows star spectra as recorded on extremely sensitive photographic film by a telescope trained on a group of stars. The bands of light are spectra of individual stars. (*Yerkes Observatory photograph*)

**6.2
the quantum
theory**

The quantum theory is concerned with the rules that govern the gain or loss of energy from an object. Planck's contribution was to see that when we deal with gain or loss of energy from objects in the atomic or subatomic size range, the rules *seem* to be different than those that apply when we are dealing with energy gain or loss from objects of ordinary dimensions. A very crude and fanciful analogy might best illustrate what is involved. Imagine a large dump truck loaded with perhaps 20 tons of fine-grained sand. Let's say that the amount of sand on the truck is measured by driving it onto a supersensitive scale that measures the weight of a 30-ton object to the nearest pound. With this as our measuring device, the gain or loss of sand from the truck might be too small to be measured. For example, a spoonful of sand might be added or deleted with no change in scale reading.

Now imagine, if you will, a tiny little truck, operated by men the size of tiny mites. For this little truck a full load of sand would consist of perhaps a dozen grains of sand of the same size as those carried by the large truck. In this microscopic world, the load on the truck can be added to or decreased only by rolling on or off one or more full grains of sand. On a scale that weighs this tiny truck, even one grain of sand, the smallest piece attainable, represents a substantial fraction of the full load and is easily measurable.

In our analogy the sand represents energy. An object of ordinary, or macroscopic, dimensions, like the dump trucks we have seen on highways, contains energy in so many tiny pieces that the gain or loss of individual pieces is completely unnoticed. On the other hand, an object of atomic dimensions, such as our imaginary little truck, contains such a small amount of energy that the gain or loss of even the smallest possible piece makes a substantial difference. The essence of Planck's quantum theory is that there *is* such a thing as a smallest allowable gain or loss of energy. Even though the amount of energy gained or lost at one time

FIGURE 6.6 Max Planck, physicist. Born in 1858 in Kiel, Germany, Planck was the son of a law professor at the University of Kiel. When he announced his intention to study physics, Planck was warned that all the major discoveries had already been made in this field. Nevertheless, Planck became a physicist, and in 1892 was named professor of physics at the University of Berlin. In 1900 he presented a paper before the Berlin Physical Society which launched the greatest intellectual revolution in the history of science. (*Library of Congress*)

may be very tiny, there is a limit to how small it may be. Planck termed the smallest allowed increment of energy gained or lost, **aquantum.** In our analogy, a single grain of sand represents a quantum of sand.

You should keep in mind that the rules regarding the gain or loss of energy are always the same, whether we are concerned with objects on the size scale of our ordinary experience or with microscopic objects. However, it is only when dealing with matter at the atomic level of size that the impact of the quantum restriction is evident. Humans, being creatures of macroscopic dimensions, had no reason to suppose that the quantum restriction existed until they devised means of observing the behavior of matter at the atomic level. The major tool for doing this at the time of Planck's work was by observation of the radiant energy absorbed or emitted by matter. An object can gain or lose energy by absorbing or emitting radiant energy. Planck assumed that the amount of energy gained or lost at the atomic level by absorption or emission of radiation had to be a whole number multiple of a constant times the frequency of the radiant energy. Let us call this amount of energy gained or lost ΔE. Then, according to Planck's theory

$$\Delta E = h\nu, \ 2h\nu, \ 3h\nu, \ \text{etc.} \tag{6.2}$$

The constant h is known as Planck's constant. It has the value of 6.625×10^{-34} joule-sec, or J-sec. The smallest increment of energy at a given frequency, $h\nu$, is called a *quantum* of energy.

SAMPLE EXERCISE 6.2

Calculate the smallest increment of energy (that is, the quantum of energy) that can be emitted or absorbed at a frequency of 5×10^{13}/sec.

Solution: We obtain the value for ΔE from Equation [6.2] by inserting the values for h, 6.625×10^{-34} J-sec, and for ν, 5×10^{13}/sec:

$$\Delta E = (6.625 \times 10^{-34} \text{ J-sec})(5 \times 10^{13}/\text{sec})$$
$$= 3.31 \times 10^{-20} \text{ J}$$

Using Equation [6.1] we can easily determine the wavelength of radiation with a frequency of 5×10^{13}/sec:

$$\lambda = c/\nu$$
$$= \frac{3.0 \times 10^8 \text{ m/sec}}{5 \times 10^{13}/\text{sec}}$$
$$= 6 \times 10^{-6} \text{ m}$$
$$= (6 \times 10^{-6} \text{ m}) \left(\frac{10^9 \text{ nm}}{\text{m}} \right)$$
$$= 6000 \text{ nm}$$

Radiation of 6000 nm wavelength lies in the infrared region of the spectrum. Planck's theory tells us that an atom or molecule emitting radiation with a frequency of 5×10^{13}/sec cannot lose energy by radiation except in multiples of 3.31×10^{-20} J. It cannot, for example, lose 4.62×10^{-20} J by radiation, because this is not a multiple of the smallest amount, or *quantum*, of energy that can be lost at this wavelength.

At this point you may be wondering about the practical applications of Planck's quantum theory. A few years after Planck presented his theory, scientists began to see its applicability to a great many experimental observations. It soon became apparent that Planck's theory had within it the seeds of a revolution in the way the physical world is viewed. Let's consider a few applications that are of special importance for chemistry.

THE PHOTOELECTRIC EFFECT

Einstein used the quantum theory to explain the photoelectric effect, the phenomenon that occurs when electrons are knocked out of a metal by radiant energy. He assumed that the radiant energy striking the metal surface is a stream of tiny energy packets. Each energy packet, called a **photon**, is a quantum of energy $h\nu$. *Photons of high frequency radiation have high energies, whereas photons of lower frequency radiation have lower energy.* When the photons are absorbed by the metal, their energy is transferred to an electron in the metal. A certain amount of energy is required for the electron to overcome the attractive forces that hold it within the metal. Otherwise it cannot escape from the metal surface, even if the light beam is quite intense. On the other hand, if a photon has sufficient energy, the electron is emitted (see Figure 6.7). If a photon has more than the minimum energy required to free an electron, the excess appears as the kinetic energy of the emitted electron.

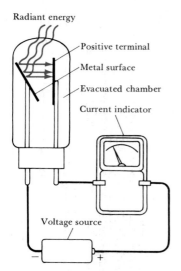

Radiant energy

Positive terminal

Metal surface

Evacuated chamber

Current indicator

Voltage source

FIGURE 6.7 The photoelectric effect. When photons of sufficiently high energy strike the metal surface in the tube, electrons are emitted from the metal. The electrons are drawn toward the other electrode, which is a positive terminal. As a result, current flows in the circuit.

In later chapters (see, for example, Section 10.6) we'll see that a model similar to Einstein's can be used to explain why certain wavelengths of light can cause chemical changes to occur, whereas others cannot.

LINE SPECTRA

The spectrum reaching us from the sun is continuous; that is, it contains a broad range of frequencies of radiant energy as shown in Figure 6.8. However, Franz Fraunhofer noticed as early as 1813 that radiation of certain frequencies was almost absent from the solar spectrum. These frequencies are of lower intensity because hydrogen and other atoms present in the sun's outer atmosphere absorb much of this radiation as it moves outward from the sun.

Not only are atoms able to absorb certain frequencies of light passing through them, but they also emit radiation when suitably excited. For example, when hydrogen is placed under reduced pressure in a tube such as that depicted in Figure 6.9, and a high voltage is applied, the hydrogen emits light. (A similar arrangement with neon gas in the tube produces the familiar red-orange glow of neon lights.) When the light coming from the hydrogen-filled tubes is separated into its monochromatic components, only certain frequencies of light are found to be present. A spectrum containing radiation of only specific wavelengths is called a **line spectrum.** The emission lines seen in the line spectrum of excited hydrogen atoms correspond exactly to many of the strong absorptions observed in the solar spectrum.

The absorptions and emissions of light by hydrogen atoms correspond to energy changes within the atoms. The fact that only certain frequencies are absorbed or emitted tells us that only certain sizes of

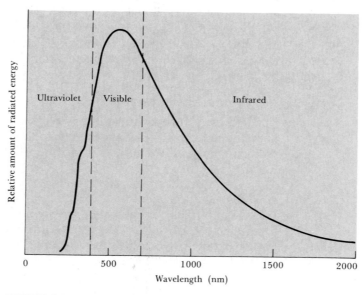

FIGURE 6.8 The intensity of solar radiation as a function of wavelength. Certain specific wavelengths are much diminished in intensity; this figure shows only the overall broad outline of the sun's emissions.

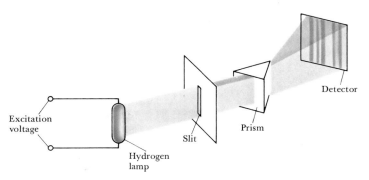

FIGURE 6.9 Emission of radiation by hydrogen atoms excited in an electrical discharge.

energy change are allowed. Other substances, such as neon and sodium vapor, also emit line spectra in a discharge tube. The spectra are much more complex than those for hydrogen. The important point, however, is that the spectra do consist of sharp lines.

Although it was not realized at the time by the scientists who gathered the data, the line spectra for atoms provide beautiful applications of the quantum theory. It remained for Neils Bohr in 1914 to propose a theory of the hydrogen atom that incorporated Planck's theory.

6.3
Bohr's theory of hydrogen

Bohr's theory of the hydrogen atom took into account two important developments that were relatively new at that time. The first was Rutherford's experiments, which established the nuclear nature of the atom (Section 2.7). The other was Einstein's work, which showed that radiant energy could be thought of as a stream of discrete bundles of energy called photons.

Bohr reasoned that the existence of the sharp line spectra meant that hydrogen atoms could absorb or emit only certain frequencies or wavelengths of light. Because the frequencies of radiation are directly related to energy, this must mean that *only certain energy changes are possible within the hydrogen atom.*

Bohr's theory consists of a series of postulates that may be summarized as follows:

1. The electron of a hydrogen atom moves about the central proton in a circular orbit. However, an electron in an atom cannot have just any energy; only orbits of certain radii, and having certain energies, are "allowed." This idea is an extension of Planck's concept of the quantization of energy.
2. In the absence of radiant energy, an electron in an atom remains indefinitely in one of the allowed energy states. When radiant energy is present, however, the atom may absorb energy. When this happens, the electron undergoes a change from one allowed energy state to another. The frequency of the radiant energy absorbed (ν) corresponds exactly to the energy difference (ΔE) between two of the allowed energies: $\Delta E = h\nu$.

We need not concern ourselves with the details of how Bohr used quantum theory to calculate the energies of the electron. The main point is that he was able to calculate a set of allowed energies. Each of these allowed energies corresponds to a circular path of different radius. In Bohr's model, each allowed orbit was assigned an integer n, known as the **principal quantum number,** that may have values from 1 to infinity. The radius of the electron orbit in these energy states varies as n^2:

$$\text{Radius} = n^2(0.53 \times 10^{-8} \text{ cm}) \qquad [6.3]$$

The number 0.53×10^{-8} cm represents the radius calculated by Bohr for the lowest energy, or most stable, orbit of the electron. Once the radii of the electron orbits are determined, it is possible to calculate the energy of an electron occupying any orbit. Bohr found that the energy of an electron occupying an orbit with principal quantum number n is proportional to $1/n^2$:

$$E \propto \left(\frac{1}{n^2}\right)$$

$$E = -R_H\left(\frac{1}{n^2}\right) \qquad [6.4]$$

The proportionality constant is called the Rydberg constant (R_H). It has the value 2.179×10^{-18} J. The orbital radii and energies are illustrated for $n = 1, n = 2$, and $n = 3$ in Figure 6.10. Another, more common, way of representing the allowed energies is shown in Figure 6.11.

The negative sign in Equation [6.4] denotes stability relative to some reference state. In other words, the more negative the value for energy, the more stable the system is. (One way to remember this convention is to think of a ball rolling around on a surface with many hills and valleys. The ball will naturally come to rest in a valley; it is most stable when its potential energy is lowest.) The reference, or zero-energy, state for the electron in hydrogen is chosen to be that in which the electron is completely separated from the nucleus. This, of course, corresponds to an infinitely large value for the principal quantum number n:

$$E = \frac{-R_H}{n^2}$$

$$= \frac{-R_H}{\infty^2} = 0$$

The energy of the electron in any other orbit is then negative relative to this reference state. The lowest energy, or most stable state, with $n = 1$, is known as the **ground state.** When the electron is in a higher energy orbit, that is, $n = 2$ or higher, the atom is said to be in an electronically **excited state.**

Energy (J)

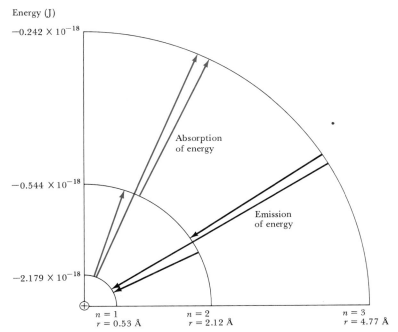

FIGURE 6.10 Radii and energies of the three lowest energy orbits in the Bohr model of hydrogen. The arrows refer to transitions of the electron from one allowed energy state to another. When the transition takes the electron from a lower to a higher energy state, absorption occurs. When the transition is from a higher to a lower energy state, emission occurs.

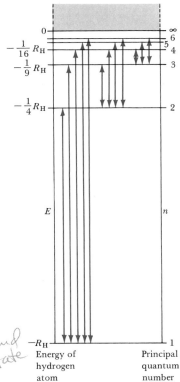

FIGURE 6.11 Energy levels in the hydrogen atom from the Bohr theory. The arrows refer to transitions of the electron from one allowed energy state to another, as described in Figure 6.10. Only the lowest six energy levels are shown.

Bohr's theory of the hydrogen atom satisfactorily explained the observed line spectra. The absorptions or emissions in the line spectra correspond to transitions of the electron from one orbit to another. Radiant energy is absorbed when the electron moves from one orbit to another having a larger radius; it requires energy to pull the electron away from the nucleus. Conversely, energy is emitted when the electron moves from a larger orbit to another having a smaller radius. The changes in energy, ΔE, are given by the expression

$$\Delta E = E_2 - E_1$$
$$= \frac{-R_H}{n_2{}^2} + \frac{R_H}{n_1{}^2}$$
$$= R_H \left(\frac{1}{n_1{}^2} - \frac{1}{n_2{}^2} \right)$$

Since $\Delta E = h\nu$,

$$\Delta E = h\nu = R_H \left(\frac{1}{n_1{}^2} - \frac{1}{n_2{}^2} \right) \qquad [6.5]$$

In this expression, n_1 and n_2 represent the quantum numbers for the initial and final states, respectively. Notice that when the final state quantum number (n_2) is larger than the initial state quantum number (n_1), the term in brackets is positive, and ΔE is positive. This means that

159

the system has absorbed a photon and thus increased in energy. The reverse applies when n_1 is larger than n_2, as happens in emission.

SAMPLE EXERCISE 6.3

In terms of Bohr's theory, explain the origin of the Fraunhofer lines in the solar spectrum.

Solution: Hydrogen atoms in the ground electronic state, present in the outermost regions of the sun, absorb solar energy at frequencies which correspond to excitation of the electrons from $n = 1$ to $n = 2$, $n = 1$ to $n = 3$, and so on. Similarly, hydrogen atoms may be present which are already in an electronically excited state, and these may absorb energy as they are excited to still higher energies, for example, $n = 2$ to $n = 3$, $n = 2$ to $n = 4$, and so on. All these transitions are represented by the upward directed arrows in Figure 6.11.

We can use Equation [6.5] to calculate the frequencies of Fraunhofer lines, as in Sample Exercise 6.4.

SAMPLE EXERCISE 6.4

Calculate the frequencies of the three hydrogen lines that correspond to the value $n_1 = 1$, and $n_2 = 2, 3,$ or 4.

Solution: We employ Equation [6.5] to find the value for ν. The value for R_H, mentioned above, is in units of J, and h is in units of J-sec. Then we obtain:

$$\nu = \frac{2.179 \times 10^{-18}\,\text{J}}{6.625 \times 10^{-34}\,\text{J-sec}}\left(\frac{1}{n_1{}^2} - \frac{1}{n_2{}^2}\right)$$

For $n_1 = 1$, and $n_2 = 2$, we obtain

$$\nu = 3.29 \times 10^{15}/\text{sec}\left(\frac{1}{1^2} - \frac{1}{2^2}\right)$$

$$= 2.47 \times 10^{15}/\text{sec}$$

Proceeding in the same way we obtain $2.92 \times 10^{15}/\text{sec}$ and $3.08 \times 10^{15}/\text{sec}$ for the other two frequencies. (You should double-check these to make sure of your ability to do the problem.)

Complete removal of the electron from a hydrogen atom, corresponding to a transition from the $n = 1$ (ground) state to the $n = \infty$ state, is known as ionization. This is represented as

$$H \longrightarrow H^+ + e^-$$

The energy required for ionization from the ground state is called the **ionization potential** (I).

SAMPLE EXERCISE 6.5

Using Equation [6.5], calculate the energy required for ionization of an electron from the ground state of the hydrogen atom.

Solution: The ionization potential may be written as the difference between the final and initial state energies. We have $n_2 = \infty$, $n_1 = 1$.

$$I = \Delta E = E_2 - E_1 = R_H\left(\frac{1}{1^2} - \frac{1}{\infty^2}\right)$$

$$= R_H(1 - 0)$$

This is just equal to R_H, 2.179×10^{-18} J.

It is often useful to express the energy on a molar basis. To do this we simply multiply by Avogadro's number:

$$I = \left(2.179 \times 10^{-18}\,\frac{\text{J}}{\text{atom}}\right)\left(6.022 \times 10^{23}\,\frac{\text{atoms}}{\text{mole}}\right)$$

$$\times \left(\frac{1\,\text{kJ}}{1000\,\text{J}}\right)$$

$$= 1312\,\text{kJ/mole}$$

Bohr's theory was very important, because it introduced the idea of quantized energy states for electrons in atoms. The theory was adequate for the hydrogen atom and for hydrogenlike ions such as He^+ and Li^{2+} that consist of a nuclear center and only one electron. But it was not adequate to account for the lines observed in the atomic spectra of any other atom or ion, except in a rather crude way. Bohr's theory is deficient because it does not take adequate account of the repulsions between electrons. We know that charged particles of the same sign repel one another. Thus in any atom containing two or more electrons, the electrons repel one another while at the same time they are attracted to the nuclear center.

Bohr's theory was eventually replaced by a new way of viewing atoms, called **quantum mechanics,** or **wave mechanics.** As we shall see, the concept of quantized energy states introduced by Bohr remains, but in addition still further applications of Planck's quantum theory enter the picture.

6.4
matter waves

In the years following Bohr's development of a theory for the hydrogen atom, the dual nature of radiant energy had become a familiar concept. Depending on the experimental circumstances, radiation might appear to have either a wavelike or particlelike character. Louis de Broglie, a young man working on his Ph.D. thesis in physics at the Sorbonne, in Paris, made a rather daring, intuitive extension of this idea. If radiant energy could under appropriate conditions behave as though it were particulate in nature, could not matter under appropriate conditions possibly show the properties of a wave? Suppose that the electron in orbit around the nucleus of a hydrogen atom could be thought of as a wave, with a characteristic wavelength, as illustrated in Figure 6.2. De Broglie suggested that the electron in its circular path about the nucleus has associated with it a particular wavelength. Then if the wave is to be a stable one, the circumference of the orbit must be a whole number of wavelengths long, as illustrated in Figure 6.12. If this were not so, the wave would partially cancel itself on each successive orbit, so that the average amplitude, or height of the wave, would be zero. But if a wave has no amplitude, it does not exist. For an orbit to be allowed, then, the wavelength of the electron must be related to the circumference of the orbit by the equation

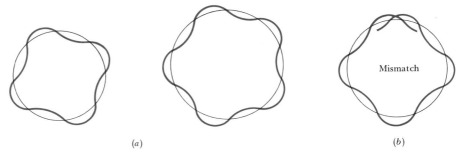

(a)   (b)

FIGURE 6.12 Schematic illustration of "interference" in the electron wave in a hydrogen atom orbit. In (a) the orbits are an integral number of wavelengths in total length. In (b) it is a nonintegral number. This is therefore a nonallowed orbit.

$$2\pi r = n\lambda \qquad [6.6]$$

where

$$n = 1, 2, 3 \ldots$$

But if this is so, what determines the characteristic wavelength of the electron? De Broglie showed that a particle of mass m and velocity v possesses a characteristic wavelength given by

$$mv = h/\lambda \qquad [6.7]$$

He used the term **matter waves** to describe the waves characteristic of material particles. If we solve Equation [6.7] for λ we can determine the sorts of characteristic wavelengths we might expect for a particle:

$$\lambda = \frac{h}{mv} \qquad [6.8]$$

Because De Broglie's hypothesis is perfectly general, any object of mass m and velocity v would give rise to a characteristic matter wave. However, it is easy to see from Equation [6.8] that the wavelength associated with an object of ordinary size, such as a golf ball, is so tiny as to be completely out of the range of any possible observation. This is not so for electrons, because their mass is so small.

SAMPLE EXERCISE 6.6

What is the characteristic wavelength of an electron with a velocity of 5.97×10^6 m/sec?

Solution: The mass of the electron is 9.107×10^{-28} g. The value of Planck's constant is 6.625×10^{-34} J-sec. ($1\ J = 1\ kg\ m^2/sec^2$)

$$\lambda = \frac{h}{mv}$$

$$\lambda = \frac{(6.625 \times 10^{-34}\ \text{J-sec})}{(9.107 \times 10^{-28}\ \text{g})(5.97 \times 10^6\ \text{m/sec})}$$

$$\times \left(\frac{1\ \text{kg m}^2/\text{sec}^2}{1\ \text{J}}\right)\left(\frac{10^3\ \text{g}}{1\ \text{kg}}\right)$$

$$= 1.22 \times 10^{-10}\ \text{m} = 0.122\ \text{nm}$$

By comparing this value with the wavelengths of electromagnetic radiations shown in Figure 6.3, we see that the characteristic wavelength is about the same as that of X rays.

Within a few years after De Broglie published his theory, the wave properties of the electron were demonstrated experimentally. Electrons were diffracted by crystals, just as X rays, which are definitely radiant energy, are diffracted. (We shall have more to say about the X-ray diffraction experiment in Chapter 11.)

The technique of electron diffraction has been highly developed. In the electron microscope, the wave characteristics of electrons are used to obtain electron diffraction pictures of tiny objects. The electron microscope is an important technique for studying surface phenomena at the very highest magnifications. An example of an electron microscope picture is shown in Figure 6.13. Pictures such as this are powerful demonstrations that tiny particles of matter can indeed behave as waves.

FIGURE 6.13 A piece of graphite photographed by means of an electron microscope (magnification is about 15 million times). The bright bands are layers of carbon atoms that are only 3.41 Å apart. (*P. A. Marsch and A. Voet, J. M. Huber Corp.*)

THE UNCERTAINTY PRINCIPLE

Discovery of the wave properties of matter raised new and interesting questions. If a subatomic particle can exhibit the properties of a wave phenomenon, is it possible to say precisely just where that particle is located? One can hardly speak of the precise location of a wave. The amplitude, or intensity, of a wave can be defined at a certain point, as illustrated in Figure 6.2, but the wave as a whole extends in space. Its location is therefore not defined precisely, at least not in the same sense that one can define the location of a particle. We might logically expect to be able to measure not only a particle's location, but also its direction and speed of motion.

However, the German physicist Werner Heisenberg* concluded that there is a fundamental limitation on just how precisely we can hope to know both the location and the energy of a particle. Just as in the case of quantum effects, the limitation becomes important only when we deal with matter at the subatomic level, that is, with masses as small as that of an electron. Heisenberg's principle is called the **uncertainty principle.** When applied to the electrons in an atom, this principle states that it is inherently impossible for us to know both the exact energy of the electron and its location in space. Thus it is not appropriate to imagine the electrons as moving in well-defined circular orbits about the nucleus, always at the same radius.

*Heisenberg, born in 1901, is one of the leading physicists of the twentieth century. He received the Nobel Prize in physics in 1932.

6.5
the quantum-mechanical
description of the atom

the electronic structures
of atoms

De Broglie's hypothesis and Heisenberg's uncertainty principle set the stage for a new and more broadly applicable theory of atomic structure. In this new approach, all attempt to define precisely the instantaneous location and energy of the electrons is abandoned. The wave nature of the electron is recognized, and its behavior is described in terms appropriate to waves.

6.5 the quantum-mechanical description of the atom

The mathematics employed to determine electron energy levels in even so simple a system as the hydrogen atom by quantum-mechanical methods are quite advanced. There is therefore no point in our attempting to present any mathematical details here. We can, however, understand the idea content of quantum mechanics with the aid of a qualitative description.

In solving the hydrogen atom problem by quantum mechanics, the problem is formulated mathematically so that the electron moving around the nucleus has wave properties. The attraction of the electron for the nucleus and the kinetic energy of motion of the electron are considered. Then, certain conditions are imposed on the possible solutions to make them physically reasonable. The result is a set of mathematical expressions that describe the allowed energy states of the electron. These expressions, usually represented by the symbol ψ, are called **wave functions.***

A wave function provides information about an electron's location in space when it is in a certain allowed energy state. The allowed energy states are the same as those predicted by the Bohr theory. However, the Bohr theory suggests that the electron is in a circular orbit of some particular radius about the nucleus. In the quantum-mechanical model for hydrogen, it is not so simple to describe the electron's location. The uncertainty principle suggests that if we know the energy of the electron with high accuracy, our knowledge of its location is very uncertain. Thus, for an individual electron, we cannot hope to specify its location around the nucleus. Rather we must be content with a kind of statistical knowledge. In the quantum-mechanical model we therefore speak of the *probability* that the electron will be in a certain region of space at a given instant. As it turns out, the square of the wave function (ψ^2) at a given point in space represents the probability that the electron will be found at that location. If the value of the wave function is summed over all the points in space around the nucleus, its value must equal one. This is true because probability of one is certainty, and it is certain that the electron must be somewhere in the space around the nucleus.

If we imagine the electron to be spread out in the entire volume element, then we can also think of the value for ψ^2 at a particular tiny volume element as an **electron density**. Regions of high electron density are regions where the electron is most likely to be found, as illustrated in Figure 6.14. The concepts of electron density and probability function

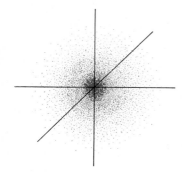

FIGURE 6.14 Electron density distribution in the ground state of the hydrogen atom.

*In mathematics, the term *function* is used to describe a relationship between two variables—for example, time and distance. Thus, the distance (d) covered by an auto moving at a constant speed (s) is a *function* of the amount of time (t) it has been in motion: $d = st$.

are more or less the same, and represent two ways of viewing the same idea.

The complete quantum-mechanical solution to the hydrogen atom problem yields a set of wave functions and a corresponding set of energies. Each wave function represents an allowed solution. The term **orbital** denotes an allowed energy state for the electron. It also refers to the probability function that defines the distribution of electron density in space. Thus an orbital has both a characteristic energy and a characteristic shape. Note that an *orbital* (quantum-mechanical model) is not the same as an *orbit* (Bohr model).

Because the space around the nucleus is three-dimensional, we might guess that the atomic wave functions would have associated with them three quantum numbers, as indeed they do. One of these, the principal quantum number (n) relates to the distance of the electron from the nucleus; it is the quantum mechanical equivalent of Bohr's principal quantum number. In the hydrogen atom, orbitals possessing the same principal quantum number (n) are of the same energy. As we shall see in the next chapter, this is not the case for atoms having many electrons.

The second quantum number is known as the azimuthal quantum number, and is given the symbol l. It defines the shape of the orbital. The possible values for l are limited by the value for n; l may have integral values from 0 to $n - 1$. Thus, for example if $n = 3$, l could have values of 0, 1, or 2. Orbitals are designated by letters corresponding to different values of l, as follows:

l	0	1	2	3	4
Designation of orbital	s	p	d	f	g

We shall consider such notation more fully in a moment; first let's consider the third quantum number.

The third quantum number, labeled m_l, is called the magnetic or orientational quantum number. It describes the orientation of the orbital in space. This quantum number may have integral values of l, $l - 1$, $l - 2$, and so on, down to $-l$. Thus, when $l = 0$, m_l must be zero. When $l = 1$, m_l can take on values of 1, 0, or -1. The possible values of the three quantum numbers through $n = 4$ are presented in Table 6.2.

SAMPLE EXERCISE 6.7

(a) What are the possible values of l when $n = 4$? (b) How many total orbitals are there for which $n = 4$?

Solution: (a) Since l may have values from 0 to $n - 1$, the possible values of l are 0, 1, 2, and 3.

(b) For $l = 0$, there is but one orbital, corresponding to $m_l = 0$. For $l = 1$ there are three orbitals, corresponding to $m_l = 1, 0$, and -1. For $l = 2$, there are five orbitals, corresponding to $m_l = 2, 1, 0, -1$, and -2. For $l = 3$, there are seven orbitals, corresponding to $m_l = 3, 2, 1, 0, -1, -2$, and -3. Thus the total number of orbitals with major quantum number $n = 4$ is 16.

A particular orbital is designated by a number corresponding to the value for the principal quantum number (n) and a letter corresponding to the value for l. Thus a $3d$ orbital corresponds to $n = 3$ and $l = 2$. As shown in Table 6.2, there are five such $3d$ orbitals corresponding to the five permitted values of m_l: 2, 1, 0, −1, and −2.

TABLE 6.2

relationship among values of n, l, and m_l through $n = 4$

n	l	ORBITAL DESIGNATION	m_l	NUMBER OF ORBITALS
1	0	$1s$	0	1
2	0	$2s$	0	1
	1	$2p$	1, 0, −1	3
3	0	$3s$	0	1
	1	$3p$	1, 0 −1	3
	2	$3d$	2, 1, 0, −1, −2	5
4	0	$4s$	0	1
	1	$4p$	1, 0, −1	3
	2	$4d$	2, 1, 0, −1, −2	5
	3	$4f$	3, 1, 2, 0, −1, −2, −3	7

SAMPLE EXERCISE 6.8

Write the designations for orbitals with the following assigned quantum numbers: (a) $n = 3$, $l = 0$; (b) $n = 5$, $l = 2$.

Solution: (a) The designation of orbitals with $l = 0$ is s; therefore $3s$.

(b) The designation for orbitals with $l = 2$ is d; therefore $5d$.

We'll consider later how we can distinguish the various possible values of m_l by means of a subscript added to the orbital symbol. First, however, we should consider how we can visualize the orbitals.

6.6
representations
of orbitals

THE s ORBITALS

The lowest energy (most stable) orbital, the one with $n = 1$, is referred to as the $1s$ orbital. The electron density for the $1s$ orbital is spherically symmetric as shown in Figure 6.14. Figures of this type, showing electron density, are one of the several ways we have to help us visualize orbitals. This figure indicates that the probability of finding the electron around the nucleus decreases as we move away from the nucleus in any direction. When the probability function (ψ^2) for the $1s$ orbital is graphed as a function of the distance from the nucleus (r), it rapidly approaches zero at large distance, as shown in Figure 6.15. This effect indicates that the electron, which is drawn toward the nucleus by electrostatic attraction, is not likely ever to get very far from the nucleus.

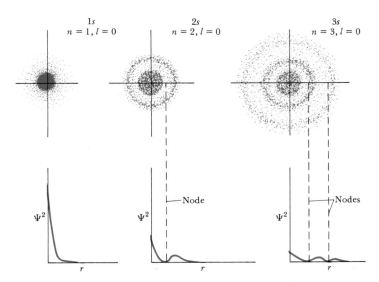

Node

Nodes

Ψ^2

Ψ^2

Ψ^2

r

r

r

FIGURE 6.15 Electron density distributions in 1s, 2s, and 3s orbitals. The lower part of the figure shows how the electron density, represented by ψ^2, varies as a function of distance from the nucleus. In the 2s and 3s orbitals, the electron density function drops to zero at certain distances from the nucleus. The spherical surfaces around the nucleus at which ψ^2 is zero are called nodes.

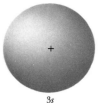

FIGURE 6.16 Contour representations of the 1s, 2s and 3s orbitals. The spherical surfaces connect points of equal value of ψ^2. The surface encloses 90 percent of the total ψ^2 for each orbital.

If we similarly consider the 2s and 3s orbitals of hydrogen we find that they are also spherically symmetric. Indeed, *all* s orbitals are spherically symmetric. The manner in which the probability function (ψ^2) varies with r for the 2s and 3s orbitals is shown in Figure 6.15. Notice that for the 2s orbital, ψ^2 goes to zero and then increases again in value before finally approaching zero at a larger value of r. The intermediate regions where ψ^2 goes to zero are called **nodal surfaces** or simply **nodes.** The number of nodes increases with increasing value for the principal quantum number (n). The 3s orbital possesses two nodes as illustrated in Figure 6.15. Notice also that as n increases the electron spends more and more of its time further from the nucleus. That is, the size of the orbital increases as n increases.

The most widely used method of representing orbitals is to display a boundary surface that encloses some substantial fraction, say 90 percent, of the total electron density for the orbital. For the s orbitals these contour representations are merely spheres. The contour or boundary surface representations of the 1s, 2s, and 3s orbitals are shown in Figure 6.16. They have the same shape, but they differ in size. The fact that there are nodes within the 2s and 3s surfaces is lost in these representations. This is not a serious disadvantage; it turns out that for most qualitative discussions the most important features of orbitals are their size and shape. These features are adequately represented by the contour diagrams.

In discussing chemical bonding it is useful to know not only the size and shape of an orbital but also whether the wave function (ψ) is positive or negative in sign at the contour surface. All s orbitals have positive values of ψ at their contour surfaces. This sign is therefore shown on the representations depicted in Figure 6.16.

167

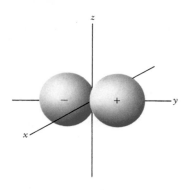

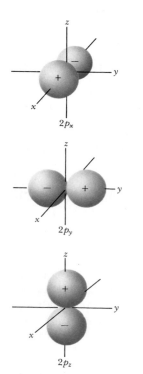

$2p_x$

$2p_y$

$2p_z$

THE p ORBITALS

When $l = 1$, the orbital is known as a p orbital. The contour surface for such orbitals is shown in Figure 6.17. The surface resembles two tangent spheres. The sign of the wave function (ψ) is different in the two lobes. This fact should not be allowed to confuse the physical interpretation of the contour surface. The probability function (ψ^2), which corresponds to electron density, is always positive. It is also useful to recall that we are making no statement of how the electron is moving within the volume outlined by this contour surface.

Whenever $l = 1$, m_l can have values of 1, 0, or -1. There are therefore three p orbitals corresponding to each principal quantum number, beginning with $n = 2$. For example, there are three 2p orbitals, three 3p orbitals, and so forth. The orbitals of a given principal quantum number have the same size and shape but differ from each other in orientation. The three 2p orbitals are shown in Figure 6.18. It is convenient to label these as the $2p_x$, $2p_y$, and $2p_z$ orbitals. The letter subscript indicates the axis along which the orbital is oriented. As it turns out, there is no necessary connection between one of these subscripts and a particular value of m_l. To explain why this is so would require discussion of material beyond the scope of an introductory text.

Just as with the s orbitals, the distance from the nucleus to the center of the electron density moves outward as we go from the 2p to 3p to 4p orbitals. In other words, orbital size increases with increase in the principal quantum number (n). In accurate contour representations of the 3p and higher p orbitals, there are small regions of electron density separating the major lobes. These details of the shapes of the p orbitals are not of major chemical importance. Most important is the general directional character of the p orbitals and the fact that the two major lobes always correspond to opposite signs of the wave functions. We shall therefore always represent the p orbitals as shown in Figure 6.18, regardless of the value for the principal quantum number (n).

THE d AND f ORBITALS

When $l = 2$, the orbital is a d orbital. There are no d orbitals with n lower than 3. This follows from the fact that l can never be larger than $n - 1$. There are five equivalent 3d orbitals corresponding to the five possible values for m_l: 2, 1, 0, -1, and -2. Likewise there are five 4d orbitals, and so forth. Just as in the case of the p orbitals, the differing values of m_l correspond to different orientations of orbitals in space. The most useful representations of the 3d orbitals are shown in Figure 6.19. Notice that four of the orbitals are of the same shape but have differing orientations. The fifth orbital, labeled the d_{z^2}, has a different shape. It is not possible to represent the five d orbitals as having the same shape in an ordinary x, y, z axis system. Although the fifth d orbital looks different, it has the same energy in an atom as any of the other four orbitals.

The representations of higher d orbitals are very much like those for the 3d. The contour representations shown in Figure 6.19 are commonly employed for all d orbitals, regardless of major quantum number.

There are seven equivalent f orbitals (for which $l = 3$) for each value of n of 4 or greater. The f orbitals are difficult to represent in three-dimensional contour diagrams. We shall have no need to concern ourselves with orbitals having values for l greater than 3.

As we shall see in a later chapter, an understanding of the number and shapes of atomic orbitals is important to a proper understanding of the molecules formed by combining atoms. *You should commit to memory the orbital representations shown in Figures 6.16, 6.18 and 6.19.*

FOR REVIEW

summary

After reading this chapter, you should have a clear understanding of the description of electrons in atoms in terms of the wave model.

In the modern view of atomic structure, electrons are distributed in the space about a massive nucleus of positive charge. A fundamental limitation on our ability to measure things prevents us from specifying precisely where the electrons are to be found. Instead we must settle for a statistical kind of description, in which we speak of the probabilities of the electron being found in a certain tiny volume surrounding a particular point in space. Although the positions of the electrons are not defined, except in this averaged sense, their energies are precisely known. Each allowed state of an electron in an atom corresponds to a particular set of values for three **quantum numbers.** Each such allowed energy state is termed an **orbital.** An orbital is described by a combination of an integer and letters, corresponding to the three values for the quantum numbers. The **principal quantum number** (n) is indicated by the integers 1, 2, 3, ... This quantum number relates most directly to the size and energy of an orbital. The **azimuthal quantum number** (l) is indicated by the letters s, p, d, f, and so on, corresponding to values of l of 0, 1, 2, 3, ... The l quantum number defines the shape of the orbital. The **magnetic** or **orientational quantum number** (m_l) describes the orientation of the orbital in space. For example, the three $3p$ orbitals are designated $3p_x$, $3p_y$ and $3p_z$, the subscript letters indicating the axis along which the orbital is oriented.

The orbitals we have been describing are essentially those of a hydrogen atom. In the atoms of other elements, there are many electrons moving about a single nucleus. These electrons are attracted by the nucleus and at the same time are repelled by one another. The resulting distribution of electron density and the energies of the allowed energy states of the electrons are thus the product of extremely complex forces. Nevertheless, as we shall see in the following chapter, the electronic structures of atoms having many electrons can be built up by the progressive addition of electrons to orbitals that are very much like those of the hydrogen atom.

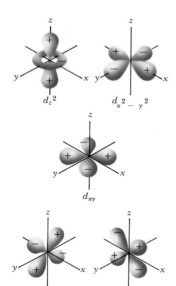

FIGURE 6.19 Contour representations of the five $3d$ orbitals.

learning goals

Having read and studied this chapter, you should be able to:

1. Describe the wave properties and characteristic speed of propagation of radiant energy (electromagnetic radiation).
2. Use the relationship $\lambda v = c$, which relates the wavelength (λ) and frequency (v) of radiant energy to its speed (c).
3. Describe how the wavelength and frequency differ in the various parts of the electromagnetic spectrum, such as the infrared, visible, and ultraviolet.

4. Explain the essential feature of Planck's quantum theory, namely, that the smallest increment, or quantum, of radiant energy of frequency ν that can be emitted or absorbed is $h\nu$, where h is Planck's constant.

5. Explain how Einstein accounted for the photoelectric effect by considering the radiant energy to be a stream of particlelike photons striking a metal surface. In other words you should be able to explain all the observations about the photoelectric effect using Einstein's model.

6. Explain what is meant by the term *spectrum* and by the expression *line spectrum* in referring to the light emitted or absorbed by an atom.

7. List the assumptions made by Bohr in his theory of the hydrogen atom. Most importantly, you should be able to explain how Bohr's theory relates to Planck's quantum theory.

8. Explain the concept of an allowed energy state and how this concept is related to the quantum theory.

9. Calculate the energy differences between any two allowed energy states of the electron in hydrogen.

10. Explain the concept of ionization potential.

11. Calculate the characteristic wavelength of a particle from a knowledge of its mass and velocity.

12. Describe the uncertainty principle and explain the limitations it places on our ability to define simultaneously the location and energy of a subatomic particle, particularly an electron.

13. Explain the concepts of *orbital, electron density,* and *probability* as used in the quantum-mechanical model of the atom.

14. Describe the three quantum numbers used to define an orbital in an atom and list the limitations placed on the values each may have.

15. Describe the correspondence between letter designations and values for the azimuthal quantum number (l).

16. Describe the shapes of s, p, and d orbitals and list the relative signs of the wave functions in different regions.

17. Explain the difference between the wave function (ψ) and the square of the wave function (ψ^2) and describe the physical significance of ψ^2.

key terms

Among the more important terms and expressions used for the first time in this chapter are the following:

The **azimuthal quantum number**, l (Section 6.5), is one of the quantum numbers that specifies an atomic orbital. The value for l defines the shape of the orbital. The values allowed for l are restricted by the value for the principal quantum number (n); l may take on integral values from 0 to $n - 1$.

The **electron density** (Section 6.5) at a particular point in space in an atom is the probability that the electron will be found in the region immediately around that point (ψ^2).

The **ionization potential** (Section 6.3) for the hydrogen atom is the energy required to lift an electron from its lowest energy, or most stable, orbit to a point infinitely far from the nucleus. In effect, this involves removing the electron from the atom.

A **line spectrum** (Section 6.2) contains radiation only of certain specific wavelengths.

Matter wave (Section 6.4) is the term applied to the wave characteristics of a subatomic particle.

A node (Section 6.6) as applied to electron density in atoms, is the locus of points (for example, a plane or a spherical surface) at which the electron density is zero. For example, the node in a $2s$ orbital, Figure 6.15, is a spherical surface.

An orbit (Section 6.3) in the Bohr theory of the hydrogen atom is any one of the allowed energy states of the electron. Each orbit corresponds to a different value for the principal quantum number (n).

An orbital (Section 6.5) represents an allowed energy state of an electron in the modern, quantum-mechanical model of the atom. The term orbital is also used to describe the spatial distribution of the electron.

The orientational quantum number, m_l (Section 6.5), describes the orientation of an orbital in space. This quantum number may have integral values of l, $l - 1$, $l - 2$, and so on, down to $-l$. For each value of l there are thus $2l + 1$ values for m_l.

A photon (Section 6.2) is a quantum, or smallest increment, of radiant energy, $h\nu$.

The principal quantum number, n (Section 6.5), is the quantum number that relates most directly to the size and energy of an atomic orbital. It may take on integer values of 1, 2, 3 . . .

A quantum (Section 6.2) is the smallest increment of radiant energy that may be absorbed or emitted. The magnitude of the quantum of radiant energy is $h\nu$.

Radiant energy, or electromagnetic radiation (Section 6.1) is a form of energy possessing wave character and being propagated through space with a characteristic speed, 3.00×10^8 m/sec.

The term spectrum (Section 6.1) refers to the distribution among various wavelengths of the light emitted or absorbed by an object. A continuous spectrum contains radiation distributed over many wavelengths.

The uncertainty principle (Section 6.4) states that there is an inherent uncertainty in the precision with which we can simultaneously specify the location and energy of a particle. It is of importance only for the lightest particles such as the electron.

A wave function (Section 6.5) is a mathematical description of an allowed energy state, or orbital, for an electron in the quantum-mechanical model for an atom.

EXERCISES

radiant energy

6.1   The mean distance of the moon from earth is 239,000 mi. How long did it take for the TV pictures from the Apollo mission to reach earth from the moon?

6.2   The equipment designed to search for evidence of biological life on Mars is expected to transmit its findings to earth from as far away as 420 million mi. How long will it take for these transmissions to reach earth? Suppose they are received at a station in California, and then immediately relayed to New York 3400 mi away. How much later will they be received in New York. than in California?

6.3   A particular AM radio station broadcasts at a frequency of 1120 kHz. Its sister FM station broadcasts at a frequency of 98.7 MHz. What are the wavelengths of the radiations from each station?

6.4   Giant radio telescopes listening for radiation from outer space have detected strong radio signals with a wavelength of 21 cm. What is the frequency of this radiation?

6.5   The red color of blood is due to a strong absorption of light centered at about 450 nm in the visible region of the spectrum. What is the frequency of light which matches this absorption?

6.6   Which of the following types of electromagnetic radiation is highest in energy? Which is lowest? (a) sunlight through a green filter; (b) radiation from an FM station; (c) radiation from an infrared heat lamp; (d) radiation from a hot oven

quantum theory

6.7 The yellow lamps that illuminate highway intersections and city streets are filled with sodium vapor that is electrically excited and then emits light. The strongest line emitted by the sodium is at a wavelength of 589.2 nm. What is the magnitude of a quantum of energy at this wavelength?

6.8 In the photoelectric effect experiment shown in Figure 6.7, why is there no current flowing when the light is off? What causes current to flow when the light is on?

[6.9] The energy required to remove an electron from the metal by a photon is called the photoelectric work function. For cesium metal the photoelectric work function is 3.05×10^{-19} J. What is the maximum wavelength of light that will cause electrons to be emitted from cesium metal?

[6.10] Photocells that are used in "electric eye" door openers and burglar alarms operate on the principle of photoelectric emission as illustrated in Figure 6.7. If the lamp used as the source of light has an effective wavelength of 540 nm, would it be satisfactory to use copper as the metal in the photocell? The photoelectric work function (that is, the energy required to remove an electron from copper metal) is 6.69×10^{-19} J.

6.11 From the results of Sample Exercise 6.5, calculate the wavelength of light that would ionize a hydrogen atom.

6.12 Explain how it is possible to tell from study of the light from a distant star that hydrogen is present in the outer atmosphere of the star.

6.13 When the light from a neon lamp is dispersed in a prism to form a spectrum, it is found that the spectrum is not continuous; that is, it does not consist of a broad range of wavelengths of light. Rather, it consists of several sharp lines. Explain in general terms why the emissions form a line spectrum rather than a continuous spectrum.

Bohr's theory

6.14 In Bohr's theory of the hydrogen atom, under what conditions does the atom absorb or emit radiant energy?

6.15 What is the energy of the electron, in Joules, when it is in the $n = 6$ state of the hydrogen atom? When the electron moves from this state to the one in which $n = 4$, is energy emitted or absorbed by the atom? Explain.

[6.16] At what frequency would you expect to find the lowest frequency line in the series of hydrogen

absorption lines for which $n_1 = 3$? What is the wavelength of this line? In what region of the spectrum does it appear?

6.17 The He$^+$ ion is isoelectronic with H; that is, it has the identical electronic structure. Would you expect the ionization potential for He$^+$ to be larger or smaller than for H? Explain.

6.18 Atoms of He differ from H in having two electrons moving about the central nucleus. How does this complicate the task of calculating the energies of these electrons? Is Bohr's theory suitable for He atoms? Explain.

matter waves

6.19 According to the de Broglie relationship, what requirement must the characteristic wavelength of the electron have in order for it to be in an allowed orbit?

6.20 Cite one or more pieces of experimental evidence that show that it is possible for particles of matter to possess wave properties.

6.21 Neutrons from an atomic reactor can be selected according to velocity and made into beams of neutrons with the same velocity. What is the characteristic wavelength of a neutron moving with a velocity of 3.22×10^3 m/sec? (Refer to Table 2.3 for required data.)

6.22 What is the characteristic wavelength of a golf ball of mass 82 g, moving with a speed of 265 km per hour?

wave functions; orbitals; quantum numbers

6.23 What is meant by the term *wave function?* In what sense does the square of the wave function have physical significance?

6.24 Give a definition of the term *orbital*. What are some of the characteristics of an orbital?

6.25 Which of the following are permissible sets of quantum numbers for an electron in an atom? (a) $n = 3, l = 1, m_l = -1$; (b) $n = 3, l = 1, m_l = 2$; (c) $n = 2, l = 2, m_l = -1$; (d) $n = 11, l = 0, m_l = 0$; (e) $n = 4, l = -2, m_l = 1$. For those combinations of orbital designations that are permissible, write the appropriate orbital designation in terms of numeral and letter (that is, $1s$, and so on).

6.26 What is the physical significance of the requirement that the square of the wave function, summed over all the space around the nucleus, must equal one?

6.27 If the value of ψ^2 for the $1s$ orbital is summed over all space would its value differ from the value of ψ^2 for the 3s orbital summed over all space? Explain.

6.28 What is the meaning of the term *node?* What is the probability of the electron's being found at a node? Explain.

6.29 Suppose that you begin at a distance far from the nucleus on the x axis, travel along the x axis, through the nucleus to a distance far on the $-x$ axis. How many nodal surfaces would you pass through for each of the following orbitals? (a) 3s; (b) $2p_x$.

6.30 What characteristic of orbitals is determined by the value for azimuthal quantum number (l)?

6.31 From inspection of Figure 6.15 describe how the most probable distance of the electron from the nucleus differs from the 1s to 2s to 3s orbitals. Can you offer a reason for this difference?

6.32 Without consulting the figures in your text, draw the contour representations for each of the following orbitals: p_x; d_{xz}; 3s; $d_{x^2-y^2}$.

6.33 How many different orbitals are there that can have each of the following designations: (a) $n = 4$; (b) $4p$; (c) $2d$; (d) $3d_{x^2-y^2}$?

6.34 What general relationship can you deduce between the principal quantum number n and the number of allowed orbitals with that value of n?

6.35 How many orbitals are there that may have the following designations: (a) $4f$; (b) $3p_x$; (c) $4d$; (d) $2p$?

6.36 Draw contour diagrams for all the atomic orbitals of principal quantum number 3.

[6.37] Of the two ions Li^{2+} or He^+, which has the lower energy electronic ground state? Explain.

general exercises

6.38 Explain or define each of the following terms: (a) orbital; (b) azimuthal quantum number; (c) characteristic wavelength; (d) node; (e) line spectrum.

6.39 How much energy is required to ionize a mole of hydrogen atoms that are in an excited state such that the electron is in the orbital $n = 3$?

[6.40] The circumference of the circular orbit of the electron in the $n = 2$ state of hydrogen in Bohr's model is 13.32×10^{-8} cm. Using the de Broglie relationship, calculate the velocity of the electron in this orbit, assuming that it has a characteristic wavelength that is just one full circumference.

6.41 Draw the shapes of each of the following orbitals: (a) p_z; (b) d_{xz}; (c) 2s.

6.42 For each of the following pairs of hydrogen orbitals, indicate which is the higher in energy, or whether they are of equal energy: (a) 2s, 3s; (b) $3p_z$, $3p_x$; (c) $3p_y$, $3d_{xy}$; (d) $3p_x$, 4s.

6.43 Calculate the wavelength of the photon that is required to cause ionization of the hydrogen atom from an excited state in which $n = 2$.

6.44 Although the sign of the wave function for an orbital may differ in different regions of space around the nucleus, the sign of ψ^2 is always positive. What is the physical meaning of ψ^2?

periodic relationships among the elements

7

In the early days of chemistry, the emphasis was on observations, on learning how chemical substances behave. However, as the quantity of chemical information grew, more attention was given to the possibilities of classifying it in useful ways. The most important product of these attempts at classification is the periodic table of the elements, which was introduced in Chapter 2. In this chapter we shall look at the structure of the periodic table in some detail. We shall also see that the table, which developed in an entirely empirical manner, is in fact intimately related to the electronic structures of the elements.

7.1

the modern periodic table

A modern version of the periodic table is shown in Figure 7.1. This same figure is also reproduced for ready reference on the inside front cover of the text. As was noted in Chapter 2, the elements in the table are arranged in the order of increasing atomic number, beginning with the lightest element, hydrogen. Elements that bear a close chemical resemblance are arranged in vertical rows. Each vertical column of elements is referred to as a group or family. Thus, for example, the elements Li, Na, K, Rb, and Cs (the alkali metals) are aligned together vertically to form a group that is labeled 1A.* Some properties of the alkali metals are shown in Table 7.1. These elements are alike in many of their chemical and physical properties. All are shiny, soft, metallic solids of relatively low

*Although the element francium, Fr, is a member of group 1A, all the isotopes of the element are radioactive. Francium has been observed in the laboratory as a product of nuclear reactions, but is present in nature in such tiny quantities that it has never been detected there.

Periodic Table of the Elements

Active metals

1A	2A

Transition metals

3B	4B	5B	6B	7B	8B			1B	2B

Nonmetals

3A	4A	5A	6A	7A	8A

1A	2A	3B	4B	5B	6B	7B	8B			1B	2B	3A	4A	5A	6A	7A	8A
1 H 1.00797																	2 He 4.0026
3 Li 6.939	4 Be 9.0122											5 B 10.811	6 C 12.01115	7 N 14.0067	8 O 15.9994	9 F 18.9984	10 Ne 20.183
11 Na 22.9898	12 Mg 24.312											13 Al 26.9815	14 Si 28.086	15 P 30.9738	16 S 32.064	17 Cl 35.453	18 Ar 39.948
19 K 39.098	20 Ca 40.08	21 Sc 44.956	22 Ti 47.90	23 V 50.942	24 Cr 51.996	25 Mn 54.9380	26 Fe 55.847	27 Co 58.9332	28 Ni 58.71	29 Cu 63.54	30 Zn 65.37	31 Ga 69.72	32 Ge 72.59	33 As 74.9216	34 Se 78.96	35 Br 79.909	36 Kr 83.80
37 Rb 85.47	38 Sr 87.62	39 Y 88.905	40 Zr 91.22	41 Nb 92.906	42 Mo 95.94	43 Tc (99)	44 Ru 101.07	45 Rh 102.905	46 Pd 106.4	47 Ag 107.870	48 Cd 112.41	49 In 114.82	50 Sn 118.69	51 Sb 121.75	52 Te 127.60	53 I 126.9044	54 Xe 131.30
55 Cs 132.905	56 Ba 137.33	57 *La 138.91	72 Hf 178.49	73 Ta 180.948	74 W 183.85	75 Re 186.2	76 Os 190.2	77 Ir 192.2	78 Pt 195.09	79 Au 196.967	80 Hg 200.59	81 Tl 204.37	82 Pb 207.19	83 Bi 208.980	84 Po (210)	85 At (210)	86 Rn (222)
87 Fr (223)	88 Ra (226)	89 †Ac (227)	104 Rf (257)	105 Ha (260)													

*Lanthanide series

58 Ce 140.12	59 Pr 140.907	60 Nd 144.24	61 Pm (147)	62 Sm 150.35	63 Eu 151.96	64 Gd 157.25	65 Tb 158.924	66 Dy 162.50	67 Ho 164.930	68 Er 167.26	69 Tm 168.934	70 Yb 173.04	71 Lu 174.97

† Actinide series

90 Th 232.038	91 Pa (231)	92 U 238.03	93 Np (237)	94 Pu (242)	95 Am (243)	96 Cm (247)	97 Bk (247)	98 Cf (249)	99 Es (254)	100 Fm (253)	101 Md (256)	102 No (253)	103 Lw (257)

FIGURE 7.1 Periodic table of the elements. The atomic weight of each element is given beneath the symbol for the element. Approximate values for radioactive elements are listed in parentheses.

density. All are very reactive chemically and readily lose one electron to form ions of +1 charge. For example, they react vigorously with water to form a metal hydroxide, with evolution of hydrogen. The general equation for this reaction is

$$2M(s) + 2H_2O(l) \longrightarrow 2MOH(aq) + H_2(g) \qquad [7.1]$$

The symbol M in this equation represents any one of the alkali metals.

At the opposite side of the table, the elements He, Ne, Ar, Kr, Xe and Rn also bear close resemblance to one another. All exist as monatomic gases, with very little or no tendency to undergo chemical reaction. They resemble one another quite closely, but they differ greatly from other elements having atomic numbers close to theirs. Although elements within each group or family are similar, there are differences that can be related to the vertical position of the element. The increasing numbers of protons, neutrons, and electrons in the atom with increasing atomic number causes systematic differences in chemical and physical properties. For example, the melting points and densities of the alkali metals (Table 7.1) vary in a fairly regular manner with increasing atomic number.

TABLE 7.1

some physical and chemical properties of the alkali metals

ELEMENT	SYMBOL	ATOMIC NUMBER	ATOMIC WEIGHT	MELTING POINT (°C)	DENSITY (g/cm³)	FORMULA OF HYDROXIDE	FORMULA OF CHLORIDE
Lithium	Li	3	6.939	181	0.53	LiOH	LiCl
Sodium	Na	11	22.9898	98	0.97	NaOH	NaCl
Potassium	K	19	39.102	63	0.86	KOH	KCl
Rubidium	Rb	37	85.47	39	1.53	RbOH	RbCl
Cesium	Cs	55	132.905	29	1.87	CsOH	CsCl

The essential feature of the periodic table is then that *when the elements are arranged in the order of increasing atomic number, similarities in physical and chemical properties are repeated at regular intervals.* Each interval corresponds to a row, or period, of the table. The elements hydrogen and helium are counted as a first short row. The second row of the table consists of eight elements, extending from lithium, atomic number 3, to neon, atomic number 10. The third row begins with sodium, atomic number 11, which has properties similar to those of lithium. It ends with argon, atomic number 18, which resembles neon. Thus, there are eight elements each in the second and third rows. The fourth row consist of 18 elements. Note in Figure 7.1 that the table is expanded to make room for the set of 10 elements that do not correspond with any of those in the second and third rows. The elements in this portion of the table are called **transition metals.**

There are three rows of 18 elements. In addition, a series of elements called the **lanthanides,** with atomic numbers ranging from 58 to 71, needs to be fitted into the sixth row. Another similar series called the **actinides,** atomic numbers 90 through 103, needs to be fitted into the seventh row.

To expand the table horizontally still further would make the table too unwieldy. The compromise generally employed is to place these two series of elements below the main body of the table, as shown in Figure 7.1.

The label attached to each group or family has a loose relationship to the electronic structures of the elements, but it is mainly just a means of identifying the groups. Note that there are both A and B groups. Elements of the same group number but different letter labels do have certain characteristics in common, but in general there is not a close connection. It is therefore very important not to confuse an element in an A group with one in a B group. For example, copper (Cu), element number 29, is like potassium (K) in being able to form a chloride of the formula CuCl, just as potassium can form KCl. In many other respects, copper is very much unlike potassium and generally bears little resemblance to it or to any other group 1A element.

7.2
historical development of the periodic table

During the earliest years of the nineteenth century, many new elements were discovered in a short period of time. By 1830 there were about 56 known elements. The identification of many new elements and the development of their descriptive chemistry naturally led to various attempts at classification. The classification process that occurred in chemistry in those years is common to the development of all science. As the quantity and variety of data increase, some means of orderly classification is sought, simply as a means of managing the large number of facts at hand. When a workable classification scheme is found, it may form the basis for development of a theory that accounts for the regularities observed. In 1869, Dmitri Mendeleev in Russia (Figure 7.2) and

FIGURE 7.2 Dmitri Mendeleev. Mendeleev rose from very poor beginnings to a position of great eminence in nineteenth-century science. He was born in Siberia, the youngest child in a family of at least 14. His mother endured great personal sacrifice to make it possible for him to enroll in a university in Saint Petersburg. Mendeleev proved to be a brilliant student in sciences and mathematics and eventually was able to study in France and Germany. He spent most of his career as a professor of chemistry in the University of Saint Petersburg. Despite his eminence as a scientist, he was often in trouble because of his liberal, unorthodox opinions. (*Library of Congress*)

	B 10.81	**C** 12.01	**N** 14.01
	Al 26.98	**Si** 28.09	**P** 30.97
Zn 65.37	?	?	**As** 74.92
Cd 112.41	**In** 114.82	**Sn** 118.69	**Sb** 121.75

FIGURE 7.3 A portion of Mendeleev's periodic table showing the symbol for the element and the modern value for atomic mass.

Lothar Meyer in Germany, working quite independently of each other, published very similar schemes for classification of the elements. Their tables of the elements were the forerunners of the modern periodic table. Mendeleev arranged the elements in order of increasing atomic weight and observed that when this is done, similar chemical and physical properties recur periodically. By arranging the elements so that those with similar characteristics are in vertical groupings, Mendeleev constructed the periodic table.

Although Meyer and Mendeleev came to essentially the same conclusion about the periodicity of properties, Mendeleev must be given credit for more vigorously advancing his ideas and stimulating much new work in chemistry. By sticking to his notion that elements of similar characteristics must be listed in groups, he was forced to leave several spaces in his table blank. For example, arsenic (As) was the element of next highest atomic weight after zinc (Zn). But its placement immediately after Zn in the table would have required that it fall under aluminum (Al). This, however, did not make sense in terms of its properties. Rather, it clearly belonged under phosphorus, as shown in Figure 7.3. This meant that in the table there were two blank spaces that Mendeleev boldly predicted would be filled by as yet undiscovered elements. He gave these elements the names eka-aluminum and eka-silicon and suggested that they might be found in nature with other members of their respective families. For example, eka-aluminum might be found in certain ores containing aluminum, because according to the periodic law it was likely to have properties similar to those of aluminum.

By noting that the properties of elements within a vertical family varied in a regular way with increasing atomic weight, Mendeleev was able to predict the properties of the unknown elements. Thus, the properties of eka-silicon should be intermediate between those of silicon (Si) and tin (Sn). In 1871 Mendeleev predicted the properties for eka-silicon, and not many years later, in 1886, the element germanium (Ge) was discovered. That the element germanium was the eka-silicon predicted by Mendeleev is shown by the data listed in Table 7.2.

SAMPLE EXERCISE 7.1

Note the densities listed below for the elements aluminum and indium. From these values and the data listed in Table 7.2 and Figure 7.3, predict the atomic weight and density for eka-aluminum (gallium). (Note: density of aluminum is 2.70 g/cm^3; density of indium is 7.30 g/cm^3.)

Solution: We would predict the atomic weight to be intermediate between the two values for the adjacent elements of higher and lower atomic weights, Zn and Ge. The average of the atomic weights of these two elements is 69.0, as compared with observed atomic weight of 69.72. We might estimate the density as intermediate between the values for Al and In, namely 5.00 g/cm³. The observed value is 5.90 g/cm³. Incidentally, the average of the densities of Zn and Ge, the adjacent elements in the horizontal row, is 6.25 g/cm³, which comes a bit closer to the observed density for Ga.

TABLE 7.2

comparison of the properties for eka-silicon predicted by Mendeleev with the known properties of the element germanium

PROPERTY	MENDELEEV'S PREDICTION FOR EKA-SILICON	OBSERVED PROPERTIES OF GERMANIUM
Appearance	Gray	Grayish-white
Atomic weight	72	72.59
Density (g/cm³)	5.5	5.35
Specific heat (cal/g-K)	0.073	0.074
Formula of oxide	XO_2	GeO_2
Density of oxide (g/cm³)	4.7	4.70
Formula of chloride	XCl_4	$GeCl_4$
Density of chloride (g/cm³)	1.9	1.84

The development of the periodic classification of the elements did much to systematize the study of chemistry. Nevertheless, many elements did not seem to fit very well into the periodic table. Only much later was it realized that the ordering of the elements should be in terms, not of atomic weight, but rather of atomic number. Once the concept of atomic number, that is, the number of protons in the nucleus of the atom and the number of electrons surrounding the nucleus, was established, all the essential basis for the modern periodic table was at hand. The modern version of Mendeleev's periodic law is thus: When the elements are listed in the order of increasing atomic number, similar chemical and physical properties recur periodically.

7.3
the periodic table and electron configuration

The atomic number corresponds not only to the number of protons in the nucleus of an atom but also to the number of electrons in that atom. Thus, the periodic table represents an ordering of the elements not only according to nuclear charge but also according to the number of electrons about the nucleus. It was natural, therefore, for chemists to search for a model that would account for the chemical properties of elements in terms of the arrangement, or configuration, of electrons in the atoms.

Two American chemists, Gilbert N. Lewis and Irving Langmuir, reasoned that if the chemical properties of the elements are repeated at intervals, then their electronic arrangements must be repeating in some way. They were led to the idea of layers, or shells, of electrons. Once a shell of electrons is filled, additional electrons must go into a new shell. They reasoned that completion of a shell should be marked by some sort of special behavior, perhaps by chemical inertness. The elements called the noble, or rare, gases, atomic numbers 2, 10, 18, 36, 54, and 86, are chemically very nonreactive. Lewis and Langmuir therefore concluded that these elements mark the completion of a shell, the attainment of a chemically stable arrangement of electrons. The chemical behavior of the other elements might then be understood in terms of a tendency by one means or another to achieve the same electronic arrangement as the stable, inert gases. For example, we saw in Section 2.9 that sodium, atomic number 11, has one electron more than the "magic number" of 10 that characterizes neon. Sodium therefore readily loses an electron on chemical reaction to form the sodium ion (Na^+), which has the inert gas arrangement.

The Lewis-Langmuir theory was very important in the development of chemistry, because it provided chemists with a model associating chemical behavior with electron configuration. In later chapters, as we develop our ideas of chemical bonding in more detail, we shall be using ideas directly derived from the theoretical work of Lewis and Langmuir. At this point, we have before us the task of understanding the basis on which the Lewis-Langmuir model rests. Why are electrons arranged in shells? Why do certain stable electronic arrangements occur periodically? To answer these questions and to account in more detail for the way in which various atomic properties vary with atomic number, we must apply the ideas of atomic orbitals and quantum numbers developed in the previous chapter.

7.4

energies of orbitals

We have seen how the allowed energy states of the electron in a hydrogen atom are characterized by three quantum numbers. When a hydrogen atom is in its lowest energy, or ground, state, the electron resides in the $1s$ orbital. The electron may be promoted into a higher energy orbital by absorption of radiation of appropriate energy. By observing the frequencies of the radiant energy absorbed by hydrogen atoms when they are excited, it is possible to determine the energies of the allowed orbitals. These energies are accounted for quite well by both the Bohr and quantum-mechanical models.

The energy level diagram for the orbitals of hydrogen and other one-electron systems such as He^+ and Li^{2+} is as shown in Figure 7.4. We use a box to represent each orbital; the orbitals are arranged in order of increasing energy. All orbitals having the same value of principal quantum number, for example the $3s$, $3p$, and $3d$, have essentially the same energy. In scientific jargon, orbitals that have the same energy are said to be *degenerate*.

The energy of any particular orbital in a one-electron system depends on the principal quantum number n and the charge of the nucleus (Z) according to the formula

$$\text{Energy} = \frac{-R_H Z^2}{n^2} \qquad [7.2]$$

R_H is the Rydberg constant. Note that when $Z = 1$, this equation becomes Equation [6.5]. Equation [7.2] tells us that in one-electron systems

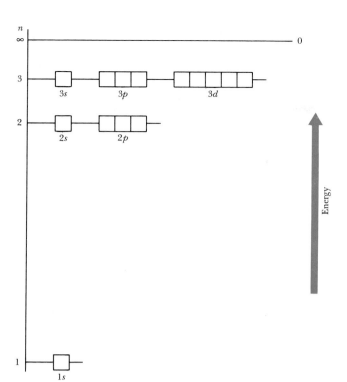

FIGURE 7.4 Orbital energy levels in the hydrogen atom and in hydrogenlike ions (those containing just one electron). Note that all orbitals with the same value for the principal quantum number, n, have the same energy. This is true only in one-electron systems.

the energy does not depend on the quantum numbers l and m_l. The dependence on Z^2 reflects the attractive force of the nucleus. An electron in a $1s$ orbital of He^+ ($Z = 2$), for example, is four times more stable than an electron in a $1s$ orbital of H ($Z = 1$). Furthermore, the $1s$ orbital in He^+ is smaller than the $1s$ orbital of H. Because the electron is more strongly attracted to the nucleus of higher charge, its average distance from the nucleus is smaller.

ATOMIC ORBITALS IN ATOMS HAVING MANY ELECTRONS

In atoms having many electrons, the electrons are acted upon not only by the attractive force of the nucleus, but also by the repulsive forces operating between electrons. We know that the electrons must continue to obey the laws of quantum mechanics, even though their behavior is more complex than in the case of the one-electron atom or ion. It has been found that the three quantum numbers we discussed previously are also useful in describing the electron arrangements of many-electron atoms. However, the relative energies of the orbitals are no longer like those shown in Figure 7.4.

In a many-electron atom, the electrons interact with one another as well as with the nucleus. Electrons that are relatively close to the nucleus shield the nucleus from electrons further out. Thus, an electron in the outer periphery of the atom does not experience the full charge of the nucleus. The **effective nuclear charge** that a particular electron experiences depends on its average distance from the nucleus as compared with the other electrons in the same atom. This idea is illustrated in Figure 7.5. At

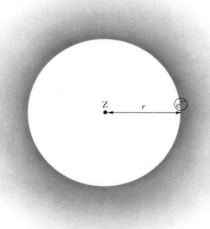

FIGURE 7.5 Shielding of the nuclear charge from an electron by other electrons within an atom. All electronic charge within the sphere of radius r shields the electron located at radius r from the nuclear charge. Electronic charge outside this radius has no effect.

some particular instant, the electron shown is at distance r from the nucleus. Assume that the averaged motion of all the other electrons results in a spherical electron distribution. All the electron density within the sphere of radius r shields the electron from the nucleus, canceling a certain amount of the nuclear charge. Of course each electron is moving about; as it moves closer to or farther away from the nucleus, the effective nuclear charge it experiences increases or decreases. However, we are interested in its *average* behavior. The average distance of the electron from the nucleus determines its energy. In this picture of a many-electron atom, we can retain the three quantum numbers, n, l, and m_l, which were employed to describe the hydrogenlike orbitals. Thus, we can continue to think of the electrons as occupying orbitals with designations such as $1s$, $2p$, $4d$, and so forth.

One result of the interactions between electrons in many-electron atoms is to cause orbitals with the same value for n but different values for l to differ in energy. To see the reason for this, consider the orbitals for which $n = 3$. The manner in which the probability function ψ^2 varies as we move outward from the nucleus differs for the $3s$, $3p$, and $3d$ orbitals. That is, the average distance from the nucleus for a $3s$ electron differs from that for a $3p$, which in turn differs from that for a $3d$. The $3s$ electron distribution extends closer to the nucleus than does the $3p$, and the $3p$ in turn extends closer than does the $3d$. As a result, the other electrons of the atom are less effective in shielding the $3s$ electrons from the nucleus. Thus the $3s$ electrons experience a larger effective nuclear charge than do the $3p$ electrons, and these in turn experience a larger effective charge than do the $3d$ electrons. The $3s$ electrons have thus a lower energy (that is, they are more stable) than the $3p$, which in turn are lower in energy than the $3d$. We find that the energy level diagram for the hydrogenlike orbitals (Figure 7.4) must be modified to that shown in Figure 7.6. This is a *qualitative* energy level diagram; the exact energies and their spacings differ from one atomic species to another. In all cases, however, the relative energies of the orbitals through $n = 3$ are as shown.

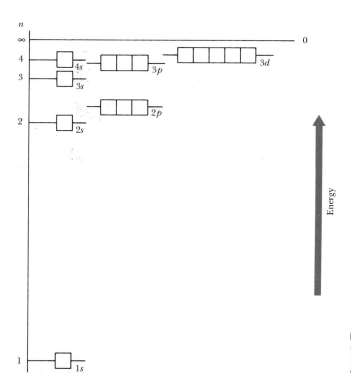

FIGURE 7.6 The ordering of orbital energy levels in many-electron atoms. Note that orbitals with the same value for the principal quantum number, n, but differing values of azimuthal quantum number, l, differ in energy.

SAMPLE EXERCISE 7.2

Based on the energy level diagram of Figure 7.6, would you expect the average distance from the nucleus of a $3d$ electron to be greater or less than that of a $2p$? Explain.

Solution: The energy of the $2p$ orbitals is considerably lower than for a $3d$. This indicates that the average attractive interaction of the $2p$ electron with the nucleus is much greater than for an electron in a $3d$ orbital. The increased attractive interaction is due to a smaller average distance of the $2p$ electron from the nucleus.

With this much background we now have an idea of the ordering of the energies of atomic orbitals in many-electron atoms. We must now consider what rules govern the placement of electrons into these orbitals. However, before the rules can be stated in their most useful form, we must learn about a fourth, and final, quantum number.

7.5
**electron spin
and the Pauli
exclusion
principle**

When the atomic spectra of atoms was first studied, certain complicating features of the spectra were noted. Lines that were at first thought to be single lines were found under high resolution to be closely spaced pairs. The only way in which these extra splittings could be accounted for was to introduce a new quantum number in addition to the three quantum numbers we have already encountered. This new quantum number is associated with the electron itself. The electron behaves as though it were spinning on its own axis and thereby acting like a tiny magnet. It was

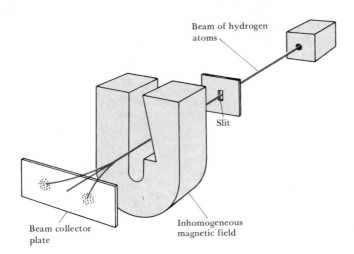

Beam of hydrogen
atoms

Slit

Beam collector
plate

Inhomogeneous
magnetic field

FIGURE 7.7 A diagrammatic illustration of the Stern-Gerlach experiment. A beam of hydrogen atoms is allowed to pass through an inhomogeneous magnetic field. Atoms in which the electron-spin quantum number, m_s is $+\frac{1}{2}$ are deflected in one direction, whereas those in which m_s is $-\frac{1}{2}$ are deflected in the other.

necessary to define an **electron-spin quantum number** (m_s), which has values of $+\frac{1}{2}$ or $-\frac{1}{2}$, corresponding to the two possible orientations of the electron spin in a magnetic field.

In 1921 Otto Stern and Walter Gerlach succeeded in actually separating a beam of atoms into two groups according to the orientation of the electron spin. Their experiment is diagramed in Figure 7.7. Let us assume that the beam of atoms is hydrogen. As the atoms pass into the region of strong magnetic field, the electron in each atom interacts with the magnetic field, causing the atom to be deflected from its straight-line path. The direction in which the atom is deflected depends on the orientation of the spin of the electron. We expect that there will be equal numbers of electrons with each of the two possible orientations, as is found to be the case. The presence of electron spin turns out to be important in determining the electronic structures of atoms. The first to recognize this was a German physicist, Wolfgang Pauli. In 1924 he spelled out what has become known as the **Pauli exclusion principle**. This principle declares that no two electrons in an atom can have the same set of four quantum numbers, n, l, m_l, and m_s. The exclusion principle provides the key to one of the great problems of chemistry—an explanation of the structure of the periodic table of the elements.

7.6

the electron configurations of the elements

We are now in a position to make use of the theoretical ideas we have been learning about in this and the previous chapter. Our goal is to understand how the electrons are arranged in each element of the periodic table. When you have mastered this material, you will be able to describe the arrangement of electrons in any element. This is an important skill, because the electron arrangements in atoms are the key to their chemical behavior.

The most stable, or ground, state of an atom will be that in which all the electrons are in the lowest possible energy states. If there were no restrictions on the possible values for the quantum numbers of the electrons, all the electrons would crowd into the $1s$ orbital, because this is lowest in energy. The Pauli principle, however, tells us that there are limits on the quantum numbers that the electrons may have. The $1s$ orbital corresponds to values $n = 1$, $l = 0$, and $m_l = 0$. The electron that occupies this orbital can have a spin quantum number (m_s) of $+\frac{1}{2}$ or $-\frac{1}{2}$. If one electron has quantum numbers $n = 1$, $l = 0$, $m_l = 0$, and $m_s = +\frac{1}{2}$, then a second electron can have $n = 1$, $l = 0$, $m_l = 0$, and $m_s = -\frac{1}{2}$. Thus, the exclusion principle requires that there can be at most *two electrons* in the $1s$ orbital, or for that matter in any *single atomic orbital*. In hydrogen there is one electron in the $1s$ orbital. In helium, atomic number 2, there are two electrons in this orbital. These electrons must have opposite values for the electron-spin quantum number (m_s). The electrons are said to be paired. Because of this pairing, the magnetic moments of the electrons effectively cancel. If the Stern-Gerlach experiment (Figure 7.7) were performed on a beam of helium atoms, no separation of the beam by the magnetic field would occur as it does with hydrogen.

A particular arrangement of electrons in the orbitals of an atom is referred to as an **electron configuration.** It is convenient to have a shorthand notation for representing the electron configuration. This is done by writing the symbol for each orbital occupied by an electron, with a superscript to indicate the number of electrons occupying that orbital. For hydrogen the electron configuration is $1s^1$; for helium it is $1s^2$. Another way of representing the electron configurations of these atoms is as follows:

$1s$

H $\boxed{\uparrow}$

He $\boxed{\uparrow\downarrow}$

Each orbital is represented by a box, and each electron by a half-arrow. A half-arrow pointing upward ($\uparrow$) represents an electron spinning in one direction ($m_s = +\frac{1}{2}$), whereas a downward half-arrow ($\downarrow$) represents an electron spinning in the opposite direction ($m_s = -\frac{1}{2}$). We shall refer to representations of this type as **orbital diagrams.**

The two electrons present in helium complete the filling of orbitals with principal quantum number $n = 1$. Helium therefore possesses a very stable electron configuration, as reflected in its chemical inertness. The electron configurations of lithium and of several elements that follow it in the periodic table are shown in Table 7.3. Recall that a maximum of two electrons can be placed in each orbital. Thus, for lithium, with three electrons, the third electron cannot enter the $1s$ orbital, but must be placed in the next most stable orbital, the $2s$. The change in principal quantum number for the third electron represents a large jump in energy, and a corresponding jump in the average distance of the electron from the nucleus. We may say that it represents the start of a new shell of electrons. As you can see by examining the periodic table, lithium represents the start of a new row of the periodic table. It is the first member of the alkali metals family (group 1A).

The element that follows lithium is beryllium; its electron configuration is $1s^22s^2$ (Table 7.3). Boron, atomic number 5, has an electron configuration $1s^22s^22p^1$. The fifth electron must be placed in a $2p$ orbital, because the $2s$ orbital is filled. Because each of the three $2p$ orbitals are of equal energy, it doesn't matter which $2p$ orbital is occupied. With the next element, carbon, we come to a new situation. We know that the sixth electron must go into a $2p$ orbital, where there is already one electron. However, does this new electron go into the $2p$ orbital that already has one electron, or into one of the others? This question is answered by **Hund's rule**, which states that electrons occupy equivalent orbitals singly to the maximum extent possible, and with their spins parallel. In the case of carbon, then, the sixth electron goes into one of the other $2p$ orbitals, and with its spin in the same orientation as the other $2p$ electron. Hund's rule is based on the fact that electrons repel one another, because they have the same electrical charge. By occupying different orbitals, the electrons remain as far as possible from one another in space, thus minimizing electron-electron repulsions. When electrons must occupy the same orbital, the repulsive interaction between the paired electrons is greater than between electrons in different, equivalent orbitals.

TABLE 7.3
electron configurations of several lighter elements

ELEMENT	ORBITAL DIAGRAM				ELECTRON CONFIGURATION
Li	↑↓	↑			$1s^22s^1$
Be	↑↓	↑↓			$1s^22s^2$
B	↑↓	↑↓	↑		$1s^22s^22p^1$
C	↑↓	↑↓	↑ ↑		$1s^22s^22p^2$
Ne	↑↓	↑↓	↑↓ ↑↓ ↑↓		$1s^22s^22p^6$
Na	↑↓	↑↓	↑↓ ↑↓ ↑↓	↑	$1s^22s^22p^63s^1$
	$1s$	$2s$	$2p$	$3s$	

Neon, the last member of the second period, has ten electrons. Two electrons fill the $1s$ orbital, two electrons fill the $2s$ orbital, and the remaining six electrons fill the $2p$ orbitals. The electron configuration is thus $1s^22s^22p^6$ (Table 7.3). In neon, all of the orbitals with $n = 2$ are filled. The filling of the $2s$ and $2p$ orbitals by the eight electrons that they can hold represents a very stable configuration. As a result, neon is chemically quite inert. Lewis and Langmuir, in their model for the electron configurations of elements, noted that the octet of electrons (eight) in the outermost shell of an atom or ion represents an especially stable arrangement.

Sodium, atomic number 11, marks the beginning of a new row of the periodic table. Sodium has a single $3s$ electron beyond the stable configuration of neon. We can abbreviate the electron configuration of sodium as follows:

Na   [Ne] $3s^1$

The symbol [Ne] represents the electron configuration of the ten electrons of neon, $1s^2 2s^2 2p^6$. Writing the electron configuration in this manner helps us focus attention on the electron arrangement at the periphery of the atom. The outer electrons are the ones largely responsible for the chemical behavior of an element. For example, we can write the electron configuration of lithium as

Li   [He] $2s^1$

By comparing this with the electron configuration for sodium, it is easy to appreciate why lithium and sodium are so similar chemically: They have the same type of outer electron configuration. All the members of the alkali metal family (group 1A) have a single s electron beyond an inner-core noble gas configuration.

SAMPLE EXERCISE 7.3

Draw the orbital diagram representation for the electron configuration of oxygen, atomic number 8.

Solution: The ordering of orbitals is as shown in Figure 7.6. Two electrons each go into the $1s$ and $2s$ orbitals. This leaves four electrons for the three $2p$ orbitals. Following Hund's rule, we put one electron into each $2p$ orbital until all three have one each. The fourth electron must then be paired up with one of the three electrons already in a $2p$ orbital, so that the correct representation is

SAMPLE EXERCISE 7.4

What is the characteristic electron configuration of the group 7A elements, the halogens?

Solution: The first member of the halogen family is fluorine, atomic number 9. (Even though hydrogen is listed in the periodic table in the same vertical row with the halogens, it is not counted as a halogen.) The abbreviated form of the electronic configuration for fluorine is

F   [He] $2s^2 2p^5$

Similarly, the abbreviated form of the electron configuration for chlorine, the second halogen, is

Cl   [Ne] $3s^2 3p^5$

From these two examples we see that the characteristic outer electron configuration of a halogen is $ns^2 np^5$, where n ranges from 2 in the case of fluorine to 5 in the case of iodine.

The rare gas element argon marks the end of the period started by sodium. The configuration for argon is $1s^2 2s^2 2p^6 3s^2 3p^6$. The element following argon in the periodic table is potassium (K), atomic number 19. In all its chemical properties, potassium is very obviously a member of the alkali metal family. The experimental facts about the properties of potassium leave no doubt that the outermost electron of this element occupies an *s* orbital. But this means that the highest energy electron has *not* gone into a 3*d* orbital, which we might naïvely have expected it to do. In this case the ordering of energy levels is such that the 4*s* orbital is lower in energy than the 3*d* (see Figure 7.6).

Following complete filling of the 4*s* orbital (this occurs in the calcium atom), the next set of equivalent orbitals to become filled is the 3*d*. (You'll find it helpful as we go along to refer often to the periodic table.) Beginning with scandium, and extending through zinc, electrons are added to the five 3*d* orbitals until they are completely filled. Thus the fourth row of the periodic table is ten elements wider than the previous rows because of the insertion of the elements known as the **transition metals.** Note the position of these ten elements in the periodic table (Figure 7.1).

In accordance with Hund's rule, electrons are added to the 3*d* orbitals singly until all five orbitals have one electron each. Additional electrons are then placed in the 3*d* orbitals with spin pairing until the shell is completely filled. The orbital diagram representations of two transition elements are as follows:

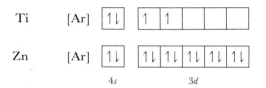

The 3*d* and 4*s* orbital energies are very close together. Occasionally, an electron may be moved from one of these types of orbital to another. For example, we might expect chromium to have the outer electron configuration $4s^2 3d^4$, but it is actually $4s^1 3d^5$. This happens because there is a special stability associated with precisely half-filled sets of equivalent orbitals. Apparently there is just enough gain in stability in arriving at this arrangement to cause the electron to move from a 4*s* to a 3*d* orbital.

Upon completion of the transition series, the 4*p* orbitals begin to be occupied, until the completed octet of outer electrons is again arrived at in krypton (Kr), atomic number 36. Krypton is another of the rare gases. Rubidium (Rb) marks the beginning of the fifth row of the periodic table. This row is in every respect like the previous one, except that the value for *n* is one greater. The sixth row of the table begins similarly to the previous one: one electron in the 6*s* orbital of cesium (Cs) and two electrons in the 6*s* orbital of barium (Ba). The next element, lanthanum (La), represents the start of the third series of transition elements. But with cerium (Ce), element 58, a new set of orbitals, the 4*f*, enter the picture. The energies of the 5*d* and 4*f* orbitals are very close. For lanthanum itself, the 5*d* orbital energy is just a little lower than the 4*f*. However, for the elements immediately following lanthanum, the 4*f* orbital energies are a little lower, so that the highest energy electrons go into the 4f orbitals.

There are seven equivalent 4f orbitals, corresponding to the seven allowed values of m_l ranging from 3 to -3. Thus it requires 14 electrons to completely fill the 4f orbitals. The 14 elements corresponding to the filling of the 4f orbitals are elements 58–71, known as the **rare earth,** or **lanthanide,** elements. In order not to make the periodic table unduly wide, the rare earth elements are set together below the other elements. The properties of the rare earth elements are all quite similar, and they occur together in nature. For many years it was virtually impossible to separate them from one another.

Following completion of the rare earth series, the third transition element series is completed, followed by filling of the 6p orbitals. This brings us to radon (Rn), heaviest of the rare gas elements. The final row of the periodic table begins as the one before it. The **actinide** elements involve completion of the 5f electron orbitals. This series consists mainly of elements not found in nature, but rather which have been synthesized in nuclear reactions.

USING THE PERIODIC TABLE TO WRITE ELECTRON CONFIGURATIONS

Our rather brief survey of electronic configurations of the elements has taken us through the entire periodic table. You will find that a familiarity with the general structure of the table will enable you to write down the electronic configuration of any element. There are a few instances in which minor shifts of an electron or two from one orbital to another occur, when orbitals have closely similar energies. We have given as one example the case of chromium, which possesses an outer electron configuration $4s^1 3d^5$, rather than the $4s^2 3d^4$ we might have expected. Another interesting case occurs with copper and its congeners,* silver and gold. In copper the configuration is found to be $4s^1 3d^{10}$. Evidently the stability associated with completing the d orbital level causes the electron to move from ns to $(n - 1)d$. There are a few other similar instances among the transition elements, lanthanides, and actinides. Although these minor departures from the expected are interesting, they are not of great chemical significance.

We've seen that the periodic table is structured so that elements with the same outer shell electron configuration are arranged vertically. The elements can be grouped also in terms of the *type* of orbital into which the electrons are placed. These different groupings are indicated by the shadings on the outline of the periodic table shown in Figure 7.8. The first two groups of elements on the left contain the **active metals,** with outermost s electrons. The **transition elements** are those for which the d orbitals are incomplete and being filled. The **representative elements** are those for which the outermost p orbitals are being filled. The **rare gas,** or noble gas, elements are those for which the octet of outermost electrons has been attained. The two series of elements for which the f orbitals are being filled are sometimes called the **inner transition elements.**

*A congener is an element in the same family or group of the periodic table as another.

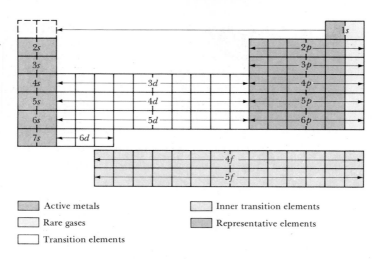

FIGURE 7.8 A block diagram of the periodic table showing the groupings of the elements according to the type of orbital being filled with electrons. The arrangement of elements in this figure is the same as in the periodic table (Figure 7.1).

SAMPLE EXERCISE 7.5

Write the electron configuration for the element bismuth, atomic number 83.

Solution: We can do this by simply moving across the periodic table one row at a time, and writing the occupancies of the orbitals corresponding to each row:

First row	$1s^2$
Second row	$2s^2 2p^6$
Third row	$3s^2 3p^6$
Fourth row	$4s^2 3d^{10} 4p^6$
Fifth row	$5s^2 4d^{10} 5p^6$
Sixth row	$6s^2 4f^{14} 5d^{10} 6p^3$

Total: $1s^2 2s^2 2p^6 3s^2 3p^6 3d^{10} 4s^2 4p^6 4d^{10} 4f^{14}$
$5s^2 5p^6 5d^{10} 6s^2 6p^3$

Note that 3 is the lowest possible value which n may have for a d orbital, and that 4 is the lowest possible value of n for an f orbital.

The total of the superscripted numbers should equal the atomic number of bismuth, 83. It does not matter a great deal precisely in which order the orbitals are listed. They may be listed, as shown above, in the order of increasing major quantum number. However, it is also possible to list them in the sequence read from the periodic table: $1s^2 2s^2 2p^6 3s^2 3p^6 4s^2 3d^{10} 4p^6 5s^2 4d^{10} 5p^6 6s^2 4f^{14} 5d^{10} 6p^3$. Writing only the outer electron configuration beyond the nearest rare gas we have [Xe] $6s^2 4f^{14} 5d^{10} 6p^3$.

The configurations of many of the heavier elements are not known for certain. The configurations must be deduced by analysis of atomic spectra. These methods are extremely complicated, especially for the heavier elements, in which the energy levels are closely spaced.

SAMPLE EXERCISE 7.6

Draw the orbital diagram representation for palladium, atomic number 46.

Solution: Palladium has ten electrons beyond the nearest rare gas configuration, characterized by

190

krypton, atomic number 36. Examining the periodic table we see that palladium is a transition element from the fifth row of the table. This means that it has in its valence shell $5s$ and $4d$ electrons. Two electrons occupy the $5s$ orbital; eight must be placed in the five $4d$ orbitals. These go in one at a time until five have been placed, then the remaining three are paired with electrons in three of the $4d$ orbitals. Thus we have:

Pd   [Kr] $\boxed{\uparrow\downarrow}$ $\boxed{\uparrow\downarrow\,|\,\uparrow\downarrow\,|\,\uparrow\downarrow\,|\,\uparrow\,|\,\uparrow}$

              $5s$ $4d$

COMPARISON OF A AND B GROUPS

In our initial survey of the periodic table we had noted that there are elements that carry the same group number but have differing letter designations. Let's examine the significance of this distinction in terms of electron configuration. As an example, locate the group 1A and 1B elements on the periodic table. In both families of group 1 elements the outer electron configuration involves a single s electron. In the case of the alkali metals (group 1A), this single s electron is outside a completed octet of electrons. In the group 1B elements, sometimes referred to as the *coinage metals,* the single electron is outside a completed set of 18 electrons. For example, in copper the single $4s$ electron is outside a completed $3s^23p^63d^{10}$ arrangement. The presence of the added ten d electrons, and of course the increase of ten in nuclear charge that goes along with them, has a profound effect on the chemical behavior of the single s electron. Thus, while the group 1A metals are extremely active chemically, the 1B metals are relatively unreactive. The situation is similar in the case of the group 2A and 2B elements, for which the outer electron configuration is ns^2.

When we compare the A and B subgroup elements with still higher group numbers the comparison becomes more tenuous. We use the convention that the B subgroup elements are those involving the transition metals. The group B elements are then those which have an electron configuration involving ns and $(n-1)d$ orbitals. The A subgroup elements, on the other hand, have electron configurations involving ns and np electrons. We expect that there will not be a great deal of similarity in the chemical properties of elements belonging to different subgroups, and indeed this is found to be the case. For example, chromium in group 6B does not have much resemblance to selenium in group 6A. However, there are some points of similarity, and the periodic table helps us to understand how they arise. Sample Exercise 7.7 provides an example.

SAMPLE EXERCISE 7.7

The elements of group 5B are capable of combining with up to five chlorines to form compounds with empirical formula MCl_5—for example, $NbCl_5$. The group 5A elements from the same rows of the periodic table are also capable of forming compounds of the empirical formula MCl_5—for example, $SbCl_5$.

Account for this similarity in behavior in terms of the electron configurations of the 5B and 5A elements.

Solution: The element niobium (Nb) has an electron configuration [Kr] $5s^24d^3$. The group 5A element in the same row, antimony (Sb), has electron configuration [Kr] $5s^2 4d^{10} 5p^3$. In forming the penta-chlorides, all five electrons above the inert gas structure in Nb are used. In Sb the five electrons used are the $5s$ and $5p$. The ten $4d$ electrons in Sb are not readily involved in chemical bonding; they behave as part of the inner core of electrons. The maximum number of electrons involved in bonding in either group 5B or 5A is thus five.

7.7
electron shells
in atoms

We have seen how the electron configuration of an atom may be built up by adding electrons to orbitals of successively higher energy, in accordance with the Pauli exclusion principle and Hund's rule. But as we build up the periodic table by increasing the nuclear charge and correspondingly increasing the number of electrons, what happens to the *total* spatial distribution of electronic charge about the nucleus? Very accurate calculations of the total electronic charge distribution in many-electron atoms can be made with the aid of large computers. These calculations show that the total electronic charge distribution does not fall off continuously away from the nucleus. Rather, it occurs in layers, corresponding to the notion of shells of electron density.

One way of illustrating these findings is shown in Figure 7.9. The quantity plotted on the vertical axis is called the radial electron density. It corresponds to the probability of the electron density being located at a particular distance out from the nucleus. The radial electron density as a function of distance from the nucleus is shown in Figure 7.9 for the first three rare gas elements.

Note that for helium there is only one maximum in the radial electron density, but there are two for neon and three for argon. Each of these maxima is due mainly to electrons in the atom which have the

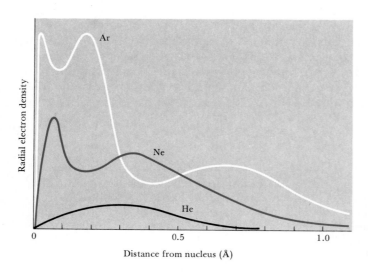

FIGURE 7.9 Radial electron density graphs for the first three rare gas elements, He, Ne, and Ar. The maxima that occur in the radial electron density correspond to electrons with the same value of principal quantum number n.

same value for the principal quantum number n. Thus, for helium the $1s$ electrons possess a maximum in radial electron density at about 0.3 Å. In argon, the maximum in the $1s$ radial electron density occurs at only 0.05 Å. The second maximum which occurs at larger radial distance is due to both the $2s$ and $2p$ electrons. The third maximum is due to $3s$ and $3p$ electrons.

The reason for the smaller radial distance of the orbital of the $1s$ electrons in the heavier atom is clear when we recall that the nuclear charge of helium is only 2, while that for argon is 18. The $1s$ electrons are the innermost electrons of the atom. The electrons of quantum number $n = 2$ and greater, present in elements beyond helium, therefore do not do much to shield the $1s$ electrons from the increasing nuclear charge. As a result, the size of the $1s$ orbital shrinks steadily as nuclear charge increases.

To summarize, the results of calculations show that in many-electron atoms the inner electrons are pulled with ever-increasing force into the region around the nucleus as the nuclear charge increases. The radial electron density shows maxima, corresponding to the traditional idea of shells of electrons, although these shells are diffuse and overlap considerably. As will be pointed out in the next section, the existence of these shells is reflected in the values observed for successive ionization potentials.

7.8
ionization
potentials

Recall from our discussion in the previous chapter that the ionization potential is the energy required to remove an electron from a gaseous atom or ion. The first ionization potential for an element is therefore that required for the process shown in Equation [7.3],

$$M(g) \longrightarrow M(g)^+ + e^- \qquad [7.3]$$

where M is a gaseous, neutral atom. The second ionization potential is then the energy for the removal of the second electron, Equation [7.4]:

$$M(g)^+ \longrightarrow M(g)^{2+} + e^- \qquad [7.4]$$

Successive ionization potentials are defined in a similar manner. The values of successive ionization potentials are known for many elements. Values for the elements sodium through argon are listed in Table 7.4.

As we might expect, each successive removal of an electron requires more energy. The reason for this is that the positive nuclear charge that provides the attractive force remains the same, whereas the number of electrons, which produce repulsive interaction, steadily decreases. For example, the electronic configuration for silicon (Si) is $1s^2 2s^2 2p^6 3s^2 3p^2$. If we look at the successive ionization potentials for silicon given in Table 7.4, we see a steady increase from 780 kJ/mole to 4350 kJ/mole for the four values of I that correspond to loss of the four electrons with principal quantum number $n = 3$. The fifth electron, however, requires considerably more energy for removal, 16,100 kJ/mole. This sharp increase in ionization potential reflects the fact that the fifth electron, which is in a $2p$ orbital, penetrates closer to the nucleus than do the $3s$ and $3p$ electrons. The $2p$ electron in silicon not only has a smaller average distance r from the nucleus, but it also experiences a larger effective nuclear charge Z because it penetrates the charge distribution of the other electrons.

The effect of change in the principal quantum number can be seen in other comparisons as well. For example, if we compare Mg and Al, we see that the energies required to remove first one and then two electrons from the two metals are not so very different. Yet the energy required to remove the third electron from Mg, 7730 kJ/mole, is much greater than the energy required to remove a third electron from Al, 2740 kJ/mole. Yet, the process is the same in both cases. That is,

$$M(g)^{2+} \longrightarrow M(g)^{3+} + e^- \qquad\qquad [7.5]$$

The difference in energies must therefore arise predominantly from the fact that a third electron removed from Mg is a $2p$, whereas the third electron removed from Al is a $3s$.

SAMPLE EXERCISE 7.8

As can be seen in Table 7.4, the energy required to remove an electron from P^{4+} is 6270 kJ/mole, as compared with 16,100 kJ/mole for removal of an electron from Si^{4+}. Account for the large difference.

Solution: The outer electron configuration of phosphorus is $3s^23p^3$. After removal of four of these electrons, the highest-energy electron remaining is a $3s$. The outer electron configuration of silicon is $3s^23p^2$. After removal of these four electrons the highest-energy electron remaining is a $2p$. It requires considerably less energy to remove the $3s$ electron, which lies largely outside the $1s^22s^22p^6$ core of electrons, than to remove an electron from the $2p$ level, as would be necessary for silicon.

TABLE 7.4

successive values of ionization potentials (I) for the elements sodium through argon (kJ/mole)[a]

ELEMENT	I_1	I_2	I_3	I_4	I_5	I_6	I_7
Na	490	4560					
Mg	735	1445	7730				
Al	580	1815	2740	11,600			
Si	780	1575	3220	4350	16,100		
P	1060	1890	2905	4950	6270	21,200	
S	1005	2260	3375	4565	6950	8490	27,000
Cl	1255	2295	3850	5160	6560	9360	11,000
Ar	1525	2665	3945	5770	7230	8780	12,000

[a]Although the ionization potentials are given here in units of kJ/mole, they are also often given in units of electron volts; 1 electron volt is equal to 96.49 kJ/mole.

These and similar ionization potential data thus support the notion that the electrons of a many-electron atom are arranged in shells or layers, in accordance with the postulates of the Lewis-Langmuir theory of the electronic structures of atoms. It also supports the idea that only the outermost electrons, those beyond the noble gas core, are involved in the sharing and transfer of electrons that gives rise to chemical change. The reason for this is that the inner electrons are too tightly bound to the nucleus to be lost from the atom or even shared with another atom.

PERIODIC TRENDS IN IONIZATION POTENTIALS

It is of interest to observe how the first ionization potentials vary with atomic number. Figure 7.10 shows a graph of I_1 versus atomic number. It is evident that there is an overall periodicity in this property. Overlooking for the moment the lesser displacements, there is a gradual increase in I_1 with atomic number for the elements in any one horizontal row. Thus the alkali metals show the lowest ionization potentials, and the rare gas elements the highest. For the elements of any one vertical group or family of elements, there is a gradual decrease in ionization potential with increasing atomic number. For example, it requires more energy to remove an electron from a lithium atom than from a potassium atom. A few simple considerations help to explain these observations. Proceeding along any one horizontal row of the table, the electrons which are added to counterbalance the increasing nuclear charge do not completely shield

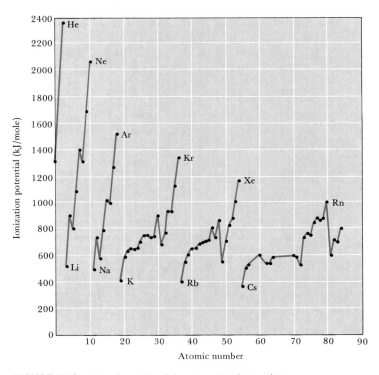

FIGURE 7.10   Ionization potential versus atomic number.

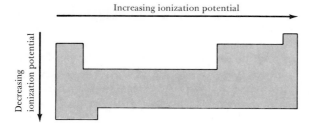

FIGURE 7.11 Variation of the first ionization potential for the elements in relation to the periodic table.

the outermost electrons from the nucleus. Thus the effective nuclear charge increases steadily. On the other hand, when we compare the elements of a vertical row, we are comparing elements with the same outer electron arrangements, but differing values for n, the principal quantum number. As n increases so also does the average distance of the electron from the nucleus. As its average distance from the nucleus increases, the electron becomes easier to remove. Therefore, if all other factors are the same, ionization potential decreases with increasing atomic radius. In terms of the periodic table, the ionization potentials vary in the manner shown in Figure 7.11.

One application of these considerations is the comparative behavior of the group 1A, 1B, 2A, and 2B elements discussed in Section 7.6. Table 7.5 lists the first ionization potentials for group 1 elements, and the first and second ionization potentials for the group 2 elements. Note that the B group metals possess generally higher ionization potentials. These higher potentials are the result mainly of a higher value for effective nuclear charge for the B group metals. This in turn is the result of incomplete shielding of the nucleus by the d electrons. The s electrons of the B group elements thus experience a larger effective nuclear charge.

**7.9
electron
affinities**

The electron affinity (E) is the energy change that occurs when an electron is added to a gaseous atom or ion. The process may be represented for a neutral atom as

$$M(g) + e^- \longrightarrow M^-(g) \tag{7.6}$$

For an ion of $1+$ charge, the equation is

$$M^+(g) + e^- \longrightarrow M(g) \tag{7.7}$$

For most neutral and for all positively charged species, energy is evolved when the electron is added; E is thus negative in sign. The process shown in Equation [7.7] is just the opposite of ionization of the neutral atom. The electron affinity thus bears a close relationship to the ionization potential; the electron affinity of a singly charged positive ion is just the negative of the ionization potential of the corresponding neutral atom. The electron affinities of neutral atoms are quite difficult to measure, and not very many values are available for the elements. Table 7.6 lists the

TABLE 7.5

comparative values of ionization potentials for group 1A, 1B, 2A, and 2B elements (kJ/mole)

1A	I_1	1B	I_1	2A	I_1	I_2	2B	I_1	I_2
K	418	Cu	859	Ca	589	1145	Zn	907	1733
Rb	403	Ag	730	Sr	549	1064	Cd	867	1630
Cs	375	Au	890	Ba	503	965	Hg	994	1805

values that are known with reasonable accuracy. A negative value for E corresponds to a process that occurs with evolution of energy. That is, the resulting negative ion is more stable than the separated atom and electron.

You will find it difficult to see much in the way of periodic trends in these values, but the comparative values among the halogens (F, Cl, Br, I) tell us some interesting things. The addition of a single electron to a halogen atom completes the octet of electrons around the nucleus and thus might be expected to lead to a fairly stable arrangement. But at the same time, addition of an electron increases the total repulsion energy of the electrons. There is thus a balance achieved between the electron-nuclear attraction that holds the added electron, and the increased electron-electron repulsions. As we proceed from fluorine to iodine the added electron is going into a p orbital of increasing major quantum number. The average distance of the electron from the nucleus steadily increases, and electron-nuclear attraction should thus steadily decrease. If this were all that is involved, fluorine would have the highest electron affinity. But the orbitals that hold the outermost electrons of the halogen are increasingly spread out as we proceed from fluorine to iodine. The electron-electron repulsions between these electrons and the added one therefore decrease with increasing atomic weight of the halogen. A lower electron-nuclear attraction is thus counterbalanced by lower electron-electron repulsion. The overall result is that the electron affinities differ very little among the halogens.

Note that addition of one electron to an oxygen atom results in evolution of energy, that is, it leads to a more stable species than the

TABLE 7.6

electron affinities of some elements

ELEMENT	ION FORMED	E(kJ/mole)
H	H^-	-72
F	F^-	-330
Cl	Cl^-	-350
Br	Br^-	-325
I	I^-	-295
O	O^-	-135
O	O^{2-}	$+710$
S	S^{2-}	$+375$

originally separated atom and electron. However, addition of a second electron to form O^{2-}, even though it results in a completed octet of electrons about the nucleus, requires energy; the sign of E is therefore positive. In this case, and for S^{2-} also, the electron-electron repulsions outweigh the electron-nuclear attractions.

7.10
atomic sizes

According to the quantum-mechanical model, an atom does not have a sharply defined boundary that determines its size. The electronic charge density in an atom simply drops off with increasing distance from the nucleus, approaching zero at large distance. However, there are techniques that make it possible to measure the distances between atoms in compounds. These can be used to develop a table of atomic radii, which should be fairly good approximate measures of the relative sizes of the atoms. For example, the distance between the centers of the Br atoms in Br_2 is 2.286 Å, making it possible to assign the Br atom a radius of 1.14 Å. Figure 7.12 graphs the atomic radii of the elements versus atomic number.

It is evident that atomic radii show periodic variation. In any one period the alkali metals are largest, and the halogens smallest. Furthermore, we find that among the elements of any one group or family of the periodic table, for example, among the alkali metals, the radius increases regularly with increasing atomic number.

We can understand these variations using the same line of reasoning employed in discussing the variations in ionization potential. Proceeding from left to right across a horizontal row of the periodic table, the effective nuclear charge experienced by the outermost electrons increases

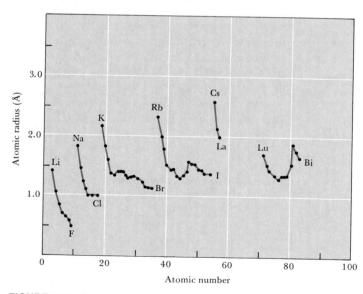

FIGURE 7.12 Atomic radii versus atomic number. The rare-gas elements are not included in this graph, because there is no simple way of relating their radii to those of the other elements on the basis of solid-state structure determinations. Gaps in the graph are due to lack of experimental data.

as a result of incomplete shielding. Thus, the orbital containing the electron is contracted. On the other hand, in any vertical row, orbital size increases with increasing value for the principal quantum number.

FOR REVIEW

summary

Our major concern in this chapter has been the periodic table of the elements. We saw that the table developed in an entirely empirical way. Chemists had begun to amass a great many facts about the elements, and there was a need for some basis on which to organize them. The elements in the periodic table are arranged so that those with similar chemical characteristics are grouped together in vertical columns called **groups** or **families**. Originally this was accomplished by aligning the elements in the order of increasing atomic weight.

When the arrangements of protons, neutrons, and electrons in atoms was understood, the periodic table took on a new significance. It then became clear that the fundamental property of an element in the periodic table is not atomic weight, but atomic number, the number of electrons surrounding the nucleus. If the chemical properties of elements are repeated as we proceed to higher atomic numbers, it must be because the repetition of some characteristic arrangement of the electrons gives rise to certain chemical properties. Lewis and Langmuir used this approach in developing the idea of shells of electrons within atoms. Special stability was associated with filling of a shell or a subshell.

The key to understanding the periodic table in terms of atomic orbitals is the **Pauli exclusion principle**, which places a limit of two on the number of electrons that may occupy any one atomic orbital. Armed with this rule, we can begin with the atomic orbitals obtained for hydrogen and hydrogenlike atoms (Chapter 6) and begin to put in electrons. Some allowance needs to be made for the added complexity of many-electron atoms. The most important change is that orbitals of the same principal quantum number (n) but differing azimuthal quantum number (l) have differing energies in a many-electron atom or ion. With the aid of the Pauli exclusion principle and knowing the order of atomic orbitals in terms of energy, it is then possible to build up the entire periodic table. The most significant point is that the periodic table can be explained in terms of the **electron configurations** of the atoms. Similarities of chemical properties arise because elements have similar arrangements of electrons in their outermost, incomplete shells. The Lewis and Langmuir idea that the electrons are arranged in layers or shells is correct.

Many properties of atoms that have chemical significance exhibit periodic character. Among the most important are the first ionization potential and atomic radii.

learning goals

Having read and studied this chapter, you should be able to:

1. Explain the basis on which the modern periodic table is constructed.
2. Define the term "group" or "family" in terms of electron configuration.
3. Describe the various blocks of elements in the periodic table in terms of the type of orbital being occupied by electrons.
4. Relate the historical development of the periodic table, with emphasis on the logical process employed by Mendeleev and Meyer in placing the elements in the table.

5. Explain the basis of the Lewis and Langmuir theory of electronic structures of atoms.

6. List the factors that determine the energy of an electron in a many-electron atom. You should be able to explain the fact that electrons with the same value of principal quantum number (n) but differing values of the azimuthal quantum number (l) possess different energies.

7. Explain the concept of effective nuclear charge as it relates to the energies of electrons in atoms.

8. State the significance of the Pauli exclusion principle and how it helps to explain the electronic structures of the elements.

9. Explain the orbital energy level diagram for many-electron atoms, as depicted in Figure 7.6.

10. Write the electron configuration for any element once you know its place in the periodic table.

11. Use the orbital diagram representation for electron configurations of atoms.

12. List the names for the blocks of elements that are based on a particular type of orbital being filled.

13. Explain the effect of increasing nuclear charge on the radial density function in many-electron atoms.

14. Account for the observed variations in first ionization potentials among the elements, as depicted in Figure 7.10. You should also be able to explain the observed changes in values of the successive ionization potentials for a given atom.

15. Explain the variation in atomic radii with atomic number. You should be able to relate this variation, as shown in Figure 7.12, to the variation in corresponding values of the first ionization potentials.

16. Explain the concept of electron affinity and its relationship to ionization potential.

key terms

Among the more important terms and definitions used for the first time in this chapter are the following:

The **active metals** (Section 7.6), groups 1A and 2A, are those in which electrons occupy only the s orbitals of the valence shell.

The **effective nuclear charge** (Section 7.4) is the charge at the nucleus experienced by an electron in a many-electron atom. This charge is not the full nuclear charge, because there is some shielding of the nuclear charge by other electrons in the atom. How much shielding occurs depends on the average distance of the electron from the nucleus, compared with the other electrons in the atom.

The **electron affinity** (Section 7.9) is the energy change that occurs when an electron is added to a gaseous atom or ion.

An **electron configuration** (Section 7.6) is a particular arrangement of electrons in the orbitals of an atom.

Electron spin (Section 7.5) is a property of the electron that makes it behave as though it were a tiny magnet. Associated with the electron spin is a **spin quantum number** (m_s), which may have values of $+\frac{1}{2}$ or $-\frac{1}{2}$.

Hund's rule (Section 7.6) states that electrons must occupy equivalent orbitals one at a time until all orbitals have at least one electron, before pairing of electrons in the orbitals occurs. Note carefully that the rule applies only to orbitals that are **degenerate**, which means that they have the same energy.

The **inner transition elements** (Section 7.6), or **lanthanides** and **actinides,** are those in which the $4f$ or $5f$ orbitals are partially occupied.

The **Pauli exclusion principle** (Section 7.5) states that no two electrons in an atom may have all four quantum numbers, n, l, m_l, and m_s, the same. As a consequence of this principle, there can be no more than two electrons in any one atomic orbital.

The **periodic law** (Section 7.1) in its modern version states that when the elements are arranged in the order of increasing atomic number, similar chemical and physical properties recur periodically.

The **representative elements** (Section 7.6) are those in which the p orbitals are partially occupied.

Transition elements (Section 7.6) are those in which the d orbitals are partially occupied.

EXERCISES

the periodic table and periodic properties

7.1 Using a handbook of chemistry or other reference source, make a table of the following properties of the group 2A elements: density, atomic weight, formula of chloride, formula of oxide, first ionization potential, second ionization potential, third ionization potential. Is there a basis in the data for classifying the elements in one group? Explain.

7.2 By examining the modern periodic table, find as many examples as you can of violations of Mendeleev's periodic law that the chemical and physical properties of the elements are periodic functions of their atomic *weights*. Why do these violations occur? State the modern version of the periodic law.

7.3 The formula for water is, of course, H_2O. Write the formulas one might expect for the corresponding hydrogen compounds of the other elements in group 6A.

7.4 Although potassium (K) has a lower atomic weight than argon (Ar), it is clear that K should follow Ar in the periodic table. What reasons can you give for why K and Ar occur in this order, rather than K first and then Ar?

energies of orbitals

7.5 If the ionization potential of H is 1312 kJ/mole, what is the ionization potential of Li^{2+}?

7.6 Consider the element scandium, atomic number 21. Make a drawing such as that shown in Figure 7.6 and place in the boxes all the electrons of the scandium atom. Which electron in scandium experiences the smallest effective nuclear charge? Which electrons experience the largest?

7.7 Which of the four quantum numbers n, l, m_l, and m_s is related to the energy of an electron in a hydrogen atom? Which are related to the energy of an electron in a many-electron atom?

7.8 Which orbital in each of the following sets is lower in energy in a many-electron atom? (a) $3p$, $5s$; (b) $2s$, $2p$; (c) $3d$, $3s$; (d) $3d$, $4s$; (e) $3d$, $4f$

7.9 How does the average distance from the nucleus of a $2s$ electron in a neon atom compare with that for a $2p$ electron? Explain.

electron spin: the Pauli principle

7.10 If you could do the Stern-Gerlach type of experiment shown in Figure 7.7 with the first four elements of the periodic table, which would give rise to a separation of beams? How would your results provide experimental evidence for the Pauli exclusion principle?

7.11 The Stern-Gerlach experiment illustrated in Figure 7.7 was first carried out using a beam of silver atoms. Draw the orbital diagram representation for the electron configuration of silver. How many unpaired electrons does this atom have?

7.12 How does the Pauli exclusion principle account for the existence of inert chemical behavior at atomic numbers 2, 10, 18, 36, and 54?

7.13 What is the total number of electrons in an atom that may have the following quantum numbers? (a) $n = 3$, $l = 1$; (b) $n = 4$, $l = 1$; (c) $n = 3$, $l = 2$, $m_l = -2$; (d) $n = 4$; (e) $n = 3$, $m_s = +\frac{1}{2}$

7.14 Make up a table for the element carbon that lists the quantum numbers for each electron in the atom.

electron configurations of the elements

7.15 What is the general electronic configuration that characterizes each of the following groups of elements? (For example, the alkali metals are characterized by ns^1.) (a) the rare gases; (b) group 1B; (c) group 5A

7.16 What changes in the electron configurations of each of the following elements are necessary for them to attain a rare gas configuration? (a) S; (b) Ca; (c) F; (d) H

7.17 The elements of group 2B commonly lose two electrons in chemical reaction, forming the 2+ ions, as for example, Zn^{2+}. Does this ion have a rare gas configuration? If not, how does it differ and why?

7.18 Identify the specific element or groups of elements that can have the following electron configurations: (a) $1s^2 2s^2 2p^6 3s^2 3p^5$; (b) outer electron configuration $ns^2(n-1)d^7$; (c) [Kr] $5s^2 4d^3$

7.19 Write out the complete electron configuration of each of the following elements: (a) B; (b) Ge; (c) Ru; (d) Rb. (For example, Li = $1s^2 2s^1$).

7.20 Indicate the number of unpaired electrons to be expected in each of the following atoms: (a) C; (b) V; (c) In, (d) Sm.

7.21 The number of unpaired electrons in the electron configuration of Cr is not what might be expected. What is the reason for this? Show an orbital diagram representation of the outer orbital electron configurations in this element.

7.22 Just as in the case of hydrogen, many-electron atoms may absorb energy with excitation of an electron to a higher energy orbital. Indicate which of the following electron configurations corresponds to an atom in its most stable (ground) state, and which to an atom in an excited state. (a) $1s^2 2s^1 2p^1$; (b) [Ar] $4s^2 3d^3 4p^1$; (c) [Ar] $4s^2 3d^{10} 4p^2$

7.23 When lithium atoms are excited they emit line spectra that are similar in many respects to those produced by excited hydrogen atoms. Which electron or electrons of the lithium atom do you suppose are responsible for the observed line spectra?

periodicity and atomic properties

7.24 The most common oxyacid of the element sulfur is sulfuric acid (H_2SO_4). Predict the formulas of readily available oxyacids of selenium and chromium. Which of these predictions would you feel most confident of? Explain.

7.25 How does atomic size vary within a particular family? Explain.

7.26 How does atomic size vary within a horizontal row of the periodic table? Explain how this variation arises.

7.27 Would you expect the rare gas element argon to show a negative electron affinity? Explain your answer.

7.28 The ionization potential for scandium (Sc) is 631 kJ/mole, and that for yttrium (Y) is 615 kJ/mole. In Sc an electron is lost from a $4s$ orbital, in Y from a $5s$. In which atom does the electron experience a higher effective nuclear charge? Explain your answer.

[7.29] Make a graph of the third ionization potentials (I_3) listed in Table 7.4, as a function of atomic number. Discuss the reason for the large value seen for Mg. Discuss also a possible reason for the minimum in I_3 at P.

7.30 How does ionization potential vary in the series of elements from potassium through krypton? Explain the observed variation in terms of effective nuclear charge.

7.31 How does ionization potential vary with vertical position among the group 5A elements? Explain the observed trend.

7.32 How is the change in radii of the atoms in the series from potassium through krypton related to the change in ionization potentials in the same series? Explain the reason for the observed trend.

7.33 The Mg^{2+} and Na^+ ions have the same number of electrons about the nucleus (ten). Which ion would you expect to have the smaller radius? Explain.

general exercises

7.34 It requires considerably more energy to ionize a $4s$ electron from zinc than from calcium. Explain.

7.35 The ionization potential of Sr is 546 kJ/mole, whereas that for Cd is 864 kJ/mole. Explain the reason for the higher value for Cd.

[7.36] The element technetium, atomic number 43, is not observed in nature because it happens to be radioactive. Assuming that it can be synthesized in a nuclear reactor in quantity, predict some of its characteristic properties. These should include electron configuration, density, melting point, and formulas of oxides and chlorides. Consult a handbook of chemistry for information on related elements.

7.37 The difference in atomic numbers between arsenic (As), an element of group 5A, and antimony (Sb), also an element of group 5A is 18. Where are these 18 electrons in the Sb atom?

7.38 Suppose that the universe were different than it is and that the electron spin quantum number could be +1, 0, or −1. Assuming that the Pauli principle still held, which atomic numbers would correspond to special chemical stability?

7.39 Notice in Figure 7.12 that the radius of lanthanum (La) is larger than that for lutetium (Lu). The atomic number of lutetium is 14 greater than that for lanthanum. The 14 electrons added to the atom in going from La to Lu occupy the $4f$ orbitals. Suggest a reason why Lu has a smaller radius than La.

7.40 What is the lowest value of principal quantum number n for which there can be a g subshell? How many electrons are required to fully occupy a g subshell? Are there any elements known in which electrons occupy the g subshell?

8 chemical bonding

In the previous two chapters you learned about the modern theory of atomic structure and how it accounts for the arrangements of electrons within the atom. This theory is quite successful in relating the configuration of electrons in an element to its position in the periodic table. However, the real test of a theory of atomic structure is that it helps us to understand the chemical properties of the elements. That is, there should be a relationship between the electron configurations of the atoms and the chemical reactions they undergo. In this chapter, we shall be concerned with the types of chemical substances formed by combinations of the elements. We shall also attempt to understand various ways that elements combine with one another in terms of the electronic structures of the atoms themselves.

As we examine the relationship between electron configuration and chemical properties, we shall find it useful to classify the chemical forces which hold atoms together in compounds into three broad groups: (1) ionic bonds, (2) covalent bonds, and (3) metallic bonds.

The term **ionic bond** refers to the electrostatic forces that exist between particles of opposite charge. As we shall see, ions may be formed from atoms by transfer of one or more electrons from one atom to another. Ionic substances generally result from the interaction of metals from the far left side of the periodic table with the nonmetallic elements from the far right side (excluding the rare gases, group 8A).

The **covalent bond** results from a sharing of electrons between two atoms. Covalent bonding is most often observed in the interactions of nonmetallic elements with one another.

Metallic bonds are found in solid metals such as copper, iron, and aluminum. In the metals, each metal atom is bonded to several neighboring atoms. The bonding electrons are relatively free to move throughout the three-dimensional structure. Metallic bonds give rise to such typical metallic properties as high electrical conductivity and luster.

The term **valence** is commonly used in discussions of both ionic and covalent bonding; it is a measure of the capacity of an element to form chemical bonds with other elements. The valence is a positive integer. In an ionic substance, the valence is the charge on a monatomic ion. For example, in MgO, the ions present are Mg^{2+} and O^{2-}. The valence of each element is thus 2. Originally, valence was also determined by the number of hydrogen atoms with which an element combined; the valence of hydrogen was set equal to 1. Thus, in H_2O, the valence of oxygen is 2. You will see when we have discussed covalent bonding in more detail that a modern and more general definition of valence in covalent substances would be expressed in terms of the number of electron pairs shared with other atoms.

8.1
ionic bonds

Deep within the earth below the city of Detroit and beneath the rolling plains of Kansas there lie enormous deposits of a white mineral, *halite*. This material, also known as sodium chloride (NaCl), familiar to everyone as common table salt, was deposited in these and other places millions of years ago, when extensive primordial seas dried upon the changing surface of the earth. Sodium chloride is the most abundant dissolved substance present in seawater and is found in human body tissues in large quantities. We find it before us on the dinner table every day. It is the most familiar example of an ionic compound.

The sodium and chlorine in solid NaCl are found to consist of a regular three-dimensional array, as shown in Figure 8.1. Many lines of experimental evidence lead to the conclusion that the sodium exists in this lattice as the positive ion Na^+ and chlorine as the negative ion Cl^-.

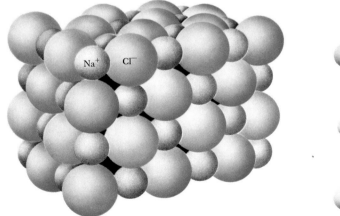
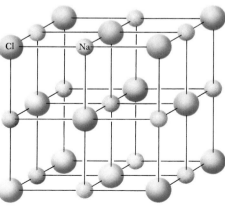

FIGURE 8.1 Two ways of representing the crystal structure of sodium chloride. The structure on the left shows the ions in their correct sizes relative to the distances between them. The larger spheres represent the chloride ions.

We might ask why sodium and chlorine react with one another to form this particular compound, and why it has this particular kind of ionic lattice. The driving force in all chemical reactions is energetic in character. Elements react with one another to form compounds when energy is released, and more stable arrangements result. The sodium chloride lattice has a great deal of stability because of the packing of the oppositely charged Na^+ and Cl^- ions together as shown in Figure 8.1. A measure of just how much stabilization results from this packing is given by the **lattice energy.** This quantity is the energy required for a mole of the solid ionic substance to be separated completely into ions far removed from one another. We can write the process as

$$NaCl(s) \longrightarrow Na^+(g) + Cl^-(g) \qquad [8.1]$$

To get a picture of this process, imagine that the lattice shown in Figure 8.1 expands from within, so that the spaces between the ions grow larger and larger, until the ions are very far apart. The energy that would be required for that to occur for a lattice containing 1 mole of Na^+ and 1 mole of Cl^- ions is the lattice energy.

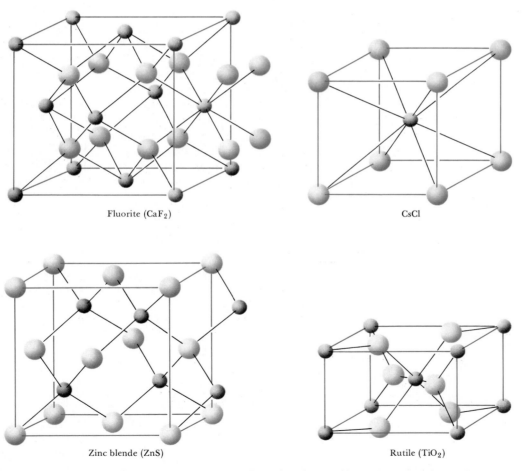

Fluorite (CaF_2)   CsCl

Zinc blende (ZnS)   Rutile (TiO_2)

FIGURE 8.2   Examples of several important types of crystal structure.

The lattice energy for NaCl(s) amounts to 785 kJ/mole. This is a very large amount of energy, and it accounts for the fact that sodium chloride is a stable, solid substance with a high melting point. We see from the structure of NaCl (Figure 8.1) that each sodium ion is surrounded by six nearest neighbor chloride ions of opposite charge. Similarly each chloride ion is surrounded by six sodium ions. This attractive force between each ion and its nearest neighbors of opposite charge provides much of the stabilizing lattice energy. Furthermore, each ion also experiences repulsive interactions with ions of like charge in the lattice and is attracted to other ions of opposite charge in addition to its nearest neighbors. The lattice energy is the result of all the electrostatic interactions, taken over the entire lattice. Ionic substances may have arrangements of ions that differ from that shown in Figure 8.1. The lattice arrangements for a few other common ionic substances are shown in Figure 8.2. In all these substances, the forces between ions lead to brittle, crystalline solids with high melting points, properties characteristic of ionic compounds.

Ionic substances are formed very readily when the elements that react together can attain rare-gas electron configurations by gain or loss of electrons. In the case of sodium chloride, for example, sodium ($1s^22s^22p^63s^1$) possesses one electron beyond the ten that mark the neon electron configuration ($1s^22s^22p^6$). Sodium loses this electron in reacting with chlorine to form NaCl. Chlorine, on the other hand, with an electron configuration of $1s^22s^22p^63s^23p^5$, is one electron short of the argon configuration ($1s^22s^22p^63s^23p^6$). It acquires this electron when it reacts with sodium to form NaCl.

Electron-dot formulas are a simple and useful way of showing the electron arrangements about atoms and the changes that occur on chemical bond formation. The electron-dot formula for an atom consists of a dot for each valence-shell electron, that is, for each electron that occupies an orbital outside the rare-gas configuration. For example, the electron-dot formulas for hydrogen, chlorine, oxygen, and sulfur are as follows:

$$\text{H}\cdot \qquad :\ddot{\text{Cl}}\cdot \qquad \cdot\ddot{\text{O}}\cdot \qquad \cdot\ddot{\text{S}}\cdot$$

Note that the electron-dot formulas for O and S are the same. This is so because the valence-shell electron configuration of the two group 6A elements is the same (ns^2np^4). Using electron-dot formulas we can represent the formation of ionic sodium chloride from sodium and chlorine as follows:

$$\text{Na} + \cdot\ddot{\text{Cl}}: \longrightarrow \text{Na}^+ + :\ddot{\text{Cl}}:^-$$

Removal of a second electron from sodium requires much more energy than removal of the first (Table 7.4). Sodium is therefore never found in an ionic substance as Na^{2+}. We find that the other elements of group 1A are also found only as the 1+ ions Li^+, K^+, Rb^+, or Cs^+ in ionic substances. Similarly, addition of a second electron to chloride ion to form a hypothetical Cl^{2-} is never observed, because there is no tendency for chloride ion to add another electron. The other group 7A elements (the halogens) are also found only as the $1-$ ions F^-, Br^-, or I^- in ionic substances.

Magnesium, an element of group 2A, also forms an ionic compound with chlorine, of composition $MgCl_2$. In this instance, the metal achieves a rare-gas configuration by loss of two electrons. As you can see by looking at Table 7.4, it requires more energy to remove the two electrons from magnesium than is required for removing just one from sodium. This energy is more than recovered, however, in the increased lattice energy of $MgCl_2$, which comes from the higher charge on the metal ion. Thus magnesium and the other group 2A metals do not form ionic compounds in which the metal has a $1+$ charge; there is no known MgCl. Similarly, the group 6A elements are found in ionic compounds as O^{2-}, S^{2-}, and so forth, in which the ion possesses a rare-gas configuration. In the formation of the ionic solid MgO, both magnesium and oxygen attain the rare-gas configuration, by transfer of two electrons:

$$Mg + :\ddot{O}: \longrightarrow Mg^{2+} + [:\ddot{O}:]^{2-}$$

SAMPLE EXERCISE 8.1

Which substance would have the higher lattice energy, NaF or MgO? Explain.

Solution: Magnesium oxide, MgO, would have the higher lattice energy. The electrostatic attraction between oppositely charged ions increases with the charge on the ion. For this reason, it would require a greater amount of energy to separate a mole of Mg^{2+} ions and a mole of O^{2-} ions to infinite separation, as compared with separating a mole of Na^+ and a mole of F^-.

SAMPLE EXERCISE 8.2

Predict the formula of the compound formed between aluminum (Al) and fluorine; between aluminum and oxygen.

Solution: Aluminum, with atomic number 13, has three electrons beyond the inert gas configuration. It might then be expected to lose three electrons to fluorine atoms. Each fluorine atom attains the rare-gas configuration by accepting one electron. The expected formula is thus AlF_3.

In forming a compound with oxygen, each aluminum is again capable of losing three electrons, each oxygen accepts two. The ratio of oxygen atoms to aluminum atoms in the final compound must therefore be $3:2$. We write the formula for aluminum oxide as Al_2O_3.

Ionic-bond theory correctly predicts the charges found on many simple ions, based on the notion that attainment of a rare-gas configuration leads to maximum stability. For some elements, however, the rule must be modified, and for others it is not applicable at all. For example, metals of group 1B (Cu, Ag, Au) are observed to occur often as the $1+$ ions (as in CuBr and AgCl). Silver possesses a $4d^{10}5s^1$ outer electron

TABLE 8.1

some transition-metal ions with their corresponding electron configurations

ION	OUTER ELECTRON CONFIGURATION
Cr^{3+}	$3d^3$
Mn^{2+}	$3d^5$
Fe^{2+}	$3d^6$
Fe^{3+}	$3d^5$
Co^{2+}	$3d^7$
Ni^{2+}	$3d^8$
Cu^+	$3d^{10}$
Cu^{2+}	$3d^9$
Mo^{2+}	$4d^4$
Re^{4+}	$5d^3$
Pd^{2+}	$4d^8$
Ag^+	$4d^{10}$
Au^+	$5d^{10}$
Au^{3+}	$5d^8$

configuration. In forming Ag^+ the $5s$ electron is lost. This leaves a completely filled shell of 18 electrons in the $n = 4$ level. Because it is a completed shell, it is somewhat like a rare-gas arrangement. Similarly, the group 2B elements most commonly are seen as the 2+ ions (Zn^{2+}, Cd^{2+}, Hg^{2+}) in ionic compounds. The valence-shell s electrons are lost in forming the ions, leaving an electronic arrangement consisting of 18 electrons in the highest occupied level.

For most of the transition metals, the attainment of a rare-gas configuration by loss of electrons is not feasible; that would require the loss of too many electrons. The outer electron configurations of these elements are either $(n - 1)d^x ns^2$ or $(n - 1)d^x ns^1$, where n is 4, 5, or 6, and x may vary from 1 to 10. In forming ions the transition metals lose the valence-shell s electrons first, then as many d electrons as are required to form an ion of particular charge. Most of the transition metals are found in more than one charge state. For example, the element chromium is found in compounds as Cr^{2+} or Cr^{3+}. There are no simple rules to tell which charge state of a transition-metal ion will exist in a particular case. Table 8.1 lists several of the more common ions of transition metals with their outer electronic structures.

The term **cation** is used to describe a positive ion, whether it is formed from a single atom or is polyatomic. Similarly, the term **anion** refers to any negatively charged ion. In polyatomic ions, two or more atoms are bound together by predominantly covalent bonds. The group of atoms as a whole then acts as a charged species in forming an ionic compound with an ion of opposite charge. Examples of polyatomic cations are the vanadyl ion, VO^{2+}, and the familiar ammonium ion, NH_4^+. However, most polyatomic ions are negatively charged. Table 2.5 lists the names and formulas of several frequently encountered polyatomic anions.

8.2

sizes of ions

The radii of ions are interesting because they help us understand better how the electrons in ions are attracted to the central nuclear charge. They are also important in many practical ways. For example, the sizes of ions are important in determining the lattice energy in an ionic solid. They also determine how easily the ions can be removed from water in water-softening devices. As another example, many metal ions are important in biological reactions. Biological systems are often very specific; they work well with one metal ion but not with another, even though it has the same charge and seems very similar. It often happens that only a small difference in ionic size is sufficient to cause one metal ion to be biologically active and another not to be so.

When ions are drawn close together, electron-electron repulsions eventually become as large as the attractive forces. The repulsive forces place a limit on the distance of closest approach. We may thus think of ions as having characteristic radii that determine their distances from other ions. From X-ray diffraction studies of ionic solids, which will be described in Chapter 11, it is possible to determine the distances between ions. Using data for a large number of structures, these distances have been analyzed to obtain a set of ionic radii.

The radii for several ions with rare-gas configurations are listed in Table 8.2. As we might expect, the radii of cations are smaller than the radii of the corresponding neutral atoms from which they are derived, as illustrated in Figure 8.3. For example, the radius of the potassium atom is 2.2 Å, whereas the radius of K^+ is 1.33 Å. Positive ions are formed by removing one or more electrons from the outermost region of the atom. Thus the most spatially extended orbitals are vacated. In addition, removal of an electron decreases the total electron-electron repulsions. On the other hand, the radii of anions are larger than those of the corresponding neutral atoms. For example, the atomic radius of Br^- is 1.96 Å, whereas for bromine atom it is 1.15 Å. Electrons added to atoms to form negative ions with the rare-gas configuration go into p orbitals that are already partially filled. The increased electron-electron repulsions caused by the additions result in a larger average distance from the nucleus. The electrons spread out in space to minimize their interactions with one another.

TABLE 8.2

radii (in Å) of ions with rare-gas electron configurations

GROUP 1A		GROUP 2A		GROUP 3A		GROUP 6A		GROUP 7A	
Li^+	0.68	Be^{2+}	0.30			O^{2-}	1.45	F^-	1.33
Na^+	0.98	Mg^{2+}	0.65			S^{2-}	1.90	Cl^-	1.81
K^+	1.33	Ca^{2+}	0.94	Al^{3+}	0.45	Se^{2-}	2.02	Br^-	1.96
Rb^+	1.48	Sr^{2+}	1.10	Sc^{3+}	0.68	Te^{2-}	2.22	I^-	2.19
Cs^+	1.67	Ba^{2+}	1.31	Y^{3+}	0.90				

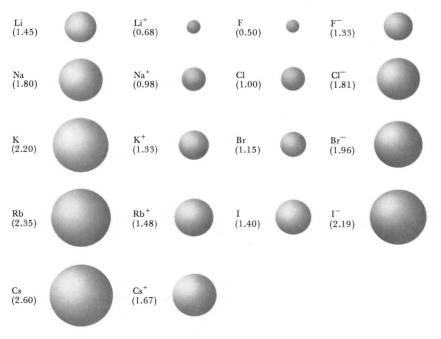

FIGURE 8.3 The relative sizes of atoms and ions.

SAMPLE EXERCISE 8.3

Based on the data given in Figure 8.3, how would you compare the effective nuclear charge experienced by a $4p$ electron in Br^-, as compared with a $4p$ electron in Br? Explain.

Solution: The effective nuclear charge experienced by a $4p$ electron in Br^- is smaller than in Br. The *actual* nuclear charge is the same in both cases. The only difference is that an extra electron has been added to Br to form Br^-. This extra electron, which goes into the vacant $4p$ orbital of Br, acts to some extent to shield the nucleus from the other $4p$ electrons. Thus, they experience a smaller effective nuclear charge. This effect, along with the increased electron-electron repulsions, causes Br^- to have a larger radius than Br.

The effects we have just described are also seen in the variation in radius in an **isoelectronic series** of ions. The term isoelectronic means that the ions possess the same number and arrangement of electrons. For example, in the series O^{2-}, F^-, Na^+, Mg^{2+}, and Al^{3+}, there are ten electrons arranged in the neon electron configuration about each nucleus. The nuclear charge in this series increases steadily in the order listed. With the number of electrons remaining constant, the radius of the ion decreases as the nuclear charge increases, attracting the electrons more strongly toward the nucleus:

		Increasing nuclear charge $\longrightarrow$		
O^{2-}	F^-	Na^+	Mg^{2+}	Al^{3+}
1.45	1.33	0.98	0.65	0.45

We saw in Section 7.10 that as the principal quantum number of an orbital increases, the average distance of the electron from the nucleus increases also. The relative radial extensions of the $1s$, $2s$, and $3s$ orbitals, shown in Figure 6.15, provide a good example. Thus, for ions of the same charge in any one family, the ionic radius increases with increasing period, that is, as the principal quantum number of the outermost occupied orbital increases.

One further comparison worth keeping in mind is the relative sizes of ions from the A and B subgroups. You may recall from the discussion in Section 7.8 that the outermost s electron of the group 1B elements experiences a higher effective nuclear charge. This happens because the ten d electrons added in going from a group 1A element to a group 1B element in the same row (for example, in going from K to Cu) do not completely shield the valence s electron from the nucleus. They also do not completely shield one another from the nucleus. As a result, not only is the atom of the group 1B element smaller, the ion formed by removal of the valence s electron is smaller also. For example, the radius of Cu^+, 0.96 Å, is less than that for the corresponding group 1A ion, K^+, radius 1.33 Å.

SAMPLE EXERCISES 8.4

The radius of Zn^{2+} is 0.74 Å, as compared with 0.99 Å for Ca^{2+}. Account for the smaller radius of Zn^{2+}.

Solution: Zinc occurs in the same horizontal row of the periodic table as Ca, but it has ten additional electrons, and a nuclear charge ten larger than for

Ca. As a result of incomplete shielding of the added nuclear charge by the added $3d$ electrons, the effective nuclear charge experienced by all the electrons in the $n = 3$ shell in Zn^{2+} is considerably greater than in Ca^{2+}. As a result, the radius of Zn^{2+} is smaller.

8.3
covalent
bonding

We have seen that ionic substances possess several characteristic properties. They are usually brittle substances with a high melting point. They are usually also crystalline, meaning that the solids have flat surfaces that make characteristic angles with one another. Ionic crystals can often be cleaved; that is, they break apart along smooth, flat surfaces. The characteristics of ionic substances result from the ionic forces that maintain the ions in a rigid, well-defined, three-dimensional arrangement such as one of those illustrated in Figures 8.1 and 8.2.

The vast majority of chemical substances do not have the characteristics of ionic materials; we need only think of water, gasoline, banana peelings, hair, antifreeze, and plastic bags as examples. Most of the substances with which we come in daily contact tend to be gases, liquids, or solids with low melting points; many vaporize readily—for example, mothball crystals. Many in their solid forms are plastic rather than rigidly crystalline—for example, paraffin or plastic bags.

For the very large class of substances that do not behave like ionic substances, a different model for the bonding between atoms is required. G. N. Lewis and Irving Langmuir developed a model for bonding in such molecules that involves the idea of bonds formed by shared electron pairs. They reasoned that an atom might acquire a rare-gas electron configuration by *sharing* electrons with other atoms. They used the term **covalence** to describe the sharing of electron pairs between atoms. A chemical bond formed by the sharing of electron pairs is called a covalent bond.

The hydrogen molecule, H_2, furnishes the simplest possible example of a covalent bond. Using electron-dot formulas, formation of the H_2 molecule by combination of two hydrogen atoms can be represented as

$$H\cdot \; + \; \cdot H \longrightarrow H\!:\!H \qquad\qquad [8.2]$$

The shared pair of electrons provide each hydrogen atom with two electrons in its valence shell (the $1s$) orbital, so that in a sense it has the electron configuration of the rare gas helium. Similarly, when two chlorine atoms combine to form the Cl_2 molecule,

$$:\!\ddot{C}l\cdot \; + \; \cdot \ddot{C}l\!: \longrightarrow :\!\ddot{C}l \; : \; \ddot{C}l\!: \qquad\qquad [8.3]$$

each chlorine atom, by sharing in the bonding electron pair, acquires eight electrons (an octet) in its valence shell, and thus achieves the rare-gas electron configuration of argon. The structures shown above for H_2 and Cl_2 are called **Lewis structures.** In writing Lewis structures, it is the usual practice to show each electron pair shared between atoms as a line, and the unshared electron pairs as pairs of dots. Thus the Lewis structures for Cl_2 and H_2O are shown as follows:

$$: \overset{..}{\underset{..}{Cl}} - \overset{..}{\underset{..}{Cl}} : \qquad H \overset{\displaystyle \overset{..}{\underset{}{O}}:}{\diagup \quad \diagdown} H$$

Note that oxygen acquires the rare-gas arrangement of an octet of electrons by sharing a pair of electrons with each of two hydrogens.

In the Lewis model the valence (or covalence) of an element is associated with the number of electron pairs shared to complete the octet of electrons. The model was especially successful in accounting for the compositions of compounds of the nonmetals, in which covalent bonding predominates. For example, consider the simple hydrides of the non-metals shown in Table 8.3. We have already accounted for the composi-

TABLE 8.3

simple hydrides of the nonmetallic elements

GROUP 4A	GROUP 5A	GROUP 6A	GROUP 7A
CH_4	NH_3	OH_2	FH
SiH_4	PH_3	SH_2	ClH
GeH_4	AsH_3	SeH_2	BrH
SnH_4	SbH_3	TeH_2	IH
PbH_4	BiH_3		

tion of H_2O. The same line of reasoning also accounts for the compositions of the other hydrides of the second row (period) of the periodic table. The Lewis structures for these compounds are as follows:

$$
\begin{array}{cccc}
\overset{\displaystyle H}{\underset{\displaystyle H}{H-\overset{|}{\underset{|}{C}}-H}} &
H-\overset{..}{\underset{\displaystyle H}{\underset{|}{N}}}-H &
:\overset{..}{\underset{\displaystyle H}{\underset{|}{O}}}-H &
:\overset{..}{\underset{..}{F}}-H
\end{array}
$$

The Lewis structures for the corresponding hydrides for elements of the following periods would be precisely the same, with only a change in the symbol for the central element. This is so because in the Lewis structure we represent only the valence-shell electrons.

SAMPLE EXERCISE 8.5

Using the Lewis-Langmuir theory and the theory of ionic bonding described earlier, explain the formulas of the following hydrides: NaH; MgH_2; AlH_3; SiH_4; PH_3; SH_2; HCl.

Solution: The hydrides of the metallic elements are ionic compounds consisting of metallic cations and the hydride ion, H^-. These ionic substances are formed by transfer of one or more electrons from the metal to hydrogen atoms. Each hydrogen atom accepts one electron to form H^-. One hydride ion is required for each electron removed from the metal to form the rare-gas configuration. The first three compounds thus correspond to the compositions Na^+H^-; $Mg^{2+}2H^-$; $Al^{3+}3H^-$. The remaining compounds are best formulated as covalent, in which an electron pair is shared between the central atom and each hydrogen. Si, an element of group 4A, requires four electrons to attain the rare-gas configuration of eight valence-shell electrons. Phosphorus requires three, sulfur two, and chlorine one. The formulas of the hydrides are in accord with the number of electrons needed.

STRUCTURES OF THE NONMETALLIC ELEMENTS

The test of a bonding model such as the Lewis-Langmuir theory comes in comparisons of experimental data with the predictions of the model. In examining covalently bonded structures, the data of principal importance are the molecular weight and the structure. The structure of the molecule involves the particular locations of atoms with respect to one another and tells us which atoms are connected to which by chemical bonds. We also obtain information about the distances between atoms. As we shall see, these give us an idea of the number of electron pairs bonding two atoms together. If we imagine a line drawn through two atoms that are bonded together, the angle made by two such lines at a particular atom is called a **bond angle.** These geometrical features, bond distances and bond angles, provide important clues about how valence electrons are arranged in molecules.

We shall not concern ourselves here with how chemists go about determining the structures of molecules. Let us simply make use of the information available to compare the known structures of the nonmetallic elements with the predictions of the Lewis-Langmuir model.

Consider first the group 7A elements, the halogens, consisting of fluorine, chlorine, bromine, iodine, and astatine (not a stable element). Both fluorine and chlorine exist as greenish-yellow gases. By measuring the densities of the gases under known conditions of pressure and temperature, it has been determined that both substances have the formula X_2, that is, two atoms per molecule. Bromine is a dark red liquid that boils at 58°C, whereas iodine is a black crystalline solid that very easily evaporates. In the vapor state, bromine and iodine also have molecular weights that correspond to two atoms of the element per molecule. Because each halogen atom possesses seven electrons in its valence shell, the sharing of one electron with another halogen atom provides for completion of the octet about each atom. The Lewis structure for the halogen diatomic molecules is thus

$$: \overset{..}{\underset{..}{X}} - \overset{..}{\underset{..}{X}} :$$

where X can be any one of the four halogens, F, Cl, Br, or I.

In the group 6A elements, there are six electrons in the valence shell of each atom. The Lewis theory predicts that each atom will share two electrons, but it does not predict whether this will occur by sharing with just one other atom or with two others. Oxygen exists in nature as the diatomic molecule O_2. The Lewis theory predicts the structure

$$: \overset{..}{O} = \overset{..}{O} :$$

As we shall see later this is *not* the correct electronic structure for O_2. It is one of the few cases in which the Lewis theory gives rise to an incorrect prediction. We must defer a discussion of the correct structure for O_2 until Chapter 9.

The element sulfur exists in several different forms. The different forms of an element are referred to as **allotropes.** The most common allotrope of sulfur at room temperature consists of an eight-membered ring of sulfur atoms, as shown in Figure 8.4. In this structure each sulfur

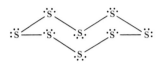

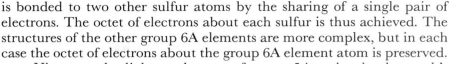

is bonded to two other sulfur atoms by the sharing of a single pair of electrons. The octet of electrons about each sulfur is thus achieved. The structures of the other group 6A elements are more complex, but in each case the octet of electrons about the group 6A element atom is preserved.

Nitrogen, the lightest element of group 5A, exists in the earth's atmosphere as the very stable N_2 molecule. According to the Lewis theory, because each nitrogen atom possesses five electrons in its valence shell, the sharing of three electron pairs is required to achieve the octet configuration:

$$:N\equiv N:$$

FIGURE 8.4 The structure of S_8 molecules as found in the most common allotropic form of sulfur at room temperature.

The bond between the nitrogen atoms is an example of a **multiple bond,** that is, a bond formed by the sharing of two or more pairs of electrons between the same two atoms.* The Lewis structure for N_2 is in complete accord with all the known properties for this molecule.

Nitrogen gas (N_2) is a nonpolar diatomic molecule with exceptionally low chemical reactivity. The low chemical reactivity results from the fact that the nitrogen-nitrogen bond energy is very high (941 kJ/mole). The study of the structure of N_2 reveals that the nitrogen atoms are separated by only 1.10 Å. The short N—N bond distance is a result of the sharing of more than one electron pair between a pair of atoms. From structure studies of many different compounds in which nitrogen atoms share one or two electron pairs, it has been learned that the average distance between bonded nitrogen atoms varies with the number of shared electron pairs:

N—N	N=N	N≡N
1.47 Å	1.24 Å	1.10 Å

As a general rule, the distance between bonded atoms decreases as the number of shared electron pairs increases.

In contrast to nitrogen, its congener phosphorus does not exist as a diatomic molecule, except at high temperatures in the vapor state. When it condenses from the vapor state it forms so-called white phosphorus, which consists of P_4 molecules. These molecules consist of a tetrahedral arrangement of phosphorus atoms, as shown in Figure 8.5. The Lewis structure for this molecule shows that each phosphorus is bonded to three other phosphorus atoms by single, shared electron pair bonds. In addition, each phosphorus atom possesses an unshared pair of electrons. The geometry of the P_4 molecule requires that the bond angle between each P—P bond be only 60°. We expect that shared electron pairs are concentrated in the region between the atoms. The structure for P_4 shown in Figure 8.5 suggests that the bonding electron pairs in this molecule will be crowded rather closely together. In fact, the P_4 molecule is quite reactive, probably because of the strain introduced by electron-electron repulsions.

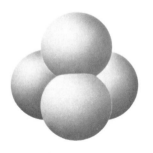

FIGURE 8.5 The structure of P_4 molecules as found in white phosphorus, a common allotropic form of the element.

*Bonds formed by the sharing of a single pair of electrons are called *single bonds*. Multiple bonds are called *double* or *triple bonds,* depending on whether two or three pairs of electrons are shared between atoms; N_2 has a triple bond.

The heavier elements of group 5A, arsenic, antimony, and bismuth, show increasingly metallic properties with increasing period. The allotropy of arsenic is somewhat similar to that of phosphorus, but for antimony and bismuth the structures of the elements are more complex and possess many of the characteristics of metals. These elements are frequently referred to as metalloids, to indicate that they are on the borderline between metals and nonmetals in their properties. The group 5A elements exemplify the rule that *in any one group, the metallic characteristics of the elements increase with increasing atomic number.* In the periodic table shown in the inside front cover and in Figure 7.1, the heavy line that runs diagonally from just below B to Po represents the approximate dividing line between metallic and nonmetallic elements.

Carbon, the lightest of the group 4A elements, exists in two major allotropic forms, graphite and diamond. In graphite, each carbon is bonded to three other carbon atoms in a plane. (We shall discuss the structure of graphite in Chapter 9.) The structure of diamond is shown in Figure 8.6. You can see that each carbon atom is surrounded by four other carbon atoms arranged at the corners of a tetrahedron. Each carbon shares a pair of electrons with each of the surrounding four carbon atoms, thus completing the octet of electrons about each atom. The stability of the diamond lattice comes from the fact that the carbon atoms are interconnected in a three-dimensional array of strong carbon-carbon single bonds. Diamond is a very hard and brittle material. In fact, industrial grade diamonds are employed in the blades of saws for the most demanding cutting jobs. The melting point of diamond, above 3500°C, is higher than that for any other element. It is exceptionally inert chemically. However, when heated to about 1800°C in the absence of air, diamond converts to graphite, the other allotropic form of the element. Diamonds also burn at about 900°C when heated in air or in oxygen. (Naturally, there is not much interest in carrying out experiments of this kind.)

Diamond

FIGURE 8.6 Structures of diamond, a major allotropic form of carbon.

The structures of elemental silicon and germanium are the same as for diamond. No graphitelike allotrope of these elements is known. Tin exists in two allotropic forms. White tin is the more metallic in character; this is the more stable form at room temperature and above. However, at lower temperatures, the element converts to a grey form that has the diamond structure. The conversion of tin from the metallic form to a grey powder was a great nuisance when tin was used for construction in cold climates. This phenomenon was referred to as "tin disease." Tin is another example of an element on the borderline between metallic and nonmetallic, with perhaps more metallic character in its properties. Lead, the heaviest element of group 4A, is quite markedly metallic in character.

8.4
drawing Lewis
structures

We have seen in the foregoing section some simple examples of Lewis structures. It is actually fairly easy to draw the Lewis structures for most compounds and ions formed from nonmetallic elements. It is a good idea to follow a regularly established procedure:

1. First write the symbols for the atoms involved so as to show which atoms are connected to which. This is often very easy; for example, in CH_4, we know that the hydrogens must be connected to carbon. But when the formula is H_3PO_3, or N_2O, or $H_4C_2O_2$, it isn't so immediately obvious which atoms are bonded to which. In these cases you must have more information before you can make an unambiguous choice. You may be asked, however, to work out various possibilities and attempt to judge which of these is the most likely.

2. Add up the number of valence-shell electrons. (Use the periodic table.) If the species involved is neutral, this amounts to just the sum of the valence electrons for all the atoms present. If the ion is negatively charged, add the number of charges; if it is positively charged, subtract the charge of the ion.

3. Draw in a single bond between each pair of atoms to be connected by bonding. Then attempt to put unshared electron pairs on each atom to meet the requirements of the octet rule. If placing all the electrons in this manner uses up all the available electrons, the structure is complete. If, on the other hand, the octet rule cannot be satisfied for each atom in this manner, then double or triple bonds must be formed with a reduction in the number of unshared electron pairs.

The rules are best explained by a few sample exercises:

SAMPLE EXERCISE 8.6

Draw the Lewis structure for phosphorus trichloride, PCl_3.

Solution: The valence-shell electron-dot formula for the two elements involved in this structure are:

$$:\overset{\cdot\cdot}{\underset{\cdot\cdot}{Cl}}\cdot \qquad \cdot\overset{\cdot\cdot}{P}\cdot$$

We see that chlorine requires one electron to complete its valence-shell octet of electrons, whereas phosphorus needs three. If each chlorine shares an electron pair with phosphorus, then the bonding of three chlorines to a central phosphorus atom will result in an octet of electrons about both chlorine and phosphorus:

$$:\overset{\cdot\cdot}{\underset{\cdot\cdot}{Cl}}-\overset{\cdot\cdot}{\underset{|}{P}}-\overset{\cdot\cdot}{\underset{\cdot\cdot}{Cl}}:$$
$$:\overset{}{\underset{\cdot\cdot}{Cl}}:$$

Alternatively, we might have proceeded using the rules above. We note that the total number of valence shell electrons to be placed is $3 \times 7 + 5 = 26$. We put in three single bonds between each Cl and P, which requires 6 electrons. The remaining ten electron pairs are then placed three on each chlorine atom and one on the phosphorus atom to give the Lewis structure shown above. Because there is an octet of electrons about each atom, the structure is reasonable.

SAMPLE EXERCISE 8.7

Draw the Lewis structure for phosphorus oxytrichloride, $POCl_3$.

Solution: The valence-shell electron arrangements for the three elements involved in this structure are

$$:\overset{\cdot\cdot}{\underset{\cdot\cdot}{Cl}}\cdot \qquad \cdot\overset{\cdot\cdot}{P}\cdot \qquad \cdot\overset{\cdot\cdot}{O}\cdot$$

The structure of $POCl_3$ involves the arrangement just completed for PCl_3, but with addition of an oxygen atom. Experimental studies of the molecule show that the oxygen atom is bonded to phosphorus. We note that oxygen has six electrons in its valence shell; it thus requires two more to achieve the octet arrangement. On the other hand, phosphorus already has a completed octet in PCl_3. We can imagine that bonding of oxygen to phosphorus occurs by "donation" of the unshared pair of electrons on phosphorus to the oxygen atom:*

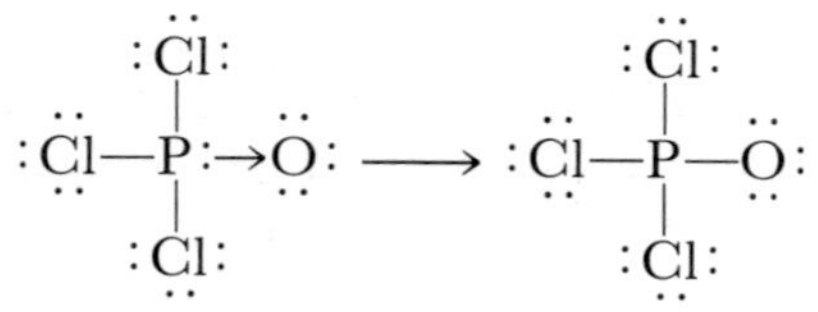

*A bond of this kind, in which one of the two atoms involved has been the source of the electron pair, is often called a *coordinate covalent*, or *dative*, bond. We should remember that electrons do

This arrangement results in an octet of electrons about both phosphorus and oxygen.

not "belong" to certain atoms. However, in accounting for the overall placement of electrons we can see that both electrons in the P—O bond can be associated originally with the phosphorus.

SAMPLE EXERCISE 8.8

The chlorate ion, ClO_3^-, consists of a central chlorine surrounded by three oxygen atoms. Draw the Lewis structure for this ion.

Solution: The valence shell electron arrangements in chlorine and oxygen are as follows:

The total number of electrons we must account for is $3 \times 6 + 7 + 1 = 26$. The extra electron is added to account for the fact that the ion is negatively charged. After putting in the single bonds and distributing the unshared electron pairs we have:

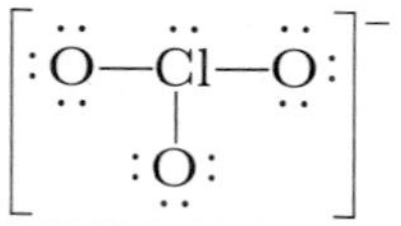

SAMPLE EXERCISE 8.9

Draw the correct Lewis structure for acetylene, C_2H_2.

Solution: In solving this problem let us attempt to predict the correct structure before we know anything about the observed arrangement of the atoms. The choice of reasonable structures is really quite limited. Because hydrogen can have at most two electrons in its valence shell orbital, the $1s$, hydrogen can be bonded via a shared electron pair to only one other atom. Thus we can rule out structures such as H—C—H—C. Either the two hydrogens are connected one each to the two carbons, or both are bonded to the same carbon. This gives us

```
    C
    C
H       H    or    H  C  C  H
```

as possible skeletal structures. The valence-shell electron arrangements for the two elements involved are:

Thus for acetylene, C_2H_2, we have a total of ten electrons to place. Putting in single bonds and unshared pairs gives us:

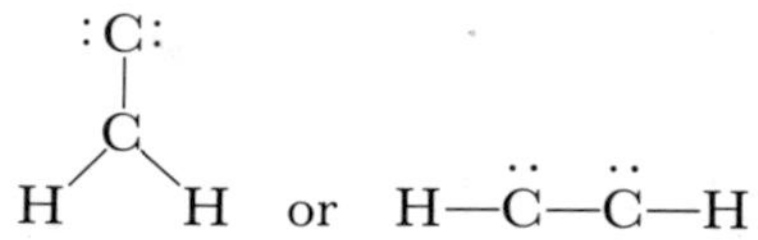

In neither case do we obtain an octet about the carbons. By using multiple bonds we can obtain the following best possible structures:

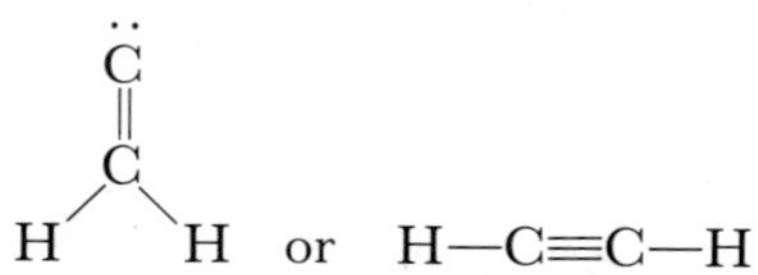

The structure on the right satisfies the octet rule for both carbon atoms, whereas that on the left does not. In addition, the structure on the right is consistent with the experimental facts about acetylene. The molecule is known to be linear, that is, all the atoms lie on a straight line. The carbon-carbon bond distance in acetylene is only 1.21 Å, as compared with distance of 1.54 Å between carbon atoms in diamond, which we have seen has carbon-carbon single bonds. The short distance indicates a triple bond. The energy of the carbon-carbon bond in acetylene is estimated to be about 832 kJ/mole, somewhat less than for the N≡N bond in N_2.

**8.5
resonance
forms**

We sometimes encounter substances in which the known arrangement of atoms is not adequately described by a single Lewis structure. The structural chemistry of nonmetallic elements affords several instances. Consider ozone, O_3, about which we shall have much to say in Chapter 10. This fascinating molecule represents an allotropic form of oxygen. We've already noted that allotropes are different forms of an element. In the examples already seen, the allotropes involved different bonding arrangements of atoms in solids. Ozone, however, involves a different number of atoms and a different bonding arrangement than does O_2; both allotropes are gaseous molecules.

Ozone is a bent molecule, with both O—O distances the same:

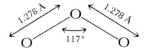

Because each oxygen atom contributes 6 valence-shell electrons, the ozone molecule has 18 valence-shell electrons. In writing the Lewis structure, we find that we must have one double bond to attain an octet of electrons about each atom:

But this structure cannot by itself be correct, because it requires that one O—O bond be different from the other, contrary to the observed structure. However, in drawing the Lewis structure we could just as easily have put the O=O bond on the left:

The two alternative Lewis structures for ozone are equivalent except for the placement of electrons. Equivalent Lewis structures of this sort are called **resonance forms**. To properly describe the structure of ozone, we must write both Lewis structures and indicate that the real molecule is described by an average of the structures suggested by the two resonance forms:

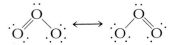

The double-headed arrow is used to indicate that the structures shown are resonance forms.

The fact that we must write more than one Lewis structure to describe a molecule or ion does not imply anything especially different about these species. You must not suppose that the molecule really exists in two or more different forms and oscillates rapidly between them. There is only one form of the molecule, that which is observed experimentally. The fact that we need to write two or more different resonance forms is simply a limitation of the use of Lewis structures in describing the electron distributions in molecules.

One rule which must be followed in writing resonance forms is that the arrangement of the nuclei must be the same in each structure. That is, the same atoms must be bonded to one another in all the structures, so that the only differences are in the arrangements of electrons.

As an additional example of resonance forms, let us draw the Lewis structure for the nitrate ion, NO_3^-, one of the most commonly encountered anions. We find that three equivalent Lewis structures are required in this instance:

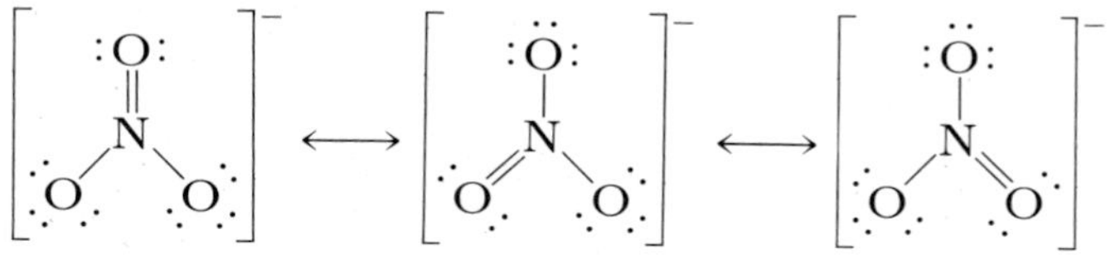

Note that the arrangement of nuclei is the same in each structure; only the placement of electrons differs. All three Lewis structures taken together adequately describe the nitrate ion, which is observed to be planar (all atoms in the same plane), with all three N—O distances equal.

In the examples of resonance structures we have seen so far, all the resonance forms have the same importance in contributing to the overall description of the molecule or ion. In the example presented in Sample Exercise 8.10, the contributing resonance structures are not all equally important.

SAMPLE EXERCISE 8.10

The molecule chlorine dioxide, ClO_2, is bent, with a central chlorine atom bound to two oxygen atoms. The two Cl—O distances are observed to be equal. Describe the ClO_2 molecule in terms of Lewis structures.

Solution: The chlorine atom has 7 valence-shell electrons, and oxygen has 6. We therefore have a total of 19 electrons ($2 \times 6 + 7$) to place in this molecule. Note that ClO_2 has an odd number of electrons. This is an unusual situation; it means that one electron must remain unpaired.

We draw the arrangement of atoms in accord with the experimental facts and put one single bond between chlorine and each oxygen:

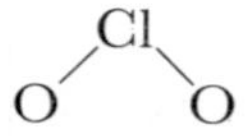

This leaves us with 15 electrons to place. We put electron pairs on the atoms to achieve an octet on each atom, or as close to it as we can get. It turns out that the odd electron must go on one or the other of the atoms, so that three possible structures result:

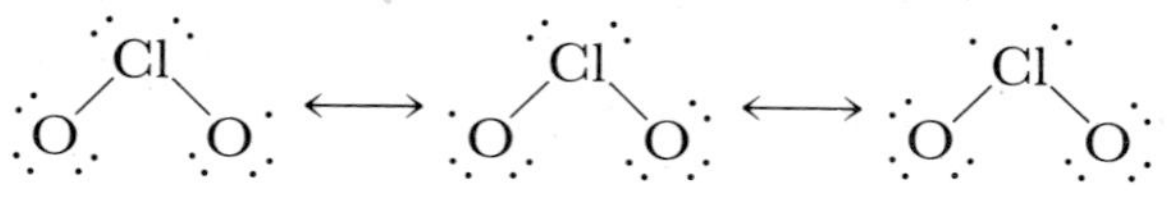

Two of these are equivalent, but the other is different, in that the odd electron is on chlorine rather than oxygen. To find out how much weight to attach to this third resonance structure in comparison with the other two, we would have to have experimental information on how the odd electron is distributed in the real ClO_2 molecule.

In the Lewis theory, attention is focused on attainment of an octet of electrons about each atom. There are many molecules and ions, however, in which the octet rule is not obeyed. As we saw in Sample Exercise 8.10, one fairly small group consists of molecules in which, as in ClO_2, there is an odd number of electrons. In the vast majority of molecules the number of electrons is even, and complete pairing of spins occurs. But in molecules such as ClO_2, NO, and NO_2, the number of electrons is odd. Obviously, under these conditions, complete pairing of all electrons is impossible.

A second possible failure to obey the octet rule comes when there are fewer than eight electrons about an atom. One example of this is boron trifluoride, BF_3, a planar molecule with three fluorine atoms around each boron, as shown in Figure 8.7. The Lewis structure, also shown in the figure, indicates that there are only six electrons about the boron. One of the valence-shell orbitals on the boron must therefore be vacant. The chemical behavior of this molecule reflects this vacancy. It is attracted to other molecules in which there is an atom with an unshared pair of electrons that can be donated to the boron. For example, it reacts with ammonia, NH_3, to form the compound BF_3NH_3:

$$
\begin{array}{c}
\text{H} \\
\text{H}\!-\!\text{N:} \\
\text{H}
\end{array}
+
\begin{array}{c}
\text{F} \\
\text{B}\!-\!\text{F} \\
\text{F}
\end{array}
\longrightarrow
\begin{array}{c}
\text{H} \qquad \text{F} \\
\text{H}\!-\!\text{N}\!-\!\text{B}\!-\!\text{F} \\
\text{H} \qquad \text{F}
\end{array}
\qquad [8.4]
$$

We shall return to a consideration of this type of reaction in discussing acid-base chemistry in Chapter 16.

A third and most numerous class of substances in which the octet rule is not obeyed is that in which there are *more* than eight electrons in the valence shell about an atom. As an example, consider PCl_5. This compound may be made by reaction of PCl_3 with chlorine gas:

$$PCl_3 + Cl_2 \longrightarrow PCl_5 \qquad [8.5]$$

When we draw the Lewis structure for this molecule we are forced to place ten electrons about the central phosphorus atom:

$$
\begin{array}{c}
\text{Cl} \\
| \quad \text{Cl} \\
\text{Cl}\!-\!\text{P} \\
| \quad \text{Cl} \\
\text{Cl}
\end{array}
$$

Among the many other examples of molecules and ions with "expanded" valence shells are PF_5, AsF_6^-, SF_4, SF_6, and ClF_3. We shall see in Chapter 9 how such structures can be accounted for by the use of additional atomic orbitals. Even without referring to these considerations, however, it is possible to see some logical pattern in their occurrences. We note that:

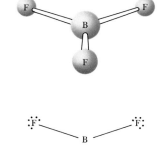

FIGURE 8.7 Geometrical structure and Lewis structure for boron trifluoride, BF_3.

1. The occurrences of expanded valence shells increase with increasing size of the central atom. There are no instances of such molecules

among nonmetals of the second row. For example, the NF_3 molecule is a stable species, but NF_5 is unknown. On the other hand, both PF_3 and PF_5 are well-characterized molecules.

2. Expanded valence shells occur most often when the central atom is bonded to the smallest and most strongly electron-attracting atoms, such as F, Cl, and O.

These observations suggest that expanded valence shells may occur when size considerations allow more atoms to crowd about a central atom than are required by the octet rule. For example, consider the pentahalides of phosphorus. Phosphorus pentafluoride, PF_5, is a very stable species. Phosphorus pentachloride, PCl_5, is reasonably stable, but dissociates in the vapor phase at $300°C$ into PCl_3 and Cl_2. Phosphorus pentabromide, PBr_5, dissociates in the vapor state even more readily; the compound PI_5 is not known. We see from these observations that the stability of the pentahalide decreases as the size of the halogen atom increases and as its electronegativity decreases.

8.7
bond polarity;
electronegativities

The electron pairs shared between two different atoms are not necessarily shared equally. We can visualize two extreme cases in the degree to which electron pairs are shared. On the one hand, we have bonding between two identical atoms, as in Cl_2 or N_2, where the electron pairs must be equally shared. At the other extreme, illustrated by NaCl, there will be essentially no sharing of electrons. We know that in this case the compound is best described as composed of Na^+ and Cl^- ions. The $3s$ electron of the Na atom is, in effect, transferred completely to chlorine. The bonds occurring in most covalent substances fall somewhere between these extremes.

The concept of **bond polarity** is useful in describing the sharing of electrons between atoms. A nonpolar bond is one in which the electrons are shared equally between two atoms. In a polar covalent bond, one of the atoms exerts a greater attraction for the electrons than the other. In the limiting case, the difference in relative abilities to attract electrons leads to formation of an ionic bond.

The tendency of an atom to attract electrons to itself in a chemical bond is referred to as **electronegativity.** We can't readily measure some single property that would give us a value for the electronegativity of an element. Instead, a numerical estimate of electronegativity must be obtained from observations of several different properties.

On theoretical grounds, the best measure of electronegativity for an atom is an average of its ionization potential (Section 7.8) and electron affinity (Section 7.9). Atoms that show high ionization potentials also show strong attraction for electrons in bonds. The attraction is even greater when the atom also shows a relatively large electron affinity. The electronegativities of metallic elements are low, because metals have low ionization potentials and low electron affinities. In contrast, nonmetals possess high electronegativities as a result of their high ionization potentials and relatively high electron affinities.

We noted in Section 7.9 that values of electron affinity are available only for a few elements. It is therefore necessary to employ other data to obtain estimates of electronegativities. Figure 8.8 shows electronegativity values for many of the elements. The numbers listed in this table are all based on a particular value chosen for one of the elements. The value of 2.5 for carbon is the most commonly accepted reference; this was the value chosen for carbon by Pauling, who first developed the concept of electronegativity. The actual value chosen for the reference is not so important; we are interested mainly in the relative values of electronegativity in comparing two elements. Notice that the most electronegative element is fluorine, with an electronegativity of 4.0. The least electronegative element, cesium, has an electronegativity of 0.79. All the other elements have electronegativities that lie between these extremes. The values listed for the transition metals are those for the $+2$ state. When the element is in a higher charge state, its electronegativity is higher.

Note that in a horizontal row of the table there is a more or less steady increase in electronegativity in moving from left to right, that is, from the most metallic to the most nonmetallic elements. Notice also that, with a few exceptions, there is an overall decrease in electronegativity with increasing atomic number in any one group of the periodic table. This is what we might expect, because we know that ionization potentials tend to decrease with increasing atomic number in a group, and electron affinities don't change very much.

Keep in mind that electronegativities are *approximate* measures of the *relative* tendencies of these elements to attract electrons to themselves in a chemical bond. The electronegativity varies with the type of chemical environment in which an element is situated. We have noted, for example, that the electronegativities of transition metals vary with their valence. Similarly, the electronegativity of chlorine in its bonding to phosphorus in PCl_3 (Sample Exercise 8.6) is likely to be different from its value in the chlorate ion, ClO_3^- (Sample Exercise 8.8), in which it is in a much different bonding situation. Variations of this sort are not so large as to render the concept of electronegativity useless, but we must avoid placing too much reliance on precise values for electronegativities. So long as we remain aware of their limitations, electronegativity values provide a useful guide to polarities in chemical bonds.

The electronegativity difference between two atoms is a measure of the polarity of the bond between them. For example, in HF the electronegativity difference between H and F is 1.8; correspondingly, this bond is more polar than in HI, where the electronegativity difference is only 0.4. Figure 8.8 reveals that Cl has a higher electronegativity than P; consequently, the shared electron pairs in PCl_3 are probably displaced slightly toward chlorine, and the PCl bonds would therefore be slightly polar. The P—F bond in PF_3 should be even more polar because the electronegativity difference between P and F is larger.

DIPOLE MOMENTS

As a result of the unequal sharing of electrons between atoms, the centers of positive and negative charge in molecules may not coincide. The molecule is then said to be polar. The degree of its polarity is measured

1A	2A	3B	4B	5B	6B	7B	8B			1B	2B	3A	4A	5A	6A	7A
																H 2.2
Li 1.0	Be 1.6											B 1.8	C 2.5	N 3.0	O 3.4	F 4.0
Na 0.93	Mg 1.3											Al 1.6	Si 1.9	P 2.2	S 2.6	Cl 3.2
K 0.82	Ca 1.0	Sc 1.4	Ti 1.5	V 1.6	Cr 1.7	Mn 1.6	Fe 1.8	Co 1.9	Ni 1.9	Cu 2.0	Zn 1.6	Ga 1.8	Ge 2.0	As 2.2	Se 2.6	Br 3.0
Rb 0.82	Sr 0.9	Y 1.2	Zr 1.3	Nb 1.6	Mo 2.2	Tc –	Ru 2.2	Rh 2.3	Pd 2.2	Ag 1.9	Cd 1.7	In 1.8	Sn 1.8	Sb 2.0	Te 2.1	I 2.7
Cs 0.79	Ba 0.9								Pt 2.3	Au 2.5	Hg 2.0	Tl 2.0	Pb 2.3	Bi 2.0	Po –	

FIGURE 8.8 Electronegativities of the elements. The values for the transition metals are those for the +2 valence state.

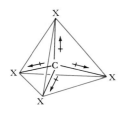

FIGURE 8.9 Structure of the CX_4 molecule.

by its **dipole moment,** μ. The dipole moment of a molecule increases as the quantity of charge which is separated increases, or as the distance between the positive and negative centers increases. For example, suppose that the shared electron pair in the molecule A—B is shifted slightly toward A, because A has the higher electronegativity. This creates an overall partial negative charge at A, and an overall partial positive charge at B. The magnitude of the dipole moment depends on the size of partial charge created at each atom, and on the distance separating them. If the difference in electronegativities between A and B were to grow larger, this would result in larger partial charges at the two atoms. The dipole moment would thus increase, even though the distance between the atoms were constant. On the other hand, if the distance separating A and B were to increase, while the partial charges on each atom remained the same, that would also result in an increase in dipole moment.

Dipole moments are normally reported in units of Debye, D. The dipole is usually represented pictorially as an arrow with a cross base, $\longmapsto$. The head of the arrow points toward the negative end of the dipole.

Table 8.4 shows the dipole moments and bond dissociation energies for the hydrogen halides. Note that the dipole moment decreases as the electronegativity difference decreases.

Although dipole moments furnish a useful indication of charge distribution, they are not always easy to interpret in terms of bond polarity. In a molecule with several chemical bonds, the overall dipole moment is the sum of all the individual bond dipole moments. But equal bond dipole moments pointing in opposite directions cancel. As a result, the dipole moment sometimes tells us nothing about bond polarities. For example, in CH_4, CCl_4, and CF_4, the carbon atom is tetrahedrally surrounded by the other four atoms to which it is bonded (Figure 8.9). Each of the individual C—X bonds has a bond dipole moment. However, this set of four individual dipoles, because of their symmetric arrangement about the central atom, add up to exactly zero. The overall dipole moment is therefore zero, no matter how polar the C—X bond may be.

TABLE 8.4

some properties of hydrogen halides

COMPOUND	ELECTRONEGATIVITY DIFFERENCE	DIPOLE MOMENT (D)	BOND ENERGY (kJ/mole)
HF	1.8	1.91	565
HCl	1.0	1.03	431
HBr	0.8	0.79	364
HI	0.5	0.38	297

SAMPLE EXERCISE 8.11

In each of the following cases, indicate the directions of the individual bond dipole moments and the overall molecular dipole moment: (a) CO_2; (b) SF_6; (c) ICl.

Solution: (a) Carbon dioxide is a linear molecule. The overall molecular dipole moment is zero because the individual C—O bond dipoles point in opposite directions. They thus cancel by symmetry. (b) Sulfur hexafluoride is a highly symmetric molecule with six equivalent S—F bonds arranged around the central sulfur:

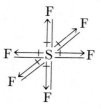

Each pair of S—F bond dipoles, directed in opposite directions, cancels, so that the overall dipole moment is zero. (c) Chlorine is a more electronegative element than iodine. The dipole moment of ICl is directed toward chlorine as the negative end:

$$I—Cl$$

$$\longmapsto$$

8.8
oxidation and reduction; oxidation numbers

The reaction between sodium and chlorine atoms to form NaCl involves transfer of an electron from sodium to chlorine. We can imagine the overall reaction as the result of two separate processes:

$$Na\cdot \longrightarrow Na^+ + e^- \qquad [8.6]$$

$$:\ddot{C}l\cdot + e^- \longrightarrow [:\ddot{C}l:]^- \qquad [8.7]$$

Each of these processes is called a **half-reaction**. Adding these two half-reactions together gives the overall formation of the ionic species from the neutral atoms:

$$Na + Cl \longrightarrow Na^+ + Cl^-$$

As another example, consider the formation of MgO from magnesium and oxygen atoms:

$$\dot{Mg} \longrightarrow Mg^{2+} + 2e^- \qquad [8.8]$$

$$\cdot \ddot{O} \cdot + 2e^- \longrightarrow [:\ddot{O}:]^{2-} \qquad [8.9]$$

$$\overline{Mg + O \longrightarrow Mg^{2+} + O^{2-}} \qquad [8.10]$$

Processes shown in Equations [8.6] and [8.8], which involve loss of electrons from the metal, are called **oxidation**. Processes shown in Equations [8.7] and [8.9], which involve gain of electrons by the nonmetal atoms, are called **reduction**. As a general definition, oxidation corresponds to loss of electrons, reduction corresponds to gain of electrons. A substance that has lost electrons is said to be oxidized; one that gains electrons is said to be reduced.

There is very little difficulty in determining what has been oxidized and what reduced when the transfer of electrons from one atom to another is complete, as in the above examples. But it is not so clear what has been oxidized and what reduced, if anything, in the following reaction:

$$H_2(g) + Cl_2(g) \longrightarrow 2HCl(g) \qquad [8.11]$$

We know that HCl is not an ionic substance; it is a gas boiling at $-84°C$, with all the properties of a polar covalent compound. In a covalently bonded substance such as HCl, the electrons are shared between atoms, not transferred as in an ionic substance. From the relative electronegativities of the elements we can usually determine that one atom claims a greater share of the electron pair than the other. But this is by no means a quantitative measure. Thus we can't really say what the net charge on a particular atom in a covalently bonded molecule or ion is.

As an aid in balancing chemical equations, and as a kind of accounting system for electrons in chemical reactions, a set of rules has been established for assigning **oxidation numbers** to atoms in molecules and ions. The oxidation number is a whole number that reflects, in a formal way only, the charge assigned to a particular atom. Insofar as possible, the assignment of oxidation numbers follows the relative electronegativities of the elements. A few simple rules are used to assign oxidation numbers:

1. The oxidation number of a substance in the elemental state is zero. Thus, for Cl_2, N_2, Na, or P_4, the oxidation number of each atom is zero.

2. In a compound, the more electronegative elements are assigned negative oxidation numbers; the less electronegative elements are assigned positive oxidation numbers. The magnitude of the oxidation number corresponds more or less to the valence, or number of shared electron-pair bonds. For example, hydrogen always has an oxidation number of 1. It is -1 when hydrogen is bonded to a less electronegative element, for example, in NaH; it is $+1$ when it is bonded to a more electronegative element. Thus, in HCl, hydrogen is assigned an oxidation number of $+1$, and chlorine an oxidation number of -1.

3. In any molecule or ion, the sum of the positive and negative oxidation numbers must equal the overall charge on the species. For example, in OF_2, fluorine is assigned an oxidation number of -1 because it is the more electronegative element. Oxygen must then have an oxidation number of $+2$. In AlO_2^-, if oxygen is assigned an oxidation number of -2, then Al must have an oxidation number of $+3$ in order that the overall charge come out -1. From this rule it follows that the oxidation number of the element in a simple ion is just equal to the charge on that ion. For example, the oxidation number of Tl in Tl^{3+} is $+3$. The term **oxidation state** is nearly synonymous with oxidation number, and we use the two expressions interchangeably. To illustrate, the oxidation number of Cr in Cr_2O_3 is $+3$. The Cr is said to be in the $+3$ oxidation state.

The periodic table provides us with many guidelines to the assignment of oxidation numbers. All the elements of group 1A have oxidation numbers of $+1$ in their compounds. There is simply no other commonly observed means of chemical bonding for these metals other than loss of an electron to form the $+1$ ion. The situation is not so clear-cut with the elements of group 1B. Although the $+1$ oxidation state is observed

(CuBr, AgCl), the $+2$ oxidation state ($CuCl_2$) and even the $+3$ oxidation state ($AuCl_3$) are also noted.

The elements of group 2A are always found in the $+2$ oxidation state—for example, Mg^{2+} and Sr^{2+}. The group 2B elements are also most commonly found in the $+2$ oxidation state, but Hg can be an exception. In Hg_2Cl_2 it is in the $+1$ oxidation state. In group 3A, the most commonly encountered element, Al, is almost always found in the $+3$ oxidation state, as in Al_2O_3 or $Al(NO_3)_3$.

The most electronegative of the nonmetals are usually always found in characteristic negative oxidation states. Fluorine, the most electronegative of the elements, is always found in the -1 oxidation state. Oxygen is nearly always in the -2 oxidation state. The only common exception to this general rule occurs in peroxides. In hydrogen peroxide, H_2O_2, if we assign hydrogen an oxidation number of $+1$, then oxygen must have an oxidation number of -1. In the peroxide ion, O_2^{2-}, and in ionic peroxides such as BaO_2, the oxygen retains an oxidation number of -1.

Using these general remarks and rules as guidelines, it should be possible for you to assign oxidation numbers to elements in most bonding situations.

SAMPLE EXERCISE 8.12

Assign oxidation numbers to the element in colored type in each of the following species: (a) **Pb**Cl_2; (b) H_2**S**O_3; (c) **Cl**O_3^-; (d) K**Mn**O_4.

Solution: (a) The chlorine in $PbCl_2$ is expected to be in the -1 oxidation state. The oxidation number for Pb is therefore $+2$. (b) Each oxygen has an oxidation number of -2, each hydrogen an oxidation number of $+1$. In order that the overall sum of oxidation numbers be zero, the oxidation number of S must be $+4$. (c) Oxygen has a greater electronegativity than chlorine (Figure 8.8), and so it is assigned a -2 oxidation number. In order that the overall ionic charge total -1, the oxidation number of Cl must equal $+5$. (d) Each oxygen in $KMnO_4$ has an oxidation number of -2; the oxidation number of K is, of course, $+1$. The overall sum of oxidation numbers must equal zero: $+1 + 4(-2) + x = 0$. The oxidation number for Mn in this compound is therefore $+7$.

It is important to keep in mind that the oxidation numbers, or oxidation states, do not correspond to real charges on the atoms, except in the special case of simple ionic substances. Nevertheless, they furnish a useful means of organizing chemical facts, especially in the case of metallic elements. Their most frequent use is in balancing chemical equations for reactions in which changes in oxidation numbers occur.

OXIDATION NUMBERS AND NOMENCLATURE

The rules for naming simple inorganic compounds were discussed in Chapter 2. This is a good point at which to review those rules. Now that the concept of oxidation number has been explained, we can add another widely used method for naming binary compounds. You learned in Section 2.10 that the name of a binary compound consists of the name of the less electronegative element first, followed by the name of the more electronegative element, modified to have an -*ide* ending. The following examples are illustrative:

BaSe	barium selenide	
ZnS	zinc sulfide	
MgH_2	magnesium hydride	

When one or the other of the elements involved has more than one possible oxidation state, the number of atoms may be included in the name as a Greek prefix (1 = mono-, 2 = di-, 3 = tri-, 4 = tetra-, 5 = penta-, 6 = hexa-, 7 = hepta-), *or* the oxidation state is indicated using a Roman numeral, as in these examples:

MnO_2	manganese dioxide	*or* manganese(IV) oxide
Mn_2O_3	dimanganese trioxide	*or* manganese(III) oxide
P_2O_5	diphosphorus pentoxide	*or* phosphorus(V) oxide
P_2O_3	diphosphorus trioxide	*or* phosphorus(III) oxide
$SnCl_4$	tin tetrachloride	*or* tin(IV) chloride
$SnCl_2$	tin dichloride	*or* tin(II) chloride
XeF_4	xenon tetrafluoride	*or* xenon(IV) fluoride
XeF_6	xenon hexafluoride	*or* xenon(VI) fluoride

OXIDATION-REDUCTION EQUATIONS

Many chemical reactions involve the transfer of electrons from one species to another. For example, we saw in Equations [8.6] and [8.7] that formation of NaCl from sodium and chlorine involves loss of an electron from the sodium atom and the acquisition of this electron by the chlorine atom. In each of the two half-reactions a change in oxidation number has occurred. In Equation [8.6], the oxidation number of sodium changes from zero to +1; in Equation [8.7], the oxidation number of chlorine has changed from zero to −1. The overall reaction that includes these two half-reactions is called an **oxidation-reduction reaction.**

The formation of NaCl from the elements sodium and chlorine is a clear example of complete transfer of an electron. In many other chemical reactions it is not so clear that a real transfer of electrons has occurred. Nevertheless, there may be a change in the oxidation numbers of two or more of the species taking part in the reaction. Such reactions are also classified as oxidation-reduction processes. We may define an oxidation-reduction reaction as one in which two or more species undergo changes in oxidation numbers. In terms of this more general definition, we can define oxidation as an increase in oxidation number, reduction as a decrease in oxidation number.

8.9
balancing oxidation-reduction equations

Many oxidation-reduction reactions, involving perhaps only two substances, are easy to balance. For example, the reaction of aluminum with oxygen to form Al_2O_3, which we might write as

$$Al + O_2 \longrightarrow Al_2O_3 \qquad [8.12]$$

can be balanced by simply recognizing that we need to have three oxygen atoms on the left for each two aluminums. Thus we could write

$$2Al + \frac{3}{2}O_2 \longrightarrow Al_2O_3 \qquad [8.13]$$

If we wish to remove the fraction from before the O_2, the entire equation can be multiplied by 2, to give

$$4Al + 3O_2 \longrightarrow 2Al_2O_3 \qquad [8.14]$$

But now let's consider a more complicated case. When lead dioxide, PbO_2, is treated with manganese nitrate, $Mn(NO_3)_2$, in the presence of sulfuric acid, H_2SO_4, lead sulfate, $PbSO_4$, precipitates out, and permanganic acid, $HMnO_4$, is left in solution. Thus the overall unbalanced reaction is

$$PbO_2 + H_2SO_4 + Mn(NO_3)_2 \longrightarrow$$
$$PbSO_4 + HNO_3 + HMnO_4 \qquad [8.15]$$

To balance this equation, we must first decide which atoms undergo a change in oxidation number. To do this, we should determine the oxidation number of each atom on both sides of the equation, using the rules set out above. However, we can make things easier for ourselves by recognizing that certain groups of atoms appear in the same form on both sides of the equation. Note the presence of sulfuric acid, H_2SO_4, on the left. We can think of this as a combination of the sulfate ion, SO_4^{2-}, with two hydrogen ions. On the right the sulfate ion appears in combination with lead, in $PbSO_4$. The sulfate ion, SO_4^{2-}, is a stable species that very often appears on both sides of a chemical equation, with no change in its overall charge. This means that all the atoms in the sulfate ion are unchanged in their oxidation numbers in going from reactants to products. Similarly, the nitrate ion, NO_3^- appears in combination with Mn^{2+} on the left, and in combination with the hydrogen ion on the right. By learning to recognize the common polyatomic ions, Table 2.5, you can greatly simplify your task of balancing many oxidation-reduction reactions.

Next we notice that hydrogen is bonded on the left side to the sulfate ion, SO_4^{2-}, and on the right to the nitrate ion, NO_3^- and the permanganate ion, MnO_4^-. In all these cases it has an oxidation number of $+1$. Thus, hydrogen also does not undergo any change in oxidation number in this reaction. The oxygen that is bonded to Pb on the left is assigned its usual oxidation number of -2. All of the oxygens appearing on the right-hand side of this equation also show an oxidation number of -2. Thus, there is no change in oxidation number of the oxygen appearing in this equation. We have now identified all the elements and groups that do *not* undergo a change in oxidation number. Those that do undergo changes are Pb and Mn. In PbO_2 the lead has an oxidation number of $+4$, because it is bonded to two oxygen atoms with oxidation numbers of -2 each. On the right, Pb is bonded to a single SO_4^{2-} ion, thus has an oxidation number of $+2$. Manganese on the left is bonded to two nitrate ions with net changes of -1 each; its oxidation number is thus $+2$. On the right, the Mn in $HMnO_4$ is calculated to have an oxidation number of $+7$, using the rules outlined above.

Our next step is to write the oxidation numbers above those species that undergo change and note the magnitude of the change in going from the left- to the right-hand side of the equation:

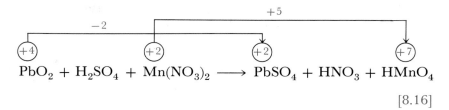

$$\quad [8.16]$$

The number above each line shows the magnitude of the change in oxidation number in going from reactants to products.

It is clear from examining these values that the reaction is an oxidation-reduction reaction in which lead is reduced (decrease in oxidation number) and manganese is oxidized (increase in oxidation number). The first principle to be considered in balancing such an equation is that *the total change in oxidation number of all the species being reduced must equal the total change in oxidation number of all the substances being oxidized.* In other words, the total number of electrons in the system must be conserved. Because reduction of each lead requires two electrons, and oxidation of each manganese releases five electrons, we must multiply the coefficients of Pb and Mn by five and two, respectively, to achieve a balance in the number of electrons gained and lost. We add these coefficients to the equation, and rewrite the total changes in oxidation numbers as follows:

$$
\begin{array}{c}
\overset{\rule{6cm}{0.4pt}\; -\,(5\times 2)=-10\;\rule{1cm}{0.4pt}\searrow}{} \\
5PbO_2 + H_2SO_4 + 2Mn(NO_3)_2 \longrightarrow 5PbSO_4 + HNO_3 + 2HMnO_4 \\
\underset{\rule{5cm}{0.4pt}\; +\,(2\times 5)=+10\;\rule{4cm}{0.4pt}\nearrow}{}
\end{array}
$$

$$\quad [8.17]$$

The next step is to balance correctly the sulfate and nitrate. We need five SO_4^{2-} to combine with the five Pb^{2+} on the right; we need four HNO_3 on the right to take care of the four NO_3^- from the left hand side. Adding these coefficients gives us

$$5PbO_2 + 5H_2SO_4 + 2Mn(NO_3)_2 \longrightarrow$$
$$5PbSO_4 + 4HNO_3 + 2HMnO_4 \qquad [8.18]$$

At this point the equation is almost, but not quite, balanced. The final step is to achieve an oxygen and hydrogen balance. We note that there are ten oxygens on the left (aside from those in the SO_4^{2-} and NO_3^- ions, which we have already accounted for), and only eight on the right. In addition, there are ten hydrogens on the left, and only six on the right. The additional four hydrogens and two oxygens are added on the right by forming two moles of water. The overall balanced equation is thus:

$$5PbO_2 + 5H_2SO_4 + 2Mn(NO_3)_2 \longrightarrow$$
$$5PbSO_4 + 4HNO_3 + 2HMnO_4 + 2H_2O \qquad [8.19]$$

In summary, the procedure for balancing an oxidation-reduction equation by the method we have outlined here is as follows:

1. Assign the oxidation numbers to all the elements in the equation.
2. Determine which elements have undergone a change in oxidation number and the number of units gained or lost in each change.
3. Use the values found in step 2 to determine the simplest ratio of moles of reactants that will lead to equal losses and gains in oxidation numbers. Add these numbers as coefficient of the reactants and products, and add the proper coefficients to other reactants and products as needed.
4. Determine the number of moles of any other remaining substances, for example, H_2O, needed to complete the balancing.

SAMPLE EXERCISE 8.13

When an orange-colored solution of potassium dichromate, $K_2Cr_2O_7$, is reacted with hydrogen chloride, HCl, chlorine gas is given off, and the violet-green color of chromium trichloride, $CrCl_3$, appears. Balance the chemical equation that describes this reaction.

Solution: We first write an unbalanced equation that describes as completely as possible what our observations tell us about the reaction.

$$K_2Cr_2O_7 + HCl \longrightarrow CrCl_3 + Cl_2$$

We then carry out the first step in the process outlined above by assigning oxidation numbers to each atom on both sides of the equation. These are written above the symbol for each element:

$$\overset{+1\ +6\ -2}{K_2Cr_2O_7} + \overset{+1-1}{HCl} \longrightarrow \overset{+3\ -1}{CrCl_3} + \overset{0}{Cl_2}$$

Remember that not all the substances in the reaction system are yet shown on both sides of the equation. However, we do have all the substances that undergo oxidation-reduction. We note that the two elements that undergo change in oxidation numbers are chromium and chlorine. We write the oxidation numbers of these two substances and the changes in oxidation numbers which occur:

$$\overset{-3}{\underset{+6 \quad -1 \quad +3 \quad 0}{K_2Cr_2O_7 + HCl \longrightarrow CrCl_3 + Cl_2}} \quad {+1}$$

To achieve a balance in the gains and loss of oxidation numbers we need three Cl for each Cr. There are already two Cr atoms on the left. We add the coefficient 2 to the Cr on the right, add a coefficient 6 to the Cl on the left and a 3 to the Cl_2 on the right to give

$$\overset{-(2 \times 3) = -6}{K_2Cr_2O_7 + 6HCl \longrightarrow 2CrCl_3 + 3Cl_2} \\ {+(6 \times 1) = +6}$$

Now we have to balance the other elements. There are K^+ and O^{2-} on the left, and none of either element on the right. No change in oxidation number occurs for either of these elements in this equation. The excess oxide O^{2-} is combined with hydrogen ions to form water on the right. The K^+ combines with whatever negative charge is left over. To do this systematically, we first add $7H_2O$ on the right to take care of the oxygen:

$$K_2Cr_2O_7 + 6HCl \longrightarrow$$
$$2CrCl_3 + 3Cl_2 + 7H_2O$$

This means that there are 14 hydrogens on the right, but only 6 on the left. We therefore add 8 more HCl on the left to supply these needed H^+. Now we have a total of 14 chloride on the left:

$$K_2Cr_2O_7 + 14HCl \longrightarrow$$
$$2CrCl_3 + 3Cl_2 + 7H_2O$$

Six of these end up on the right in the form of Cl_2,

and another 6 as $CrCl_3$. The remaining two are combined with the K^+ to form KCl. Our completely balanced equation is thus

$$K_2Cr_2O_7 + 14HCl \longrightarrow$$
$$2CrCl_3 + 3Cl_2 + 7H_2O + 2KCl$$

The method we have used here for balancing oxidation-reduction equations is sometimes called the oxidation-number method. Another equally popular method, the method of half-reactions, is best appreciated after we have discussed the nature of ionic solutions and electrochemical processes. The oxidation-number method we have used here is quite adequate for balancing any oxidation-reduction equation, no matter how complex.

FOR REVIEW

summary

In this chapter, we have dealt with the interactions between atoms that lead to formation of chemical bonds. **Ionic bonding** results from the complete transfer of electrons from one atom to another, with formation of a three-dimensional lattice of charged particles. The stabilities of ionic substances result from the powerful electrostatic attractive forces between an ion and all the surrounding ions of opposite charge.

Covalent bonding results from the sharing of electrons between atoms. The rules that govern this sharing are based on the stability of the rare-gas electron configuration. We can represent shared electron-pair structures of molecules by means of Lewis structures, which show the sharing of electron pairs between atoms. It sometimes happens that a single Lewis structure is inadequate to represent a particular molecule, but that an average of two or more Lewis structures does form a satisfactory representation. In these cases, the Lewis structures are referred to as **resonance forms.**

It is important to recognize that even in covalent bonding, electrons may not be shared equally between two atoms. **Electronegativity** is a measure of the ability of an atom to compete with other atoms for the electrons shared between them. Highly electronegative elements strongly attract electrons. The electronegativities of the elements, which show a regular periodic relationship, are an important guide to chemical behavior. We shall be using the concept of electronegativity often throughout the text.

One application of electronegativity is in the assignment of **oxidation numbers,** formal whole-number charges assigned to atoms in molecules and ions. Although the oxidation numbers do not represent the real charges on atoms except in simple ionic substances, they are of great value in helping us to organize chemical facts and are an aid in the balancing of reactions and in the naming of compounds.

Oxidation may be defined as a process in which an atom undergoes an increase in oxidation number. **Reduction** is a process in which an element undergoes a decrease in oxidation number. In an **oxidation-reduction reaction,** both oxidation and reduction occur in such a manner as to balance the total increases and decreases in oxidation numbers.

learning goals

Having read and studied this chapter, you should be able to:

1. Describe the origin of the energy terms that lead to stabilization of ionic lattices.
2. Describe the more commonly observed arrangements of ions in lattices, as illustrated in Figures 8.1 and 8.2.
3. Predict on the basis of the periodic table the probable formulas of ionic substances formed between common metals and nonmetals.
4. List the commonly observed oxidation states of the more frequently encountered transition metals such as Fe, Cu, Ni, and so on.
5. Describe the effects of gain or loss of electrons on atomic radii in producing ionic radii.
6. Explain the concept of an isoelectronic series and the origin of ionic radius changes within such a series.
7. Describe the basis of the Lewis-Langmuir theory and predict the covalency of common nonmetallic elements from their position in the periodic table.
8. Give the geometrical structures, and be able to draw Lewis structures, for the lighter nonmetallic elements.
9. Write the Lewis structures for molecules and ions containing covalent bonds, using the periodic table.
10. Give the reasons for writing more than one Lewis structure to represent the structures of certain substances.
11. Explain what is meant by the term bond polarity.
12. Explain the significance of electronegativity and in a general way relate the electronegativity of an element to its position in the periodic table.
13. Explain what is meant by the dipole moment of a molecule, and what relationship, if any, it may bear to bond polarities.
14. Assign oxidation numbers to atoms in molecules and ions.
15. Give the meaning of the terms oxidation, reduction, and oxidation-reduction reactions.
16. Balance an oxidation-reduction equation using the method of oxidation numbers.
17. Assign acceptable names to simple inorganic compounds and ions.

key terms

Among the more important terms and expressions used for the first time in this chapter are the following:

An **allotrope** (Section 8.3) is one of the forms of an element, when that element is capable of existing in more than one form. For example, O_2 and O_3 (ozone) are allotropic forms of oxygen.

Bond polarity (Section 8.7) is a measure of the difference in ability of the two atoms in a chemical bond to attract electrons.

A **cation** (Section 8.1) is a positively charged ion; an **anion** is negatively charged.

A **covalent bond** (Section 8.3) is a bond formed between two or more atoms by a sharing of electrons.

The **dipole moment** (Section 8.7) of a molecule is a measure of the separation between the centers of positive and negative charges in polar molecules.

Electronegativity (Section 8.7) is a measure of the ability of an atom that is bonded to another atom to attract electrons to itself.

A **half-reaction** (Section 8.8) is half of an overall oxidation-reduction reaction, corresponding to either the oxidation half or the reduction half.

An **ionic bond** (Section 8.1) is a bond formed on the basis of the electrostatic forces that exist between oppositely charged species in solid lattices made up of ions. The ions are formed from atoms by transfer of one or more electrons.

A **Lewis structure** (Section 8.3) is a representation of covalent bonding in a molecule. Covalently shared electron pairs are shown as lines, and unshared electron pairs are shown as a pair of dots. Only the valence-shell electrons are shown.

Oxidation (Section 8.8) is the half of an oxidation-reduction process that corresponds to an increase in oxidation number.

Oxidation number (Section 8.8) is a whole number assigned to an atom in a molecule or ion that corresponds to the formal charge assigned to that atom.

Reduction (Section 8.8) is the half of an oxidation-reduction process that corresponds to a decrease in oxidation number.

The term **rare-gas configuration** (Section 8.1) refers to an outer, or valence-shell, electron configuration, ns^2np^6.

Resonance forms (Section 8.5) are individual Lewis structures in cases where two or more Lewis structures are equally good descriptions of a single molecule. The resonance structures in such an instance are "averaged" to give a correct description of the real molecule.

Valence (Introduction) may be defined as the capacity of an atom for entering into chemical combination with other atoms. Ionic valence is equal to the number of electrons gained or lost in forming the ionic species. Covalence is equal to the number of electrons from an atom that are involved in shared electron-pair bonds with other atoms.

EXERGISES

ionic bonds

8.1 What are some of the observable characteristics that distinguish ionic substances from covalently bonded substances?

8.2 By comparing the properties of KBr with Br_2, provide some arguments for why it is proper to view KBr as an ionic substance. (You should consult a handbook of chemistry for data on the two substances).

8.3 The lattice energy of NaF is 916 kJ/mole, whereas that for KCl is only 715 kJ/mole. Account for the difference.

8.4 Predict the formula of the ionic compound formed between each of the following pairs of elements and write a balanced chemical equation for the formation of the compound: (a) Sr and Br; (b) Be and O; (c) Al and S; (d) Zn and F.

8.5 Which of the following substances would you expect to be a stable substance? In each case give a reason based on electron configuration. (a) CaO; (b) $AlCl_2$; (c) TiF_5; (d) VF_5; (e) Rb_2O; (f) MnF_2

8.6 Write the outer electron configurations of each of the following metallic ions: (a) Mn^{4+}; (b) Cu^+; (c) Au^+; (d) Co^{3+}.

8.7 Write the expected formulas for each of the following compounds (refer to Table 2.5 as necessary): (a) gallium chloride; (b) titanium fluoride; (c) ammonium sulfide; (d) calcium phosphate; (e) aluminum nitrate; (f) potassium permanganate.

8.8 Complete and balance each of the following equations:

(a) $Ag(s) + O_2(g) \longrightarrow$

(b) $Tl(s) + Cl_2(g) \longrightarrow$

(c) $Ba(s) + F_2(g) \longrightarrow$

(d) $Al(s) + O_2(g) \longrightarrow$

(e) $Mg(s) + I_2(g) \longrightarrow$

(f) $Li(s) + P_4(g) \longrightarrow$

(g) $Sc(s) + Cl_2(g) \longrightarrow$

8.9 Which of the following ions possess a rare gas configuration? (a) Ca^{2+}; (b) In^+; (c) Ga^{3+}; (d) Sc^{3+}; (e) Se^{2-}

8.10 Explain each of the following trends in lattice

energies: (a) $MgO > KCl$; (b) $MgO > SrS$; (c) $NaF > NaCl > NaBr$.

sizes of ions

8.11 How does the ionic radius vary in the series of ions Cl^-, K^+, Ca^{2+}? Explain the variation.

8.12 Which of the following species in each case would be expected to have the larger electron affinity? (Review Section 7.9 if you've forgotten what electron affinity is). (a) K^+ or Li^+; (b) Cl^- or K^+

8.13 In the $2+$ ions of the fourth row of the periodic table, extending from Ca^{2+} to Zn^{2+}, how do the ionic radii and the effective nuclear charge seen by the highest energy electrons vary? What relationship is there between these two quantities?

8.14 In the series of ions O^{2-}, S^{2-}, Se^{2-}, and Te^{2-}, why does the ionic radius continuously increase?

8.15 Is the series of ions Rb^+, Sr^{2+}, Y^{3+} an isoelectronic series? Why does the ionic radius vary as it does in this series?

covalent bonding

8.16 How does the covalence shown by an atom relate to the electron configurations of rare gases? Explain with the aid of a specific example.

8.17 Describe the difference between covalence and ionic valence.

8.18 Draw the Lewis dot formulas like those shown in Equation [8.3] for the formation of: (a) N_2 from two N atoms; (b) Na_2 from two Na atoms; (c) HBr from H plus Br.

Lewis structures

8.19 Draw Lewis structures for the stable hydrides of Pb, Sb, Te, and I.

8.20 Draw reasonable Lewis structures for each of the following known substances: N_2; N_2H_2; N_2H_4; NH_3.

8.21 Draw Lewis structures for each of the following forms of nonmetallic elements: S_6; P_4; Cl_2.

8.22 Draw the Lewis structures for all the members of the oxychloride anion series, ClO^-, ClO_2^-, ClO_3^-, and ClO_4^-.

8.23 Draw the Lewis structures for the series of compounds H_2S_2, H_4P_2, and H_6Si_2, containing in each case a single bond between the heavy nonmetallic atoms.

resonance structures

8.24 The O—O bond distance in hydrogen peroxide, H_2O_2, is 1.48 Å. The O—O bond in hydrogen peroxide consists of a single pair of shared electrons.

In ozone the distance between the central oxygen and either of the other oxygens is 1.28 Å. How are these relative values explained in terms of resonance theory?

8.25 Draw Lewis structures with all appropriate resonance forms, for each of the following: (a) SO_2; (b) NO_2^-; (c) HClO.

8.26 Why is it necessary to use resonance structures to describe adequately the structure of nitrate ion, NO_3^-, using Lewis structures?

exceptions to the octet rule

8.27 Describe what is meant by the octet rule. In what ways can the distribution of electrons in a compound deviate from the octet rule? Give an example in each case.

8.28 Which of the following elements is most likely to show a departure from the octet rule in its covalent compounds: (a) C; (b) F; (c) Sb? Explain your answer.

8.29 SF_6 is a very stable substance that is used as an electrically insulating gas in high-voltage transformers. Yet SCl_6 is quite unstable and has never been isolated. Explain the reasons for this difference in stabilities.

bond polarities; electronegativities

8.30 In each of the following pairs of compounds, indicate which has the more polar bonds; (a) BF_3, CCl_4; (b) LiI, KF; (c) $GeBr_4$, $GeCl_4$; (d) OF_2, BeF_2; (d) ZnS, MgO.

8.31 Why does the electronegativity of the halogens decrease with increasing atomic number?

8.32 Explain why the electronegativities of the elements increase in a more or less regular way from left to right in each horizontal row of the periodic table.

8.33 In the following series of compounds, which metal would you expect to have the highest actual positive charge: (a) ZnS; (b) SrO; (c) $CdBr_2$? Explain.

8.34 Account for the fact that CCl_4 has no dipole moment, but $CHCl_3$ does.

8.35 Assuming that the bond angles around the central carbon are tetrahedral in all cases, arrange the following compounds in the expected order of increasing polarity: (a) CF_3Cl; (b) CF_4; (c) CF_2Cl_2; (d) CF_2H_2; (e) CH_4.

oxidation numbers; oxidation-reduction

8.36 How does the oxidation number differ from the charge on an atom? Why don't we use charges on atoms instead of oxidation numbers?

8.37 Indicate the oxidation numbers most commonly associated with each of the following elements or groups of elements and give an example in each case (in some instances it may be appropriate to indicate more than one common oxidation state, with an example of each): (a) the group 7A elements; (b) the group 3A elements; (c) sulfur; (d) cadmium; (e) the alkali metals; (f) phosphorus.

8.38 Assign oxidation numbers to the atoms in each of the following molecules and ions: (a) NO; (b) SiF_4; (c) $KClO_3$; (d) MnO_2; (e) N_2H_4; (f) CaF_2; (g) Na_2HPO_4; (h) CH_2O.

8.39 Identify which of the following balanced chemical equations are oxidation-reduction equations.

(a) $Mg + 2H_2O \longrightarrow Mg(OH)_2 + H_2$
(b) $BaCl_2 + H_2SO_4 \longrightarrow BaSO_4 + 2HCl$
(c) $Br_2 + C_2H_4 \longrightarrow C_2H_4Br_2$
(d) $2KClO_3 \longrightarrow 2KCl + 3O_2$

8.40 Combine each of the two oxidation half-reactions shown below with each of the two reduction half-reactions shown to produce four balanced oxidation-reduction reactions.

$Ca \longrightarrow Ca^{2+} + 2e^-$
$K \longrightarrow K^+ + e^-$
$Br_2 + 2e^- \longrightarrow 2Br^-$
$O_2 + 4e^- \longrightarrow 2O^{2-}$

balancing oxidation-reduction equations

8.41 In each of the following balanced oxidation-reduction equations, identify those elements that undergo changes in oxidation number and indicate the magnitude of the change in each case.

(a) $3H_2S + 2HNO_3 \longrightarrow 3S + 2NO + 4H_2O$
(b) $5H_2SO_3 + 2MnO_4^- \longrightarrow$
$\qquad 5SO_4^{2-} + 2Mn^{2+} + 4H^+ + 3H_2O$
(c) $2CrO_2^- + 3ClO^- + 2OH^- \longrightarrow$
$\qquad 2CrO_4^{2-} + 3Cl^- + H_2O$
(d) $2Cu^{2+} + 2H_2O \longrightarrow 2Cu + O_2 + 4H^+$
(e) $2KOH + Cl_2 \longrightarrow KCl + KClO + H_2O$
(f) $BaSO_4(s) + C(s) \longrightarrow BaS(s) + 4CO$
(g) $5PbO_2(s) + 2Mn^{2+} + 5SO_4^{2-} + 4H^+ \longrightarrow$
$\qquad 5PbSO_4(s) + 2MnO_4^- + 2H_2O$
(h) $CH_4 + 2O_2 \longrightarrow CO_2 + 2H_2O$

8.42 Balance each of the following reactions:

(a) $H_2SO_3 + HIO_3 \longrightarrow H_2SO_4 + I_2 + H_2O$
(b) $PbO_2 + HBr \longrightarrow PbBr_2 + Br_2 + H_2O$
(c) $BaSO_4 + C \longrightarrow BaS + CO$

(d) $FeSO_4 + KClO_3 + H_2SO_4 \longrightarrow$
$\qquad Fe_2(SO_4)_3 + KCl + H_2O$
(e) $H_3AsO_4 + Zn + HNO_3 \longrightarrow$
$\qquad AsH_3 + Zn(NO_3)_2 + H_2O$
(f) $HIO_3 + HI \longrightarrow I_2 + H_2O$
(g) $Ag_2S(s) + HNO_3 \longrightarrow$
$\qquad AgNO_3 + NO + S + H_2O$
(h) $KMnO_4(s) \longrightarrow$
$\qquad K_2MnO_4(s) + MnO_2(s) + O_2$

naming inorganic compounds

8.43 Indicate the name of each of the following ions: (a) Ti^{2+}; (b) Cr^{2+}; (c) CrO_4^{2-}; (d) Ga^{3+}; (e) SO_3^{2-}; (f) MnO_4^{2-}; (g) Ca^{2+}; (h) IO_2^-.

8.44 Name each of the following compounds, giving two different names for each: (a) CrF_3; (b) $TiCl_3$; (c) $Fe_2(SO_4)_3$; (d) CdO; (e) PbO_2.

8.45 The formula for phosphoric acid is H_3PO_4. Name each of the following compounds: (a) $Ca_3(PO_4)_2$; (b) H_3PO_2; (c) $AlPO_3$.

8.46 Name each of the following: (a) $KClO_2$; (b) $LiBrO_3$; (c) OF_2; (d) K_2MnO_4; (e) $Fe(ClO_2)_2$.

8.47 Name each of the following, using two different naming methods: (a) NF_3; (b) N_2F_2; (c) N_2F_4.

8.48 Name each of the following binary compounds using two different names for each: (a) Cr_2O_6; (b) NbF_5; (c) SeO_2; (d) TiF_3; (e) AgF_2; (f) P_2Cl_4.

general exercises

[8.49] For each of the following word descriptions, write balanced chemical equations to describe what happens. Draw the Lewis structures of each molecule shown in italics. (a) Lithium nitride reacts with water to form *ammonia* and lithium hydroxide. (b) Copper metal reacts with nitric acid to form copper(II) nitrate and *nitrogen(IV) oxide*. (c) Hypophosphorous acid reacts with water to form phosphorous acid, and liberating hydrogen. (d) *Phosphorous acid* reacts with *hydrogen peroxide* to form *phosphoric acid* and water. (e) (Four balanced equations needed). Nitrogen and hydrogen gas are combined at high perssure and temperature to form ammonia. The ammonia is then oxidized with the aid of a catalyst to form *nitric oxide*. This oxide reacts with oxygen to produce nitrogen dioxide. Nitrogen dioxide reacts with water to form nitric acid and nitrous acid.

8.50 Draw the Lewis structure for the cyanamide ion, CN_2^{2-}. (Hint: Think of a familiar neutral molecule with which CN_2^{2-} is isoelectronic).

8.51 Draw suitable Lewis structures for each of the following anions (resonance forms may be required):

(a) CN^-; (b) PO_4^{3-}; (c) CO_3^{2-}; (d) $C_2H_3O_2^-$ (that is, CH_3COO^-).

8.52 Magnesia, MgO, used in making bricks for lining of high-temperature furnaces, melts at a very high temperature. By comparison, NaF could not be used for furnace bricks, because it melts at a temperature that is much too low. Suggest an explanation for the big difference in properties.

8.53 The ionic radius of Sr^{2+} is 1.13 Å, whereas that for Cd^{2+} is 0.97 Å. Account for the difference.

[8.54] For the alkali metal atoms and ions in Figure 8.3, make up a little table of the difference in radius between the atom and positive ion for each element. To what is the difference due? Does the difference change as the atomic number of the alkali metal changes? Can you account for any variations you note?

[8.55] Potassium permanganate solutions are widely used to analyze for the concentrations of easily oxidized species in solution. To standardize the permanganate solution, the usual practice is to titrate a solution of known concentration of oxalic acid dissolved in sulfuric acid solution. The unbalanced equation for the reaction is

$$KMnO_4 + H_2C_2O_4 + H_2SO_4 \longrightarrow$$
$$K_2SO_4 + MnSO_4 + CO_2 + H_2O$$

Balance this equation. Suppose that 0.465 g of pure oxalic acid were dissolved in dilute sulfuric acid and the solution adjusted to precisely 0.500 liter volume. Then 50 ml of this solution was titrated using a potassium permanganate solution of unknown concentration. It required 37.5 ml of permanganate solution to titrate (that is, to completely react with) all the oxalic acid. What is the concentration of the permanganate solution?

8.56 Bacterial decay of organic matter under anaerobic conditions (absence of air) can produce hydrides of the nonmetallic elements. In this manner methane, CH_4, hydrogen sulfide, H_2S, and phosphine, PH_3, are formed. (Phosphine is very reactive and produces a glow when it burns in air. The "will-o-the-wisp," a glow seen in marshes on a dark night, is probably due to the oxidation of PH_3 released to the air.) Write balanced equations showing the oxidation of each of these hydrides by molecular oxygen. Assume in each case that the highest possible oxidation state is attained by the nonmetal.

geometries of molecules; molecular orbitals

9

It is important for us to realize that molecules occupy three dimensions and have particular shapes determined by the bonding arrangements of the atoms. The shape of a molecule may be important in determining its chemical and physical properties. For example, consider the compounds methane, CH_4, and water, H_2O. You might suppose that two substances with such nearly similar formulas would have similar properties. Yet methane boils at $-161°C$, and water boils at $100°C$. We'll see in this chapter that much of this great difference in boiling points can be traced to the different shapes of the two molecules. Molecular shape is especially important in biochemical reactions, in which only molecules that are of a certain shape and size may be able to enter into a reaction. For example, small changes in the shapes and sizes of drug molecules may alter activity or may reduce the toxic side effects of a drug that is otherwise beneficial. Those properties of a molecule related to its shape and size are referred to as **steric effects**.

By the geometrical structure of a molecule, we mean the three-dimensional arrangement of the atoms. We can imagine that the atoms, represented as spheres, are connected by straight lines, as in the SF_6 molecule, shown in Figure 9.1. The term **bond distance** refers to the distance between two bonded atoms along this straight line. The term **bond angle** refers to the angle that is made by two of these lines at a given atom. It is possible to determine bond-distance and bond-angle information about a molecule from various experimental techniques. We won't discuss these techniques here, but you should simply accept the fact that by using them the chemist has acquired a great deal of infor-

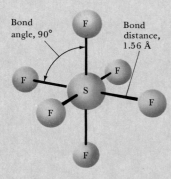

Bond angle, 90°

Bond distance, 1.56 Å

F F F S F F F

FIGURE 9.1 The geometrical structure of the SF_6 molecule, showing bond-distance and bond-angle information.

9.1

the valence-shell
electron-pair repulsion
(VSEPR) model

geometries of molecules;
molecular orbitals

mation about molecular geometries. Our concern is with how these geometries can be explained.

In Chapter 8, we employed the simple and empirical Lewis model to account for the formulas of covalent compounds. The Lewis structures that we draw are not geometrical figures. They simply describe the bonding connections between atoms in a two-dimensional representation. For example, the Lewis structure for methane, CH_4, does not tell us whether the molecule is planar or tetrahedral. To develop a model that explains three-dimensional structure, we must take into account the electrostatic interactions between electron pairs.

**9.1
the
valence-shell
electron-pair
repulsion
(VSEPR)
model**

We have seen that an atom is bonded to other atoms in a molecule by electron pairs that occupy its valence-shell atomic orbitals. These electron pairs are attracted to the central nucleus by the electron-nucleus attractive force. At the same time, though, they are repelled by the other electron pairs about the central atom. As a result, the electron pairs remain as far apart from one another as possible, while maintaining their distance from the central nucleus.

You will remember from Chapter 8 that the most commonly observed electronic arrangement about an atom in covalent compounds, especially for the nonmetallic elements, is an octet, or four pairs, of electrons. We might then ask how four electron pairs could arrange themselves about a central atom so as to minimize electron-pair repulsions, while maintaining their distance from the central nucleus. Two possibilities are shown in Figure 9.2. As it turns out, the tetrahedral arrangement is the one that has the electron pairs as far apart as possible for a given distance from the central nucleus. (This can be demonstrated using some rather lengthy trigonometric calculations, but we won't go through these.) We therefore consider that an octet of electrons about an atom consists of four electron pairs, with each pair occupying a region of space directed toward the vertex of a tetrahedron. If each of these four electron pairs bonds the central atom to another atom, for example as in methane, then the observed arrangement of bonded atoms about the central atom is tetrahedral.

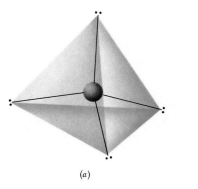
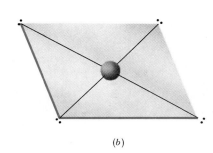

(a)   (b)

FIGURE 9.2 Two possible arrangements of four electron pairs around a central atom: (a) tetrahedral; (b) square planar. For a given distance of the electron pairs from the central atom, the repulsions are lower for the tetrahedral arrangement than for any other, such as the square planar.

Extending this idea we can formulate Table 9.1, which lists differing numbers of electron pairs in the valence shell and the geometry that corresponds to maximum separation of the pairs. Our model correctly predicts the structures of molecules with expanded valence shells; PF_5 and PCl_5 are observed, for example, to be trigonal bipyramids, as shown in Table 9.1. Similarly, SF_6 is found to be octahedral, as predicted.

The model for molecular geometries we have been discussing is often called the **valence-shell electron-pair repulsion (VSEPR) model**. It is

TABLE 9.1

geometries of electron-pair distributions about a central atom as a function of the number of electron pairs

NUMBER OF ELECTRON PAIRS	GEOMETRICAL ARRANGEMENT	PREDICTED BOND ANGLES	EXAMPLE
2	Linear	180°	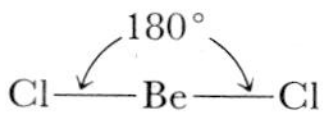
3	Trigonal planar	120°	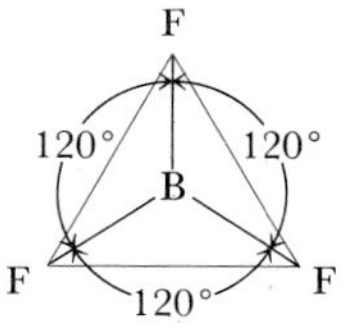
4	Tetrahedral	109.5°	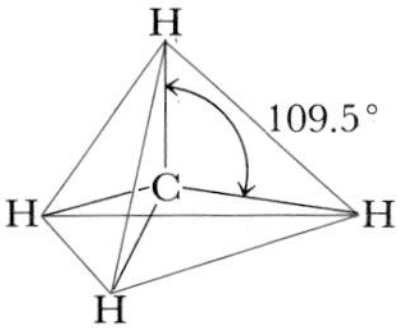
5	Trigonal bipyramidal	120° 90°	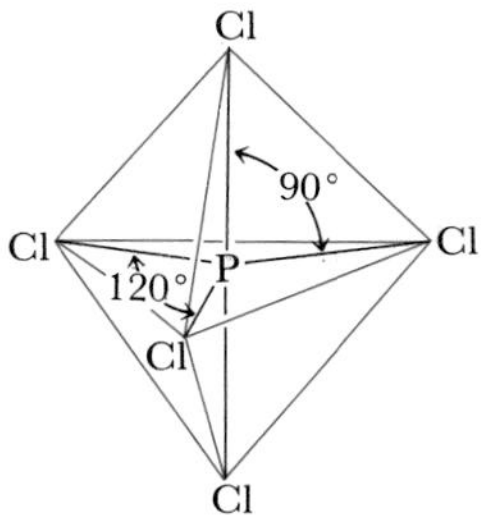
6	Octahedral	90°	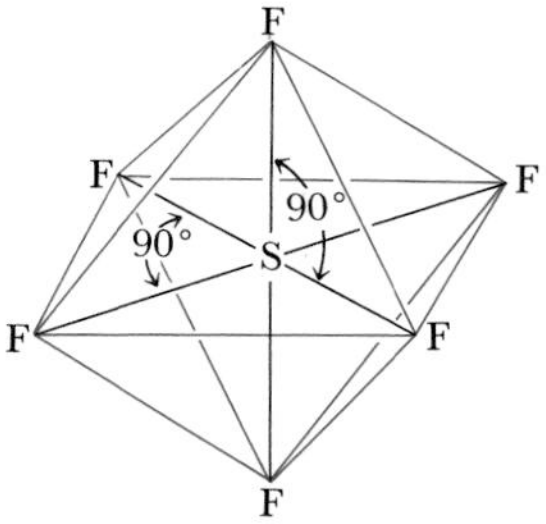

241

9.1

the valence-shell
electron-pair repulsion
(VSEPR) model

geometries of molecules;
molecular orbitals

based on the idea that the arrangement of bonded atoms around a central atom is determined by the repulsions between the electron pairs around that central atom. Let's consider some additional results based on this model.

It often happens that the electron pairs about a central atom are not all shared with other atoms. For example, remember that in the Lewis structure for water there are two unshared electron pairs about the oxygen:

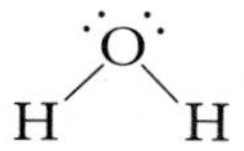

$$\ddot{O}$$
$$H \qquad H$$

These unshared electron pairs are as important as are shared pairs in determining the structure. However, when the structure of a molecule is studied experimentally, only the positions of the atoms are observed, not the positions of electrons. The presence of unshared electron pairs must therefore be inferred from what we see of the structures of molecules. In

TABLE 9.2

distribution between shared and unshared electron pairs in hydrides of the second-row nonmetals

COMPOUND	NUMBER OF SHARED PAIRS	NUMBER OF UNSHARED PAIRS	STRUCTURE
CH_4	4	0	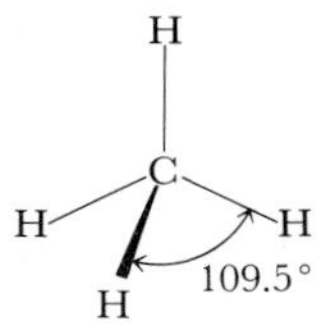
NH_3	3	1	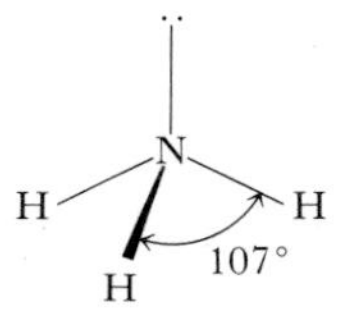
H_2O	2	2	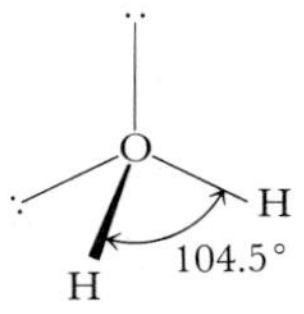
HF	1	3	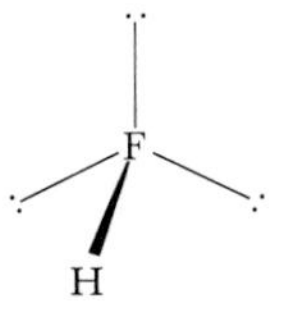

the series of molecules CH_4, NH_3, H_2O, HF, there is in each case an octet of electrons about the central atom. However, as shown in Table 9.2, the number of shared and unshared pairs of electrons varies. By taking account of the unshared electron pair, the VSEPR model predicts that CH_4 will be tetrahedral, NH_3 pyramidal, and H_2O bent. The VSEPR model tells us nothing about HF, since the structure is determined by only a bond distance.

It is possible to extend the VSEPR model a little further. We can imagine that each electron pair around a central atom has a certain volume requirement, but that the volume requirement of an unshared pair of electrons is larger than for a pair that is shared between two atoms. This is so because the shared pair is held by the attractive forces of two nuclear centers. This should have the effect of contracting the spatial distribution of the electrons. If we use this idea we conclude that the HNH angle in NH_3 should be a little less than the tetrahedral angle of $109.5°$, found in CH_4, and indeed it is, as shown in Table 9.2. Similarly, the HOH angle in H_2O is reduced from the tetrahedral angle. In the third row hydrides, the corresponding angles in PH_3 and H_2S are reduced even more.

Let's apply the idea that unshared electron pairs have a larger volume requirement to molecules with five electron pairs in the valence shell. In the trigonal bipyramid, shown in Figure 9.3, electron pairs in the axial and radial positions are not equivalent. In the axial position, an electron pair is repelled by three other electron pairs at $90°$ angles from it. For an electron pair in the radial position, there are only two such $90°$ interactions and two others much further removed, at $120°$. As a result, the repulsions from other electron pairs are larger for a pair located in an axial position than for a pair located in a radial position. Thus we predict that electron pairs with larger volume requirements will go into the radial locations.

The molecule SF_4 represents an example of a molecule with four shared electron pairs and one unshared pair. In the alternative VSEPR structures, shown in Figure 9.4, the unshared electron pair could be in either an axial or radial location. From the observed geometrical structure, also shown in Figure 9.4, we can infer that the unshared electron pair resides in a radial position, as predicted. Note that the "axial" S—F bonds are slightly bent back, suggesting that the unshared electron pair, with its greater volume requirement, is pushing them back.

Using the idea that unshared electron pairs are important in determining the observed geometry, we can predict molecular geometries for various numbers of shared and unshared electron pairs about a central atom. Table 9.2 outlines these possibilities for the case of four electron pairs. In Table 9.3 are listed the geometries to be expected for four, five, or six electron pairs. The guiding rule in arriving at these predictions is that the unshared electron pairs have a larger volume requirement than shared electron pairs, and that this determines in some cases where they are placed. Some of the possible arrangements have been omitted from the table because there are no known molecules with those arrangements.

The VSEPR model is very useful in predicting the structures of molecules. The predictions of the model are basically contained in Tables

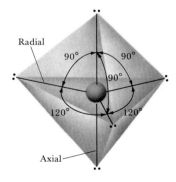

FIGURE 9.3 Trigonal bipyramidal arrangement of five electron pairs about a central atom. The repulsions experienced by the axial pairs are greater. For this reason, unshared electron pairs tend to occupy the radial positions.

9.1

the valence-shell
electron-pair repulsion
(VSEPR) model

geometries of molecules;
molecular orbitals

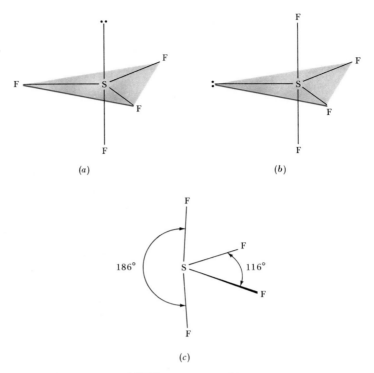

FIGURE 9.4 Alternative VSEPR structures for SF_4, and the observed structure. (*a*) Unshared pair in axial position. (*b*) Unshared pair in radial position. (*c*) Observed structure.

9.1 and 9.3. When the VSEPR model is applied to predicting the structures of molecules or ions containing multiple bonds, one additional rule is required: *A multiple bond affects molecular geometry as though it were a single electron pair.*

To use the VSEPR model to predict geometry, you should proceed in the following way:

1. Count the number of valence electrons and write out the Lewis structure for the molecule or ion.
2. From the Lewis structure determine the number of unshared electron pairs and number of bonds to other atoms. (Remember to count multiple bonds as a single bond.)
3. Use Tables 9.1 and 9.3 to predict the geometrical arrangement.

SAMPLE EXERCISE 9.1

Using the VSEPR model, predict the geometrical structures of the following: (a) $SnCl_3^-$; (b) PO_4^{3-}; (c) CO_2; (d) ethylene, C_2H_4.

Solution: (a) In $SnCl_3^-$ the central tin is bonded to three chlorines. Tin possesses four valence electrons (group 4A); each chlorine contributes one, and there is an extra electron providing the negative charge. Thus there are eight electrons in the valence shell of tin. The Lewis structure for the ion is as follows:

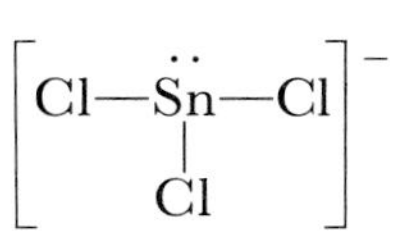

$$\left[\begin{array}{c} \ddot{C}l-\overset{..}{Sn}-Cl \\ | \\ Cl \end{array} \right]^{-}$$

The four electron pairs should thus be disposed at the corners of a tetrahedron. One of the corners is occupied by the unshared electron pair. The geometrical arrangement of the atoms is thus pyramidal:

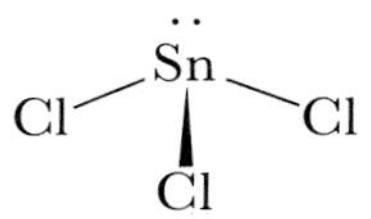

$$\overset{\displaystyle \overset{..}{Sn}}{Cl \diagdown \underset{\displaystyle |}{} \diagup Cl} \\ Cl$$

(b) The Lewis structure for the phosphate ion is

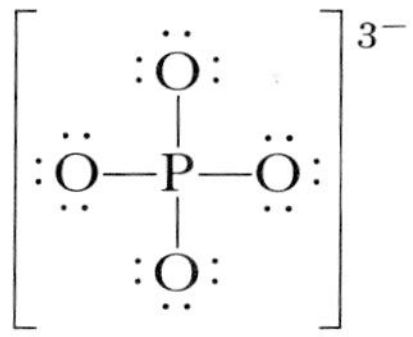

$$\left[\begin{array}{c} :\ddot{O}: \\ | \\ :\ddot{O}-P-\ddot{O}: \\ | \\ :\ddot{O}: \end{array} \right]^{3-}$$

There are thus four electron pairs in the valence shell of phosphorus; the ion should be tetrahedral.

(c) The Lewis structure for carbon dioxide is

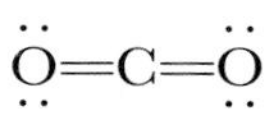

$$\ddot{O}=C=\ddot{O}$$

The rule that each multiple bond behaves as a single electron pair means that, in effect, we have two electron pairs in the valence shell. The molecule should therefore be linear (Table 9.1). The CO_2 molecule is in fact observed to be linear.

(d) The Lewis structure for ethylene is

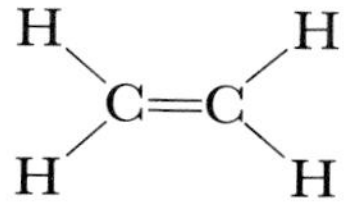

$$\underset{H}{\overset{H}{\diagdown}} C = C \underset{H}{\overset{H}{\diagup}}$$

The multiple bond between carbon atoms is treated as a single electron pair, so we have effectively three electron pairs in the valence shell. The three bonds about each carbon are distributed in a plane, with 120° bond angles (Table 9.1). Ethylene is thus predicted to be planar with approximately 120° bond angles, as observed:

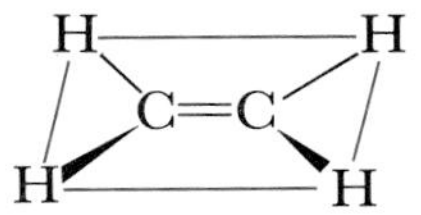

TABLE 9.3

relationship between numbers of shared and unshared electron pairs and observed molecular shapes

NUMBER OF ELECTRON PAIRS ABOUT CENTRAL ATOM	GEOMETRICAL ARRANGEMENT OF ELECTRON PAIRS	NUMBER OF BONDING PAIRS	OBSERVED MOLECULAR SHAPE		EXAMPLE	FORMULA
4	Tetrahedral	4	Tetrahedral	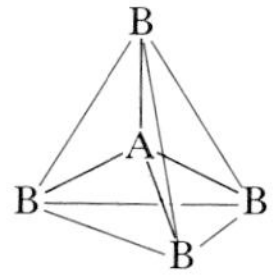	CH_4	AB_4
		3	Trigonal pyramidal	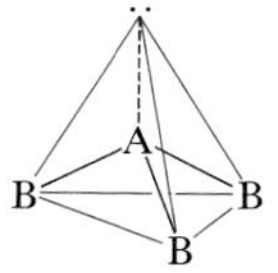	NH_3	AB_3
		2	Bent	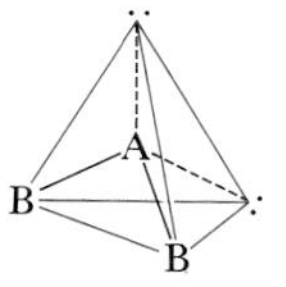	OH_2	AB_2
		1	Linear	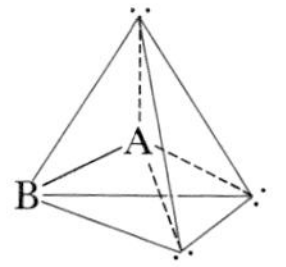	FH	AB

TABLE 9.3 (*Cont.*)

NUMBER OF ELECTRON PAIRS ABOUT CENTRAL ATOM	GEOMETRICAL ARRANGEMENT OF ELECTRON PAIRS	NUMBER OF BONDING PAIRS	OBSERVED MOLECULAR SHAPE		EXAMPLE	FORMULA
5	Trigonal bipyramidal	5	Trigonal bipyramidal		PCl_5	AB_5
		4	Seesaw		SF_4	AB_4
		3	T-shaped		ClF_3	AB_3
		2	Linear		XeF_2	AB_2
6	Octahedral	6	Octahedral		SF_6	AB_6
		5	Square pyramidal		BrF_5	AB_5
		4	Square planar		XeF_4	AB_4

9.2
hybrid orbitals and molecular shape

The VSEPR theory provides a simple model for predicting the shapes of molecules. However, it does not explain why bonds exist between atoms. The shared and unshared pairs are taken as given; the model is used simply to deduce the shape of the molecule or ion. In developing a theory of covalent bonding, chemists have also approached the problem from another direction. Suppose we take the formula and geometrical structure of the molecule as given. How can we account for the observed geometries in terms of the atomic orbitals used by the atoms in forming bonds to one another?

In the Lewis-Langmuir theory, covalent bonding occurs when atoms share electrons. But sharing of electrons can occur only if an atomic orbital of one atom shares the same region of space as an atomic orbital of another. The shared electron pair then occupies an orbital that consists in part of the atomic orbitals contributed from each atom. Orbitals that occupy the same region of space are said to *overlap*. This idea is illustrated in Figure 9.5, which shows the coming together of two hydrogen atoms to form the H_2 molecule. The overlap of the $1s$ orbital on one H atom with that on the other increases as the atoms draw near. The observed bond distance is a compromise between increased overlap of the atomic orbitals, which draws the atoms together, and the nuclear-nuclear repulsion, which forces them apart. We'll see as we go along that there are some rules that determine effective overlap of orbitals in more complicated cases.

In extending this view of bond formation to polyatomic molecules, we want a model that localizes, as much as possible, the valence electron pairs in the regions between atoms. The problem is to construct such a model, starting with the valence-shell atomic orbitals of the atoms. To see how this might be accomplished, let's consider methane, CH_4. This molecule has a tetrahedral arrangement of hydrogen atoms around a central carbon, as shown in Table 9.2. The valence shell orbitals on carbon available for bonding are the $2s$ and three $2p$ orbitals. Each hydrogen atom has, of course, a single $1s$ orbital. The three $2p$ orbitals of carbon point along the x, y, and z axes (Figure 6.18), and the $2s$ orbital is spherically symmetrical. It is therefore not possible to employ these four orbitals individually to make the four equivalent bonds between carbon and hydrogen. But if they are not appropriate by themselves for forming the four bonds to hydrogen, perhaps some combination of them is more satisfactory.

We can imagine a stepwise procedure of getting a carbon atom initially in its ground state ready for bonding to four hydrogen atoms.

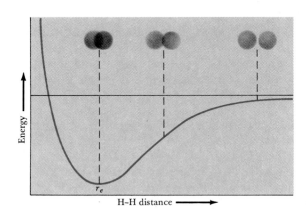

FIGURE 9.5 Formation of the stable H_2 molecule by the overlap of two hydrogen $1s$ orbitals. The minimum in the energy surface, at r_e, represents the equilibrium H—H bond distance in H_2.

The electronic configuration of the ground state carbon atom, in orbital diagram representation, is as follows:

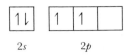

Notice that a carbon atom in its ground state could at most form two bonds to two other atoms, since there are just two unpaired electrons available. To obtain the capability for forming four bonds, one of the $2s$ electrons must be "promoted" to the vacant $2p$ orbital:

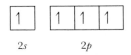

Because the $2p$ orbital is of higher energy than the $2s$ this promotion costs energy. Next, the mathematical functions that describe the $2s$ and $2p$ orbitals can be combined with one another. The combination yields four equivalent orbitals, called **hybrid orbitals,** which point toward the corners of a tetrahedron. Each one of these hybrid orbitals consists of a certain amount of the $2s$ orbital and a certain amount of the $2p$ orbitals. Because the four hybrid orbitals are made up of one $2s$ and three $2p$ orbitals, they are labeled sp^3 hybrids. When the $2s$ and $2p$ orbitals are mixed so as to give the four hybrid orbitals we obtain:

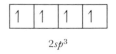

Formation of the sp^3 hybrid orbital set is shown in Figure 9.6. The carbon is now "prepared" to form four equivalent bonds with tetrahedral geometry. The interaction of the four sp^3 hybrid orbitals from carbon with four hydrogen atoms to form methane, CH_4, is shown in Figure 9.7.

The first of the steps outlined above, promotion of the $2s$ electron, requires energy. The second does not, since it merely means an averaging to give four equivalent orbitals, each with energy equal to the average of the original set. The overall process does require energy. The reason it occurs is that the system more than recovers this energy in bond formation. As we've seen, a ground-state carbon atom could at most form two bonds to hydrogen, using the two electrons in its $2p$ orbitals. With promotion and hybridization, four bonds are possible. Furthermore, hybridization permits greater orbital overlap, because the hybrid orbitals extend out further from the central nucleus.

In molecules such as NH_3 and H_2O, the central atom has about it four electron pairs in an approximately tetrahedral arrangement. The orbitals used by the central atom can be thought of as sp^3 hybrid orbitals. In NH_3 one of the sp^3 hybrid orbitals contains the unshared electron pair,

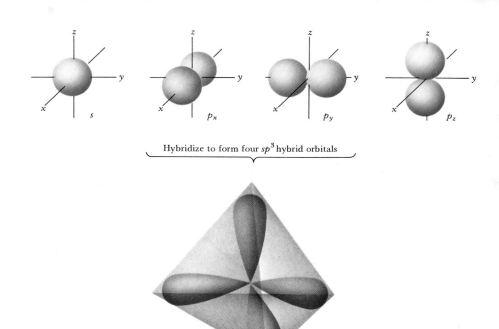

FIGURE 9.6 Formation of sp^3 hybrid orbitals from a set of one s and three p orbitals.

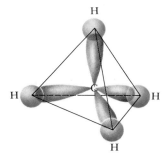

FIGURE 9.7 Formation of methane by overlap of four hydrogen atom $1s$ orbitals with the four sp^3 hybrid orbitals of a central carbon atom.

the other three are employed in the bonds to hydrogen. In H_2O two of the orbitals contain unshared pairs, two are employed in bonds to hydrogen. In these examples the hybrid orbitals employed by the central atom are not pure sp^3 hybrids, since the bond angles about the central atom are not exactly the 109.5° tetrahedral angle, as shown in Table 9.2. In the mathematical formulation of hybrid orbitals, it is possible for the contribution of the s orbital to be a little greater in one hybrid orbital, a little smaller in another, in order to make the angles between the hybrid orbitals come as close as possible to the observed bond angles. We shall make no attempt to follow through with these refinements, but shall instead employ just the idealized equivalent hybrid set.

Various combinations of atomic orbital sets can be mixed, or hybridized, to obtain different geometries of orbitals about a central atom. Table 9.4 shows several of the more important hybrid orbital combinations and the geometries to which they correspond. The hybrid orbital sets are illustrated in Figure 9.8. Note that expanded valence shells about atoms, that is, those in which there are more than eight electrons in the valence shell, can be accommodated by mixing in d orbitals along with s and p.

SAMPLE EXERCISE 9.2

The molecule BeH_2 is known to be linear. Account for the bonding in BeH_2 in terms of the hybrid orbitals employed by Be in bonding to the two hydrogen atoms.

Solution: The orbital diagram for the Be atom is as follows:

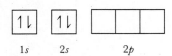

The Be atom in its ground state is incapable of forming bonds with other atoms, because all the electrons are paired. However, suppose one of the electrons is promoted to the $2p$ orbital:

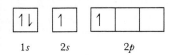

and the $2s$ and one of the $2p$ orbitals are mixed to form two sp hybrid orbitals:

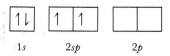

The electrons in the two sp hybrid orbitals can form shared electron pair bonds with two hydrogen atoms. The sp hybrid orbitals are directed at $180°$ angles from one another, Figure 9.8, so the BeH_2 molecule is linear.

The purpose in formulating hybrid orbitals is to provide a convenient model in which we can imagine the electrons to be localized in the region between two atoms. The picture of hybrid orbitals has limited predictive value; that is, we cannot say in advance that in NH_3 the nitrogen uses essentially sp^3 hybrid orbitals. Once given the molecular geometry, however, we can employ the concept of hybridization to describe the atomic orbitals employed by the central atom in bonding.

SAMPLE EXERCISE 9.3

From a knowledge of the geometry about the central atom in each of the following, indicate the hybridization of orbitals employed: (a) $TeCl_2$—bent; (b) SF_4—see Figure 9.4; (c) PH_4^+—tetrahedral.

Solution: (a) Tellurium has six valence electrons. By covalent bonding with two chlorine atoms, it acquires an octet of electrons. This set of four electron pairs will occupy four roughly equivalent orbitals, so that we might suppose that the hybridization would be approximately sp^3. Two of the hybrid orbitals contain unshared electron pairs, the other two contain the pairs shared with Cl:

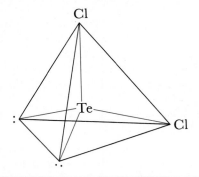

(b) The valence orbitals of the sulfur in SF_4 contain ten electrons, six from sulfur and one from each of the four fluorines. With an expanded octet of ten electrons, the use of a d orbital on the sulfur is indicated. This leads us to dsp^3 hybridization, which is that of a trigonal bipyramid, as shown in Figure 9.8. Four of the five electron pairs about the sulfur in SF_4 are involved in bonding to fluorine. The remaining one is used to contain the unshared electron pair. The observed structure of SF_4 is shown in Figure 9.4.

(c) The phosphorus atom of PH_4^+ contributes five valence electrons to bonding. Each of the four hydrogen atoms contributes one, for a total of nine electrons. But the overall charge on the ion is $1+$, because one electron has been removed. There are therefore eight electrons about P in PH_4^+ to be distributed among four equivalent bonds. The structure of PH_4^+ is therefore expected to be tetrahedral, with sp^3 hybrid orbitals employed by phosphorus.

TABLE 9.4

geometrical arrangements characteristic of hybrid orbitals

ATOMIC ORBITAL SET	HYBRID ORBITAL SET	GEOMETRICAL ARRANGEMENT	EXAMPLES
s,p	sp	Linear (180° angle)	$Be(CH_3)_2$, $HgCl_2$
s,p,p	sp^2	Trigonal planar (120° angles)	BF_3
s,p,p,p	sp^3	Tetrahedral (109.5° angles)	CH_4, $AsCl_4^+$, $TiCl_4$
d,s,p,p	dsp^2*	Square planar (90° angles)	$PdBr_4^{2-}$
d,s,p,p,p	dsp^3*	Trigonal bipyramidal (120° and 90° angles)	PF_5
d,d,s,p,p,p	d^2sp^3*	Octahedral (90° angles)	SF_6, $SbCl_6^-$

*Depending on the particular element involved, the d orbital that mixes with the s and p may be of major quantum number one lower, or of the same major quantum number. This has no effect on the geometrical characteristics of the resulting hybrid set.

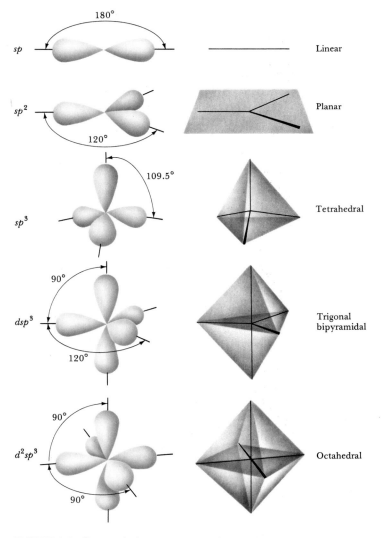

FIGURE 9.8 Geometrical arrangements characteristic of hybrid orbital sets.

The concept of hybridization may be applied also to molecules containing multiple bonds. For example, we have seen (Sample Exercise 9.1) that ethylene possesses a carbon-carbon double bond and is a planar molecule. The planar arrangement of three bonds about each carbon suggests that the hybrid orbital set it uses to bond to the other carbon and the two hydrogens is sp^2 (Table 9.4). Because the valence orbitals on carbon consist of a $2s$ and *three* $2p$ orbitals, one $2p$ orbital remains unused after forming the sp^2 hybrid set. This is illustrated in Figure 9.9. We notice that the carbon atoms of ethylene are bonded together through overlap of sp^2 hybrid orbitals. The sign of the wave function in the orbitals is positive in each case. Looking along the C—C bond axis of the molecule, the bond is completely symmetrical. Bonds of this kind are called σ (sigma) bonds. The bonds between carbon and hydrogen are also σ bonds. Because the hydrogen uses a $1s$ orbital that is spherical, each bond is symmetrical with respect to that particular C—H bond axis.

The formation of the σ bonds gives three electron pairs about carbon in the sp^2 hybrid orbitals, and an electron on each carbon in the $2p$ orbital that is perpendicular to the plane of the molecule. The p orbitals shown on the two carbon atoms in Figure 9.9 can overlap with one another in a sideways fashion. By means of this overlap a second C—C covalent bond is formed, as shown in Figure 9.10. Each carbon has achieved an octet of electrons. The second C—C bond differs from the first, because it is not symmetric about the C—C bond axis. Above the plane of the molecule the orbital is positive, and below it is negative. This kind of bond, in which the sign of the orbital changes about the bond axis, is called a π (pi) bond.

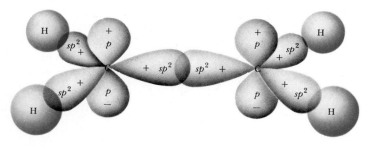

FIGURE 9.9 Hybridization of carbon orbitals in ethylene. The σ bond framework, formed from sp^2 hybrid orbitals on the carbon atoms, determines the observed geometrical structure of the molecule.

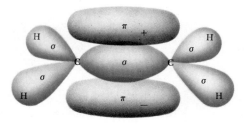

FIGURE 9.10 Formation of the π bond in ethylene by overlap of the $2p$ orbitals on each carbon atom. Note that the centers of charge density in the π bond are above and below the bond axis, whereas in σ bonds the centers of charge density lie on the bond axes.

It is useful to consider the hybridization about carbon in ethylene in terms of an orbital diagram:

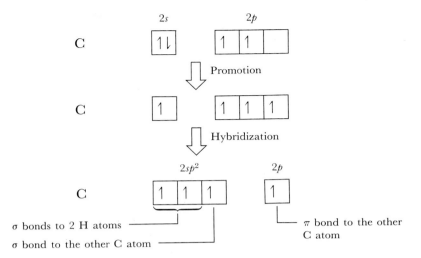

The first step in our imaginary process of preparing the carbon atom for bonding is again the promotion of one of the $2s$ electrons to the vacant $2p$ orbital. We then form an sp^2 hybrid orbital set from the $2s$ and two of the $2p$ orbitals. The orbitals of the hybrid set are used in bonding as indicated above. By electron pairing in each orbital, carbon gains an octet of electrons in its valence shell orbitals. Because two pairs of electrons are shared between carbon atoms in ethylene, as compared with only one pair in an ordinary C—C single bond, you might suppose that the C—C bond distance would be shorter in ethylene than, for example, in ethane, C_2H_6. This is indeed the case as the following comparison shows:

Ethylene
C—C distance = 1.34 Å

Ethane
C—C distance = 1.54 Å

According to the hybrid orbital picture, the double bond between carbons in ethylene consists of one σ and one π bond. This picture of the bonding in ethylene is in good accord with its chemical properties. For example, ethylene reacts with bromine, Br_2, to form dibromoethane:

$$\text{Ethylene} + Br_2 \longrightarrow \text{Dibromoethane} \qquad [9.1]$$

In this reaction the double bond is opened, and a bromine adds to each carbon. The C—C π bond is converted to two C—Br σ bonds.

253

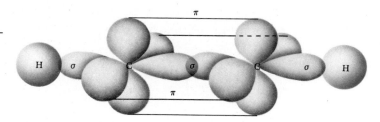

FIGURE 9.11 Formation of two π bonds in acetylene from the overlap of two sets of carbon $2p$ orbitals.

Acetylene, C_2H_2, is a linear molecule. The carbon in this instance may be visualized as using sp hybrid orbitals to bond to the other carbon and to hydrogen. This leaves two valence p orbitals at right angles to the axis of the sp hybrid set (Figure 9.11). These overlap to form a pair of π bonds. We've seen (Sample Exercise 8.9) that the Lewis structure for acetylene requires a carbon-carbon triple bond. This triple bond can be thought of as formed from one σ and two π bonds.

DELOCALIZED ORBITALS

In terms of the history of chemistry, benzene, C_6H_6, is an extremely prominent substance. For many years the relationship between the structure and composition of this molecule remained a mystery. Finally, August Kekulé, in 1865, devised a cyclic structure for the molecule. A fundamental understanding of how the electrons are distributed was not possible, however, until the development of modern chemical-bonding theory.

Benzene consists of a cyclic arrangement of six carbon atoms, with a single hydrogen attached to each carbon. The molecule is planar and has six perfectly equal C—C and C—H bonds. All the bond angles around each carbon are 120°. The equivalence of the C—C bonds can be accounted for by writing two Lewis structures and saying that the benzene molecule is intermediate between each of these forms:

Any one C—C bond is a double bond in one of the resonance structures, and a single bond in the other structure. Thus each C—C bond is somewhere between a single and double bond in character. The C—C distance in benzene is 1.395 Å, intermediate between the values given above for the C—C single bond in ethane (1.54 Å) and the C=C double bond in ethylene (1.34 Å).

To describe benzene in terms of hybridization of carbon orbitals, we follow the procedure of setting up a hybrid orbital set consistent with the skeletal structure for the molecule. Since each carbon is surrounded by three atoms at 120° angles in a plane, the appropriate hybrid set (Table 9.2) is sp^2, as shown in Figure 9.12(a). This leaves a p orbital on each carbon perpendicular to the plane of the benzene ring. The situation is very much like that in ethylene, except that now we have six carbon $2p$ orbitals, in a cyclic arrangement. The six carbon $2p$ atomic orbitals interact with one another to form π orbitals. Each of the $2p$ orbitals overlaps with two others, one on each adjacent carbon atom, to form a kind of doughnut above and below the plane of the benzene ring, as illustrated in Figure 9.12(b). The sign of the wave function is positive in one doughnut, and negative in the other.

Because there is one electron from each $2p$ orbital, there is a total of three electron pairs in the π orbitals formed in this manner. The electrons in the π orbitals of benzene are said to be *delocalized* in the sense that any one electron is free to move around the entire circle of carbon atoms. This delocalization of the π electrons gives benzene a special stability. For example, this substance does not react readily with bromine, as does ethylene. Benzene and many related molecules are referred to as **aromatics.*** It is usual to represent aromatic molecules by leaving off the hydrogens attached to carbon and showing only the carbon-carbon framework. The presence of π electrons may be shown by using one of the Lewis structures or by placing a circle in the center of the carbon ring. Several aromatic compounds are depicted in Figure 9.13.

---

*The name aromatics arose because some of the compounds related to benzene possess a spicy fragrance. However, benzene and many other aromatics do not have a pleasant smell.

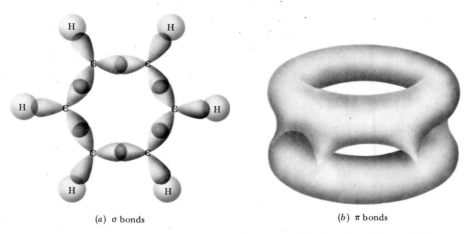

(a) σ bonds   (b) π bonds

FIGURE 9.12 The σ and π bond networks in benzene, C_6H_6. (a) The σ bonds all lie in the molecular plane and are formed from carbon sp^2 hybrid orbitals. (b) The π bond network is formed from overlap of a $2p$ orbital on a carbon with the $2p$ orbitals of each of its neighbors. Six π molecular orbitals result.

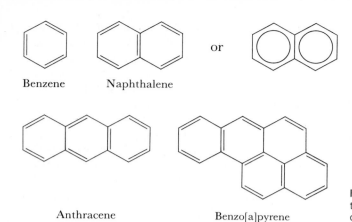

or

Benzene   Naphthalene

Anthracene   Benzo[a]pyrene

FIGURE 9.13 Several aromatic organic structures. Each vertex represents a carbon atom. A hydrogen is attached to each carbon that has only three bonds in the structures as shown.

Many of the aromatic hydrocarbons are carcinogenic, that is, cancer causing. Benzo[a]pyrene,* shown in Figure 9.13, is among the most potent carcinogens. It is found in significant quantities in urban atmospheres and, most especially, in cigarette smoke. It is regarded as a major cause of lung cancer. The major sources of benzo[a]pyrene in urban atmospheres are coal-burning power plants, engines that burn gasoline and diesel fuel, and incinerators used for refuse disposal. Studies carried out in Britain show that for nonsmokers, the incidence of lung cancer among urban dwellers is substantially higher than for rural populations with the same age profile. In both groups the incidence of lung cancer is much higher for cigarette smokers than for nonsmokers.

In the compounds shown in Figure 9.13, the delocalization of π electrons can be continued from one ring to another in the larger molecules by building up one hexagonal arrangement of carbon atoms onto another. The limiting case of this occurs in graphite. The structure of graphite is shown in Figure 9.14. The carbon atoms are arranged in layers; each carbon atom is surrounded by three other carbon atoms, all at the same distance of separation. The distance between adjacent carbon atoms in the plane is 1.42 Å, intermediate between the values for C—C

*The symbol [a] in the name for benzo[a]pyrene is used to designate the manner in which the six-membered rings are connected together.

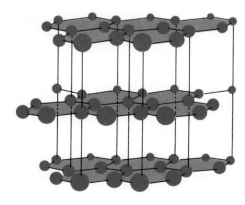

FIGURE 9.14 The structure of graphite.

single (1.54 Å) and C=C double bonds (1.34 Å). The distance between adjacent layers is 3.41 Å, too great a distance for a true bond to exist. Graphite occurs in grey, shiny plates and is usually found in masses of thin, easily separated sheets (see Figure 9.15). The layers easily slide past one another when rubbed, giving the substance a greasy feel. Graphite has been used as a lubricant and in making the "lead" in pencils.

The properties of graphite are due to the nature of the bonding between carbon atoms. Because the individual layers in graphite are not directly bonded together, they slide past one another easily and are readily separated. We can understand the bonding within each layer by supposing that each carbon is bonded to its three neighbors by sp^2 hybrid σ bonds. The remaining $2p$ orbital is employed in π bonding to the same three atoms, as illustrated in Figure 9.16. But because those three neigh-

FIGURE 9.15 A piece of graphite photographed with an electron microscope (magnification about 15 million times). The bright bands are layers of carbon atoms that are only 3.41 Å apart. (*P. A. Marsch and A. Voet, J. M. Huber Corp.*)

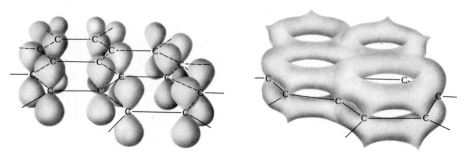

FIGURE 9.16 Formation of the π bonds in graphite.

boring atoms are also involved in π bonding to two other carbons, the network of π bonds extends over essentially the entire plane. The electrons that occupy the π bonds are free to move from one bond to the next. Thus graphite represents an extension to an entire plane of the kind of delocalization we saw in benzene. In benzene the π electrons are free to move about the circumference of the ring. In graphite they are free to move over the entire plane. Because of this freedom of motion, graphite is a good conductor of heat and electricity in directions along the planes of carbon atoms. (If you have ever taken apart a flashlight battery, you know that the central electrode in the battery is made of graphite.) Although graphite readily conducts heat and electricity along the planes, it is an insulator in the direction normal to (perpendicular to) the planes. This is so because there is no means by which electrons can move easily from one plane to the other.

HYBRID ORBITALS AND MOLECULAR GEOMETRIES

On the basis of all the examples we've seen, we can formulate a few general conclusions that are helpful in using the concept of hybrid orbitals to discuss molecular structures.

1. Every pair of bonded atoms shares one or more pairs of electrons. In every bond at least one pair of electrons is localized in the space between the atoms, in a σ bond. The appropriate set of hybrid orbitals used to form the σ bonds between an atom and its neighbors is determined by the observed geometry of the molecule. The relationship between hybrid orbital set and geometry about an atom is given in Table 9.4.
2. The electrons in σ bonds are localized in the region between two bonded atoms and do not make a significant contribution to the bonding between any other two atoms.
3. When atoms share more than one pair of electrons, the additional pairs are in π bonds. The centers of charge density in a π bond lie above and below the bond axis.
4. π bonds may extend over more than two bonded atoms. Electrons in π bonds that extend over more than two atoms are said to be delocalized.

SAMPLE EXERCISE 9.4

Formaldehyde, which is a planar molecule, has the following Lewis structure:

$$H \atop H \diagdown C=O \;\; ..$$

The bond angles about the carbon atom are nearly $120°$. Describe the bonding in formaldehyde in terms of an appropriate set of hybrid orbitals at the carbon atom.

Solution: The presence of three bonds in a plane about a central atom suggests sp^2 hybrid orbitals for the σ bonds (Table 9.4). There remains a $2p$ orbital on carbon, perpendicular to the plane of the three σ bonds. This orbital overlaps with a similarly oriented orbital on oxygen to form a π bond between carbon and oxygen (see Figure 9.17).

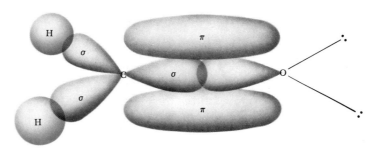

FIGURE 9.17 Formation of σ and π bonds in formaldehyde.

**9.4
molecular
orbitals**

The models for covalent bonding and molecular geometries discussed in this and the previous chapter are very useful. They provide a nice way of relating the formulas and structures of molecules to the electron configurations of the atoms involved. For example, we can understand why methane has the formula CH_4, and why the arrangement of C—H bonds about the central carbon is tetrahedral. But in all of this discussion, we have sidestepped a rather important question: Why do atoms combine to form covalent bonds in the first place? The answer to this question has to be expressed in terms of energy.

We have seen that electrons in atoms exist in allowed energy states called atomic orbitals. The quantum theory tells us that, in a similar way, electrons exist in molecules in allowed energy states that are called **molecular orbitals.** Because molecules are more complex than atoms, it is no surprise that molecular orbitals are more complex than atomic orbitals. As a very useful approximate description, we can think of them as formed by combining atomic orbitals. We have already done this in considering hybridization. For example, in methane, CH_4, we imagine that the bonding electrons are in orbitals formed by combining each sp^3 hybrid orbital from carbon with a $1s$ orbital from a hydrogen atom. But there is more involved in the combining of atomic orbitals than we have so far considered. We must examine the formation of molecular orbitals from atomic orbitals in more detail to understand why molecules are more stable or less stable than the separated atoms.

There are various rules and restrictions for how atomic orbitals can be combined:

1. When a set of atomic orbitals is combined to form molecular orbitals, the number of molecular orbitals formed is equal to the number of atomic orbitals in the set.

2. The average energy of the molecular orbitals formed by combining a set of atomic orbitals is approximately equal to the average energy of the atomic orbitals. However, some of the molecular orbital energies are lower than the energies of the starting atomic orbitals, while others are higher.

3. The Pauli principle is obeyed for molecular orbitals just as for atomic orbitals; there can be at most two electrons in each molecular orbital, and they must have their spins paired.

4. Atomic orbitals combine most effectively with other atomic orbitals of comparable energy.

5. The effectiveness with which two atomic orbitals combine is proportional to their overlap with one another. Orbitals that have zero overlap thus cannot combine at all.

6. When a molecular orbital is formed by overlap of two nonequivalent atomic orbitals, the bonding molecular orbital contains a greater contribution from the atomic orbital of lower energy. Conversely, the antibonding molecular orbital contains a greater contribution from the higher energy atomic orbital. (Bonding and antibonding orbitals will be explained shortly.)

Let us now illustrate these rules by considering the molecular orbital descriptions of simple diatomic molecules. We begin with two hydrogen atoms coming together to form H_2, as illustrated in Figure 9.5. The $1s$ orbital on each H atom interacts with the $1s$ orbital on the other atom to form two molecular orbitals (rule 1). These are illustrated in the energy level diagram of Figure 9.18. One of the combinations leads to a lowering in energy; this is the **bonding orbital,** labeled σ. The other leads to a destabilization; this is referred to as the **antibonding orbital,** labeled σ^* (rule 2). Each hydrogen atom contributes one electron to the bonding;

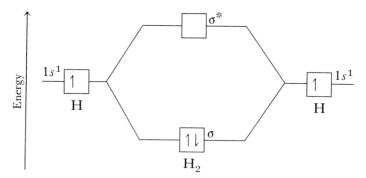

FIGURE 9.18 Energy-level diagram for the molecular orbitals in the H_2 molecule.

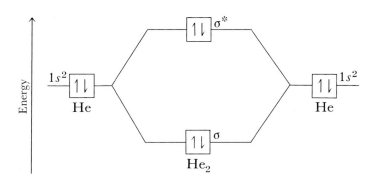

$1s^2$ He

σ^*

σ

He$_2$

$1s^2$ He

FIGURE 9.19 Energy-level diagram for the molecular orbitals in the He$_2$ molecule.

thus there are two electrons to be placed in the molecular orbitals formed from the atomic orbitals. Here we use the Pauli exclusion principle (rule 3). The two electrons in H$_2$ are placed in the lower-energy molecular orbital. Because the two $1s$ orbitals forming this molecular orbital are of the same energy, they combine effectively (rule 4). This means that the bonding σ orbital is substantially lower in energy than the $1s$ orbitals that make up the molecular orbital. In other words, H$_2$ is more stable than two separate H atoms.

Now let's consider the coming together of two helium atoms to form the He$_2$ molecule. Again, we have two $1s$ orbitals interacting to form a bonding σ and antibonding σ^* pair of molecular orbitals. In this case, however, each helium atom contributes two electrons, so that the total number to be placed is four. As Figure 9.19 shows, two are placed in the σ orbital, and the other two must be placed in the σ^* orbital. Using rule 2 we recognize that the σ^* orbital is destabilized to the same extent that the σ orbital is stabilized. Thus the two electrons that must go into the σ^* orbital destabilize the He$_2$ molecule just as much as the two electrons in the σ orbital stabilize it. We predict from this model that there is no net stabilization in He$_2$. Laboratory studies have shown that there is no significant tendency for two helium atoms to bond together to form He$_2$.

From these two examples of the molecular-orbital model, we see that H$_2$ is stable because electrons can be accommodated in bonding molecular orbitals. The He$_2$ molecule, on the other hand, is not stable because there are as many electrons in antibonding orbitals as in bonding orbitals.

Recall from our earlier discussion of atomic orbitals (Section 6.6) that the wave function, the mathematical expression that describes the orbital, has either a positive or negative sign, depending on the type of orbital involved and the region of space. The wave functions for the outer regions of s orbitals are always positive. However, when we combine atomic orbitals to form molecular orbitals, we can combine them with the same sign or with different signs. In the bonding molecular orbital, the wave functions are combined with the same sign, whereas in the antibonding orbital they are combined with opposite signs.

Figure 9.20 shows a contour diagram of the wave function of the σ

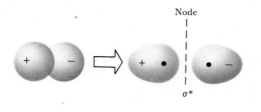

FIGURE 9.20 Contour diagram of the wave functions for the σ and σ* molecular orbitals in H_2. Note the presence of a nodal plane midway between the atoms in the σ* antibonding orbital.

and σ* molecular orbitals. The wave function in the σ orbital has the same sign everywhere in the region between the two nuclei. In the σ* orbital, on the other hand, the wave function changes sign at the midpoint between the two atoms. That is, there is a node in the wave function at this point.

It is easy to see from the figure how the contour diagrams shown originate. The molecular orbitals are obtained by combination of the 1s atomic orbitals on the two atoms. When these two atomic orbitals are combined with the same sign, they overlap to give a molecular orbital that has the same sign everywhere, that is, a σ orbital. When the atomic orbitals overlap with opposite signs, the wave functions cancel in the middle region, and two lobes of opposite sign are developed, thus forming the σ* orbital. In this case, most of the wave function amplitude is outside the region of the nuclei.

Remember that the electron density distribution is given by the square of the wave function (Section 6.5). From Figure 9.20 we can readily see why the electrons in the bonding orbital really are bonding. They concentrate in the region between the positive nuclei, shielding them from one another and maximizing the attractive interactions between nuclei and electrons. On the other hand, electrons in the antibonding molecular orbital are concentrated in the region in which they attract the nuclei *away* from one another, thus disrupting the molecule.

SAMPLE EXERCISE 9.5

Would you expect the polyatomic ion He_2^+ to be stable relative to a separated He atom and He^+ ion? Explain.

Solution: The energy level diagram for this system is shown in Figure 9.21. Two helium 1s orbitals interact to form bonding and antibonding molecular orbitals. In He_2^+ we have a total of three electrons. Two of these are placed in the bonding orbital, the third in the antibonding orbital, as shown in Figure 9.21.

The stability gained from two electrons in the bonding orbital is greater than the destabilization due to one electron in the antibonding orbital. The He_2^+ molecular ion is therefore predicted to be stable relative to its dissociation products. It has been shown in laboratory studies that He_2^+ is stable relative to separated He and He^+.

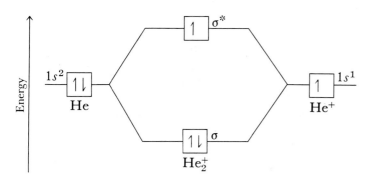

FIGURE 9.21 Energy-level diagram for the molecular orbitals in He_2^+.

In considering the formation of Li_2 from two Li atoms, a new factor enters the picture. The lithium atom has the electron configuration $1s^22s$. In combining two lithium atoms to form the Li_2 molecule we must consider the possible interactions of the $1s$ orbital on one lithium atom with the $2s$ orbital on the other. Such an interaction is theoretically possible, but rule 4 tells us that it will not be of importance. The energy difference between the $1s$ and $2s$ orbitals is simply too great for a strong interaction. As a result, the energy-level diagram for the Li_2 molecule is as shown in Figure 9.22. Notice that the molecular orbitals are labeled with subscripts to indicate the set of atomic orbitals from which they are formed. The electrons in the σ_{1s} and σ_{1s}^* orbitals make no net contribu-

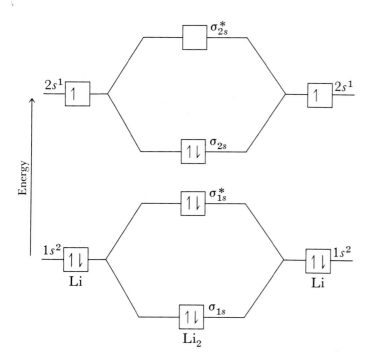

FIGURE 9.22 Energy-level diagram for the Li_2 molecule.

263

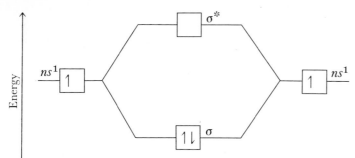

FIGURE 9.23 Energy-level diagram for alkali-metal diatomic molecules.

tion to the bonding. Their average energy is just the energy of the $1s$ electrons in the isolated Li atoms. All of the bonding in the Li_2 molecule is thus due to the $2s$ electrons. This example illustrates the general rule that *filled atomic subshells do not contribute to bonding in molecule formation.* This rule means that whenever an atom has a completed s, p, or d level, the electrons in those atomic orbitals do not contribute to bond formation. It also means that all filled shells do not contribute to bonding. We need only consider the valence-shell electrons. This conclusion is equivalent to the assumption we make when we draw Lewis structures that show only the valence-shell electrons.

Using this rule, we can account for the bonding in the other alkali-metal diatomic molecules. The valence-shell part of the molecular-orbital diagram for Na_2 or K_2 would look just like the valence-shell part for Li_2 (see Figure 9.23). In Figure 9.23, n represents the major quantum number of the valence shell s orbital. There is no need to show any of the electrons below the valence shell, because they do not contribute to bonding.

Although Li_2 and the other alkali metals are similar to H_2 in terms of the energy-level diagram, the change in major quantum number is of great importance. Table 9.5 shows the bond-dissociation energies of H_2 and the alkali-metal diatomic molecules through Cs_2. (The alkali metals are not normally in the form of diatomic molecules, but they can be studied as diatomic molecules in the gas phase at high temperatures.) The bond-dissociation energy decreases steadily in this series. The major factor responsible for this effect is a decrease in the extent to which the valence-shell s orbitals overlap (rule 5). The inner-shell electrons set a limit on how close the nuclei can draw together, because of repulsions between filled shells. In addition, the s orbitals become more spatially extended with an increase in major quantum number. The overall result is a decrease in the extent of overlap of the s orbitals with increasing major quantum number for the diatomic molecules of Table 9.5.

SAMPLE EXERCISE 9.6

Beryllium, the fourth element of the periodic table, does not form a stable diatomic molecule. Provide a reason for this in terms of molecular orbital formation.

Solution: The electron configuration for Be is $1s^2 2s^2$. The energy-level diagram for Be_2 involves interactions of the $1s$ and $2s$ orbitals, just as for Li_2, as shown above. The four valence-shell electrons in Be_2, however, completely fill both σ_{2s} and σ_{2s}^* orbitals, to leave a net bonding of zero.

TABLE 9.5

bond-dissociation energies for H_2 and the alkali metals in diatomic molecules

MOLECULE	MAJOR QUANTUM NUMBER OF s ORBITAL	BOND DISTANCE (Å)	DISSOCIATION ENERGY (kJ/mole)
H_2	1	0.75	430
Li_2	2	2.67	100
Na_2	3	3.08	72.4
K_2	4	3.92	49.4
Rb_2	5	4.2	45.2
Cs_2	6	4.7	42.6

OVERLAPS OF p AND s ATOMIC ORBITALS

When we consider the formation of diatomic molecules for the elements beyond beryllium, we need to take account of the ways in which the p orbitals on the atoms may combine with one another and with s orbitals. In some instances the overlap is zero, for reasons of symmetry. This means that the shapes and orientations of the orbitals and the signs of the wave functions are such that the overlap in one region of space is exactly cancelled by an equal overlap of opposite sign in another. Figure 9.24 shows the contour diagrams for the molecular orbitals formed by combining s and p atomic orbitals in all the possible ways that lead to nonzero overlap. These combinations are of two kinds. When the orbitals that make up the combination are symmetric about the bond axis, the molecular orbital formed is of the σ type. When the orbitals that make up the combination change sign about the bond axis, the molecular orbital formed from the atomic orbitals is labeled π. (You will remember that we used these same designations in describing bonds formed from overlaps of hybrid atomic orbitals, Section 9.3.)

For the combinations that give nonzero overlap, both a bonding and antibonding combination may be obtained. Note that the electron density in the bonding orbitals tends to be concentrated in the region between the nuclei, whereas in the antibonding orbital it is concentrated in the regions in back of the nuclei.

9.5 molecular-orbital diagrams for diatomic molecules

We are now ready to consider the molecular-orbital energies in a diatomic molecule formed from two atoms of an element from the second row of the periodic table. Without specifying which element is involved, let us imagine that two atoms of a second-row element come together to form a molecule. The atomic orbitals on each atom interact with those on the other atom, in accordance with the rules described above. From

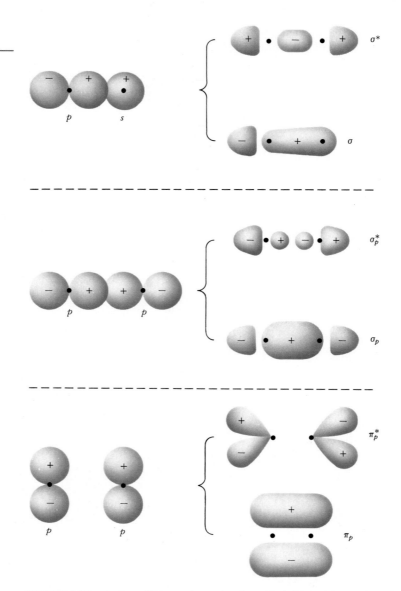

FIGURE 9.24 Contour diagrams for molecular orbitals formed by overlap of s and p orbitals. These diagrams are designed to illustrate the general shapes of bonding and antibonding orbitals and are not accurate representations.

these interactions, a set of molecular orbitals results, as shown in Figure 9.25. We might expect that the $2s$ orbitals would interact exclusively with one another and the $2p$ orbitals with other $2p$ orbitals. Because there are two possible ways in which the $2p$ orbitals can interact, there are both σ and π molecular orbitals formed in this case. The only complication in this simple picture comes from the fact that the σ_{2s} and σ_{2p} orbitals can overlap with one another, as can the σ^*_{2s} and σ^*_{2p}. As a result, the σ_{2p} and σ^*_{2p} orbitals, formed from the $2p$ orbitals, are pushed a little upward in energy; the σ_{2s} and σ^*_{2s} orbitals, formed from the $2s$ orbitals, are pushed downward. These changes in energy result from a certain amount of

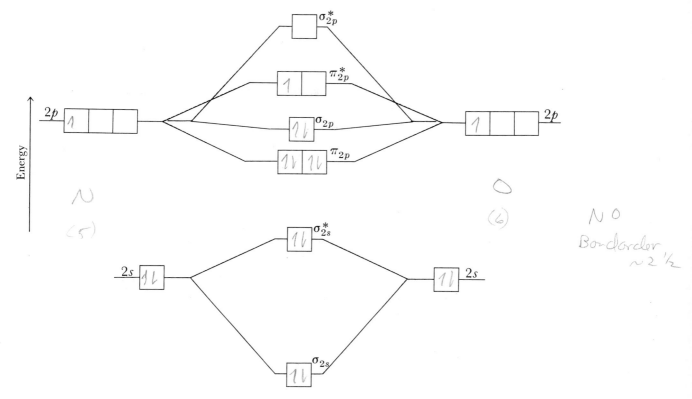

FIGURE 9.25 General energy-level diagram for molecular orbitals of second-row diatomic molecules.

mixing of the σ_{2s} and σ_{2p} orbitals, and of the σ_{2s}^* and σ_{2p}^* orbitals. Note also that the energy splitting between bonding and antibonding orbitals is larger for the σ_{2p} orbitals than it is for the π_{2p} orbitals. This occurs because the overlap of p orbitals is greater when they are oriented along the axis with respect to one another in σ fashion than when they are oriented in π fashion (rule 5).

Using this basic energy-level diagram we can readily deduce the electronic configuration of any of the second-row diatomic molecules. Some of these molecules are familiar substances, for example, O_2 or N_2. Others are known only in the vapor state at high temperature or under other unusual conditions. In all cases, however, the electronic structures are known from experiments. **Bond order** in these molecules is defined as the excess of bonding electron pairs over antibonding pairs. Thus in Li_2, with two valence electrons, the bond order is one. The molecule Be_2, with four valence shell electrons, should have a net bond order of zero, because the four electrons are placed in the σ_{2s} and σ_{2s}^* orbitals. However, B_2 should be stable, with a net bonding pair, for a bond order of one. As shown in Table 9.6, the two bonding electrons occupy the pair of π orbitals of equal energy. Hund's rule operates here just as for electrons in atomic orbitals: Electrons occupy equivalent orbitals singly and with spins parallel until all equivalent orbitals have been occupied. Thus B_2 is predicted to have two unpaired electrons, in agreement with experimen-

TABLE 9.6

electronic configuration and some experimental data for several second-row diatomic molecules

	B_2	C_2	N_2	O_2	F_2	Ne
σ^*_{2p}	☐	☐	☐	☐	☐	↑↓
π^*_{2p}	☐ ☐	☐ ☐	☐ ☐	↑ ↑	↑↓ ↑↓	↑↓ ↑↓
σ_{2p}	☐	☐	↑↓	↑↓	↑↓	↑↓
π_{2p}	↑ ↑	↑↓ ↑↓	↑↓ ↑↓	↑↓ ↑↓	↑↓ ↑↓	↑↓ ↑↓
σ^*_{2s}	↑↓	↑↓	↑↓	↑↓	↑↓	↑↓
σ_{2s}	↑↓	↑↓	↑↓	↑↓	↑↓	↑↓
Bond order	One	Two	Three	Two	One	Zero
Bond-dissociation energy (kJ/mole)	290	620	941	495	155	
Bond distance (Å)	1.59	1.31	1.10	1.21	1.43	
Ionization potential (kJ/mole)	—	1150	1495	1205	1700	

(handwritten annotations: "paramagnetic" with arrow pointing to O₂ π^*_{2p}; "bond order"; "$\sigma_{2p} - \sigma^*_{2p}$")

tal results. Pairing of the π electrons comes in C_2, which has a net bond order of two. With nitrogen, the bond order reaches its maximum, three. In view of this high bond order, the exceptional stability of the N_2 molecule and its high bond-dissociation energy are understandable.

The next molecule in our series, O_2, is especially interesting. We have a total of twelve valence electrons to place. As shown in Table 9.6, the last two of these must go into the π^*_{2p} orbitals with spins parallel. Thus the bond order in O_2 is two. Because of the unpaired electrons, molecular oxygen is paramagnetic. (A paramagnetic substance is one that is drawn into a magnetic field.) Oxygen is the only commonly available, simple molecule with this property. The paramagnetism of O_2 can be demonstrated by observing the effect of a magnetic field on a tube containing liquid oxygen, as shown in Figure 9.26. When the field is applied, the tube is moved laterally as the sample is drawn into the magnetic field. The prediction of the paramagnetism of O_2 is a most elegant achievement of molecular-orbital theory. The simple Lewis theory (Section 8.3) does not account for the paramagnetism, and there is no other bonding model for O_2 in which this property is explained so naturally.

In the next molecule in our series, F_2, the electrons in the π^*_{2p} orbitals are paired; the net bond order is one. In the hypothetical Ne_2 molecule, all valence molecular orbitals would be filled, for a net bond order of zero. The Ne_2 molecule is nonexistent. It is interesting to compare the electronic structures of the diatomic molecules with their observable properties. One of the most important measures of bond strength is the **bond-dissociation energy**—that is, the energy required to

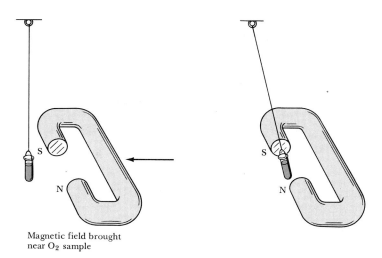

Magnetic field brought
near O_2 sample

FIGURE 9.26 Illustration of an experiment that shows the paramagnetic character of O_2. When the magnetic field is applied, the sample of liquid O_2 is drawn into the field.

separate the two atoms of the molecule to a very large distance from one another. Another important characteristic of a bond is the distance separating the bonded atoms. We have seen in previous discussions (Section 8.3) that bond distances are smaller in the case of multiple bonds. A third property of the diatomic molecules that relates to bonding is the **ionization potential**—that is, the energy required to remove the highest energy electron from the molecule.

The bond-dissociation energies, bond distances, and ionization potentials for the diatomic molecules are listed in Table 9.6. Note that the bond-dissociation energies of molecules with the same bond order are not the same. This should not surprise us, because overlaps differ, and many other factors contribute to the total energy of the molecule. Still, it is roughly true that bond-dissociation energy increases with bond order. Similarly, bond length decreases with increasing bond order.

In addition to the neutral diatomic molecules listed in Table 9.6 there are several known diatomic ions, such as N_2^+ and O_2^+. These species can be produced in the gas phase and their properties studied, even though they are too reactive to permit isolation. We'll see in Chapter 10 that N_2^+ and O_2^+ are important components of the earth's upper atmosphere. It is possible, using the energy-level diagram for molecular orbitals, to predict some of the properties of these ions.

SAMPLE EXERCISE 9.7

Predict the following properties of O_2^+: (a) number of unpaired electrons, (b) bond order, (c) bond-dissociation energy, and (d) bond length.

Solution: O_2^+ has one electron less than O_2. The electron removed from O_2 to form O_2^+ is one of the two unpaired π^* electrons. O_2^+ should therefore have just one unpaired electron left. Because the electron removed has come from an antibonding orbital, the bond order in O_2^+ is larger than for O_2; it is in fact intermediate between O_2 and N_2. Counting a single electron as contributing a bond order of $\frac{1}{2}$, the bond order is $2\frac{1}{2}$. The bond-dissociation energy and bond length should be about midway between that for O_2

and N_2; say 720 kJ/mole and 1.15 Å, respectively. The observed properties of O_2^+ are: Number of unpaired electrons, one; bond length, 1.123 Å; dissociation energy, 625 kJ/mole.

In principle, the energy-level diagram shown in Figure 9.25 is applicable to the diatomic molecules of the third- and higher-row elements. Except for the halogens, however, in which only a single bond is possible, the formation of diatomic molecules among the other elements is not the most stable form of bonding. The molecules P_2 and S_2 have been observed in the high-temperature vapors of these elements, but at lower temperatures other forms of the element are more stable (Section 8.3). Since P—P and S—S single bonds are quite stable, the major reason for the relative instability of the diatomic molecules in these two cases seems to be that their π bonds are not very strong.

POLAR COVALENT MOLECULAR ORBITALS

The concept of molecular orbitals can also be used to understand molecule formation between two different atoms. When the atoms are not very different, the energy-level diagram of Figure 9.25 can be used with some modification. The energy levels of one of the atoms are shifted with respect to the other. Such a diagram should be applicable to such molecules as NO, CO, CN, and so forth.

When the two atoms forming a molecule are appreciably different, extensive changes need to be made in the energy-level diagram. In many instances, it is possible to make simplifying assumptions that work quite well. For example, consider HF, in which the two atoms have quite different electronegativities. This is a simple case because hydrogen has only a single orbital for interaction with the orbitals of fluorine. The energy of the hydrogen $1s$ orbital is higher than that of the highest valence-shell orbital of fluorine, the $2p$, as shown in Figure 9.27. It is *much* higher than the fluorine $2s$ orbital. Thus, we need consider only the interaction of the hydrogen $1s$ orbital with the $2p$. Of the three $2p$ orbitals, only one is of correct symmetry to interact with the hydrogen $1s$.

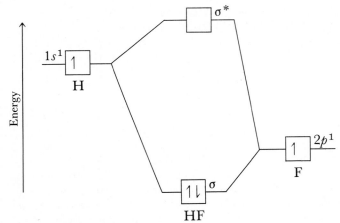

FIGURE 9.27 Energy-level diagram for HF molecular orbitals.

The interaction leads to a bonding and antibonding pair of orbitals, as shown in Figure 9.27. The other orbitals of fluorine are relatively unperturbed by molecule formation.

The bonding orbital occupied by a pair of electrons in HF is a polar covalent bond. This means that the contributions of the two atomic orbitals to the molecular orbitals are not the same. In the bonding orbital, the fluorine $2p$ orbital is more important than the hydrogen $1s$. As a result, the bonding electron pair is shifted more toward the fluorine atom than toward hydrogen. In the antibonding orbital, the contribution of the hydrogen $1s$ orbital is greater than that of the fluorine $2p$. Hydrogen fluoride provides an example of rule 6 in the list of rules for formation of molecular orbitals.

9.6 metallic bonding

Anyone who has handled a length of copper wire or an iron bolt or has observed a piece of freshly cut sodium knows that metals possess distinct physical properties. The three metals just mentioned differ from one another greatly in physical characteristics, and yet their properties are similar. In addition to their characteristic luster, metals are most distinctively characterized by their high electrical and heat conductivities. When an electrical potential is applied across a length of metal, current flows. This current flow occurs without any displacement of metal atoms. It is therefore due to the flow of electrons through the metal. Metallic conduction is characterized by a small resistance that increases with increase in temperature.

The characteristic feel of metals is due to their high heat conductivities. The electrical and thermal (heat) conductivities vary in the same manner from one metal to another. For example, silver and copper, which possess the highest electrical conductivities, also possess the highest thermal conductivities. This suggests that the two types of conductivity have the same origin.

A freshly prepared metal surface is characterized by its lustrous appearance. In most metals, light of all wavelengths is reflected. The color in metals such as gold and copper is due to absorption of light from the blue, or high-energy, region of the visible spectrum.

Most metals are **malleable,** which means that they can be hammered into sheets, and **ductile,** which means that they can be drawn into wires. These properties indicate that the atoms of the metallic lattice are capable of slipping with respect to one another. This is not a characteristic of ionic solids and the crystals of most covalent compounds. These types of solids are typically brittle and fracture easily along certain planes.

The metals form solid structures in which the atoms are arranged as close-packed spheres or in some similar packing arrangement. In such structures, as depicted in Figure 9.28, each atom is in contact with several nearest neighbors. For example, in copper, which possesses a close-packing arrangement called cubic close packing, each copper is in contact with 12 other copper atoms. None of the metal atoms possess sufficient valence electrons to form a localized electron-pair bond with each of these neighbors. For example, magnesium has only two valence electrons,

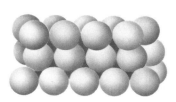

FIGURE 9.28 Close-packing arrangements of spherical objects. A packing of this kind, or a very similar arrangement, characterizes nearly all metallic structures.

yet it is surrounded by 12 neighboring magnesium atoms. If each atom is to share its bonding electrons with all its neighbors, these electrons must be able to move from one bonding region to another.

In discussing the structures of molecules such as benzene, we saw that electrons can in some cases be delocalized, or distributed over several nuclear centers. This happens when the atomic orbitals on a single atom are able to interact with atomic orbitals on more than one other atom. In graphite (Section 9.3), the electrons are delocalized over an entire plane. It is useful to think of the bonding in metals in a similar way. The valence atomic orbitals on one metal atom overlap with those on several nearest neighbors. These in turn overlap with the atomic orbitals of still other atoms. The interactions of all the valence orbitals of the metal atoms with the valence orbitals of adjacent metal atoms gives rise to a huge number of very closely spaced molecular orbitals that extend over the entire metal lattice. The energy separations between these metal orbitals is so tiny that for all practical purposes it is possible to think of them as forming a continuous band of allowed energy states, as shown in Figure 9.29. The electrons available do not completely fill this band. One can think of the energy band as a partially filled container for electrons. The imcomplete filling of the allowed energy levels gives rise to characteristic metallic properties. The electrons near the top of the occupied levels require very little energy input to be "promoted" to a still higher unoccupied orbital. Under the influence of any source of excitation energy, such as an applied electrical potential or an input of thermal energy, electrons are excited to previously vacant levels and are thus freed to move through the lattice, giving rise to electrical or thermal conductivity.

The description we have given of the electronic structures of metals has been drastically simplified. In transition metals, in which s, p, and d metal orbitals may overlap, the band structure is quite complex. In all these elements, however, the characteristic metallic properties of high electrical and thermal conductivities, luster, malleability, and ductility are found to varying degrees.

Metallic bonding is found also in **alloys**, materials that have metallic properties and that are composed of more than one element. Similar metals may form alloys in which the two metallic elements are homogeneously mixed throughout the solid, but the composition may be varied by altering the relative amounts of the two elements. Gold and silver or copper and nickel are examples of pairs of metals that exhibit

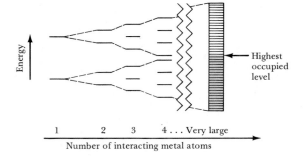

FIGURE 9.29 Schematic illustration of the interactions of atomic metal orbitals to form the delocalized orbitals of the metal lattice. The two atomic orbitals on each metal atom in this example might represent an s and p orbital. The main point is the formation of a very large number of molecular orbitals with very closely spaced energy levels. The number of electrons available does not completely fill these orbitals.

this type of alloy formation. Another very important type of alloy is one in which much smaller atoms, usually of a nonmetallic element, are located throughout the metallic lattice. In this type of alloy, called **interstitial,** the smaller atoms are found in the spaces between the larger spherical atoms of the metallic lattice. The interstitial atoms participate in bonding to neighboring atoms. The presence of the extra bonds provided by the interstitial atoms causes the metal lattice to become harder, stronger, and less ductile. For example, iron containing less than 3 percent carbon is much harder than pure iron and has a much higher tensile strength and other desirable physical properties. High-carbon steels, containing between 0.5 and 0.9 percent carbon, are used to make cutting tools and springs. Alloys of this kind differ from ordinary chemical compounds in that the composition is not fixed. The ratio of nonmetallic to metallic element may vary over a wide range, imparting a variety of specific physical and chemical properties to the materials.

FOR REVIEW

summary

In this chapter, we've applied the basic principles of chemical bonding to several important areas of chemical structure and behavior. The three-dimensional structures of molecules are determined by the distances between bonded atoms and by the directions of chemical bonds with respect to one another around a particular atom. The **valence-shell electron-pair repulsion (VSEPR) model** explains these relative directions in terms of the repulsions that exist between electron pairs. According to this model, electron pairs around an atom orient themselves so as to minimize electrostatic repulsions; that is, they remain as far apart as possible. By recognizing that unshared electron pairs take up more space (exert greater repulsive forces) than shared electron pairs, it is possible to account for the departures of bond angles from the idealized values and to explain many other aspects of molecular structure.

The Lewis-Langmuir model for covalent bonding introduced in Chapter 8 can be extended to account very nicely for the geometrical properties of molecules. We can imagine that the atoms in a molecule are bonded to one another by electron pairs that occupy pairs of overlapping atomic orbitals. The extent to which the atomic orbitals share the same region of space, called overlap, is important in determining the amount of stability that results from bond formation. The bonds directed along the internuclear axes are called σ **bonds.** It is possible to formulate orbitals on an atom that are directed toward each of the other atoms surrounding it by forming **hybrid orbitals.** These orbitals are made up of mixtures of the familiar s, p, and d atomic orbitals. Depending on the particular number of other atoms bonded to an atom and their arrangement in space, a particular set of hybrid orbitals can be formulated that has the necessary directional characteristics.

In addition to the σ **orbitals,** which determine the geometry of the bonding around a particular atom, there may be also π **bonds** constructed from remaining atomic orbitals. Thus double bonds, consisting of a σ and a π bond, or a triple bond, consisting of a σ and two π bonds, may be formed. In some molecules the π bonds may extend, or be delocalized, over several atoms. Delocalization of the π electrons in a cyclic structure, such as in benzene, or throughout a plane, as in graphite, leads to a special stability.

The coming together of atoms to form molecules may be viewed also as the coming together of atomic orbitals to form molecular orbitals. Atomic orbitals may combine with one another in various ways. The rules for combining atomic orbitals on atoms to form molecular orbitals allow us to account very well for the observed properties of the diatomic molecules formed by the first several elements of the periodic table. The molecular-orbital model is particularly impressive in explaining the fact that the O_2 molecule contains two unpaired electrons.

Metals consist of a three-dimensional arrangement of metallic atoms, usually of a type known as close packing. This means that each metal atom is in contact with the maximum number of other atoms, and the spheres occupy the volume as efficiently as possible. The valence electrons of the metal atoms are shared among all the atoms of the lattice and are relatively free to move within the structure.

learning goals

Having read and studied this chapter, you should be able to:

1. Relate the number of electron pairs in the valence shell of an atom in a molecule to the geometrical arrangement around that atom.
2. Explain why unshared electron pairs exert a greater repulsive interaction on other pairs than do shared electron pairs.
3. Predict the geometrical structure of a molecule or ion from its Lewis structure.
4. Explain the concept of hybridization and its relationship to geometrical structure.
5. Assign a hybridization to the valence orbitals of an atom in a molecule, knowing the number and geometrical arrangement of the atoms to which it is bonded.
6. Formulate the bonding in a molecule in terms of hybrid σ orbitals and π orbitals, from its Lewis structure.
7. Explain the concept of delocalization in π bonds.
8. Explain the concept of orbital overlap and the reason why overlap may in some cases be zero because of symmetry.
9. Describe how molecular orbitals are formed by overlap of atomic orbitals.
10. Explain the relationship between bonding and antibonding molecular orbitals.
11. Construct the molecular-orbital energy-level diagram for a diatomic molecule or ion built from elements of the first or second row and predict the bond order and number of unpaired electrons.
12. Describe the geometrical arrangement of atoms characteristic of metallic elements.
13. Explain how metallic bonding differs from typical covalent bonding found in nonmetallic structures.
14. Account in a qualitative way for the characteristic properties of metals in terms of their electronic structures.

key terms

Among the more important terms and expressions used for the first time in this chapter are the following:

Alloys (Section 9.6) are substances composed of more than one element and possessing metallic properties.

Aromatic molecules (Section 9.3) are compounds of carbon and hydrogen (other elements may also be present) that are characterized by a planar, cyclic arrangement of carbon atoms linked by both σ and π bonds.

Bond order (Section 9.5) is expressed as the number of bonding electron pairs shared between two atoms, less the number of antibonding electron pairs.

Hybridization (Section 9.2) refers to the mixing of different types of atomic orbitals to produce a set of equivalent hybrid orbitals.

Metallic bonds (Section 9.6), which exist in solid metals such as copper or aluminum, involve the sharing of valence electrons among all the atoms of the metallic lattice. Metallic bonding arises because there are very many more metal-metal contacts than there are electron pairs to be shared between the atoms. The valence electrons are free to move throughout the metallic lattice.

A **molecular orbital** (Section 9.4) is an allowed state for an electron in a molecule. A molecular orbital is entirely analogous to an atomic orbital, which is an allowed state for an electron in an atom.

The term **overlap** (Section 9.4) refers to the extent to which atomic orbitals on different atoms share the same region of space to form a molecular orbital. When overlap is large, a strong bond may be formed.

Paramagnetism (Section 9.5) is a property that a substance may possess if it contains one or more unpaired electrons. A paramagnetic substance is drawn into a magnetic field.

The **valence-shell electron-pair repulsion (VSEPR) model** (Section 9.1) accounts for the geometrical arrangements of shared and unshared electron pairs around a central atom in terms of the repulsions between electron pairs.

EXERCISES

the VSEPR model

9.1 What factor is of primary importance in determining the arrangement of electron pairs about a central atom?

9.2 Explain how the structures indicated for the following molecules are explained in terms of the VSEPR model: (a) PCl_3—pyramidal; (b) SiF_4—tetrahedral; (c) $SbCl_5$—trigonal bipyramidal.

9.3 Explain the variation in HMH angles: CH_4, HCH angle = 109.5°; NH_3, HNH angle = 107°; H_2O, HOH angle = 104.5°.

9.4 Write the Lewis structure and predict the geometrical structure for each of the following: (a) NF_3; (b) $AlCl_4^-$; (c) $POCl_3$; (d) CH_2Cl_2; (e) $TiCl_4$.

[9.5] Assuming that each of the following species is covalently bonded, which would you expect to be linear? (a) NH_2^-; (b) H_2S; (c) C_2H_2; (d) OF_2

9.6 Predict the shape at the central atom for each of the following molecules or ions: (a) NO_2^-; (b) BrF_3; (c) SO_4^{2-}; (d) $SnCl_6^{2-}$.

9.7 The bond angle at the central atom in NF_3 is 103°, whereas in BF_3 it is 120°. What single factor accounts for the difference in bond angles?

[9.8] The three species NO_2^+, NO_2, and NO_2^- all have the nitrogen atom in the middle and bonded to two oxygen atoms. The O—N—O bond angle in the three species is 180°, 134°, and 115°, respectively. Explain this variation in bond angle.

hybrid orbitals

9.9 Why does the promotion of an electron from the $2s$ to the $2p$ orbital of carbon require energy?

9.10 How is the energy required to form hybrid orbitals from an atom in its ground state recovered?

9.11 Explain what is meant by *overlap* of atomic orbitals. What conditions are necessary for overlap to occur?

9.12 Indicate the type of hybrid orbital set that can be formed from each of the following sets of atomic orbitals and describe the geometry of the set: (a) $2s$, $2p_x$; (b) $4s$, $4p_x$, $4p_y$, $4p_z$; (c) $5s$, $5p_x$, $5p_y$, $5p_z$, $4d_{x^2-y^2}$, $4d_{z^2}$.

[9.13] In the series of N—N bonded compounds N_2, N_2H_2, and N_2H_4, describe the hybridization at the nitrogen in each case and the overall geometry of the molecule.

9.14 Predict the approximate hybridization at the central atom in each of the following: (a) PF_3; (b) $SiCl_4$; (c) ICl_4^+; (d) AsF_5.

[9.15] From the following set of molecules and ions, select pairs in which the hybridization at the central atom is the same: (a) NH_4^+; (b) C_2H_4; (c) $Si(CH_3)_4$; (d) PCl_5; (e) C_6H_6; (f) $SeCl_4$.

[9.16] The structure of caffeine, a stimulant of the central nervous system found in coffee, is shown in Figure 9.30. Indicate the number of carbon atoms

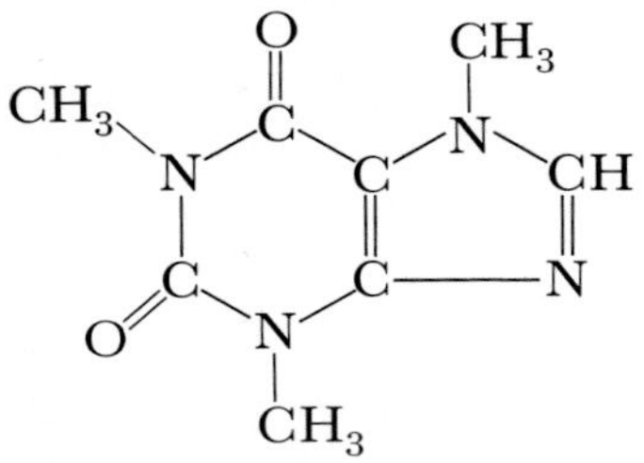

FIGURE 9.30 The structure of caffeine.

possessing approximate sp, sp^2, and sp^3 hybridizations. Do the same for the nitrogen atoms. (Note that, in keeping with the usual practice, the unshared pairs on N and O atoms are not shown.)

9.17 When bromine is added to ethylene, an immediate reaction occurs. However, when it is added to benzene, no reaction is observed. Account for the difference.

9.18 Graphite rods are employed as electrodes in dry cells and in many electrochemical processes in industry. Explain how this material is capable of conducting an electric current.

9.19 Molecules that have the same molecular formula, but different structures, are called *isomers*. Draw the three isomers of dichloroethylene, $C_2H_2Cl_2$. Which have a dipole moment?

9.20 The carbon-carbon bond distance in acetylene is 1.20 Å. How does this compare with the C—C bond distances in ethane and ethylene? Explain the trend in C—C bond distances in this series of substances.

9.21 Using an orbital diagram, show the formation of hybrid orbitals at carbon in the acetylene molecule and indicate with a label the type of bond formed by each orbital of the carbon (see Figure 9.9).

9.22 How many electrons occupy π orbitals in each of the following molecules: (a) ethylene; (b) acetylene; (c) N_2; (d) naphthalene?

molecular orbitals

9.23 Why is the bonding orbital formed from two atomic orbitals lower in energy than the starting atomic orbitals?

9.24 In each of the following pairs, which molecule or ion is the more stable relative to dissociation? Explain. (a) Li_2, Li_2^+; (b) O_2^+, N_2; (c) Be_2, Be_2^+

9.25 How would you expect the overlap of valence-shell s orbitals to vary in the series Li_2, Na_2, K_2? Explain.

9.26 For each of the following molecules, indicate the bond order and list the σ- and π-bond components: (a) N_2; (b) HCl; (c) CN.

9.27 Draw pictorial representations of all the valence-shell molecular orbitals occupied by electrons in the C_2 molecule.

9.28 Draw a general orbital energy-level diagram for the bonding orbitals of the diatomic molecules of group 3A.

9.29 Compare the Lewis model and molecular-orbital model for the bonding in O_2. How do they differ? Which model is in better accord with the experimental facts?

9.30 Which of the following molecules or ions would you expect to be paramagnetic: (a) LiH; (b) O_2^+; (c) SCl; (d) CO? Explain.

9.31 Predict the number of unpaired electrons, bond-dissociation energy, and bond distance for NO^+. Explain the reasons behind your predictions. Would you expect the ionization potential of NO^+ to be greater or less than that for N_2? Explain.

metallic bonding

9.32 Are the gaseous diatomic molecules Li_2 and Al_2 metallic in character? Explain.

9.33 The melting point and density of the three metals K, Ca, and Sc vary as follows:

ELEMENT	MELTING POINT (°C)	DENSITY (g/cm³)
K	62	0.84
Ca	810	1.6
Sc	1200	2.5

Suggest a reason for the observed variation in properties.

9.34 What experimental evidence can you offer to show that the valence electrons in a metal are not localized between pairs of metallic atoms, as in ordinary covalent compounds?

[9.35] When tin and lead are mixed together as molten metals and allowed to cool, the two elements separate out as almost pure, separate, solid phases. On the other hand, gold and silver under the same conditions solidify together as a single solid phase with each metal dissolved in the other. What does this suggest about the relative sizes of gold and silver atoms as compared with tin and lead atoms? Explain in terms of the characteristics of metal structures.

general exercises

9.36 Give a careful explanation for each of the following statements: (a) The Li_2 molecule is stable relative to its separated atoms. (b) Bond distance decreases with increasing bond order between a pair of bonded atoms. (c) The number of atomic orbitals that is mixed to form a hybrid orbital set is equal to the number of hybrid orbitals formed. (d) The H—P—H angle in PH_3 is less than the tetrahedral angle of 109.5°. (e) Acetylene reacts with two moles of bromine to form tetrabromoethane.

[**9.37**] Adding nitrogen to many metals makes them very brittle and hard and gives them a high melting point. Suggest a reason for this. (Note: The nitrogen is added as NH_3 or a similar substance, at high temperature, and is found in the metal as individual N atoms).

9.38 Draw one of the Lewis structures for anthracene, showing all C—C and C—H bonds.

9.39 Account for the geometry about the central atom in each of the following structures on the basis of VSEPR theory and indicate the hybrid orbital set that is consistent with the geometry predicted by the VSEPR model: (a) PF_3; (b) BF_4^-; (c) PCl_6^-; (d) $TiCl_6^{2-}$; (e) C_2F_4; (f) H_3O^+; (g) ClF_3; (h) SeO_4^{2-}.

9.40 The carbon-oxygen bond lengths in the series of molecules, CH_3OH, H_2CO, CO are 1.43 Å, 1.23 Å, and 1.13 Å, respectively. Write the Lewis structure for each molecule. How does the C—O bond order vary in this series? Is this variation consistent with the observed variation in carbon-oxygen bond length? Explain.

9.41 In terms of hybrid orbitals, explain why PF_5 is a reasonably stable molecule, but NF_5 is not a known compound.

9.42 Graph the C—C bond distance versus bond order for the compounds ethane, ethylene, acetylene (Sample Exercise 8.9), and benzene. Draw a smooth curve through the data. Explain the variation observed.

chemistry of the atmosphere

10

In the previous chapters of this text we have dealt for the most part with principles that govern the chemical and physical behavior of matter. We are now in a position to apply these principles to an understanding of the world in which we live. We start in this chapter with a fairly detailed look at the chemistry of earth's atmosphere. We shall be concerned for the most part with quite simple molecules whose structures and mode of bonding we have treated in some detail. We shall begin the study of atmospheric chemistry by considering the outermost region of the atmosphere; we shall then work our way downward to the earth's surface. In a sense, we might think of ourselves as creatures from another world, visiting earth for the first time and learning about it as we proceed toward the surface of the planet.

10.1
the outer regions

In Chapter 5 we saw that the atmosphere can be divided into various regions on the basis of the temperature profile (see Figure 5.9). Those regions, and the temperatures typical of various altitudes, are also shown in Figure 10.1. This figure also conveys other information about the characteristics of the atmosphere that will be of interest in later discussions. It is important as you read this chapter to keep in mind a proper perspective of altitude. The troposphere, in which all our weather occurs, extends to only about 12 km. The highest flying commercial jet aircraft fly at about 10 to 12 km. The U.S. version of the supersonic transport plane (SST) was designed to fly at about 20 km. Therefore, the region of the atmosphere at around 100 km, for example, is far above any ordinary

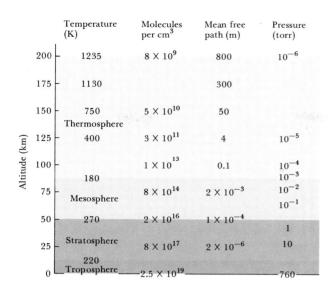

Altitude (km)		Temperature (K)	Molecules per cm³	Mean free path (m)	Pressure (torr)
200		1235	8×10^9	800	10^{-6}
175		1130		300	
150		750	5×10^{10}	50	
	Thermosphere				
125		400	3×10^{11}	4	10^{-5}
100			1×10^{13}	0.1	10^{-4}
		180			10^{-3}
75			8×10^{14}	2×10^{-3}	10^{-2}
	Mesosphere				10^{-1}
50		270	2×10^{16}	1×10^{-4}	
					1
25	Stratosphere		8×10^{17}	2×10^{-6}	10
		220			
0	Troposphere		2.5×10^{19}		760

FIGURE 10.1 Regions of the atmosphere, showing the relationship between altitude and temperature, pressure, number of molecules per unit volume, and mean free path of molecules.

human reach, and far removed from the atmospheric changes we call weather. Nevertheless, the chemical processes occurring at these high elevations are very important for our well-being, as we shall see later in this chapter.

A very sharp decrease in pressure with altitude (see also Figure 5.10) is reflected in a corresponding decrease in the number of molecules per unit volume. As the number of molecules per unit volume decreases, the mean free path, that is, the average distance traveled by a molecule between collisions (Section 5.11), increases, as Figure 10.1 shows. The very steep drop in pressure with altitude means that most of the earth's atmosphere is concentrated near the earth's surface. Thus, the troposphere, which extends to only 12 km altitude, contains about 80 percent of the total atmospheric mass; 99 percent of the total atmosphere is in the region below 30 km.

Although the upper reaches of the atmosphere contain only a small fraction of the atmospheric mass, the upper atmosphere plays an important role in determining the conditions of life at the earth's surface. These upper layers form the outer bastion of defense against the hail of radiation and high-energy particles with which the planet is continually bombarded. In absorbing these assaults, the molecules and atoms of the atmosphere undergo chemical change.

Up to an altitude of about 90 km, the average molecular weight of the atmosphere remains about the same as that at sea level, 28.96 amu. Above about 90 km the average molecular weight decreases as shown in Figure 10.2. This sharp decline means, of course, that the composition of the atmosphere is not constant. The changing composition has two causes. The first is called **diffusive separation;** the lightest molecules and atoms experience the least gravitational attraction to earth. Over a long time period, they drift to the top of the atmosphere. The element helium, which is a *very* minor constituent of the atmosphere at sea level (Table 5.1), is a major constituent in the region between 500 and 1000 km. (Remember, however, that there are very few atoms or molecules of *any* kind at these elevations.)

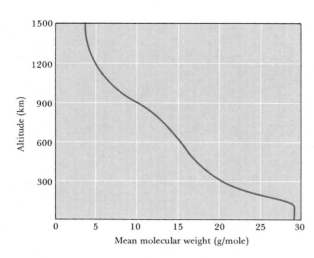

FIGURE 10.2 Mean molecular weight of the atmosphere as a function of altitude.

The second cause of the changing composition of the atmosphere is chemical in origin. Electromagnetic radiation and high-energy particles from the sun bombard the atmosphere and are absorbed. The energy absorptions that occur cause dissociation and ionization of atoms and molecules. For example, oxygen molecules are caused to dissociate into oxygen atoms:

$$O_2 + h\nu \longrightarrow 2O \qquad [10.1]$$

(The symbol $h\nu$ as used here denotes that the chemical process shown results from absorption of a photon.) As processes of this type become sufficiently extensive, the average molecular weight of the atmosphere decreases. For example, a gas consisting of O_2 molecules has a molecular weight of 32 g/mole. A gas that consisted of just O atoms would have a molecular weight of 16 g/mole. A gas consisting of a mixture of O_2 and O atoms has an average molecular weight of between 32 and 16 g/mole. We must now examine in more detail the nature of the chemical reactions that occur as a result of the absorption of radiation.

**10.2
photodissociation**

The sun emits radiant energy over a wide range of wavelengths, as shown in Figure 6.8. The shorter-wavelength, higher-energy radiations in the ultraviolet and extreme ultraviolet range of the spectrum are sufficiently energetic to cause chemical changes. We have already seen, in Section 6.2, that electromagnetic radiation can be pictured as a stream of photons. The energy of each photon is given by the relationship $E = h\nu$, where h is Planck's constant and ν is the frequency of the radiation. For a chemical change to occur when radiation falls on the earth's atmosphere, two conditions must be met. First, there must be photons with energy at least as large as that required to break a chemical bond, remove an electron, or otherwise accomplish whatever chemical process is being considered. Second, molecules must absorb these photons. This requirement means that the energy of the photons is converted into some other form of energy within the molecule.

SAMPLE EXERCISE 10.1

What is the frequency of radiation that has a wavelength of 310 nm? What is the energy of a photon with this wavelength?

Solution: Recall from Section 6.1 that the product of frequency and wavelength of radiation equals the velocity of light:

$$\nu\lambda = c = 3 \times 10^8 \text{ m/sec}$$

Therefore, rearranging this equation we have

$$\nu = \frac{c}{\lambda} = \left(\frac{3 \times 10^8 \text{ m/sec}}{310 \text{ nm}}\right)\left(\frac{10^9 \text{ nm}}{\text{m}}\right)$$

$$= 9.68 \times 10^{14}/\text{sec}$$

The energy of the photon is given by $E = h\nu$:

$$E = (6.625 \times 10^{-34} \text{ J-sec})(9.68 \times 10^{14}/\text{sec})$$
$$= 64.1 \times 10^{-20} \text{ J}$$

We are often concerned with energy quantities such as bond energies, expressed on a molar basis. To obtain an appropriate comparison we must multiply the energy of the photon by Avogadro's number, $N = 6.022 \times 10^{23}$. Thus the molar energy equivalence of the photon of wavelength 310 nm is

$$E = \left(64.1 \times 10^{-20} \frac{\text{J}}{\text{photon}}\right)$$
$$\times \left(\frac{6.022 \times 10^{23}}{\text{mole}} \text{ photons}\right)$$
$$= 386 \text{ kJ/mole}$$

You might compare this value with some of the bond-dissociation energies listed for diatomic molecules in Table 9.6.

One of the most important processes occurring in the upper atmosphere is dissociation of the oxygen molecule as a result of absorption of a photon (**photodissociation**), as shown in Equation [10.1]. The minimum energy required to cause this change is determined by the dissociation energy of O_2, 495 kJ/mole. By working in the opposite direction from that taken in Sample Exercise 10.1, we can determine the longest-wavelength photon having sufficient energy to dissociate the O_2 molecule.

SAMPLE EXERCISE 10.2

What is the wavelength of a photon corresponding to a molar bond-dissociation energy of 495 kJ/mole?

Solution: We must first calculate the energy required on a per molecule basis, and then determine the wavelength of a photon with that energy:

$$\left(495 \times 10^3 \frac{\text{J}}{\text{mole}}\right)\left(\frac{1 \text{ mole}}{6.022 \times 10^{23} \text{ molecules}}\right)$$

$$= 82.2 \times 10^{-20} \frac{\text{J}}{\text{molecule}}$$

From $E = h\nu$ we have

$$\nu = \frac{E}{h} = \left(\frac{82.2 \times 10^{-20} \text{ J}}{6.625 \times 10^{-34} \text{ J-sec}}\right)$$
$$= 1.24 \times 10^{15}/\text{sec}$$

The wavelength is then

$$\lambda = \frac{c}{\nu} = \left(\frac{3.00 \times 10^8 \text{ m/sec}}{1.24 \times 10^{15}/\text{sec}}\right)\left(\frac{10^9 \text{ nm}}{1 \text{ m}}\right)$$
$$= 242 \text{ nm}$$

The calculations in Sample Exercise 10.2 tell us that any photon of wavelength *shorter* than 242 nm will have sufficient energy to dissociate the O_2 molecule. (Remember that shorter wavelength means higher energy!)

The second condition that must be met before dissociation actually occurs is that the photon must be absorbed by O_2. Fortunately for us, O_2

absorbs much of the high-energy, short-wavelength radiation from the solar spectrum before it reaches the lower atmosphere. As it does so, atomic oxygen (O) is formed. The oxygen composition of the atmosphere as a function of altitude is illustrated in Figure 10.3. At higher elevations the dissociation of O_2 is very extensive: at 400 km only 1 percent of the oxygen is in the form of O_2; the other 99 percent is in the form of atomic oxygen. At 130 km, O_2 and O are just about equally abundant. Below this elevation O_2 is more abundant than O.

Recall from our discussion of the electronic structures of diatomic molecules that the bond-dissociation energy of N_2 is very high (Table 9.6). Thus, only photons of very short wavelength possess sufficient energy to cause dissociation of this molecule. Furthermore, N_2 does not readily absorb photons, even when they do possess sufficient energy. The overall result is that very little atomic nitrogen is formed in the upper atmosphere by dissociation of N_2.

One of the most interesting photochemical processes in the atmosphere is photodissociation of water. The partial pressure of water in the atmosphere is quite appreciable near the surface, but decreases very rapidly with increasing altitude. The water-vapor level in the stratosphere at 30 km is only about three parts per million (3 ppm).* Although it has not been possible to measure the water-vapor level at still higher altitudes, it seems quite certain that it is never more than a few parts per million. The amount of water that finds its way to the top of the

*Parts per million is a very commonly used unit for expressing the relative concentrations of trace constituents. A concentration of one part per million means that in a million parts of total atmosphere, one part by volume is due to the trace constituent. From Dalton's law of partial pressures and Avogadro's hypothesis (Section 5.7), this also means that of a million total molecules, one molecule on the average is due to the trace constituent.

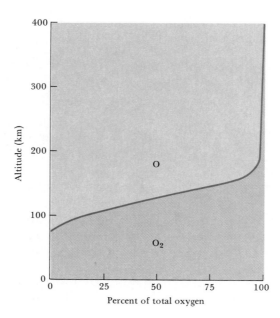

FIGURE 10.3 Oxygen composition of the atmosphere as a function of altitude.

atmosphere from the surface is thus very small. Once in the upper atmosphere, however, water undergoes photodissociation:

$$H_2O + h\nu \longrightarrow H + OH$$
$$OH + h\nu \longrightarrow H + O$$

[10.2]

Some scientists think that in the early stages of the earth's history, when it had no oxygen atmosphere, photodissociation of water was the means by which the planet acquired an oxygen atmosphere.

The oxygen formed in the photodissociation of water eventually diffuses to a lower elevation; there it undergoes chemical reactions to form O_2, NO, and so on. The hydrogen atoms also do this, but a significant fraction of them undergo a different fate. We have already noted that the atmosphere as a whole is acted on by the earth's gravitational field. Hydrogen atoms, the least massive of all atomic or molecular particles, experience the least gravitational force. At the same time, however, they possess the same average energy of motion as the other atoms or molecules in the same region of the atmosphere. Diffusive separation thus occurs; in the outermost regions of the atmosphere, called the **exosphere,** hydrogen and helium are the most abundant species. Because of their low masses, both hydrogen and helium are lost from the atmosphere to outer space.

10.3 ionization processes

In 1901 Guglielmo Marconi carried out a sensational experiment, by receiving in St. John's, Newfoundland, a radio signal transmitted from Land's End, England, some 2900 km away. Because radio waves were thought to travel in straight lines, it had been assumed that radio communication over large distances on earth would be impossible. The fact that Marconi's experiment was successful suggested that the earth's atmosphere in some way substantially affected radio-wave propagation. His discovery led to intensive study of the upper atmosphere. In about 1924 the existence of electrons in the upper atmosphere was established by experimental studies. These electrons cause the reflection of radio waves back to earth and thus make possible long-distance communication. The electron concentration in the atmosphere under more or less typical conditions is shown in Figure 10.4. Most radio-wave reflection occurs from electrons in the region between 100 and 250 km in altitude.

For each electron present in the upper atmosphere, there is a corresponding positively charged ion. The electrons in the upper atmosphere are a result of ionization of molecules, which is caused mainly by solar radiation (**photoionization**), and to a lesser extent by high-energy electrons and protons that stream into the earth's atmosphere from the sun. We shall concern ourselves here only with ionization caused by radiation. For photoionization to occur, a photon must be absorbed by the molecule, and this photon must have enough energy to cause removal of the highest-energy electron. Some of the more important ionization processes occurring in the upper atmosphere, that is, above about 90 km, are shown in Table 10.1, along with the ionization potentials and λ_{max},

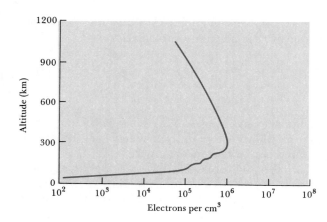

FIGURE 10.4 Electron concentration in the atmosphere as a function of altitude.

the maximum wavelength of a photon capable of causing ionization. Photons with energies sufficient to cause ionization have wavelengths in the short, or high-energy, region of the ultraviolet. These wavelengths are completely filtered out of the radiation reaching earth as a result of absorption by the upper atmosphere.

REACTIONS OF ATMOSPHERIC IONS

The molecular ions formed by photoionization are very reactive species. Without any additional input of energy from solar radiation, they react very rapidly on contact with a variety of charged species and neutral molecules. Reactions of this kind are called **thermal reactions;** they occur spontaneously under the temperature and pressure conditions of the reactants. The thermal reactions of the molecular ions are of three basic types. As we discuss these reactions, bear in mind that only processes that are exothermic, that is, those in which heat is evolved, need to be considered. Endothermic processes, those that require an input of energy, take place so slowly as to be of no interest in atmospheric chemistry.

Dissociative recombination. One of the obvious possibilities for reaction of a molecular ion is recombination with an electron to yield the neutral molecule. A great deal of energy is released when the electron recombines with the molecular ion; in fact, this amount of energy is equal to the ionization potential of the neutral molecule, because recombination is just the reverse of ionization. Unless there is a way to transfer this excess energy—for example, by collision with another molecule—the excess

TABLE 10.1

ionization processes, ionization potentials, and maximum wavelength of a photon capable of causing ionization

PROCESS	IONIZATION POTENTIALS (kJ/mole)	λ_{max} (nm)
$N_2 + h\nu \longrightarrow N_2^+ + e^-$	1495	80.1
$O_2 + h\nu \longrightarrow O_2^+ + e^-$	1205	99.3
$O + h\nu \longrightarrow O^+ + e^-$	1313	91.2
$NO + h\nu \longrightarrow NO^+ + e^-$	890	134.5

energy causes dissociation of the molecule. The likelihood of getting rid of the excess energy by collision with another molecule is very small in the upper atmosphere. The density of molecules is relatively low, and collisions do not occur frequently. As a result, essentially all the recombinations of electrons with molecular ions result in dissociation:

$$N_2^+ + e^- \longrightarrow N + N \qquad [10.3]$$

$$O_2^+ + e^- \longrightarrow O + O \qquad [10.4]$$

$$NO^+ + e^- \longrightarrow N + O \qquad [10.5]$$

A reaction of this type is referred to as **dissociative recombination.** It is the principal source of the atomic nitrogen present in the upper atmosphere.

Charge transfer. When a molecular ion undergoes collision with a neutral species, there may be transfer of an electron. For example, when an N_2^+ ion encounters an O_2 molecule, an electron is transferred from O_2 to N_2^+:

$$N_2^+ + O_2 \longrightarrow N_2 + O_2^+ \qquad [10.6]$$

For such a charge-transfer reaction to occur, the ionization potential of the molecule losing the electron must be less than the ionization potential of the molecule formed after transfer occurs. For example, the ionization potential of O_2 is less than that of N_2. When this condition is met, the reaction is exothermic, as illustrated in Figure 10.5. The excess energy of reaction is carried off in the kinetic energy of the molecules. On the basis of the data on ionization potentials in Table 10.1, the reaction shown in Equation [10.6] and the following reactions should be exothermic:

$$O^+ + O_2 \longrightarrow O + O_2^+ \qquad [10.7]$$

$$O_2^+ + NO \longrightarrow O_2 + NO^+ \qquad [10.8]$$

$$N_2^+ + NO \longrightarrow N_2 + NO^+ \qquad [10.9]$$

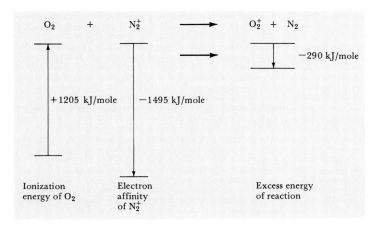

FIGURE 10.5 Illustration of the energy relations in a charge-transfer reaction. Recall (Section 7.9) that the electron affinity of a positive ion (N_2^+) is just the ionization potential, with reversed sign, of the corresponding neutral molecule (N_2).

SAMPLE EXERCISE 10.3

Using Hess's law (Section 4.4) and the data in Table 10.1, calculate the energy change in the reaction shown in Equation [10.7].

Solution: We can write Equation [10.7] as the sum of two processes, for each of which the heat is given in Table 10.1 (recall that when we reverse the direction of a process, we reverse the sign of the energy change):

$$O^+ + e^- \longrightarrow O \qquad\qquad E = -1313 \text{ kJ/mole}$$
$$\underline{O_2 \longrightarrow O_2^+ + e^- \qquad\qquad E = +1205 \text{ kJ/mole}}$$
$$O^+ + O_2 \longrightarrow O + O_2^+ \qquad E = -108 \text{ kJ/mole}$$

The negative sign for E denotes an exothermic process.

Because the N_2 molecule has the highest ionization potential of any of the species present in the upper atmosphere, N_2^+ is capable of undergoing charge transfer with every molecule with which it collides. Charge-transfer reactions are extremely rapid. As a result, although N_2^+ is produced extensively by photoionization, there is very little N_2^+ present in the upper atmosphere. These gas-phase charge-transfer reactions represent the simplest examples of a very important type of oxidation-reduction reaction known as **electron transfer.** Reactions in which an electron is transferred from one chemical species to another are of great importance in all areas of chemistry, including biochemistry. We shall encounter other examples of this type of reaction in later chapters.

Atom transfer. In the simple charge-transfer reaction, all chemical bonds remain intact. An electron is simply transferred from one species to another. There is a class of reactions, however, in which an atom is transferred. The two important reactions of this type occurring in the upper atmosphere are

$$O^+ + N_2 \longrightarrow NO^+ + N \qquad\qquad [10.10]$$
$$N_2^+ + O \longrightarrow NO^+ + N \qquad\qquad [10.11]$$

As before, these reactions are of significance because they are exothermic and proceed very readily. Notice that the product of both reactions is NO^+. Because the ionization potential of NO is lower than that of the other abundant species present in the upper atmosphere (Table 10.1), once the NO^+ ion is formed it does not become neutralized to any extent by charge-transfer reaction. The only means for removal of the ion is the dissociative-recombination reaction, Equation [10.5]. As a result, even though the neutral NO molecule is only a minor constituent of the upper atmosphere, present to the extent of about one part per million in total concentration, NO^+ is the most abundant ion in the upper atmosphere.

At this point, let's pause for another look at Figure 10.1, to see where we have been. All of the chemical processes we have discussed so far are important in the atmosphere above about 100 km. Note again that this is far above the troposphere, the only part of the atmosphere with which we have any direct contact. Yet if it were not for the absorptions of short-wavelength solar radiation in the upper atmosphere, life on earth as we know it would be impossible. However, not all the short-wavelength radiation is filtered out by N_2, O_2, and NO. To learn how some of the remaining short-wavelength radiation is filtered out, we must focus our attention on a lower region of the atmosphere.

10.4
ozone in the upper atmosphere

At an elevation of about 90 km, most of the short-wavelength solar radiation capable of causing ionization has been absorbed. As a result, the concentration of ions and electrons drops off very rapidly at about this elevation, as shown in Figure 10.4. Radiation capable of causing dissociation of the O_2 molecule remains sufficiently intense, however, so that photodissociation of O_2, Equation [10.1], remains important down to 30 km. The chemical processes that occur in the region below about 90 km following photodissociation of O_2 are very different than processes that occur at higher elevations. In the mesosphere and stratosphere the concentration of O_2 is much greater than that of atomic oxygen (Figure 10.3). Thus, when O atoms are formed in the mesosphere and stratosphere, they undergo frequent collisions with O_2 molecules. These collisions lead to formation of ozone, O_3:

$$O + O_2 \longrightarrow O_3^* \qquad\qquad [10.12]$$

The asterisk over the O_3 denotes that the ozone molecule contains an excess of energy. Reaction of O with O_2 to form O_3 results in release of 105 kJ/mole. This energy must be gotten rid of by the O_3 molecule in a very short time, or else it will simply fly apart again into O_2 and O. This decomposition is shown in Equation [10.13] as the reverse of the process by which O_3 is formed. The double arrows, $\rightleftharpoons$, indicate that the reaction is reversible; that is, that it may occur in either direction. The energy-rich O_3 molecule can get rid of the excess energy by colliding with another atom or molecule and transferring some of the excess energy to it. Let us represent the atom or molecule undergoing the collision as M. (Nearly always M is O_2 or N_2, because these are the most abundant molecules.) The transfer of energy can then be represented as in the second reaction, Equation [10.14]:

$$O + O_2 \rightleftharpoons O_3^* \qquad\qquad [10.13]$$
$$\underline{O_3^* + M \longrightarrow O_3 + M^* \qquad\qquad [10.14]}$$
$$O + O_2 + M \longrightarrow O_3 + M^* \qquad\qquad [10.15]$$

The reaction in Equation [10.14] competes with the reverse version of Equation [10.13]. When collisions are not very frequent, the reverse reaction in Equation [10.13] wins out; most of the O_3^* molecules formed fall apart into O and O_2 before they undergo a stabilizing collision. However, when the number of collisions per unit time is high, formation of O_3 via Equation [10.14] is favored. Because the number of molecules per unit volume increases rapidly with decreasing elevation, the frequency of stabilizing collisions is greater at the lower elevations. However, at much lower altitudes, the solar radiation energetic enough to produce dissociation of O_2 becomes largely absorbed. The overall result of the opposing factors is a maximum in the rate of production of ozone at about 50 km altitude. The ozone molecule, once formed, does not stay around very long; ozone itself is capable of absorbing solar radiation, with the result that it is decomposed into O_2 and O. Because the energy required for this process is only 105 kJ/mole, photons of wavelength shorter than 1140 nm are sufficiently energetic. The strongest and most important absorptions are of photons with wavelengths from about 200 to 310 nm. Radiation in this wavelength range is not strongly absorbed by any species other than ozone. If it were not for the layer of ozone in the stratosphere, therefore, these short-wavelength, high-energy photons would penetrate to the earth's surface. Plant and animal life as we know it could not survive in the presence of this high-energy radiation. The "ozone shield" is thus essential for our continued well-being.

The photodecomposition of ozone reverses the reaction leading to its formation. We thus have a cyclic process of ozone formation and decomposition, summarized as follows:

$$O_2 + h\nu \longrightarrow O + O$$

$$O + O_2 + M \longrightarrow O_3 + M^* \quad \text{(heat released)}$$

$$O_3 + h\nu \longrightarrow O_2 + O$$

$$O + O + M \longrightarrow O_2 + M \quad \text{(heat released)}$$

The overall result of this cycle is that ultraviolet radiation from the sun is converted into heat energy. The ozone cycle in the stratosphere is responsible for the temperature rise that reaches its maximum at the stratopause, as illustrated in Figure 5.9.

The scheme described above for the life and death of ozone molecules accounts for some but not all of the known facts about the ozone layer. Many chemical reactions involving substances other than just oxygen are involved. In addition, the effects of turbulence and winds in mixing up the stratosphere must be considered. A very complicated picture results. It is quite certain, however, that the oxides of nitrogen are important in the ozone cycle.

Nitric oxide, NO, and its close relative nitrogen dioxide, NO_2, are present in the stratosphere in low concentrations. Ozone reacts with NO to form NO_2 and O_2; then NO_2 reacts with atomic oxygen to regenerate NO and form O_2. The NO is then ready again to react with O_3. The overall reaction involving NO is simply:

$$O_3 + NO \longrightarrow NO_2 + O_2$$
$$\underline{NO_2 + O \longrightarrow NO + O_2}$$
$$O_3 + O \longrightarrow 2O_2 \qquad\qquad \text{[10.16]}$$

We see from this sequence of reactions that NO serves the function of increasing the rate of decomposition of O_3. There is no net change in the chemical state of the NO. We have here a very simple example of a catalyst, a substance that has the effect of increasing the rate of a chemical reaction, without itself undergoing a net chemical change.

The overall result of ozone formation and removal reactions, coupled with atmospheric turbulence and other factors, is to produce an ozone profile in the upper atmosphere as shown in Figure 10.6.

Supersonic transport (SST) planes have been the subject of intense controversy during the past several years because of their possible effect on the ozone layer. In any internal-combustion engine, the temperatures attained may be sufficiently high so that there is some reaction of the nitrogen and oxygen of the atmosphere to form NO:

$$N_2 + O_2 \longrightarrow 2NO \qquad\qquad [10.17]$$

This reaction is quite endothermic at ordinary temperatures, but it does proceed at a measurable rate at high temperatures such as those encountered in intensely hot flames. Some NO is therefore produced in the engines of an SST. The SST planes are designed to fly at stratospheric altitudes (Figure 10.6). Thus, the NO produced in the engines of the SST's would be released to the stratosphere at an altitude where there is a fairly high concentration of ozone. Attempts have been made to estimate how much additional NO would be contributed to the atmosphere by a fleet of SSTs, flying a certain number of hours each day. The conclusion of many experts is that enough NO would be produced to significantly decrease the total ozone in the stratosphere, especially over those regions of the earth that are directly under the flight paths of the planes. Thus, the SST flights would result in an increased amount of

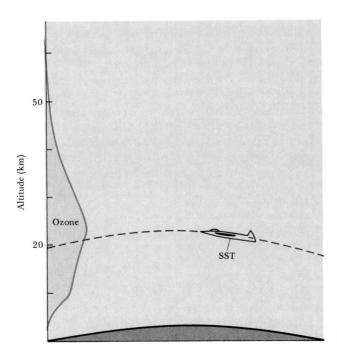

FIGURE 10.6   The ozone profile of the atmosphere.

high-energy ultraviolet radiation at sea level. Among the detrimental effects of a higher level of ultraviolet radiation would be an increased incidence of skin cancer among those receiving exposure to the sun's rays.

Recently a new source of destruction of the ozone layer has been recognized. **Chlorofluoromethanes**, principally CF_2Cl_2 and $CFCl_3$, are widely used as aerosol propellants and as refrigerant gases. The atmospheric concentrations of these substances have been steadily increasing. They are quite inert chemically; there seems to be no relatively rapid chemical process that removes chlorofluoromethanes from the lower atmosphere. The lifetimes of these molecules in the atmosphere are thus controlled by the rate at which they diffuse into the stratosphere and become subject to the action of ultraviolet light. The action of high-energy light with wavelengths in the range of 190 to 225 nm results in **photolysis,** or light-induced rupture, of a carbon-chlorine bond:

$$CF_xCl_{4-x} + h\nu \longrightarrow CF_xCl_{3-x} + Cl \qquad [10.18]$$

There may be further photochemical breakdown of the CF_xCl_{3-x} fragment. Calculations suggest that the rate of chlorine-atom formation will be maximized at an altitude of about 30 km. The atomic chlorine produced by photolysis is capable of rapid reaction with ozone to form chlorine oxide and molecular oxygen. Chlorine oxide is capable of reaction with atomic oxygen to reform atomic chlorine:

$$
\begin{aligned}
Cl + O_3 &\longrightarrow ClO + O_2 \qquad &[10.19]\\
ClO + O &\longrightarrow Cl + O_2 \qquad &[10.20]\\
\hline
O_3 + O &\longrightarrow 2O_2 \qquad &[10.16]
\end{aligned}
$$

This pair of reactions is analogous to those involving nitric oxide to produce the net reaction shown in Equation [10.16]. In both cases the original species is regenerated. The overall result is reaction of ozone with atomic oxygen to form molecular oxygen. Many uncertainties are involved in any quantitative estimate of how much ozone destruction there might be from this man-made source. Because rates of diffusion of molecules into the stratosphere from the earth's surface are likely to be very slow, it may be several decades before the full impact of the chlorofluoromethanes is felt. However, assuming that production of these compounds continues, at some time in the future the rate of ozone destruction from atomic chlorine could outstrip the rates of destruction from natural sources. This particular example illustrates very well how serious the consequences of human activities may become. The chlorofluoromethanes have been regarded as relatively harmless chemicals. In the long run, however, they may prove to be very dangerous indeed.

**10.5
chemistry
of the
troposphere**
In the previous sections, we have described the photodissociation of oxygen and ozone. These two processes, and, to a lesser extent, other photodissociations and photoionizations, result in essentially complete absorption of all solar radiation of less than about 300 nm wavelength at the altitude of the tropopause. Because the major constituents of the

atmosphere do not interact with radiation of wavelength longer than 300 nm, the photochemical reactions that occur in the troposphere are entirely those of minor atmospheric constituents. Many of the minor constituents occur to only a slight extent in the natural environment, but exhibit much higher local concentrations in certain areas as a result of human activities. Table 10.2 is a summary of information on several minor atmospheric constituents. Some of these are of interest for their role as air pollutants. We shall discuss here the most important characteristics of a few of these minor constituents and their chemical role as air pollutants.

TABLE 10.2

sources and typical concentrations of several minor atmospheric constituents

MINOR CONSTITUENT	SOURCES	TYPICAL CONCENTRATIONS
Carbon dioxide (CO_2)	Decomposition of organic matter; release from the oceans; fossil-fuel combustion	320 ppm throughout troposphere
Carbon monoxide (CO)	Decomposition of organic matter; industrial processes; fuel combustion	0.05 ppm in nonpolluted air; 1 to 50 ppm in urban traffic areas
Methane (CH_4)	Decomposition of organic matter; natural-gas seepage	1 to 2 ppm throughout troposphere
Nitric oxide (NO)	Electrical discharges; internal-combustion engines; combustion of organic matter	0.01 ppm in nonpolluted air; 0.2 ppm in smog atmospheres
Ozone (O_3)	Electrical discharges; diffusion from stratosphere; photochemical smog	0 to 0.01 ppm in nonpolluted air; 0.5 ppm in photochemical smog
Sulfur dioxide (SO_2)	Volcanic gases; forest fires; bacterial action; fossil-fuel combustion; industrial processes (roasting of ores, and so on)	0 to 0.01 ppm in nonpolluted air; 0.1 to 2 ppm in polluted urban environment

SULFUR COMPOUNDS

Sulfur-containing compounds are present to some extent in the natural, unpolluted atmosphere. They originate in the bacterial decay of organic matter, in volcanic gases, and from other sources listed in Table 10.2. Some scientists think that a certain amount of sulfur dioxide may also originate in the oceans. The concentration of sulfur-containing compounds in the atmosphere as a result of distribution of material from natural sources is very small in comparison with the concentrations that may build up in urban and industrial environments as a result of human activities. Sulfur compounds, chiefly sulfur dioxide, SO_2, are among the most unpleasant and harmful of the common pollutant gases. Table 10.3 shows the concentrations of several pollutant gases in a *typical* urban environment (not one that is particularly affected by smog). According to these data, the level of sulfur dioxide would be 0.08 ppm or higher about half the time. This concentration is considerably lower than that of other pollutants, notably carbon monoxide. Nevertheless, sulfur dioxide is regarded as the most serious health hazard among the pollutants shown,

TABLE 10.3

concentrations of atmospheric pollutants likely to be exceeded about 50 percent of the time in a typical urban atmosphere

POLLUTANT	CONCENTRATION (ppm)
Carbon monoxide	10
Hydrocarbons	3
Sulfur dioxide	0.08
Nitrogen oxides	0.05
Total oxidants (ozone and others)	0.02

especially for persons with respiratory difficulties. Studies of the medical case histories of large population segments in urban environments have shown clearly that those living in the most heavily polluted parts of cities have higher levels of respiratory diseases and shorter life expectancies. One industrial process that may produce very high local levels of SO_2 is the roasting, or smelting, of ores. By this process a metal sulfide is oxidized, driving off SO_2, as in the following example:

$$2ZnS + 3O_2 \longrightarrow 2ZnO + 2SO_2 \qquad [10.21]$$

Smelting operations account for about 8 percent of the total SO_2 released in the United States. About 80 percent of the SO_2 generated comes from the combustion of coal and oil. The extent to which SO_2 emissions are a problem in the burning of fossil fuels depends on the level of sulfur concentration in the coal or oil. Oil that is burned in the power plants of electrical generating stations is the nonvolatile residue that remains after the low boiling fractions have been distilled off. Some oil, such as that from the Middle East, is relatively low in sulfur, whereas Venezuelan oil is relatively high. Because of concern about SO_2 pollution, low-sulfur oil is in greater demand and is consequently more expensive.

Coals vary considerably in their sulfur content. Much of the coal lying in beds east of the Mississippi is relatively high in sulfur content, up to 6 percent by weight. Much of the coal lying in the western states has a lower sulfur content. (This coal, however, also has a lower heat content per unit weight of coal, so that the difference in sulfur content on the basis of a unit amount of heat produced is not as large as is often assumed.)

Altogether, more than 30 million tons of SO_2 are released into the atmosphere in the United States each year. This material does a great deal of damage to both property and human health. Not all the damage, however, is caused by SO_2 itself; it is likely, in fact, that SO_3, formed by oxidation of SO_2, is the major culprit.

Sulfur dioxide is not readily oxidized in clean dry air. It is, however, very rapidly oxidized to sulfur trioxide, SO_3, by O_2 in the presence of metal oxide dust particles. Studies show that the reaction between SO_2 and O_2 occurs on the surface of a dust particle. In some way, the surface of the metal oxide promotes the reaction. The surface of the metal oxide thus serves as a **heterogeneous catalyst.** The term *heterogeneous* refers to the fact that the reaction occurs on a surface, whereas ordinarily it would have occurred in the gas phase. The term *catalyst* indicates that the surface of the metal oxide increases the rate of reaction over the rate of homogeneous reaction. (There is more on heterogeneous catalysis in Chapter 13.) Very finely divided solids suspended in air are referred to as **particulate matter**. The levels of particulate matter are generally quite high in polluted air. In the stack gases issuing from power plants, for example, there is a considerable quantity of metal-containing fly ash. It is reasonable to suppose from the studies with metal oxides that these particles serve as catalytic surfaces for oxidation of SO_2.

A second pathway for SO_2 oxidation is via fog or cloud droplets; it is known that oxidation of SO_2 dissolved in water is quite rapid. Finally, there is a possibility for photochemical oxidation of SO_2. To understand how this might occur, we must pause to consider the possibilities for electronic transitions in molecules that are analogous to those that occur in an atom such as the hydrogen atom.

We have already observed that the solar radiation reaching the troposphere contains no wavelengths shorter than about 300 nm. A photon of 300 nm wavelength corresponds to an energy of about 400 kJ/mole. The energy required to dissociate an SO_2 molecule into SO and O is 565 kJ/mole. Ionization of SO_2 requires photons of even higher energy. But it is possible for processes other than ionization or bond rupture to occur when a photon is absorbed by a molecule or atom. In discussing the electronic structures of hydrogen and other atoms, we noted (Section 6.3) that electrons may be excited from one allowed energy state to another. Such an excitation requires photons that possess an energy precisely equal to the difference in energy between the two states. Electrons in molecules also exist in allowed energy states, called molecular orbitals, just as electrons in atoms exist in atomic orbitals. And just as in atoms, there is an entire galaxy of allowed energy states of higher energy than the highest one that is occupied. Certain selection rules govern the probability that a molecule will be able to undergo a particular electronic transition in which an electron is promoted from an occupied orbital to one of higher energy. These rules and all the details regarding them are beyond the scope of this text. It is sufficient for our purposes to know that such a transition may occur, and that a molecule is then considered to be in an *excited state*. So long as the molecule remains in this excited state, it is more reactive than is a molecule in the normal ground state; thus, such a molecule is more likely to enter into chemical reaction. Considerably less energy is required to cause electronic excitation in this manner as compared with dissociation or ionization. The photochemical oxidation of SO_2 may thus be written as a two-step process:

$$SO_2 + h\nu \longrightarrow SO_2^{\#} \qquad\qquad [10.22]$$

$$SO_2^{\#} + O_2 \longrightarrow SO_3 + O \qquad\qquad [10.23]$$

In this scheme $SO_2^{\#}$ denotes electronically excited SO_2. The details of the process by which SO_3 is formed are still not clear; the reactions shown are simply an indication of the overall effect, which is to produce SO_3.

It is evident from the above discussion that SO_2 may be oxidized to SO_3 by any of several pathways, depending on the particular nature of the atmosphere. Once SO_3 is formed it dissolves in water droplets, forming sulfuric acid, H_2SO_4:

$$SO_3 + H_2O \longrightarrow H_2SO_4 \qquad [10.24]$$

The acidic droplets thus formed are a menace to health and are strongly corrosive when they come in contact with metals, paints, and similar substances. An example of this corrosive effect is shown in Figure 10.7.

In some areas, where the atmosphere also contains ammonia, NH_3, an acid-base reaction may occur, producing ammonium hydrogen sulfate, $NH_4(HSO_4)$, or ammonium sulfate, $(NH_4)_2SO_4$:

$$NH_3 + H_2SO_4 \longrightarrow NH_4(HSO_4) \qquad [10.25]$$

$$NH_4(HSO_4) + NH_3 \longrightarrow (NH_4)_2SO_4 \qquad [10.26]$$

The thick haze that overlays many heavily industrial areas consists largely of a dispersal of ammonium sulfate formed in this manner.

Aside from the damage to human health, billions of dollars each year are lost as a result of corrosion resulting from SO_2 pollution. Obviously we all want this noxious gas removed from the environment. But removal of sulfur from coal or oil is difficult and, therefore, expensive. Rather than attempt to remove sulfur from fuel before it is burned, the sulfur dioxide formed when the fuel is combusted may be removed.

FIGURE 10.7 Notre Dame Cathedral, Paris. The extensive decay of the stone in several areas of this cathedral is due to the presence of SO_2 in the atmosphere. The SO_2 reacts with limestone, $CaCO_3$, and calcium sulfate, $CaSO_4$, is eventually formed, leading to a powdering or blistering of the surface. Wind and weather certainly contribute to the decay of buildings and statuary, but the major culprit is the chemical process that destroys the limestone.

There are many possible ways of doing this. One way involves blowing powdered limestone, $CaCO_3$, into the combustion chamber. The carbonate (limestone) is decomposed into lime (CaO) and carbon dioxide:

$$CaCO_3 \longrightarrow CaO + CO_2 \qquad [10.27]$$

The lime then reacts with SO_2 to form calcium sulfite:

$$CaO + SO_2 \longrightarrow CaSO_3 \qquad [10.28]$$

Only about half the SO_2 is removed by contact with the dry solid. It is necessary to "scrub" the furnace gas with an aqueous suspension of lime to remove the $CaSO_3$ formed, and to remove any unreacted SO_2. This process, which is illustrated in Figure 10.8, is difficult to engineer, reduces the heat effectiveness of the fuel, and leaves an enormous solid waste disposal problem. An electric power plant that would serve the needs of a population of about 150,000 people would produce about 160,000 tons per year of solid waste if it were equipped with the purification system just described. This is three times the normal fly-ash waste from a plant of this size. Various schemes may be employed to recover elemental sulfur or some other industrially useful chemical from the SO_2, but as yet no process has been found sufficiently attractive from an economic point of view to warrant large-scale development. Pollution by sulfur dioxide remains a major problem and will probably continue to remain so for some time.

NITROGEN OXIDES; PHOTOCHEMICAL SMOG

The atmospheric chemistry of the nitrogen oxides is interesting because these substances are components of smog, a phenomenon with which

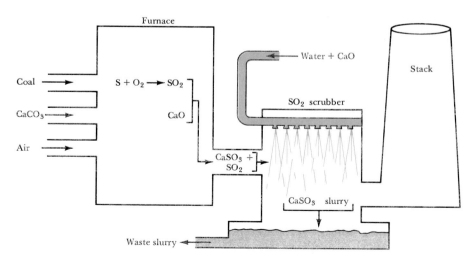

FIGURE 10.8 A common method for removing SO_2 from combusted fuel. Powdered limestone decomposes into CaO, which reacts with SO_2 to form $CaSO_3$. The $CaSO_3$ and any unreacted SO_2 enter a purification chamber where a shower of CaO and water precipitates the $CaSO_3$ into a watery residue called slurry and converts the remaining SO_2 to $CaSO_3$.

city dwellers are all too familiar. The term **smog** refers to a particularly unpleasant condition of pollution in certain urban environments, which occurs when weather conditions produce a relatively stagnant air mass. The smog made famous by Los Angeles, but now common in many other urban areas as well, is more accurately described as a **photochemical smog,** because photochemical processes play an essential role in its formation.

Nitric oxide, NO, is formed in small quantities in the cylinders of internal-combustion engines via the direct combination of nitrogen with oxygen, as described by Equation [10.17]. Prior to installation of control measures, typical emission levels of NO_x were 4 g per mile. (The x is either 1 or 2; both NO and NO_2 are formed, though NO predominates.) The most recent auto-emission standards call for NO_x emission levels of less than 1 g per mile.

Nitric oxide is oxidized very slowly in air. A certain amount of NO_2 is present, however; this is formed either directly in the automobile engine or by slow oxidation of NO. The dissociation of NO_2 into NO and O requires 304 kJ/mole. This requirement corresponds to a photon wavelength of 393 nm. In sunlight, NO_2 undergoes dissociation to NO and O:

$$NO_2 + h\nu \longrightarrow NO + O \qquad [10.29]$$

The atomic oxygen formed undergoes several possible reactions. One of these is with the abundant molecular oxygen to form ozone, as described earlier:

$$O + O_2 + M \longrightarrow O_3 + M^* \qquad [10.15]$$

Ozone is capable of rapidly oxidizing NO to NO_2, a reaction we have also seen earlier in connection with Equation [10.16]:

$$O_3 + NO \longrightarrow NO_2 + O_2 \qquad [10.30]$$

To see the significance of these reactions for smog formation, we must look at Figure 10.9, which shows the time dependence of the concentrations of various smog components. For the moment look at just the curves for the nitrogen oxides. In the early morning hours, the NO_2 concentration is low. As auto traffic builds up, and NO is formed, oxidation by ozone, Equation [10.30], takes over. Note that the ozone level does not noticeably increase during this period.

In addition to nitrogen oxides and carbon monoxide, an automobile engine also emits as pollutants unburned and partially burned hydrocarbons, compounds made up entirely of carbon and hydrogen. A typical engine without emission controls emits about 10–15 g of such organic compounds per mile. The newest standards require that hydrocarbon emissions be less than 1 g per mile. Table 10.4 shows typical concentrations of trace constituents in photochemical smog. The most important organic compounds in this list for smog formation are olefins and aldehydes. An **olefin** is an organic compound, a type of hydrocarbon, containing a double bond between carbon atoms. Ethylene, C_2H_4, is the

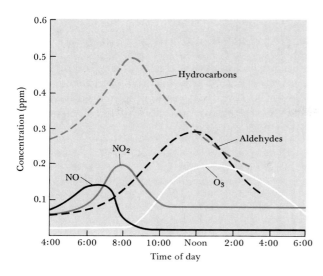

FIGURE 10.9 Concentration of smog components as a function of time of day. (*After P. A. Leighton, "Photochemistry of Air Pollution." Academic Press, 1961*)

simplest member of the series. **Aldehydes** are compounds containing a carbon-oxygen double bond on a carbon atom at the end of a hydrocarbon chain. Formaldehyde, acetaldehyde, and propionaldehyde are examples:

$$
\underset{\text{Formaldehyde}}{\overset{\displaystyle O}{\overset{\|}{\text{H}-\text{C}-\text{H}}}}
\qquad
\underset{\text{Acetaldehyde}}{\overset{\displaystyle O}{\overset{\|}{\text{CH}_3-\text{C}-\text{H}}}}
\qquad
\underset{\text{Propionaldehyde}}{\overset{\displaystyle O}{\overset{\|}{\text{CH}_3-\text{CH}_2-\text{C}-\text{H}}}}
$$

Note that in Figure 10.9 the hydrocarbon levels decrease as aldehyde levels increase. This occurs because some of the atomic oxygen produced in photodissociation of NO_2 reacts with organic compounds, eventually

TABLE 10.4

typical concentrations of trace pollutants during a photochemical smog (levels are subject to wide variation from one situation to another)

CONSTITUENT	CONCENTRATION (ppm)
NO_x	0.2
NH_3	0.02
CO	40
O_3	0.5
CH_4	2
C_2H_4	0.5
Higher olefins[a]	0.25
C_2H_2 (acetylene)	0.25
Aldehydes	0.6
SO_2	0.2

[a]Higher olefins are compounds that contain a hydrocarbon chain attached to one of the carbons of the C=C bond.

producing aldehydes through a complex series of reactions. Ozone is also capable of reaction with olefins to eventually yield aldehydes. Many of the compounds formed by reaction of atomic oxygen and ozone with organic compounds are **free radicals,** molecular fragments that contain an unpaired electron. They are very reactive and lead to a complex chemistry in the polluted atmosphere. One group of molecules formed in all this is the peroxyacylnitrates (PAN), especially unpleasant substances that cause eye irritation and breathing difficulties:

$$\underset{\displaystyle R-\overset{\displaystyle \overset{\textstyle O}{\|}}{C}-O-O-NO_2}{}$$

In this diagram, R represents an organic group such as CH_3, C_6H_5, and so on.

During the afternoon hours, the smog atmosphere contains relatively high concentrations of ozone, as shown in Figure 10.9. Much of the nitrogen oxides have been converted into PAN and several other related compounds. In addition, nitric acid is formed by dissolution of NO_2 in water droplets if these are present. Besides the gas phase components, an **aerosol** (a dispersion of tiny droplets in the air) usually forms and reduces visibility markedly. This aerosol consists mainly of organic compounds produced in the smog atmosphere. The entire mixture is thoroughly unpleasant and harmful to health.

Reduction or elimination of smog requires that the essential ingredients for its formation be removed from automobile exhaust. The stricter emission standards in effect since 1975 are designed to reduce drastically the levels of two of the major ingredients of smog—NO_x and hydrocarbons. Whether the means for meeting these standards now in use will be successful remains to be seen. Emission-control systems are not notably successful in poorly maintained autos.

CARBON MONOXIDE

In terms of total mass, carbon monoxide is the most important of all the pollutant gases. The level of CO present in fresh, nonpolluted air is small, probably on the order of 0.05 to 0.1 ppm. The estimated total amount of CO in the earth's atmosphere is about 5.2×10^{14} g. In the United States alone, however, about 1×10^{14} g of CO are produced each year. The CO is formed mostly in the incomplete combustion of fossil fuels. The major sources of CO in the United States are automobile and power-plant emissions.

The total amount of CO generated by man each year is estimated to be about 30 percent of that present in the atmosphere. However, the atmospheric level as a whole has not increased to the extent that these figures would imply. This suggests that there is a *sink* for CO, that is, a process that consumes it. Although CO is susceptible to oxidation to CO_2 by atmospheric oxygen, the reaction is extremely slow. The major pathway for removal of CO from the air over landmasses seems to be consumption by microorganisms in the soil. In addition, some of the gaseous CO may dissolve in the oceans. Very little is known about this process; in

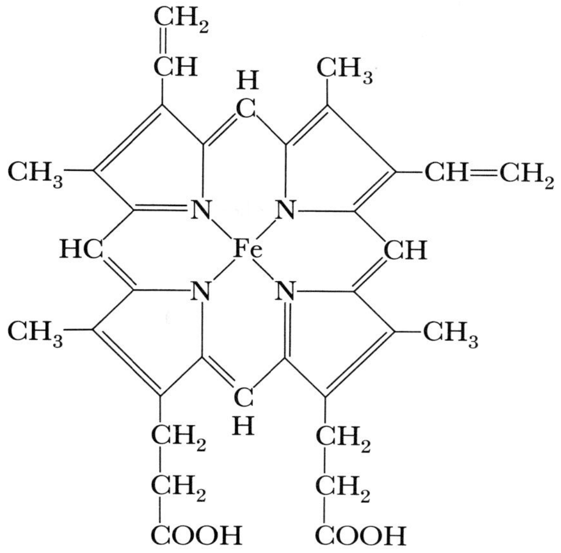

CH₂ ... the structure diagram:

$$CH_2$$
$$\|$$
$$CH \quad \underset{C}{H} \quad CH_3$$

CH₃—

—CH=CH₂

HC — Fe — CH

CH₃—

—CH₃

CH₂   H   CH₂
CH₂   CH₂
COOH   COOH

FIGURE 10.10 The structure of the heme molecule.

fact, it is not clear whether the oceans serve as a source or sink of CO. Finally, CO may be removed from the lower atmosphere by diffusion into the stratosphere, where it is removed by reactions with more reactive molecules and atoms present there. Experts estimate that the average residence time of CO in the atmosphere is about 6 months. Thus, the averaged concentration of CO in the total atmosphere is not large. However, this compound is a serious pollution hazard in two special situations: (1) in urban centers where traffic density is high and (2) in the cigarettes smoked by large numbers of Americans.

Carbon monoxide is a relatively unreactive molecule, and it might be supposed that it would not be a health hazard. It does have the unusual ability, however, of binding very strongly to **hemoglobin,** the iron-containing protein that is responsible for oxygen transport in the blood. Hemoglobin consists of four protein chains loosely held together in a cluster. Each chain has within its folds a heme molecule. The structure of heme is shown in Figure 10.10. The important characteristics of heme are that the iron is situated in the center of a plane of coordinating nitrogen atoms. An oxygen molecule reacts with the iron atom to form a species called **oxyhemoglobin.** Under appropriate conditions the oxygen is released from the iron. The equilibrium between hemoglobin and oxy-hemoglobin may be illustrated graphically as shown in Figure 10.11. Oxygen is picked up by hemoglobin in the lungs, and released in the tissues where it is needed for cell metabolism, that is, for the chemical processes occurring in the cell.

Carbon monoxide also happens to bind very strongly to the iron in hemoglobin. The complex is called **carboxyhemoglobin,** and is represented as COHb. The affinity of human hemoglobin for CO is about 210 times greater than for O_2. As a result, a relatively small quantity of CO can inactivate a substantial fraction of the hemoglobin in the blood for oxygen transport. For example, a person breathing air that contains only 0.1 percent of CO takes in enough CO after a few hours of breathing to

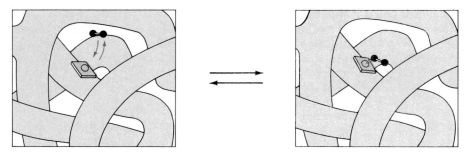

FIGURE 10.11 The equilibrium between hemoglobin and oxyhemoglobin.

convert up to 60 percent of the hemoglobin into COHb, thus reducing the blood's normal oxygen-carrying capacity by 60 percent.

Considerable evidence has accumulated that chronic exposure to CO in polluted air for a long period of time may have adverse health effects. Under normal conditions, a nonsmoker breathing unpolluted air has about 0.3–0.5 percent carboxyhemoglobin, COHb, in the bloodstream. This small amount arises mainly from the production of small amounts of CO in the course of normal body chemistry, and from the small amount of CO present in clean air. Exposure to higher concentrations of CO causes the COHb level to increase. This doesn't happen instantly, but requires several hours. Similarly, when the CO level is suddenly decreased, it requires several hours for the COHb concentration to level off at a lower value. Table 10.5 shows the percentages of COHb in blood that are typical of various groups of people. It is interesting to note that the federal Clean Air Act of 1971 was designed to control CO emissions so that the COHb percentage would remain below 1.5 percent in active nonsmokers.

The CO concentration in city traffic often reaches 50 ppm and may go as high as 140 ppm in traffic jams. Persons who work in areas with high traffic density, such as policemen, guards in traffic tunnels, and cab drivers, show abnormally high percentages of COHb as compared with the population as a whole. The most serious source of carbon monoxide poisoning, however, comes from cigarette smoking. The inhaled smoke from cigarettes contains about 400 ppm of CO. The effect of smoking on COHb percentage is evident from the data in Table 10.5. A study of a group of San Francisco longshoremen presented further proof of the

TABLE 10.5

carboxyhemoglobin (COHb) percentages in the blood of persons under various conditions

	COHb (%)
Continuous exposure, 10 ppm CO	2.0
Continuous exposure, 30 ppm CO	5.0
Nonsmokers, Chicago (1970)	2.0
Smokers, Chicago (1970)	5.8
Nonsmokers, Milwaukee (1969–1971)	1.1
Smokers, Milwaukee (1969–1971)	5.0

dramatic relationship between smoking and COHb percentage. Non-smokers in the group averaged 1.3 percent COHb; light smokers (less than half a pack per day) averaged 3.0 percent; moderate smokers averaged 4.7 percent; and heavy smokers (two packs or more per day) averaged 6.2 percent COHb.

There is widespread evidence that chronic exposure to CO impairs performance on standardized tests. Thus, it is most definitely not a good idea to smoke heavily before and during a test. In addition, motor performance is also impaired by high COHb percentages. For example, there is evidence that drivers responsible for traffic accidents have, on the average, higher than normal percentages of COHb in their blood. As has been mentioned, a chronically high level of COHb means that a certain fraction of the hemoglobin in the blood is not available for oxygen transport. This in turn means that the heart must work that much harder to ensure an adequate supply of oxygen. It is not surprising, therefore, that many medical researchers believe that chronic exposure to CO is a contributing factor in heart disease and in heart attacks.

WATER VAPOR, CARBON DIOXIDE, AND CLIMATE

We have seen how the atmosphere makes life as we know it possible on earth by screening out harmful short-wavelength radiation. In addition, the atmosphere is essential in maintaining a reasonably uniform and moderate temperature on the surface of the planet. The two atmospheric components of major importance in maintenance of the earth's surface temperature are carbon dioxide and water.

About 71 percent of all the solar radiation that strikes the outer atmosphere is eventually absorbed by the planet. The remainder is reflected back into space. The temperature of the planet as seen from outer space is determined by the amount of absorbed energy. The maximum in intensity of solar radiation occurs in the visible portion of the spectrum (see Figure 6.8). Because the earth's atmosphere is reasonably transparent in this region of the spectrum, most absorption of solar radiation takes place at the earth's surface. Thus, although the absorption of short-wavelength photons in the upper atmosphere is essential for the continued existence of living things, the absorption does not represent a large fraction of all the energy absorbed by the planet.

The earth is in overall thermal balance with its surroundings. This means that the planet radiates energy into space at a rate equal to the rate at which it absorbs energy from the sun. The sun is a very hot body with a temperature of about 6000 K. As seen from outer space the earth is relatively cold, with a temperature of about 254 K. The distribution of wavelengths in the radiation emitted from an object is determined by its temperature. The radiation emitted by relatively cold objects is in the low-energy, or long-wavelength, region of the spectrum. This means that the maximum in the wavelength of radiation from the earth is in the far infrared region, around 12,000 nm (Figure 6.3). But although the troposphere is transparent to visible light, it is not at all transparent to infrared radiation. Figure 10.12 shows the distribution of radiation from the earth's surface and, on the same scale, the wavelengths absorbed by water vapor and carbon dioxide. Clearly, these atmospheric gases absorb much of the outgoing radiation from the earth's surface. It is indeed fortunate for us that they do so; they serve to maintain a livably uniform tempera-

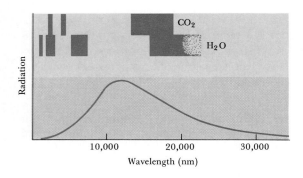

FIGURE 10.12 Long-wavelength radiation from earth, as compared with the absorption of infrared radiation by carbon dioxide and water.

ture at the surface by holding in, as it were, the infrared radiation from the surface.

The partial pressure of water vapor in the atmosphere varies greatly from place to place and time to time, but, in general, it is highest near the surface and drops off very sharply with increased elevation. Carbon dioxide, by contrast, is uniformly distributed throughout the atmosphere, at a concentration of about 320 ppm. Because water vapor absorbs infrared radiation so strongly, it plays the major role in maintaining the atmospheric temperature at night, when the surface is emitting radiation into space and not receiving energy from the sun. In very dry desert climates, where the water vapor concentration is unusually low, it may be extremely hot during the day, but very cold at night. In the absence of an extensive layer of water vapor to absorb and then radiate back part of the infrared radiation, the surface loses this radiation into space and cools off very rapidly.

Carbon dioxide plays a secondary, but very important, role in maintaining the surface temperature. John Tyndall, in 1861, was the first to suggest that it might be possible to account for the earth's past climatic history in terms of variations in the concentration of carbon dioxide in the atmosphere. Periods of unusually high temperatures, the so-called interglacial periods, could be accounted for in terms of unusually high CO_2 levels. The glacial periods, on the other hand, could have been caused by unusually low levels of carbon dioxide. It is now generally recognized that climate is the result of extraordinarily complex, interacting factors, and it seems unlikely that any one factor could have been responsible for drastic climatic change. Nevertheless, interest in the possible climatic effects of changing carbon dioxide levels has increased of late, as a result of new information on the impact of human activities on the atmosphere.

The worldwide combustion of fossil fuels, principally coal and oil, on a prodigious scale in the modern era has materially increased the carbon dioxide level of the atmosphere. From measurements such as those graphed in Figure 10.13 it is clear that the CO_2 concentration in the atmosphere is steadily increasing. From a knowledge of the infrared-absorbing characteristics of CO_2 and water, and using a theoretical model for the atmosphere, it has been estimated that if the CO_2 level were to double from its present level, the average surface temperature of the planet would increase 2.3°C. On the basis of present and expected

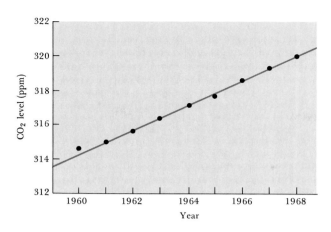

FIGURE 10.13 Annual mean CO_2 level in the atmosphere, 1960–1968. (*After B. Bolin and W. Bischof,* Tellus, 22, *431, 1970*)

future rates of fossil-fuel use, about 70 percent of the present known reserves of coal and essentially all the oil will have been consumed by about 2050. This amount of fuel consumption should just about double the atmospheric CO_2 level. If the calculated effect of a doubling of CO_2 level on surface temperature is correct, this means that the earth's temperature will be 2.3°C higher within 75 years. Such a small change may seem insignificant, but it is not. Major changes in global climate could result from a temperature change of this or even smaller magnitude. Because so many factors go into determining climate, it is not possible to predict with certainty precisely what changes will occur. It is clear, however, that humanity has acquired the potential, by changing the CO_2 concentration in the atmosphere, for substantially altering the climate of the planet. Unfortunately, if it should turn out for the worst, as seems altogether likely, there is little or nothing we can presently visualize that could be done about it. The continued high rate of combustion of fossil fuels is therefore a matter for long-range concern.

FOR REVIEW

summary

In this chapter we've examined chemical processes of major importance occurring throughout the atmosphere. We've seen that in the upper reaches of the atmosphere, only the simplest chemical species can survive the bombardment of highly energetic particles and radiation from the sun. The average molecular weight of the atmosphere at high elevations is lower than at the earth's surface, because the lightest atoms and molecules diffuse upward and because of **photodissociation.**

Absorption of radiation may also lead to ionization. Molecular ions formed by **photoionization** may recombine with electrons and thus form again the neutral molecule. However, the energy of recombination is so high that the molecule exists only momentarily before dissociating. Reaction of an electron with a molecular ion is thus referred to as **dissociative recombination.** Molecular ions may also react by removing an electron from another neutral molecule, in a **charge-transfer reaction.** Another possible reaction pathway is **atom transfer.**

Ozone is produced in the mesosphere and stratosphere as a result of reaction of atomic oxygen with O_2. Ozone is itself decomposed by absorption of a photon or by reaction with an active species such as NO. Human activities could result in the addition to the stratosphere of atomic chlorine, which is capable of reacting with ozone in a catalytic cycle to convert ozone to O_2. A marked reduction in the ozone level in the upper atmosphere would have serious adverse consequences, because the ozone layer filters out certain wavelengths of ultraviolet light that are not taken out by any other atmospheric component.

In the troposphere, the lower atmosphere in which we live, the chemistry of trace atmospheric components is of major importance. Many of these minor components are pollutants, which degrade the quality of the environment and present a health hazard in proportion to their concentration in the atmosphere. Sulfur dioxide is one of the more noxious and prevalent pollutants. It is oxidized in air to form sulfur trioxide, which upon dissolving in water forms sulfuric acid. To control this source of pollution, it is necessary to prevent SO_2 from escaping from industrial operations in which it is formed. One method for doing this involves reacting the SO_2 with CaO to form calcium sulfite, $CaSO_3$.

Photochemical smog is a complex mixture of components in which both nitrogen oxides and ozone play important roles. The smog components are generated mainly in automobile engines, and smog control consists largely in controlling emissions from automobiles.

Carbon monoxide is found in high concentrations in the exhaust of automobile engines and in cigarette smoke. This compound is a health hazard because of its ability to form a strong bond with hemoglobin and thus reduce the capacity of blood for oxygen transfer from the lungs.

Carbon dioxide and water vapor are the only components of the atmosphere that strongly absorb infrared radiation. The level of carbon dioxide in the atmosphere is thus of importance in determining worldwide climate. As a result of the extensive combustion of fossil fuels (coal, oil, and natural gas) the carbon dioxide level of the atmosphere is steadily increasing. Burning of a large fraction of the known reserves of fossil fuels could increase the CO_2 concentration in the atmosphere sufficiently to produce marked, and almost certainly undesirable, changes in the global climate.

learning goals

Having read and studied this chapter, you should be able to:

1. List the names of the regions of the earth's atmosphere and the approximate altitude interval for each.

2. Explain why the mean molecular weight of the atmosphere decreases above about 120 km.

3. Explain what is meant by the term photodissociation and calculate the maximum wavelength of a photon that is energetically capable of producing photodissociation, given the dissociation energy of the bond to be broken in the process.

4. Describe the way in which the element oxygen is distributed between the atomic and molecular form as a function of altitude and explain the reasons for this distribution.

5. Explain what is meant by photoionization and relate the energy requirement for photodissociation to the ionization potential of the species undergoing ionization.

6. Explain the process known as dissociative recombination and explain why it occurs.

7. Explain the conditions under which reaction between a molecular ion and a neutral molecule lead to charge transfer.

8. Write atom-transfer reactions that account for the presence of NO^+ in the upper atmosphere.

9. Explain the presence of ozone in the mesosphere and stratosphere in terms of appropriate chemical reactions.

10. Describe how nitrogen oxides or atomic chlorine function in the stratosphere as ozone-removal agents.

11. Explain how atomic chlorine might appear in the stratosphere as a product of the chlorofluoromethanes.

12. List the names and chemical formulas of the more important pollutant substances present in the troposphere and in urban atmospheres.

13. List the major sources of sulfur dioxide as an atmospheric pollutant and the various means by which SO_2 may be oxidized to SO_3.

14. List the more important reactions of nitrogen oxides and ozone that occur in smog formation.

15. Explain why carbon monoxide constitutes a health hazard.

16. Explain why the concentration of carbon dioxide in the troposphere has an effect on the average temperature at the earth's surface.

key terms

Among the more important terms and expressions used for the first time in this chapter are the following:

In an **atom-transfer reaction** (Section 10.3) an atom is transferred from one species to another.

Carboxyhemoglobin (Section 10.5) is a complex formed between carbon monoxide and hemoglobin, in which CO is bound to the iron atom.

A **catalyst** (Section 10.4) is a substance that affects the rate of a chemical reaction, but does not itself undergo a net, overall chemical change.

A **charge-transfer reaction** (Section 10.3) is one in which an electron is transferred from a neutral atom or molecule to a positively charged atom or molecule.

Chlorofluoromethanes (Section 10.4) are compounds of the general formula CF_xCl_{4-x}, $x = 1$, 2, or 3, used as propellent gases in aerosol spray cans and in refrigeration units.

Dissociative recombination (Section 10.3) is the process by which an electron recombines with an ionized molecule. The energy of recombination is released by rupture of a chemical bond in the neutral species.

Hemoglobin (Section 10.5) is an iron-containing protein responsible for oxygen transport in the blood.

A **heterogeneous catalyst** (Section 10.5) is a substance that catalyzes a reaction occurring in another phase. For example, a solid substance might act as a heterogeneous catalyst for a gas-phase reaction.

Photochemical smog (Section 10.5) is a complex mixture of undesirable substances produced by the action of sunlight on an urban atmosphere polluted with automobile emissions. The major starting ingredients are nitrogen oxides and organic substances, notably olefins and aldehydes.

Photodissociation (Section 10.2) refers to the breaking of a molecule into two or more neutral fragments as a result of absorption of light.

Photoionization (Section 10.3) refers to the removal of an electron from an atom or molecule by absorption of light.

10.1 Name the regions of the atmosphere in the order of increasing altitude and indicate the altitude interval for each.

10.2 Which region of the atmosphere contains the largest total number of gas molecules? Explain the reason for your answer.

[10.3] In terms of the chemical processes described in this chapter, explain why the temperature of the atmosphere increases dramatically above about 90 km.

10.4 Helium constitutes about 50 percent of the atmosphere at an elevation of 600 km. At that altitude the atmospheric pressure is 3×10^{-9} torr, and the average temperature is 1600 K. At the surface of the earth, helium is present in the atmosphere to the extent of 5.2 ppm. Is the concentration of helium higher at the surface or at an altitude of 600 km?

10.5 What factors are responsible for the fact that the mean molecular weight of the atmosphere decreases above about 150 km?

[10.6] The average molecular weight of the atmosphere near the surface of the earth is determined mainly by the percentages of N_2 and O_2. Assume that the atmosphere contains 80 percent N_2 and 20 percent O_2; calculate the mean molecular weight. Now assume that all the O_2 in this atmosphere is dissociated into oxygen atoms. What would be the mean molecular weight of the atmosphere in this case?

10.7 What two requirements must be met in order that photodissociation occur?

10.8 The dissociation energy of a carbon-chlorine bond is typically about 325 kJ/mole. What is the maximum wavelength of a photon that can cause C—Cl bond dissociation?

10.9 The dissociation energy of a typical carbon-chlorine bond is about 325 kJ/mole, while the dissociation energy of a typical carbon-fluorine bond is about 480 kJ/mole. What is the maximum wavelength of light that can cause bond dissociation in each case?

10.10 Write chemical equations that describe the photodissociation of each of the following molecules: (a) O_2; (b) N_2; (c) H_2O; (d) CF_2Cl_2; (e) NO.

[10.11] Suppose that it were possible to provide the moon with several "instant oceans" by filling several of the large maria (lunar "seas") with water. What do you think would happen over a long period of time to the water-vapor atmosphere established on the moon? Describe several stages of the changes that would lead to whatever you think would be the final result.

10.12 Atomic nitrogen can be produced in the upper atmosphere by (a) photoionization of N_2; (b) atom transfer between N_2 and O^+; (c) dissociative recombination of NO^+; (d) atom transfer between N_2^+ and O. Write balanced chemical equations for each process.

10.13 The concentration of electrons in the region of about 120 km altitude is much greater during the daytime than at night. Explain this in terms of the source and eventual fate of the electrons.

[10.14] Calculate the overall heat absorbed or evolved in each of the following changes:

(a) $O_2^+ + NO \longrightarrow NO^+ + O_2$
(b) $N_2^+ + e^- \longrightarrow N + N$
(c) $O_2^+ + N \longrightarrow NO^+ + O$
(d) $N + NO \longrightarrow N_2 + O$

A needed bond-dissociation energy, in addition to the energies given in Table 9.6 and Table 10.1, is: NO, 682 kJ/mole.

chemistry of the stratosphere

10.15 Explain what is meant by the term stabilizing collision. How critical is the nature of the molecule M involved in such a process? Explain.

[10.16] In terms of the energy requirements, explain why photodissociation of oxygen is more important than photoionization of oxygen at altitudes below about 90 km.

10.17 Why is the presence of ozone in the stratosphere and mesosphere of significance to us? Explain the possible effects of a substantial reduction in the total quantity of ozone.

10.18 Explain by means of balanced chemical equations how nitrogen oxides act as catalysts for the destruction of stratospheric ozone.

10.19 Beginning with the intact chlorofluoromethane CF_2Cl_2, write equations showing how a catalytic effect for destruction of ozone may be established in the stratosphere.

10.20 It has been suggested that the chlorofluoromethanes may not diffuse into the stratosphere, but instead may be transported to the polar regions and there frozen out in polar ice. Look up the melting and boiling points of CF_2Cl_2 and $CFCl_3$ in a handbook of chemistry. Does this seem like a reasonable possibility, assuming that there is an atmospheric transport mechanism?

[10.21] Explain how two factors working in opposite directions cause a maximum in the fractional abundance of ozone in the vicinity of about 30 km. (We might put this question in a slightly different form: The rate of formation of ozone is highest at 50 km, and yet the maximum in ozone concentration occurs at a lower elevation, 30 km. Why is this so?)

chemistry of the troposphere

10.22 For each of the following gases, make a list of known or possible naturally occurring sources: (a) CO; (b) SO_2; (c) CH_4; (d) NO.

10.23 Assuming an SO_2 concentration of 0.1 ppm over an urban area of 200 sq mi, and assuming the SO_2 to be evenly distributed to an elevation of 1000 ft, calculate the total mass of SO_2 present.

10.24 In a recent study carried out in Sweden, it was found that in a particular location far from industrial activity the content of sulfate in rainfall amounted to 150 mg/m² per year. Calculate the total mass of sulfate falling in a square mile per year.

10.25 Assuming an overall efficiency of about 10 percent, how much calcium carbonate would be required to remove the SO_2 formed in burning a ton of oil containing 1.5 percent sulfur?

10.26 Indicate three different pathways by which SO_2 can be oxidized in air. What is the significance of the oxidation of SO_2 as an environmental problem?

10.27 Write balanced chemical equations for each of the following processes: (a) formation of nitric oxide in an automobile cyclinder; (b) oxidation of nitric oxide by ozone; (c) formation of ammonium sulfate in a polluted industrial atmosphere; (d) oxidation of carbon monoxide in a catalytic muffler; (e) formation of ozone in an urban atmosphere.

10.28 From Figure 10.9 we see that the concentration of hydrocarbons in an urban atmosphere increases to a maximum rather early in the day and steadily decreases thereafter. What is the reason for this behavior?

10.29 Explain why continuous exposure to low concentrations of carbon monoxide in the air constitutes a health hazard.

[10.30] We have noted that the affinity of carbon monoxide for hemoglobin is about 210 times that of O_2. Assume that a person is inhaling air that contains 100 ppm of CO. If all the hemoglobin leaving the lungs carries either oxygen or carbon monoxide, calculate the fraction in the form of carboxyhemoglobin.

10.31 Indicate the major source or sources of high carbon monoxide concentrations in air and discuss what might be done to lower carbon monoxide concentrations.

10.32 Except for two substances, the components of the earth's atmosphere are transparent to long-wavelength, infrared radiation. What are these two components? In what way does the absorption of infrared radiation affect the earth's climate? Explain how increased levels of infrared-absorbing substances in the atmosphere could lead to a higher average surface temperature.

general exercises

10.33 It has recently been pointed out that today there may be increased amounts of NO in the troposphere as compared with the past because of massive use of nitrogen-containing compounds in fertilizers. Assuming that NO can eventually diffuse into the stratosphere, what role might it play in affecting the conditions of life on earth? Explain.

10.34 Compare typical concentrations of CO, SO_2, and NO in nonpolluted air (Table 10.2) and urban air (Table 10.3) and indicate in each case at least one possible source of higher values in Table 10.3.

10.35 Describe a process that is a major source of each of the following species, writing an equation or equations where appropriate: (a) atomic nitrogen at 150 km altitude; (b) atomic hydrogen at 150 km altitude; (c) O_3 at 50 km altitude; (d) NO^+ at 120 km altitude; (e) sulfur trioxide at 500 m altitude; (f) NO_2 at 500 m altitude over a city in daytime.

10.36 Give one example each of photoionization and photodissociation processes occurring in the upper atmosphere that help to prevent high-energy solar radiation from reaching the earth's surface.

10.37 Experiments have been performed in which metals such as sodium or barium have been released into the atmosphere at altitudes of about 120 km. Assuming that the metals are present in the atomic form, what reactions would you expect to occur with the ionic species present? Explain.

[10.38] From Equation [5.12] and Figure 10.1, calculate the average time that an O_2 molecule formed between two oxygen atoms would have to exist at 175 km altitude before undergoing a collision. Do the same calculation for an O_2 molecule at 75 km. How do the results help to explain the presence of atomic oxygen at 175 km and its relatively low concentration at 75 km? (Hint: See Equations [10.13] through [10.15] for a similar case.)

liquids, solids, and intermolecular forces

11

The physical properties of a substance depend to a large extent on its state, or phase. We have seen that matter exists in three states: gas, liquid, and solid (Section 2.1). Some properties characteristic of each state are given in Table 11.1. We discussed the physical properties of gases in Chapter 5. We now turn our attention to the other two states: liquids and solids. We shall consider the general properties of these states, seeking to account for them using the kinetic-molecular theory introduced in Chapter 5. One of the key concepts to be discussed in this chapter is that of intermolecular forces, the attractive forces responsible for holding molecules together in liquids and solids.

11.1
a molecular model for liquids and solids

The kinetic-molecular theory describes gases as being composed of widely separated molecules undergoing rapid, random motion. Temperature is a measure of the average kinetic energy of the particles; thus, as a gas is cooled, its molecules move more slowly. When a gas is cooled sufficiently, it condenses to form a liquid. We visualize this condensation as occurring when the kinetic energy of the molecules drops below their average energy of attraction. When this situation occurs, the attractive forces cause the molecules to stick together when they collide, thereby forming the liquid. The liquid is visualized as consisting of particles clustered closely together. Because the particles are close together, liquids are not compressible like gases; furthermore, they have definite volumes that are independent of the size of their container.

We also visualize the particles as being able to move freely about the volume of the liquid. This idea allows us to explain why liquids can be

TABLE 11.1

some characteristic properties of the states of matter

STATE	PROPERTY
Gas	1. Assumes both the volume and shape of container. 2. Is compressible. 3. Diffusion occurs rapidly. 4. Flows readily.
Liquid	1. Assumes the shape of the portion of the container it occupies. 2. Does not expand to fill container. 3. Is virtually incompressible. 4. Diffusion occurs slowly through a liquid. 5. Flows readily.
Solid	1. Retains its own shape and volume. 2. Is virtually incompressible. 3. Diffusion is extremely slow through solids. 4. Does not flow.

SAMPLE EXERCISE 11.1

Using the kinetic-molecular model for gases and liquids just described, explain why diffusion occurs more rapidly in a gas than in a liquid.

Solution: The particles of a gas are farther apart than the particles of a liquid and move a greater distance between collisions than do the liquid particles. This condition results in more rapid diffusion for gases. The situation may be likened to moving across a nearly empty room toward a door versus moving toward the door through a crowd. It takes more time to cover the required distance in the second case.

poured and why they assume the shape of their container. As in the case of gases, the average kinetic energy of the molecules is proportional to temperature.

When a liquid is cooled sufficiently, it may freeze, thus forming a solid. We visualize this solidification as occurring when the kinetic energy of the particles is low enough to permit the attractive forces between particles to virtually lock them in place. Like liquids, solids are not very compressible because the particles are close together. Because the particles are not free to undergo long-range movement, solids are rigid. The kinetic-molecular model for gases, liquids, and solids is summarized schematically in Figure 11.1. With this model in mind, let us now examine the properties of liquids and solids more closely.

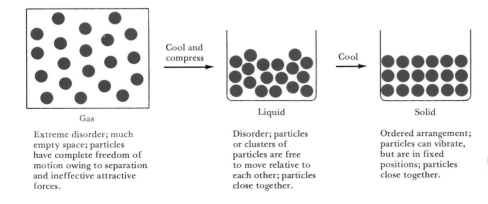

Gas

Extreme disorder; much empty space; particles have complete freedom of motion owing to separation and ineffective attractive forces.

Liquid

Disorder; particles or clusters of particles are free to move relative to each other; particles close together.

Solid

Ordered arrangement; particles can vibrate, but are in fixed positions; particles close together.

FIGURE 11.1 A molecular-level comparison of gases, liquids, and solids.

When a liquid is placed in an open container, it slowly escapes into the gas phase, eventually leaving the container empty. This phenomenon, known as **vaporization** or **evaporation,** can be explained in terms of the kinetic-molecular model. A particle near the liquid's surface can escape into the gas phase whenever it gains sufficient kinetic energy through collisions to overcome its attraction to other particles surrounding it in the liquid. The weaker the attractive forces, the more readily vaporization occurs. When two liquids are compared, the one that evaporates more readily is described as being the more **volatile.** For example, alcohol (ethanol) is more volatile than water. The rate of vaporization increases with increasing temperature. Figure 11.2 provides a graphic explanation for this behavior. In that figure the distribution of kinetic energies of the particles at the surface of the liquid is compared at two temperatures with the attractive forces at the surface. Notice that the distribution curves are like those shown earlier for gases (Figure 5.12); average kinetic energy increases with increasing temperature. Therefore as the temperature of a liquid is increased, there is an increase in the number of particles having sufficient kinetic energy to escape from the surface.

Loss of particles with high kinetic energy causes the average kinetic energy of the particles remaining in the liquid to decrease, just as the removal of the highest grades on an examination lowers the class average. The temperature of the remaining liquid is thereby lowered, because temperature reflects the *average* kinetic energy of the particles. Anyone getting out of a swimming pool, particularly on a windy day, has experienced the cooling that accompanies evaporation. Cooling caused by evaporation is easily detected if the liquid is in an insulated container and cannot readily absorb energy from its surroundings. Otherwise, the liquid simply abstracts energy from its surroundings and thereby remains at essentially the same temperature throughout the vaporization process.

SAMPLE EXERCISE 11.2

Why does vaporization occur more rapidly in a breeze than in still air?

Solution: When a particle escapes from the liquid surface, it does not move very far before it begins to undergo collisions with the molecules in the air. The resultant random motion of the particle may bring it back in contact with the liquid, and it will be recaptured. The net rate of vaporization depends on the rate at which the particles escape the liquid minus the rate at which they are recaptured. When a breeze is blowing, the particles that have escaped the surface are blown away. This reduces the likelihood of the particles colliding with the liquid surface and thereby increases the net rate of vaporization.

The quantity of energy required to convert a liquid to a gas at a constant temperature is known as the **heat of vaporization.** The molar heats of vaporization (ΔH_{vap}) of some common liquids at their boiling points are summarized in Table 11.2.

The rate of metabolic heat generation is such that an adult human's body temperature would rise 1 to 30°F per hour (depending on the level of activity) if the heat were not dissipated. This rise would result in a heat stroke at 106° and death at 110 to 112°F. Because heat does not flow from colder to hotter ob-

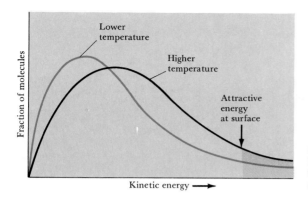

FIGURE 11.2 Kinetic energy distribution for the surface molecules of a hypothetical liquid compared to the intermolecular attractive forces at the liquid's surface. The shaded areas represent the fraction of molecules with sufficient kinetic energy to escape from the surface.

jects, cooling by simple conductive heat loss from the body is not possible when the outside temperature is greater than body temperature. Our bodies rely, for cooling, on the heat absorbed by evaporation of water; we are cooled by sweating. Thus, one of the significant roles played by water in our bodies is that of a cooling agent. We may note that water is particularly well suited for this role; it takes more energy to vaporize a gram of water than to vaporize an equal mass of any known liquid substance. Comparing water and mercury (Table 11.2) we have, for water:

$$\left(40.7 \; \frac{kJ}{mole}\right)\left(\frac{1 \; mole}{18.0 \; g}\right) = 2.26 \; kJ/g$$

For mercury we have:

$$\left(59.3 \; \frac{kJ}{mole}\right)\left(\frac{1 \; mole}{200.5 \; g}\right) = 0.296 \; kJ/g$$

It is also interesting that when the surrounding air is humid and contains many water molecules, evaporation of water is slower, and we feel more uncomfortable.

VAPOR PRESSURE

If a liquid is placed in a closed container, the particles entering the vapor phase cannot escape. In their random motion many particles strike the liquid and are recaptured. Thus two processes occur simultaneously: evaporation (entry of particles into the vapor phase) and condensation (entry of particles into the liquid from the vapor phase). The rate of condensation increases as the number of particles in the vapor phase increases. The rate of condensation and the rate of evaporation are compared qualitatively as a function of time in Figure 11.3. When the rates of these two processes become equal, the number of particles in the vapor phase becomes stabilized. This steady state, in which there is a balance between the dynamic processes of evaporation and condensation,

TABLE 11.2

heats of vaporization of some common substances at their boiling points

SUBSTANCE	FORMULA	ΔH_{vap} (kJ/mole)	BOILING POINT (°C)
Benzene	C_6H_6	30.8	80.2°
Ethanol	C_2H_5OH	39.2	78.3°
Ether	$C_2H_5OC_2H_5$	26.0	34.6°
Mercury	Hg	59.3	356.9°
Methane	CH_4	10.4	−164°
Water	H_2O	40.7	100°

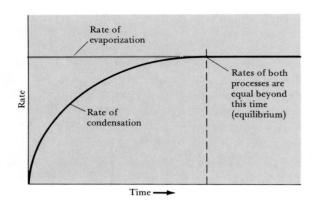

FIGURE 11.3 A comparison of the rates of evaporation and condensation of a liquid as a function of time as the system approaches equilibrium.

is an example of a **dynamic equilibrium.** This equilibrium condition, which represents a situation in which two opposing processes take place at the same rate, can be represented symbolically with double arrows:

$$\text{Liquid} \rightleftharpoons \text{gas}$$

When equilibrium is reached, there is no further change in the number of particles in the vapor. To all outward appearances vaporization seems to cease. If the pressure exerted by the vapor is monitored as shown in Figure 11.4, it is found to increase until equilibrium is reached and thereafter remains constant. A graph of vapor pressure plotted as a function of time would look like that shown in Figure 11.5. The magnitude of the **equilibrium vapor pressure** depends on the attractive forces between particles of the liquid and on the temperature of the liquid. We shall hereafter adopt the common procedure of referring to the equilibrium vapor pressure simply as the vapor pressure of the liquid. The variation in vapor pressure with temperature for several liquids is shown in Figure 11.6. Vapor pressure increases with increasing temperature.

BOILING POINTS

A liquid is said to boil when vapor bubbles form in the interior of the liquid; this condition occurs when the vapor pressure equals the external pressure acting on the liquid's surface. Consequently, the boiling point of

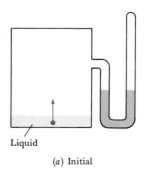

Liquid

(a) Initial

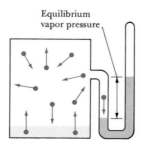

Equilibrium
vapor pressure

(b) At equilibrium

FIGURE 11.4 Schematic representation of the evaporation of a liquid in a closed container. At equilibrium the rate of condensation and the rate of evaporation are equal. This produces a stable vapor pressure, one that does not change with time so long as the temperature remains constant.

a liquid depends on pressure. From Figure 11.6 it can be seen that the boiling point of water at 1 atm pressure (760 torr) is 100°C. The boiling point of a liquid at 1 atm pressure is called its **normal boiling point.** At 650 torr water boils at 96°C.

SAMPLE EXERCISE 11.3

What is the boiling point of ethanol at 500 torr?

Solution: From Figure 11.6 it can be seen that the boiling point must be about 70°C, as compared to the normal boiling point of 78.3°C.

The fact that boiling points vary with pressure is utilized in the pressure cooker. The time required to cook food depends on the temperature. As long as water is present, the maximum temperature of the food is the boiling point of water. Pressure cookers are sealed and allow steam to escape only when it exceeds a predetermined pressure; the pressure above the water can therefore increase above atmospheric pressure. The higher pressure causes water to boil at a higher temperature thereby allowing the food to get hotter. It therefore cooks more rapidly. The effect of pressure on boiling point also explains why it takes longer to cook food at higher elevations than at sea level; at higher altitudes the atmospheric pressure is lower, and water boils at a lower temperature.

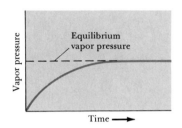

FIGURE 11.5 Vapor pressure of a liquid in a sealed container as a function of time.

VISCOSITY

Some liquids literally flow like molasses, whereas others flow quite easily. The resistance of liquids to flow is referred to as their **viscosity.** The larger the viscosity, the more slowly the liquid flows. Molasses is a viscous liquid; water is not. Motor oils are given ratings (20 wt, 30 wt, and so on) based on viscosity. Viscosity is related to the ease with which individual molecules can move by other molecules. Viscosity thus depends on the

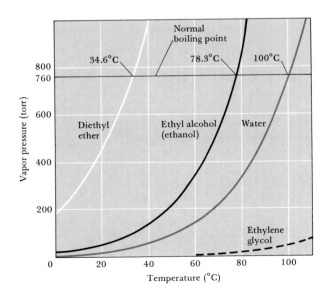

FIGURE 11.6 Equilibrium vapor pressures of four common liquids shown as a function of temperature. The temperature at which the vapor pressure is 760 torr is the normal boiling point of the liquid.

313

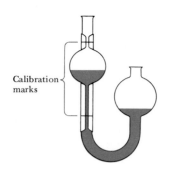

Calibration marks

FIGURE 11.7 An Ostwald viscometer, a device that is commonly used to measure viscosity.

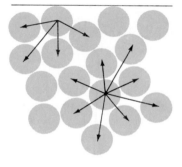

FIGURE 11.8 A molecular-level view of the unbalanced intermolecular forces on the surface of a liquid that result in surface tension.

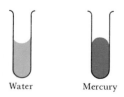

Water   Mercury

FIGURE 11.9 A comparison between the shape of the meniscus of water in a glass tube and the shape of the mercury meniscus.

attractive forces between molecules and on whether there are structural features that permit the molecules to become entangled. One way to determine viscosity is by measuring the time it takes for a given amount of liquid to flow through a small-diameter tube under the influence of gravity. A common device used to measure viscosity is shown in Figure 11.7. Viscosity decreases with increasing temperature, because at higher temperatures the attractive forces between molecules are less important in comparison with the increased kinetic energies of the particles. The viscosities of some common liquids are listed in Table 11.3.

SURFACE TENSION

Liquids have a tendency to assume a minimum surface area. This minimum is achieved when the liquid has a spherical shape. For example, when water is placed on a waxy surface, it "beads up," forming distorted spheres. The state of minimum energy is the state of minimum surface area; energy must therefore be supplied to increase the surface area. The energy required to increase the surface area of a liquid by a unit amount is known as its **surface tension**. The origin of surface tension is an imbalance of forces at the surface of the liquid, as shown in Figure 11.8. There is a net inward pull on the surface that contracts the surface and makes the liquid behave almost as if it had a skin. This effect permits a carefully placed needle to float on the surface of water or some insects to "walk" on water even though their densities are greater than that of water.

The forces between like molecules that are reflected in a substance's vapor pressure, boiling point, heat of vaporization, viscosity, and surface tension are called **cohesive**. The forces between unlike substances, such as water and glass, are called **adhesive**. In a glass tube, the adhesive forces between water and glass are sufficiently strong relative to the cohesive forces for water to form a concave-upward surface (Figure 11.9). Such a curved surface on a liquid is known as a **meniscus**. For mercury, cohesive forces are greater than the adhesive forces with glass; mercury does not adhere to glass and a concave-downward meniscus is observed (Figure 11.9).

When a liquid adheres to or wets the walls of a tube as water adheres to glass, the liquid is drawn up the tube as shown in the case of

TABLE 11.3

viscosities and surface tensions of some common liquids at 20°C

SUBSTANCE	FORMULA	VISCOSITY (kg/m-sec)	SURFACE TENSION, (J/m²)
Benzene	C_6H_6	0.65×10^{-3}	2.89×10^{-2}
Ethanol	C_2H_5OH	1.20×10^{-3}	2.23×10^{-2}
Ether	$C_2H_5OC_2H_5$	0.23×10^{-3}	1.70×10^{-2}
Glycerin	$C_3H_8O_3$	1490×10^{-3}	6.34×10^{-2}
Mercury	Hg	1.55×10^{-3}	46×10^{-2}
Water	H_2O	1.00×10^{-3}	7.26×10^{-2}

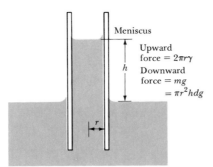

FIGURE 11.10 Capillary rise is used to determine the surface tension of a liquid.

water in Figure 11.9. This phenomenon is known as **capillary action.** Wetting of the walls of a tube tends to increase the surface area of the liquid. Surface tension tends to reduce this area, and consequently the liquid is drawn up the tube, until the upward force caused by the adhesion with the walls and the cohesion of the liquid is balanced by the downward force caused by the pull of gravity on the liquid. It is not intuitively obvious, but the upward force is approximately $2\pi r\gamma$, where r is the radius of the tube and γ is the surface tension of the liquid. The downward force is just the mass of the column, m, times the acceleration due to gravity, g ($g = 9.807$ m/sec^2). The liquid mass is the volume of the column, $\pi r^2 h$, where h is the height of the column, times the density of the liquid, d. Therefore

$$\text{Upward force} = \text{downward force}$$
$$2\pi r\gamma = \pi r^2 hdg$$
$$\gamma = \tfrac{1}{2}rhdg \qquad\qquad [11.1]$$

Consequently, surface tension can be determined from the height to which a liquid rises in a tube of known radius. This relationship is shown in Figure 11.10. The surface tensions of several common liquids are given in Table 11.3. The surface tension of 7.26×10^{-2} J/m^2 for water indicates that an energy of 7.26×10^{-2} J must be supplied to increase the surface area of a given amount of water by 1 m^2.

11.3
intermolecular
forces

We have noted that some properties of liquids—for example, their boiling points—reflect the magnitudes of the forces between particles in the liquid. If the particles are molecules, these forces are termed **intermolecular forces.** When water is vaporized, the bonds that are broken are the *inter*molecular bonds between molecules, not the *intra*molecular bonds within the molecules:

It requires 920 kJ to overcome the intramolecular forces and break a mole of gaseous water into hydrogen and oxygen atoms. In contrast, only 40.7 kJ are required to overcome intermolecular forces and convert a mole of water to steam. Generally, the stronger the intermolecular bonds, the higher the molar heat of vaporization, boiling point, surface tension, and viscosity of the substance.

If we examine an extended list of substances and their physical properties, the following generalizations concerning intermolecular forces emerge:

1. For molecules of approximately equal molecular weight, intermolecular forces increase with increasing polarity. Comparison of the boiling points of CH_4 and NH_3 or Br_2 and ICl (Table 11.4) reflect this fact.

2. As we saw in Section 5.11, when molecules have approximately equal polarity, attractive forces increase with increasing molecular weight. Another example of this trend is seen by comparing the boiling points of CH_4, SiH_4, GeH_4, and SnH_4, all of which are nonpolar. This comparison is made in Table 11.5.

3. The attractive forces between molecules in which hydrogen is bonded to a small electronegative atom such as N, O, or F are much higher than those in other compounds of similar molecular weight and polarity. This effect is seen in the series H_2O, H_2S, H_2Se, and H_2Te in which H_2O has an anomalously high boiling point as seen in Table 11.5 and in Figure 11.11. A similar trend is seen in the series NH_3, PH_3, AsH_3, and SbH_3 and in the series HF, HCl, HBr, and HI. In each case the first member of the series (associated with O, N, or F) has an abnormally high boiling point, whereas the boiling points of the other members of each family increase with increasing molecular weight.

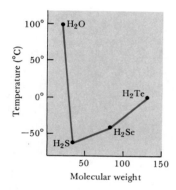

FIGURE 11.11 Normal boiling points as a function of molecular weight through a simple series of compounds, the hydrides of Group 6A.

TABLE 11.4

boiling points as a function of dipole moments

COMPOUND	FORMULA	MOLECULAR WEIGHT	BOILING POINT (°C)	DIPOLE MOMENT (D)
Methane	CH_4	16	-164	0
Ammonia	NH_3	17	-75	1.49
Bromine	Br_2	160	59	0
Iodine chloride	ICl	162	97	0.65

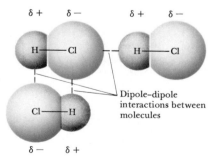

Dipole-dipole interactions between molecules

FIGURE 11.12 Dipole-dipole interactions between polar HCl molecules. The Greek letter δ (delta) followed by a positive or negative sign means a partial positive or a partial negative charge.

TABLE 11.5

comparison of the boiling points of the hydrides of group 4A and group 6A

GROUP 4A		GROUP 6A	
compound	boiling point (°C)	compound	boiling point (°C)
CH_4	−164	H_2O	100
SiH_4	−112	H_2S	−61
GeH_4	−90	H_2Se	−41
SnH_4	−52	H_2Te	−2

Keeping these generalizations in mind, let us now consider the various types of forces between molecules that account for them.

DIPOLE-DIPOLE ATTRACTIONS

The higher attractive forces between polar molecules compared with nonpolar ones of approximately equal molecular weight are readily explained in terms of the attractions between dipoles. Recall that polar molecules are those to which a positive and a negative end can be assigned owing to molecular shape and to charge separation caused by unequal sharing of electrons (Section 8.7). A simple example is HCl ($\mu = 1.03$ D) shown in Figure 11.12. The positive hydrogen end of one molecule experiences an electrostatic attraction to the negatively charged chlorine end of another molecule. Because the charges of the ends of a dipole are not as great as the charges of the ions of an ionic compound like NaCl, these attractive forces are not as great in magnitude as those of ionic materials. Furthermore, these forces are not as effective at large distances as those between ions. The energy of interaction between two ions varies as $1/r$, whereas that between two dipoles varies as $1/r^3$. Therefore, as the distance between two ions is doubled the energy of interaction decreases to half its original magnitude. However, as the distance between two dipoles doubles, the energy of interaction decreases to $1/2^3 = 1/8$ its original value. Consequently, dipole-dipole forces operate over much shorter distances than do ion-ion forces.

LONDON DISPERSION FORCES

The type of force just described cannot exist between nonpolar particles such as helium atoms. Yet some type of attraction must be present; otherwise it would be impossible to liquefy such nonpolar materials. If we were able to take a series of instantaneous pictures of a helium atom, we would find that in several of these the two electrons and the nucleus would be arranged in such a fashion that an instantaneous dipole results. This dipole can induce a similar dipole in an adjacent atom (Figure 11.13). The motions of the electrons in the adjacent atoms are no longer completely independent but become somewhat synchronized. These instantaneous induced dipole attractions are known as **London dispersion forces,** named after Fritz London, a German physicist, who first derived the quantum-mechanical theory for this attraction in 1930. The strength of these attractive forces depends on how easily distorted or polarized the

317

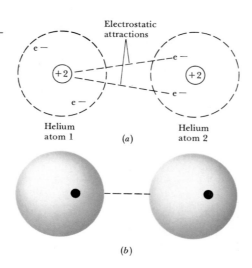

FIGURE 11.13 Two schematic representations of an instantaneous dipole on two adjacent helium atoms, showing the electrostatic attraction between them.

electron cloud is. In general, the larger the molecule, the farther its electrons are from the nuclei and consequently the greater its polarizability. Therefore, the magnitude of these London dispersion forces increases with an increase in molecular size. Because molecular size and weight generally parallel each other, it is often suggested that the magnitude of these forces increases with increasing molecular weight. This generalization is the one that we suggested as we examined trends in intermolecular forces. However, it applies only so long as molecules of reasonably similar shape are compared. For example, *n*-pentane and neopentane, which are shown in Figure 11.14, have the same molecular formula, C_5H_{12}, yet the boiling point of *n*-pentane is 27°C higher than that of neopentane. The difference can be traced to the different shapes

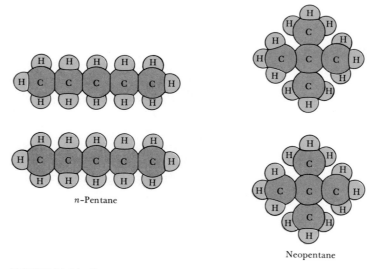

FIGURE 11.14 Illustration of the effect of molecular shape on intermolecular attraction. The boiling point of *n*-pentane is 36.2°C, whereas that of neopentane is 9.5°C.

of the two molecules. The overall attraction between molecules is greater in the case of n-pentane* because there are more sites of interaction; the molecules are able to come in contact over the entire length of the molecule. In the case of neopentane, less contact is possible between molecules.

It is also noteworthy that these induced dipole forces contribute to the attractive forces between all molecules, even polar ones. Thus in the series H_2S, H_2Se, and H_2Te (Figure 11.11 and Table 11.5), boiling points increase with increasing molecular weight. Dispersion forces operate only over very short distances, and so the molecules must be really close in order for the forces to have any effect. The energy of interaction varies as $1/r^6$ so that doubling the distance between particles decreases the energy by a factor of $2^6 = 64$.

London dispersion forces are often referred to as **van der Waals forces,** although this term is also used in a more general sense to describe all intermolecular attractions. We shall adopt this more general usage in this text.

HYDROGEN BONDS

We have noted the unusual boiling points of H_2O, NH_3, and HF. The unusually strong intermolecular interaction that gives rise to this property is referred to as **hydrogen bonding.** This important type of intermolecular attraction can be viewed as largely a dipole-dipole attraction. It is found in systems where a hydrogen atom bound to an oxygen, nitrogen, or fluorine atom on one molecule can interact with an oxygen, nitrogen, or fluorine atom on another. When hydrogen ($EN = 2.2$)† is bonded to the small, highly electronegative fluorine ($EN = 4.0$), oxygen ($EN = 3.4$), or nitrogen ($EN = 3.0$) atoms, the resulting bond is quite polar, and the electrons are bound more closely to the more electronegative atom. The proton is left rather deficient in surrounding electron density. It therefore experiences an attractive interaction with the lone pairs of electrons on electronegative atoms of nearby molecules. In HF, where hydrogen bonding is especially strong, the hydrogen behaves almost as if it were simultaneously bound to two fluorine atoms (Figure 11.15).

Hydrogen bonds are extremely important in biological systems. For

*The n in n-pentane is an abbreviation for the word "normal." A normal hydrocarbon is one whose carbon atoms are arranged in a straight chain.

†We use here EN as an abbreviation for electronegativity (see Section 8.7).

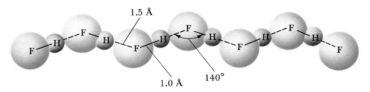

FIGURE 11.15 The structure of crystalline HF showing the hydrogen bonding between HF molecules. The 140° H—F—H angle is probably determined in part by the directionality of the lone-pair electrons in HF and in part by crystal packing forces.

example, the $>$N—H and $>$C$=$O groups found in proteins are able to form a hydrogen bond between them:

Hydrogen bond

$$>N—H \text{-----} :\ddot{O}=C<$$

As we shall see later, hydrogen-bond interactions play a significant role in determining the structures of proteins. Hydrogen bonds are also responsible for many of the properties, such as high heat of vaporization and high boiling point, that make water such a unique substance. Although hydrogen bonds in H_2O, NH_3, and HF represent the strongest known intermolecular attractions, even in these cases they amount to less than 10 percent of the strengths of typical covalent bonds.

SAMPLE EXERCISE 11.4

Which of the following substances is most likely to exist as a gas at room temperature and normal atmospheric pressures: P_4O_{10}, Cl_2, AgCl, or I_2?

Solution: In essence, the question asks which substance has the weakest intermolecular attractive forces, because the weaker these forces, the more likely the substance is to exist as a gas at any given temperature and pressure. We should therefore select Cl_2, because this is both a nonpolar molecule and also has the lowest molecular weight. In fact Cl_2 does exist as a gas at room temperature and normal atmospheric pressure, whereas the others are solids. Of the other substances, AgCl is least likely to be a gas because it exists as Ag^+ and Cl^- ions with very strong ionic bonds holding the ions within the solid.

SAMPLE EXERCISE 11.5

List the substances $BaCl_2$, H_2, CO, HF, and Ne in order of increasing boiling points.

Solution: The boiling point reflects the attractive forces in the liquid. These are stronger for ionic substances than for molecular ones, and so $BaCl_2$ has the highest boiling point. The intermolecular forces of the remaining substances depend on molecular weight, polarity, and hydrogen bonding. The other molecular weights are H_2 (2), CO (28), HF (20), and Ne (20). The boiling point of H_2 should be the lowest, because it is nonpolar and has the lowest molecular weight. The molecular weights of CO, HF, and Ne are roughly the same. HF has hydrogen bonding, and so it has the highest boiling point of the three. CO, which is slightly polar and has the highest molecular weight, is next. Ne, which is nonpolar, comes last of these three. The predicted boiling points are therefore

$$H_2 < Ne < CO < HF < BaCl_2$$

The actual normal boiling points are H_2 (20 K), Ne (27 K), CO (83 K), HF (293 K) and $BaCl_2$ (1813 K).

As the size of an object decreases, gravitational forces have a less significant impact on its behavior. In discussing objects on a molecular level, we generally ignore gravitational forces altogether because they are relatively insignificant compared to molecular forces. The contrast between large and small objects, reflecting the relative importance of gravitational and molecular forces, is illustrated by comparing the world of humans with that of flies. A human cannot walk on a ceiling because of the force of gravity on his body. However, a fly can, not because it has suction cups for feet, but because gravitational forces are less important than the molecular forces of adhesion between its feet and the ceiling.

**11.4
crystalline
solids**

Solids are rigid; they cannot be poured like liquids or compressed like gases. Solids such as quartz or diamond possess highly regular crystalline shapes or cleavage planes. These facts suggest a regular atomic arrangement within the solid. Indeed, a statement of this regular arrangement is sometimes included as part of the definition of solids. For our purposes, we shall divide materials that we would ordinarily call solids because of their rigidity into two groups: crystalline solids and amorphous solids. **Crystalline solids** are characterized by a regular three-dimensional arrangement of atoms. **Amorphous solids** lack this regular atomic-level organization. Familiar materials of the second type include substances such as rubber and glass, which are composed of large or complicated molecules. In some texts, amorphous solids are referred to as supercooled liquids because they have the molecular disorder of liquids. In fact, glass is capable of flowing, as revealed by a careful examination of the window panes of very old houses. The panes are thicker at the bottom than at the top because the glass has flowed under the continued influence of the force of gravity. In this section our focus is on crystalline solids. Amorphous solids will be discussed in Section 11.5.

X-RAY STUDY OF CRYSTAL STRUCTURE

Much of what we know about the internal molecular-level regularity of crystalline solids is revealed by their interaction with X rays. X rays, you will recall, are electromagnetic waves of short wavelength and high energy (Section 6.1). They were first discovered in 1895 by the German physicist Wilhelm Roentgen. However, it was not until 1913 that they were used to determine the location of atoms within a crystalline solid. In that year, the Englishman W. L. Bragg and his father, W. H. Bragg, determined the location of the zinc and sulfur atoms in a ZnS crystal from a mathematical analysis of the X-ray diffraction pattern of this substance. This analysis followed the suggestion of the German physicist Max van Laue, who in 1912 had suggested that crystals should diffract X rays. Diffraction is the term used to describe the way light is spread out as it passes through a narrow opening as shown in Figure 11.16. If the light passes through two slits whose separation is on the same order of magnitude as the wavelength of the light, a series of light and dark bands

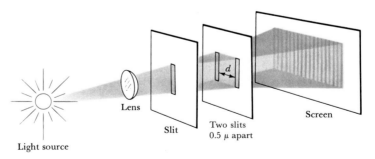

Light source   Lens   Slit   Two slits 0.5 μ apart   Screen

FIGURE 11.16 The diffraction pattern produced by light passing through two narrow slits whose separation, *d*, is on the order of the wavelength of the light.

known as a diffraction pattern is produced. This pattern is shown in Figure 11.16.

You can see this phenomenon if you look between your thumb and forefinger at a monochromatic light source and then bring your fingers close together. Just before they touch, an area consisting of a series of closely spaced lines is evident. The light areas result from **constructive interference** of the scattered light waves. Constructive interference occurs when waves are in phase and consequently reinforce each other; that is to say, the crests and troughs of the waves coincide, thus resulting in a wave of the same wavelength as the original waves but with greater amplitude as shown in Figure 11.17. The dark areas result from **destructive interference,** produced by out-of-phase waves that cancel each other as shown in Figure 11.17.

X rays impinging on a solid interact with electrons and become scattered,* producing diffraction patterns such as that shown for NaCl in Figure 11.18. These patterns can be interpreted in terms of the regular arrangement of the atoms within the solid. The diffraction can be visualized as resulting from the scattering of the X-ray waves by atoms in parallel crystal planes as shown in Figure 11.19. The incoming X rays are in phase at AB. Wave ACA' is scattered or reflected by an atom in the first layer of the solid, whereas wave BEB' is reflected by an atom in the second layer. If these two waves are to be in phase at $A'B'$, the extra distance covered by BEB' must be a whole-number multiple of the wavelength, λ. In Figure 11.19, the extra distance, DEF, is 2λ. Now

---

*Each atom absorbs some of the energy of the X rays and then reemits it in all directions, thereby scattering the X rays.

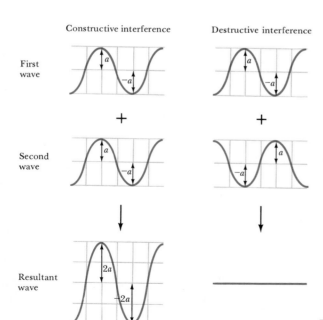

FIGURE 11.17 Constructive and destructive interference of waves.

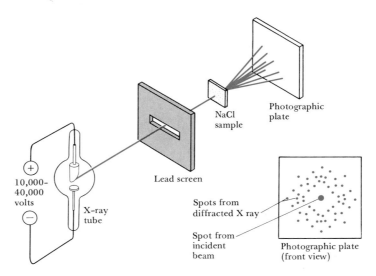

FIGURE 11.18 The Laue X-ray diffraction pattern for NaCl and the experimental method by which it is obtained.

notice that the triangle CDE is a right triangle. Using trigonometry, it can be shown that the distance DE is $d \sin \theta$, where d is the distance between the planes and θ is the angle between the incoming wave and the plane. Because the distance DEF is twice DE, we have

$$DEF = 2\lambda = 2d \sin \theta$$

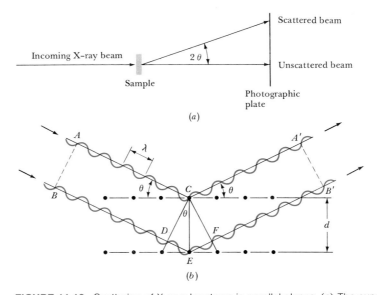

FIGURE 11.19 Scattering of X rays by atoms in parallel planes. (a) The experimental arrangement with the X-ray beam scattered at an angle of 2θ from the unscattered beam. (b) The atomic-level view of the atoms and waves. The incoming X rays of wavelength λ are diffracted by atoms, which are represented as dots. The atoms are arranged in planes that are separated by a distance d. The incoming X rays make an angle of θ with the planes and are scattered at an angle of 2θ from the unscattered beam, which is not shown in (b).

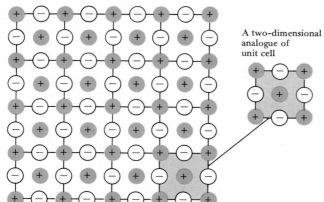

A two-dimensional
analogue of
unit cell

FIGURE 11.20 A cross-section of the space lattice of NaCl, showing the repeating pattern of its unit cell.

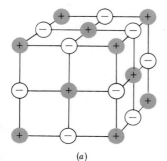

(a)

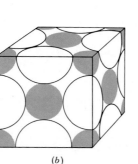

(b)

FIGURE 11.21 Two representations of the unit cell of sodium chloride. In (a) the ions are shown as point charges. In (b) we get a better feel for the relative sizes of the Na$^+$ and Cl$^-$ ions and how ions are shared between unit cells.

It can be shown that the general equation for constructive interference is

$$n\lambda = 2d \sin \theta, \quad \text{where } n = 1, 2, 3, 4 \ldots . \qquad [11.2]$$

This relationship, known as the **Bragg law,** allows determination of the spacing between planes from the known wavelength of the light and experimentally determined values of θ.

CRYSTAL LATTICES AND UNIT CELLS

The order characteristic of crystalline solids allows us to convey a picture of an entire crystal by looking at only a small part of it. The three-dimensional arrangement of particles in space is called the **crystal lattice** or **space lattice.** That portion of the space lattice that is consistent with the compound formula and generates the entire crystal by simple displacement in three dimensions is called the **unit cell.** A simple two-dimensional example is afforded by a sheet of wallpaper. The smallest area that contains the repeating pattern of the sheet is the "unit cell" for the pattern. The unit cell may be considered the "fundamental building block" of the crystal lattice because the lattice can be constructed by stacking unit cells. The NaCl lattice, which we have discussed on previous occasions, provides a convenient example. A two-dimensional portion, representing a slice through the lattice and showing the Na$^+$ and Cl$^-$ ions simply as point charges, is shown in Figure 11.20. Two representations of the unit cell of NaCl are given in Figure 11.21. As shown in Figure 11.21(b), the particles at the corners, edges, and faces of the unit cell do not lie wholly within it. Sample Exercise 11.6 amplifies on this point.

SAMPLE EXERCISE 11.6

Determine the net number of Na$^+$ and Cl$^-$ ions in the NaCl unit cell (Figure 11.21).

Solution: There is $\frac{1}{8}$ of a Na$^+$ on each corner (each Na$^+$ on a corner is shared by eight cubes which intersect at that point). There is $\frac{1}{2}$ of a Na$^+$ on each face; $\frac{1}{4}$ of a Cl$^-$ on each edge; and a whole Cl$^-$ in the center of the cube. Because this may be hard for you to visualize, the sharing of particles on the corners and faces of a unit cell is shown in Figure 11.22. For

NaCl we therefore have the following:

$$Na^+: \quad (\tfrac{1}{8}\ Na^+ \text{ per corner})(8 \text{ corners}) = 1\ Na^+$$
$$(\tfrac{1}{2}\ Na^+ \text{ per face})(6 \text{ faces}) = 3\ Na^+$$
$$Cl^-: \quad (\tfrac{1}{4}\ Cl^- \text{ per edge})(12 \text{ edges}) = 3\ Cl^-$$
$$(1\ Cl^- \text{ per center})(1 \text{ center}) = 1\ Cl^-$$

Thus the unit cell contains $4\ Na^+$ and $4\ Cl^-$. This result agrees with the compound's stoichiometry: one Na^+ for each Cl^-.

SAMPLE EXERCISE 11.7

If the unit cell of NaCl is 5.64 Å on an edge, calculate the density of NaCl.

Solution: The volume of the unit cell is $(5.64\ \text{Å})^3$. Because each unit cell contains $4\ Na^+$ and $4\ Cl^-$ (Sample Exercise 11.6), its mass is

$$4\ (23.0\ \text{amu}) + 4\ (35.5\ \text{amu}) = 234.0\ \text{amu}$$

The density is mass/volume:

$$\text{Density} = \frac{234.0\ \text{amu}}{(5.64\ \text{Å})^3}\left(\frac{1\ \text{g}}{6.02 \times 10^{23}\ \text{amu}}\right)$$
$$\times \left(\frac{1\ \text{Å}}{10^{-8}\ \text{cm}}\right)^3$$
$$= 2.17\ \text{g/cm}^3$$

Any crystalline solid can be described in terms of one of seven types of unit cells. These unit cells are shown in Figure 11.23(*a*). The size and shape of the unit cell is determined by the lengths of the edges of the unit cell (*a*, *b*, and *c*) and by the angles α, β, and γ, which are the angles at which these edges intersect each other. Particles are located at the corners of these cells and may also be located at other special positions within the cell. Consider the cubic cell, for example. When particles are located only at the corners of this cell, the unit cell is described as **simple cubic**—see Figure 11.23(*b*). When particles are located at the corners and at the center of the unit cell, the cell is known as **body-centered cubic**. A third type of cubic cell has particles located at each corner, as well as at the center of each face; this is known as a **face-centered cubic** cell.

CLOSE PACKING

We frequently visualize the unit cells of crystals in terms of points connected by imaginary lines. The points represent atoms, ions, or molecules, and the lines help us visualize the symmetry of the crystal. It is important to remember, however, that the particles of a solid actually take up much more of the space of the crystal lattice than these representations show. The unit cell of NaCl shown in Figure 11.21(*b*) and the

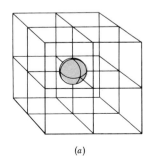

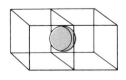

FIGURE 11.22 Particles at the corners and faces of unit cells are shared by other unit cells. Here we see the sharing of a corner atom by eight unit cells (*a*) and the sharing of a face-centered atom by two unit cells (*b*).

(*a*)   (*b*)

unit cell of CH_4 shown in Figure 11.24 provide a more realistic picture of how the particles actually pack within the solid. Notice in Figure 11.24 that the CH_4 molecules can be approximated as spheres. Many crystals, particularly those of metals, can be visualized as consisting of equal-sized spheres packed together in space.

> "There are several ways," Dr. Breed said to me, "in which certain liquids can crystallize—can freeze—several ways in which their atoms can stack and lock in an orderly, rigid way."
>
> That old man with spotted hands invited me to think of the several ways in which cannonballs might be stacked on a courthouse lawn, of the several ways in which oranges might be packed into a crate.
>
> "So it is with atoms in crystals, too; and two different crystals of the same substance can have quite different physical properties." *

*Excerpt from *Cat's Cradle,* by Kurt Vonnegut, Jr. Copyright © 1963 by Kurt Vonnegut, Jr. Reprinted by permission of Delacorte Press/Seymour Lawrence and Donald C. Farber for Kurt Vonnegut, Jr.

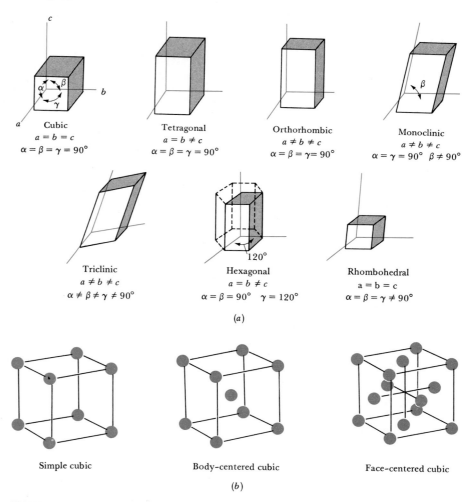

Cubic
$a = b = c$
$\alpha = \beta = \gamma = 90°$

Tetragonal
$a = b \neq c$
$\alpha = \beta = \gamma = 90°$

Orthorhombic
$a \neq b \neq c$
$\alpha = \beta = \gamma = 90°$

Monoclinic
$a \neq b \neq c$
$\alpha = \gamma = 90°$ $\beta \neq 90°$

Triclinic
$a \neq b \neq c$
$\alpha \neq \beta \neq \gamma \neq 90°$

Hexagonal
$a = b \neq c$
$\alpha = \beta = 90°$ $\gamma = 120°$

Rhombohedral
$a = b = c$
$\alpha = \beta = \gamma \neq 90°$

(a)

Simple cubic   Body-centered cubic   Face-centered cubic

(b)

FIGURE 11.23 (a) The seven classes of unit cells that can be used to describe the crystal lattices of all crystalline solids. (b) Three types of cubic cells.

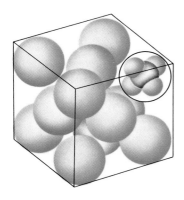

FIGURE 11.24 The unit cell of solid methane. Each large sphere represents a CH_4 molecule as shown at the upper right.

It is instructive to consider how equal-sized spheres can pack most efficiently (that is, with the minimum amount of empty space). Such arrangements, referred to as closest-packed structures, are quite common. They were mentioned briefly in our earlier discussion of the bonding in metals, in Section 9.6. The most efficient arrangement of a layer of equal-sized spheres is shown in Figure 11.25(*a*). Each sphere is surrounded by six others in the layer. The most efficient arrangement of the spheres in a second layer is in the depressions of the first layer, labeled *A* in Figure 11.25(*a*). This is shown in Figure 11.25(*b*). The spheres of the third layer sit in depressions in the second layer. However, there are two types of depressions, and they lead to two different structures. If the spheres of the third layer are placed immediately above those of the first layer, in the positions labeled *B* in Figure 11.25(*b*), the structure known as the **hexagonal close-packed structure** results—Figure 11.25(*c*). If we consider more than three layers of spheres, the arrangement of the layers in this structure can be represented as ABABAB. . . . The second type of close-packed structure results if the third-layer spheres are placed in the positions labeled *C* in Figure 11.25(*b*). The resultant structure is known as the **cubic close-packed structure**—Figure 11.25(*d*). In this case the stacking sequence can be represented as ABCABC. . . . Although it is not obvious from Figure 11.25, the cubic-close-packed arrangement of spheres has a face-centered cubic unit cell. In Figure 11.26, the close-packed structures are viewed from a perspective that permits the unit cells to be seen more clearly. In both of the close-packed structures, each sphere has twelve nearest neighbors (that is, a **coordination number** of twelve).

When unequal-sized spheres are packed in a lattice, the large

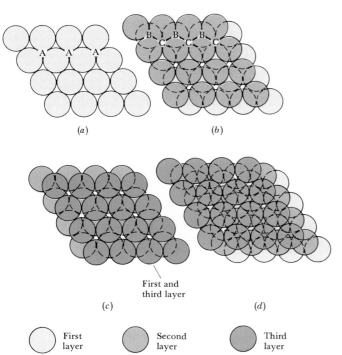

(a)

(b)

First and third layer

(c)

(d)

First layer

Second layer

Third layer

FIGURE 11.25 The closest packing of equal-sized spheres. (*a*) One layer. (*b*) Two superimposed layers. (*c*) Three superimposed layers in hexagonal close-packed arrangement. (*d*) Three superimposed layers in cubic close-packed arrangement.

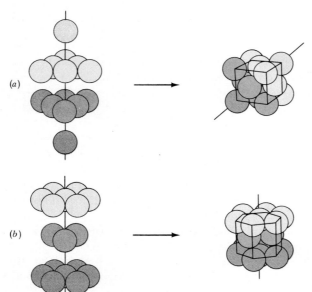

FIGURE 11.26 The stacking of spheres by means of (a) cubic close-packed structure and (b) hexagonal close-packed structure, drawn so as to show the unit cell of each lattice.

particles sometimes assume one of the close-packed arrangements with small particles occupying the holes between the large spheres. For example, in Li_2O the oxide ions assume a cubic close-packed structure, whereas the Li^+ ions occupy small cavities that exist between oxide ions. We shall consider this particular view of crystals further in Chapter 22 when we discuss the structures of silicates, the primary structural materials of our planet.

CRYSTAL DEFECTS

Although we have generally pictured crystalline solids as being composed of perfectly ordered arrays of particles, this is merely a useful abstraction like that of an ideal gas. Real crystals contain imperfections whose number and type can play an important role in determining the properties of the solid. For example, a crystal lattice that has many sites where particles are missing (vacancies) can be more readily deformed than a perfect crystal lattice of the same substance. It is not hard to appreciate that structural imperfections can occur readily. In forming a solid, the structural units can be thought of as a mob that has been required to assume a military drill formation. The faster crystal formation occurs, the greater the chance for defects. Some of the types of defects that can occur and their role in determining the properties of substances will be considered in Chapter 22.

BONDING IN SOLIDS

The properties of solids are determined not only by the arrangement of the particles in the lattice, but also by the types of forces that exist between the particles. In Table 11.6, solids are organized according to the types of particles within the lattice and the types of forces that hold these particles together. It is important to study this table carefully because it

TABLE 11.6

crystal classifications

CRYSTAL CLASSIFICATION	FORM OF UNIT PARTICLES	FORCES BETWEEN PARTICLES	PROPERTIES	EXAMPLES
Atomic	Atoms	London dispersion forces	Soft, very low melting point, poor thermal and electrical conductors	Rare gases—Ar, Kr
Molecular	Polar or non-polar molecules	Van der Waals forces (London dispersion, dipole-dipole forces, hydrogen bonds)	Fairly soft, low to moderately high melting point, poor thermal and electrical conductors	Methane, CH_4; sugar, $C_{12}H_{22}O_{11}$; dry ice, CO_2
Ionic	Positive and negative ions	Electrostatic attraction	Hard and brittle, high melting point, poor thermal and electrical conductors	Typical salts—for example, NaCl, $CaCl_2$.
Covalent (network)	Atoms that are connected in covalent-bond network	Covalent bond	Very hard, very high melting point, poor thermal and electrical conductors	Diamond, quartz
Metallic	Cations in electron cloud	Metallic bond	Soft to very hard, low to very high melting point, excellent thermal and electrical conductors, malleable and ductile	All metallic elements—for example, Cu, Fe, Al

summarizes a great deal of information about solids, relating atomic-level rearrangements and forces to the bulk properties of the crystal.

Solid argon and methane are examples of **atomic solids** and **molecular solids,** respectively. Because the interparticle forces are of the weak van der Waals type, such solids normally have relatively low melting points and exhibit trends in melting points that are similar to those discussed for the boiling points of molecular substances. Most substances that are gases or liquids at room temperature form molecular solids at low temperature.

Water forms a molecular crystal when it freezes to produce ice. Normally, the solid form of a substance is more dense than the liquid. However, in the case of water, the solid form is less dense than the liquid. This unusual property has some important consequences. First of all, it permits ice to float on water. When ice forms in cold weather, it covers the top of the water, thereby insulating the water below. If ice were more dense than water, ice forming at the top of a lake would fall to the bottom, and the lake could freeze solid. Most aquatic life could not survive under these circumstances. The expansion of water upon freezing is also what causes water pipes to break in freezing weather. The low density of ice compared to water can be understood in terms of hydrogen-bonding interactions between water molecules. The interactions in the liquid are random. However, when water freezes, the molecules assume the ordered arrangement shown in Figure 11.27. This structure, which extends in all directions in space, permits the maximum number of hydrogen-bonding interactions between the H_2O molecules. Because the structure has large hexagonal holes, ice is more open and less dense

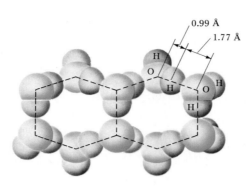

FIGURE 11.27 The arrangement of water molecules in ice. Each hydrogen atom on one water molecule is oriented toward a nonbonding pair of electrons on an adjacent water molecule. Distances between the centers of the bonded atoms are shown.

than the liquid. Application of pressure depresses the melting point of ice because the pressure causes the solid to revert to the more dense liquid. In the case of H_2S, which has almost no hydrogen bonding, the molecules do not assume an open network when they crystallize. Instead, H_2S crystallizes to form a cubic close-packed lattice. Correspondingly, the solid form is more dense than the liquid form, and increasing pressure increases the melting point.

Ionic solids have higher melting points than do atomic and molecular solids, because electrostatic bonds are stronger than van der Waals bonds. Ionic solids are also harder and fracture when struck rather than simply deforming.

In **covalent** or **network solids** the lattice units are joined by covalent bonds. Such materials are consequently much stronger than molecular solids. Diamond, whose structure and bonding was discussed in Section 8.3, is an example of this type of lattice (see Figure 8.6). Quartz (see Figure 11.28) is another example.

The bonding in **metals** is unlike that of the previous types. Each

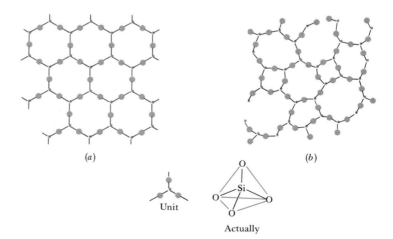

FIGURE 11.28 A schematic comparison of (*a*) crystalline SiO_2 (quartz) and (*b*) amorphous SiO_2 (quartz glass). The small dots represent silicon atoms; the large ones represent oxygen atoms. The structure is actually three-dimensional and not planar as drawn. The unit shown as the basic building block (silicon and three oxygens) actually has four oxygens, the fourth coming out of the plane of the paper and capable of bonding to other silicon atoms.

atom in a metallic lattice typically has eight to twelve atoms adjacent to it, reflecting a close-packing arrangement. The bonding is too strong to be of the van der Waals type, and yet there are not enough valence electrons for ordinary covalent bonds between the atoms. An explanation for the bonding, which is known as band theory and which is essentially an extension of molecular-orbital theory, was discussed in Section 9.6. This model allows us to picture metals as composed of metal atoms held together by electrons that are distributed throughout all the spaces between the atoms. The electrons are free to move through the orbitals that extend over the entire metal. However, they maintain a uniform average distribution. The properties and structures of metals will be examined more closely in Chapter 22.

11.5
amorphous
solids

As noted in Section 11.4, not all solids have a regular arrangement of particles. For example, the material obtained when quartz, SiO_2, is melted and then rapidly cooled is amorphous. Quartz has a three-dimensional structure like that of diamond (Section 8.3). When quartz is melted (at approximately 1600°C) it becomes a viscous, tacky liquid. Although the silicon-oxygen network remains largely intact, many Si—O bonds are broken, and the orderliness of the quartz is lost. If the melt is rapidly cooled, the atoms are unable to return to their orderly arrangement and an **amorphous solid** known as quartz glass or silica glass results. The lack of molecular-level regularity in this glass is represented in Figure 11.28, where its structure is compared with that of quartz.

Even when a solid lacks any long-range order, small regions of regularity known as **crystallites** may exist. Their existence permits description of these solids in terms of the degree of crystallinity. Such crystallites are often found, for example, in synthetic **polymers,** large molecules composed of many molecular parts fused together.

The simplest synthetic polymer is polyethylene, a waxy-feeling substance used in packaging films, wire insulation, and molded articles. Polyethylene is formed by causing ethylene, C_2H_4, molecules to fuse together to form chains. Typically 700–2000 C_2H_4 molecules combine to form a chain containing 1400–4000 carbon atoms:

Ethylene   Polyethylene

The regular layering of these chains produces crystallites as shown in Figure 11.29.

The properties of a plastic* like polyethylene are determined by at

---

*Although the term plastic has come to mean a certain type of synthetic material, the term is most precisely used to describe any material that changes shape when a force is exerted and maintains its distortion upon removal of the force. Materials like rubber that distort but return to their original shape when the force is removed are called elastomers.

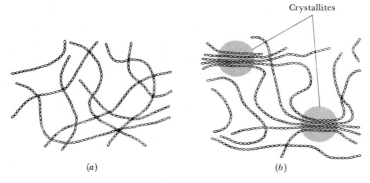

FIGURE 11.29 (a) An amorphous polymer, compared with (b) a polymer containing regions of crystallinity, or crystallites.

least three factors: (1) the length of the polymer chain; (2) the degree of crystallinity; and (3) the extent of bonding between chains. As the length of a polymer chain increases, intermolecular attractive forces increase, thus making the polymer mechanically stronger and harder. Mechanical strength and hardness also increase as the degree of crystallinity increases. The regular arrangement of chains in crystallites permits closer approach of the molecules, thereby increasing intermolecular attractions.

The effect of the degree of crystallinity on the properties of polyethylene can be seen in Table 11.7. The degree of crystallinity depends on the conditions of the polymerization.

To soften a plastic or make it more pliable a substance with low molecular weight known as a **plasticizer** may be added during the course of fabrication. The plasticizer molecules occupy positions between polymer strands, thereby interfering with intermolecular forces between chains and lowering the degree of crystallinity. The oily film that develops on the insides of the windows of new cars left standing in the sun is due in part to loss of plasticizers from the plastics in the cars' interiors. Continual loss of the plasticizer leaves the plastics brittle, causing them eventually to crack.

In contrast to the weakening of bonds between chains by use of a plasticizer, the bonds can be strengthened by replacing the intermolecular bonds with covalent bonds between the chains. These bonds are known as cross-links. The process of vulcanizing rubber, which increases its rigidity, involves cross-linking of polymer chains.

TABLE 11.7

properties of polyethylene as a function of crystallinity

| | DEGREE OF CRYSTALLINITY | | | | |
	55%	62%	70%	77%	85%
Melting point (°C)	109	116	125	130	133
Density (g/cm^3)	0.92	0.93	0.94	0.95	0.96
Stiffness[a]	25	47	75	120	165
Yield stress[a]	1700	2500	3300	4200	5100

[a]These tests reflect the increased mechanical strength of the polymer with increased crystallinity. The physical units for the stiffness test are psi $\times$ 10^{-3} (psi = pounds per square inch), while those for the yield stress test are psi. Discussion of the exact meaning and significance of these tests is beyond the scope of this text.

When solids are converted to liquids, intermediate phases possessing some of the molecular order characteristic of crystalline solids are sometimes obtained. These intermediate phases are known as **liquid crystals,** because they possess some of the attributes of both liquids and solids; for example, liquid crystals exhibit the optical properties of solids and the fluidity of liquids. Three types of molecular arrangements, known as *smectic, nematic,* and *cholesteric* liquid crystals, have been identified. These are shown in Figure 11.30.

Several unsymmetrical molecules that have approximately rodlike shapes, such as *para*-azoxyanisole, belong to the nematic category. The *para*-azoxyanisole molecule has the following form:*

$$\text{CH}_3\text{O} \overline{\bigcirc} \text{N} \overset{\overset{\displaystyle O}{|}}{=} \text{N} \overline{\bigcirc} \text{OCH}_3$$

para-Azoxyanisole

The molecules in **nematic liquid crystals** are parallel to each other but can slide or roll relative to each other, and their centers are oriented randomly.

Smectic liquid crystals show greater order. Not only are the molecules parallel to each other, but they are arranged in layers that are generally one molecule thick. Because the layers are able to slide readily over each other, liquid crystals of this type are usually turbid and viscous.

Cholesteric liquid crystals are so named because the majority of the molecules of this category are derivatives of cholesterol. The molecules are arranged in thin layers, and the orientation of molecules in one layer is twisted relative to that in the next layer, producing a helical (spiral) pattern throughout the crystal.

*The hexagon symbol is used to represent benzene and compounds derived from it (see Sections 9.3 and 24.1).

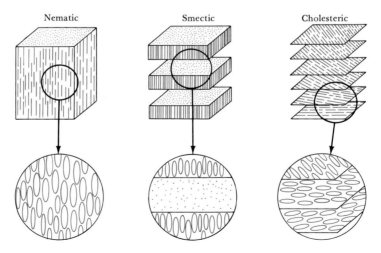

FIGURE 11.30 Nematic, smectic, and cholesteric liquid crystals.

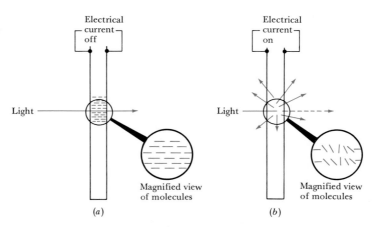

FIGURE 11.31 Light passes readily through a properly oriented nematic liquid crystal (*a*), but is scattered when the molecules are disordered by the application of an electrical signal (*b*).

There has been much recent interest in liquid crystals because of their potential use in electro-optical devices such as display boards on clocks, wristwatches, and calculators, and electronically tuned optical filters. A thin layer of liquid crystal is placed between two electrodes, at least one of which is transparent. For example, the electrodes may be made from glass having a thin conductive coating of tin oxide. In the absence of an electrical signal across the electrodes, the molecules are uniformly oriented as represented schematically in Figure 11.31. In this configuration light passes readily through the cell. Application of an electrical signal across the electrodes causes the liquid to become opaque, probably because of a disruption in the organization of the molecules.

11.7
phase changes

Earlier in this chapter (Section 11.2), we considered the liquid-vapor phase change. Before we close this chapter, it is useful to extend our discussion of phase changes to include solids. Just as liquids have char-

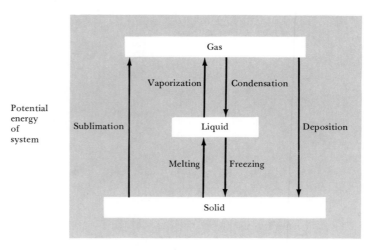

FIGURE 11.32 The possible phase changes and the terms associated with them.

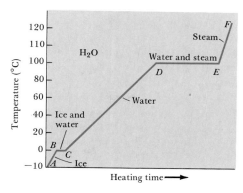

FIGURE 11.33 The heating curve for water with heat added at a constant rate. The rate of climb (slope) of the portions *AB*, *CD*, and *EF* depends on the specific heats of ice (2.092 J/g-°C), water (4.184 J/g-°C), and water vapor (1.841 J/g-°C), respectively. (The specific heat of a substance is the energy required to increase the temperature of 1 gram of that substance by 1°C.) The lengths of the level segments *BC* and *DE* are related to the heat of fusion (6.02 kJ/mole) and heat of vaporization (40.67 kJ/mole), respectively.

acteristic boiling points and heats of vaporization associated with the liquid-to-gas phase change, solids have characteristic **melting points** and **heats of fusion** associated with the solid-to-liquid phase change. The melting point of a solid is the same as the **freezing point** of its liquid. Under appropriate conditions, solids can also be transformed directly to the gas phase without passing through the liquid phase. The solid-to-gas phase change is known as **sublimation,** whereas the reverse process is often called **deposition.** The various phase changes are summarized in Figure 11.32.

HEATING AND COOLING CURVES

When a substance is heated, its temperature increases unless it is undergoing a phase change such as from solid to liquid. During the time a pure substance is melting or boiling its temperature remains constant. In Figure 11.33 the temperature of a sample of water (starting as ice) is plotted as a function of time as heat is added at a uniform rate. The resultant diagram is known as a **heating curve.** We understand the increasing temperature as indicating that added energy is used to increase the average kinetic energy of the particles. At the melting and boiling points temperature remains constant because the added energy is used to overcome attractive forces between particles; it increases the potential energy of the system. In an impure or amorphous solid the melting point may vary over a limited range, and the heating curve shows a slight increase in temperature during melting as shown in Figure 11.34. For example, Plexiglas melts over the approximate range of 119° to 130°C.

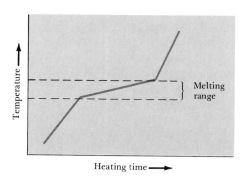

FIGURE 11.34 Variation in temperature during the melting of an impure or amorphous solid.

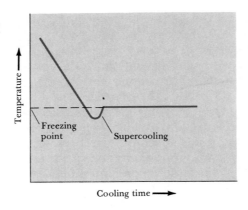

FIGURE 11.35 A portion of the cooling curve of a liquid showing supercooling.

This effect is caused by the lack of regularity in intermolecular forces throughout the solid.

The heating curve is retraced backwards when a substance is cooled, producing a **cooling curve.** Upon cooling, the temperature of a liquid may temporarily fall below its freezing point as shown in Figure 11.35. This phenomenon is known as **supercooling** and represents the time lag associated with changing molecules from their random organization to the orderly arrangement of a crystalline solid.

PHASE DIAGRAMS

The conditions for equilibria between the various phases of a substance can be represented simultaneously on a single graph known as a **phase diagram.** The phase diagram of water is shown in Figure 11.36; those of other substances are qualitatively similar. The conditions of equilibrium exist only along the lines that separate the areas labeled solid, liquid, and gas. For example, the liquid and vapor phases of H_2O are in equilibrium at a water vapor pressure of 0.57 atm (433 torr) and 85°C, point A. The line through points C and D represents the vapor pressure of liquid water, while the line through points D and E represents the vapor pressure of ice. The line through B and D represents the conditions of pressure and temperature at which liquid water is in equilibrium with ice. As we noted in Section 11.4, the melting point of water decreases with increas-

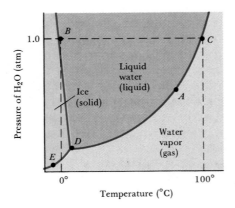

FIGURE 11.36 The phase diagram of H_2O showing the normal melting point, the normal boiling point, and the triple point. The axes are not drawn exactly to scale, and so the triple point should actually be almost on the horizontal axis, a fraction above 0°C.

ing pressure. Therefore, the line through points B and D slopes to the left as pressure increases. For most substances, the melting point increases with increasing temperature and the line representing the solid-liquid equilibrium would slope to the right. Point D, which represents the single point on the diagram where all three phases are in equilibrium with each other, is known as the **triple point**. For water this point is at 0.0098 °C and 0.006 atm (4.58 torr). The areas between the equilibrium lines, the gas, liquid, and solid states, represent nonequilibrium conditions under which a single phase is present. For example, at 0.5 atm and 60 °C, water exists entirely as a liquid. Point B represents the temperature at which ice melts when the external pressure is 1 atm; this is known as the normal melting point of the ice. Point C represents the normal boiling point of water.

To understand the utility of the diagram, consider what happens to water if its temperature is held at 85 °C while the water vapor pressure is increased from 0.5 atm to 1.0 atm. As Figure 11.36 shows, water exists entirely as a gas at 0.5 atm and 85 °C. When the pressure reaches 0.57 atm, point A, the water vapor condenses and an equilibrium between liquid and vapor is established. As the pressure is increased above this the water is converted entirely to the liquid form.

Consider now what happens to H_2O when the pressure is maintained at 0.57 atm and the temperature increased from -10 °C to 70 °C. At -10° the H_2O exists as a solid. At just above 0 °C (0.005 °C) it melts, and a solid-liquid equilibrium exists. Above this temperature, the water exists entirely as a liquid until the temperature reaches 85 °C. At that temperature a liquid-gas equilibrium is established. Above 85 °C the H_2O exists entirely as a gas.

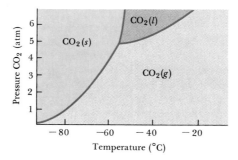

FIGURE 11.37 The phase diagram of CO_2.

SAMPLE EXERCISE 11.8

Referring to the phase diagram for CO_2 (Figure 11.37), determine the phase in which CO_2 exists at a CO_2 pressure of 3 atm and a temperature of -80 °C.

Solution: The CO_2 exists as a solid under these conditions since the point falls in the area marked s. Since this point does not fall on a line, it does not represent an equilibrium condition.

SAMPLE EXERCISE 11.9

What is the lowest pressure at which solid CO_2 is able to melt? (Refer to Figure 11.37; the triple point of CO_2 is -57 °C and 5.2 atm.)

Solution: The lowest pressure is 5.2 atm; below this the CO_2 sublimes. It is for this reason that solid CO_2 (Dry Ice) is such a convenient coolant. It sublimes rather than melts as it absorbs sufficient energy at ordinary pressures.

In order for ice to sublime, its temperature must be below 0.0098°C (the triple-point temperature), and the water vapor pressure must be below its equilibrium vapor pressure. These conditions are employed in the preparation of freeze-dried foods. The food is frozen and then introduced into a vacuum chamber. Water is thereby removed from the food through sublimation.

FOR REVIEW

summary

In this chapter we've considered the general properties of liquids and solids and discussed them in terms of the kinetic-molecular model. For example, we noted that liquids **evaporate** and have characteristic **heats of vaporization** and **vapor pressures**. The vapor pressure is the pressure exerted by the vapor when it is in **equilibrium** with the liquid. When the vapor pressure equals the external pressure, the liquid boils. The **normal boiling point** of a liquid is the temperature at which the vapor pressure equals 1 atm. Two other properties of liquids that were discussed are **viscosity** and **surface tension**. Among the properties of solids that were discussed are **melting points** and **heats of fusion**.

The properties of liquids and solids are due in large measure to the intermolecular forces that exist between particles. Three types of intermolecular forces were discussed: **dipole-dipole**, **induced-dipole** (*London-dispersion*), and **hydrogen-bonding** forces.

The properties of solids also depend on the arrangement of particles within the solid. Much of what we know about the structures of solids comes from X-ray studies. We noted that X rays interact with the particles of a solid and are thereby scattered, thus producing **diffraction patterns**. The Bragg law, $n\lambda = 2d \sin \theta$, gives the conditions for constructive interference of a series of parallel waves scattered by parallel layers of particles. **Crystalline solids** have regular arrangements of particles, known as **crystal lattices,** that can be described in terms of **unit cells.** It is also convenient in many instances to consider the structures of solids in terms of the packing of spheres. Two closest-packed structures for equal-sized spheres were discussed, the **hexagonal close-packed** structure and the **cubic close-packed** structure. The cubic close-packed structure has a **face-centered cubic** unit cell. **Amorphous solids** lack long-range molecular organization, but may possess small regions of regularity known as **crystallites**.

Whereas the regularity of crystalline solids extends in all three dimensions in space, some substances can exist in phases in which the regularity extends in only one or two dimensions. These phases possess some properties of liquids and some of solids and are known as **liquid crystals**.

We have also seen that the temperature changes that accompany the heating or cooling of a substance can be represented graphically in terms of **heating curves** or **cooling curves**. Equilibria between the different physical states for a substance can be displayed graphically in the form of **phase diagrams**.

learning goals

Having read and studied this chapter, you should be able to:

1. Distinguish between gases, liquids, and solids on a molecular level.
2. Use the kinetic-molecular theory to explain the properties of each phase, such as melting point, boiling point, heat of fusion, heat of vaporization, viscosity, surface tension, and vapor pressure.
3. Explain the relation between pressure, vapor pressure, temperature, and boiling point.

4. Recognize and describe the types of intermolecular forces, their relative strengths, and the effects of distance on each.

5. Predict relative boiling points (and other physical properties) of substances, having been given their molecular weights and structural formulas.

6. Determine the number of particles in a unit cell and use this to calculate the density of the substance.

7. Predict the type of solid (atomic, molecular, ionic, covalent network, or metallic) formed by a substance and predict its general properties.

8. Distinguish between crystalline and amorphous solids.

9. Describe the structures of liquid crystals.

10. Draw a phase diagram of a substance given appropriate data.

11. Use a phase diagram to predict what phases are present at any given temperature and pressure.

key terms

Among the more important terms and expressions used for the first time in this chapter are the following:

An **amorphous solid** (Section 11.5) is a solid that lacks a regular and long-range pattern to its molecular arrangement.

The **boiling point** (Section 11.2) of a liquid is the temperature at which its vapor pressure equals the external pressure. The **normal boiling point** is the temperature at which the liquid boils when the external pressure is 1 atm (that is, the temperature at which the vapor pressure of the liquid is 1 atm).

Capillary action (Section 11.2) is the term used to describe the process by which a liquid rises in a tube because of a combination of adhesion with the walls of the tube and cohesion between liquid particles.

A **crystalline solid** (Section 11.4) (or simply a **crystal**) is a solid whose internal arrangement of atoms, molecules, or ions shows a regular repetition in any direction through the solid.

A **dynamic equilibrium** (Section 11.2) is a state of balance in which opposing processes occur at the same rate.

Hydrogen bonds (Section 11.3) are intermolecular attractions between molecules containing hydrogen bonded to oxygen, nitrogen, or fluorine.

London dispersion forces (Section 11.3) are intermolecular forces resulting from attractions between induced dipoles.

The **melting point** (Section 11.7) of a solid (or the **freezing point** of a liquid) is the temperature at which solid and liquid phases coexist in equilibrium. The **normal melting point** is the melting point at 1 atm pressure.

A **meniscus** (Section 11.2) is the curved upper surface of a liquid column.

Surface tension (Section 11.2) is the intermolecular, cohesive attraction that causes a liquid surface to become as small as possible. Surface tension makes a liquid behave as though it had a thin skin over its surface.

The **triple point** (Section 11.7) of a substance is the temperature at which solid, liquid, and gas phases coexist in equilibrium.

A **unit cell** (Section 11.4) is the smallest portion of a crystal that reproduces the structure of the entire crystal when repeated in different directions in space. It is the repeating unit or "building block" of the crystal lattice.

Van der Waals forces (Section 11.3) are weak intermolecular attractions arising from dipole-dipole or induced-dipole attractions.

Vapor pressure (Section 11.2) is the pressure exerted by a vapor in equilibrium with its liquid or solid phase.

Viscosity (Section 11.2) is a measure of the resistance of fluids to flow.

EXERCISES

properties of liquids

11.1 Why does your arm feel cool after it is rubbed with alcohol?

11.2 Explain why water stored in porous clay pots can remain cool even in hot areas.

11.3 In 1958 two mountain climbers on Mt. Washington in New Hampshire froze to death even though the temperature never fell below 2°C (the wind, however, averaged 65 mph). Explain how the climbers could have frozen under these conditions.

11.4 Why is the equilibrium between a liquid and its vapor referred to as dynamic?

11.5 Explain, on a molecular level, why the vapor pressure of water at 20°C is lower than that of alcohol at the same temperature.

11.6 Explain how each of the following affects the vapor pressure of a liquid: (a) surface area; (b) temperature; (c) intermolecular attractive forces; (d) volume of liquid.

[11.7] The vapor pressure of water as a function of temperature is tabulated below.

TEMPERATURE (°C)	VAPOR PRESSURE (torr)
0	4.6
10	9.2
20	17.5
30	31.8
50	92.5
75	289.1
100	760.0

Plot the logarithm of the vapor pressure versus $1/T$, where T is the absolute temperature. What conclusions can be reached regarding the relation between vapor pressure and temperature?

11.8 Using Figure 11.6, predict the boiling point of (a) ethanol at 300 torr and (b) water at 200 torr.

11.9 Why do you think a pressure cooker would be more popular in Denver than in San Francisco?

11.10 Explain why the boiling point of a substance is more pressure-dependent than its melting point.

11.11 Why can a needle, which is much more dense than water, float on the surface of water?

11.12 What is the surface tension of a liquid whose density is 0.88 g/cm^3 if it rises a distance of 4.5 cm in a capillary tube whose inside radius is 0.040 cm?

11.13 A liquid whose density is 0.79 g/cm^3 rises 1.15 cm in a capillary tube whose inside radius is 0.050 cm. Calculate the surface tension of this liquid.

Which of the liquids listed in Table 11.2 could this substance be?

11.14 The surface tension of water is given in Table 11.2. Using this value, calculate how far water will rise in a glass capillary whose inside radius is 0.50 mm.

intermolecular forces

11.15 Predict the type of intermolecular forces found in each of the following substances: (a) CCl_4; (b) CH_3OH; (c) Ne; (d) NH_3; (e) H_2O_2; (f) O_3; (g) CO_2; (h) HCl; (i) Cu.

11.16 Indicate whether the following properties increase in magnitude, decrease in magnitude, or remain unaffected by an increase in the strength of intermolecular forces: (a) vapor pressure; (b) normal boiling point; (c) normal melting point; (d) surface tension; (e) viscosity; (f) heat of fusion; (g) heat of vaporization; (h) molecular weight.

11.17 List the following substances in order of increasing boiling point: (a) CH_4; (b) CI_4; (c) CCl_4; (d) CBr_4; (e) CF_4.

11.18 In each of the following pairs, which substance would you expect to have the higher boiling point? Briefly justify your answer. (a) O_2 and N_2; (b) HF and HCl

11.19 Which of the following compounds would you expect to exhibit hydrogen bonding: (a) PH_3; (b) CH_4; (c) H_2O; (d) HBr; (e) HF?

11.20 In the following pairs of substances, indicate which member will have the highest heat of vaporization. Briefly justify your choices. (a) He or Kr; (b) CH_3OH or CH_3OCH_3; (c) F_2 or Cl_2

11.21 Why is it preferable to use liquid propane (C_3H_8) as a bottled gas instead of liquid butane (C_4H_{10}) in cold climates?

11.22 Rationalize the following observations: (a) the boiling point of $CH_3CH_2CH_3$ ($-42.2°C$) is higher than that of CH_3CH_3 ($-88.6°C$); (b) the boiling point of $CH_3CH_2CH_2CH_3$ ($-0.5°C$) is higher than that of CH_3CHCH_3 ($-12°C$)
$$\underset{\overset{|}{CH_3}}{}$$

solids and liquid crystals

11.23 Describe the attractive forces that must be weakened when each of the following substances is melted: (a) $CaCl_2$; (b) CO_2; (c) copper; (d) xenon; (e) quartz; (f) O_2; (g) water.

11.24 Rationalize the following observations: (a) quartz (SiO_2) has a much higher melting point than CO_2; (b) NaCl is harder than sodium metal; (c) BeO has a higher melting point than LiF.

11.25 Calculate the number of particles in the face-centered cubic and body-centered cubic unit cells.

11.26 What X-ray wavelength would be defracted at an angle of 8.40° if the interplanar spacing is 2.00 Å?

11.27 The first-order ($n = 1$) reflection of X rays of a wavelength of 1.54 Å is found at an angle of 10.5°. What is the distance between the planes causing this diffraction?

11.28 Krypton crystallizes in a face-centered cubic lattice in which the edge length of the cell is 5.59 Å. Calculate the density of solid krypton.

11.29 Lithium crystallizes in a body-centered cubic lattice with a unit cell whose edge length is 3.51 Å. Calculate the density of lithium.

11.30 Copper crystallizes in a face-centered cubic lattice with a density of 8.93 g/cm³. What is the length of the edge of the unit cell?

[11.31] If a substance is composed of atoms arranged in a simple cubic unit cell, determine the fraction of the volume of the unit cell that is occupied by atoms, assuming the atoms on each corner touch each other.

11.32 Irradiation of polyethylene with X rays introduces cross-linking into the material. What effect would this have on the properties of the polyethylene?

11.33 What is the simplest formula of the following polymer?

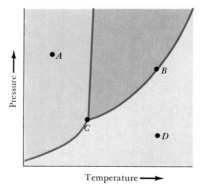

Would you expect this substance to form a crystalline or amorphous solid? Briefly justify your answer.

11.34 Which of the following molecules would you expect could form a liquid crystal? Briefly justify your answer.

$$H_3C-\underset{\underset{CH_3}{|}}{\overset{\overset{CH_3}{|}}{C}}-O-H \quad \text{or}$$

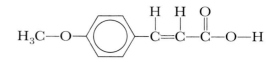

phase diagrams and heating curves

11.35 Explain why it's possible to boil water in a paper cup without burning the cup.

11.36 Two pans of boiling water are located side-by-side on a stove. One is boiling rapidly while the other is boiling slowly. Is the water in the two pans at different temperatures? Explain.

11.37 Roughly sketch the heating curve for ethanol given the following information on this substance: normal melting point = −117.3°C; normal boiling point = 78.5°C; heat of fusion = 4.79 kJ/mole; heat of vaporization = 39.3 kJ/mole; heat capacity of solid form = 44 J/mole-°C; heat capacity of liquid form = 112 J/mole-°C; and heat capacity of gaseous form = 79 J/mole-°C.

11.38 How much energy is required to convert 50.0 g of ice at −10.0° to steam at 105°C (see caption to Figure 11.33).

11.39 How much energy is required to convert 50.0 g of ethanol at −10.0° to gaseous ethanol at 105°C (see data given in Question 11.37).

11.40 Krypton has a normal boiling point of −152°C, a normal melting point of −157°C, and a triple point of −169°C and 133 torr. Roughly sketch the phase diagram for krypton, showing the three points given above and indicating the areas in which each phase is stable.

11.41 Consider the phase diagram in Figure 11.38. State which phase or phases are present at each point.

FIGURE 11.38 Phase diagram.

general exercises

11.42 Compare the following so as to bring out the salient differences between them: (a) boiling point and normal boiling point; (b) crystalline solid and amorphous solid; (c) melting point and triple point; (d) covalent and molecular solid; (e) London dispersion forces and van der Waals forces; (f) intermolecular forces and intramolecular forces; (g) space lattice and unit cell; (h) liquid and liquid crystal; (i) evaporation and sublimation; (j) a liquid at $25\,°C$ and the same liquid at $50\,°C$; (k) crystal and crystallite; (l) hexagonal closest-packed structure and cubic closest-packed structure.

11.43 Using kinetic-molecular theory, explain the following observations: (a) gases are more compressible than liquids or solids; (b) solids are rigid whereas gases and liquids are fluid; (c) most liquids contract in volume when they freeze.

11.44 Explain what happens to the temperature of a solid as it is heated at a constant rate just past its melting point? What is happening on a molecular level?

11.45 Rationalize the observation that the heat of vaporization of a substance is generally much greater than the heat of fusion of the same substance.

11.46 Identify the class of solid (Table 11.6) formed by each of the following substances: (a) bronze; (b) C_6H_6 (benzene); (c) graphite; (d) KBr; (e) polyethylene.

11.47 Explain why H_2S melts at $-83\,°C$, whereas ice melts at $0\,°C$.

11.48 Which member of each of the following pairs would you expect to have the higher melting point? Briefly justify your answer. (a) $CaCl_2$ or CCl_4; (b) I_2 or Br_2

11.49 The oxygen atoms in ice assume the same arrangement as the carbon atoms in diamond. Explain why diamond is so much harder and more dense than ice.

11.50 Which will rise further in a glass capillary: water or C_2H_5OH (ethanol)? Explain briefly.

[11.51] One of the crystal forms of iron is face-centered cubic. (a) If the iron atom has a radius of $1.26\,Å$, and the iron atoms "touch" each other across the diagonal of each face, calculate the length of the unit cell. (b) Calculate the density of iron using the unit cell dimension calculated in part (a) of this question.

[11.52] Platinum has a density of $21.5\,g/cm^3$. How many atoms are in a unit cell if it is cubic with an edge length of $3.914\,Å$. Is the unit cell simple cubic, body-centered cubic, or face-centered cubic?

[11.53] KCl has the same structure as NaCl. The length of the unit cell is $6.28\,Å$. The density of KCl is $1.984\,g/cm^3$ and its formula weight is 74.55. Using this information, calculate Avogadro's number.

11.54 What is the radius of a glass capillary in which a capillary depression of $3.0\,mm$ is observed for mercury at $0\,°C$. The density of mercury at $0\,°C$ is $13.59\,g/cm^3$.

11.55 Chlorine has a triple point at $-103\,°C$ and 8.9 torr; its normal boiling point is $-34.6\,°C$ and its solid form is denser than the liquid form at 1 atm pressure. Make a rough sketch of the phase diagram of chlorine.

solutions

12

Very few of the materials that we encounter in everyday life are pure substances; most are mixtures. Many of these mixtures are homogeneous; that is, their components are uniformly intermingled on a molecular level. We have seen that homogeneous mixtures are called solutions (Sections 2.2 and 3.11). Examples of solutions abound in the world around us. The air we breathe is a homogeneous mixture of several gaseous substances. The oceans are a solution of many dissolved substances in water. The fluids that run through our bodies are solutions, carrying a great variety of essential nutrients, salts, and so forth.

Solutions may be gaseous, liquid, or solid; examples of each kind are given in Table 12.1.* Recall that the solvent is the component whose phase is retained when the solution forms (Section 3.11); if all components are in the same phase, the one in greatest amount is called the solvent. Other components are called solutes. Liquid solutions are the most common, and it is on this type of solution that we focus our attention in this chapter. We shall examine the solution process, the factors that determine the amount of solute that can dissolve in a given quantity of solvent, and the general properties of solutions. At the end of the chapter we shall consider a type of mixture, known as a colloid, that is on the borderline between heterogeneous mixtures and solutions. Before we examine these topics, however, it is useful to discuss ways of describing the concentrations of solutions, that is, the amount of solute dissolved in a given quantity of solvent or solution.

*In order for a liquid or solid solute to form a gaseous solution, the liquid or solid must itself become a gas. Therefore, although three types of liquid solutions and three types of solid solutions are given in Table 12.1, only one type of gaseous solution is listed.

TABLE 12.1

examples of solutions

STATE OF SOLUTION	STATE OF SOLVENT	STATE OF SOLUTE	EXAMPLE
Gas	Gas	Gas	Air
Liquid	Liquid	Gas	Oxygen in water
Liquid	Liquid	Liquid	Alcohol in water
Liquid	Liquid	Solid	Salt in water
Solid	Solid	Gas	Hydrogen in platinum
Solid	Solid	Liquid	Mercury in silver
Solid	Solid	Solid	Silver in gold (certain alloys)

12.1

ways of expressing concentration

The concentration of a solution can be expressed either qualitatively or quantitatively. The terms dilute and concentrated are used to qualitatively describe a solution. A solution with a relatively small concentration of solute is said to be **dilute;** one with a large concentration is said to be **concentrated.**

Several quantitative expressions of concentration are employed in chemistry. We discussed one of these, **molarity** (M), in Section 3.11. We defined molarity as the number of moles of solute in a liter of solution, Equation [12.1]:

$$\text{Molarity} = \frac{\text{moles solute}}{\text{volume of soln in liters}} \qquad [12.1]$$

One of the simplest ways of expressing concentration is in terms of the **weight percentage.** The weight percentage of a component of a solution is given by Equation [12.2]. If a solution of hydrochloric acid contains 36 percent HCl by weight, it would have 36 g of HCl for each 100 g of solution.

$$\text{Wt \% of component} = \frac{\text{mass of component in soln}}{\text{total mass of soln}} \times 100 \quad [12.2]$$

SAMPLE EXERCISE 12.1

A solution is made containing 6.9 g of $NaHCO_3$ per 100 g of water. What is the weight percentage of solute in this solution?

Solution:

$$\text{Wt \% of solute} = \frac{\text{mass solute}}{\text{mass soln}} \times 100$$

$$= \frac{6.9 \text{ g}}{6.9 \text{ g} + 100 \text{ g}} \times 100 = 6.5\%$$

Notice that the mass of solution is the sum of the mass of solvent and the mass of solute. The weight percentage of solvent in this solution is $(100 - 6.5)\% = 93.5\%$.

The **mole fraction** of a component of a solution is given by Equation [12.3].

$$\text{Mole fraction of component} = \frac{\text{moles component}}{\text{total moles of all components}} \quad [12.3]$$

This concept was used in Chapter 5 when we discussed Dalton's law of partial pressures. The symbol X is used for mole fraction with a subscript to indicate the component on which attention is being focused. For example, the mole fraction of HCl in a hydrochloric acid solution can be represented as X_{HCl}. The mole fractions of all components of a solution will add to 1.

SAMPLE EXERCISE 12.2

Calculate the mole fraction of HCl in a solution of hydrochloric acid containing 36 percent HCl by weight.

Solution: Assume that there are 100 g of solution (you can verify for yourself that assuming any other quantity will not change the result, though it can make the arithmetic more difficult). The solution therefore contains 36 g of HCl and 64 g of H_2O.

$$\text{Moles HCl} = (36 \text{ g HCl})\left(\frac{1 \text{ mole HCl}}{36.5 \text{ g HCl}}\right)$$

$$= 0.99 \text{ mole HCl}$$

$$\text{Moles H}_2\text{O} = (64 \text{ g H}_2\text{O})\left(\frac{1 \text{ mole H}_2\text{O}}{18 \text{ g H}_2\text{O}}\right)$$

$$= 3.6 \text{ moles H}_2\text{O}$$

$$X_{HCl} = \frac{\text{moles HCl}}{\text{moles H}_2\text{O} + \text{moles HCl}}$$

$$= \frac{0.99}{3.6 + 0.99} = \frac{0.99}{4.6} = 0.22$$

The **molality** (m) of a solution is defined as the number of moles of solute in a kilogram of solvent:

$$\text{Molality} = \frac{\text{moles solute}}{\text{mass of solvent in kg}} \quad [12.4]$$

Notice the difference between molarity and molality. These two ways of expressing concentration are similar enough to be easily confused. Molality is defined in terms of the mass of solvent, whereas molarity is defined in terms of the volume of solution. A 1.50 molal (written 1.50 m) solution contains 1.50 moles of solute for every kilogram of solvent.

SAMPLE EXERCISE 12.3

What is the molality of a solution made by dissolving 5.0 g of toluene (C_7H_8) in 225 g of benzene (C_6H_6)?

Solution:

$$\text{Molality} = \frac{\text{moles C}_7\text{H}_8}{\text{kg C}_6\text{H}_6} =$$

$$\left(\frac{5.0 \text{ g C}_7\text{H}_8}{225 \text{ g C}_6\text{H}_6}\right)\left(\frac{1 \text{ mole C}_7\text{H}_8}{92.0 \text{ g C}_7\text{H}_8}\right)\left(\frac{1000 \text{ g C}_6\text{H}_6}{1 \text{ kg C}_6\text{H}_6}\right)$$

Convert grams C_7H_8 to moles   Convert grams C_6H_6 to kg

$$= 0.24 \ m$$

In Sample Exercise 12.4 we see, by way of example, how the various ways of quantitatively expressing concentration are related.

SAMPLE EXERCISE 12.4

Given that the density of a solution of 5.0 g of toluene and 225 g of benzene (Sample Exercise 12.3) is 0.876 g/ml, calculate the concentration of the solution in (a) molarity, (b) mole fraction of solute, and (c) weight percentage of solute.

Solution: (a) The total mass of the solution is equal to the mass of the solvent plus the mass of the solute:

$$\text{mass soln} = 5.0 \text{ g} + 225 \text{ g} = 230 \text{ g}$$

The density of the solution is used to convert the mass of the solution to its volume:

$$\text{Milliliters soln} = (230 \text{ g})\left(\frac{1 \text{ ml}}{0.876 \text{ g}}\right) = 263 \text{ ml}$$

Density must be known in order to interconvert molarity and molality, because one is on a mass basis whereas the other is on a volume basis. The number of moles of solute must be known to calculate either molarity or molality:

$$\text{Moles C}_7\text{H}_8 = (5.0 \text{ g C}_7\text{H}_8)\left(\frac{1 \text{ mole C}_7\text{H}_8}{92.0 \text{ g C}_7\text{H}_8}\right)$$

$$= 0.054 \text{ mole}$$

Molarity is moles of solute per liter of solution:

$$\text{Molarity} = \frac{\text{moles C}_7\text{H}_8}{\text{liter soln}}$$

$$= \left(\frac{0.054 \text{ mole C}_7\text{H}_8}{263 \text{ ml soln}}\right)\left(\frac{1000 \text{ ml soln}}{1 \text{ l. soln}}\right)$$

$$= 0.21 \text{ } M$$

Compare this molarity with the molality calculated in Sample Exercise 12.3.

(b) The mole fraction of solute is calculated as follows:

$$\text{Moles C}_7\text{H}_8 = (5.0 \text{ g C}_7\text{H}_8)\left(\frac{1 \text{ mole C}_7\text{H}_8}{92.0 \text{ g C}_7\text{H}_8}\right)$$

$$= 0.054 \text{ mole}$$

$$\text{Moles C}_6\text{H}_6 = (225 \text{ g C}_6\text{H}_6)\left(\frac{1 \text{ mole C}_6\text{H}_6}{78.0 \text{ g C}_6\text{H}_6}\right)$$

$$= 2.88 \text{ moles}$$

$$X_{\text{C}_7\text{H}_8} = \frac{0.054 \text{ mole}}{0.054 \text{ mole} + 2.88 \text{ moles}} = \frac{0.054}{2.93}$$

$$= 0.018$$

(c) The weight percentage of solute is calculated as follows:

$$\text{Wt \% C}_7\text{H}_8 = \frac{5.0 \text{ g C}_7\text{H}_8}{5.0 \text{ g C}_7\text{H}_8 + 225 \text{ g C}_6\text{H}_6} \times 100$$

$$= \frac{5.0 \text{ g}}{230 \text{ g}} \times 100 = 2.2\%$$

The concentration units that we have just discussed are the ones that we shall use through the rest of this chapter and elsewhere in this text. However, there is one further concentration expression that may be encountered in other places, **normality.** Normality (abbreviated N) is defined as the number of **equivalents** of solute per liter of solution. An equivalent is defined according to the type of reaction being examined. For acid-base reactions, an equivalent of an acid is the quantity that supplies one mole of H^+; an equivalent of a base is the quantity reacting with one mole of H^+. In an oxidation-reduction reaction, an equivalent is the quantity of substance that gains or loses one mole of electrons. The masses of one equivalent of several substances are given in Table 12.2. An equivalent is always defined in such a way that 1 equivalent of reagent A will react with 1 equivalent of reagent B. For example, in an

TABLE 12.2
equivalent-mass relationships

REACTANT	PRODUCT	REACTION TYPE	MASS OF 1 MOLE OF REACTANT (g)	MASS OF 1 EQUIVALENT OF REACTANT (g)
$KMnO_4$	Mn^{2+}	Reduction (5 e^-)	158.0	$158.0/5 = 31.6$
$KMnO_4$	MnO_2	Reduction (3 e^-)	158.0	$158.0/3 = 52.7$
$Na_2C_2O_4$	CO_2	Oxidation (2 e^-)	134.0	$134.0/2 = 67.0$
H_2SO_4	SO_4^{2-}	Acid (2 H^+)	98.0	$98.0/2 = 49.0$
$Al(OH)_3$	Al^{3+}	Base (3 OH^-)	78.0	$78.0/3 = 26.0$

oxidation-reduction reaction, 31.6 g of $KMnO_4$ is stoichiometrically equivalent to 67.0 g of $Na_2C_2O_4$ (refer to Table 12.2). Likewise, in an acid-base reaction, 49.0 g of H_2SO_4 is stoichiometrically equivalent to 26.0 g of $Al(OH)_3$.

Where $KMnO_4$ is reduced to Mn^{2+}, thereby gaining five electrons, we have

$$1 \text{ mole } KMnO_4 = 5 \text{ equivalents of } KMnO_4$$

Therefore, if 1 mole of $KMnO_4$ is dissolved in sufficient water to form 1 l. of solution, the concentration of the solution can be expressed as either 1 M or 5 N. Normality is always a whole-number multiple of molarity. In oxidation-reduction reactions, the whole number is the number of electrons gained or lost by one formula unit of the substance. In acid-base reactions the whole number is the number of H^+ or OH^- available in a formula unit of the substance.

SAMPLE EXERCISE 12.5

What is the molarity and normality of a solution of H_2SO_4 made by dissolving 5.00 g of H_2SO_4 in enough water to make 200 ml of solution? The H_2SO_4 is used as an acid, forming SO_4^{2-}.

Solution: The molecular weight of H_2SO_4 is 98.0 amu. The molarity is

Molarity =

$$\left(\frac{5.00 \text{ g } H_2SO_4}{200 \text{ ml soln}}\right)\left(\frac{1000 \text{ ml soln}}{1 \text{ l. soln}}\right)\left(\frac{1 \text{ mole } H_2SO_4}{98.0 \text{ g } H_2SO_4}\right)$$

$$= 0.255 \frac{\text{moles } H_2SO_4}{\text{l. soln}} = 0.255 \ M$$

Because H_2SO_4 has two available hydrogen ions, there are 2 chemical equivalents (equiv) in a mole and the normality is twice the molarity, 0.510 N:

$$\text{Normality} = \left(\frac{5.00 \text{ g } H_2SO_4}{200 \text{ ml soln}}\right)\left(\frac{1000 \text{ ml soln}}{1 \text{ l. soln}}\right)$$

$$\times \left(\frac{2 \text{ equiv } H_2SO_4}{1 \text{ mole } H_2SO_4}\right)\left(\frac{1 \text{ mole } H_2SO_4}{98.0 \text{ g } H_2SO_4}\right)$$

$$= 0.510 \frac{\text{equiv } H_2SO_4}{\text{l. soln}} = 0.510 \ N$$

12.2

the solution process

Sometimes a substance is said to dissolve when a chemical reaction leads to a homogeneous mixture. For example, zinc metal can be dissolved by hydrochloric acid. The following reaction occurs:

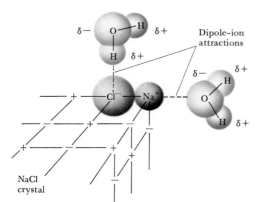

NaCl
crystal

FIGURE 12.1   Interactions between H_2O molecules and the Na^+ and Cl^- ions of an NaCl crystal.

$$Zn(s) + 2HCl(aq) \longrightarrow H_2(g) + ZnCl_2(aq) \qquad [12.5]$$

In this instance the chemical form of the substance being dissolved is changed. If the solution is evaporated to dryness, $Zn(s)$ is not recovered as such; instead $ZnCl_2(s)$ is recovered. In contrast, when $NaCl(s)$ is dissolved in water, it can be recovered by evaporation of its solution to dryness. Our focus throughout this chapter is on solutions from which the solute can be recovered unchanged from the solution.

What happens when a substance like NaCl dissolves in water? When NaCl is added to water, the water molecules orient themselves on the surface of the NaCl crystals as shown in Figure 12.1. The positive end of the water dipole is oriented towards the Cl^- ions, while the negative end of the water dipole is oriented toward the Na^+ ions. The dipole-ion attractions between water molecules and Na^+ and Cl^- ions are sufficiently strong to pull these ions from their positions in the crystal.* Once

*Notice that the corner Na^+ ion is held in the crystal by only three adjacent Cl^- ions. In contrast, an Na^+ ion on the edge of the crystal has four nearby Cl^- ions, and a Na^+ ion in the interior of the crystal has six surrounding Cl^- ions. The corner Na^+ ion is therefore particularly vulnerable to removal from the crystal. Once this Na^+ ion has been removed, adjacent Cl^- ions are similarly exposed and are therefore removed more easily than before.

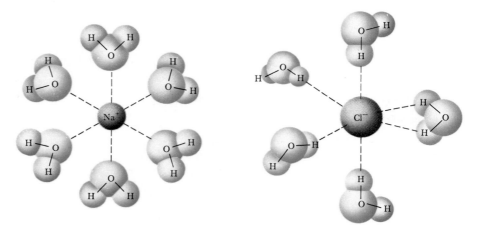

FIGURE 12.2   Hydrated Na^+ and Cl^- ions. The negative ends of the water dipole point toward the positive ion. The positive ends of the water dipole point toward the negative ion. We do not know whether one or both positive hydrogens touch the negative ion.

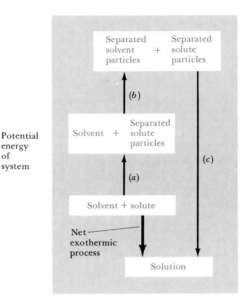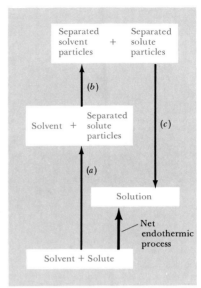

FIGURE 12.3 Analysis of the energy changes accompanying the solution process: (*a*) energy required to separate solute particles; (*b*) energy required to separate solvent particles; (*c*) energy released when solute and solvent particles interact with each other. The figure on the left shows a net exothermic heat of solution, whereas the one on the right shows a net endothermic heat of solution.

removed from the crystal, the Na^+ and Cl^- ions are surrounded by water molecules as shown in Figure 12.2. Such interactions between solute and solvent molecules is known as **solvation.** When the solvent is water it is known as **hydration.**

Sodium chloride dissolves in water because the water molecules interact with the Na^+ and Cl^- ions sufficiently strongly to overcome the attraction between Na^+ and Cl^- ions in the crystal. To form a solution, the water molecules must also separate to make room for the ions. Therefore, we can visualize three types of interaction that take place in the solution process: (1) solute-solute interactions; (2) solute-solvent interactions; and (3) solvent-solvent interactions. The energy changes accompanying each of these interactions are shown in Figure 12.3. Energy is required to overcome solute-solute and solvent-solvent attractions, while energy is released when solute and solvent interact with one another. As shown in Figure 12.3, the net solution process can be either exothermic or endothermic. For example, ammonium nitrate, NH_4NO_3, dissolves by an endothermic reaction ($\Delta H = 26.4 \text{ kJ/mole}$). It has consequently been used to make instant ice packs, which are used in athletics. The solid NH_4NO_3 is placed inside a thin-walled plastic bag. This in turn, is sealed inside a thicker-walled bag together with some water. The small bag can be broken by kneading the larger bag. The resultant solution gets quite cold.

SOLUBILITY

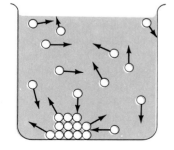

FIGURE 12.4 Movement of solute particles in solvent containing excess solute. Both dissolution and crystallization occur.

As the solution process proceeds and the concentration of solute particles in solution increases, the chances of their colliding with the surface of the solid increases (Figure 12.4). Such a collision may result in the solute

349

particle becoming attached to the solid. This process, which is the opposite of the solution process, is called **crystallization.** Thus, two opposing processes may occur in a solution in contact with undissolved solute. This situation is represented in Equation [12.6] by use of a double arrow

$$\text{Solute} + \text{solvent} \underset{\text{crystallize}}{\overset{\text{dissolve}}{\rightleftharpoons}} \text{solution} \qquad [12.6]$$

When the rates of these opposing processes become equal, there will be no further net increase in the amount of solute in solution. This is an example of a **dynamic equilibrium,** similar to that discussed in Section 11.2, where the processes of evaporation and condensation were considered. A solution that is in equilibrium with undissolved solute is said to be **saturated.** Additional solute will not dissolve if added to a saturated solution. The amount of solute needed to form a saturated solution in a given quantity of solvent is known as the **solubility** of that solute. For example, the solubility of NaCl in water at 0°C is 35.7 g per 100 ml of water. This is the maximum amount of NaCl that can be dissolved in water to give a stable, equilibrium solution at that temperature. If less solute is added than the equilibrium amount, the solution is said to be **unsaturated.** In some cases it is possible to prepare solutions that contain more solute than the equilibrium amount. Such solutions, which are said to be **supersaturated,** are unstable, and under the proper conditions solute will crystallize from them to give saturated solutions. Sodium acetate, $NaC_2H_3O_2$, forms supersaturated solutions very easily. The solubility of this substance at 0°C is 119 g per 100 ml of H_2O. It is more soluble at higher temperatures. If a hot solution containing more than 119 g of $NaC_2H_3O_2$ per 100 ml of H_2O is slowly cooled to 0°C, the excess solute remains dissolved; the resultant solution is supersaturated. Addition of a small crystal of $NaC_2H_3O_2$ gives a surface on which crystallization can start, and crystallization occurs until the concentration of the solution drops to the saturation level.

The solubility of a substance in a solvent is promoted when the solute particles interact strongly with the solvent molecules. Thus, for example, sodium chloride is quite soluble in water but does not dissolve to any significant extent in gasoline or similar organic liquids. The polar water molecules interact strongly with the Na^+ and Cl^- ions of NaCl, whereas nonpolar organic liquids cannot do so. But there is another aspect to solubility in addition to the strength of interaction between solute and solvent. The formation of a solution between carbon tetrachloride, CCl_4, and hexane, C_6H_{14}, is *spontaneous* just like that between sodium chloride and water; that is the solution forms of its own accord, without any extra input of energy from the surroundings. Yet the interactions between CCl_4 and C_6H_{14} molecules are not substantially different than those between CCl_4 molecules themselves or between C_6H_{14} molecules themselves.

To understand more clearly the distinction between these two examples, let's consider a seemingly abstract question: What are the essential attributes of any spontaneous process? If the process that we are considering is a book falling to the floor, the answer to our general question seems simple. The book is acted upon by the force of gravity. At its initial height, the book has a potential energy higher than that of the book on the floor. Unless restrained, the book falls, thereby losing energy. This leads us to the first basic principle identifying spontaneous processes and influencing their direction: *Processes in which the energy content of the system decreases tend to occur spontaneously.* Spontaneous processes tend to be exothermic. Change occurs in the direction that leads to a lower energy content. The second basic principle is this: *Processes in which the disorder of the system increases tend to occur spontaneously.* A familiar example is your room. It spontaneously gets messier and more disorganized. It gets straightened and organized only when energy is provided by you, your servant, relative, or roommate.

Molecules can increase their disorder by moving in a larger volume, thereby being more widely dispersed. Molecules of different types can increase their disorder by mixing. Both of these processes occur spontaneously unless the molecules are restrained by sufficiently strong intermolecular forces or by physical barriers. Thus, because of the strong bonds holding the sodium and chloride ions together, sodium chloride does not spontaneously dissolve in gasoline. On the other hand, gases spontaneously expand unless restrained by their containers; in this case intermolecular forces are too weak to restrain the molecules. We shall discuss spontaneous processes again in Chapter 15; at that time we shall consider the balance between the tendency toward lower energy and toward increased disorder in greater detail. For the moment we need to be aware that the formation of a solution is always favored by the increase in disorder that accompanies mixing. Consequently, a solution will form unless solute-solute or solvent-solvent interactions are too strong relative to the solute-solvent interactions.

12.3 effect of solute and solvent on solubility

The interplay of the interactions between particles in the solution process is best appreciated by examining specific examples. Consider octane, C_8H_{18}, and water, two substances that do not form a solution. Because C_8H_{18} molecules are nonpolar, the forces of attraction that exist between them are relatively weak London dispersion forces. In contrast, very strong hydrogen bonds exist between water molecules. Attractive forces between polar and nonpolar molecules, such as those between H_2O and C_8H_{18} are much weaker than hydrogen bonds. The H_2O–C_8H_{18} interactions are insufficient to overcome the strong H_2O–H_2O interactions; consequently, the C_8H_{18} molecules cannot penetrate between the H_2O molecules. Furthermore, the increased disorder that would accompany mixing is not sufficient to overcome the attractions between water molecules. For the same reasons, water does not form solutions with other hydrocarbons (compounds of carbon and hydrogen). For example, oil, which is a mixture of hydrocarbons, does not form a solution with water. Because oil is less dense than water it spreads over the surface of water forming an oil slick.

In contrast to the behavior of hydrocarbons and water, water mixes in all proportions with ethanol, CH_3CH_2OH.* Pairs of liquids that mix in all proportions are said to be **miscible,** while liquids that do not mix

*This is the common alcohol of alcoholic beverages. The name alcohol is a family name for a class of compounds having an OH group attached to a carbon atom. There are many different alcohols, some of which are listed in Table 12.3.

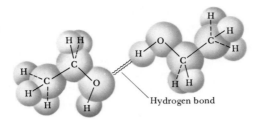

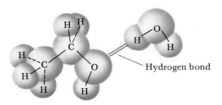

FIGURE 12.5 Hydrogen-bonding interactions between ethanol molecules and between water and ethanol molecules.

are said to be **immiscible**. The C_2H_5OH molecules are able to form hydrogen bonds with water molecules as well as with each other. This is shown in Figure 12.5. Because of this hydrogen-bonding ability, the solute-solute, solvent-solvent, and solute-solvent forces are not appreciably different within a mixture of C_2H_5OH and H_2O. There is no significant change in the environment of the molecules as they are mixed. The increase in disorder accompanying mixing therefore plays a significant role in the formation of the solution.

The number of carbon atoms in an alcohol affects its solubility in water as shown in Table 12.3. As the length of the carbon chain increases,

TABLE 12.3

solubilities of some alcohols in water

ALCOHOL	SOLUBILITY IN H_2O (moles/100 g H_2O at 20°C)
CH_3OH (methanol)	∞ [a]
CH_3CH_2OH (ethanol)	∞
$CH_3CH_2CH_2OH$ (propanol)	∞
$CH_3(CH_2)_3OH$ (butanol)	0.11
$CH_3(CH_2)_4OH$ (pentanol)	0.030
$CH_3(CH_2)_5OH$ (hexanol)	0.0058
$CH_3(CH_2)_6OH$ (heptanol)	0.0008

[a] Infinity symbol indicates that there is no real limit to the solubility of this alcohol in water.

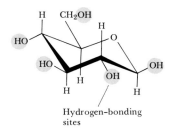

H CH₂OH
HO   H
              O
HO   OH
   H   OH
   H   H

Hydrogen-bonding
sites

FIGURE 12.6 Structure of glucose. Color spheres indicate sites capable of hydrogen bonding with water.

the OH group becomes an ever smaller part of the molecule and the molecule becomes more like a hydrocarbon. The solubility of the alcohol decreases correspondingly. If the number of OH groups along the carbon chain increases, more solute-water hydrogen bonding is possible and solubility generally increases. In the case of glucose, $C_6H_{12}O_6$, there are five OH groups on a six-carbon framework that make the molecule very soluble in water (83 g dissolve in 100 ml of water at 17.5°C). The glucose molecule is shown in Figure 12.6.

Examination of pairs of substances such as those listed in the preceding paragraphs has led to an important generalization: *Substances with similar intermolecular attractive forces tend to be soluble in one another.* This generalization is often simply stated as *"likes dissolve likes."* Nonpolar substances are soluble in nonpolar solvents, whereas ionic and polar solutes are soluble in polar solvents. Network solids like diamond and quartz are not soluble in either polar or nonpolar solvents because of the strong bonding forces within the solid.

SAMPLE EXERCISE 12.6

Predict whether each of the following substances is more likely to dissolve in carbon tetrachloride, CCl_4, which is a nonpolar solvent, or in water: C_7H_{16}, $NaHCO_3$, HCl, I_2.

Solution: Both C_7H_{16} and I_2 are nonpolar. We would therefore predict that they would be more soluble in CCl_4 than in H_2O. On the other hand, $NaHCO_3$ is ionic and HCl is polar covalent. Water would be a better solvent than CCl_4 for these two substances.

Vitamins B and C are water soluble; the A, D, E, and K vitamins are soluble in nonpolar solvents and in the fatty tissue of the body (which is nonpolar). Because of their water solubility, vitamins B and C are not stored to any appreciable extent in the body, and so foods containing these vitamins should be included in the daily diet. In contrast, the fat-soluble vitamins are stored in sufficient quantities to keep vitamin-deficiency diseases from appearing even after a person has subsisted for a long period on a vitamin-deficient diet. With the ready availability of vitamin supplements, cases of hypervitaminosis, illness

caused by an excessive amount of vitamins, are now being seen by physicians in this country. Because only the fat-soluble vitamins are stored, they are the ones for which true hypervitaminosis has been observed. The different solubility patterns of the water-soluble vitamins and the fat-soluble ones can be rationalized in terms of the structures of the molecules. The chemical structure of vitamin A (retinol) is shown below, whereas that of vitamin C (ascorbic acid) is on the next page.

Note that the vitamin A molecule is an alcohol with a very long carbon chain. It is nearly nonpolar, and

Vitamin A

because the OH group is such a small part of the molecule, the molecule mainly resembles a hydrocarbon. The situation is like that of the long-chain alcohols listed in Table 12.3. In contrast, the vitamin C molecule is smaller and has more OH groups that can form hydrogen bonds with water. It is somewhat like the glucose example discussed above.

Vitamin C

12.4 effect of temperature and pressure on solubility

The solubility of a solute depends not only on the nature of the solute and the nature of the solvent, but also on the temperature. In the case of gases, the solubility also depends on pressure.

The effect of temperature on solubility depends on the enthalpy change for the solution process. For solutes that dissolve with endothermic heats of solution, solubility increases with increasing temperature. For the exothermic processes, solubility decreases as temperature is increased. These temperature effects can be understood in terms of **LeChatelier's principle.** This principle, which we shall examine more closely in Section 14.3, can be stated as follows: Any attempt to change the conditions of a system at equilibrium causes the equilibrium to shift in the direction that partially offsets the change. Thus if we have a solution in equilibrium with undissolved solute and then increase its temperature, the equilibrium shifts in the direction that absorbs heat. If the solution process is endothermic, as represented in Equation [12.7], an increase in temperature will cause more solute to dissolve.

$$\text{Solute} + \text{solvent} + \text{heat} \rightleftharpoons \text{solution} \qquad [12.7]$$

(Increase in T shifts equilibrium $\longrightarrow$)

The effect of temperature on the water solubilities of several solids is shown in Figure 12.7. The solution process for most ionic solids in water is endothermic. Consequently, for most solids solubility increases with increasing temperature.

The solution process is exothermic for most gases dissolving in water. Consequently solubility decreases with increasing temperature. If a glass of cold tap water is warmed, bubbles of air are seen on the side of the glass. Similarly, a carbonated beverage like soda pop goes "flat" as it is allowed to warm; as the temperature of the solution increases, CO_2 escapes from the solution. The decreased solubility of O_2 in water as temperature increases is one of the effects of the thermal pollution of water. The effect is particularly serious in deep lakes, because warm water is less dense than cold water. It therefore tends to remain on top of the cold water, at the surface. This situation impedes the dissolving of oxygen into the deeper layers, thus stifling the respiration of all aquatic life needing oxygen. Fish may suffocate and die in these circumstances.

PRESSURE EFFECTS

The solubilities of gases are increased by increasing pressure, whereas the solubilities of liquids and solids are not appreciably affected. The effect of

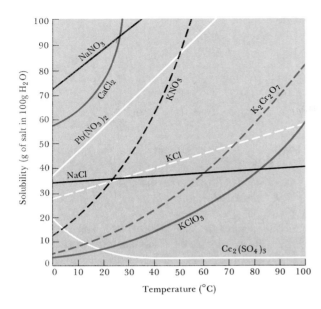

FIGURE 12.7 The solubilities of several common ionic solids shown as a function of temperature.

pressure on solubility can be understood in terms of LeChatelier's principle. An increase in pressure will shift an equilibrium in the direction that assumes the smallest volume. Because gases occupy much larger volumes than their solutions, the solubilities of gases increase with increased pressure. On the other hand, volume does not change appreciably when a solid or liquid dissolves. Consequently pressure has a much smaller effect on the solubilities of solids or liquids than on the solubilities of gases.

The solubility of a gas, C_g, is directly proportional to the partial pressure, P_g, of the gas above the solution. This relationship is given by Equation [12.8], where k is the proportionality constant between C_g and P_g.

$$C_g = kP_g \qquad [12.8]$$

This relationship is known as **Henry's law.** The solubility of pure nitrogen in water at 25°C and a partial pressure of 0.78 atm (this is the partial pressure of N_2 in air at 1.0 atm) is 5.3×10^{-3} mole/l. If the partial pressure is doubled, to 1.56 atm, the solubility of N_2 in water is doubled, to 1.06×10^{-2} mole/l.

The effect of pressure on solubility is utilized in the production of carbonated beverages like champagne, beer, and soda pop. These are bottled under a carbon dioxide pressure slightly greater than 1 atm. When the bottles are opened to the air, the partial pressure of CO_2 above the solution is decreased, and CO_2 bubbles out of the solution.

Deep-sea divers rely on compressed air for their oxygen supply. According to Henry's law, the solubilities of gases increase with pressure. If a diver is suddenly exposed to atmospheric pressure, where the solubility of gases is less, bubbles form in the blood stream and in other fluids of the body. These bubbles affect nerve impulses and give rise to the disease known as "the bends," or decompression sickness. Nitrogen is the main problem because it has the highest partial pressure in air and because it can be removed only through the respiratory system. Oxygen is consumed in metabolism. Substitution of helium for nitrogen minimizes this effect, because helium has a much lower solubility in biological fluids than does N_2. Cousteau's divers on *Conshelf III* used a mixture of 98 percent helium and 2 percent oxygen. At the high pressures (10 atm) experienced by the divers, this percentage of oxygen gives an oxygen partial pressure of about 0.2 atm, which is the partial pressure in normal air at 1 atm. If the oxygen partial pressures become too great, the urge to breathe is reduced, CO_2 is not removed from the body, and this leads to CO_2 poisoning.

12.5 electrolyte solutions

Some substances exist in aqueous solution as ions while others do not. Sodium chloride, an ionic solid, dissolves in water, giving hydrated Na^+ and Cl^- ions that are able to move randomly throughout the solution. A solution labeled as 1.0 M NaCl actually contains 1.0 mole of Na^+ and 1.0 mole of Cl^- per liter of solution. On the other hand, when glucose, $C_6H_{12}O_6$, and ethanol, C_2H_5OH, dissolve, they are not fragmented into ions but maintain their molecular structure in solution. Substances like NaCl that form ions in aqueous solution are called **electrolytes**, whereas those like $C_6H_{12}O_6$ and C_2H_5OH that do not are known as **nonelectrolytes**. Electrolytes can be recognized experimentally by their ability to conduct an electric current. The qualitative abilities of solutions to conduct an electric current can be determined using a device such as that shown in Figure 12.8. Ions in the solution complete the electric circuit and thereby permit the light bulb to glow. There are three common types of electrolytes: acids, bases, and salts (ionic solids).

Electrolytes that dissociate completely in aqueous solution are called **strong electrolytes**, whereas those that dissociate partly are called **weak electrolytes**. Acids and bases that are strong electrolytes are known as **strong acids** and **strong bases**. Those acids and bases that are only partially ionized in solution are known as **weak acids** and **weak bases**. The terms weak acid and weak base have no relation to how reactive the acid or base is. Hydrofluoric acid, HF, is a weak electrolyte and is therefore called a weak acid. A 0.1 M solution of HF is 8 percent ionized. However, HF is a very reactive acid that vigorously attacks many substances, including glass.

The following generalizations are useful in recognizing which substances are strong electrolytes and which weak:

1. *Acids:* The common strong acids are HCl, HBr, HI, HNO_3, and H_2SO_4. Nearly all other acids are weak electrolytes.
2. *Bases:* The common strong bases are the hydroxides of the alkali metals and the alkaline earths (except Be). NH_3 is a weak electrolyte.
3. *Salts:* Nearly all common salts are strong electrolytes.

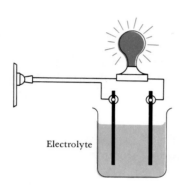

Electrolyte

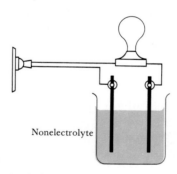

Nonelectrolyte

FIGURE 12.8 A simple device that can be used to distinguish between electrolyte and nonelectrolyte solutions.

SAMPLE EXERCISE 12.7

Classify each of the following substances as non-electrolyte, weak electrolyte, or strong electrolyte: $CaCl_2$, HNO_3, CH_3OH (methanol), $HC_2H_3O_2$ (acetic acid), KOH, H_2O_2.

Solution: Only one of the substances, $CaCl_2$, is a salt. It is a strong electrolyte. Two of the substances, HNO_3 and $HC_2H_3O_2$, are acids: HNO_3 is a common strong acid (strong electrolyte); because $HC_2H_3O_2$ is not a common strong acid, our best guess would be that it is a weak acid (weak electrolyte). This is correct. There is one base, KOH. It is one of the common strong bases (a strong electrolyte) because it is a hydroxide of an alkali metal. The remaining compounds, CH_3OH and H_2O_2, are neither acids, bases, nor salts. They are nonelectrolytes.

SAMPLE EXERCISE 12.8

What is the concentration of all species in a 0.1 M solution of $Ba(NO_3)_2$?

Solution: $Ba(NO_3)_2$ is a salt and as such we expect it to be a strong electrolyte. It ionizes into Ba^{2+} and the polyatomic nitrate ion, NO_3^-. A 0.1 M solution of $Ba(NO_3)_2$ is 0.1 M in Ba^{2+} and 0.2 M in NO_3^- (since 1 mole of $Ba(NO_3)_2$ supplies 1 mole of Ba^{2+} and 2 moles of NO_3^-).

12.6
metathesis reactions

When aqueous solutions of $AgNO_3$ and NaCl are mixed, AgCl separates from the solution as a solid; it is not soluble in water. We say that the AgCl **precipitates** from the solution. Sodium nitrate, $NaNO_3$, remains in solution. The equation for the reaction is as follows:

$$AgNO_3(aq) + NaCl(aq) \longrightarrow AgCl(s) + NaNO_3(aq) \qquad [12.9]$$

Such reactions, in which positive ions (**cations**) and negative ions (**anions**) exchange partners, are known as **metathesis reactions**. We can gain a clearer understanding of such reactions if we remember that ionic compounds like $AgNO_3$ and NaCl are strong electrolytes; they exist in aqueous solution as ions. The possible interactions between these ions are shown by arrows below:

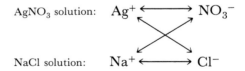

One of these possible combinations, AgCl, is insoluble in water. The reaction can be represented as shown in Equation [12.10]. This equation is known as an **ionic equation**; it shows the form of the reactants and products in solution:

$$Ag^+(aq) + NO_3^-(aq) + Na^+(aq) + Cl^-(aq) \longrightarrow$$
$$AgCl(s) + Na^+(aq) + NO_3^-(aq) \qquad [12.10]$$

Equations such as [12.9] are known as **molecular equations**. Notice that in ionic equation [12.10], Na^+ and NO_3^- ions occur on both sides of the arrow. Such ions, which go through a chemical reaction unchanged, are known as **spectator ions.** If the spectator ions are not shown, the **net ionic equation** results:

$$Ag^+(aq) + Cl^-(aq) \longrightarrow AgCl(s) \qquad [12.11]$$

In order to predict whether precipitation will occur when two solutions are mixed, we must have more knowledge of solubilities than presented so far in this text. The rule "likes dissolve likes" allows us to predict correctly that a substance like AgCl is more likely to be soluble in water than in a nonpolar solvent. However, it doesn't permit us to predict the extent to which AgCl will dissolve in water. In fact, only 1.1×10^{-5} moles of AgCl dissolves in a liter of water at 25°C. For all practical purposes, the compound is insoluble. In our discussions, any substance whose solubility is less than 0.01 mole per liter at 25°C will be referred to as being insoluble. Those substances whose solubilities are 0.01 to 0.1 M will be considered as moderately soluble, and those whose solubilities are greater than 0.1 M will be labeled as being soluble. We shall consider solubility from a quantitative point of view in Chapter 17, but the following are some qualitative generalizations concerning the solubilities of common ionic compounds in water:

1. Salts of the alkali metals are *soluble.*
2. Ammonium (NH_4^+) salts are *soluble.*
3. Salts containing nitrate (NO_3^-), chlorate (ClO_3^-), perchlorate (ClO_4^-), and acetate ($C_2H_3O_2^-$) are *soluble.*
4. All chlorides (Cl^-), bromides (Br^-), and iodides (I^-) are *soluble* except for those of Pb^{2+}, Hg_2^{2+}, and Ag^+, which are *insoluble.*
5. All sulfates (SO_4^{2-}) are *soluble* except for those of Sr^{2+}, Ba^{2+}, Hg_2^{2+}, Hg^{2+}, and Pb^{2+}, which are *insoluble.* The sulfate salts of Ca^{2+} and Ag^+ are *moderately soluble.*
6. All hydroxides (OH^-) are *insoluble* except for those of the alkali metals, which are *soluble,* and the hydroxides of Ca^{2+}, Ba^{2+}, and Sr^{2+}, which are *moderately soluble.*
7. All sulfites (SO_3^{2-}), carbonates (CO_3^{2-}), chromates (CrO_4^{2-}), and phosphates (PO_4^{2-}) are *insoluble* except for those of NH_4^+ and the alkali metals, which are *soluble.*
8. All sulfides (S^{2-}) are *insoluble* except for those of NH_4^+, the alkali metals, and the alkaline earths, which are *soluble.*

SAMPLE EXERCISE 12.9

Write balanced molecular, ionic, and net ionic equations for reactions (if any) that occur when solutions of the following compounds are mixed: (a) $BaCl_2$ and Na_2SO_4; (b) KCl and Na_2SO_4.

Solution: (a) Both $BaCl_2$ and Na_2SO_4 are soluble and ionize giving Ba^{2+}, Cl^-, Na^+, and SO_4^{2-} ions. The possible metathesis products are $BaSO_4$ and

NaCl. The $BaSO_4$ is insoluble according to rule 5 in the list just given. Therefore the molecular equation is

$$BaCl_2(aq) + Na_2SO_4(aq) \longrightarrow$$
$$BaSO_4(s) + 2NaCl(aq)$$

The ionic equation is

$$Ba^{2+}(aq) + 2Cl^-(aq) + 2Na^+(aq) + SO_4^{2-}(aq) \longrightarrow$$
$$BaSO_4(s) + 2Na^+(aq) + 2Cl^-(aq)$$

The net ionic equation is

$$Ba^{2+}(aq) + SO_4^{2-}(aq) \longrightarrow BaSO_4(s)$$

(b) Both reactants are soluble and ionize in solution. There are no possible insoluble salts resulting from metathesis. Both $NaCl$ and K_2SO_4 are soluble. Therefore, there is no reaction.

In the preceding paragraphs and in Sample Exercise 12.9, we have seen that metathesis reactions occur whenever a precipitate forms in solution. It is sometimes said that the formation of an insoluble substance is one of the "driving forces" for metathesis reactions. The formation of a gaseous substance and the formation of a nonelectrolyte are also "driving forces" for metathesis reactions. Some common examples of such reactions are as follows:

1. Carbonates reacting with acids:

$$CO_3^{2-}(aq) + 2H^+(aq) \longrightarrow CO_2(g) + H_2O(l) \qquad [12.12]$$

2. Sulfites reacting with acids:

$$SO_3^{2-}(aq) + 2H^+(aq) \longrightarrow SO_2(g) + H_2O(l) \qquad [12.13]$$

3. Sulfides reacting with acids:

$$S^{2-}(aq) + 2H^+(aq) \longrightarrow H_2S(g) \qquad [12.14]$$

4. Bases reacting with acids:

$$H^+(aq) + OH^-(aq) \longrightarrow H_2O(l) \qquad [12.15]$$

SAMPLE EXERCISE 12.10

Write balanced molecular and net ionic equations for any reactions that occur when the following compounds are mixed: (a) $Na_2CO_3(aq) + HCl(aq)$; (b) $NiS(s) + HCl(aq)$; (c) $NaOH(aq) + H_3PO_4(aq)$.

Solution: (a) This reaction is of the type of Equation [12.12].

Molecular: $Na_2CO_3(aq) + 2\,HCl(aq) \longrightarrow$
$$2NaCl(aq) + CO_2(g) + H_2O(l)$$

Net ionic: $CO_3^{2-}(aq) + 2H^+(aq) \longrightarrow$
$$CO_2(g) + H_2O(l)$$

(b) This reaction is of the type of Equation [12.14], except that the sulfide is insoluble. Most sulfides will react with acids even though they are insoluble. Similarly, solid carbonates and hydroxides will generally react with acids.

Molecular: $NiS(s) + 2HCl(aq) \longrightarrow$
$$NiCl_2(aq) + H_2S(g)$$

Net ionic: $NiS(s) + 2H^+(aq) \longrightarrow$
$$Ni^{2+}(aq) + H_2S(g)$$

Notice that in writing the net ionic equation, both solids and gases are included. The only species that goes through the reaction unchanged is the Cl^- ion.

(c) This is the familiar acid-base reaction summarized by Equation (12.15).

Molecular: $3NaOH(aq) + H_3PO_4(aq) \longrightarrow$
$$3H_2O(l) + Na_3PO_4(aq)$$

Net ionic: $H^+(aq) + OH^-(aq) \longrightarrow H_2O(l)$

12.7

colligative
properties

A number of physical properties of solutions depend on the number of solute particles present in solution but not on the identity of the solute. Such properties are called colligative properties. They include vapor-pressure lowering, boiling-point elevation, freezing-point depression, and osmotic pressure.

VAPOR-PRESSURE LOWERING

If a nonvolatile solute is added to a solvent, the vapor pressure of the solvent is decreased. For example, sugar and salt both lower the vapor pressure of water, thereby causing water to evaporate more slowly. Experiments have shown that the vapor-pressure lowering depends on the concentration of the solute particles. For example, 1.0 mole of a nonelectrolyte such as glucose and 0.5 mole of NaCl will produce essentially the same vapor-pressure lowering in a given quantity of water. There will be 1.0 mole of particles in both solutions, because 0.5 mole of NaCl dissolves giving 0.5 mole Na^+ and 0.5 mole Cl^-. Quantitatively, the vapor pressure of solutions containing nonvolatile solutes is given by Raoult's law, Equation [12.16], where P_A is the vapor pressure of the solution, X_A is the mole fraction of the solvent, and $P_A°$ is the vapor pressure of the pure solvent:

$$P_A = X_A P_A° \qquad [12.16]$$

For example, the vapor pressure of water is 17.5 torr at 20°C. Imagine holding the temperature constant while adding glucose, $C_6H_{12}O_6$, to the water so that the resulting solution has $X_{H_2O} = 0.80$ and $X_{C_6H_{12}O_6} = 0.20$. According to Equation [12.16], the vapor pressure of water over the solution will be 14 torr:

$$P_{H_2O} = (0.80)(17.5 \text{ torr}) = 14 \text{ torr}$$

The vapor-pressure lowering (VPL) is given by Equation [12.17]

$$\text{VPL} = X_B P_A° \qquad [12.17]$$

where X_B is the mole fraction of solute particles. In our example, the vapor pressure will be lowered by 3.5 torr:

$$\text{VPL} = (0.20)(17.5 \text{ torr}) = 3.5 \text{ torr}$$

It is easy to explain this effect qualitatively. At a given temperature, the average kinetic energy of all particles in the solution is the same. Essentially the same number of particles are at the surface in both the solution and pure solvent. However, in the case of the solution, only a fraction of these particles are volatile and have sufficient kinetic energy to escape from the solution.

SAMPLE EXERCISE 12.11

Calculate the vapor-pressure lowering caused by the addition of 100 g of sucrose, $C_{12}H_{22}O_{11}$, to 1000 g of water if the vapor pressure of the pure water at 25°C is 23.8 torr.

Solution:

$$X_{C_{12}H_{22}O_{11}} = \frac{\text{moles } C_{12}H_{22}O_{11}}{\text{moles } H_2O + \text{moles } C_{12}H_{22}O_{11}}$$

$$\text{Moles } C_{12}H_{22}O_{11} = (100 \text{ g}) \left(\frac{1 \text{ mole } C_{12}H_{22}O_{11}}{342 \text{ g } C_{12}H_{22}O_{11}} \right)$$

$$= 0.292 \text{ mole}$$

$$\text{Moles } H_2O = (1000 \text{ g}) \left(\frac{1 \text{ mole } H_2O}{18.0 \text{ g } H_2O} \right)$$

$$= 55.5 \text{ moles}$$

$$X_{C_{12}H_{22}O_{11}} = \frac{0.292}{55.5 + 0.292} = \frac{0.292}{55.8}$$

$$\text{VPL} = \left(\frac{0.292}{55.8} \right)(23.8 \text{ torr}) = 0.125 \text{ torr}$$

BOILING-POINT ELEVATION

We saw in Section 11.2 that the vapor pressure of a liquid increases with increasing temperature; it boils when its vapor pressure is the same as the external pressure over the liquid. Because nonvolatile solutes lower the vapor pressure of the solution, a higher temperature is required to cause the solution to boil. This relationship between the vapor pressure of the liquid and its boiling point is shown in Figure 12.9. The increase in boiling point, ΔT_b (relative to the boiling point of the pure solvent), is directly proportional to the molality of the solute as shown in Equation [12.18].

$$\Delta T_b = K_b m \qquad\qquad [12.18]$$

The magnitude of K_b, which is called the **molal boiling-point-elevation constant,** depends on the solvent. Some typical values for several common solvents are given in Table 12.4.

For water, K_b is $0.52°C/m$; therefore, a $1\ m$ aqueous solution of $C_6H_{12}O_6$, or any other aqueous solution that is $1\ m$ in nonvolatile solute particles, will boil at a temperature $0.52°C$ higher than pure water.

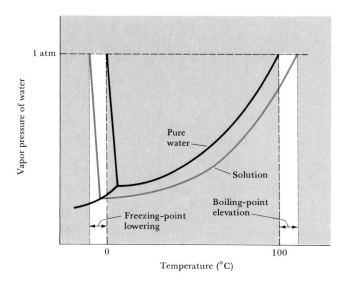

FIGURE 12.9 The effect of addition of a nonvolatile solute on the phase diagram of water. The vapor pressure is lowered, resulting in a decrease in the normal freezing point and an increase in the normal boiling point of water.

TABLE 12.4
molal boiling-point-elevation and freezing-point-depression constants

SOLVENT	NORMAL BOILING POINT (°C)	$K_b(°C/m)$	NORMAL FREEZING POINT (°C)	$K_f(°C/m)$
Water, H_2O	100.0	0.52	0.0	1.86
Benzene, C_6H_6	80.1	2.53	5.5	5.12
Carbon tetrachloride, CCl_4	76.8	5.02	−22.3	29.8
Ethanol, C_2H_5OH	78.4	1.22	−114.6	1.99
Chloroform, $HCCl_3$	61.2	3.63	−63.5	4.68

FREEZING-POINT DEPRESSION

The reduction in vapor pressure produced by a nonvolatile solute also causes a lowering of the freezing point of the solution, as seen in Figure 12.9. This effect arises because the solute is not normally soluble in the solid phase of the solvent. For example, pure ice almost always separates when aqueous solutions freeze. Consequently, the vapor pressure of the solid is unaffected by the solute. As shown in Figure 12.9, the freezing point of the solution is at a lower temperature than that for pure water. The line that represents the equilibrium between solid and liquid therefore lies to the left of the corresponding line for pure water.

Like the boiling-point elevation, the decrease in freezing point, ΔT_f, is directly proportional to the molality of the solute:

$$\Delta T_f = K_f m \qquad [12.19]$$

The values of K_f, **the molal freezing-point-depression constant,** for several common solvents are given in Table 12.4. For water K_f is 1.86°C/m; therefore a 0.5 m aqueous solution of NaCl or any other aqueous solution that is 1 m in nonvolatile solute particles will freeze 1.86°C lower than pure water. The freezing-point lowering caused by solutes explains the use of antifreeze (Sample Exercise 12.12) and the use of calcium chloride, $CaCl_2$, to melt ice on roads during winter.

SAMPLE EXERCISE 12.12

Calculate the freezing point and the boiling point of a solution of 100 g of ethylene glycol ($C_2H_6O_2$) in 900 g of H_2O.

Solution:

$$\text{Molality} = \frac{\text{moles } C_2H_6O_2}{\text{kg } H_2O} =$$

$$\left(\frac{100 \text{ g } C_2H_6O_2}{900 \text{ g } H_2O}\right)\left(\frac{1 \text{ mole } C_2H_6O_2}{62.0 \text{ g } C_2H_6O_2}\right)\left(\frac{1000 \text{ g } H_2O}{1 \text{ kg } H_2O}\right)$$

$$= 1.79 \text{ } m$$

$$\Delta T_f = K_f m = \left(1.86 \frac{°C}{m}\right)(1.79 \text{ } m) = 3.33°C$$

Freezing point = (normal f.p. of solvent) − ΔT_f
$$= 0.00°C − 3.33°C = −3.33°C$$

$$\Delta T_b = K_b m = \left(0.52 \frac{°C}{m}\right)(1.79 \text{ } m) = 0.93°C$$

Boiling point = (normal b.p. of solvent) + ΔT_b
$$= 100.00°C + 0.93°C$$
$$= 100.93°C$$

Ethylene glycol is the main component of antifreeze. It protects the cooling system of the car from freezing by lowering the freezing point of water. Because it also raises the boiling point of water, it is also useful in the summer to prevent what antifreeze advertisements call "boil-over."

List the following solutions in order of their expected freezing points: 0.05 m CaCl$_2$; 0.15 m NaCl; 0.10 m HCl; 0.05 m HC$_2$H$_3$O$_2$; 0.10 m C$_{12}$H$_{22}$O$_{11}$.

Solution: First notice that CaCl$_2$, NaCl, and HCl are strong electrolytes, HC$_2$H$_3$O$_2$ is a weak electrolyte, and C$_{12}$H$_{22}$O$_{11}$ is a nonelectrolyte. The molality of each solution in total particles is as follows:

0.05 m CaCl$_2$ (0.15 m in particles)

0.15 m NaCl   (0.30 m in particles)

0.10 m HCl   (0.20 m in particles)

0.05 m HC$_2$H$_3$O$_2$ (between 0.05 and 0.10 m in particles)

0.10 m C$_{12}$H$_{22}$O$_{11}$ (0.10 m in particles)

The freezing points are expected to run from 0.15 m NaCl (lowest freezing point), to 0.10 m HCl to 0.05 m CaCl$_2$ to 0.10 m C$_{12}$H$_{22}$O$_{11}$ to 0.05 HC$_2$H$_3$O$_2$ (highest freezing point).

Because NaCl is a strong electrolyte, the expected freezing-point depression of a 0.100 m aqueous solution of NaCl is (0.200 m) (0.186°C/m) = 0.372°C. However, the measured value is slightly less than this, 0.348°C. Although strong electrolytes like NaCl give nearly the effect expected from the number of ions present in solution, the effect is always slightly less than that of a corresponding number of nonelectrolyte particles. This is due to the residual electrostatic attraction between ions, even when separated by solvent molecules. This attraction makes solutions of electrolytes behave as though their ion concentrations were less than they actually are.

OSMOSIS

Certain materials, including many membranes in biological systems, are semipermeable, meaning that they allow some particles to pass through, but not others. **Osmosis** is the net movement of solvent molecules but not solute particles through a semipermeable membrane from a more dilute solution into a more concentrated one. Consider two solutions separated by a semipermeable membrane as shown in Figure 12.10. The net migration of solvent occurs from the right arm to the left arm of the apparatus as if the solutions were trying to equalize their concentrations. Pressure can be applied to the left arm of the apparatus, as shown in Figure 12.11, to prevent the osmosis. The pressure required to stop osmosis from pure solvent into a solution is known as the **osmotic pressure**, π, of the solution. The osmotic pressure is related to concentration as shown in Equation [12.20]

$$\pi = MRT \qquad [12.20]$$

where M is molarity, R is the ideal gas constant, and T is the temperature on the Kelvin scale.

If two solutions of identical osmotic pressure are separated by a semipermeable membrane, no osmosis will occur. The two solutions are said to be **isotonic**. If one solution is of lower osmotic pressure, it is described as being **hypotonic** with respect to the more concentrated solution. The more concentrated solution is said to be **hypertonic** with respect to the dilute solution.

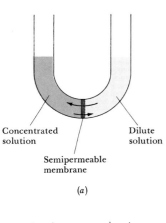

Concentrated solution

Dilute solution

Semipermeable membrane

(a)

(b)

FIGURE 12.10 Osmosis: (a) net movement of solvent from the solution with low solute concentration into the solution with high solute concentration; (b) osmosis stops when the column of solution on the left becomes high enough to exert sufficient pressure to stop the osmosis.

Osmosis plays a very important role in living systems. For example, the membranes of red blood cells are semipermeable. Placement of red blood cells in a solution that is hypertonic relative to the intracellular solution (the solution within the cells) causes water to move out of the cell as shown in Figure 12.12. This causes the cell to shrivel, a process known as **crenation.** Placement of the cell in a solution that is hypotonic relative to the intracellular fluid causes water to move into the cell. This causes rupturing of the cell, a process known as **hemolysis.** Persons needing replacement of body fluids or nutrients who cannot be fed orally are administered solutions by intravenous (or IV) infusion, meaning slow addition to the veins. To prevent crenation or hemolysis of red blood cells, the IV solutions must be isotonic with the intracellular fluids of the cells.

SAMPLE EXERCISE 12.14

The average osmotic pressure of blood is 7.7 atm at 25°C. What concentration of glucose, $C_6H_{12}O_6$, will be isotonic with blood?

Solution:

$$\pi = MRT$$

$$M = \frac{\pi}{RT} = \frac{7.7 \text{ atm}}{\left(0.082 \dfrac{\text{l.-atm}}{\text{K-mole}}\right)(298 \text{ K})}$$

$$= 0.31 \ M$$

In clinical situations, the concentrations of solutions are generally expressed in terms of weight percentages. The weight percentage of a 0.31 M solution of glucose is 5.3 percent. The concentration of NaCl that is isotonic with blood is 0.16 M since NaCl ionizes to form two particles, Na^+ and Cl^- (a 0.155 M solution of NaCl is 0.310 M in particles). A 0.16 M solution of NaCl is 0.9 percent in NaCl. Such a solution is known as a physiological saline solution.

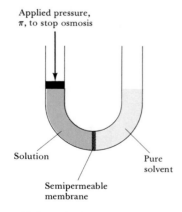

Applied pressure, π, to stop osmosis

Solution

Pure solvent

Semipermeable membrane

FIGURE 12.11 Applied pressure on the left of apparatus stops net movement of solvent from the right side of the semipermeable membrane. This applied pressure is known as the osmotic pressure of the solution.

A wide variety of interesting examples of osmosis can be drawn. A cucumber placed in concentrated brine loses water via osmosis and shrivels into a pickle. A carrot that has become limp because of water loss to the atmosphere can be placed in water. Water moves into the carrot through osmosis, making it once again firm. People eating a lot of salty food experience water retention in tissue cells and intercellular spaces because of osmosis. The resultant swelling or puffiness is called edema. Movement of water from soil into plant roots and subsequently into the upper portions of the plant is due at least in part to osmosis. The preservation of meat by salting and of fruit by adding sugar protects against bacterial action. Through the process of osmosis, a bacterium on salted meat or candied fruit loses water, shrivels, and dies.

In osmosis, water moves from an area of high water concentration (low solute concentration) into an area of low water concentration (high solute concentration). Such movement of a substance from an area where its concentration is high to an area where it is low is spontaneous. Biological cells transport not only water, but other select materials through their membrane walls. This permits entry of nutrients and allows for disposal of waste materials. In some cases, substances must be moved from an area of low concentration to one of high concentration. This movement is called **active transport.** It is not spontaneous but requires expenditure of energy by the cell.

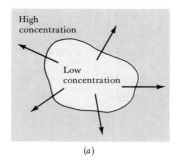
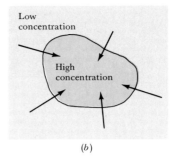

(a)   (b)

FIGURE 12.12 Osmosis through the semipermeable membrane of a cell: (a) crenation caused by movement of water from the cell; (b) hemolysis caused by movement of water into the cell.

DETERMINATION OF MOLECULAR WEIGHT

The colligative properties of solutions provide an important means of experimentally determining molecular weights. Any of the four colligative properties could be used to determine molecular weight. The procedures are illustrated in Sample Exercises 12.15 and 12.16.

SAMPLE EXERCISE 12.15

A solution of an unknown nonvolatile nonelectrolyte was prepared by dissolving 0.250 g in 40.0 g of CCl_4. The normal boiling point of the resultant solution was increased by 0.357°C. Calculate the molecular weight of the solute.

Solution: Using Equation [12.18] we have

$$\text{Molality} = \frac{\Delta T_b}{K_b} = \frac{0.357°C}{5.02°C/m} = 0.0711 \ m$$

Thus the solution contains 0.0711 moles of solute per kilogram of solvent. The solution was prepared from 0.250 g solute and 40.0 g of solvent. The number of grams of solute in a kilogram of solvent is therefore

$$\frac{\text{Grams solute}}{\text{kg } CCl_4} = \left(\frac{0.250 \text{ g solute}}{40.0 \text{ g } CCl_4}\right)\left(\frac{1000 \text{ g } CCl_4}{1 \text{ kg } CCl_4}\right)$$

$$= \frac{6.25 \text{ g solute}}{1 \text{ kg } CCl_4}$$

Notice that a kilogram of solvent contains 6.25 g, which from the ΔT_b measurement must be 0.0711 m. Therefore

$$0.0711 \text{ mole} = 6.25 \text{ g}$$

$$1 \text{ mole} = \frac{6.25 \text{ g}}{0.0711} = 87.9 \text{ g}$$

Therefore

$$\text{MW} = 87.9 \text{ amu}$$

SAMPLE EXERCISE 12.16

The osmotic pressure of 0.200 g of hemoglobin in 20.0 ml of solution is 2.88 torr at 25°C. Calculate the molecular weight of hemoglobin.

Solution:

$$\pi = MRT$$

$$M = \frac{\pi}{RT} = \frac{2.88 \text{ torr}}{\left(0.082 \frac{\text{l.-atm}}{\text{K-mole}}\right)(298 \text{ K})}\left(\frac{1 \text{ atm}}{760 \text{ torr}}\right)$$

$$= 1.55 \times 10^{-4} \frac{\text{mole}}{\text{liter}}$$

A liter of solution contains

$$\left(\frac{0.200 \text{ g solute}}{20.0 \text{ ml soln}}\right)\left(\frac{1000 \text{ ml soln}}{1 \text{ l. soln}}\right) = 10.0 \text{ g}$$

Therefore

$$1.55 \times 10^{-4} \text{ mole} = 10.0 \text{ g}$$

$$1 \text{ mole} = \frac{10.0 \text{ g}}{1.55 \times 10^{-4}} = 6.45 \times 10^4 \text{ g}$$

Therefore

$$\text{MW} = 64,500 \text{ amu}$$

Osmotic pressure is particularly useful for measuring the molecular weights of high molecular weight substances.

TABLE 12.5
types of colloids

PHASE OF COLLOID	DISPERSING (SOLVENTLIKE) SUBSTANCE	DISPERSED (SOLUTELIKE) SUBSTANCE	COLLOID TYPE	EXAMPLE
Gas	Gas	Gas	—	None (all are solutions)
Gas	Gas	Liquid	Aerosol	Fog
Gas	Gas	Solid	Aerosol	Smoke
Liquid	Liquid	Gas	Foam	Whipped cream
Liquid	Liquid	Liquid	Emulsion	Milk
Liquid	Liquid	Solid	Sol	Paint
Solid	Solid	Gas	Solid foam	Marshmallow
Solid	Solid	Liquid	Solid emulsion	Butter
Solid	Solid	Solid	Solid sol	Ruby glass

12.8
colloids

When finely divided clay particles are dispersed through water, they do not remain suspended but eventually settle out of the water. The dispersed clay particles are much larger than molecules, consisting of many thousands or even millions of atoms. In contrast, the dispersed particles of a solution are of molecular size. Between these extremes is the situation in which dispersed particles are larger than molecules, but not so large that the components of the mixture separate under the influence of gravity. These intermediate types of dispersions or suspensions are called **colloidal dispersions** or simply **colloids**. Thus colloids are on the dividing line between solutions and heterogeneous mixtures. Like solutions, colloids can be either gases, liquids, or solids. Examples of each are listed in Table 12.5.

The size of the dispersed particle is the property used to classify a mixture as a colloid. Colloid particles range in diameter from approximately 10 Å to 2000 Å, whereas solute particles are smaller. The colloid particle may consist of many atoms, ions, or molecules, or may even be a single giant molecule. For example, the hemoglobin molecule, which carries oxygen in blood, has molecular dimensions of $65 \times 55 \times 50$ Å and a molecular weight of 64,500 amu. Colloids can be made either by breaking up large pieces of matter or by clustering small particles into colloidal size.

Even though colloid particles are so small that the dispersion appears uniform even under a microscope, they are large enough to scatter light very effectively. Consequently, colloids appear cloudy or

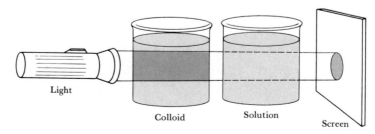

FIGURE 12.13 The Tyndall effect. Light is scattered by the colloidal dispersion but not by the solution.

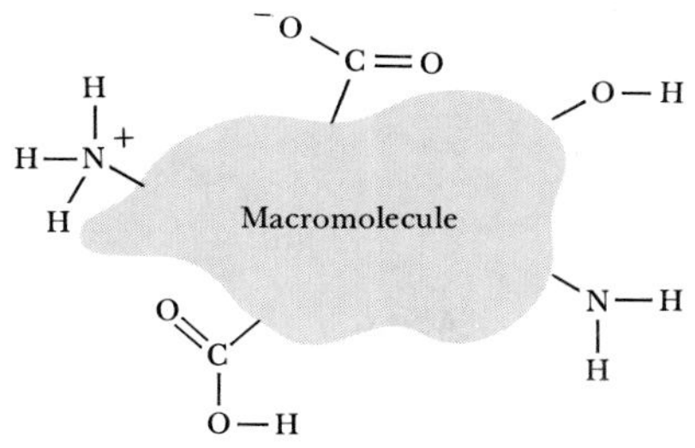

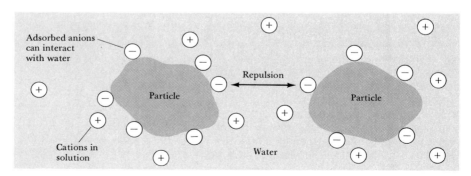

FIGURE 12.14 Examples of hydrophilic groups at the surface of a giant molecule (macromolecule) that help keep the macromolecule suspended in water.

opaque. Furthermore, because they scatter light, a light beam can be seen as it passes through a colloidal suspension as shown in Figure 12.13. This scattering of light by colloidal particles is known as the **Tyndall effect**. Such light scattering makes it possible to see the light beam coming from the projection housing in a smoke-filled theatre or the light beam from an automobile on a dusty dirt road.

HYDROPHILIC AND HYDROPHOBIC COLLOIDS

The most important colloids are those in which the dispersing medium is water. Such colloids are frequently referred to as **hydrophilic** (water loving) or **hydrophobic** (water hating). Hydrophilic colloids are most like the solutions that we have previously examined. In the human body, the extremely large molecules that make up such important substances as enzymes and antibodies are kept in suspension by interaction with surrounding water molecules. The molecules fold so that polar or charged groups can interact with water molecules at the periphery of the molecules. These hydrophilic groups generally contain oxygen or nitrogen. Some examples are shown in Figure 12.14.

Hydrophobic colloids can be prepared in water only if they are stabilized in some way. Otherwise, their natural lack of affinity for water causes them to separate from the water. Hydrophobic colloids can be stabilized by adsorption of ions on their surface as shown in Figure 12.15. These adsorbed ions can interact with water, thereby stabilizing the colloid. At the same time the mutual repulsion between colloid particles with adsorbed ions of the same charge keeps the particles from colliding and thereby getting larger. Hydrophobic colloids can also be stabilized by the presence of other hydrophilic groups on their surfaces. For example, small droplets of oil are hydrophobic. They do not remain suspended

FIGURE 12.15 Schematic representation of the stabilization of a hydrophobic colloid by adsorbed ions.

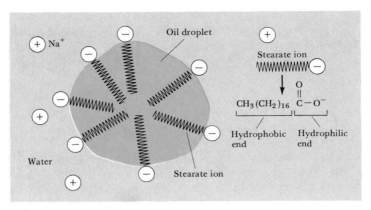

FIGURE 12.16 Stabilization of an emulsion of oil in water by stearate ions.

in water; instead they separate, forming an oil slick on the surface of the water. Addition of sodium stearate, whose structure is

$$\underset{\underset{\text{Hydrophobic end}}{\uparrow}}{CH_3(CH_2)_{16}} \overset{\overset{\displaystyle O}{\parallel}}{\underset{\underset{\text{Hydrophilic end}}{\uparrow}}{CO^-Na^+}}$$

Sodium stearate

or any similar substance having one end that is hydrophilic (polar or charged) and one that is hydrophobic (nonpolar), will stabilize a suspension of oil in water as shown in Figure 12.16. The hydrophobic ends of the stearate ions interact with the oil droplet, whereas the hydrophilic ends point out toward the water with which they interact.

The cleansing action of a soap or detergent is due to its ability to stabilize hydrophobic colloids. These cleansing agents contain substances such as sodium stearate. The action of such molecules in removing dirt from a piece of cloth is illustrated in Figure 12.17.

REMOVAL OF COLLOID PARTICLES

It is often desirable to remove colloidal particles from a dispersing medium, as in the removal of smoke from stacks or butter from milk. This removal, which cannot be accomplished by simple filtration, is sometimes referred to as **coagulation.** It is normally accomplished in one of two ways: (1) by heating, or (2) by adding an electrolyte to the mixture. When a suspension is heated, the increased molecular motion of the particles increases collisions between particles. Whenever they stick together following collision, the colloid particle increases in size. As the particle increases in size, it eventually settles out of solution. The effect of added electrolytes is to neutralize the charge of ions adsorbed on the surface of hydrophobic colloids. The effect of electrolytes is seen in the depositing of suspended clay in a river as it mixes with salt water. This results in the formation of river deltas wherever rivers empty into oceans or other salty bodies of water.

Semipermeable membranes can be used to separate ions from colloidal particles; the ions can pass through the membrane, whereas the

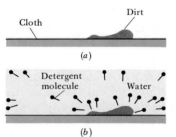

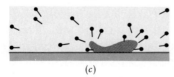

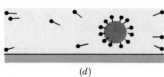

FIGURE 12.17 Schematic representation of the action of detergent ions on dirt. (a) Dirt on a piece of cloth. (b) Detergent molecules in water attach their hydrophobic ends to the dirt and (c) begin to lift it. (d) Detergent surrounds the dirt and holds it in suspension so that it can be washed away.

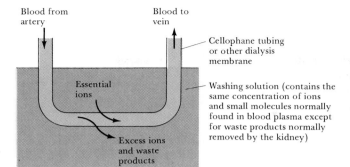

Blood from artery

Blood to vein

Cellophane tubing or other dialysis membrane

Essential ions

Washing solution (contains the same concentration of ions and small molecules normally found in blood plasma except for waste products normally removed by the kidney)

Excess ions and waste products

FIGURE 12.18 An illustration of dialysis as applied in the operation of an artificial kidney machine. A dialysis membrane permits small particles to pass from an area of high concentration to one of low concentration. Colloid-sized particles, such as hemoglobin, do not pass through the membrane.

colloid particles cannot. This type of separation is known as **dialysis.** This process is used in the purification of blood in artificial kidney machines (see Figure 12.18).

In the genetic disease known as sickle-cell anemia, hemoglobin molecules are abnormal, in that they have lower solubility, especially in the unoxygenated form. Consequently as much as 85 percent of the hemoglobin in the red blood cells crystallizes from solution. This distorts the cells into a sickle shape as shown in Figure 12.19. These clog the capillaries, thus causing gradual deterioration of the vital organs. The disease is hereditary, and if both parents carry the defective genes, it is likely that the child will possess only abnormal hemoglobin. Such children seldom survive more than a few years after birth. The reason for the insolubility of hemoglobin in sickle-cell anemia can be traced to a change in one part of the molecule. The normal group of atoms shown in the next column is replaced at one place in the hemoglobin molecule by the abnormal combination.

$$-CH_2-CH_2-\overset{\overset{\displaystyle O}{\|}}{C}-OH \qquad -\underset{\underset{\displaystyle CH_3}{|}}{CH}-CH_3$$

Normal   Abnormal

The normal group of atoms has a polar group that can interact with water, thereby enhancing the solubility of the hemoglobin. The abnormal group of atoms is nonpolar (hydrophobic), and its presence leads to the aggregation of this defective form of hemoglobin into particles too large to remain suspended in biological fluids.

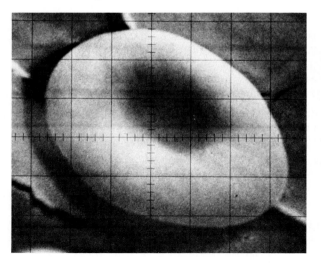

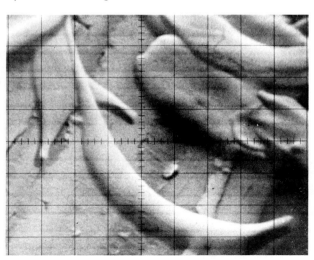

FIGURE 12.19 Electron micrograph showing a normal red blood cell (left; magnification 10,000 times) and sickled red blood cells (right; magnification 5,000 times). (*Courtesy Philips Electronic Instruments, Inc.*)

summary

Solutions are homogeneous mixtures of atoms, ions, or molecules. The relative amounts of **solute** and **solvent** in a solution can be described qualitatively (**dilute** or **concentrated** solutions) or quantitatively (**weight percentage, molarity, molality, normality,** and **mole fraction** were discussed).

The extent to which a solute will dissolve in a particular solvent depends on the relative magnitudes of solute-solute, solute-solvent, and solvent-solvent attractive forces as well as on the changes in disorder accompanying mixing. The rule "likes dissolve likes" was found to be useful in rationalizing solubilities. It is possible to change the solubility of a solute by changing temperature and pressure. If the solution process is endothermic, an increase in temperature promotes solubility. With a gas, an increase in pressure promotes solubility. These effects can be understood in terms of **LeChatelier's principle.**

Substances that exist in solution as ions are called **electrolytes.** The positive ions are known as **cations** while the negative ones are called **anions.** Those substances that are completely ionized in solution are called **strong electrolytes. Metathesis reactions** occur between electrolytes if an insoluble substance, a gas, or a nonelectrolyte can form. **Net ionic equations** focus attention on the particular species that actually undergo some change during the reaction.

The presence of a solute in a solvent lowers the vapor pressure and the freezing point and increases the boiling point of the solvent. These changes together with osmotic pressure are **colligative properties,** depending on the number of particles in solution. Colligative properties can be used to calculate the molecular weights of nonvolatile nonelectrolytes.

True solutions can be differentiated from **colloids** on the basis of particle size. A number of properties of colloids were examined.

learning goals

Having read and studied this chapter, you should be able to:

1. Define molarity, molality, mole fraction, normality, and weight percentage and calculate concentrations in any of these concentration units.
2. Convert concentration in one concentration unit into any other (given the density of the solution where necessary).
3. Give definitions of the qualitative terms used to describe solutions: dilute, concentrated, saturated, unsaturated, and supersaturated.
4. Describe the solution process, including the bonds made and broken when a substance dissolves.
5. Describe the role of disorder in the solution process.
6. Rationalize the solubilities of substances in various solvents in terms of their molecular structures and intermolecular forces.
7. Discuss the effects of pressure and temperature on solubilities.
8. Predict which substances are electrolytes and which are nonelectrolytes; predict which electrolytes are strong electrolytes and which are weak.
9. Predict the course of metathesis reactions and write net ionic equations for these reactions.
10. Predict the effect of solute concentration on vapor pressure, freezing point, boiling point, and osmotic pressure.
11. Explain the difference between the magnitude of changes in colligative properties caused by electrolytes compared to those caused by nonelectrolytes.

12. Calculate the molecular weight of a solute using colligative properties.

13. Describe how a colloid differs from a true solution.

key terms

Among the more important terms and expressions used for the first time in this chapter are the following:

Colloids (Section 12.8) are particles that are larger than normal molecules, but that are nevertheless small enough to remain suspended in a dispersing medium indefinitely.

Dialysis (Section 12.8) is the separation of small solute particles from colloid particles by means of a semipermeable membrane.

An **electrolyte** (Section 12.5) is a solute that gives a solution containing ions; such a solution will conduct an electrical current. **Strong electrolytes** (strong acids, strong bases, and most common salts) are completely ionized in solution. **Weak electrolytes** are partially ionized in solution.

LeChatelier's principle (Section 12.4) states that any attempt to change the conditions of a system at equilibrium results in a shift in the position of the equilibrium in the direction that tends to offset the change.

Molality (Section 12.1) is the concentration of a solution expressed as moles of solute per kilogram of solvent; abbreviated m.

Mole fraction (Section 12.1) is the ratio of the number of moles of one component of a solution to the total number of moles of all substances present in the solution; abbreviated X, with a subscript identifying the component.

A **net ionic equation** (Section 12.6) is an equation for a reaction involving ions that is obtained by eliminating **spectator ions** (those ions that go through the reaction unchanged and appear on both sides of the overall ionic equation).

Normality (Section 12.1) is the concentration of a solution expressed as equivalents of solute per liter of solution; abbreviated N. An **equivalent** is defined according to the type of reaction being considered; one equivalent of one reactant will react with one equivalent of a second reactant.

Osmosis (Section 12.7) is the net movement of solvent through a semipermeable membrane toward the solution with greatest solute concentration. The **osmotic pressure** of a solution is the pressure that must be applied to a solution to stop osmosis from pure solvent into the solution.

A **saturated solution** (Section 12.2) is a solution in which undissolved solute and dissolved solute are in equilibrium. The **solubility** of a substance is the amount of solute that dissolves to form a saturated solution. Solutions containing less solute than this are said to be **unsaturated,** whereas those containing more are said to be **supersaturated.**

Solvation (Section 12.2) is the clustering of solvent molecules around a solute particle. When the solvent is water, this clustering is known as **hydration.**

The **Tyndall effect** (Section 12.8) describes the visible path of a light beam passing through a colloidal dispersion caused by scattering of light by the colloid particles.

The **weight percentage** (Section 12.1) of a solution is the number of grams of solute it contains in each 100 grams of solution.

EXERCISES

concentrations of solutions

12.1 Calculate the number of moles of solute present in each of the following solutions: (a) 100 ml of 0.25 M H_2SO_4; (b) 100 g of 0.25 m HCl; (c) 100 g of 0.50 percent NaCl.

12.2 Describe how you would prepare each of the following solutions, starting with solid $AgNO_3$: (a) 200 ml of 0.300 M solution of $AgNO_3$; (b) 500 g of 0.300 m solution of $AgNO_3$; (c) 150 g of 0.30 percent $AgNO_3$; (d) an aqueous $AgNO_3$ solution for which $X_{AgNO_3} = 0.03$.

12.3 What mass of $CaCl_2$ is dissolved in each 100 g of water in each of the following solutions: (a) a 0.10 m solution; (b) a solution with $X_{CaCl_2} = 0.10$; (c) a 0.10 percent $CaCl_2$ solution.

12.4 Calculate the molarity of each of the following solutions: (a) one that contains 10.0 g of H_2SO_4 in 500 ml of solution; (b) a 3.0 N solution of H_3PO_4 (used to form PO_4^{3-}); (c) one made by diluting 5.0 ml of 3.0 M H_2SO_4 to a total volume of 25.0 ml.

12.5 In a certain reaction, Sn^{4+} is reduced to Sn^{2+}. (a) What is the normality of a 0.20 M solution of Sn^{4+}? (b) What is the molarity of a 0.20 N solution of Sn^{4+}?

12.6 Calculate the mole fraction of benzene in a solution made by dissolving 3.0 g of benzene, C_6H_6, in 7.0 g of cyclohexane, C_6H_{12}. Calculate the weight percentage and molality of the benzene in this solution. If the density of the solution is 0.81 g/ml, calculate the molarity of the benzene in the solution.

12.7 The concentration of H_2SO_4 in a bottle labeled "concentrated sulfuric acid" is 18 M. The solution has a density of 1.84 g/ml. What is the mole fraction and weight percentage of H_2SO_4 in this solution?

solubility and the solution process

12.8 Do you think that it's proper to say that zinc dissolves in hydrochloric acid? Explain your answer.

12.9 Why are solutions usually used in carrying out chemical reactions?

12.10 Which ion is likely to be more tightly hydrated, Na^+ or Mg^{2+}?

12.11 Which member of the following pairs has the greater disorder? (a) water or steam; (b) $NaCl(s)$ or $NaCl(aq)$; (c) $CO_2(g)$ and $H_2O(g)$ or $C_{12}H_{22}O_{11}(s)$

12.12 Which member of the following pairs of substances will be more soluble in water? (a) $CO_2(g)$ or $H_2(g)$; (b) $CH_3OCH_3(l)$ or $CH_3OH(l)$; (c) $AlCl_3(s)$ or $CCl_4(l)$

12.13 DDT has the following structure:

This substance has a solubility of only about 10^{-6} g per liter of water. However, it concentrates in the fatty tissues of animals. Rationalize this behavior.

12.14 Why is Br_2 more soluble than I_2 in CCl_4?

12.15 Which of the following substances are soluble in water? (a) KCl; (b) $AgCl$; (c) $Fe(NO_3)_2$; (d) FeS; (e) $NaOH$; (f) $Ba(OH)_2$; (g) $CaSO_4$; (h) $Ca(NO_3)_2$

12.16 Explain why seawater contains so much of the alkali metal and alkaline earth ions.

12.17 The partial pressure of CO_2 in air at sea level is 0.239 torr. If the solubility of CO_2 at this partial pressure and 25°C is 3.4×10^{-2} mole/l., calculate the constant, k, in Henry's law. What is the solubility of CO_2 at $P_{CO_2} = 0.010$ atm?

12.18 The heat of solution of $NaHCO_3(s)$ is $\Delta H = -18.0$ kJ/mole. Will the solubility of $NaHCO_3$ increase or decrease with increasing temperature?

12.19 Apply LeChatelier's principle to the evaporation of liquids:

$$Liquid + heat \rightleftharpoons vapor$$

Which way does the equilibrium shift with (a) increasing temperature; (b) increasing pressure?

electrolytes

12.20 Classify each of the following substances as a nonelectrolyte, weak electrolyte, or strong electrolyte: (a) H_2SO_4; (b) $Ca(OH)_2$; (c) NH_3; (d) $C_{12}H_{22}O_{11}$; (e) HCN; (f) $NaNO_3$; (g) O_2; (h) H_2SO_3; (i) $BaCl_2$. Identify those that are acids, bases, or salts.

12.21 What is the molarity of K^+ in a solution made by dissolving 1.50 g KCl and 3.20 g K_2SO_4 together in enough water to make 500 ml of solution?

12.22 Complete and balance the following equations. If there is no reaction, write *N.R.* (All reagents are in aqueous solution.)

(a) $HCl + NaHCO_3 \longrightarrow$

(b) $HCl + AgCl \longrightarrow$

(c) $HCl + Al(OH)_3 \longrightarrow$

(d) $HCl + KBr \longrightarrow$

(e) $H_2S + Cu(NO_3)_2 \longrightarrow$

12.23 Write balanced net ionic equations for the following reactions:

(a) $NH_4NO_3(aq) + NaOH(aq) \longrightarrow$
    $NH_3(g) + H_2O(l) + NaNO_3(aq)$

(b) $MgCl_2(aq) + NaOH(aq) \longrightarrow$
$$Mg(OH)_2(s) + NaCl(aq)$$

12.24 The capacity of natural water to neutralize acids is an important property. This neutralizing ability is due to the presence not only of OH^-, but also of HCO_3^- and CO_3^{2-} ions. How do these neutralize acids?

12.25 Write balanced net ionic equations for the reactions that occur when aqueous solutions of the following substances are mixed; (a) calcium acetate and sodium carbonate; (b) ammonium sulfate and lead(II) nitrate; (c) potassium hydroxide and nickel(II) nitrate.

12.26 Two students are arguing over whether it would be better to prepare $CaCl_2$ from $CaCO_3$ and HCl or from $CaSO_4$ and NaCl. Which do you think would be better? Why?

colligative properties

12.27 Why does fresh water evaporate faster than seawater under the same conditions?

12.28 The vapor pressure of benzene, C_6H_6, at 25°C is 95 torr. How much is the vapor pressure lowered if 5.0 g of a nonvolatile substance, C_6H_6O, is added to 45.0 g of benzene?

12.29 Arrange the following aqueous solutions in order of their increasing vapor pressures: (a) 0.1 m NaCl; (b) 0.1 m CH_3OH; (c) 0.1 m $CaCl_2$; (d) 0.2 m KCl.

12.30 Explain how NaCl or $CaCl_2$ works to melt ice on highways.

12.31 What are the freezing and boiling points of a solution made by dissolving 10.0 g of glucose, $C_6H_{12}O_6$, in 40.0 g of H_2O?

[12.32] How much ethylene glycol, $C_2H_6O_2$, is needed to protect the cooling system of an automobile to -10°C if the capacity of the cooling system is 4 gal? (The density of ethylene glycol is 1.12 g/cm^3. Assume the volume of the solution is the volume of the solute plus the volume of the solvent.)

12.33 A 0.01 m solution of acetic acid, $HC_2H_3O_2$, freezes at -0.0194°C. Rationalize this freezing point.

12.34 The freezing point of a 0.010 M solution of H_2SO_4 is 0.0047°C. Show that this is consistent with 100 percent ionization of the acid according to the reaction

$$H_2SO_4 \longrightarrow H^+ + HSO_4^-$$

and 50 percent ionization of the resultant HSO_4^-

$$HSO_4^- \longrightarrow H^+ + SO_4^{2-}$$

12.35 Why does salty soil kill plants?

[12.36] Seawater contains the following ions: Cl^- (0.57 m), SO_4^{2-} (0.029 m), HCO_3^- (0.002 m), Na^+ (0.49 m), Mg^{2+} (0.055 m), K^+ (0.011 m), and Ca^{2+} (0.011 m). Calculate the approximate freezing point, boiling point, and osmotic pressure of seawater. (The density of sea water is 1.024 g/ml.)

12.37 Which of the following solutions exhibits the largest osmotic pressure? (a) 0.1 M $C_{12}H_{22}O_{11}$; (b) 0.1 M NaCl; (c) 0.1 M $HC_2H_3O_2$

12.38 What is the osmotic pressure of 1.0 g of aspirin ($C_9H_8O_4$) in 100 ml of solution at 25°C?

12.39 A solution containing 0.50 g/l. of the hormone insulin exhibits an osmotic pressure of 2.54 torr at 20°C. What is the approximate molecular weight of insulin?

12.40 A solution of 3.0 g of the artificial sweetener saccharin in 100 g of ethanol has a boiling point of 78.7°C. The normal boiling point of pure ethanol is 78.5°C. If K_b for ethanol is 1.22°C/m, what is the approximate molecular weight of saccharin?

colloids

12.41 Explain why highly saline water flowing into a reservoir will hasten the sedimentation of silt, thereby shortening the lifetime of the reservoir.

12.42 Explain why milk curdles when poured into an acidic solution (cottage cheese is made commercially by treating milk with hydrochloric acid).

12.43 Explain why dialysis stabilizes some colloidal dispersions.

12.44 Explain why colloids are opaque (like milk) or translucent, whereas solutions are transparent.

12.45 Explain the cleansing action of soap.

12.46 In one method for analyzing solutions for Cl^-, the Cl^- is precipitated as AgCl upon the addition of $AgNO_3$. The AgCl is then collected, dried, and weighed. During this procedure, prior to filtration, the solution is heated for a short time, then cooled. Suggest a possible explanation for this heating.

general exercises

12.47 Distinguish between: (a) molarity and molality; (b) solution and emulsion; (c) hydrophilic and hydrophobic; (d) saturated and supersaturated; (e) electrolyte and nonelectrolyte; (f) strong and weak acid; (g) osmosis and dialysis.

12.48 Would water serve equally well as a solvent if it had a linear structure? Explain.

12.49 What are the expected freezing and boiling points of a solution made by dissolving 5.0 g of NaCl in 25.0 g of water?

[12.50] Acetic acid, $HC_2H_3O_2$, is moderately soluble in benzene. However, it is reported that the acetic acid molecules exist in benzene as dimers (pairs) rather than as individual molecules. Describe how you would proceed experimentally to check the correctness of this report.

12.51 The solution of hydrogen peroxide that is commonly used as a disinfectant contains 3.0 percent H_2O_2 by weight. This aqueous solution has a density of 1.0 g/ml. What is the molality, mole fraction, and molarity of H_2O_2 in this solution?

12.52 Write balanced net ionic equations for the following reactions: (a) $BaCl_2$ and Na_2SO_4 solutions are mixed; (b) hydrochloric acid is added to magnesium metal; (c) nitric acid is added to calcium carbonate.

12.53 Consider the following equilibrium:

$$HCl(g) + H_2O \rightleftharpoons H_3O^+(aq) + Cl^-(aq) + energy$$

Which way will this equilibrium shift if (a) the temperature is decreased; (b) the pressure is increased?

12.54 How many grams of aluminum metal can be "dissolved" by 20.0 ml of 3.00 M HCl?

$$2Al(s) + 6H^+(aq) \longrightarrow 2Al^{3+}(aq) + 3H_2(g)$$

12.55 Progesterone, a female hormone, contains 9.5 percent H, 10.2 percent O, and 80.3 percent C. Calculate its empirical formula. A solution of 0.100 g of progesterone in 5.00 g of benzene freezes at 5.18°C. What is the molecular weight and molecular formula of progesterone?

rates of chemical reactions

13

Chemistry is by its very nature concerned with change. Substances with well-defined properties are converted by chemical reactions into other materials with different properties. Chemists want to know which new substances are formed from a given set of starting reactants. However, it is equally important to know how rapidly chemical reactions occur, and to understand the factors that control their speeds. For example, what factors are important in determining how rapidly foods spoil? What determines the rate at which steel rusts? How does one design a rapidly setting material for dental fillings? What factors control the rate at which fuel burns in an auto engine, and how does burning rate determine the pollutant content of the engine exhaust?

The area of chemistry concerned with the speeds, or rates, at which chemical reactions occur is called kinetics. In this chapter we learn how to express and determine the rates at which reactions occur. We shall also learn how reaction rates are affected by variables such as concentration, temperature, and the presence of catalysts.

13.1
reaction rate

When butyl chloride is placed in water, it reacts to form butyl alcohol and hydrochloric acid, according to the equation

$$C_4H_9Cl + H_2O \longrightarrow C_4H_9OH + HCl \qquad [13.1]$$

Suppose that we begin with a 0.1 M solution of butyl chloride in water and observe the concentration of the chloride as a function of time. We

rates of chemical reactions

TABLE 13.1

rate data for reaction of C_4H_9Cl with water

TIME (sec)	$[C_4H_9Cl]^a$ (M)	RATE (M/sec)
0	0.100	
50	0.0905	1.90×10^{-4}
100	0.0820	1.70×10^{-4}
150	0.0741	1.58×10^{-4}
200	0.0671	1.40×10^{-4}
300	0.0549	1.22×10^{-4}
400	0.0448	1.01×10^{-4}
500	0.0368	0.80×10^{-4}
800	0.0200	0.56×10^{-4}
10,000	0	

^{a}The brackets indicate concentration of C_4H_9Cl.

might then collect data as shown in the first two columns of Table 13.1. We might also make a graph of these data, by plotting the concentration of the chloride on the vertical axis versus time on the horizontal axis, as illustrated in Figure 13.1.

From these experimental data we can determine the rate, or speed, with which butyl chloride reacts with water. The rate of a reaction is expressed as the change in concentration of a reactant molecule with time. For example, in the reaction of butyl chloride with water, the rate is given by the ratio of a change in concentration of butyl chloride to a corresponding change in time:

$$\text{Rate} = \frac{\text{decrease in concentration of } C_4H_9Cl}{\text{time interval}} = \frac{-\Delta[C_4H_9Cl]}{\Delta t}$$

[13.2]

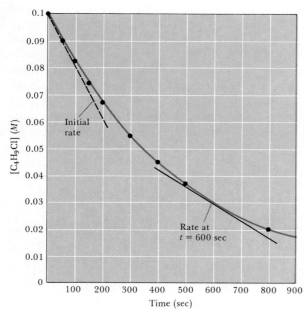

FIGURE 13.1 Concentration of butyl chloride as a function of time. The reaction rate at any time is given by the slope of the tangent to the curve at that time.

Here $\Delta[C_4H_9Cl]$ represents a small change in concentration of C_4H_9Cl during the small interval Δt.* The term $\Delta[C_4H_9Cl]$ is expressed in units of moles/liter (M); time is normally in seconds, and so rate has the units of M/sec. The negative sign before the right-hand term indicates that the concentration of butyl chloride is decreasing with time.

The rate could also be expressed in terms of the appearance of a product. For example, the rate of the reaction shown in Equation [13.1] is also given by

$$\text{Rate} = \frac{\Delta[C_4H_9OH]}{\Delta t} \qquad [13.3]$$

The rate of formation of butyl alcohol is precisely the same as the rate of disappearance of the butyl chloride; one molecule of alcohol is formed for each molecule of chloride that reacts.

The rate of a chemical reaction is not constant, but changes continuously with time. As reactants are consumed, the reaction rate normally decreases. At some point the reaction stops; that is, there is no longer any change in concentrations with time. To show how reaction rate changes with time, let us calculate the rate of the reaction shown in Equation [13.1] at various times, using the data in Table 13.1. During the first time interval of 50 sec, the chloride concentration decreased from 0.100 M to 0.0905 M. The rate for this interval is given by inserting the numerical quantities into Equation (13.2):

$$\text{Rate} = \frac{(0.100 - 0.0905)\ M}{50\ \text{sec}} = 1.90 \times 10^{-4}\ M/\text{sec}$$

Similarly we might calculate the rates for all the other time intervals, as shown in the third column of Table 13.1. Note that the rate steadily decreases as the reaction proceeds.

We could also obtain the rate of the reaction at any time from the graph of concentration versus time, Figure 13.1. The rate at a particular time is given by the slope of a straight line tangent to the curve at the time of interest. As an example, the rate of reaction at time = 600 sec is equal to the slope of the line segment shown in Figure 13.1. The **initial rate** corresponds to the slope of the curve at zero time, shown as the dotted line in Figure 13.1. The rate decreases from this initial rate to a value of zero at long times, when the reactant has been consumed. Sample Exercise 13.1 shows how to obtain the numerical value of the rate by calculating the slope of the line.

*Throughout the text we shall use the convention that square brackets around a quantity indicate the concentration of that quantity.

SAMPLE EXERCISE 13.1

From Figure 13.1, estimate the initial rate of reaction of C_4H_9Cl with water.

Solution: The initial rate, given by the slope of the dotted line in Figure 13.1, can be determined by noting the concentrations of C_4H_9Cl at two different time points on this line. Let us use 0 and 200 sec. Figure 13.2 (a detailed view of a portion of Figure 13.1) shows how we read from the line the change in concentration which corresponds to the 200-sec time

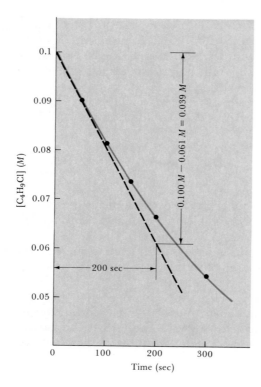

FIGURE 13.2 Portion of curve showing concentration of butyl chloride as a function of time.

interval. The initial rate, determined from the slope, is then the change in concentration divided by the time interval:

$$\text{Initial rate} = \frac{(0.100 - 0.061)M}{200 \text{ sec}}$$

$$= 1.95 \times 10^{-4} \ M/\text{sec}$$

13.2
dependence of reaction rate on concentrations

Ammonium and nitrite ions react in aqueous solution to form nitrogen gas and water, according to the equation

$$NH_4^+ + NO_2^- \longrightarrow N_2 + 2H_2O \qquad [13.4]$$

If we were to study the rate of this reaction, we might do so by measuring the concentration of ammonium or nitrite ion as a function of time. In addition, because one of the products of the reaction is a gas that is not very soluble in water, we could also measure the extent of reaction by measuring the volume of nitrogen gas produced as the reaction proceeds; let us call this quantity V_{N_2}. The rate of reaction is given by any one of the following expressions:

$$\text{Rate} = \frac{-\Delta[NH_4^+]}{\Delta t} = \frac{-\Delta[NO_2^-]}{\Delta t} = \frac{\Delta V_{N_2}}{\Delta t} \qquad [13.5]$$

By measuring one of these quantities as a function of time, we can determine the rates of reaction for a series of solutions in which the starting concentrations of the reactants are allowed to vary. Table 13.2

TABLE 13.2
rate data for the reaction of ammonium and nitrite ions in water at 25°C

INITIAL NO_2^- CONCENTRATION (M)	INITIAL NH_4^+ CONCENTRATION (M)	OBSERVED INITIAL RATE (M/sec)
0.0100	0.200	5.4×10^{-7}
0.0200	0.200	10.8×10^{-7}
0.0400	0.200	21.5×10^{-7}
0.0600	0.200	32.3×10^{-7}
0.200	0.0202	10.8×10^{-7}
0.200	0.0404	21.6×10^{-7}
0.200	0.0606	32.4×10^{-7}
0.200	0.0808	43.3×10^{-7}

shows a list of initial reaction rates for a series of solutions with varying initial concentrations of reactants. By examining these data you can see that the initial reaction rate depends on the initial concentrations of the reactants.

The manner in which the rate of a reaction varies with reactant concentrations is expressed in the **rate law,** a mathematical expression between the rate and some function of concentrations. We expect that the rate law for the reaction shown in Equation [13.4] will be of the form:

$$\text{Rate} = k[NH_4^+]^m[NO_2^-]^n \qquad [13.6]$$

The **rate constant,** k, expresses the proportionality between reaction rate and the concentrations of reactants that appear in the rate law. For any given reaction at a given temperature, k is a constant; its value does not change with changes in the concentrations of reactants. We can employ the initial rate data shown in Table 13.2 to determine the values of the constants m and n in Equation [13.6]. Notice that in the first four solutions listed, the initial concentration of ammonium ion is constant. This means that for these solutions the expression $k[NH_4^+]_i^m$ in Equation [13.6] (the subscript i stands for initial concentration) is constant. Any variation in initial rate for these four solutions must therefore be due to variation in the term $[NO_2^-]_i^n$. Now we notice that when the initial nitrite concentration doubles, the rate doubles; when it goes up by a factor of approximately four, as in the third solution, the initial rate increases by a factor of about four. Because the initial rate is proportional to the initial concentration of nitrite, the value of n in Equation [13.5] must be 1. Similarly, the last four entries in Table 13.2 show that when $[NO_2^-]_i$ is held constant and $[NH_4^+]_i$ allowed to vary, the initial rate varies in direct proportion to $[NH_4^+]_i$; that is, m in Equation [13.6] must also be 1. From these experimental observations we can conclude that the rate law for the reaction shown in Equation [13.4] is

$$\text{Rate} = k[NH_4^+][NO_2^-] \qquad [13.7]$$

In a rate law such as that in Equation [13.6], the overall **reaction order** is the sum of the powers to which all the reactant concentrations appearing in the rate expression are raised. The order with respect to each reactant is simply the exponent to which the concentration of that particular reactant is raised. In Equation [13.7] the concentration of NH_4^+ is raised to the first power; we say that the reaction is first order with respect to ammonium ion. Similarly, it is first order with respect to nitrite ion, NO_2^-. The overall order of the reaction is two.

SAMPLE EXERCISE 13.2

Suppose that the rate law for the reaction $A \longrightarrow B$ has been found to be of the form

$$\text{Rate} = k[A]^m$$

From the following data, determine the overall order of the reaction, and the order with respect to A.

INITIAL CONCENTRATION OF A (M)	INITIAL RATE (M/sec)
0.05	3×10^{-5}
0.10	12×10^{-5}
0.20	49×10^{-5}

Solution: If the reaction were first order in $A(m = 1)$, the initial rate would increase in direct proportion to $[A]_i$. But in fact the initial rate increases much more rapidly than this. When $[A]_i$ is doubled, the initial rate increases by a factor of four; when $[A]_i$ increases fourfold, the initial rate increases by a factor of about 16. These results tell us that m in the rate law is two. That is, the initial rate increases as the square of the initial concentration of A. The overall order of the reaction, and the order with respect to A, is thus two.

In the rate law Equation [13.7] the order of each reactant is the same as the coefficient which appears before it in the balanced equation for the reaction, Equation [13.4]. In general, however, *there is no necessary connection between the order of a reactant in the rate expression and its coefficient in the balanced equation.* The following examples of balanced equations and the experimentally determined rate law illustrate this very well:

$$2N_2O_5(g) \longrightarrow 4NO_2(g) + O_2(g) \tag{13.8}$$
$$\text{Rate} = k[N_2O_5]$$

$$CHCl_3(g) + Cl_2(g) \longrightarrow CCl_4(g) + HCl(g) \tag{13.9}$$
$$\text{Rate} = k[CHCl_3][Cl_2]^{1/2}$$

SAMPLE EXERCISE 13.3

What is the overall reaction order in the rate laws for the reactions shown in Equations [13.8] and [13.9]?

Solution: The overall reaction order in reaction Equation [13.8] is 1; in reaction Equation [13.9] it is $\frac{3}{2}$. (Note from the latter example that reaction order need not be an integer.)

The rate law for a reaction must be determined by experiments in which the concentrations of reactants are varied in a systematic way and the effects of such variations on rate are noted. Several different techniques are available for doing this. For example, the rate law for reaction

Equation [13.4] was determined by varying initial concentrations and noting the effects on initial rates. In other systems the mathematical analysis of data required to obtain the rate law may become very complex. Let us look at a couple of the simplest types of kinetics systems, beginning with first-order reactions.

13.3
**first-order
reactions** In a first-order reaction, the rate depends on the concentration of a single reactant, raised to the first power. As a simple example of such a reaction, consider the rearrangement of methyl isonitrile to acetonitrile:

$$H_3C-N\equiv C: \longrightarrow H_3C-C\equiv N: \qquad [13.10]$$

Methyl isonitrile   Acetonitrile

In this reaction the molecule undergoes an intramolecular rearrangement. The reaction is first order with respect to isonitrile (we use R to represent the methyl group, CH_3):

$$\text{Rate} = \frac{-\Delta[RNC]}{\Delta t} = k[RNC] \qquad [13.11]$$

Notice the units in this equation; [RNC] has units of molarity, and t is in units of seconds. Thus,

$$\text{Rate} = \frac{(M)}{(\text{sec})} = k(M)$$

Rearranging,

$$k = \frac{(M)}{(\text{sec})}\frac{1}{(M)} = \frac{1}{\text{sec}} = \text{sec}^{-1}$$

Equation [13.11] can be rearranged to yield

$$\frac{\Delta[RNC]}{[RNC]} = -k\,(\Delta t) \qquad [13.12]$$

By using calculus this expression can be integrated.* When this is done a very useful expression results:

$$\log\left(\frac{[RNC]_0}{[RNC]_t}\right) = \frac{kt}{2.30} \qquad [13.13]$$

In Equation [13.13] $[RNC]_0$ is the concentration of isonitrile at the start of the reaction, and $[RNC]_t$ is the concentration at time t. Using the properties of logarithms (Appendix A.2), the equation can be rearranged a little more to give

*There is no need in this course for you to be able to perform calculus operations; you need only remember the result, Equation [13.13].

$$\log [RNC]_0 - \log [RNC]_t = \frac{kt}{2.30}$$

$$\log [RNC]_t = \frac{-kt}{2.30} + \log [RNC]_0 \qquad [13.14]$$

In general, for a reactant species A,

$$\log [A]_t = \left(\frac{-k}{2.30}\right)t + \log [A]_0 \qquad [13.15]$$

You may recognize this as the equation of a straight line in the form

$$y = ax + b \qquad [13.16]$$

In this form a is the slope of the linear relationship between x and y, and b is the intercept on the y axis when $x = 0$. In applying this to Equation [13.15], $\log [A]_t$ is one variable, t is the other; $(-k/2.30)$ is the slope of the line and $\log [A]_0$ is the intercept at time $t = 0$.

SAMPLE EXERCISE 13.4

In the rearrangement of methyl isonitrile in the gas phase at 198.9°C, the data shown below were obtained for the concentration of CH_3NC as a function of time. Determine whether the rearrangement reaction does indeed fit a first-order rate law; if it does, determine a value for the first-order rate constant.

TIME (sec)	PRESSURE, CH_3NC (torr)
0	150
1,000	142
2,000	135
5,000	115
10,000	88.8
15,000	68.2
20,000	52.4
30,000	31.0

Solution: The first point to realize in attacking this problem is that pressure is a legitimate unit of concentration for a gas, because the number of molecules per unit volume is directly proportional to the pressure. To test for first-order behavior, we graph the log of the isonitrile pressure on the vertical axis against the time on the horizontal axis, as shown in Figure 13.3. We see that this graph does indeed yield a straight line, indicating that the rate data obey Equation [13.15]. To obtain the value for k we must determine the slope of the line. We choose two well-separated points on the line drawn through the data and read off the corresponding coordinates.

$$\text{Slope} = \frac{\Delta \log P}{\Delta t} = \frac{2.105 - 1.765}{18,300 - 3000}$$

$$= -2.22 \times 10^{-5}/\text{sec}$$

Because the slope equals $-k/2.30$,

$$k = -(2.30)(-2.22 \times 10^{-5}) = 5.11 \times 10^{-5}/\text{sec}$$

From a knowledge of the rate constant, it is possible to calculate the concentration of reactant or product at any given time, when the initial concentration is known, as shown in Sample Exercise 13.5

SAMPLE EXERCISE 13.5

Assume that at time $t = 0$ the pressure of methyl isonitrile is 120 torr. Employing the first-order rate

constant calculated in Sample Exercise 13.4, determine the pressure of methyl isonitrile at $t = 7.4 \times 10^3$ sec.

Solution: We can use Equation [13.15]:

$$\log [RNC]_t = \frac{-5.11 \times 10^{-5}/\text{sec}}{2.30}(7400 \text{ sec})$$
$$+ \log 120$$
$$= -0.164 + 2.079$$
$$= 1.915$$

Taking the antilog we obtain:

$$[RNC] = 82 \text{ torr}$$

HALF-LIFE

The **half-life** of a reaction, $t_{1/2}$, is the time required for the concentration of the reactant to decrease to halfway between its initial and final values. In the cases we'll be considering, the final concentration is zero. To obtain an expression for $t_{1/2}$ for a first-order reaction, we begin with Equation [13.13] and employ X as a general symbol for the reactant. The half-life corresponds to the time when $X_t = \frac{1}{2}(X_0)$. Inserting these quantities into the equation, we have

$$\log \frac{[X]_0}{\frac{1}{2}[X]_0} = \left(\frac{k}{2.30}\right)t_{1/2} \qquad [13.17]$$

$$\log 2 = \left(\frac{k}{2.30}\right)t_{1/2} \qquad [13.18]$$

$$t_{1/2} = \frac{(2.30)\log 2}{k} \qquad [13.19]$$

$$= \frac{0.693}{k} \qquad [13.20]$$

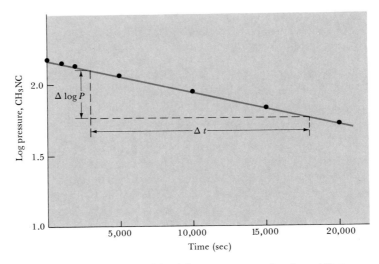

FIGURE 13.3 Log of methyl isonitrile pressure as a function of time.

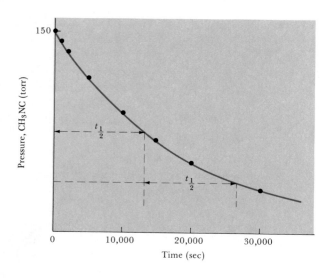

FIGURE 13.4 Pressure of methyl isonitrile as a function of time. Two successive half-lives of the rearrangement reaction, Equation [13.10], are shown.

Notice that $t_{1/2}$ is independent of the initial concentration of reactant. This result tells us that if we measure reactant concentration at *any* time in the course of a first-order reaction, the concentration of reactant will be half of that measured value at a time $0.693/k$ later. The concept of half-life is widely used in describing radioactive decay. This application is discussed in detail in Section 20.3.

The data from Sample Exercise 13.4 for the first-order rearrangement of methyl isonitrile are graphed in Figure 13.4. The first half-life is shown at 13,320 sec. At a time 13,320 sec later, the isonitrile concentration has decreased to $\frac{1}{2}$ of $\frac{1}{2}$, or $\frac{1}{4}$ the original concentration. *It is a characteristic of a first-order reaction that the concentration of the reactant decreases by factors of $\frac{1}{2}$ in a series of regularly spaced time intervals.*

SAMPLE EXERCISE 13.6

From the half-life determined from Figure 13.4, calculate the pressure of methyl isonitrile remaining after 53,280 sec, assuming an initial pressure of 150 torr.

Solution: The half-life for the reaction is shown above to be 13,320 sec. The time 53,280 sec is thus 53,280/13,320, or precisely four half-lives. The factor by which the concentration of isonitrile has diminished in this period of time is thus $(\frac{1}{2})^4$, or 0.0625. The pressure of remaining isonitrile is therefore

$$(0.0625)(150) = 9.4 \text{ torr}$$

**13.4
higher-order
reactions**

As the order of a reaction increases, the complexity of the mathematics required to obtain the rate constant from the rate data increases as well. The expression for a second-order reaction that is second order in just one reactant is

$$\text{Rate} = k[\text{X}]^2 \qquad\qquad [13.21]$$

If the expression is rearranged to yield an equation in k, we obtain

$$k = \frac{\text{rate}}{[\text{X}]^2} \qquad [13.22]$$

As we have seen, rate ordinarily has units of M/sec or moles/(l. sec). The concentration of X will ordinarily be expressed in molarity, M. Inserting these units into Equation [13.22], we have:

$$
\begin{aligned}
k &= \frac{(M/\text{sec})}{(M)^2} \\
&= M^{-1}/\text{sec} \qquad [13.23] \\
&= \text{l.}/\text{mole-sec}
\end{aligned}
$$

In these units a second-order rate constant for a typical reaction occurring at a conveniently measurable rate in solution would be on the order of 10^{-5}–10^{-6} l./mole-sec. The half-life for a second-order reaction of the form of Equation [13.21] is given by the expression $t_{1/2} = 1/k[\text{X}]_0$. It is *not* independent of the initial concentration of reactant, as is $t_{1/2}$ for a first-order reaction.

Second-order reactions may also be first order in each of two reactants:

$$\text{Rate} = k[\text{X}][\text{Y}] \qquad [13.24]$$

The second-order rate constant has the same units in this case as described above, l./(mole-sec). Second-order rate laws as in Equation [13.24] are very commonly observed for reactions occurring in solution. For example, we showed in Section 13.2 that the reaction between NH_4^+ and NO_2^-, Equation [13.4], has a rate law of this form—see Equation [13.7].

Reactions of order higher than second are not common. It is very common, however, to obtain rate laws that are quite complex, as illustrated by Equation [13.9].

13.5

the temperature dependence of reaction rates

The rates of most chemical reactions increase as the temperature rises. We see examples of this generalization in many biological processes around us. The rate at which grass grows and the metabolic activity of the common housefly are both greater in warm weather than in the cold of winter. As another example, food cooks more rapidly in boiling water than in merely hot water. However, it is dangerous to place too much emphasis on the apparent rates at which biological systems operate as a function of temperature, because they are very complex and are adapted through evolutionary development to operate optimally in a narrow temperature range. To obtain a clear understanding of how temperature affects reaction rates, we must examine simple reaction systems. As an example, let us consider the reaction about which we have already learned quite a bit, the first-order rearrangement of methyl isonitrile, Equation [13.10]. Figure 13.5 shows the experimentally determined rate constant for this reaction as a function of temperature. It is evident that

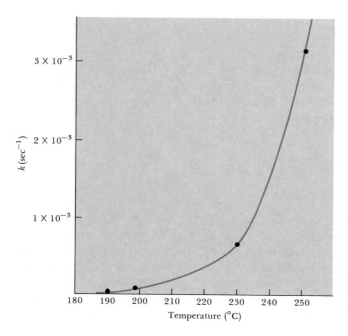

the rate of the reaction increases rapidly with temperature. Furthermore, the increase is nonlinear.

As we seek an explanation for this behavior, perhaps the first question to ask is, why do *any* reactions go slowly? What keeps reactions from simply occurring immediately? If the methyl isonitrile molecules are going eventually to rearrange into acetonitrile, why don't they all do it at once?

ACTIVATION ENERGY

We know from the kinetic-molecular theory of gases that with increasing temperature the average energy of the gas molecules increases. The fact that the rate of the methyl isonitrile reaction increases with increasing temperature suggests that perhaps the rearrangement is related to the kinetic energies of the molecules. Svante Arrhenius suggested in 1888 that before reaction can occur, a certain minimum amount of energy must be available to "propel" the molecules from one chemical state into another. The situation is rather like that shown in Figure 13.6. The boulder will

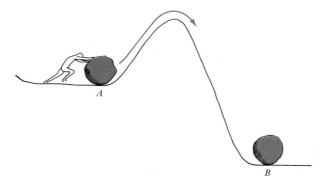

FIGURE 13.6 Illustration of the potential-energy profile for a boulder. The boulder must be moved over the energy barrier before it can come to rest in the lower energy location, *B*.

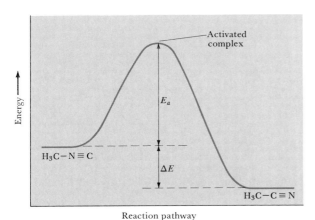

FIGURE 13.7 Energy profile for the rearrangement of methyl iso-nitrile. The molecule must surmount the activation-energy barrier before it can form the product, acetonitrile.

be in a lower (or more stable) potential-energy state in valley B than in valley A. Before it can come to rest in B, however, it must acquire the energy to overcome the barrier blocking its passage from the one state into the other. In the same way, molecules may require a certain minimum energy to overcome the forces that tend to keep them as they are, if they are to form the new chemical bonds that will result in a different arrangement. In our methyl isonitrile example, we might imagine that for rearrangement to occur, the N≡C portion of the molecule must turn over:

$$H_3C-N\equiv C: \longrightarrow \left[H_3C \cdots \overset{\cdot\cdot}{\underset{\cdot\cdot}{\overset{C}{\underset{N}{\vert\vert\vert}}}} \right] \longrightarrow H_3C-C\equiv N:$$

Even though the bonding may be more stable in the product acetonitrile than in the starting compound, energy is required to force the molecule through the relatively unstable intermediate state to the final result. The energy of the molecule as it proceeds along this reaction pathway is shown in Figure 13.7. Arrhenius called the energy barrier between the starting molecule and the highest energy along the reaction pathway the **activation energy,** E_a. The particular arrangement of atoms that has the maximum energy is often called the **activated complex.**

Energy is transferred between molecules through collisions. Thus, within a certain period of time, any particular isonitrile molecule might acquire enough energy to overcome the energy barrier and be converted into acetonitrile. At any given temperature only a small fraction of collisions will occur with sufficient energy to overcome the barrier to reaction. However, as shown in Figure 5.12, the distribution of molecular speeds is more spread out toward higher values when the gas is at a higher temperature. The distribution of kinetic energies of molecules changes in a similar way, as shown in Figure 13.8. This graph shows that at the higher temperature a larger fraction of molecules possesses the minimum energy needed for reaction.

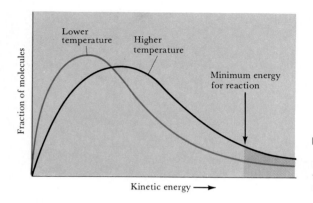

FIGURE 13.8 The distribution of kinetic energies in a sample of gas molecules at two different temperatures. At the higher temperature a larger number of molecules have higher energies. Thus, a larger fraction in any one instant will have more than the minimum energy required for reaction.

Similar considerations apply to other reactions, such as bimolecular reactions in which two molecules or ions come together and react; the activation energy required for reaction to occur must be present in the molecules as they collide.

It is rather easy to see that not every collision leads to reaction when two molecules or atoms react. Molecules in the gas phase at ordinary temperatures and pressures collide with one another about 10^{10} times per second. Consider the reaction of hydrogen with iodine to form hydrogen iodide:

$$H_2 + I_2 \longrightarrow 2HI \qquad [13.25]$$

In this reaction the rate law is

$$Rate = k[H_2][I_2]$$

If every collision between an H_2 and I_2 molecule in a mixture of these gases at ordinary temperature and pressures resulted in reaction, the reaction would be over in much less than a second. Instead, at room temperature the reaction proceeds very slowly. Obviously, every collision does not result in reaction. In fact, we can calculate that only about one in every 10^{13} collisions is effective. This tells us that the energy barrier to reaction of hydrogen and iodine is quite substantial. Only a *very* small fraction of the collisions occur with enough energy to carry the molecules over the energy barrier to the products. As the temperature increases, the fraction of collisions that are sufficiently energetic increases. With each 10°C rise in temperature, the rate of this reaction triples.

THE ARRHENIUS EQUATION

Arrhenius noted that the increase in rate with increasing temperature for most reactions is nonlinear, as in the example shown in Figure 13.5. He found that most reaction-rate data obeyed the equation

$$k = Ae^{-E_a/RT} \qquad [13.26]$$

In this equation, called the **Arrhenius equation,** e is the base of the natural logarithm scale (see Appendix A.2). The term E_a is the activation energy, which we have already defined; R is the gas constant and T, of course, is absolute temperature; A is constant, or nearly so, as temperature is varied. It is called the **frequency factor.**

Equation [13.26] can be put into a more convenient form by taking the natural log of both sides (see Appendix A.2 if you need review on the use of logs):

$$\ln (k) = \ln (Ae^{-E_a/RT})$$
$$\ln k = \ln A - E_a/RT \qquad [13.27]$$

and then converting natural logs to logs base 10:

$$\log k = \log A - \frac{E_a}{2.30\ RT} \qquad [13.28]$$

This expression has the form of a straight line, in which one variable is $\log k$, and the other is $1/T$. The slope of the line is given by $-E_a/2.30\ R$; the intercept, at $(1/T) = 0$, is $\log k = \log A$.

SAMPLE EXERCISE 13.7

The following table shows the rate constant for rearrangement of methyl isonitrile at various temperatures (these are the data that are graphed in Figure 13.5).

TEMPERATURE (°C)	$k(\text{sec}^{-1})$
189.7	2.52×10^{-5}
198.9	5.25×10^{-5}
230.3	6.30×10^{-4}
251.2	3.16×10^{-3}

From these data calculate the activation energy for the reaction.

Solution: We must first convert temperatures to the absolute temperature scale, K. We then take the inverse of these temperatures, and obtain the corresponding log values for k. This gives us the following table:

T(K)	$1/T$(K)	LOG k
462.7	2.160×10^{-3}	-4.60
471.9	2.118×10^{-3}	-4.29
503.3	1.986×10^{-3}	-3.20
524.2	1.907×10^{-3}	-2.50

A graph of these data results in a straight line, as shown in Figure 13.9. The data points shown in the graph lie very close to the best straight line through all four points. The slope of the line is obtained by choosing two well-separated points, as shown, and reading off the coordinates of each:

$$\text{Slope} = \frac{-2.45 - (-4.45)}{.00190 - .00214} = 8330 \qquad [13.29]$$

The numerator in Equation [13.29] has no units, because logs have no units. The denominator has the units of $1/T$, that is, $1/\text{K}$. Thus, the overall units for the slope are K. The slope is equal to $-E_a/2.30\ R$. We want the value for the molar gas constant R in J/K-mole (Table 5.2), which is 8.317. Thus we obtain

$$\text{Slope} = \frac{-E_a}{2.30\ R}$$
$$E_a = -(\text{slope})(2.30\ R)$$
$$= -(-8330\ \text{K})(2.30)\left(8.32\frac{\text{J}}{\text{K-mole}}\right)$$
$$\times \left(\frac{1\ \text{kJ}}{1000\ \text{J}}\right)$$
$$= 159\ \text{kJ/mole}$$

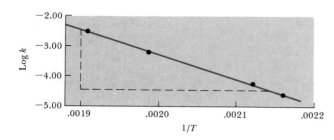

FIGURE 13.9 Log of the rate constant for rearrangement of methyl isonitrile as a function of $1/T$. The linear relationship is predicted from the Arrhenius equation.

13.6

reaction mechanisms

A knowledge of how the rate of a chemical reaction depends on concentrations, temperature, and pressure has many practical applications. For example, it is necessary to have such information to design a chemical plant to produce a certain desired substance. To understand how a certain pollutant chemical such as a herbicide behaves when released into the environment, it is necessary to know how it reacts under various conditions in nature, and at what speeds. Aside from such practical reasons for studying the kinetics of chemical reactions, there is also a more fundamental goal. A knowledge of the rate law and the activation energy for a chemical reaction can be of help in arriving at a reaction **mechanism,** a detailed picture of how the reaction occurs. The mechanism describes the order in which bonds are broken and formed and the changes in relative positions of atoms in the course of the reaction. The formulation of detailed mechanisms for chemical reactions presents one of the great challenges of chemistry. Once a reaction mechanism has been devised, the chemist can use it to predict new reaction possibilities and to test them by additional experiments.

As an example of the kind of thinking that goes into arriving at a reaction mechanism, let's consider the following simple reaction,

$$O^+ + NO \longrightarrow NO^+ + O \qquad [13.30]$$

which occurs in the upper atmosphere. The overall reaction is strongly exothermic (see Table 10.1). The reaction is first order in O^+ and first order in NO:

$$\text{Rate} = \frac{\Delta[O]}{\Delta t} = k[O^+][NO] \qquad [13.31]$$

The activation energy is quite small, and so the overall reaction energy profile for the course of the reaction looks as shown in Figure 13.10. Now this is just about as far as kinetics information can take us. We can assume from this that the activated complex involves a coming together of the O^+ and NO, but there remain some important unanswered questions.

It is possible to imagine two quite different pathways for the reaction (see Section 10.3). These alternative pathways are illustrated in Figure 13.11. In the **atom-transfer pathway,** the incoming O^+ approaches the NO molecule on the side opposite the oxygen bound to nitrogen and essentially bumps off the oxygen atom from the other side. This possibility can

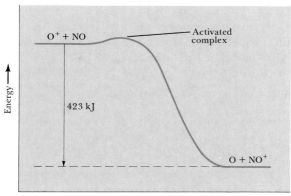

FIGURE 13.10 Energy profile for reaction of O^+ with NO.

Reaction pathway

be tested in a laboratory experiment by generating the O^+ from oxygen which is highly enriched in the rare oxygen isotope ^{18}O. Because this isotope is present to the extent of only 0.2 percent in nature, the NO contains only 0.2 percent $N^{18}O$. If the $^{18}O^+$ which comes up to the NO molecule ends up on the NO^+, the overall result of reaction would be as follows:

$$^{18}O^+ + NO \longrightarrow N^{18}O^+ + O \qquad [13.32]$$

The other possibility for the reaction mechanism is an **electron-transfer pathway,** in which an electron jumps from NO to O^+, and the two entities then move apart. If this pathway is followed, the overall result is as follows:

$$^{18}O^+ + NO \longrightarrow NO^+ + {}^{18}O \qquad [13.33]$$

By measuring the location of ^{18}O in the reaction products, the two mechanisms can be distinguished. The results of laboratory studies show that the product NO^+ incorporates no ^{18}O. Thus, reaction proceeds according to the electron transfer pathway.

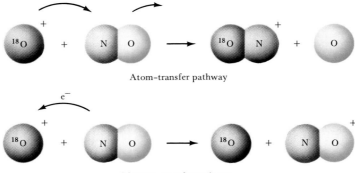

FIGURE 13.11 Alternative pathways for the reaction $O^+ + NO \longrightarrow O + NO^+$. The results of experiments using ^{18}O labeling show that the reaction proceeds according to the electron-transfer pathway.

Not all reactions proceed from reactants to products in a single step. In reactions that proceed through two or more steps, the overall rate of reaction is determined by the rate of the slowest step. The rate law for a reaction reflects all the steps that precede the slow step, or **rate-determining step.** To illustrate, consider the gas phase reaction of chlorine with chloroform:

$$CHCl_3(g) + Cl_2(g) \longrightarrow CCl_4(g) + HCl(g) \qquad [13.9]$$
$$Rate = k[CHCl_3][Cl_2]^{1/2}$$

Because the rate law does not correspond to the overall equation for the reaction, we know that this reaction cannot proceed by a simple coming together of $CHCl_3$ and Cl_2 molecules. A mechanism that is consistent with the observed rate law involves a dissociation of Cl_2 molecules to give Cl atoms:

$$Cl_2 \rightleftharpoons 2Cl \qquad [13.34]$$

The rate-determining step is postulated to be reaction of chlorine atoms with $CHCl_3$:

$$Cl + CHCl_3 \longrightarrow HCl + CCl_3 \qquad (slow) \qquad [13.35]$$

A *rapid* combination of CCl_3 with chlorine atoms follows:

$$Cl + CCl_3 \longrightarrow CCl_4 \qquad [13.36]$$

The rate of appearance of product is determined by the rate of the slowest step, Equation [13.35]:*

$$Rate = \frac{\Delta[HCl]}{\Delta t} = k_2[Cl][CHCl_3] \qquad [13.37]$$

But what is the concentration of chlorine atoms? If the chlorine atoms are in equilibrium with Cl_2 molecules, the concentration of chlorine atoms can be expressed as:

$$[Cl] = K_{eq}^{1/2}[Cl_2]^{1/2}$$

(We have had to anticipate a bit here from Chapter 14. We need only note that K_{eq} is constant. Thus, the concentration of chlorine atoms is proportional to the square root of the chlorine molecule concentration.) Substituting this into the rate expression, Equation [13.37], we have

$$\frac{\Delta[HCl]}{\Delta t} = k_2 K_{eq}^{1/2}[Cl_2]^{1/2}[CHCl_3] \qquad [13.38]$$

---

*The subscript 2 on k indicates that k is the rate constant for the second step in the set of reactions that describes the overall reaction.

The two constants in this equation can be lumped together as one constant, giving us:

$$\frac{\Delta[\text{HCl}]}{\Delta t} = k[\text{CHCl}_3][\text{Cl}_2]^{1/2} \qquad [13.39]$$

The final step in the reaction, Equation [13.36], occurs rapidly, so that the CCl_4 molecule appears just as rapidly as CCl_3 is formed via Equation [13.35]. In this example, the fact that the reaction is only $\frac{1}{2}$ order in Cl_2 provides a valuable clue as to the probable mechanism of the reaction.

13.7 catalysis A **catalyst** is a substance that acts to change the speed of a chemical reaction without itself undergoing a permanent change in the process. Nearly all catalysts increase reaction rates. Catalysts are very common; most reactions occurring in the human body, the atmosphere, the oceans, or in industrial chemical processes are affected by catalysts.

If you have been exposed to chemical laboratory work, it is likely that you have carried out the reaction in which oxygen is produced by heating of potassium chlorate, KClO_3:

$$2\text{KClO}_3(s) \xrightarrow{\Delta} 2\text{KCl}(s) + 3\text{O}_2(g) \qquad [13.40]$$

In the absence of a catalyst, KClO_3 does not readily decompose in this manner, even on strong heating. However, mixing black manganese dioxide, MnO_2, with the KClO_3 before heating causes the reaction to occur much more readily. The MnO_2 can be recovered largely unchanged from this reaction, so it is clear that the overall chemical process is still the same. Thus, MnO_2 acts as a catalyst for decomposition of KClO_3. As another example, we know that a cube of sugar, when dissolved in water at 37°C, does not undergo oxidation at a significant rate. The sugar could be recovered essentially unchanged from the solution after several days. Yet when sugar is ingested into the human body it is rapidly oxidized and soon ends up mostly as carbon dioxide and water:

$$\text{C}_{12}\text{H}_{22}\text{O}_{11} + 12\text{O}_2 \longrightarrow 12\text{CO}_2 + 11\text{H}_2\text{O} \qquad [13.41]$$

The oxidation of sugar in the biochemical system has been greatly speeded up by the presence of one or more catalysts. These catalysts are **enzymes,** protein molecules that act to catalyze specific biochemical reactions. (Enzymes are discussed at length in Chapter 25.)

Much industrial chemical research is devoted to the search for new and more effective catalysts for reactions of commercial importance. Extensive research efforts also are devoted to finding means of inhibiting or removing certain catalysts that promote undesirable reactions, such as those involved in corrosion of metals, aging, and tooth decay.

HOMOGENEOUS CATALYSIS

A catalyst that is present in the same phase as the components of a chemical reaction is known as a **homogeneous catalyst**. For example, a homogeneous catalyst for a reaction occurring in solution would itself be dissolved in the solution.

In Chapter 10 we considered a simple example of homogeneous catalysis, the action of NO in promoting the decomposition of ozone, O_3. The NO acts as a catalyst by reacting with O_3 to form NO_2 and O_2. The NO_2 thus formed then reacts with atomic oxygen present in the stratosphere to reform NO and yield O_2 as the other product. The sequence of reactions and the overall result are as follows:

$$NO + O_3 \longrightarrow NO_2 + O_2$$
$$NO_2 + O \longrightarrow NO + O_2$$
$$\overline{O_3 + O \longrightarrow 2O_2}$$

In this example, NO acts as a catalyst for O_3 decomposition because it speeds up the rate of the overall reaction without itself undergoing any net, or overall, chemical change.

As another example, hydrogen peroxide, H_2O_2, when dissolved in water undergoes slow decomposition, leading to oxygen and water:

$$2H_2O_2 \longrightarrow 2H_2O + O_2 \tag{13.42}$$

This reaction has a very strong tendency to occur in the direction shown. By this we mean that eventually essentially all the hydrogen peroxide present in solution will have decomposed to oxygen and water. In the absence of a catalyst, however, the reaction occurs at an extremely slow rate. Many different substances are capable of catalyzing the reaction; among these is bromine, Br_2. The bromine reacts with hydrogen peroxide in acidic solution, forming bromide ion and liberating oxygen:

$$Br_2 + H_2O_2 \longrightarrow 2Br^- + 2H^+ + O_2 \tag{13.43}$$

We recognize this as a simple oxidation-reduction reaction (Chapter 8) in which bromine is reduced to bromide ion, and the oxygen of hydrogen peroxide is oxidized from the -1 oxidation state in H_2O_2 to the 0 oxidation state of O_2. If this reaction were all that were involved, bromine would not be a catalyst, because it undergoes chemical change in this reaction. It happens that hydrogen peroxide reacts with bromide ion in acidic solution to form bromine:

$$2Br^- + H_2O_2 + 2H^+ \longrightarrow Br_2 + 2H_2O \tag{13.44}$$

In this oxidation-reduction reaction, bromide ion is oxidized by hydrogen peroxide to bromine; the oxygen of hydrogen peroxide is reduced, forming water. The overall sum of reaction Equations [13.43] and [13.44] is just Equation [13.42]. (Add these two reactions together yourself to make sure you see that reaction Equation [13.42] results.) We see that bromine

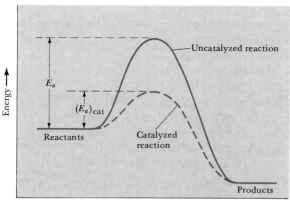

FIGURE 13.12 Energy profile for a catalyzed and uncatalyzed reaction. The catalyst functions in this example to lower the activation energy for reaction. Notice that the energies of reactants and products are unchanged by the catalyst.

is indeed a catalyst in the reaction, because it speeds the overall reaction without itself undergoing any net, or overall, change.

On the basis of the Arrhenius expression for a chemical reaction, Equation [13.26], the rate constant k is determined by the activation energy E_a and the frequency factor A. A catalyst may affect the rate of reaction by altering the value for either E_a or A. Because E_a appears in the exponential term, the most dramatic catalytic effects come from lowering of E_a. As a general rule, *a catalyst lowers the overall activation energy for chemical reaction.* The lowering of E_a by a catalyst is shown schematically in Figure 13.12.

A catalyst usually lowers the overall activation energy for reaction by providing a completely different pathway for reaction. The two examples given above involve a reversible, cyclic reaction of the catalyst with the reactants. For example, in the decomposition of hydrogen peroxide, two successive reactions of H_2O_2, with bromine and then with bromide, are involved. Because these two reactions together serve as a catalytic pathway for hydrogen peroxide decomposition, *both* of these reactions must have significantly lower activation energies than the uncatalyzed decomposition, as shown schematically in Figure 13.13.

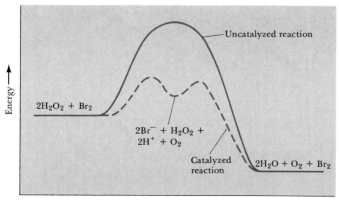

FIGURE 13.13 Energy profile for the uncatalyzed decomposition of hydrogen peroxide, and for the reaction as catalyzed by Br_2. The catalyzed reaction involves two successive steps, each of which has a lower activation energy than the uncatalyzed reaction.

395

The catalysts we have so far discussed (and this includes enzymes, which are discussed in more detail in Chapter 25) are homogeneous catalysts. However, a great many reactions are catalyzed by substances that exist in a different phase from the reactants.

HETEROGENEOUS CATALYSIS

A **heterogeneous catalyst** exists in a different phase from the reactant molecules. For example, a reaction between molecules in the gas phase might be catalyzed by a finely divided metal oxide. In the absence of a catalyst, the reaction would occur slowly in the gas phase. However, when the catalyst is present, the reaction occurs more rapidly on the surface of the solid catalyst.

We saw in Section 10.5 an example of heterogeneous catalysis, in the oxidation of SO_2 to SO_3. Many industrially important reactions occurring in the gas phase are catalyzed by solid surfaces. Reactions occurring in solution may also be catalyzed by solids. Heterogeneous catalysts are often composed of finely divided metal or metal oxides. Because the catalyzed reaction occurs on the surface, special methods are often used to prepare catalysts so that they have very large surface areas.

The initial step in heterogeneous catalysis is usually **adsorption** of reactants. As noted in Chapter 2, the term *ad*sorption should be distinguished from *ab*sorption. Adsorption refers to binding of molecules to a surface, whereas absorption refers to uptake of molecules into the interior of another substance. Adsorption occurs because the atoms or ions at the surface of a solid are extremely reactive. Unlike their counterparts in the interior of the substance, they have unfulfilled valence requirements. The unused bonding capability of surface atoms or ions may be utilized to bond molecules from the gas or solution phase to the surface of the solid. In practice, not all the atoms or ions of the surface are reactive; various impurities may be adsorbed at the surface, and these may occupy many potential reaction sites and block further reaction. The places where reacting molecules may become adsorbed are called **active sites**. The number of active sites per unit amount of catalyst depends on the nature of the catalyst, on its method of preparation, and on its treatment before use.

As an example of heterogeneous catalysis, consider the hydrogenation of ethylene to form ethane:

$$\underset{\text{Ethylene}}{\overset{\text{H}\qquad\text{H}}{\underset{\text{H}\qquad\text{H}}{\text{C}=\text{C}}}} + \text{H}_2 \longrightarrow \underset{\text{Ethane}}{\overset{\text{H}\qquad\text{H}}{\underset{\text{H}\qquad\text{H}}{\text{H}-\text{C}-\text{C}-\text{H}}}} \qquad [13.45]$$

In the absence of a catalyst, this reaction does not occur at all readily. However, in the presence of a very finely divided metal such as nickel, palladium, or platinum, the reaction occurs rather easily at room temperature, under a few hundred atmospheres of hydrogen pressure. The mechanism by which reaction occurs is shown diagrammatically in Figure 13.14. Both ethylene and hydrogen are adsorbed at the metal

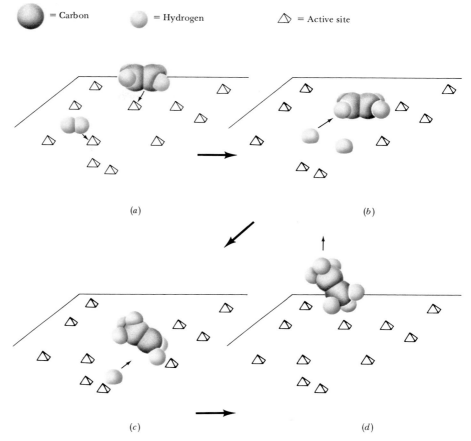

= Carbon   = Hydrogen   △ = Active site

(a)   (b)

(c)   (d)

FIGURE 13.14 Mechanism for reaction of ethylene with hydrogen on a catalytic surface. (a) The hydrogen and ethylene are adsorbed at the metal surface. (b) The H—H bond is broken to give adsorbed hydrogen atoms. (c) These migrate to the adsorbed ethylene and bond to the carbon atoms. (d) As C—H bonds are formed, the adsorption of the molecule to the metal surface is decreased, and ethane is released.

surface, Figure 13.14(a). The adsorption of hydrogen results in breaking of the H—H bond and formation of two M—H bonds, where M represents the metal surface, Figure 13.14(b). The hydrogen atoms are relatively free to move about the surface. When they encounter an adsorbed ethylene, the hydrogen may become bound to the carbon, Figure 13.14(c). The carbon thus acquires four σ bonds about it, which reduces its tendency to remain adsorbed at the metal. When the other carbon also acquires a hydrogen, the ethane molecule is released from the surface, Figure 13.14(d). The active site is ready to adsorb another ethylene molecule, and thus begin the cycle again.

CATALYSIS AND AIR POLLUTION

Heterogeneous catalysis plays a major role in the fight against urban air pollution. Two components of automobile exhausts that are involved in the formation of photochemical smog are nitrogen oxides and unburned hydrocarbons of various types (Section 10.5). In addition, automobile

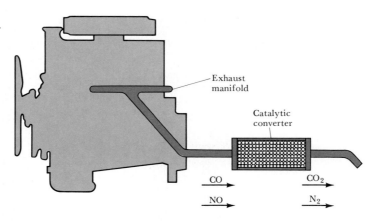

FIGURE 13.15   Illustration of the arrangement and functions of a catalytic converter.

exhausts may contain considerable quantities of carbon monoxide. Even with the most careful attention to engine design and fuel characteristics, it is not possible under normal driving conditions to reduce the contents of these pollutants to an acceptable level in the exhaust gases coming from the engine. It is therefore necessary somehow to remove them from the exhaust gases before they are vented to the air. This removal is accomplished in the catalytic converter.

The catalytic converter, illustrated in Figure 13.15, must perform two distinct functions: (1) oxidation of CO and unburned hydrocarbons to carbon dioxide and water and (2) reduction of nitrogen oxides to nitrogen gas:

$$CO, \text{ hydrocarbons } (C_xH_y) \xrightarrow{O_2} CO_2 + H_2O$$

$$NO, NO_2 \longrightarrow N_2$$

These two functions require two distinctly different catalysts. The development of a successful catalyst system represents a very difficult challenge. The catalysts must be effective over a wide range of operating temperatures; they must continue to be active in spite of the poisoning action of various gasoline additives emitted along with the exhaust; they must be physically rugged enough to withstand gas turbulence and the mechanical shocks of driving under various conditions for thousands of miles.

Catalysts that promote the combustion of CO and hydrocarbons are, in general, the transition-metal oxides and noble metals such as platinum. As an example, a mixture of two different metal oxides such as CuO and Cr_2O_3 might be used. These materials are supported on a structure, Figure 13.16, which allows the best possible contact between the flowing exhaust gas and the catalyst surface. Either bead or honeycomb structures made from alumina, Al_2O_3, and impregnated with the catalyst may be employed. Such catalysts operate by first adsorbing

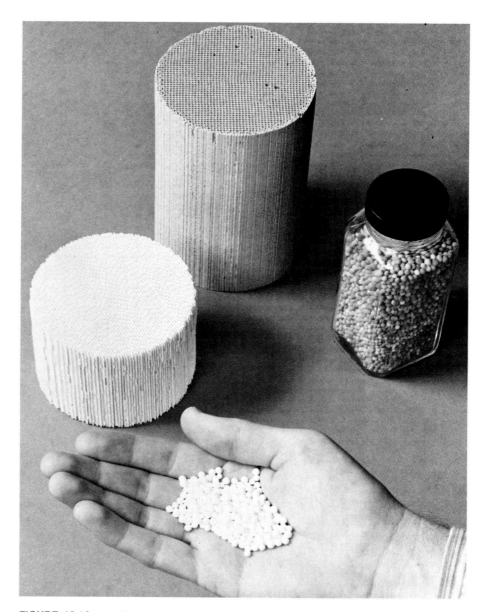

FIGURE 13.16 Catalyst support materials used in an automobile catalytic converter. Some catalytic converters use the honeycomb, or monolith, type of support, whereas others use beads that are packed in the converter in layers. In either case, the support material is impregnated with the catalyst, in such a manner that the catalyst will remain on the support structure during use. (*General Motors Corporation*)

oxygen gas, also present in the exhaust gas. This adsorption weakens the O—O bond in O_2, so that oxygen atoms are in effect available for reaction with adsorbed CO to form CO_2. Hydrocarbon oxidation probably proceeds somewhat similarly, with the hydrocarbons first being adsorbed by rupture of a C—H bond.

Reduction of nitrogen oxides is favored thermodynamically. That is, the decomposition of NO, for example, to yield N_2 and O_2 is favored, but the reaction is extremely slow. A catalyst is therefore necessary. The most effective catalysts are transition-metal oxides, and noble metals, the same kinds of materials that catalyze the oxidation of CO and hydrocarbons. The catalysts that are most effective in one reaction, however, are usually much less effective in the other. It is therefore necessary to have two different catalytic components.

As an illustration of the difficulties faced in the design of a catalytic converter, one of the frequent products of nitrogen oxide reduction in exhausts is ammonia. This comes about because water is present in the exhaust gases, and it undergoes reaction with CO in the presence of catalyst to yield H_2:

$$H_2O + CO \longrightarrow H_2 + CO_2 \qquad [13.46]$$

The hydrogen in turn is adsorbed on the catalyst and reacts with NO to form either N_2 or NH_3:

$$2NO + 2H_2 \longrightarrow N_2 + 2H_2O \qquad [13.47]$$
$$2NO + 5H_2 \longrightarrow 2NH_3 + 2H_2O \qquad [13.48]$$

Ammonia itself would, of course, be an undesirable pollutant. Furthermore, it is capable of reacting with oxygen on the catalyst surface to form NO and H_2O, putting us right back where we started. Obviously, therefore, the catalyst system for reduction of nitrogen oxides must not lead to much ammonia formation.

An additional problem that has recently come to light is the possibility of oxidation of SO_2 to SO_3 in the catalytic converter. Gasolines contain small quantities of sulfur that ordinarily appear in the exhaust gases as SO_2. In the catalytic converter, the oxidation of SO_2 to SO_3 is catalyzed. Because SO_3 dissolves in water to form highly corrosive sulfuric acid (Section 10.5), the appearance of SO_3 in the exhaust gases could cause corrosion and health problems in the vicinity of heavy traffic areas.

The activity of catalytic converters decreases with use, because of loss of active catalyst, cracking and fractures due to repeated heating and cooling, and poisoning of the catalysts. One of the most active poisons is the lead that comes from the tetramethyl lead, $Pb(CH_3)_4$, or tetraethyl lead, $Pb(C_2H_5)_4$ added to gasoline. Because of the severe catalyst poisoning that results from use of leaded fuels, the 1975 and later model cars have been engineered to discourage the use of leaded gas.

FOR REVIEW

summary

In this chapter we've learned how to express in a quantitative way the relationship between the rate of a chemical reaction and the concentrations of reactants. The **rate constant** for a reaction is a proportionality constant between the reaction rate and the appropriate function of reactant concentrations. That appropriate function must be determined by experiments in which the concentrations of reactants are varied, and the effect on reaction rate noted. The equation which relates the reaction rate to the rate constant times some function of reactant concentrations is called the **rate law.** In a first-order reaction, the reaction rate is proportional to the concentration of a single reactant, raised to the first power. In such a reaction, a graph of $\log [X]_t$, where $[X]_t$ is the concentration of reactant X at time t, versus time yields a straight line of slope $-k/2.3$, where k is the first-order rate constant. For a first-order reaction the **half-life,** $t_{\frac{1}{2}}$ is given by $0.693/k$. Reactions more complex than first-order ones yield similar but more complex expressions for rate constant and half-life.

In coming together to form products, reactants require a certain minimum amount of energy, called the **activation energy.** When the activation energy is appreciable, only a very small fraction of all collisions between reactants provides the energy required to surmount the activation-energy barrier and form the products of the reaction. From the manner in which the rate constant for a reaction varies with temperature, it is possible to determine the magnitude of the activation energy for a reaction.

From a knowledge of the rate law for a reaction, and with other information such as the activation energy, results of isotope labeling experiments, and so on, a detailed picture of how the reaction proceeds may be formulated. Such a description of the reaction is called a **mechanism.**

The activation-energy barrier to formation of products from reactants can be lowered by use of a **catalyst,** a substance that increases the rate of reaction without itself undergoing a net chemical change. Catalysts may be either **homogeneous,** that is, in the same phase with the reactants, or **heterogeneous,** in a separate phase. Heterogeneous catalysts are particularly important in large-scale industrial chemical processes, and in applications such as the catalytic converter of an automobile.

learning goals

Having read and studied this chapter, you should be able to:

1. Express the rate of a given reaction in terms of the variation in concentration of a reactant or product substance with time.
2. Estimate reaction rate at a given time from a knowledge of the variation in concentration of a reactant or product substance with time.
3. Explain the meaning of the term rate constant and state the units associated with rate constants for first- and second-order reactions.
4. Use Equation [13.15] to obtain the value of the rate constant from data for a first-order reaction.
5. Explain the concept of reaction half-life and describe the relationship between half-life and rate constant for a first-order reaction.
6. Explain the concept of activation energy and how it relates to the variation of reaction rate with temperature.
7. Determine the activation energy for a reaction from a knowledge of how the rate constant varies with temperature.
8. Describe the effect of a catalyst on the energy requirements for a reaction.
9. Relate the factors that are important in determining the activity of a heterogeneous catalyst.
10. Explain the functions of a catalytic converter in an automobile exhaust system.

key terms

Among the more important terms and expressions used for the first time in this chapter are the following:

The **activated complex** (Section 13.5) is the particular arrangement of reactant and product molecules at the point of maximum energy in the rate-determining step of a reaction.

Activation energy (Section 13.5), E_a, is the minimum energy that must be supplied by the reactants in a chemical reaction to overcome the barrier to formation of products.

The **Arrhenius equation** (Section 13.5) relates the rate constant for a reaction to the product of a constant (the frequency factor) and an exponential term containing the activation energy: $k = Ae^{-E_a/RT}$.

A **catalyst** (Section 13.7) is a substance that acts to change the speed of a chemical reaction without itself undergoing a permanent change in the process.

A **catalytic converter** (Section 13.7), or catalytic muffler, is a catalyst system built into the exhaust system of an automobile for the purpose of removing CO, NO, and unburned hydrocarbons from the exhaust.

An **enzyme** (Section 13.7) is a protein molecule that acts to catalyze specific biochemical reactions.

A **first-order reaction** (Section 13.3) is one in which the reaction rate is proportional to the concentration of a single reactant, raised to the first power.

The **half-life** (Section 13.3) of a reaction is the time required for the concentration of a reactant substance to decrease to halfway between its initial and final values.

Heterogeneous catalyst (Section 13.7) is a catalyst that is in a different phase from that of the reactant substances.

A **homogeneous catalyst** (Section 13.7) is a catalyst that is in the same phase as the reactant substances.

A **mechanism** (Section 13.6) for a chemical reaction is a detailed picture, or model, of how the reaction occurs; that is, the order in which bonds are broken and formed, and the changes in relative positions of the atoms as reaction proceeds.

The **rate constant** (Section 13.2) is a constant of proportionality between the reaction rate and the concentrations of reactants that appear in the rate law.

The **rate-determining step** (Section 13.6) in a chemical reaction is the slowest step in a reaction that proceeds via a series of steps from reactants to products.

A **rate law** (Section 13.2) is an equation in which the reaction rate is set equal to a mathematical expression involving the concentrations of reactants (and sometimes of products also).

The **reaction order** (Section 13.2) is the sum of the powers to which all the reactants appearing in the rate expression are raised.

Reaction rate (Section 13.1) is defined in terms of the decrease in concentration of a reactant molecule or the increase in concentration of a product molecule with time.

EXERCISES

reaction rates, rate laws and rate constants

13.1 Provide a short definition of each of the following terms: (a) reaction order; (b) reaction rate; (c) rate constant; (d) initial rate; (e) half-life.

13.2 For the reaction of hydrogen with iodine in the gas phase to form HI, Equation [13.25], express the rate of reaction in terms of the concentration of I_2; in terms of HI. In each case, what are the units in which rate is expressed?

13.3 The rate law for the reaction

$$NO + O_3 \longrightarrow NO_2 + O_2$$

is second order, first order each in NO and O_3. Write the complete rate expression. Assuming that all products and reactants are in the gas phase and their concentrations are measured in units of moles/liter, what are the units of the rate constant?

13.4 In the gas phase reaction of chloroform with chlorine, Equation [13.9], with rate law as given, how does the rate of reaction vary with: (a) a doubling of the concentration of $CHCl_3$; (b) a doubling in the concentration of Cl_2? What effect do each of these two changes have on the rate constant?

13.5 The first-order decomposition of N_2O_5, Equation [13.8], was studied at a particular temperature and found to have a rate constant of 6.2×10^{-4}/sec. What is the half-life for the decomposition under these conditions?

[13.6] Determine from the data in Table 13.1 whether the reaction of butyl chloride with water is a first-order reaction. If it is, calculate the first-order rate constant and half-life.

[13.7] Suppose you were assigned the task of determining the rate of decomposition of N_2O_5, Equation [13.8]. Assuming the decomposition is to be studied under conditions of constant pressure, make a sketch of a possible experimental apparatus and indicate the procedure to be used in obtaining data.

13.8 The rearrangement of methyl isonitrile was studied in the gas phase at 215°C, and the following data were obtained:

TIME (sec)	PRESSURE, CH_3NC (torr)
0	502
2000	335
5000	180
8000	95.5
12,000	41.7
15,000	22.4

From these data determine the value for the first order rate constant for rearrangement.

13.9 From the value of first-order rate constant calculated in the preceding question, or by graphical means, determine: (a) the half-life for rearrangement; (b) the initial rate of rearrangement, at time $t = 0$, in units of torr/sec. Also calculate the initial rate of rearrangement in units of M/sec.

13.10 Suppose that the rearrangement of methyl isonitrile were to be studied at 215°, as in question 13.8, but beginning with an initial pressure of 1004 torr. How would this affect the value for the first-order rate constant? How would it affect the half-life for the reaction?

13.11 The first-order rate constant for hydrolysis of a certain insecticide in water at 20°C is 1.45 yr^{-1}. A quantity of this insecticide is washed into a lake in June, leading to an overall concentration of 5×10^{-7} g/cm^3 water. Assuming that the effective temperature of the lake for a period of one year is 20°C, what is the concentration of the insecticide in June of the following year? Five years after the initial entry into the lake?

13.12 Hydrogen sulfide, H_2S, is a common and troublesome pollutant in industrial waste waters. One pathway for removal is oxidation by dissolved oxygen:

$$2H_2S + O_2 \longrightarrow 2S + 2H_2O$$

Assuming that this is a second-order reaction, first order in each reactant, how would the rate of removal of H_2S depend on the concentration of dissolved oxygen? Suppose that the concentration of dissolved H_2S in polluted water is 5×10^{-6} M. The concentration of dissolved oxygen in water in equilibrium with air is about 2.6×10^{-4} M. If the second-order rate constant for the oxidation of H_2S is 4×10^{-5} l./mole-sec, what is the initial rate at which the H_2S is oxidized?

13.13 The reaction of Cl_2 with H_2S in water

$$Cl_2 + H_2S \longrightarrow S + 2HCl$$

is first order in each reactant. The second-order rate constant at 28°C has the value 3.5×10^{-2} l./mole-sec. If at a given time the concentration of H_2S is 5×10^{-5} M, and that of Cl_2 is 0.05 M, what is the rate of formation of HCl?

effects of temperature; activation energy

13.14 The reaction

$$F + H_2 \longrightarrow HF + H$$

has been studied in the gas phase at various temperatures and has been found to have an activation energy of 22 kJ/mole. The overall change in energy in the reaction, ΔE, is -130 kJ/mole. Draw a diagram of the energy profile of the system as a function of reaction coordinate.

13.15 Suppose that two reactions, call them A and B, are studied at 25°C, and reaction B is found to proceed more rapidly. Under the same conditions of concentration, reaction A is found to occur more rapidly than B at 45°C. What can be said about the relative activation energies of the two reactions?

13.16 The first-order rate constant for hydrolysis of a particular organic chloride compound in water varies with temperature as follows:

TEMPERATURE (K)	RATE CONSTANT (sec^{-1})
300	1.0×10^{-5}
320	5.0×10^{-5}
340	2.0×10^{-4}
355	5.0×10^{-4}

From these data calculate the activation energy in units of kJ/mole.

[13.17] The rate at which fireflies flash varies with the temperature of the insect, because a certain rate-controlling reaction proceeds at speeds that are temperature dependent. The period of the firefly flash was found to be 16.3 sec at 21.0°C, and 13.0 sec at 27.8°C. What is the activation energy for the reaction that controls the rate of firefly flash?

reaction mechanisms; catalysis

13.18 The decomposition of hydrogen peroxide, Equation [13.44], is catalyzed by iodide ion. The catalyzed reaction is thought to proceed by a two-step mechanism,

$$H_2O_2 + I^- \longrightarrow H_2O + IO^- \quad \text{(slow)}$$
$$IO^- + H_2O_2 \longrightarrow H_2O + O_2 + I^- \quad \text{(fast)}$$

Make a sketch of the energy profile for the un-catalyzed decomposition of hydrogen peroxide and for the catalyzed reaction that is consistent with the mechanism described above.

13.19 The gas-phase reaction of chlorine with carbon monoxide to form phosgene

$$Cl_2 + CO \longrightarrow COCl_2$$

obeys the rate law

$$\text{Rate} = \frac{\Delta[COCl_2]}{\Delta t} = k[Cl_2]^{3/2}[CO]$$

A mechanism involving the following series of steps is consistent with the rate law:

$$Cl_2 \rightleftharpoons 2Cl$$
$$Cl + CO \rightleftharpoons COCl$$
$$COCl + Cl_2 \rightleftharpoons COCl_2 + Cl$$

Assuming that this mechanism is correct, which of the above steps is the slow, or rate-determining, step? Explain.

13.20 The rate law for the reaction

$$NO + O_3 \longrightarrow NO_2 + O_2$$

is of the form

$$\text{Rate} = k[NO][O_3]$$

Which of the following mechanisms is consistent with this rate law? Explain.

(a)
$$NO \longrightarrow N + O \quad \text{(slow)}$$
$$N + O_3 \longrightarrow NO_2 + O \quad \text{(fast)}$$
$$O + O \longrightarrow O_2 \quad \text{(fast)}$$

(b)
$$NO + O_3 \longrightarrow NO_2 + O_2$$

(c)
$$O_3 \longrightarrow O_2 + O \quad \text{(slow)}$$
$$O + NO \longrightarrow NO_2 \quad \text{(fast)}$$

13.21 The activity of a heterogeneous catalyst is highly dependent on its method of preparation and prior treatment. Explain why this is so.

13.22 Dust explosions occur when a combustible material in finely divided form is ignited. Explain why an explosion results, by considering the hetero-geneous character of the reaction and its exother-micity.

13.23 Activated charcoal, a finely divided and specially prepared form of carbon, is often used in filters for circulating air in public places such as restaurants in order to remove odors. Explain in molecular terms how such a filtering action might occur.

13.24 When D_2 is reacted with ethylene, C_2H_4, in the presence of a finely divided catalyst, ethane with two deuteriums, CH_2D—CH_2D, is formed. (Deuterium, D, is an isotope of hydrogen of mass 2.) There is very little ethane formed in which two deuteriums are bound to one carbon, for example, CH_3—CHD_2. Explain why this is so in terms of the sequence of steps involved in hydrogenation.

13.25 In general, it is found that olefins or acetylenes are more readily adsorbed on heterogeneous catalysts than are saturated organic compounds. For example, ethylene, C_2H_4, and acetylene, C_2H_2, are more readily adsorbed than ethane, C_2H_6. Provide a reasonable explanation for this in terms of the bonding in these compounds (Section 9.3) and the nature of the adsorption process.

general exercises

13.26 (a) The oxidation of ammonia to form nitric oxide and water,

$$4NH_3(g) + 5O_2(g) \longrightarrow 4NO(g) + 6H_2O(g)$$

is a highly exothermic reaction, yet ammonia does not readily oxidize in air at room temperature. Explain this in terms of the energy requirements for the reaction. (b) When a hot platinum wire is inserted into a mixture of ammonia gas and oxygen, it continues to glow. What effect does the platinum have, and why does it continue to glow?

13.27 From your reading of this chapter, make a list of the materials that find use as heterogeneous catalysts in industrial processes, catalytic converters, and so forth. Which elements of the periodic table are most commonly involved? What characteristic electronic structure do these elements have in common? Can you suggest a reason why a particular kind of electronic structure (or the availability of certain kinds of atomic orbitals) might be related to catalyst activity?

13.28 A particular reaction is found to have the following rate law:

$$\text{Rate} = k[\text{A}]^2[\text{B}]$$

Which terms in this rate law are changed by each of the following changes: (a) the concentration of A is doubled; (b) a catalyst is added; (c) the concentration of A is increased by a factor of 2, whereas the concentration of B is decreased by a factor of 4; (d) temperature is increased.

13.29 In the first-order decomposition of N_2O_5, Equation [13.8], the pressure of N_2O_5 as a function of time was found to be as follows:

TIME (sec)	PRESSURE, N_2O_5 (torr)
0	86.1
200	72.8
400	62.4
600	52.7
800	44.7
1000	38.4

From these data determine the value of the first-order rate constant.

13.30 In each of the following rate laws, what is the overall reaction order? What is the order in Cl_2?

(a) $\text{Rate} = k[C_2H_4][Cl_2]^{1/2}$

(b) $\text{Rate} = k[C_6H_6][AlCl_3]/[Cl_2]$

13.31 When natural gas issues from a burner, and no flame is applied, the gas does not burn. When a flame is applied, and the gas is caused to burn, it continues to burn so long as the gas stream remains on. Explain these familiar observations in terms of the enthalpy change in the reaction and the concept of activation energy.

13.32 The energy profile for the reaction

$$O^+ + NO \longrightarrow NO^+ + O$$

is shown in Figure 13.10. Would you expect the rate constant for this reaction to vary a great deal with temperature? Explain.

[13.33] In a second-order reaction with the rate law $k[X][Y]$, the concentration of one of the reactants, say X, could be much larger than for Y. Under these conditions, [X] does not change appreciably during reaction, and is essentially a constant. The rate thus may be written as $k'[Y]$, where k' is called the pseudo–first-order rate constant; $k' = k[X]$. The oxidation of H_2S in water under the conditions described in question 13.13 is a pseudo–first-order reaction, because O_2 is present in excess. What is the value for the pseudo–first-order rate constant? What is the half-life for oxidation of H_2S under these conditions?

13.34 One of the rules of thumb sometimes given to beginning chemistry students is that reaction rates double for each 10°C rise in temperature. Calculate the activation energy that corresponds to a doubling of rate in the interval from 20° to 30°C.

13.35 The reaction of iodide ion with peroxydisulfate ion proceeds in water at 30°C according to the equation

$$S_2O_8{}^{2-} + 2I^- \longrightarrow 2SO_4{}^{2-} + I_2$$

In a series of experiments the initial concentrations of the reagents were varied, and the initial rates of formation of I_2 observed:

INITIAL $S_2O_8{}^{2-}$ CONCENTRATION (M)	INITIAL I^- CONCENTRATION (M)	INITIAL RATE OF FORMATION OF I_2 (M/sec)
1.5×10^{-3}	1.0×10^{-2}	0.75×10^{-6}
3.0×10^{-3}	1.0×10^{-2}	1.50×10^{-6}
1.5×10^{-3}	3.0×10^{-2}	2.25×10^{-6}
3.0×10^{-3}	2.0×10^{-2}	3.00×10^{-6}

From these data determine the rate law for the reaction. Calculate the rate constant for the reaction at 30°C.

13.36 The oxidation of NO by O_3, a reaction of importance in formation of photochemical smog (Section 10.5), is first order in each of the reactants. The second-order rate constant at 28°C is 1.5×10^7 l./mole-sec. If the concentrations of NO and O_3 are each 2×10^{-8} moles/l., what is the rate of oxidation of NO?

14 chemical equilibrium

It often happens that chemical reactions do not proceed to completion. By this we mean that a mixture of reactants is not entirely converted into products. Instead, after a time, reactant concentrations no longer decrease. The reaction system at this point consists of a mixture of reactant and product substances. A chemical system in such a condition is said to be in a state of **chemical equilibrium.** We have already encountered instances of simple equilibrium conditions. For example, in a closed container, the vapor above a liquid achieves an equilibrium with the liquid phase (Section 11.2). The rate at which molecules escape from the liquid to the gas phase equals the rate at which molecules of the gas phase strike the surface and become part of the liquid. As another example, solid sodium chloride may be in equilibrium with the ions dissolved in water (Section 12.2). The rate at which ions leave the solid surface equals the rate at which other ions are removed from the liquid to become part of the solid. We know from these examples that equilibrium is not a static condition in which nothing is happening. Rather, it is dynamic; opposing processes are occurring at equal rates. In this chapter we shall examine the principles on which the idea of equilibrium is based. To demonstrate the importance of equilibrium considerations and to illustrate the most basic concepts involved, we begin with a discussion of the Haber process for synthesizing ammonia.

14.1
the Haber process

Of all the chemical reactions that man has learned to carry out and control for his own purposes, the synthesis of ammonia from hydrogen and atmospheric nitrogen is perhaps the most important. This is especially the case in the present world situation, with food shortages becoming more critical each year. The growth of plant matter requires a substantial store of nitrogen in the soil, in a form usable by plants. The quantity of food required to feed the ever-increasing human population far exceeds that which could be produced if we relied solely on naturally available nitrogen in the soil. Huge quantities of fertilizer rich in nitrogen are required to sustain the high-yield agriculture practiced in the United States today. Great amounts of fertilizer must be used throughout the world if the gap between food supplies and human needs is to be narrowed. The only widely available source of nitrogen is the N_2 present in the atmosphere. The problem thus becomes one of "fixing" atmospheric N_2, that is, converting it to a form that plants can use. This process is called **fixation**.

The N_2 molecule is exceptionally unreactive. Its lack of reactivity is due in large measure to the strong triple bond between the nitrogen atoms (Section 8.3). Because this bond is so strong, there is very little tendency for the molecule to engage in chemical reactions that will result in loss of the $N \equiv N$ triple bond and formation of other bonds to nitrogen. For this reason, the process of fixation is not easy to achieve. In nature the fixation of N_2 is carried out by a special group of nitrogen-fixing bacteria that grow on the roots of certain plants, for example, clover or alfalfa. Here we are interested in only one particular fixation reaction, the Haber process.

Fritz Haber, a German chemist, investigated the energy relations in the reaction between nitrogen and hydrogen and convinced himself that it should be possible to form ammonia in a reasonable yield from these two starting substances. The chemical reaction involved is

$$N_2(g) + 3H_2(g) \rightleftharpoons 2NH_3(g) \qquad [14.1]$$

Haber's research was of great interest to the German chemical industry. Germany was preparing for World War I, and nitrogen compounds figured heavily in the manufacture of explosives. Without a synthetic source of these nitrogen compounds Germany would be greatly handicapped. By 1913, Haber had designed a process that worked, and man began for the first time to produce ammonia on a large scale from atmospheric nitrogen. In the following year, World War I began.

From these unhappy beginnings as a major factor in international warfare the Haber process has become the world's principal source of fixed nitrogen. It is estimated that 16 million tons of ammonia were formed via the Haber process in the United States in 1975.* The ammonia produced in the Haber process can be applied directly to the soil. It may also be converted into ammonium salts, for example, ammonium sulfate, $(NH_4)_2SO_4$, or ammonium hydrogen phosphate, $(NH_4)_2HPO_4$, which are then used as fertilizers.

In designing the process that bears his name, Haber had to face two separate questions. First, is there a catalyst that will allow the reaction to occur at a reasonable speed under practically attainable conditions?

---

*The industrial fixation of nitrogen, chiefly by the Haber process, now accounts for more than 30 percent of all the nitrogen fixed on the planet.

After much long and difficult searching, Haber found a suitable catalyst; we shall return to this aspect of his work in a later section. Second, assuming that a catalyst can be found, to what extent will nitrogen be converted into ammonia? Let us now consider this latter question, which relates to chemical equilibrium.

14.2
the equilibrium constant

The Haber process consists of putting together N_2 and H_2 in a high-pressure tank at a total pressure of several hundred atmospheres, in the presence of a catalyst, and at a temperature of a few hundred degrees centigrade. Under these conditions, the two gases react to form ammonia. But the reaction does not lead to complete consumption of the N_2 and H_2. Rather, at some point the reaction apparently stops, with all three components of the reaction mixture present at the same time. The manner in which the concentration of H_2, N_2, and NH_3 vary with time in this situation is shown in Figure 14.1(a). The condition of the system in which all concentrations have achieved steady values is referred to as chemical equilibrium. The relative amounts of N_2, H_2, and NH_3 present at equilibrium do not depend on the amounts of catalyst present. However, they do depend on the relative amounts of H_2 and N_2 with which the reaction was begun. Furthermore, if only ammonia is placed into the tank under the usual reaction conditions, at equilibrium there is again a mixture of N_2, H_2, and NH_3. The variations in concentrations as a function of time for this situation are shown in Figure 14.1(b). By comparing the two parts of Figure 14.1, we can see that at equilibrium the relative concentrations of H_2, N_2, and NH_3 are the same, regardless of whether the starting mixture was a $3:1$ mixture of H_2 and N_2 or pure NH_3. Equilibrium is thus a condition of the system that can be approached from both directions. This suggests to us that equilibrium is *not* a static condition. On the contrary, both the forward reaction in Equation [14.1], in which ammonia is formed, and the reverse reaction in which H_2 and N_2 are formed from ammonia, are going on at precisely the same rates. Thus the net rate of change in the system is zero. The fact that it is possible to arrive at the same equilibrium condition from two different starting points tells us also that we have a true equilibrium. It is very important to distinguish this from another condition that is often called **metastable equilibrium.** Suppose we place a mixture of N_2, H_2, and

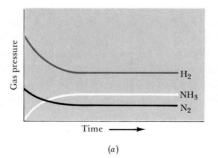
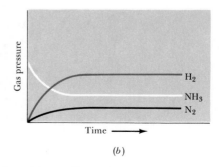

(a) (b)

FIGURE 14.1 Variation in gas pressures in formation of the equilibrium $N_2 + 3H_2 \rightleftharpoons 2NH_3$. ($a$) The equilibrium is approached beginning with H_2 and N_2 in the ratio $3:1$. (b) The equilibrium is approached beginning with NH_3.

NH_3 together in a reaction vessel in which there is no catalyst. After observing that the concentrations of the reactants were not changing with time, we might be tempted to conclude that the definition of equilibrium has been met; the rates of the forward and reverse reactions are equal. But they seem to be equal only because they are practically zero. The true equilibrium condition is reached only by increasing the temperature to speed up reaction or by introducing a catalyst.

If we were systematically to change the relative amounts of N_2, H_2, and NH_3 in the starting mixture of gases, and then analyze the gas mixtures at equilibrium, it would be possible to determine what sort of "law" governs the equilibrium state. Studies of this kind were carried out on other chemical systems by chemists in the nineteenth century, long before Haber's work. In 1864, Cato Maximilian Guldberg and Peter Waage proposed their **law of mass action.** This law expresses the relative concentrations of reactants and products at equilibrium in terms of a quantity called the equilibrium constant. Suppose we have the general reaction

$$j A + k B \rightleftharpoons p R + q S \qquad [14.2]$$

where A, B, R, and S are the chemical species involved, and j, k, p, and q are their coefficients in the balanced chemical equation. According to the law of mass action, the equilibrium condition is expressed by the equation

$$K_c = \frac{[R]^p [S]^q}{[A]^j [B]^k} \qquad [14.3]$$

where K_c is a constant, called the **equilibrium constant,** and the square brackets signify the *concentration* of the species within the brackets. The law of mass action applies only to a system that has attained equilibrium. In general, the equilibrium constant is given by the concentrations of all reaction products multiplied together, each raised to the power of its coefficient in the balanced equation, divided by the concentrations of all reactants multiplied together, each raised to the power of its coefficient in the balanced equation. The equilibrium constant is a true constant. Its value at any given temperature does not depend on the initial concentrations of reactants and products. It also does not matter whether there are other substances present, so long as they do not consume a reactant or product through chemical reaction. The value of the equilibrium constant does, however, vary with temperature.

As an illustration of the law of mass action, consider the gas phase equilibrium between dinitrogen tetroxide and nitrogen dioxide:

$$N_2O_4(g) \rightleftharpoons 2NO_2(g) \qquad [14.4]$$

Because NO_2 is a dark brown gas, and N_2O_4 is colorless, the amount of NO_2 in the mixture can be measured by measuring the intensity of the brown color of the gas mixture.

Following the rule given above, the equilibrium constant for reaction Equation [14.4] is given by the concentration of NO_2 raised to the second power, divided by the concentration of N_2O_4 to the first power:

$$K_c = \frac{[NO_2]^2}{[N_2O_4]} \qquad [14.5]$$

Suppose that to determine the numerical value for K_c, and to verify that it is indeed constant as the concentrations of NO_2 and N_2O_4 change, three samples of NO_2 were placed in sealed glass vessels. In addition a sample of N_2O_4 was placed in a fourth vessel. The vessels were allowed to remain at $100°C$ until no further change in the color of the gas was noted. The mixture of gases was then analyzed to determine the concentrations of both NO_2 and N_2O_4. The results are given in Table 14.1.

To evaluate the equilibrium constant, K_c, the equilibrium concentrations are inserted into the equilibrium-constant expression, Equation [14.5]. For example, using the first set of data,

$$[NO_2] = 0.0172 \, M \quad [N_2O_4] = 0.00140 \, M$$

$$K_c = \frac{[NO_2]^2}{[N_2O_4]} = \frac{(0.0172 \, M)^2}{(0.00140 \, M)} = 0.211 \, M$$

Proceeding in the same way, the values of K_c for the other samples were calculated, as listed in Table 14.1. Note that the value for K_c is essentially constant, even though the initial concentrations vary. Furthermore, the results of experiment 4 show that equilibrium can be attained beginning with N_2O_4 as well as with NO_2. That is, equilibrium can be approached from either direction. Finally, you should note from this example that K_c has units. Concentration appears as a squared term in the numerator, and to the first power in the denominator. The overall units of K are thus molarity. It is important to keep in mind the concentration units in which the equilibrium constant is expressed. In most cases, when the units of concentration are changed, the numerical value for K_c changes as well. Normally units of molarity are used for reactions occurring in solution. For gas phase reactions, the units used are either molarity or atmospheres of pressure.

TABLE 14.1

initial and equilibrium concentrations (molarities) of NO_2 and N_2O_4 in the gas phase at 100°C

EXPERIMENT	INITIAL N_2O_4 CONCENTRATION	INITIAL NO_2 CONCENTRATION	EQUILIBRIUM N_2O_4 CONCENTRATION	EQUILIBRIUM NO_2 CONCENTRATION	K_c (M)
1	0.0	0.0200	0.00140	0.0172	0.211
2	0.0	0.0300	0.00280	0.0243	0.211
3	0.0	0.0400	0.00452	0.0310	0.213
4	0.0400	0.0	0.0134	0.0532	0.211

SAMPLE EXERCISE 14.1.

Applying the law of mass action to the ammonia synthesis system, Equation [14.1], write the expression for the equilibrium constant.

Solution: Ammonia, NH_3, is the sole product of the reaction. It appears in the balanced equation with a coefficient of 2. Thus according to the rule given for writing the equilibrium constant, the concentration of ammonia, raised to the second power, appears in the numerator. The reactants are hydrogen and ni-

trogen. Because the coefficient of hydrogen in the balanced equation is 3, the concentration of hydrogen is raised to the third power; similarly, because the coefficient of N_2 is 1, the concentration of N_2 is raised to the first power. The product of concentration terms for the reactants appears in the denominator. Thus, the equilibrium-constant expression is:

$$K_c = \frac{[NH_3]^2}{[H_2]^3[N_2]} \qquad [14.6]$$

SAMPLE EXERCISE 14.2

Write the equilibrium-constant expression for the reaction $2NH_3 \rightleftharpoons N_2 + 3H_2$.

Solution: This reaction is just the reverse of that shown in Equation [14.1]. The substances that were reactants in that reaction are now products, and vice versa. Again following the rule for writing the equilibrium constant we obtain:

$$K_c = \frac{[N_2][H_2]^3}{[NH_3]^2} \qquad [14.7]$$

Note that this is just the reciprocal of the equilibrium-constant expression for reaction Equation [14.1], as shown in Sample Exercise 14.1. It is a general rule that the equilibrium constant expression for a reaction written in one direction is the reciprocal of the expression for the reverse reaction. This also means that the numerical value of the equilibrium constant for a reaction written in one direction is the reciprocal of the value of the equilibrium constant for the reaction written in the reverse direction.

One of the first tasks confronting Haber and his co-workers when they set to work on the problem of ammonia synthesis was the determination of the numerical value of the equilibrium constant for reaction Equation [14.1]. If K_c in Equation [14.6] is very small, this means that at equilibrium the numerator in Equation [14.6] is small relative to the denominator. That is, the amount of ammonia formed relative to the amounts of N_2 and H_2 present initially would be small. This situation can be described by saying that the equilibrium in reaction Equation [14.1] lies to the left, that is, toward the reactant side. Clearly, if the equilibrium were too far to the left, it would not be possible to develop a satisfactory synthesis for ammonia.

SAMPLE EXERCISE 14.3

In one of their experiments, Haber and co-workers initially introduced a mixture of hydrogen and nitrogen and allowed the system to attain chemical equilibrium at 472°C, with a total pressure of 10 atm. The equilibrium mixture of gases under these conditions consisted of 73.76 percent H_2, 24.58 percent N_2, and 1.66 percent NH_3. From these data, calculate the equilibrium constant of Equation [14.1].

Solution: Because the total pressure in the vessel at equilibrium is known, we can readily determine the partial pressure of each component of the mixture by applying Dalton's law (Section 5.5). Using the percentages given above, the partial pressure of each gas is:

$$P_{H_2} = (0.7376)(10 \text{ atm}) = 7.376 \text{ atm}$$
$$P_{N_2} = (0.2458)(10 \text{ atm}) = 2.458 \text{ atm}$$

$$P_{NH_3} = (0.0166)(10 \text{ atm}) = 0.166 \text{ atm}$$

Inserting each of these into the equilibrium-constant expression, Equation (14.6), we obtain:

$$K_c = \frac{P_{NH_3}{}^2}{P_{H_2}{}^3 \times P_{N_2}}$$

$$K_c = \frac{(0.166 \text{ atm})^2}{(7.38 \text{ atm})^3(2.46 \text{ atm})}$$

$$= 2.78 \times 10^{-5} \text{ atm}^{-2}$$

Notice that in the present example, the concentrations of gases are expressed in units of atmospheres pressure.* If the concentrations were converted to moles/liter, the equilibrium constant would have a different numerical value.

Knowing the value for the equilibrium constant, Haber was able to calculate the relative amount of ammonia formed under various conditions of pressure at any temperature for which he knew the equilibrium constant. The results of some of these calculations are shown in Table 14.2 for the case in which the starting pressures of H_2 and N_2 are in the ratio 3:1. Notice that as the total pressure in the system increases at any one temperature, the percentage conversion to ammonia increases.

Thus, even though a change in pressure at constant temperature does not change the value for the equilibrium constant, the *position* of equilibrium, that is, the relative amounts of reactants and products, is affected by a change in pressure. The shift of the ammonia equilibrium with change in pressure is an example of the operation of LeChatelier's principle.

**14.3
LeChatelier's
principle**

A system at equilibrium is in a dynamic state; the forward and reverse processes are occurring at equal rates, and the system is in a state of balance. An alteration in the conditions of the system may cause the state of balance to be disturbed. If this occurs, the equilibrium is shifted until a new state of balance is attained.

The manner in which a change in conditions affects a system that is in chemical equilibrium can be deduced by applying a principle put forth by Henri LeChatelier in 1884: *If a system at equilibrium is subjected to a change in conditions that disturbs its equilibrium, the system proceeds toward a new equilibrium state in such a direction as to partially offset the change in*

*We can relate pressure to molarity by rearranging the ideal gas equation to the form $P = (n/V)RT$. At a given temperature RT is a constant; thus, pressure is proportional to n/V, that is, moles/liter. The concentration, expressed in moles/liter, n/V, is equal to the pressure, expressed in atmospheres times $(1/RT)$.

TABLE 14.2

effect of temperature and total pressure on the percentage of ammonia present at equilibrium, beginning with a 3:1 H_2/N_2 mixture

TEMPERATURE (°C)	TOTAL PRESSURE (atm)			
	200	300	400	500
400	38.7	47.8	54.9	60.6
450	27.4	35.9	42.9	48.8
500	18.9	26.0	32.2	37.8
600	8.8	12.9	16.9	20.8

conditions. **LeChatelier's principle** allows us to make qualitative predictions about the response of a system at equilibrium to a perturbing change.

The position of equilibrium may be changed by a change in total pressure, by a change in any of the reactant or product concentrations, or by a change in temperature. It is well to remember that of these various perturbing changes, only a change in temperature causes a change in the numerical value of the equilibrium constant. Sample Exercise 14.4 illustrates the effect of a change in total pressure by decrease in the total volume.

SAMPLE EXERCISE 14.4

Apply LeChatelier's principle to determine whether an increase in pressure will increase or decrease the equilibrium amount of ammonia in the reaction:

$$N_2(g) + 3H_2(g) \rightleftharpoons 2NH_3(g) \qquad [14.1]$$

Solution: If the total pressure is increased by reducing the volume of the system, the system responds, according to LeChatelier, by shifting the equilibrium in a direction to reduce the pressure. Because there are only two moles of gas on the right-hand side in Equation [14.1], and four moles of gas on the left, a shift toward ammonia leads to fewer molecules of gas, and thus a lower pressure. An increase in total pressure thus leads to more ammonia, as shown in Table 14.2. It is important to keep in mind that a change in total pressure does not change the value for K_c, so long as temperature is maintained constant. Only the relative amounts of reactants and products may be changed, so as to maintain K_c constant.

LeChatelier's principle also allows us to predict the change that would result from addition of reactant or product to a system at equilibrium. For example, addition of hydrogen to an equilibrium mixture of H_2, N_2, and NH_3 would cause the system to shift in such a way as to reduce the hydrogen pressure toward its original value. This can occur only if the equilibrium shown in Equation [14.1] is shifted in the direction of forming more NH_3. At the same time, the quantity of N_2 would also be reduced slightly. Addition of more N_2 to an equilibrium system would similarly cause a shift in the direction of forming more ammonia. On the other hand, LeChatelier's principle tells us that if we add NH_3 to the system at equilibrium, the shift will be in such a direction as to reduce the NH_3 concentration toward its original value; that is, some of the added ammonia will decompose to form N_2 and H_2.

As another example of LeChatelier's principle, consider the gas phase reaction

$$H_2(g) + CO_2(g) \rightleftharpoons H_2O(g) + CO(g) \qquad [14.8]$$

The position of equilibrium in this reaction is not influenced by a change in total pressure. Because there are two moles of gas on each side of the equation, there is no change in the total pressure as the reaction is shifted in either the forward or reverse direction. On the other hand, if $H_2(g)$ were added to the system at equilibrium, the position of equilibrium would shift so as to partially undo the effect of increased H_2 pressure. That is, it would shift to the right, forming more $H_2O(g)$ and $CO(g)$. In a similar way, addition of $CO(g)$ to the reaction at equilibrium would result in increased quantities of $H_2(g)$ and $CO_2(g)$.

TABLE 14.3

percent dissociation of CO$_2$ into CO and O$_2$ as a function of temperature

TEMPERATURE (K)	PERCENT DISSOCIATION
1500	0.048
2000	2.05
2500	17.6
3000	54.8

**14.4
temperature
dependence of
the equilibrium
constant**

Consider the decomposition of carbon dioxide into carbon monoxide and oxygen:

$$2CO_2(g) \rightleftharpoons 2CO(g) + O_2(g) \qquad [14.9]$$

From a knowledge of the heats of formation of $CO_2(g)$ and $CO(g)$, we can conclude that the forward reaction is highly endothermic; using the values of ΔH_f° from Table 4.1, ΔH° for the overall reaction is calculated to be $+566$ kJ/mole. At room temperature and thereabouts, CO_2 has no observable tendency to dissociate into CO and O$_2$. However, at high temperatures the equilibrium shifts to the right as shown in Table 14.3. Clearly, the equilibrium constant for reaction Equation [14.9] is very dependent on temperature and increases with increasing temperature.

    Almost every equilibrium constant changes in value with change in temperature. The equilibrium constants for exothermic reactions, that is, those in which heat is evolved, decrease with increase in temperature. By contrast, equilibrium constants for endothermic reactions increase with increase in temperature.

    We can deduce the rules for the temperature dependence of the equilibrium constant by applying LeChatelier's principle. When heat is added to a system by increasing the temperature, the equilibrium should shift in such a direction as to partially undo the effect of the added heat. It shifts, therefore, in the direction in which heat is absorbed. If a reaction is exothermic in the forward direction, it must be endothermic in the reverse direction. Thus, when heat is added to an equilibrium system that is exothermic in the forward direction, the equilibrium shifts in the reverse direction, in the direction of reactants. In summary, the rule is that *when heat is added at constant pressure to an equilibrium system, the equilibrium shifts in the direction that absorbs heat.* Conversely, if heat is removed from an equilibrium system, the equilibrium shifts in the direction that evolves heat.

SAMPLE EXERCISE 14.5

The reaction

$$N_2O_4(g) \rightleftharpoons 2NO_2(g)$$

is endothermic; the standard enthalpy change for the reaction in the forward direction is $\Delta H^\circ =$ $+58.0$ kJ/mole. Predict the manner in which the value of the equilibrium constant changes with an increase in temperature.

Solution: The fact that the reaction is endothermic

in the forward direction means that heat is absorbed when N_2O_4 dissociates into two NO_2 molecules. According to LeChatelier's principle, the equilibrium will shift on increase in temperature in such a direction as to absorb heat. In the present example, that means a shift in the forward direction. The equilibrium constant for the reaction will therefore increase with an increase in temperature.

Let us apply these considerations to the Haber process for synthesis of ammonia, Equation [14.1]. It was important for Haber to know how the equilibrium constant for this reaction varies with temperature. He determined K_c over a range of temperatures, to obtain the values listed in Table 14.4. Notice that K_c changes very markedly with change in temperature, and that it is larger at lower temperatures. This is a matter of great practical importance. To form ammonia at a reasonable rate, higher temperatures are required. Yet at higher temperatures, the equilibrium constant is smaller, so the percentage conversion to ammonia is smaller. To compensate for this, higher pressures are needed, because high pressure favors ammonia formation.

SAMPLE EXERCISE 14.6

Using the standard heat of formation data in Table 4.1, determine the enthalpy change for the reaction

$$N_2(g) + 3H_2(g) \rightleftharpoons 2NH_3(g)$$

From this determine how the equilibrium constant for the reaction, Equation [14.6], should change with temperature.

Solution: Recall that the standard enthalpy change for a reaction is given by the standard molar enthalpies of the reactants, each multiplied by its coefficient in the balanced chemical equation, less the same quantities for the reactants. ΔH_f° for $NH_3(g)$ at 25°C is -46.19 kJ/mole. The ΔH_f° values for $H_2(g)$ and $N_2(g)$ are zero by definition, because the enthalpies of formation of the elements in their normal states at 25°C are defined as zero (Section 4.6). Because 2 moles of NH_3 are formed, the kJ/mole total enthalpy change is thus

$$2 \text{ moles}(-46.19 \text{ kJ/mole}) - 0 = -92.38 \text{ kJ}$$

The reaction in the forward direction is exothermic. An increase in temperature should therefore cause the reaction to shift in the *reverse* direction, that is, in the direction of less NH_3, and more N_2 and H_2. This is what occurs, as reflected in the values for K_c presented in Table 14.4.

TABLE 14.4

variation in K_c for the equilibrium $N_2 + 3H_2 \rightleftharpoons 2NH_3$ as a function of temperature

TEMPERATURE (°C)	K_c (atm^{-2})
300	4.34×10^{-3}
400	1.64×10^{-4}
450	4.51×10^{-5}
500	1.45×10^{-5}
550	5.38×10^{-6}
600	2.25×10^{-6}

14.5

the relationship between chemical equilibrium and chemical kinetics

It is very important to be clear on the distinction between a system at equilibrium and the rate at which the system approaches equilibrium in the first place. Imagine that we have a system initially containing only reactant molecules, and no products. As the system begins to change, the only reaction occurring is formation of products. As products accumulate, however, the reverse reaction also begins to occur. In many systems, this reverse reaction is so slow that it is completely unimportant. The system then just keeps on changing until essentially all the reactants are converted to products. Reactions of this sort are said to proceed to completion. Even though the reverse reaction, conversion of products to reactants, is possible in principle, it is not observed.

In many other reactions, on the other hand, the reverse reaction occurs more rapidly. After a time, reactant molecules are being formed just as rapidly as they are themselves reacting to form products. When the rates of the two opposing processes are equal, the overall rate of change in the system is zero, and we have chemical equilibrium. A true chemical equilibrium always involves a balancing of equal and opposite rate processes. This is true regardless of the particular mechanism or pathway by which the reaction proceeds. Figure 14.2 shows an energy profile for the reaction between reactants A and B and products C and D:

$$A + B \rightleftharpoons C + D \qquad\qquad [14.10]$$

At equilibrium, the rate at which reactants cross the energy barrier to form products equals the rate at which products cross the energy barrier in the reverse direction to form reactants. To make our argument simple, let's suppose that the rate law for the reaction in each direction is simply second order, so that

$$r_f = \text{rate of forward reaction} = k_f[A][B]$$
$$r_r = \text{rate of reverse reaction} = k_r[C][D]$$

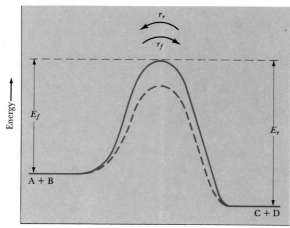

Progress of reaction

FIGURE 14.2 Schematic illustration of chemical equilibrium in the reaction A + B $\rightleftharpoons$ C + D. When equilibrium is attained, the rate of the forward reaction, r_f, equals the rate of the reverse reaction, r_r. The dashed line refers to the energy profile for a catalyzed reaction in which the activation energy is lowered. The rates of forward and reverse reactions in the catalyzed reaction are increased to the same degree.

At equilibrium these two rate processes must be equal:

$$k_f[\text{A}][\text{B}] = k_r[\text{C}][\text{D}]$$

If we rearrange this equation we obtain

$$\frac{k_f}{k_r} = \frac{[\text{C}][\text{D}]}{[\text{A}][\text{B}]} = K_c \qquad [14.11]$$

Thus the equilibrium constant is just the ratio of rate constants for the forward and reverse reactions. Recall that the rate of a reaction decreases as the height of the energy barrier increases (Section 13.5). The barrier for the forward reaction (E_f) in Figure 14.2 is lower than for the reverse reaction (E_r). Thus, k_f must be larger than k_r, so that K_c should be a large number. This is in keeping with the fact that the energies of the products are lower than the energies of the reactants.

THE EFFECTS OF CATALYSTS

Suppose that we add a catalyst to the reaction system described in Equation [14.10], so that the energy barrier to reaction is lowered, as shown by the dashed line in Figure 14.2. The rates of *both* the forward and reverse reactions are increased by the presence of a catalyst. In fact, the two rate constants are affected to precisely the same degree. In other words, it is impossible for a catalyst to lower the barrier for just the forward reaction and not the reverse reaction. Because the forward and reverse reaction rate constants are affected to exactly the same degree, their ratio is unchanged. We therefore have the rule that *a catalyst may change the rate of approach to equilibrium, but it does not change the value of the equilibrium constant.*

The rate at which a reaction approaches equilibrium is an important practical consideration. As an example, let us again consider the synthesis of ammonia from N_2 and H_2. In designing a process for ammonia synthesis, Haber had to deal with a rather serious problem. He wished to synthesize ammonia at the lowest temperature possible, consistent with a reasonable reaction rate. But in the absence of a catalyst, hydrogen and nitrogen do not react with one another at a significant rate either at room temperature or even at much higher temperatures. On the other hand, Haber had to cope with a rapid decrease in equilibrium constant with increasing temperature as shown in Table 14.4. At temperatures sufficiently high to give a satisfactory reaction rate, the amount of ammonia formed was too small. The solution to this dilemma was to develop a catalyst that would produce a reasonably rapid approach to equilibrium, at a sufficiently low temperature so that the equilibrium constant was still reasonably large. The development of a catalyst thus became the focus of Haber's research efforts.

After trying different substances to see which would be most effective, Haber finally settled on iron mixed with metal oxides. Variants of the original catalyst formulations are still employed. With these catalysts it is possible to obtain a reasonably rapid approach to equilibrium at

14.5

the relationship
between chemical equilibrium
and chemical kinetics

chemical equilibrium

temperatures around 400–500°C, and with gas pressures of 200–600 atm. The high pressures are needed to obtain a satisfactory degree of conversion at equilibrium. You can see from Table 14.2 that if an improved catalyst could be found, one that would lead to sufficiently rapid reaction at temperatures lower than 400–500°C, it would be possible to obtain the same degree of equilibrium conversion at much lower pressures. This would result in a great savings in the cost of equipment for ammonia synthesis. In view of the growing need for nitrogen as fertilizer, the fixation of nitrogen is a process of ever-increasing importance, worthy of additional research effort.

The formation of NO from N_2 and O_2 provides another interesting example of the practical importance of changes in equilibrium constant and reaction rate with temperature. The balanced equation is

$$N_2 + O_2 \rightleftharpoons 2NO \qquad [14.12]$$

The standard heat of formation of NO at 25°C is 90.4 kJ/mole. The standard heats of formation of $N_2(g)$ and $O_2(g)$ are, of course, zero, because these elements are in their normal states as gases at 25°C. Thus, the overall heat of reaction is

$$2 \text{ moles}(90.4 \text{ kJ/mole}) - 0 = 180.8 \text{ kJ}$$

The reaction is endothermic, that is, heat is absorbed when NO is formed from the elements. By applying LeChatelier's principle, we deduce that an increase in temperature will shift the equilibrium in the direction of more NO. The equilibrium constant for formation of NO from the elements at 300 K is only about 10^{-15}. On the other hand, at a much higher temperature, about 2400 K, the equilibrium constant is much larger, about 0.05. The manner in which K_c for reaction Equation [14.12] varies with temperature is shown in Figure 14.3.*

This graph helps to explain why NO is a pollution problem. In the cylinder of a modern high-compression auto engine, the temperatures

*It is necessary to use a log scale for K_c in Figure 14.3 because the values of K_c vary over such a huge range.

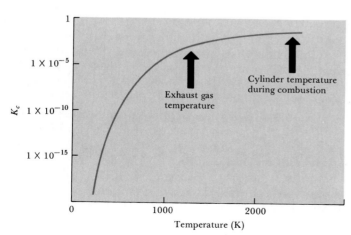

FIGURE 14.3 Variation of the equilibrium constant for the reaction $N_2 + O_2 \rightleftharpoons 2NO$ as a function of temperature.

during the fuel-burning part of the cycle may be on the order of 2400 K. Because there is a fairly large excess of air in the cylinder, these conditions provide an opportunity for the formation of some NO. After the combustion, however, the gases are quickly cooled. As the temperature drops the equilibrium of Equation [14.12] shifts strongly to the left, that is, in the direction of N_2 and O_2. But the lower temperatures also mean that the rate of the reaction also is decreased. The NO formed at high temperatures is essentially "frozen" in that form as the gas cools.

The gases exhausting from the cylinder are still quite hot, perhaps 1200 K. At this temperature, as shown in Figure 14.3, the equilibrium constant for formation of NO is much smaller. However, the rate of conversion of NO to N_2 and O_2 is too slow to permit much loss of NO before the gases are cooled still further. Getting the NO out of the exhaust gases, described in Chapter 10, depends on finding a catalyst that will work at the temperatures of the exhaust gases, and that will cause conversion of NO into something harmless. If a catalyst could be found which would convert the NO back into N_2 and O_2 the equilibrium would be sufficiently favorable. It has not proved possible to find a catalyst capable of withstanding the grueling conditions found in automobile exhaust systems that can catalyze the conversion of NO into N_2 and O_2. Instead, the catalysts used are designed to catalyze reaction of NO with H_2 or CO, as described in Section 13.7.

14.6 heterogeneous equilibria Many equilibria of importance, such as in the hydrogen-nitrogen-ammonia system, involve substances all in the same phase. Such equilibria are said to be **homogeneous**. On the other hand, it is possible for the substances in equilibrium to be in different phases, giving rise to **heterogeneous equilibria**. As an example, consider the decomposition of calcium carbonate:

$$CaCO_3(s) \rightarrow CaO(s) + CO_2(g) \qquad [14.13]$$

This system involves a gas in equilibrium with two solids. If we write the equilibrium constant expression for this process in the usual way we obtain:

$$K_c = \frac{[CaO][CO_2]}{[CaCO_3]} \qquad [14.14]$$

This example presents us with a problem we have not encountered previously: How do we express the concentration of a solid substance? Because the calcium carbonate or calcium oxide are present as pure solids, their "concentration" is a constant. The number of moles per liter of either of these solids is not changed, whether we have a large amount of solid present or just a little. The concentration of a pure substance, liquid or solid, can be expressed as the density divided by the molecular weight:

$$\frac{\text{Density}}{\text{MW}} = \frac{\text{g/cm}^3}{\text{g/mole}} = \frac{\text{moles}}{\text{cm}^3}$$

419

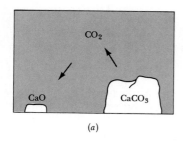

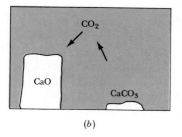

FIGURE 14.4 Heterogeneous equilibrium in $CaCO_3$ decomposition. Assuming the temperature in each case to be the same, the equilibrium pressure of CO_2 is the same in (a) as in (b).

The density of a pure liquid or solid is a constant at any given temperature, and in fact changes very little with temperature. Thus, the effective concentration of a pure liquid or solid is a constant. The equilibrium-constant expression for Equation [14.13] then simplifies to

$$K_c = \frac{[CO_2](\text{constant } 1)}{(\text{constant } 2)}$$

where constant 1 is the concentration of CaO, and constant 2 is the concentration of $CaCO_3$. Moving the constants to the left-hand side of the equation, we have

$$K'_c = K_c \frac{(\text{constant } 2)}{(\text{constant } 1)} = P_{CO_2} \qquad [14.15]$$

As a practical matter, the overall effect is the same as if we set the concentrations of both solids equal to one in the equilibrium-constant expression.

Equation [14.15] tells us that it doesn't matter at all how much $CaCO_3$ or CaO are present, so long as there is some of each in the system. As shown in Figure 14.4, we would have the same pressure of CO_2 in the system when we have a large excess of CaO as when we have an excess of $CaCO_3$. On the other hand, if one of the three ingredients is missing, we cannot have an equilibrium.

SAMPLE EXERCISE 14.7

Each of the mixtures listed below was placed into a closed container and allowed to stand. Which of these mixtures are capable of attaining the equilibrium expressed by Equation [14.13]? (a) pure $CaCO_3$; (b) CaO and a pressure of CO_2 greater than the value of K_c; (c) some $CaCO_3$ and a pressure of CO_2 greater than the value of K_c; (d) $CaCO_3$ and CaO

Solution: Equilibrium can be reached in all cases except (c). In (a), $CaCO_3$ simply decomposes until the equilibrium pressure of CO_2 is attained. In (b), CO_2 combines with the CaO present until its pressure decreases to the equilibrium value. In (c), equilibrium can't be attained, because there is no way in which the CO_2 pressure can decrease so as to attain its equilibrium value. In (d), the situation is essentially the same as in (a); $CaCO_3$ decomposes until equilibrium is attained. The presence of CaO initially makes no difference.

The physical equilibria described in Chapters 11 and 12, in our discussion of solids, liquids, and solutions, furnish examples of heterogeneous equilibria. For example, the equilibrium between a liquid substance A and its vapor, depicted in Figure 11.4, is described by the following equilibrium constant:

$$K_c = \frac{[A]_g}{[A]_l} \qquad [14.16]$$

In this example, the concentration of the vapor can be expressed in terms of pressure, as before. The concentration of a liquid, however, is constant;

there are a certain number of moles per unit volume in the liquid, and that number is constant. Thus the equilibrium-constant expression for the equilibrium between a pure liquid and its vapor can be written as

$$K = K_c[A]_l = [A]_g = P_A \qquad [14.17]$$

The numerical value of K at any temperature is equal to the vapor pressure of the liquid at that temperature. A similar equilibrium-constant expression would apply to the equilibrium between a solid and its vapor.

In all the cases of heterogeneous equilibria we have discussed so far, the value for the equilibrium constant changes with temperature in the manner predicted by LeChatelier's principle. The decomposition of calcium carbonate is an endothermic process; that is, heat is absorbed in forming CaO and CO_2 from $CaCO_3$. Thus, addition of further heat causes the equilibrium to shift in the direction of more CaO and CO_2.

SAMPLE EXERCISE 14.8

Using LeChatelier's principle, predict how the vapor pressure of liquid carbon tetrachloride, CCl_4, will change with temperature. The equilibrium of interest is

$$CCl_4(l) \rightleftharpoons CCl_4(g)$$

and the corresponding equilibrium constant is

$$K_c = P_{CCl_4}$$

Solution: Vaporization of a liquid is an endothermic process; the heat required is the heat of vaporization. By LeChatelier's principle, the equilibrium should shift in the direction of higher vapor pressure with increase in temperature, because this is the direction in which heat is absorbed. This, of course, is what does occur, as shown by vapor pressure versus temperature curves such as those illustrated in Figure 11.6.

When we consider the equilibria between two solid phases, or between a liquid and solid phase, the situation is not so obvious. For example, let's consider the equilibrium between solid and liquid water:

$$H_2O(s) \rightleftharpoons H_2O(l) \qquad [14.18]$$

We know that below 0°C this equilibrium lies to the left, and that above 0° it lies to the right. The interesting and significant thing is that it lies *completely* to the right or left. If you set a tray of ice cubes in the freezer at $-10°C$ you find that at equilibrium there is no liquid water present; *all* the water is in the form of ice. Similarly, if you move that tray to the lower part of the refrigerator, at 5°C, you find that at equilibrium *all* the water is in the form of liquid. For a given pressure, there is only one temperature at which liquid and solid water can exist in equilibrium. At 1 atm pressure, that temperature is 0°C. A mixture of water and ice remains in equilibrium indefinitely at 0°C, as long as there are no additions or removals of heat. But at any other temperature, even just a little different, the equilibrium shifts completely to one side or the other.

We can see why this happens by writing the equilibrium constant for Equation [14.18]:

$$K_c = \frac{[H_2O]_l}{[H_2O]_s} \qquad [14.19]$$

We have already indicated that the concentrations of pure liquid or solid substances are constants. Thus, as long as both water and ice are present together, regardless of the relative amounts, the ratio $[H_2O]_l/[H_2O]_s$ is a constant. This situation corresponds to equilibrium only if K_c equals $[H_2O]_l/[H_2O]_s$. But there is only one temperature at which this condition is met, and that is $0°C$. Because the melting of ice is an endothermic process, K_c is larger than $[H_2O]_l/[H_2O]_s$ above $0°C$, and smaller below it. We can see, then, that the equilibrium condition can't be satisfied if both liquid water and ice are present, except at exactly $0°C$. At any temperature above $0°$, the equilibrium is driven to complete melting, and below $0°$ to complete freezing.

ADSORPTION

We saw in Chapter 13 that many chemical reactions are catalyzed by the presence of a solid surface. For example, Michael Faraday discovered in 1833 that the reaction

$$2H_2(g) + O_2(g) \rightleftharpoons 2H_2O(g)$$

is catalyzed by a platinum metal surface. For such catalysis to occur, the H_2 and O_2 molecules must be adsorbed on the metal surface. An equilibrium is established between H_2 and O_2 molecules on the surface and those in the gas phase. Examples of such equilibria are common in heterogeneous catalysis. Figure 14.5 shows a schematic diagram of a system in which a gaseous substance is adsorbed on a solid surface. We can describe the equilibrium involved as

$$A(g) \rightleftharpoons A(ad) \qquad [14.20]$$

where $A(ad)$ refers to the substances adsorbed on the solid surface. The equilibrium constant expression is given by the expression

$$K_c = \frac{[A]_{ad}}{[A]_g} \qquad [14.21]$$

To use this equilibrium-constant expression, we must have a means for expressing the concentration of adsorbed molecules. One of the problems

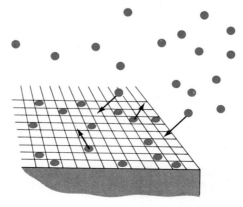

FIGURE 14.5 Schematic illustration of the equilibrium between the gas phase and adsorbed state. Each little square on the surface represents a possible adsorption site. Equilibrium involves a dynamic process, with molecules leaving and becoming adsorbed at equal rates.

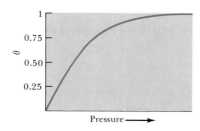

FIGURE 14.6 Variation in the fractional surface coverage, θ, with pressure of the gas being adsorbed. At high pressures, the fractional coverage approaches 1, because all sites on the surface are occupied.

in dealing with adsorption is that the concentration of adsorbed species can't increase indefinitely. There are only so many sites on the solid surface at which adsorption may occur. When these sites are all occupied, or nearly so, an increase in the pressure of the gas above the surface will not increase the concentration of adsorbed species.

The concentration of adsorbed substances can be expressed in terms of the **fractional surface coverage, θ,** the fraction of all the possible adsorption sites on the surface that are occupied. When θ is small, it increases in direct proportion to the pressure. For example, θ might be 0.03 for a gas pressure of 2 atm; we would expect then that it would be 0.06 for a pressure of 4 atm. But when θ becomes rather large, it no longer increases in direct proportion to pressure. In fact, at high pressures, θ approaches a limiting value of 1. The manner in which θ varies with pressure is shown in Figure 14.6. A curve of this form is described by the equation

$$\theta = \frac{bP}{1 + bP} \qquad [14.22]$$

The quantity b is called the **adsorption coefficient;** P represents the gas pressure. Notice that when the quantity bP is much less than 1, θ is approximately equal to bP. This represents the condition of low fractional coverage, in which the fraction of surface covered is directly proportional to pressure. Adsorption of gases is extremely important in the field of heterogeneous catalysis. The ability of a catalyst to function is often limited by the number of sites available for reaction to occur. Every effort is made, therefore, to produce catalysts that have the maximum possible surface area.

SAMPLE EXERCISE 14.9

An automobile catalytic muffler system is found to have a 0.74 fractional surface coverage of NO when the NO pressure over the catalyst is 1 torr. What will be the fractional coverage of the surface under driving conditions, when the NO pressure is typically 0.03 torr, assuming no other species compete for the surface sites?

Solution: We must first use Equation [14.22] to find the value for b, given that $P = 1$ torr and $\theta = 0.74$:

$$\theta = \frac{b(1 \text{ torr})}{1 + b(1 \text{ torr})} = 0.74$$

Solving for b we obtain 2.85 torr^{-1}. Using this value in Equation [14.22] with $P = 0.03$ torr^{-1}

$$\theta = \frac{2.85 \text{ torr}^{-1}(0.03 \text{ torr})}{1 + (2.85 \text{ torr}^{-1})(0.03 \text{ torr})}$$

$$= 0.079$$

In practice, many substances, particularly O_2, CO, and NO, all compete for sites on the catalyst surface.

summary

We've seen that it is possible to describe any chemical system represented by a balanced chemical equation in terms of an equilibrium constant, K_c, by applying the **law of mass action**. The numerical value of the equilibrium constant is not affected by changes in relative concentrations of any of the reacting substances, by the presence of catalysts, or by pressure changes. Only temperature changes affect the value of K_c. The application of **LeChatelier's principle** enables us to predict how changes made in an equilibrium system will shift the relative amounts of reactants and products. The rules for writing the equilibrium constants for heterogeneous equilibria have been described, and heterogeneous equilibria involving adsorption have been discussed.

learning goals

Having read and studied this chapter, you should be able to:

1. Write the equilibrium-constant expression for a balanced chemical reaction, whether heterogeneous or homogeneous.
2. Numerically evaluate K_c from a knowledge of the equilibrium concentrations of reactants and products.
3. Explain how the relative equilibrium quantities of reactants and products are shifted by changes in temperature, pressure, or the concentrations of any substances involved in the equilibrium.
4. Explain how the change in equilibrium constant with change in temperature is related to the heat change in the reaction.
5. Describe the effect of a catalyst on the value for the equilibrium constant.
6. Explain how to express the concentrations of pure substances in evaluating equilibrium-constant expressions.
7. Deduce when the appropriate conditions have been achieved for attainment of equilibrium in heterogeneous systems.
8. Explain the concepts involved in adsorption of substances to solid surfaces, in particular the notion of fractional surface coverage.

key terms

Among the more important terms and expressions used for the first time in this chapter are the following:

Chemical equilibrium (Section 14.2) is a state of a chemical system in which the rate of formation of products equals the rate of formation of reactants from products.

Fractional surface coverage, θ (Section 14.6), is the fraction of all sites on a solid surface that are occupied by adsorbed molecules.

The **Haber process** (Section 14.1) refers to the catalyst system and conditions of temperature and pressure developed by Fritz Haber and co-workers for the formation of NH_3 from H_2 and N_2.

Heterogeneous equilibrium (Section 14.6) refers to the equilibrium state established between substances in two or more different phases, for example, between a gas and solid, or between a solid and liquid.

Homogeneous equilibrium (Section 14.6) is the state of equilibrium established between reactant and product substances all in the same phase, for example, all gases, or all dissolved in solution.

The **law of mass action** (Section 14.2) provides the rules according to which the equilibrium constant is expressed in terms of the concentrations of reactants and products, in accordance with the balanced chemical equation for the reaction.

LeChatelier's principle (Section 14.3) tells us that when we bring to bear some disturbing influence on a system at chemical equilibrium, the relative concentrations of reactants and products will shift so as to partially undo the effects of the disturbance.

EXERCISES

the equilibrium constant

14.1 Write the expression for the equilibrium constant for each of the following reactions:

(a) $Cl_2(g) + PCl_3(g) \rightleftharpoons PCl_5(g)$

(b) $2SO_2(g) + O_2(g) \rightleftharpoons 2SO_3(g)$

(c) $P_4(g) + 5O_2(g) \rightleftharpoons P_4O_{10}(g)$

(d) $Br_2(g) + C_2H_4(g) \rightleftharpoons C_2H_4Br_2(g)$

14.2 Write the expressions for the equilibrium constant for each of the following reactions. Assuming that the concentrations of all species are expressed in units of moles/liter, what are the units of K_c in each case?

(a) $2H_2S(g) + 3O_2(g) \rightleftharpoons 2H_2O(g) + 2SO_2(g)$

(b) $I_2(g) + Cl_2(g) \rightleftharpoons 2ICl(g)$

(c) $2NO_2(g) \rightleftharpoons 2NO(g) + O_2(g)$

14.3 The compound 2-butene

$$CH_3—CH=CH—CH_3$$

is capable of existing in two forms, called the *cis-* and *trans-*, which have geometrical structures as shown:

cis-Butene   *trans*-Butene

Interconversion between these forms is ordinarily slow at room temperature, but may be catalyzed. The equilibrium constant for the above interconversion is 3.4. Suppose that $t_{1/2}$ for the first-order conversion of *cis-* $\rightarrow$ *trans-* is 0.60 hr, and that we begin with pure *cis-* compound at a pressure of 20 torr. Draw a diagram similar to Figure 14.1, showing with reasonable accuracy how the pressures of *cis-* and *trans*-butenes change with time.

14.4 Potassium chloride, KCl, is dissolved in hot water to form a saturated solution. Some of the solution is slowly cooled to room temperature; no precipitate forms. A tiny crystal of KCl is added, and very quickly a large quantity of solid KCl forms. At which points during this entire process does a true equilibrium exist? When is the system in a *metastable* state?

14.5 Suppose that the system

$$PCl_5(g) \rightleftharpoons PCl_3(g) + Cl_2(g)$$

is at equilibrium. The concentrations of substances at 250°C are

$PCl_5(g) = 0.158$ moles/l.

$PCl_3(g) = 0.081$ moles/l.

$Cl_2(g) = 0.081$ moles/l.

Calculate the equilibrium constant at 250°C. What units does it have?

14.6 The equilibrium constant for the reaction

$$I_2(g) + Br_2(g) \rightleftharpoons 2IBr(g)$$

has the value 280 at 150°C. Suppose that a quantity of IBr is placed in a closed reaction vessel and the system allowed to come to equilibrium. When equilibrium is attained, the pressure of IBr is 0.20 atm. What are the pressures of $I_2(g)$ and $Br_2(g)$ at this point?

14.7 As shown in Table 14.4, K_c for the equilibrium

$$N_2(g) + 3H_2(g) \rightleftharpoons 2NH_3(g)$$

is 4.51×10^{-5} atm^{-2} at 450°C, when concentrations are expressed in atmospheres pressure. Each of the following mixtures may or may not be at equilibrium at 450°C. Indicate in each case the direction (toward product or toward reactant?) in which the equilibrium must shift to achieve equilibrium: (a) 150 atm NH_3, 20 atm N_2, 500 atm H_2; (b) 30 atm NH_3, 600 atm H_2, no N_2; (c) 32 atm NH_3, 69 atm H_2, 69 atm N_2; (d) 100 atm NH_3, 1 atm H_2, 5 atm N_2.

14.8 From the data given in the preceding question, what is the value of K_c for the reaction

$$2NH_3(g) \rightleftharpoons 3H_2(g) + N_2(g)$$

at 450°C?

LeChatelier's principle

14.9 Write the expression for the equilibrium constant for the reaction

$$I_2(g) + Br_2(g) \rightleftharpoons 2IBr(g)$$

425

What are the units of this equilibrium constant? If the reaction is at equilibrium, what would be the effect of (a) adding additional IBr; (b) increasing the total pressure on the system by decreasing the volume; (c) adding additional I_2.

14.10 Suppose that some $PCl_5(g)$ could be removed from the equilibrium system described in question 14.5. What effect would this have on the equilibrium constant? On the position of equilibrium?

14.11 What would be the effect on the position of equilibrium of the reaction

$$PCl_5(g) \rightleftharpoons PCl_3(g) + Cl_2(g)$$

of (a) adding $Cl_2(g)$; (b) adding $PCl_3(g)$; (c) decreasing pressure by increasing the volume of the system; (d) increasing temperature (the reaction is endothermic in the forward direction).

[**14.12**] As we all know, ice is less dense at $0°C$ than liquid water. Apply LeChatelier's principle to determine how the melting temperature of ice will vary with an increase in pressure.

14.13 Write the equilibrium-constant expression for the reaction system

$$2NOBr(g) \rightleftharpoons 2NO(g) + Br_2(g)$$

How would each of the following changes affect the position of equilibrium: (a) decrease in total pressure by increase in volume; (b) increase in $NO(g)$; (c) increase in $Br_2(g)$; (d) increase in total pressure by addition of argon?

14.14 Henry's law (Section 12.5) states that the solubility of a gas in a liquid is directly proportional to the pressure of that gas over the liquid. Write the equilibrium involved in the form of an equilibrium-constant expression. Discuss Henry's law as an application of LeChatelier's principle.

**temperature dependence
of the equilibrium constant**

[**14.15**] Write the equilibrium-constant expression for the equilibrium

$$C(s) + CO_2(g) \rightleftharpoons 2CO(g)$$

The table below shows the relative percentages of $CO_2(g)$ and $CO(g)$ at a total pressure of 1 atm for several temperatures. Calculate the value of K_c at each temperature. Is the reaction exothermic or endothermic? Explain?

TEMPERATURE (°C)	PERCENT CO_2	PERCENT CO
850	6.23	93.77
950	1.32	98.68
1050	0.37	99.63
1200	0.06	99.94

14.16 The solubility of $CuSO_4$ in water, in grams $CuSO_4$ per 100 g H_2O, varies with temperature as follows:

TEMPERATURE (°C)	SOLUBILITY
15	19.3
25	22.3
30	25.5
50	33.6
60	39.0
80	53.5

Is the dissolving of $CuSO_4$ an exothermic or endothermic process? Explain.

14.17 How would you expect the equilibrium constant for each of the following reactions to vary with temperature? Explain the reasoning behind your answer in each case.

(a) $2H_2O(g) + CaO(s) + SO_2(g) \rightleftharpoons$
$\qquad CaSO_3 \cdot 2H_2O(s); \Delta H° = -79.2 \text{ kJ}$
(b) $H_2O(g) \rightleftharpoons H_2(g) + \frac{1}{2}O_2(g); \Delta H° = +242 \text{ kJ}$
(c) $H_2(g) + I_2(g) \rightleftharpoons 2HI(g); \Delta H° = -10.3 \text{ kJ}$

chemical equilibrium and chemical kinetics

14.18 For the reaction

$$2SO_{2(g)} + O_{2(g)} \longrightarrow 2SO_{3(g)}$$

the standard enthalpy change is

$$\Delta H° = -196.6 \text{ kJ}$$

The activation energy for the uncatalyzed reaction is about 160 kJ/mole. Sketch a rough drawing to scale of the energy profile for this reaction, as in Figure 14.2. Sulfur dioxide is produced in the cylinder of an auto engine by oxidation of the small quantity of sulfur present in gas. The catalytic mufflers installed in cars since 1975 result in conversion of a large portion of this SO_2 to SO_3 (Section 13.7). Using a dotted line, sketch on your figure an energy profile which might apply for SO_2 oxidation in a catalytic muffler.

14.19 In the reaction

$$NO(g) + O_3(g) \rightleftharpoons NO_2(g) + O_2(g)$$

The rate law for the forward reaction is

$$r_f = k_f[NO][O_3]$$

and for the reverse reaction

$$r_r = k_r[NO_2][O_2]$$

Using these expressions, write the equilibrium condition in terms of opposing rates, and show how K_c relates to k_f and k_r.

[**14.20**] The equilibrium constant for the reaction

$$2H_2O(g) \rightleftharpoons O_2(g) + 2H_2(g)$$

is much smaller than for the reaction

$$2NH_3(g) \rightleftharpoons N_2(g) + 3H_2(g)$$

Discuss the major reason for this in terms of bond energies.

effects of catalysts on equilibrium

14.21 Suppose you worked at the U.S. Patent Office and a patent application came across your desk in which it was claimed that a newly developed catalyst was much superior to the Haber catalyst for ammonia synthesis, because the catalyst led to much greater equilibrium conversion of N_2 and H_2 into NH_3 than the Haber catalyst under the same conditions. What would be your response?

14.22 At 1200 K, the approximate temperature of the automobile exhaust gases (Figure 14.3), the equilibrium constant for the reaction

$$2CO_2(g) \rightleftharpoons 2CO(g) + O_2(g)$$

is about 1×10^{-13} atm. Assuming that the exhaust gas (total pressure 1 atm) contains 0.2 percent CO by volume, 12 percent CO_2, and 3 percent O_2, is the system at equilibrium with respect to the above reaction? Based on your conclusion, would the CO concentration in the exhaust be lowered or increased by a catalyst that speeded up the above reaction?

14.23 Which if any of the following statements is false? For those which are, discuss the sense in which they are incorrect. (a) If a catalyst increases the rate of a forward reaction by a factor of 1000 over the uncatalyzed rate, it increases the rate of the reverse reaction by a factor of 1000 also. (b) A catalyst can promote the formation of product in some reactions by inhibiting the reverse reaction in an equilibrium. (c) Although heterogeneous catalysts must affect the rates of both forward and reverse reactions to an equal extent, homogeneous catalysts can be made to affect the rate of just the forward or just the reverse step.

heterogeneous equilibria

14.24 Which of the following chemical systems represent homogeneous equilibrium and which heterogeneous?

    (a) $H_2(g) + Cl_2(g) \rightleftharpoons 2HCl(g)$
    (b) $NH_4Cl(s) \rightleftharpoons NH_3(g) + HCl(g)$
    (c) $PCl_5(g) \rightleftharpoons PCl_3(g) + Cl_2(g)$
    (d) N_2O_4 (solution) $\rightleftharpoons 2NO_2$ (solution)

14.25 Write the expression for the equilibrium constant for each of the chemical systems listed in the preceding question. Indicate clearly any concentrations that can be combined with the equilibrium constant.

14.26 Which of the following equilibria would undergo a shift in the position of equilibrium by application of increased pressure?

    (a) $ZnSO_3(s) \rightleftharpoons ZnO(s) + SO_2(g)$
    (b) $3Fe(s) + 4H_2O(g) \rightleftharpoons Fe_3O_4(s) + 4H_2(g)$
    (c) $NH_4Cl(s) \rightleftharpoons NH_3(g) + HCl(g)$

14.27 NiO is to be reduced to nickel metal in an industrial process by use of the reaction

$$NiO(s) + CO(g) \rightleftharpoons Ni(s) + CO_2(g)$$

At 1500 K the equilibrium constant for the reaction is 700. If a CO pressure of 300 torr is to be employed in the furnace, and total pressure never exceeds 760 torr, will reduction occur?

14.28 Application of high pressure to most liquid substances causes the melting temperature to increase. What does this tell us about the relative densities of the liquid and solid forms of the substance? Explain.

[14.29] Offer reasonable explanations for the following observations: (a) Ammonia decomposition at high temperature is catalyzed by a tungsten surface. The rate of decomposition is found to be independent of the ammonia pressure. (b) The decomposition of H_2S at high temperature is catalyzed by finely divided palladium metal. The decomposition is retarded by the presence of N_2 gas.

14.30 The adsorption of a gas on a surface is normally an exothermic process. How would you expect the value for the adsorption coefficient b in Equation [14.22] to vary with temperature? Explain.

general exercises

[14.31] The equilibrium constant for the reaction

$$H_2(g) + I_2(g) \rightleftharpoons 2HI(g)$$

has the value 54.6 at 699 K. In one set of equilibrium conditions, the concentrations of gaseous species were as follows:

$$[H_2] = 0.00456\ M$$
$$[I_2] = 0.00738\ M$$
$$[HI] = 0.0135\ M$$

What are the units of K_c? Suppose the concentration of H_2 is suddenly increased to 0.0100 M. When the system has returned to equilibrium, what are the concentrations of all three species?

14.32 Consider the reaction

$$CO(g) + 2H_2(g) \rightleftharpoons CH_3OH(l)$$

Using the thermochemical data in Table 4.1, determine whether the equilibrium constant for this reaction increases or decreases with increasing temperature. Assuming equal pressures of CO and H_2, how would the extent of conversion of the gas mixture to methanol (CH_3OH) vary with total pressure?

14.33 Methanol, CH_3OH, is synthesized from CO and H_2 as shown in the preceding question, using 300–400 atm and temperatures of 200–300°C. What advantages would there be in using a more effective catalyst for this reaction?

14.34 By studying the adsorption of O_2 gas on a nickel surface, the value of the adsorption coefficient was determined to be 0.450 torr^{-1}. Using this value, calculate the fractional surface coverage of the nickel metal when the O_2 pressure is 6.00 torr.

14.35 Suppose there is a region in outer space where initially the hydrogen molecule concentration is 10^2 molecules/cm^3, the N_2 concentration is 1 molecule/cm^3, and the temperature is 100 K. At this temperature, K for the reaction

$$N_2 + 3H_2 \rightleftharpoons 2NH_3$$

is approximately 6×10^{37} atm^{-2}. Assuming that equilibrium is attained, is a significant fraction of the N_2 converted to NH_3?

free energy, entropy, and equilibrium

15

Our discussion of the Haber process in Chapter 14 should have convinced you there are very important practical considerations involved in the study of chemical equilibrium. The two big questions about any chemical reaction are: How far toward completion does the reaction proceed? How rapidly does the reaction approach equilibrium? We get the answer to the first question from a knowledge of the equilibrium constant. We learn about the second from a study of the reaction rate.

If we want to carry out a reaction that has a favorable equilibrium constant, that is, one that lies far to the right, we need only worry about getting the reaction rate into a convenient range. This may not be easy, but hard work and ingenious research might eventually lead to a catalyst that can speed up a slow reaction or to a means of controlling a reaction that is too rapid. Haber's discovery of a suitable catalyst for ammonia synthesis provides an excellent example of successful research. But suppose the equilibrium constant is unfavorable? If Haber had decided to attempt to fix nitrogen by reacting N_2 and O_2 at 400–500°C, instead of reacting N_2 with H_2, he would never have developed a successful process. The reason is that the equilibrium constant for reaction of N_2 with O_2 to form NO is so small at 400–500°C (see Figure 14.3) that the amount of NO present at equilibrium would have been too small for practical purposes. Even if one had a catalyst that would enable rapid reaction of N_2 with O_2 to form NO, the practical consequences would have been nil. There is therefore a need for some way to predict in advance whether a reaction can proceed to any significant extent before coming to equilibrium.

We have seen that chemical equilibrium represents a dynamic state. Things are happening in a system at equilibrium. However, there is no overall change as time passes, because processes that have opposite effects are going on at equal rates. We would like to have a deeper understanding of what factors control the position of equilibrium. This understanding comes through the laws of thermodynamics, which give us the basis for quantitative energy relationships.

The first law of thermodynamics (Chapter 4) is a statement of our experience that energy is conserved.* By this we mean that energy is neither created nor destroyed in processes such as the falling of a brick to the ground, the melting of an ice cube, or a chemical reaction. Energy flows from one part of nature to another, or is converted from one form into another, but the total remains constant. Once we specify a particular process or change, the first law helps us to balance the books, so to speak, on the heat released, work done, and so forth. But it says nothing about whether the process of change we specify can in fact occur. That question is encompassed by the second law of thermodynamics. Before we can state the second law in a usable form, we must define some terms.

If a reaction under a given set of conditions is capable in principle of proceeding in the forward direction to a substantial extent, it is said to be **spontaneous**. Note that we are not now discussing the rate of the reaction. A spontaneous reaction might occur very slowly. For example a mixture of $H_2(g)$ and $Cl_2(g)$ may remain together for centuries with no apparent change. After initiation by a spark or a ray of light, however, reaction to form HCl occurs with explosive speed. The point is that when a spontaneous reaction has arrived at equilibrium, however long that takes, the reactants have been largely converted into products. Thus formation of ammonia from N_2 and H_2 at 400–500°C is a spontaneous process, but formation of NO from N_2 and O_2 at these temperatures is not. The qualitative idea is illustrated in Figure 15.1.

*If we want to include nuclear reactions, we must say that mass-energy is conserved. In ordinary chemical processes, however, the mass changes involved are entirely negligible, and it is correct to say simply that total energy is conserved.

SAMPLE EXERCISE 15.1

Which of the following reaction is spontaneous?

(a) $2H_2(g) + O_2(g) \longrightarrow 2H_2O(g)$

(b) $N_2(g) \longrightarrow 2N(g)$

(c) $4Fe(s) + 3O_2(g) \longrightarrow 2Fe_2O_3(s)$

Solution: Reaction (a) we know to be spontaneous, because when the reaction occurs, it proceeds far to the right as written. However, because the reaction has a high activation energy, it proceeds very slowly unless it is ignited.

Reaction (b) is not spontaneous. The N_2 molecule is one of the most stable molecules known, and the dissociation into nitrogen atoms is completely negligible at ordinary temperatures.

Reaction (c) is spontaneous. We recognize the reaction as the rusting of iron, a process that occurs readily at room temperature. Here again, the rate of reaction varies greatly, depending on conditions. An automobile without corrosion protection might oxidize to Fe_2O_3 very slowly in a dry climate. The same car driven in a wintry climate where salt is applied to the streets would corrode much more rapidly. In either case the products of the reaction are the same, and the spontaneity of the reaction is the same.

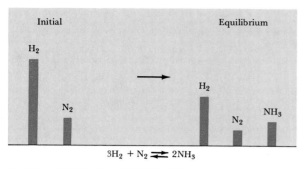

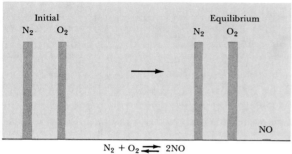

FIGURE 15.1 Illustration of the shift in concentrations in a spontaneous process, formation of ammonia from N_2 and H_2, as compared with a nonspontaneous process, formation of NO from N_2 and O_2. In a spontaneous process, a substantial change in concentrations of reactants and products occurs. In a nonspontaneous process, the changes are very small.

Nature abounds with examples of spontaneous processes. For example, if you hold a brick in your hand and let it go, the brick falls to the floor. Water placed in the freezer compartment of the refrigerator converts to ice. A shiny iron nail left outdoors turns to rust. Every one of these processes is spontaneous. In each example we can imagine a reverse process: The brick leaves the floor and moves up into your hand; ice cubes in a freezer at $-10°C$ melt; a crumbling bit of rust converts into a shiny iron nail. It is inconceivable that any of these occurrences is spontaneous. If we saw a film in which these things happened, we would conclude that the film was being run backward. Our years of observing mother nature at work have impressed us with a simple rule that is really one way of stating the **second law of thermodynamics:** *Processes that are spontaneous in one direction are not spontaneous in the reverse direction.*

But what makes a process spontaneous? We considered this question briefly in Section 12.2; let's consider it again, in more detail. When a brick falls to the floor, it loses potential energy. This potential energy is first converted into kinetic energy, the energy of motion of the brick. When the brick hits the floor, its kinetic energy is converted into heat. The overall result of the brick's fall is thus a conversion of potential energy into heat. The brick possessed potential energy because of its position relative to the floor in relation to the earth's gravitational force. In a similar way, a chemical substance may also possess potential energy relative to other substances because of the arrangements of nuclei and electrons. When these arrangements change, energy may be released. For example, when propane (bottled gas) burns to form carbon dioxide and water the reaction is strongly exothermic:

$$C_3H_8(g) + 5O_2(g) \longrightarrow$$
$$3CO_2(g) + 4H_2O(l) \qquad \Delta H° = -2202 \text{ kJ/mole} \qquad [15.1]$$

The rearrangements in space of nuclei and electrons in going from propane and oxygen to carbon dioxide and water lead to a lower chemical potential energy, so heat is evolved.

As a rough rule, reactions that are exothermic are generally spontaneous. The converse is also usually true; endothermic reactions are generally nonspontaneous. However, there are exceptions. Whether heat is absorbed or evolved is not the only criterion that determines whether a process is spontaneous.

15.2 spontaneity and entropy change

A bit of thinking brings to mind several processes that are spontaneous even though they are not exothermic. For example, consider an ideal gas confined at 1 atm pressure to a 1-l. flask, as shown in Figure 15.2. The flask is connected via a closed stopcock to another 1-l. flask that is evacuated. Now suppose the stopcock is opened. Is there any doubt about what would happen? We intuitively recognize that the gas would expand into the second flask until the pressure were equally distributed in both flasks, at 0.5 atm. In the course of expanding from the 1-l. flask into the larger volume, the ideal gas neither absorbs nor emits heat. Nevertheless, the process is spontaneous. The reverse process, in which the gas that is evenly distributed between the two flasks suddenly moves entirely into one of the flasks, leaving the other vacant, is inconceivable. Yet this is also a process that would involve no emission or absorption of heat. It is evident that some other factor than heat emitted or absorbed is important in making the process of gas expansion spontaneous.

As another example, consider the melting of ice cubes at room temperature. The process

$$H_2O(s) \longrightarrow H_2O(l) \qquad [15.2]$$

at 27°C is highly spontaneous, as we all know. Yet this is an endothermic change. The melting of ice above 0° thus represents an example of a spontaneous, endothermic process.

A similar type of process, discussed in Chapter 12, is the endothermic dissolving of many salts in water. If we add solid potassium chloride, KCl, to a glass of water at room temperature and stir, we can feel the solution growing colder as the salt dissolves. Thus the process that we can write as

$$KCl(s) + H_2O(l) \longrightarrow KCl(aq) \qquad [15.3]$$

is endothermic, and yet spontaneous.

The three processes just described have something in common that accounts for the fact that they are spontaneous. In each instance, the products of the process are in a more random or disordered state than the reactants. Let's consider each case in turn.

When we have a gas confined to a 1-l. volume as in Figure 15.2(a), we can specify the location of each and every gas molecule as being in that liter of space. After the gas has expanded, we can't be sure which gas molecules are at any one instant in the original volume, and which are on the other side. We must therefore say that the location of each and every

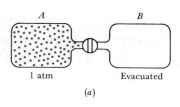

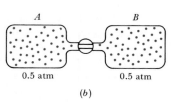

FIGURE 15.2 Expansion of an ideal gas into an evacuated space. In (a), flask A holds an ideal gas at 1 atm pressure, whereas flask B is evacuated. In (b), the stopcock connecting the flasks has been opened. The ideal gas expands to occupy both flasks A and B at a pressure of 0.5 atm.

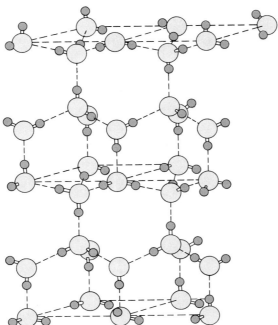

FIGURE 15.3 The structure of ice.

gas molecule is specified as being in the entire 2-l. space. In other words, the gas molecules, because they can be anywhere within a 2-l. space, are more randomized than when they were confined to a 1-l. space.

The molecules of water that make up an ice crystal are held rigidly in place in the ice crystal lattice (Figure 15.3). When the ice melts, the water molecules are free to move about with respect to one another and to turn over. Thus, in liquid water the individual water molecules are more randomly distributed than in the solid. The highly ordered solid structure is replaced by the highly disordered liquid structure.

A similar situation applies when KCl dissolves in water, although here we must be a little careful not to take too much for granted. In solid KCl, the K^+ and Cl^- are in a highly ordered, crystalline state. When the solid dissolves, the ions are free to move about in the water. They are obviously in a much more random and disordered state than before. At the same time, though, water molecules are held around the ions, as water of hydration (Section 12.3), as illustrated in Figure 15.4. These water molecules, are in a *more* ordered state than before, because they are confined to the immediate environment of the ions. Thus the dissolving of a salt involves both ordering and disordering processes. It happens that the disordering processes are dominant, so the overall effect is an increase in disorder upon dissolving a salt in water.

SAMPLE EXERCISE 15.2

Indicate in each of the following cases whether the process shown results in an increase or decrease in randomness or disorder.

(a) $4Fe(s) + 3O_2(g) \longrightarrow 2Fe_2O_3(s)$

(b) $Ag(aq)^+ + Cl^-(aq) \longrightarrow AgCl(s)$

(c) $H_2O(l) \longrightarrow H_2O(g)$

Solution: Process (a) results in a decrease in randomness, because a gas is converted into part of a solid lattice. The units of the solid oxide lattice are much more highly ordered and confined to specific locations than are the molecules of a gas. (Note that this reaction is spontaneous even though there is an overall decrease in randomness. This is so because the reaction is highly exothermic. The combined effects of enthalpy change and change in randomness are discussed later in Section 15.5.)

Process (b) also represents a decrease in randomness, because the ions that are free to move about the volume of the solution form a solid lattice in which they are confined to highly regular locations.

Process (c) occurs with an increase in randomness or disorder, because the gaseous water molecules are distributed throughout a much larger volume than in the liquid state.

As we might guess from these examples, *spontaneity is associated with an increase in disorder of a system.* The randomness or disorder is expressed as the **entropy,** S. By means of thermodynamics, it is possible to express the entropy in quantitative terms. A change in enthalpy, or heat content (ΔH), and in entropy (ΔS) is associated with every process, regardless of whether it is a physical or chemical change:

$$H_{\text{initial}} \longrightarrow H_{\text{final}} \qquad \Delta H = H_{\text{final}} - H_{\text{initial}}$$

$$S_{\text{initial}} \longrightarrow S_{\text{final}} \qquad \Delta S = S_{\text{final}} - S_{\text{initial}}$$

Entropy is thus like enthalpy, in that it is a state function (Section 4.5). The change in entropy for a process depends only on the initial and final

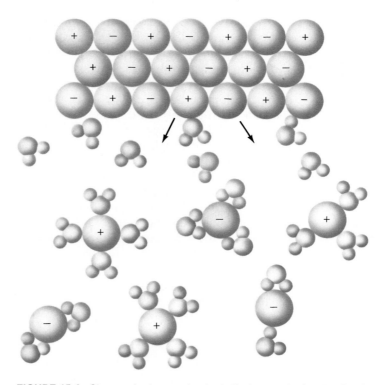

FIGURE 15.4 Changes in degree of order in the ions and solvent molecules on dissolving of an ionic solid in water. The ions become more randomized and the water molecules that hydrate the ions become less randomized.

states, and not on the particular pathway by which we get from one state to another.

The enthalpy change in a chemical reaction is often easily measured in a calorimeter, as described in Section 4.7. There is no comparable easy means for measuring the change in entropy. It has been possible, however, to determine the entropies of substances by various types of measurements. Entropies are expressed in units of joules per degree Kelvin per mole: J/K-mole.

15.3 a molecular interpretation of entropy

We shall discuss a little later how the entropy change in a chemical reaction can be calculated from tabulated entropy values, but let us first examine the *notion* of entropy in more detail.

Entropy is very closely related to the idea of probability. For any particular system that interests us, such as a collection of gas molecules or a solution, the entropy associated with a particular state of that system can be defined. For example, the gas sample in Figure 15.2 is shown in two distinctly different states. In the one, all the gas is confined to a 1-l. volume. In the other, it is distributed throughout the 2-l. volume. The gas possesses different entropy values in the two states. The gas is in a state of greater randomness—that is, it possesses higher entropy—when it is distributed throughout a 2-l. volume than when it is confined to a 1-l. volume.

An increase in volume is not the only way by which the entropy of a system can be increased. The various forms of energy that a system may contain also provide means for increased randomness or higher entropy. For example, in a gas, molecules possess kinetic energy, that is, energy of motion. In addition, if the molecules consist of more than one atom, they may also possess vibrational and rotational energy. In vibrational motion, the atoms move periodically toward and away from one another, much as a tuning fork vibrates about its equilibrium shape. Figure 15.5 shows the vibrational motions possible for the water molecule. In addition,

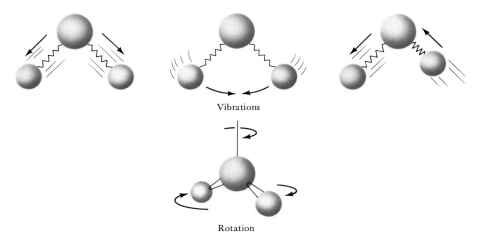

Vibrations

Rotation

FIGURE 15.5 Examples of vibrational and rotational motion, as illustrated for the water molecule. Vibrational motions involve periodic displacements of the atoms with respect to one another, a phenomenon similar to the vibrations of the arms of a tuning fork. Rotational motions involve the spinning of a molecule about an axis.

molecules may possess rotational energy, as though they were spinning like a top. The rotational motion of the water molecule is also illustrated in Figure 15.5.

As a rough analogy to the idea of storing these various forms of energy in atoms and molecules, suppose that you have a tennis ball, a top, and a tuning fork. If it were not for friction, you could throw the tennis ball into an empty room, and it would continue to bounce around the room indefinitely. Similarly, you could set the tuning fork vibrating and the top spinning. You would have stored energy in the forms of kinetic, vibrational, and rotational motions. If you were ingenious enough you could later recover the energy in other forms.

To see what this has to do with entropy, let's imagine that we begin with a pure substance that forms a perfect crystalline lattice at the lowest temperature possible, absolute zero. At this stage, none of the kinds of energy that we have been talking about are present. The individual atoms and molecules are as well defined in position and in terms of energy as they can ever be. Let us say that the entropy of our substance at this point is zero. As the temperature is raised, the units of the solid lattice begin to acquire energy. In a crystalline solid, the molecules or atoms that occupy the lattice places are constrained to remain more or less in place. Nevertheless they may store energy in the form of vibrational motion about their lattice positions. Instead of all the molecules necessarily being in the lowest possible energy state, there is a kind of expansion in the number of possible energies that the lattice atoms or molecules may have. This increase in possible energy states is not unlike the expansion of the gas illustrated in Figure 15.2. The entropy of the gas increases on expansion because the volume element in which the gas molecules move is larger. The entropy of the lattice increases with temperature because the number of possible energy states in which the molecules or atoms are distributed is larger.

Let's suppose that at some temperature a phase change occurs, converting the substance from one solid form to another. This means that the arrangement of the lattice units changes in some way, possibly so that the lattice is less regular.* This type of phase change occurs sharply at one temperature, just as do other types of phase changes, for example, from a solid to a liquid. When the change occurs, there is a change in entropy, because the two lattice arrangements do not have precisely the same degrees of randomness.

Figure 15.6 shows the variation in entropy with temperature for our sample. Note that the change in S with temperature is gradual up to the solid-state phase change and that there is then a sharp increase in S at that temperature. At temperatures above the phase change, the entropy increases with increasing temperature, up to the melting point of the solid.

When the solid melts, the units of the lattice are no longer confined to specific locations relative to other units, but are free to move about the entire volume of the unit. This added freedom of motion for the individual molecules adds greatly to the entropy content of the substance. At the temperature of melting, we therefore see a large increase in entropy

*As an example of a solid-state phase change, grey tin converts at 13°C to another solid form called white tin. White tin is stable above the transition temperature, grey tin below it. White tin has a higher entropy than grey tin.

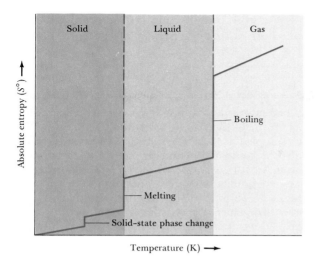

FIGURE 15.6 Entropy changes that occur as the temperature of a substance rises from absolute zero.

content. After all the solid has melted, the temperature again increases, and with it the entropy.

SAMPLE EXERCISE 15.3

Figure 15.6 shows that the entropy of a liquid increases as its temperature increases. What factors are responsible for the increase in entropy?

Solution: The average kinetic energy of the molecules in a liquid increases with temperature. As temperature increases, more molecules at any given instant possess higher energies. This "expansion" in the energies that the molecules possess is measured by the increase in entropy. The increase in S with increasing temperature results from increased energy of motion of all kinds within the liquid.

At the boiling point of the liquid there is again a big increase in entropy. The increase in this case results largely from the increased volume in which the molecules may be found. This is intuitively in line with our earlier ideas about entropy, because an increase in volume means an increase in randomness. It is less likely that a given molecule will be found in a given volume element when there is a great expansion from the liquid to the gaseous state.

As the gas is heated, the entropy increases steadily, because more and more energy is being stored in the gas molecules. The distribution of molecular speeds is spread out toward higher values, as illustrated in Figure 5.12. Again, the idea of an expansion in the range of energies in which molecules may be found helps us remember that increased average energy means increased entropy.

If we could carry out measurements that would give us data to construct a graph such as that in Figure 15.6, we would know the absolute entropy at any temperature for a substance of interest. The absolute entropy is referred to absolute zero as the starting point. **Absolute entropy values** (usually written as $S°$) are available in various tabulations of thermodynamic properties.* Although we shall have few occasions to

*For example, you can find such values in the *Handbook of Chemistry and Physics,* published by the Chemical Rubber Publishing Co., or in Lang's *Handbook of Chemistry and Physics.*

use the tabulated values, it is useful in making qualitative predictions to have a feeling for how the absolute entropy depends on structure, physical state, and so on.

SAMPLE EXERCISE 15.4

For each of the following pairs of substances, indicate which has the higher absolute entropy and suggest a reason: (a) 1 mole $NaCl(s)$ and 1 mole $HCl(g)$ at 25°C; (b) 2 moles $HCl(g)$ and 1 mole $HCl(g)$ at 25°C; (c) 1 mole $HCl(g)$ and 1 mole $Ar(g)$ at 25°C; (d) 1 mole $N_2(s)$ at 24 K and 1 mole $N_2(g)$ at 298 K.

Solution: (a) Gaseous HCl has the higher entropy per mole, because it has acquired a high degree of randomness as a result of being in the gaseous state.

(b) The sample containing 2 moles of HCl has twice the entropy of the sample containing 1 mole. (c) The HCl sample has the higher entropy, because the HCl molecule is capable of storing energy in more ways than is Ar. It may rotate, or the H—Cl distance may change periodically in a vibrational motion. (d) The gaseous N_2 sample has the higher entropy, because the entropy increases resulting from melting and then boiling of N_2 are included in its total entropy content.

15.4
entropy changes in chemical reactions

So far we have focused our attention on getting a feeling for the entropy content of substances and on how that entropy content changes with conditions. But we are mainly interested in knowing how the entropy content of a *system* changes when a chemical reaction takes place. Let us use what we have already learned to guess at whether the entropy change in a chemical reaction is positive or negative.

The entropy change in a chemical reaction is given by the sum of the entropies of the products less the sum of entropies of reactants. Thus, in the overall reaction

$$aA + bB + cC \rightleftharpoons pP + qQ + \cdots \qquad [15.4]$$

The total entropy change is given by

$$\Delta S° = (pS_P° + qS_Q° + \cdots) - (aS_A° + bS_B° + \cdots) \qquad [15.5]$$

In other words we sum the absolute entropies of all the products, multiplying each by the coefficient of the product in the balanced equation, and then subtract the same sort of sum of entropies of the reactants. Thus, if we have available a table of absolute entropy values (Appendix E), it would be no problem to calculate the standard entropy change in a chemical reaction. We shall not concern ourselves with making such quantitative calculations. Instead, let's see if we can simply predict the sign of the entropy change that occurs.

SAMPLE EXERCISE 15.5

Predict whether the entropy change of the system in each of the following reactions is positive or negative.

(a) $H_2O(l) \longrightarrow H_2O(g)$ (at 25°C)
(b) $CaCO_3(s) \longrightarrow CaO(s) + CO_2(g)$
(c) $N_2(g) + 3H_2(g) \longrightarrow 2NH_3(g)$

(d) $N_2(g) + O_2(g) \longrightarrow 2NO(g)$
(e) $Ag(aq) + Cl^-(aq) \longrightarrow AgCl(s)$

Solution: (a) The entropy change in this process is positive, because the single substance involved is going from the liquid to the gaseous state. We have already seen that the phase change from liquid to gas results in an increase in entropy (Figure 15.6). The value for $\Delta S°$ in this case is $+118$ J/K.*

(b) The entropy change here is positive, because a solid is converted into a solid and a gas. Gaseous substances generally possess more entropy than solids, and so whenever the products contain more moles of gas than the reactants, the entropy change is probably positive. The value for $\Delta S°$ in this case is $+160.5$ J/K.

(c) The entropy change in formation of ammonia from nitrogen and hydrogen is negative, because there are fewer moles of gas in the products than in the reactants. The value for $\Delta S°$ in this case is -197.5 J/K.

(d) This represents a case in which the entropy change will be small, because the same number of moles of gas is involved in the reactant and products. The value for $\Delta S°$ in this case is $+24.4$ J/K.

(e) The entropy change in the precipitation of a salt from solution is negative. The ions in solution are free to move about the entire volume of the liquid, whereas in the solid lattice they are more confined. The value for $\Delta S°$ in this case is -32.9 J/K.

*The values of $\Delta S°$ given in these solutions apply for the number of moles of substances indicated in the balanced equations.

In all of our discussion of entropy changes so far, we have concentrated our attention on the system. But a complete accounting of all entropy changes must also take account of entropy change in the surroundings as well. As an example, consider the oxidation of iron to $Fe_2O_3(s)$:

$$4Fe(s) + 3O_2(g) \longrightarrow 2Fe_2O_3(s)$$

As discussed in Sample Exercise 15.2, this chemical process results in a decrease in the degree of randomness; that is, we would expect ΔS for the process to be negative. But when the process occurs, some change also occurs in the surroundings. For example, the reaction is exothermic. Even though the reaction may occur very slowly, heat is evolved and is absorbed by the surroundings. In fact, the change that occurs in the surroundings causes an increase in the entropy of the surroundings that is larger than the decrease that occurs in the system itself. It is possible to show from the second law of thermodynamics that *every spontaneous process leads to a net overall increase in the entropy of the universe*. To put this in slightly different terms, any chemical process or physical change that occurs at a measurable rate causes an overall increase in total entropy, regardless of whether the entropy change in the system itself is positive or negative. There is always a corresponding change in the surroundings that ensures a net overall positive entropy change.

The consequences of this all-embracing statement are quite profound. We humans, for example, are very complex, highly organized, and well-ordered systems. We have a very low entropy content as compared with the same amount of carbon dioxide, water, and several other simple chemicals into which our bodies might be decomposed. But all of the thousands of chemical reactions necessary to produce one adult human have caused a very large increase in entropy of the rest of the universe. Thus the overall entropy change necessary to form and maintain a human, or for that matter any other living system, is positive.

In a similar way, the human activities that produce such an impressive ordering of the world around us—formation of copper metal from a widely dispersed copper ore; production from sand of silicon used in transistors; production of the paper on which this book is printed from trees—have along the way used up a great deal of energy that has been converted, in a sense, to disorder—coal and oil burned to form CO_2 and H_2O; a sulfide ore roasted to form SO_2 that pollutes the atmosphere; radioactive wastes scattered in the environment. Modern human society is, in effect, using up its limited storehouse of energy-rich materials in its headlong rush to exploit technology.

In recent years a few social scientists have begun to appreciate the importance of thermodynamic considerations in the economic laws that must eventually rule human activities. A leading economic theorist, Nicholas Georgescu-Roegen, published in 1971 a book entitled *The Entropy Law and the Economic Process*. He argues that the human race must eventually learn to live within the bounds of the energy supply that reaches earth daily from the sun, because it will soon have exhausted the supply of readily available energy of other sorts.

It is evident that the spontaneity of a process is determined by both the enthalpy and entropy changes. As we've seen, exothermic processes, such as the burning of propane, tend to be spontaneous. On the other hand, processes that result in an increase in entropy, such as the dissolving of a salt, tend also to be spontaneous. But there are many processes in nature in which the enthalpy and entropy changes work in opposite directions. The one tends to make a process spontaneous, whereas the other tends to make it nonspontaneous. For example, the rusting of iron in air to form Fe_2O_3 is a spontaneous process. Yet, as noted in Sample Exercise 15.2, in the reaction

$$4Fe(s) + 3O_2(g) \longrightarrow 2Fe_2O_3(s)$$

the entropy of the reaction system decreases. In this instance the strongly exothermic nature of the reaction overrides the unfavorable entropy change.

We need a new function that combines the two state functions enthalpy and entropy and that provides a criterion of whether a given change is spontaneous or nonspontaneous. This function is called the **free energy** (or sometimes the Gibbs free energy), G. The free energy is related to the enthalpy and entropy by the expression

$$G = H - TS \qquad [15.6]$$

where T is the absolute temperature.

Because G is related to two other functions that are state functions, it must itself be a state function. For a process occurring *at constant temperature and pressure,* the change in free energy is given by the expression

$$\Delta G = \Delta H - T\Delta S \qquad [15.7]$$

The free-energy function is extremely valuable because it tells us directly whether a process is spontaneous or nonspontaneous. The rule is: *For a process occurring at constant temperature and pressure, ΔG for a spontaneous process is negative. Conversely, ΔG for a nonspontaneous process is positive.*

To see how the free-energy function works, let's return to the synthesis of ammonia from hydrogen and nitrogen. Imagine that we have a certain number of moles of nitrogen, and three times that number of moles of hydrogen, under pressure in a reaction vessel at some temperature, as illustrated in Figure 15.7. The reaction vessel is fitted with a piston so that we can maintain constant pressure, by letting the piston out or pushing it in, as necessary. Now suppose that the reaction between nitrogen and hydrogen begins. Because the balanced equation is

$$N_2 + 3H_2 \rightleftharpoons 2NH_3$$

the formation of ammonia results in a smaller number of gas molecules. To maintain constant pressure, the piston would have to be pushed in as the reaction proceeds.

We know from our earlier discussion that the formation of ammonia will not be complete. At equilibrium there will be a mixture of NH_3, N_2, and H_2. We are interested in knowing what happens to the free energy of

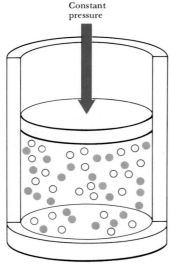

Constant
pressure

O = Hydrogen

● = Nitrogen

● = NH₃

FIGURE 15.7 Constant-pressure reaction chamber for the reaction $N_2 + 3H_2 \rightleftharpoons 2NH_3$. As the reaction proceeds, the piston is pushed in to maintain constant pressure in the chamber. The vessel is kept at constant temperature throughout.

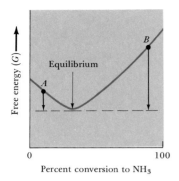

Free energy (G)

Equilibrium

A

B

0   100

Percent conversion to NH₃

FIGURE 15.8 Variation in the free-energy function, G, in the N_2, H_2, NH_3 system as a function of the percent conversion to NH_3 at constant temperature and pressure.

this system as the reaction proceeds. Figure 15.8 shows a graph of the free energy of the entire gas system as a function of the percent conversion to ammonia. At the outset, the free energy of the total gas system is the sum of the free energies of 1 mole of nitrogen and 3 moles of hydrogen. These free energies would depend on the precise temperature and pressure in the vessel. As the reaction proceeds, ammonia is formed at the expense of H_2 and N_2. The total free energy decreases with increasing ammonia formation until a minimum is reached. With still further ammonia formation the free energy increases again.

The minimum in the value for G represents the position of equilibrium in the N_2, H_2, NH_3 system. The exact position of this minimum in terms of percent conversion depends on the conditions of temperature and pressure. The percent conversion of reactants to ammonia shifts with changing conditions in accordance with LeChatelier's principle. The main idea to derive from Figure 15.8 is simply that the position of equilibrium represents a minimum in the relationship between free energy and extent of reaction. Any reaction system that is not at equilibrium is capable of proceeding spontaneously toward the equilibrium position.

The quantity ΔG for a process that takes us from one set of conditions to another is simply the difference in free energies between the two conditions. For example, in Figure 15.8 the point B represents a condition in which there is a great deal of NH_3 present and very little N_2 and H_2. The difference in free energy between that condition and equilibrium is negative; that is, there is a lowering in free energy in going from the conditions represented by point B to the equilibrium condition. This in turn means that this change in conditions at constant temperature and pressure is spontaneous. Similarly, a change in the system from point A to the equilibrium condition also represents a negative change in ΔG. This process is therefore also spontaneous. We see from this that the equilibrium condition can be approached by spontaneous change from either direction.

We have noted that free energy is a state function. This means that it is possible to tabulate standard free energies of formation for substances, just as it is possible to tabulate standard enthalpies of formation. It is important to remember that standard values for these functions imply a particular set of conditions, or standard states. The standard state for gaseous substances is 1 atm pressure, and (usually) 25°C. For solid substances the standard state is the pure solid at 25°; for liquids, the pure liquid. For substances in solution, the standard state is normally a concentration about 1 M; in accurate work it may be necessary to make certain corrections, but we need not worry about these. Just as for the standard heats of formation, the free energies of elements in their standard states are arbitrarily set to zero. This arbitrary choice of reference point has no effect on the quantity in which we are really interested, namely the *difference* in free energy between reactants and products. The rules about standard states are summarized in Table 15.1. A table of standard free energies of formation is to be found in Appendix E.

The standard free energies of formation for substances are useful in calculating the standard free-energy change in a chemical process. For the general reaction

TABLE 15.1

**conventions used in establishing standard
free-energy values**

STATE OF MATTER	STANDARD STATE
Solid	Pure solid
Liquid	Pure liquid
Gas	1 atm pressure[a]
Solution	Usually 1 molar[b]
Elements	Standard free energy of formation of the element in its normal state is defined as zero.

[a] Neglecting nonideal gas behavior.
[b] Neglecting nonideality of solutions.

$$aA + bB + \cdots \longrightarrow pP + qQ + \cdots \qquad [15.8]$$

the standard free-energy change is

$$\Delta G^\circ = (p\Delta G_f^\circ(P) + q\Delta G_f^\circ(Q) + \cdots) - (a\Delta G_f^\circ(A) + b\Delta G_f^\circ(B) + \cdots)$$
$$[15.9]$$

In this expression $\Delta G_f^\circ(P)$ represents the standard free energy of formation of product P, and all the other ΔG° have similar meanings. Stated verbally, the standard free-energy change for a reaction equals the sum of the standard free-energy values per mole of each product, each multiplied by the corresponding coefficient in the balanced equation, less the same sort of sum for the reactants.

What use can be made of this standard free-energy change for a chemical reaction? The quantity ΔG° tells us whether a mixture of reactants and products, each present under standard conditions, would spontaneously react in the forward direction to produce more product (ΔG° negative) or in the reverse direction to form more reactants (ΔG° positive). Because standard free-energy values are readily available for a large number of substances, the standard free-energy change is easy to calculate for any reaction system of interest.

SAMPLE EXERCISE 15.6

Determine the standard free-energy change for the reaction

$$N_2 + 3H_2 \rightleftharpoons 2NH_3$$

Solution: Using Appendix E we find that the standard free energies for the three substances of interest are as follows:

$N_2(g)$: $\Delta G_f^\circ = 0.0$ $H_2(g)$: $\Delta G_f^\circ = 0.0$
$NH_3(g)$: $\Delta G_f^\circ = -16.66$ kJ/mole

The standard free-energy change for the reaction of interest is

$$\Delta G^\circ = 2\Delta G_f^\circ(NH_3) - 3\Delta G_f^\circ(H_2) - \Delta G_f^\circ(N_2)$$

Inserting numerical quantities we obtain:

$$\Delta G^\circ = -33.31 \text{ kJ}$$

The fact that ΔG° is negative tells us that a mixture of H_2, N_2, and NH_3 at 25°C, each present at a pressure of 1 atm, would react spontaneously to form more ammonia. (Remember, however, that this says nothing about the rate at which the reaction occurs.)

Determine the standard free energy change for the reaction

$$2NH_3 \rightleftharpoons N_2 + 3H_2. \qquad [15.10]$$

Solution: This reaction is just the reverse of that given in the previous sample exercise. By proceeding just as above we obtain for the overall $\Delta G°$ the value $+33.31$ kJ/mole. The reason the sign of $\Delta G°$ is reversed whereas its magnitude remains the same is, of course, that the reactants have now become the products, and vice versa. It is a useful rule to remember that *when a reaction is reversed, the sign of the standard free-energy change is reversed*. Since $\Delta G°$ for the reaction in Equation [15.10] is positive at 25°C with all reactants present under standard conditions, the reaction is not capable of proceeding spontaneously in the forward direction. Rather, it proceeds spontaneously in the *reverse* direction, as we deduced in the previous sample exercise.

In applying the concept of free-energy change, it is often necessary to know the free-energy functions for substances at temperatures other than 25°C and under conditions other than the standard ones. Such applications involve a considerably more detailed treatment of thermodynamics than we are able to give in this book. We can, however, do a surprising number of interesting things with just the standard free-energy values at 25°C. Suppose we wish to know whether a proposed reaction is capable of proceeding to a reasonable degree to give products. If we calculate $\Delta G°$ for the reaction and find that it is fairly large and negative, as in Sample Exercise 15.6, this tells us that the mixture of reactants and products, all at standard conditions, is capable of reacting far in the direction of forming products. On the other hand, if we obtain a positive value for $\Delta G°$, the mixture of reactants and products will react so as to form reactants.

Normally we don't start a reaction with a mixture of reactants and products, and in fact it may not even be possible from a practical point of view to get such a mixture. Nevertheless, the $\Delta G°$ values are important, because they tell us the direction in which such a hypothetical mixture would move in approaching equilibrium. If, as we have mentioned, the $\Delta G°$ value is large and positive, the equilibrium is far in the direction of reactants. This in turn means that if we start with just reactants, almost no product will be formed.

Calculate the standard free-energy change for the reaction

$$N_2 + O_2 \rightleftharpoons 2NO$$

What does the value obtained for $\Delta G°$ tell us about the degree to which formation of NO from N_2 and O_2 is spontaneous at 25°C?

Solution: First we calculate $\Delta G°$:

$$\Delta G° = 2\Delta G_f°(NO) - \Delta G_f°(N_2) - \Delta G_f°(O_2)$$

From Appendix E we find $\Delta G_f°(NO)$ to be $+86.71$ kJ/mole. The $\Delta G_f°$ values for the elements are, of course, zero. Then

$$\Delta G° = 2 \text{ moles NO}\left(\frac{86.71 \text{ kJ}}{\text{mole NO}}\right) - O - O$$

$$= 173.4 \text{ kJ}$$

With such a large positive value for $\Delta G°$, it follows that the equilibrium mixture of N_2, O_2, and NO lies very far in the direction of N_2 and O_2. We classify this reaction qualitatively as nonspontaneous, that is, as not proceeding to any significant degree to form products. This conclusion is shown visually in Figure 15.1.

It is worthwhile to examine Equation [15.7] closely to see how the free-energy function depends on both the enthalpy and entropy changes for a given process. If it were not for entropy effects, all exothermic reactions, those in which ΔH is negative, would be spontaneous. The entropy contribution, represented by the quantity $-T\Delta S$, may increase or decrease the tendency of the reaction to proceed spontaneously. When ΔS is positive, meaning that the final state is more random or disordered than the initial state, the term $-T\Delta S$ makes a negative contribution to ΔG, that is, it increases the tendency of the reaction to occur spontaneously. When ΔS is negative, however, the term $-T\Delta S$ decreases the tendency of the reaction to occur spontaneously.

When ΔH and $-T\Delta S$ are of opposite sign, the relative importance of the two terms determines whether ΔG is negative or positive. In these instances, temperature is an important consideration. Both ΔH and ΔS are in principle capable of changing with temperature. In practice, however, the changes that occur are not very large unless very large temperature changes are involved. The only quantity in the equation

$$\Delta G = \Delta H - T\Delta S$$

that changes markedly with temperature is therefore $-T\Delta S$. At high temperatures the entropy term becomes relatively more important.

Various possible situations for the relative signs of ΔH and ΔS are shown in Table 15.2, with a few examples of each. By applying the

TABLE 15.2

characteristics of the free-energy function for various possible values of ΔH and ΔS

ΔH	ΔS	CHARACTERISTICS OF ΔG	EXAMPLES
−	+	Negative at all temperatures. Process spontaneous at all temperatures.	$2NCl_3(g) \longrightarrow N_2(g) + 3Cl_2(g)$ $2O_3(g) \longrightarrow 3O_2(g)$
+	−	Positive at all temperatures. Process proceeds spontaneously in reverse direction at all temperatures.	$3O_2(g) \longrightarrow 2O_3(g)$ $N_2(g) + 3Cl_2(g) \longrightarrow 2NCl_3(g)$
−	−	Negative at low temperatures; positive at high temperatures. Process spontaneous at low temperatures; tends to reverse at high temperatures.	$I(g) + I(g) \longrightarrow I_2(g)$ $PCl_3(g) + Cl_2(g) \longrightarrow PCl_5(g)$ $CaO(s) + CO_2(g) \longrightarrow CaCO_3(s)$
+	+	Positive at low temperatures; negative at high temperatures. Process spontaneous in reverse direction at low temperatures; spontaneous in forward direction at high temperatures.	$CaCO_3(s) \longrightarrow CaO(s) + CO_2(g)$ $PCl_5(g) \longrightarrow PCl_3(g) + Cl_2(g)$ $I_2(g) \longrightarrow I(g) + I(g)$

concepts we have developed for predicting entropy changes, it is often possible to predict how ΔG will change with change in temperature.

SAMPLE EXERCISE 15.9

Predict the direction in which ΔG for the equilibrium $N_2(g) + 3H_2(g) \rightleftharpoons 2NH_3(g)$ will change with increase in temperature.

Solution: In Sample Exercise 15.5 we saw that the change in $\Delta S°$ for the equilibrium of interest is negative. This means that the term $-T\Delta S°$ is positive, and grows larger with increasing temperature. The standard free-energy change, $\Delta G°$, is the sum of the negative quantity $\Delta H°$ and the positive quantity, $-T\Delta S°$. Because only the latter grows larger with increasing temperature, $\Delta G°$ grows less negative.

15.7 free energy and the equilibrium constant

The standard free-energy change, $\Delta G°$, is a measure of how far a mixture of all reactants and products, each in its standard state, will shift in the direction of either more reactants or more products, as it approaches equilibrium. A mixture of all reactants and products in their standard states may be considered rather approximately as halfway between pure reactants and pure products. If $\Delta G°$ for this mixture is negative, then the minimum in G, which represents equilibrium, lies in the direction of products. If $\Delta G°$ is positive, this minimum lies in the direction of reactants. From this we might guess that there is a relationship between $\Delta G°$ and the magnitude of the equilibrium constant for the reaction. This is indeed the case; the standard free-energy change, $\Delta G°$, for a process is related to the equilibrium constant K_c by the expression

$$\Delta G° = -RT \ln K_c \qquad [15.11]$$

In the more familiar log base ten units,

$$\Delta G° = -2.30 \, RT \log K_c \qquad [15.12]$$

The quantity R is the molar gas constant, 8.317 J/K-mole, and T is given in degrees Kelvin.

From Equation [15.12] we can readily see that if $\Delta G°$ is negative, $\log K_c$ must be positive. A positive value for $\log K_c$ means that $K_c > 1$. On the other hand if $\Delta G°$ is positive, $\log K_c$ is negative, which means that $K_c < 1$. To summarize:

$$\Delta G° \text{ negative} \qquad K_c > 1$$
$$\Delta G° \text{ zero} \qquad K_c = 1$$
$$\Delta G° \text{ positive} \qquad K_c < 1$$

It is possible from a knowledge of $\Delta G°$ for a reaction to calculate the value for the equilibrium constant, using Equation [15.12]. Some care is necessary, however, in the matter of units. When dealing with gases, the concentrations of reactants should be expressed in units of atmospheres. The concentrations of solids are 1 if they are pure solids; the concentrations of pure liquids are 1 if they are pure liquids; if a mixture of liquids is involved, the concentration of each is expressed as mole fraction. For substances in solution, concentrations in moles per liter are appropriate.

SAMPLE EXERCISE 15.10

From standard free energies of formation, calculate the equilibrium constant for the reaction

$$N_2 + 3H_2 \rightleftharpoons 2NH_3$$

at 25°C.

Solution: The equilibrium constant for this reaction is written as

$$K_c = \frac{(NH_3)^2}{(N_2)(H_2)^3}$$

where the gas concentrations are expressed in atmospheres pressure. The standard free-energy change for the reaction was determined in Sample Exercise 15.6 to be −33.31 kJ. Inserting this into Equation [15.12] we obtain

$$-33,310 = -2.30(8.317 \text{ J/K-mole})(298 \text{ K}) \log K_c$$
$$\log K_c = 5.83$$

Taking the antilog,

$$K_c = 6.7 \times 10^5 \text{ atm}^{-2}$$

This is a large equilibrium constant. Compare its magnitude with the equilibrium constants at higher temperature, as listed in Table 14.4. If a catalyst could be found that would permit reasonably rapid reaction of N_2 with H_2 at room temperature, high pressures would not be required to force the equilibrium toward NH_3.

The free energy has one further interesting property that comes from the fact that it is related to the degree of spontaneity of a process. Any process that occurs spontaneously can be utilized for the performance of useful work, at least in principle. For example, the falling of water in a waterfall is certainly a spontaneous process; it is also one from which it is possible to extract work, by causing the water to turn the blades of a turbine as it falls. Similarly, the burning of gasoline in the cylinders of a car leads to production of useful work in the motion of the car. How much work is extracted from a particular process depends on how it is carried out. For example, we might burn a liter of gasoline in an open container and extract no useful work at all. In an automobile engine, the overall efficiency for production of work is low, perhaps 20 percent. If the gasoline were caused to react with oxygen under other, more favorable conditions, the amount of work we could extract would be higher. In practice, we never achieve the maximum amount of work possible from a theoretical point of view. However, it is useful to know what the maximum possible work derivable from a process is, so that we might have a measure of our success in extracting work from processes of practical importance. Thermodynamics tells us that *the maximum possible work that can be derived from a spontaneous process occurring at constant temperature and pressure is equal to the free-energy change.*

For processes that are not spontaneous, the free-energy change is a measure of the minimum amount of work that must be done to cause the process to occur. In actual cases, more than the theoretical minimum amount of work must be done, because of inefficiencies in the way in which the change is caused to occur.

Free-energy considerations are very important in thinking about many nonspontaneous reactions we might wish to carry out for our own purposes, or that occur in nature. For example, we might wish to extract a metal from an ore. If we look at a reaction such as

$$Cu_2S(s) \longrightarrow 2Cu(s) + S(s)$$
$$\Delta G° = +86.2 \text{ kJ}; \Delta H° = +79.5 \text{ kJ} \qquad [15.13]$$

we find that it is endothermic and highly nonspontaneous. Clearly, then, we cannot hope to obtain copper metal from Cu_2S merely by trying to catalyze the reaction shown in Equation [15.13]. Instead, we must "do work" on the reaction in some way, in order to force it to occur as we wish. We might do this by coupling the reaction we've written with another reaction, so that we arrive at an overall reaction that *is* spontaneous. Consider, for example, the reaction

$$S(s) + O_2(g) \longrightarrow SO_2(g)$$
$$\Delta G° = -300.1 \text{ kJ}; \Delta H° = -296.9 \text{ kJ} \qquad [15.14]$$

This is an exothermic, spontaneous reaction. Adding reaction Equations [15.13] and [15.14], we obtain:

$$Cu_2S \longrightarrow 2Cu(s) + S(s)$$
$$\Delta G° = \quad 86.2 \text{ kJ}; \Delta H° = \quad 79.5 \text{ kJ}$$
$$S(s) + O_2(g) \longrightarrow SO_2(g)$$
$$\Delta G° = -300.1 \text{ kJ}; \Delta H° = -296.9 \text{ kJ}$$
$$\overline{Cu_2S + O_2(g) \longrightarrow Cu(s) + SO_2(g)}$$
$$\Delta G° = -213.9 \text{ kJ}; \Delta H° = -217.4 \text{ kJ}$$

The free-energy change for the overall reaction is the sum of the free-energy changes of the two reactions, and similarly for the overall enthalpy change. Because the negative free-energy change for the second reaction is larger than the positive free-energy change for the first, the overall reaction has a large and negative standard free-energy change.

The coupling of two or more reactions together to cause a nonspontaneous chemical process to occur is very important in biochemical systems. Many of the reactions that are absolutely essential to the maintenance of life do not occur within the human body spontaneously. These necessary reactions are made to occur, however, by coupling them with reactions that are spontaneous and energy releasing. The energy releases that accompany the metabolism of foodstuffs provide the primary source of necessary free energy. For example, a compound such as glucose, $C_6H_{12}O_6$, is oxidized in the body, and a substantial amount of energy is released.

$$C_6H_{12}O_6(s) + 6O_2(g) \longrightarrow 6CO_2(g) + 6H_2O(l)$$
$$\Delta G° = -2880 \text{ kJ}; \Delta H° = -2800 \text{ kJ} \qquad [15.15]$$

This energy is employed to "do work" in the body. However, some means is necessary to couple, or connect, the energy released by glucose oxidation to the reactions that require energy. One means of accomplishing this is shown graphically in Figure 15.9. Adenosine triphosphate (ATP) is a

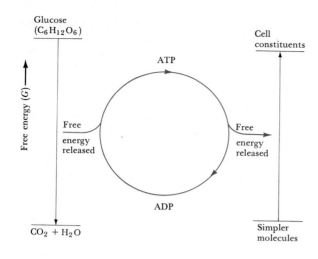

Glucose
$(C_6H_{12}O_6)$

Free energy $(G) \longrightarrow$

Free
energy
released

ATP

ADP

Cell
constituents

Free
energy
released

Simpler
molecules

$CO_2 + H_2O$

FIGURE 15.9 Schematic representation of a part of the free-energy changes that occur in cell metabolism.

high-energy molecule. When ATP is converted to a lower-energy molecule, adenosine diphosphate (ADP), energy to drive other chemical reactions becomes available. The energy released in glucose oxidation is used in part to reconvert ADP back to ATP. Thus, the ATP-ADP interconversions act as a means of storing energy and releasing it to drive needed reactions. The coupling of reactions so that the free energy released in one may be utilized by another requires particular enzymes as catalysts. In the chapter dealing with the biosphere, we shall have occasion to examine the energy relationships in living systems in more detail.

FOR REVIEW

summary

In this chapter we have seen that it is possible to understand the position of a chemical equilibrium in terms of the thermodynamic functions enthalpy and entropy. The **enthalpy change**, ΔH, in a process is a measure of the potential energy changes occurring. The **entropy change**, ΔS, is a measure of the change in randomness, of the disorder present in products as compared with reactants. These changes in randomness are not always obvious, but in a general way they can be associated with the number of different ways in which the system can be distributed among possible energy states or spatial arrangements. It is useful to employ an intuitive idea of randomness, or disorder, to characterize the entropy changes. When a system changes in such a way that it becomes less regular or less confined, the entropy increases.

The position of equilibrium is determined by both the enthalpic and entropic changes that occur. The **free-energy function**, G, combines the two state functions enthalpy and entropy to provide a new state function that relates directly to equilibrium.

learning goals

Having read and studied this chapter, you should be able to:

1. Define the term spontaneity and apply it in identifying spontaneous processes.

2. Define entropy, explain its relationship to the other thermodynamic state functions (enthalpy and free energy), and describe in a qualitative way how and why the entropy of a substance changes with temperature and when a phase change occurs.

3. Qualitatively predict whether the entropy change in a given process is near zero, is positive, or is negative.

4. Explain how the sign of the free-energy change, ΔG, determines whether a process is spontaneous in the forward direction.

5. Explain why chemical equilibrium represents a minimum in the free-energy function.

6. Calculate the standard free-energy change at constant temperature and pressure for a process from tabulated values of the standard free energies of reactants and products.

7. List the usual conventions regarding standard states, in setting the values for standard free energies.

8. Predict how ΔG will change with change in temperature.

9. Calculate the equilibrium constant for a reaction from a knowledge of the standard free-energy change.

key terms

Among the more important terms and expressions used for the first time in this chapter are the following:

Absolute entropy, $S°$ (Section 15.3), refers to the entropy of a particular substance at some temperature and in some particular state, referenced to a zero entropy for the pure solid substance at a temperature of absolute zero.

Entropy (Section 15.2) is a thermodynamic function associated with the number of different, equivalent energy states or spatial arrangements in which a system may be found. It is a thermodynamic state function, which means that once we specify the conditions for a system, that is, the temperature, pressure, and so on, the entropy is defined.

Free energy (Section 15.5) is a thermodynamic state function that combines enthalpy and entropy, in the form $G = H - TS$. For a change occurring at constant temperature and pressure, the change in free energy is $\Delta G = \Delta H - T\Delta S$. The free-energy function is intimately related to equilibrium. When ΔG for a process is negative, the process is spontaneous; that is, it is capable of proceeding in the indicated direction. The **standard free-energy change,** $\Delta G°$, is related to the equilibrium constant, K_c, by the expression, $\Delta G° = -RT \ln K_c$.

The **second law of thermodynamics** (Section 15.1) is a statement of our experience that there is a direction to the way things occur in nature: When a process occurs spontaneously in one direction, it is nonspontaneous in the reverse direction. It is possible to state the second law in many different forms, but they all relate back to the same idea about spontaneity.

A **spontaneous process** (Section 15.1) is one that is capable of proceeding in a given direction, as written or described. A process may be characterized as spontaneous even though it does not actually occur at a measurable rate under a particular set of conditions. For example, the reaction of gaseous H_2 and Cl_2 at room temperature in the dark does not occur at a measurable rate. Yet the process $H_2 + Cl_2 \rightleftharpoons 2HCl$ is spontaneous, because it is capable of proceeding to essentially complete formation of HCl once the reaction is started.

Vibrational and rotational (Section 15.3) energies refer to the storing of energies in molecules in the form of vibrational motions between the atoms of the molecules or rotational motions of the molecule as a whole.

EXERCISES

spontaneous processes

15.1 A nineteenth-century chemist, Marcellin Berthelot, suggested that all chemical processes that proceed spontaneously are exothermic. Is this correct? If you think not, can you offer any counterexamples?

15.2 When a system loses potential energy (for example, a waterfall, building collapse, and so on), that energy must, by the law of conservation of energy, appear in some other form. Make a list of some examples of potential energy and give other forms of energy into which it can be converted. Try to think of examples other than simple mechanical potential energy.

15.3 Suppose we begin with two flasks connected by a closed stopcock as in Figure 15.2. One flask contains substance A, the other substance B, both ideal gases. Then the stopcock is opened. What is the equilibrium state of the system? What changes have occurred in ΔG, ΔH, and ΔS?

15.4 Suppose you worked in the patent office and received a patent application from a person who claimed invention of a catalyst that vastly improves the yield of NO from N_2 and O_2. According to the patent, the catalyst promotes the rate of the forward reaction

$$N_2 + O_2 \longrightarrow 2NO$$

but blocks the reverse reaction

$$2NO \longrightarrow N_2 + O_2.$$

What is your response? Explain.

15.5 Based on your general chemical knowledge, which of the following processes would you classify as spontaneous?

(a) $C(s) + O_2(g) \longrightarrow CO_2(g)$
(b) $2Na(s) + Cl_2(g) \longrightarrow 2NaCl(s)$
(c) $Al_2O_3(s) \longrightarrow 2Al(s) + \frac{3}{2}O_2(g)$
(d) $2N(g) \longrightarrow N_2(g)$
(e) $2H_2O(g) + CO_2(g) \longrightarrow CH_4(g) + 2O_2(g)$

entropy changes

15.6 For several years, farmers in the Midwest used Dieldrin, an insecticide, on their fields. This compound is now found in municipal water samples taken from Lake Michigan and the Mississippi and Missouri rivers. Discuss the distribution of Dieldrin in terms of entropy change and spontaneity.

15.7 When equal volumes of water and ethyl alcohol are stirred together at room temperature, they form a homogeneous mixture. Equal volumes of water and carbon tetrachloride, on the other hand, are essentially immiscible. Assuming equal quantities of substances in both cases, compare the entropy changes that occur in each system on mixing.

15.8 For each of the processes in Sample Exercise 15.5, indicate whether the change in entropy is expected to be positive, negative, or near zero.

15.9 Indicate for each of the following kinds of changes whether the change in entropy is expected to be positive, negative, or near zero. Indicate also whether ΔH is expected to be positive, negative, or near zero: (a) a liquid solidifies; (b) the temperature of a solid is increased by 25 °C; (c) plastic sulfur, a rubbery form of sulfur made by pouring molten sulfur into cold water, slowly forms crystalline rhombic sulfur; (d) carbon tetrachloride evaporates from a beaker; (e) iodine vapor crystallizes on a surface.

15.10 In each of the following pairs, indicate which substance you would expect to possess the larger absolute entropy: (a) 1 mole $Ar(g)$ at 25 K and 1 mole $Ar(g)$ at 25 °C; (b) 1 mole $H_2O(g)$ at 80 °C and 1 mole $H_2O(l)$ at 80 °C; (c) 1 mole $KClO(s)$ at 25 °C and 1 mole $KCl(s)$ mixed with $\frac{1}{2}$ moles $O_2(g)$ at 25 °C; (d) 1 mole $HCl(g)$ at 25 °C and 1 mole $CCl_4(g)$ at 25 °C. Give a reason for your choice in each case.

15.11 Indicate whether in each of the following reactions, you would expect the entropy change to be positive or negative. Explain.

(a) $CaH_2(s) + H_2O(l) \longrightarrow$
    $\qquad Ca(OH)_2(s) + H_2(g)$
(b) $Ag_2O(s) \longrightarrow 2Ag(s) + \frac{1}{2}O_2(g)$
(c) $6H_2(g) + P_4(g) \longrightarrow 4PH_3(g)$
(d) $2Ni(s) + O_2(g) \longrightarrow 2NiO(s)$

free energy and equilibrium

15.12 The free-energy function, G, is said to be a state function. What is the significance of this statement?

15.13 How does the change in the free energy for a system relate to the spontaneity or nonspontaneity of a given process? How does it relate to the rate at which the process occurs?

15.14 For each of the following reactions at 25 °C, use Appendix E to determine the value for $\Delta G°$. Indicate which reactions proceed substantially to form products.

(a) $CO(g) + 2H_2(g) \rightleftharpoons CH_3OH(g)$

(b) $2HCl(g) + Br_2(g) \rightleftharpoons 2HBr(g) + Cl_2(g)$

(c) $2SO_2(g) + O_2(g) \rightleftharpoons 2SO_3(g)$

(d) $2N_2O(g) \rightleftharpoons 2N_2(g) + O_2(g)$

[15.15] The reaction

$$SO_2(g) + 2H_2S(g) \longrightarrow 3S(s) + 2H_2O(g)$$

is the basis of one suggested method for removal of SO_2 from power-plant stack gases. The standard molar free energies at $25°C$ for all substances involved are: $SO_2(g)$, -300.4 kJ; $H_2O(g)$, -228.6 kJ; $S(s)$, 0 kJ; $H_2S(g)$, -33.0 kJ. (a) Is this reaction at least in principle a feasible method for removal of SO_2? Calculate the value of the equilibrium constant for this reaction as written. (b) Assuming 25 torr H_2O vapor pressure and adjusting conditions so that P_{SO_2} equals P_{H_2S}, calculate the equilibrium SO_2 pressure in this system. (Remember to use gas pressures in atmospheres.) (c) The reaction as written is exothermic. Would you expect the process to be more effective or less effective at higher temperatures?

15.16 If any of the following statements are untrue, reword the statement so that it is true. (a) At constant temperature and pressure, a system proceeds spontaneously to the condition of minimum enthalpy. (b) Reactions for which both ΔH and ΔS are positive tend to go to completion at high temperatures. (c) At constant temperature and pressure, equilibrium corresponds to the condition of minimum free energy. (d) The free-energy change for an exothermic reaction is always negative. (e) The free energy change for a process is directly proportional to the amount of substance taking part in the process.

free energy and temperature

15.17 On the basis of the $\Delta H°$ and $\Delta S°$ values given for each reaction below, indicate (a) which are spontaneous at $25°C$ and (b) which can be expected to be spontaneous at high temperatures.

	$\Delta H°$ (kJ)	$\Delta S°$ (J/K)
(a) $KClO_3(s) \longrightarrow$		
$KCl(s) + \frac{3}{2}O_2(g)$	-44.7	$+59.1$
(b) $2Al(s) + 3Cl_2(g) \longrightarrow$		
$2AlCl_3(s)$	-332	-93.4
(c) $NOCl(g) \longrightarrow$		
$NO(g) + \frac{1}{2}Cl_2(s)$	$+37.6$	$+58.5$

15.18 The most widely used process at present for removal of SO_2 from stack gases in coal-burning power plants involves the reaction

$$CaO(s) + SO_2(g) \longrightarrow CaSO_3(s)$$

How would you expect the equilibrium constant for this process to vary with temperature? Based on your considerations, would you advocate putting the CaO into the system in the furnace where the $SO_2(g)$ is formed and temperatures are highest, or just before the gases generated by the power plant are vented to the stack?

free energy and the equilibrium constant

15.19 Using the data of question 14.15, calculate the standard free-energy change, $\Delta G°$, for the following reaction at $1050°C$:

$$C(s) + CO_2(g) \rightleftharpoons 2CO(g)$$

15.20 Using the data of Appendix E, determine the standard free-energy change and equilibrium constant for the reaction

$$2NO(g) + 3H_2O(g) \longrightarrow 2NH_3(g) + \frac{5}{2}O_2(g)$$

From your results, does it appear that this reaction could be of importance in removal of NO from an auto exhaust in a catalytic muffler at about $1000°C$? (There is quite a lot of $H_2O(g)$ in the exhaust.)

15.21 The reaction

$$2CO_2(g) \rightleftharpoons 2CO(g) + O_2(g)$$

is discussed in Section 14.4. From the data given there, how does the free-energy change for the reaction as written vary with an increase in temperature? Explain.

15.22 From the following absolute molar entropies at $25°C$:

$KBrO_3(s)$	$S° = 149.2$ J/K-mole
$KBr(s)$	$S° = 96.4$ J/K-mole
$O_2(g)$	$S° = 205.0$ J/K-mole

calculate $\Delta S°$ for the reaction

$$KBrO_3(s) \longrightarrow KBr(s) + \frac{3}{2}O_2(g)$$

Is the value you calculate in accord with your qualitative expectations?

Using the following values for $\Delta H°$, calculate the standard free-energy change and the equilibrium constant for the above reaction at $25°C$.

$KBrO_3(s)$	$\Delta H° = -332.2$ kJ/mole
$KBr(s)$	$\Delta H° = -392.0$ kJ/mole

Is the reaction spontaneous as written? Write the actual equilibrium constant expression. What are the units in which the concentrations of the reactants and products should be expressed?

15.23 In photosynthesis in plants the primary reaction which occurs is essentially the reverse of reaction Equation [15.15]. How do you account for the fact that such a highly nonspontaneous process occurs?

15.24 The reaction between molten sodium metal and chlorine gas,

$$2Na(l) + Cl_2(g) \longrightarrow 2NaCl(s)$$

is a highly exothermic, spontaneous process. Yet by applying a voltage across electrodes in molten NaCl, the reaction can be made to reverse:

$$2NaCl(l) \xrightarrow{\text{electricity}} 2Na(l) + Cl_2(g)$$

Is this a violation of the second law of thermodynamics? Explain.

15.25 Which of the following processes are spontaneous: (a) the freezing of water at $+5°C$; (b) formation of a solution when ethyl alcohol is added to water; (c) formation of ammonia from H_2 and N_2 at room temperature and 300 atm pressure.

15.26 What distinction must be made between the spontaneity of a reaction and the rate at which it proceeds?

15.27 Using Appendix E, compare the absolute entropy values at 298 K for the substances in each of the following pairs and explain the origin of the difference in $S°$ values in each case: (a) $Br_2(g)$ and $Br_2(l)$; (b) $Al(s)$ and $Cl_2(g)$; (c) $O_2(g)$ and $O_3(g)$; (d) $I(g)$ and $I_2(g)$.

15.28 It has been said that many pollution problems are fundamentally entropy problems. Entropy changes occur when a substance is released to the environment, and energy is required to reverse the dispersal processes. Discuss the significance of these comments, using DDT in the environment as an example.

15.29 The formation of methyl alcohol, CH_3OH, from CO and H_2 in the gas phase has been proposed as a means of using two common by-products of coal gasification. Calculate the standard free-energy change for the formation of CH_3OH, based on a balanced equation. Is the process feasible? Also calculate the absolute entropy change and predict whether the reaction will be more spontaneous or less spontaneous at higher temperatures. What effect would increased pressure have on the extent to which methanol is formed?

acids and bases

16

One of the most important classifications of substances is in terms of acid and base properties. From the earliest days of experimental chemistry it was recognized that certain substances, called acids, possess a sour taste and are capable of dissolving active metals such as zinc. Acids also cause vegetable dyes to turn a characteristic color; for example, litmus, which is obtained from certain lichens (a composite plant made up of an alga and a fungus), turns red on contact with acids. Like acids, bases possess a set of characteristic properties that can be used to identify them. Whereas acids have a sour taste (the sour taste of lemons is due to the presence of citric acid in lemon juice), bases have a characteristic bitter taste. Bases also feel slippery to the touch. Like acids, bases also cause litmus to change its color, but, whereas acids cause litmus to turn red, bases cause it to turn blue. Bases also react with many dissolved metal salts to form precipitates.

    The fact that all acids and all bases show certain characteristic chemical properties suggests that there must be an essential feature common to all the members of each class. In this chapter, we shall learn what those features of acids and bases are. We shall see how the properties of acidic and basic solutions can be understood in terms of our current understanding of structure and bonding. We shall see also that the properties of acids and bases, as we generally encounter them, are very much dependent on the fact that water is the solvent in which those properties are observed. To appreciate how really remarkable an aqueous (water) acid solution is, let's begin by considering a very common chemical reagent, hydrochloric acid.

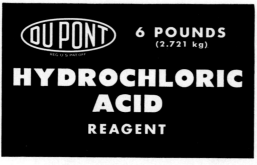

FIGURE 16.1 Reproduction of a label from a bottle of reagent grade, concentrated hydrochloric acid (*Courtesy of E. I. du Pont de Nemours & Co.*).

16.1
properties of acidic solutions

Figure 16.1 is a reproduction of the label from a bottle of reagent grade* concentrated hydrochloric acid. Although this label might at first seem to be rather dull subject matter, it contains fascinating and even amazing data. Consider first some properties of HCl. It is a gaseous substance, formed in commerce by the controlled reaction of hydrogen and chlorine:

$$H_2(g) + Cl_2(g) \longrightarrow 2HCl(g)$$

Hydrogen chloride liquefies under 1 atm pressure at $-84°C$, and freezes at $-112°$. Liquid HCl is a difficult and most unpleasant substance to work with, but some enterprising and persistent persons have learned that the liquid is a very poor conductor of electricity. This suggests that there are very few ions present in the liquid. Dry HCl gas is soluble to only a limited extent in a dry organic solvent such as benzene. The solutions do not conduct electric current; addition of an active metal such as zinc produces no evident chemical reaction.

With these facts in mind let us now examine the label shown in Figure 16.1. We note that concentrated aqueous hydrochloric acid consists of 37 percent by weight of hydrogen chloride; HCl is obviously very soluble in water. At 15°C, one volume of water dissolves up to 450 volumes of dry HCl gas at 1 atm pressure! This high solubility is nicely shown by the hydrogen chloride fountain, a popular lecture demonstration, illustrated in Figure 16.2. The molarity of concentrated hydrochloric acid solution is about 12.0 *M*, as shown in Sample Exercise 16.1. To have some appreciation for just how remarkable this solubility is, you should know that argon, which has essentially the same molecular weight as HCl, is soluble in water at 15°C and 1 atm pressure to the extent of only 0.002 *M*.

*The term reagent grade denotes that a substance is of high purity.

From the data shown in Figure 16.1, calculate the molarity of a concentrated hydrochloric acid solution.

Solution: We see that the specific gravity, or density, of the solution is 1.19 g/cm^3. Using this and the stated weight percentage, we have:

$$\left(\frac{1.19 \text{ g sol}}{1 \text{ cm}^3}\right)\left(\frac{1000 \text{ cm}^3}{\text{liter}}\right)\left(\frac{0.37 \text{ g HCl}}{1 \text{ g soln}}\right)$$

$$\times \left(\frac{1 \text{ mole HCl}}{36.5 \text{ g HCl}}\right) = \frac{12 \text{ moles HCl}}{\text{liter soln}}$$

$$= 12\ M$$

In contrast with the properties of a solution of HCl in dry benzene, an aqueous HCl solution of the same concentration reacts readily with zinc, with evolution of H_2 gas. Furthermore, the aqueous solutions strongly conduct an electrical current, indicating the presence of ions. Svante Arrhenius, who first correctly interpreted the nature of solutions of electrolytes (Section 12.5), realized that solutions that could be classified as strongly acidic *were also strongly conducting*. He noted also that HCl, HNO_3, and H_2SO_4, all of which are capable of forming strongly acidic solutions in water, share in common the fact that they contain hydrogen capable of producing an excess of H^+ ions in water. Solutions of these same substances also all react with zinc metal, with evolution of hydrogen gas. He therefore defined an acid as any substance capable of producing an excess of H^+ ions in water. Arrhenius's definition of an acid is quite reasonable, but it doesn't tell us much about what actually happens when an acid dissolves in water. To appreciate the nature of acidic solutions, and thus understand more clearly why acids behave as they do, we must look more closely into the question of how and why a substance such as HCl reacts with water to form ions.

NATURE OF THE HYDRATED PROTON

What is it about water that causes a molecule such as HCl to come apart, or dissociate, into H^+ and Cl^- ions? This question was addressed very briefly in Section 12.5; let's now consider it in more detail. The first and most obvious point to be made is that water is a polar liquid. Water molecules have dipole moments (Section 8.7), and the liquid has a high dielectric constant. This means that it has a high capacity to maintain separated charges. There must be more than this involved, however, because liquid HCl is also a polar liquid, and we've already noted that it shows only a very low electrical conductance. The key to the behavior of water lies in the unshared electron pairs on the oxygen of the water molecule; these are capable of bonding to the hydrogen ion in the following manner:

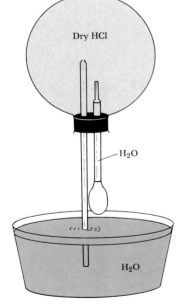

FIGURE 16.2 The hydrogen chloride fountain. The large flask is filled with dry hydrogen chloride gas at 1 atm pressure. When a small amount of water is introduced into the flask by squeezing the medicine dropper, the HCl dissolves in it very rapidly. This causes the pressure in the flask to decrease, and water is forced up the glass tube. The spray of water continues until the flask is almost completely filled.

$$\text{Cl—H} \text{---} :\overset{\cdot\cdot}{\text{O}}\text{—H} \longrightarrow \text{Cl}^- + \left[\text{H—}\overset{\cdot\cdot}{\text{O}}\text{—H}\right]^+ \qquad [16.1]$$

Here we've shown a reaction occurring between a single HCl molecule and a single water molecule. Transfer of a proton from HCl to the water molecule results in formation of H_3O^+, called the **hydronium ion**. You might note that this ion is isoelectronic (that is, it has the identical electronic structure) with ammonia, NH_3.

455

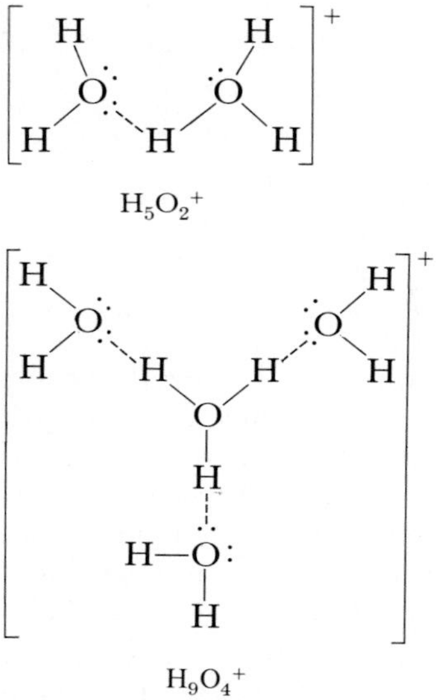

$H_5O_2^+$

$H_9O_4^+$

FIGURE 16.3 Two possible forms for the proton in water, in addition to H_3O^+. Experimental evidence indicates the existence of both these species.

The situation in water is actually much more complex than Equation [16.1] suggests. We've learned (Section 11.3) that hydrogen bonds exist throughout liquid water. The existence of this hydrogen-bond network is responsible for many of the special properties of water, for example, its high polarity and high melting and boiling points. Much research has been devoted to learning how H^+ ions fit into the complex structure of liquid water. Experimental studies show that, in part, the H^+ ions must exist as hydronium ions. In fact, it is possible to isolate salts of the form $H_3O^+Cl^-$, $H_3O^+ClO_4^-$, and others, in which there is clearly an H_3O^+ ion in the solid lattice. But just as water molecules are strongly hydrogen bonded to one another, the H_3O^+ ion in solution is hydrogen bonded to other water molecules. Thus, ions such as the two shown in Figure 16.3 are possible and have been shown to form.

We must conclude from these observations that no single species can adequately represent the proton in solution. We use the symbol $H^+(aq)$ to represent the hydrated, or aquated, hydrogen ion, that is, the hydrogen ion surrounded by solvent water. As you learn more about the properties of acidic solutions, do not lose sight of the important fact that *acidic solutions are formed by a chemical reaction in which an acid transfers a proton to water*. We write the reaction of HCl with water to form an acidic, conducting solution as shown in Equation [16.2]:

$$HCl(g) \xrightarrow{H_2O} H^+(aq) + Cl^-(aq) \qquad [16.2]$$

The chloride ion formed in the reaction with water is also hydrated (surrounded by water molecules), though not in the same way the proton is. The hydration of ions is expressed in equations by the symbol (aq) following the symbol for the ion.

16.2 Brønsted-Lowry theory of acids and bases

The nature of the reaction between an acid and water as we have just described it was first appreciated by J. Brønsted and T. M. Lowry. In 1923 they independently put forward essentially the same model for acid-base behavior in water and similar solvents. Brønsted and Lowry recognized that acid-base behavior in such solvents could be described in terms of the ability of substances to transfer protons. In the Brønsted-Lowry model, an *acid is any substance capable of donating or giving up a proton*, and a *base is any substance capable of accepting a proton*. In these terms, when HCl dissolves in water, it acts as an acid in donating a proton to the solvent. The solvent itself then acts as a base in accepting a proton from HCl. To see the implications of this idea, let us now look at a wide range of substances we call acids and note the variations in the extent to which they behave as acids.

STRONG AND WEAK ACIDS IN WATER

In terms of the Brønsted-Lowry concept, a strong acid is any substance that reacts completely with water to form $H^+(aq)$, and a weak acid is a substance that only partially so reacts. You will recall from Chapter 12 that the number of strongly acidic substances is not very large. Among

TABLE 16.1

some acids that are strong acids in water

ACID	FORMULA
Hydrochloric	HCl
Hydrobromic	HBr
Hydroiodic	HI
Nitric	HNO_3
Perchloric	$HClO_4$
Chromic	H_2CrO_4
Permanganic	$HMnO_4$
Sulfuric	H_2SO_4

the acids commonly available in the laboratory, HCl, H_2SO_4, and HNO_3 are strong. Table 16.1 lists these and several other less commonly available strong acids. We can consider the aqueous solutions of all these substances to consist entirely of ions, with no significant concentration of neutral solute molecules remaining.

For a great many other substances that show acidic properties on dissolving in water, the reaction with water is incomplete. For example, hydrofluoric acid, HF, reacts with water to only a slight extent. Thus, it is classified as a weak acid (weak electrolyte). We can write the reaction with water as an equilibrium:

$$HF(aq) \xrightleftharpoons{H_2O} H^+(aq) + F^-(aq) \qquad [16.3]$$

In a dilute solution of HF in water, say 0.2 M, about 95 percent of the HF is present in the form of the nonionized molecules. Obviously in this case the ability of the acid to transfer a proton to the solvent water is much lower than in the case of HCl.

In the Brønsted-Lowry model, HF acts as an acid in transferring a proton to solvent water, which is the base, or proton acceptor. But reaction Equation [16.3] is reversible. In the reverse reaction, fluoride ion acts as a proton acceptor in taking the proton from solvent water to form neutral HF. Thus, in the Brønsted-Lowry sense, fluoride ion is a base. It must be a much stronger base than Cl^-, because there is no tendency for reaction Equation [16.2] to occur in the reverse direction. Because HCl is completely ionized in aqueous solution, we can conclude that chloride ion, Cl^-, is a weak base; that is, it has a very weak tendency to act as a proton acceptor. By contrast, fluoride ion is a relatively strong base. Because only a small fraction of HF molecules present in solution are ionized, F^- must be a relatively stronger base than solvent water.

What these examples have shown is that in the Brønsted-Lowry model every acid has associated with it a **conjugate base** that is formed from the acid by loss of a proton. The conjugate base of HCl is Cl^-; the conjugate base of HF is F^-. Furthermore, we've seen that the stronger an acid, the weaker its conjugate base. We'll return shortly to a closer look at the relationships between acids and their conjugate bases.

Hydrocyanic acid, HCN, is ionized in water by the reaction

$$HCN(aq) \underset{}{\overset{H_2O}{\rightleftharpoons}} H^+(aq) + CN^-(aq)$$

to a lesser extent than is an HF solution of the same concentration. What is the conjugate base of HCN? Is it a stronger or weaker base than F^-?

Solution: The conjugate base of HCN is CN^-, the ion that remains after a proton has been lost to the solvent. Because HCN dissociates to a lesser extent than HF, this means that the tendency of the reverse reaction to occur is greater. In the reverse reaction the conjugate base, CN^-, accepts a proton from the solvent. In other words, CN^- is a stronger conjugate base than is F^-.

16.3 acid-dissociation constants

As we've noted, most substances that are acidic in water are actually weak acids. The extent to which an acid ionizes in aqueous medium can be expressed by the equilibrium constant for the ionization reaction. In general we can represent any acid by the symbol HX (or XH), where X^- is the formula for the conjugate base that remains when the proton ionizes. The ionization equilibrium is then given by Equation [16.4]:

$$HX(aq) \rightleftharpoons H^+(aq) + X^-(aq) \tag{16.4}$$

The corresponding equilibrium-constant expression is*

$$K_a = \frac{[H^+][X^-]}{[HX]} \tag{16.5}$$

where K_a is called the **acid-dissociation constant**. Table 16.2 shows the names, structures, and values of K_a for several weak acids. Note that many weak acids are compounds composed largely of carbon and hydrogen. Generally speaking, hydrogen atoms bound to carbon are not ionized in an aqueous medium. The ionizable hydrogens are in most instances bound to oxygen. The smaller the value for K_a, the weaker the acid. For example, phenol is the weakest acid listed in Table 16.2.

From the value for K_a it is possible to calculate the concentration of $H^+(aq)$ in a solution of a weak acid. For example, consider acetic acid, $HC_2H_3O_2$, the substance that gives the characteristic odor and acidic properties to vinegar. Let us calculate the concentration of $H^+(aq)$ in a 0.10 M solution of acetic acid.

The first step in solving any equilibrium problem is to write the equation for the equilibrium reaction. The ionization equilibrium for acetic acid can be written as

$$HC_2H_3O_2 \rightleftharpoons H^+(aq) + C_2H_3O_2^-(aq) \tag{16.6}$$

The second step is to write the equilibrium-constant expression and

*This is just a reminder that square brackets around the symbol for a chemical species means the concentration of that species. For all species that are in aqueous solution the concentration unit employed is moles/liter, M. To keep the expressions as simple as possible, we will drop the (aq) labels in writing equilibrium-constant expressions. We have also dropped the (aq) labels from all species in solution except ions, which are the species that interact most strongly with the solvent water.

TABLE 16.2

some weak acids in water[a]

ACID	FORMULA	LEWIS STRUCTURE	CONJUGATE BASE	K_a (AT 25°C)
Hydrofluoric	HF	H—F	F^-	6.8×10^{-4}
Hydrocyanic	HCN	H—C≡N	$C≡N^-$	4.9×10^{-10}
Acetic	$HC_2H_3O_2$	H—O—C(=O)—C(H)(H)—H	$C_2H_3O_2{}^-$	1.8×10^{-5}
Formic	$HCHO_2$	H—O—C(=O)—H	$CHO_2{}^-$	1.7×10^{-4}
Benzoic	$HC_7H_5O_5$	H—O—C(=O)—C_6H_5	$C_7H_5O_2{}^-$	6.5×10^{-5}
Nitrous	HNO_2	H—O—N=O	$NO_2{}^-$	4.5×10^{-4}
Carbonic	H_2CO_3	H—O—C(=O)—O—H	$HCO_3{}^-$	4.3×10^{-7}
Phenol	HOC_6H_5	H—O—C_6H_5	$C_6H_5O^-$	1.3×10^{-10}
Phosphoric	H_3PO_4	(HO)—P(=O)(O—H)—O—H	$H_2PO_4{}^-$	7.5×10^{-3}
Ascorbic (vitamin C)	$HC_6H_7O_6$	(ring structure)	$C_6H_7O_6{}^-$	8.0×10^{-5}

[a] In cases where hydrogen is present in more than one chemical environment in the molecule, the one that ionizes is shown in color.

the value for the equilibrium constant, if that is known. From Table 16.2 we have $K_a = 1.8 \times 10^{-5}$. Thus, we can write the following:

$$K_a = \frac{[H^+][C_2H_3O_2{}^-]}{[HC_2H_3O_2]} = 1.8 \times 10^{-5} \qquad [16.7]$$

As the third step, we need to express the concentrations that make up the equilibrium-constant expression. This can be done with a little accounting, which can conveniently be done below the expression for the equilibrium:

	$HC_2H_3O_2$	$\rightleftharpoons$	$H^+(aq)$	+	$C_2H_3O_2{}^-(aq)$
Initial concentration:	0.10 M		0 M		0 M
Equilibrium concentration:	$(0.10 - x)$ M		x M		x M

459

Because we seek to find a value for $[H^+]$, let us call this quantity x. The concentration of acetic acid before any of it dissociates is $0.10\ M$. The equation for the equilibrium tells us that for each molecule of $HC_2H_3O_2$ that dissociates, one $H^+(aq)$ and one $C_2H_3O_2^-(aq)$ are formed. Thus, if x moles per liter of $H^+(aq)$ are formed at equilibrium, x moles per liter of $C_2H_3O_2^-(aq)$ must also have formed, and x moles per liter of $HC_2H_3O_2$ must have been dissociated. This gives rise to the equilibrium concentrations shown above.

As the fourth step of the problem, we need to substitute the equilibrium concentrations into the equilibrium-constant expression. The substitution gives the following equation:

$$K_a = \frac{[H^+][C_2H_3O_2^-]}{[HC_2H_3O_2]} = \frac{(x)(x)}{(0.10 - x)} \qquad [16.8]$$

Because this equation has only one unknown it can be solved using algebra. However, the solution is a little tedious, because it requires use of the quadratic formula (Appendix A.3). By taking account of what is actually occurring in the solution, we can make things simpler for ourselves. Because the value of K_a is small, we might guess that x will be quite small. (In other words, perhaps only a small fraction of the $HC_2H_3O_2$ is actually ionized.) Indeed, if we solve the problem using the quadratic formula we find that $x = 1.3 \times 10^{-3}\ M$. Now you know that if a small number is subtracted from a much larger one, the result is approximately equal to the larger number. In our example we have

$$(0.10 - x) = (0.10 - 0.0013) \simeq 0.10 \qquad [16.9]$$

We can therefore make the approximation of ignoring x relative to 0.10 in the denominator of Equation [16.8]. This leads us to the following simplified expression:

$$K_a = \frac{(x)(x)}{(0.10)} = 1.8 \times 10^{-5} \qquad [16.10]$$

Solving for x we have

$$x^2 = (0.10)(1.8 \times 10^{-5}) = 1.8 \times 10^{-6}$$
$$x = \sqrt{1.8 \times 10^{-6}} = 1.3 \times 10^{-3}\ M = [H^+] \qquad [16.11]$$

From the value calculated for x we see that our simplifying approximation is quite reasonable. This type of approximation can be used whenever conditions in solution are such that only a small fraction of acid ionizes. As a general rule, if the quantity x, which is subtracted from the initial concentration of the acid, is more than about 5 percent of the initial value, it is best to use the quadratic formula. In cases of doubt, assume that the approximation is valid and solve for x in the simplified equation. Compare this approximate value of x with the initial concentraction of acid. If it is more than about 5 percent as large, the problem should be reworked using the quadratic formula. For example, if the initial concentration of acid were $0.05\ M$, and x turned out in a given case to be $0.0016\ M$, then

$$\left(\frac{0.0016\ M}{0.05\ M}\right)(100) = 3.2\%$$

In this case, the approximate solution gives a sufficiently accurate answer for our purposes.

SAMPLE EXERCISE 16.3

Calculate the concentration of $H^+(aq)$ in a $0.2\ M$ solution of HCN (refer to Table 16.2 for value of K_a).

Solution: Proceeding as in the example worked out above, we write:

$$HCN \rightleftharpoons H^+(aq) + CN^-(aq)$$

$$K_a = \frac{[H^+][CN^-]}{[HCN]} = 4.9 \times 10^{-10}$$

Let $x = [H^+]$ at equilibrium. Then we have the following concentrations:

$$HCN \rightleftharpoons H^+(aq) + CN^-(aq)$$

Initial concentration:	$0.20\ M$	$0\ M$	$0\ M$
Equilibrium concentration:	$(0.20 - x)\ M$	$x\ M$	$x\ M$

Substituting into the equilibrium constant expression:

$$K_a = \frac{(x)(x)}{(0.20 - x)} = 4.9 \times 10^{-10}$$

We next make the simplifying approximation that x, the amount of acid that dissociates, is small in comparison with the initial concentration of acid; that is,

$$(0.20 - x) \simeq 0.20.$$

Thus

$$\frac{x^2}{0.20} = 4.9 \times 10^{-10}$$

Solving for x we have

$$x^2 = (0.20)(4.9 \times 10^{-10})$$
$$= 0.98 \times 10^{-10}$$
$$x = \sqrt{0.98 \times 10^{-10}}$$
$$= 0.99 \times 10^{-5} = 9.9 \times 10^{-6} = [H^+]$$

The result obtained in Sample Exercise 16.3 is typical of the behavior of weak acids; the concentration of $H^+(aq)$ is only a small fraction of the concentration of the acid in solution. Thus, those properties of the acid solution that relate directly to the concentration of $H^+(aq)$, such as electrical conductivity or rate of reaction with an active metal, are much less in evidence for a solution of a weak acid than for a solution of a strong acid. Figure 16.4 illustrates an experiment often carried out in the chemistry laboratory to demonstrate the difference in concentration of $H^+(aq)$ in weak and strong acid solutions of the same concentration. The rate of reaction with the active metal is much faster for the solution of a strong acid. Reactions in which the rate depends on $H^+(aq)$ are common. For reactions occurring in homogeneous solution, and for which the rate law is known, the dependence of rate on acid strength can be estimated quantitatively, as shown in Sample Exercise 16.4.

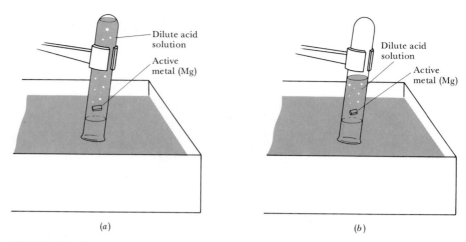

(a)   (b)

FIGURE 16.4 Demonstration of the relative rates of reaction of two acid solutions of the same concentration with an active metal. The solution in (a) is that of a weak acid, in (b) that of a strong acid. Reaction produces $H_2(g)$, which collects in the tube. From the relative amounts of gas collected in the two tubes after a period of time, it is evident that reaction is faster in (b). This indicates that even though the concentrations of acid are the same in the two tubes, the concentration of H^+ (aq) is much greater in (b).

SAMPLE EXERCISE 16.4

Experimental studies show that the rate of a particular chemical reaction is first order in $H^+(aq)$. Based on the results obtained in the calculation carried out earlier, what ratio of initial rates would you expect for the reaction run in 0.10 M HCl as compared with 0.10 M $HC_2H_3O_2$ solution, with all other conditions the same?

Solution: In a 0.10 M solution of HCl, the concentration of $H^+(aq)$ is essentially 0.10 M, because HCl is a strong acid. In a 0.10 M solution of acetic acid, the concentration of $H^+(aq)$ is calculated to be 0.0013 M. Thus in the HCl solution, the concentration of $H^+(aq)$ is 0.1/0.0013, or 77 times higher. If the reaction rate is proportional to $[H^+]$, the initial reaction rate will be 77 times faster in the solution of the strong acid.

Figure 16.5 illustrates an experiment in which the electrical conductivity of an HCl solution is compared with the conductivity of an HF solution. The conductivity of the solution of the strong acid increases approximately in proportion to the concentration. This is what one would expect; because all the acid molecules ionize, the concentration of ions in solution is directly proportional to the concentration of acid. The conductivity of the solution of the weak acid is very much less than that for a strong acid and does not vary linearly with the acid concentration. The nonlinearity of the graph arises from the fact that the percentage of acid ionized varies with the acid concentration. This is illustrated in Sample Exercise 16.5.

SAMPLE EXERCISE 16.5

Calculate the percentage of HF molecules ionized in a 0.10 M HF solution; in a 0.01 M HF solution.

Solution: The equilibrium reaction and equilibrium concentrations can be written as follows:

	HF $\rightleftharpoons$	$H^+(aq)$	+ $F^-(aq)$
Initial concentration:	0.10 M	0 M	0 M

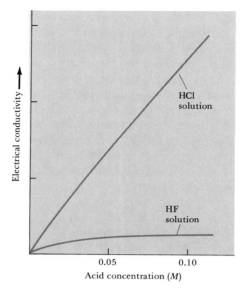

FIGURE 16.5 Electrical conductivity versus concentration for solutions of HCl, a strong acid, and HF, a weak acid. The conductivity of the HCl solution is not completely linear with concentration because of attractive forces between the ions at higher concentrations (Section 12.7). The conductivity for the HF solution is quite nonlinear with concentration and much lower than for HCl because only a fraction of the HF molecules ionize. It is nonlinear with concentration because the fraction of molecules ionizing changes with concentration.

Equilibrium
concentration: $(0.10 - x)\,M$ $x\,M$ $x\,M$

The equilibrium-constant expression is as follows:

$$K_a = \frac{[H^+][F^-]}{[HF]} = \frac{(x)(x)}{(0.10 - x)} = 6.8 \times 10^{-4}$$

We might be tempted to try solving this equation using the same approximation used in earlier examples, that is, by neglecting the concentration of acid that ionizes in comparison with the initial concentration (by neglecting x in comparison with 0.10). However K_a is large enough in this case to make that a poor approximation. We must therefore rearrange our equation, and write it in standard quadratic form:

$$x^2 = (0.10 - x)(6.8 \times 10^{-4})$$
$$= 6.8 \times 10^{-5} - (6.8 \times 10^{-4})x$$
$$x^2 + (6.8 \times 10^{-4})x - 6.8 \times 10^{-5} = 0$$

Solving this equation using the quadratic formula gives

$$x = [H^+] = [F^-] = 7.9 \times 10^{-3}\,M$$

(You should solve this problem using the quadratic formula in order to satisfy yourself about the answer and to be sure that you know how to do a problem of this sort. You might also see what answer you would get by making the simplifying approximation of neglecting x with respect to 0.10.)

From our result we can calculate the percent of molecules ionized:

Percent
ionized $= \left(\dfrac{\text{concentration ionized}}{\text{original concentration}}\right)(100)$

$$= \left(\frac{7.9 \times 10^{-3}\,M}{0.10\,M}\right)(100) = 7.9\%$$

Proceeding similarly for the 0.01 M solution we have

$$\frac{x^2}{(0.01 - x)} = 6.8 \times 10^{-4}$$

Solving the resultant quadratic expression we obtain

$$x = [H^+] = [F^-] = 2.3 \times 10^{-3}\,M$$

The percentage of molecules ionized is

$$\left(\frac{0.00234}{0.0100}\right)(100) = 23.4\%$$

Notice that in diluting the solution by a factor of 10, the percentage of molecules ionized increases by a factor of 3. We could have arrived at this conclusion qualitatively by applying LeChatelier's principle (Section 14.3) to the equilibrium represented in Equation [16.3]. There are more "particles" or reaction components on the right side of the equation. Dilution causes the reaction to shift in the direction of the larger number of particles, because this counters the effect of the decreasing concentration of particles.

The lower concentration of ions in solutions of weak acids is evident from the colligative properties of the solutions (Section 12.7). A solution of HCl causes a lowering of the freezing point or an increase in the boiling point of water comparable to that brought about by an NaCl solution of the same concentration. On the other hand, a solution of HCN or acetic acid has about the same effect as that of a sugar solution of the same concentration. This is so because only a tiny fraction of the weak acid molecules is ionized. The total number of solute particles is therefore nearly equal to the number of neutral molecules dissolved. Thus the freezing-point or boiling-point change caused by a solution of a weak acid is only about half that caused by a solution of a strong acid of the same concentration.

**16.4
autoionization
of water**

Solutions of salts and of strong acids in water are highly conducting. We might suppose that in water free of any dissolved substances the conductivity would be zero, because there would be no source of ions for carrying current. However, even the purest water shows a trace of electrical conductivity. Even in pure water, then, there remains a small concentration of ions. The ions arise because water is capable of acting as a proton donor and proton acceptor toward itself. The process by which this occurs is called **autoionization:**

$$H-\ddot{O}: + H-\ddot{O}: \rightleftharpoons \left[H-\underset{\underset{H}{|}}{\overset{..}{O}}-H^+ \right] + :\ddot{O}-H^- \qquad [16.12]$$

This reaction amounts to a spontaneous ionization of the solvent. It occurs only to a very small extent. At room temperature, only about one out of every 10^8 molecules is in the ionic form at any one instant. We know that water is a strongly hydrogen-bonded liquid. Thus, the hydrogen of one molecule may be attracted to an unshared pair of electrons on the oxygen of an adjacent molecule. Occasionally the hydrogen will transfer to the other molecule. Perhaps simultaneously there will be a transfer of H^+ from some other molecule to the oxygen which is losing a hydrogen ion. As a result of these ready transfers of H^+ from one molecule to another, there is, on the average, a certain very small fraction of molecules in the ionized form. No one molecule remains in that condition for long; the equilibria are extremely rapid. It has been found that, on the average, a proton transfers from one molecule to another in water at the rate of about 1000 times per second.

By comparing the reaction shown in Equation [16.12] with the ionization of an acid, as in reaction Equation [16.1], we see that one water molecule is acting as an acid, transferring a proton to the solvent and generating a conjugate base, in this case OH^-. We can thus rewrite Equation [16.12] in the same form as that for ionization of a weak acid:

$$H_2O \rightleftharpoons H^+(aq) + OH^-(aq) \qquad [16.13]$$

The equilibrium-constant expression for this reaction is

$$K_c = \frac{[\text{H}^+][\text{OH}^-]}{[\text{H}_2\text{O}]} \qquad [16.14]$$

The term in the denominator of this expression represents the concentration of water in pure water. We can calculate this concentration as follows:

$$\left(\frac{1000 \text{ g H}_2\text{O}}{1 \text{ liter H}_2\text{O}}\right)\left(\frac{1 \text{ mole H}_2\text{O}}{18.0 \text{ g H}_2\text{O}}\right) = \frac{55.5 \text{ moles H}_2\text{O}}{1 \text{ liter H}_2\text{O}}$$

Because the concentration of H_2O is essentially the same for any dilute solution, we can rewrite Equation [16.14] so as to define a new constant:

$$K_c[\text{H}_2\text{O}] = K_c(55.5) = K_w = [\text{H}^+][\text{OH}^-] \qquad [16.15]$$

This very important expression is called the **ion-product constant** for water. From careful measurements of conductivities in pure water, it has been determined that K_w has the value 1×10^{-14} at $25\,°\text{C}$. This is an important equilibrium constant; you should memorize it:

$$K_w = 1 \times 10^{-14} = [\text{H}^+][\text{OH}^-] \qquad [16.16]$$

The ion-product constant for water is a true constant; Equation [16.16] remains a valid expression even when the H^+ and OH^- concentrations are not equal. In acid solutions, $[\text{H}^+]$ is much greater than $[\text{OH}^-]$. In base solutions, the reverse is true. Only in neutral solutions are the two quantities exactly equal. We'll see in Section 16.9 that there is a very convenient and widely used way of expressing the relative concentrations of $\text{H}^+(aq)$ and $\text{OH}^-(aq)$. For now you should remember that:

$[\text{H}^+] > [\text{OH}^-]$ in acid solutions

$[\text{H}^+] < [\text{OH}^-]$ in basic solutions

$[\text{H}^+] = [\text{OH}^-]$ in neutral solutions

SAMPLE EXERCISE 16.6

Calculate the values of $[\text{H}^+]$ and $[\text{OH}^-]$ in a neutral solution.

Solution: By definition, in a neutral solution, $[\text{H}^+]$ equals $[\text{OH}^-]$. Let us call the concentration of each of these species in neutral solution x. Using Equation [16.16], we have

$$[\text{H}^+][\text{OH}^-] = (x)(x) = 1 \times 10^{-14}$$
$$x^2 = 1 \times 10^{-14}$$
$$x = 1 \times 10^{-7} = [\text{H}^+] = [\text{OH}^-]$$

In an acid solution, $[\text{H}^+]$ is greater than $1 \times 10^{-7}\,M$; in a basic solution it is less than $1 \times 10^{-7}\,M$.

SAMPLE EXERCISE 16.7

Calculate the concentration of $\text{H}^+(aq)$ in a solution in which $[\text{OH}^-]$ is $0.01\,M$; in which $[\text{OH}^-]$ is $2 \times 10^{-9}\,M$.

Solution: Because the ion-product constant requires that $[\text{H}^+][\text{OH}^-] = 1 \times 10^{-14}$, we have

$$[H^+][OH^-] = 1 \times 10^{-14}$$
$$[H^+][0.01] = 1 \times 10^{-14}$$
$$[H^+] = (1 \times 10^{-14})/(1 \times 10^{-2})$$
$$= 1 \times 10^{-12}$$

This solution is basic, because $[H^+] < [OH^-]$.

In the second instance,
$$[H^+][2 \times 10^{-9}] = 1 \times 10^{-14}$$
$$[H^+] = (1 \times 10^{-14})/(2 \times 10^{-9})$$
$$= 5 \times 10^{-6}$$

This solution is acidic, because $[H^+] > [OH^-]$.

16.5
solutions
of bases

An acid solution, as we've learned, is one that contains an excess of H^+ ions, and a basic solution one that contains an excess of OH^-. Arrhenius defined a base as a substance capable of furnishing OH^- ions to the solution. Thus the soluble metal hydroxides NaOH and KOH are capable of forming strongly basic solutions, because they are capable of furnishing a high concentration of OH^- in water. These compounds are ionic in nature, and contain OH^- ions in the solid lattice. They dissolve in water as would any other ionic substance (Section 12.5).

Only a few metal hydroxides are soluble in water; NaOH and KOH are the only commonly available ones. By contrast, $Mg(OH)_2$ is soluble in water to the extent of only 0.011 g/l. solution at room temperature.

Aside from the metal hydroxides, most of which are strong electrolytes, a great many other substances produce basic solutions on dissolving in water. For the most part, these substances do not contain hydroxide. Ammonia, NH_3, is a common example of such a base. We can best appreciate how ammonia functions by considering its properties in terms of the Brønsted-Lowry theory of acids and bases. Recall that in the Brønsted-Lowry theory, a base is a substance capable of accepting a proton. We might say then that if an acid in water is capable of furnishing H^+ to the solvent, a base should be a substance capable of taking H^+ from solvent.

The basic character of ammonia is due to a reaction with water, in which water acts as a proton donor toward ammonia. In the process, hydroxide ion is generated:

$$NH_3 + H_2O \rightleftharpoons NH_4^+(aq) + OH^-(aq) \qquad [16.17]$$

The reaction does not proceed very far to the right. Thus, ammonia is classified as a weak base in water.

It is useful to consider this reaction further in terms of the Brønsted-Lowry model. In the forward reaction, ammonia is the base, solvent water the acid. The products of the reaction are ammonium ion, which we can call the conjugate acid of the base ammonia, and hydroxide ion, which we can call the conjugate base of the acid, water. Thus, in the reverse reaction the acid, NH_4^+, transfers a proton to the base, OH^-, forming NH_3 and H_2O:

$$NH_3 + H_2O \rightleftharpoons NH_4^+(aq) + OH^-(aq)$$

Base 1   Acid 1   Acid 2   Base 2

According to the Brønsted-Lowry model, every acid has associated with it a conjugate base, and every base has associated with it a conjugate acid.

We've said that ammonia is a weak base in water, because the reaction shown in Equation [16.17] occurs to only a slight extent. We can put this in quantitative terms by writing the equilibrium constant expression for reaction Equation [16.17]:

$$K_c = \frac{[NH_4^+][OH^-]}{[NH_3][H_2O]}$$

[16.18]

In this expression we again encounter the concentration of water as a term in the equilibrium. Because the concentration of water is essentially constant, even when moderate concentrations of other substances are dissolved in the water, the $[H_2O]$ term is incorporated into the equilibrium constant, giving

$$K_c[H_2O] = K_b = \frac{[NH_4^+][OH^-]}{[NH_3]}$$

[16.19]

where K_b is called the **base-dissociation constant**, by analogy with the acid-dissociation constant, Equation [16.5]. Table 16.3 lists the names, structures, and K_b values for several weak bases in water. The base-dissociation equilibria for a few of these bases are listed in Table 16.4.

TABLE 16.3

some weak bases in water

BASE	FORMULA	LEWIS STRUCTURE	CONJUGATE ACID	K_b (AT 25°C)
Ammonia	NH_3	H—N̈—H | H	NH_4^+	1.8×10^{-5}
Pyridine	C_5H_5N	 (ring)N:	$C_5H_5NH^+$	1.7×10^{-9}
Hydroxylamine	H_2NOH	H—N̈—OH | H	H_3NOH^+	1.1×10^{-8}
Methylamine	NH_2CH_3	H—N̈—CH$_3$ | H	$NH_3CH_3^+$	4.4×10^{-4}
Nicotine	$C_{10}H_{14}N_2$	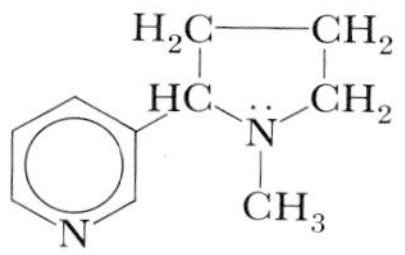 (structure)	$HC_{10}H_{14}N_2^+$	7×10^{-7} 1.4×10^{-11}
Hydrosulfide ion	HS^-	$[H—\ddot{S}:]^-$	H_2S	1.0×10^{-7}
Carbonate ion	CO_3^{2-}	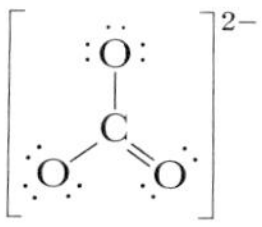 (structure)$^{2-}$	HCO_3^-	2.1×10^{-4}
Hypochlorite	ClO^-	$[:\ddot{Cl}—\ddot{O}:]^-$	$HClO$	3.1×10^{-7}

TABLE 16.4

base-dissociation equilibria in water

BASE 1		ACID 1		ACID 2		BASE 2
NH_3 Ammonia	$+$	H_2O Water	$\rightleftharpoons$	NH_4^+ Ammonium ion	$+$	OH^- Hydroxide ion
C_5H_5N Pyridine	$+$	H_2O	$\rightleftharpoons$	$C_5H_5NH^+$ Pyridinium ion	$+$	OH^-
H_2NOH Hydroxylamine	$+$	H_2O	$\rightleftharpoons$	H_3NOH^+ Hydroxyl ammonium ion	$+$	OH^-
HS^- Hydrosulfide ion	$+$	H_2O	$\rightleftharpoons$	H_2S Hydrogen sulfide	$+$	OH^-

Notice that a few of the bases listed are anions that are derived from acids by loss of a proton. For example, the hypochlorite ion is derived from (in other words, it is the conjugate base of) hypochlorous acid, HClO.

SAMPLE EXERCISE 16.8

Calculate the concentration of OH^- in a 0.15 M solution of NH_3.

Solution: We use essentially the same procedure here as used in solving problems involving the dissociation of acids. The first step is to write the equilibrium expression and the corresponding equilibrium-constant expression:

$$NH_3 + H_2O \rightleftharpoons NH_4^+(aq) + OH^-(aq)$$

$$K_b = \frac{[NH_4^+][OH^-]}{[NH_3]} = 1.8 \times 10^{-5}$$

We then tabulate the equilibrium concentrations involved in the equilibrium:

$$NH_3 + H_2O \rightleftharpoons NH_4^+(aq) + OH^-(aq)$$

Initial concentration:	0.15 M	0 M	0 M
Equilibrium concentration:	$(0.15 - x)\,M$	$x\,M$	$x\,M$

(Notice that we ignore the concentration of H_2O, because this is not involved in the equilibrium-constant expression.) Inserting these quantities into the equilibrium-constant expression gives the following:

$$K_b = \frac{[NH_4^+][OH^-]}{[NH_3]} = \frac{(x)(x)}{(0.15 - x)} = 1.8 \times 10^{-5}$$

Because K_b is small we can neglect the small amount of NH_3 that reacts with water, as compared with the total NH_3 concentration; that is, we can neglect x in comparison with 0.15 M. Then we have

$$\frac{x^2}{0.15} = 1.8 \times 10^{-5}$$

$$x^2 = (0.15)(1.8 \times 10^{-5}) = 0.27 \times 10^{-5}$$

$$x = \sqrt{2.7 \times 10^{-6}} = 1.6 \times 10^{-3}\,M = [OH^-]$$

Notice that the value obtained for x is only about 1 percent of the NH_3 concentration, 0.15 M. Therefore our neglect of x in comparison with 0.15 is justified.

It is possible to show that ammonia is a weak base in water from several lines of evidence similar to those used to distinguish between strong and weak acids. For example, the conductivity of a 0.1 M solution of NH_3 is very much lower than that of a 0.1 M solution of a strong base such as NaOH. Furthermore, reactions catalyzed by OH$^-$ are much slower in the presence of NH$_3$ than in the presence of an equivalent concentration of NaOH.

16.6
conjugate
acid-base
relationships

The equilibria shown in Table 16.4 provide examples of something that exists in every reaction of an acid or base with solvent—namely, an equilibrium between a base and its conjugate acid (or between an acid and its conjugate base). We've seen from a qualitative view that if an acid is strong, its conjugate base is relatively weak. If the acid is weak, its conjugate base is relatively strong. We can make use of these qualitative considerations in practical ways, as illustrated in Sample Exercise 16.9.

SAMPLE EXERCISE 16.9

A 0.1 M solution of NaCl is essentially neutral. What do you predict for the acidity or basicity of a 0.1 M solution of NH$_4$Cl? Write a balanced equation to support your answer.

Solution: To answer this question, we must ask about the acidity or basicity of the two ions that make up NH$_4$Cl. We know that Cl$^-$ is the conjugate base of the strong acid HCl and that it is therefore a weak base. The NH$_4$$^+$ species, however, is the conjugate

acid of the weak base NH$_3$, and it should therefore be a moderately strong acid. The overall result is that a solution of NH$_4$Cl should be acidic in character. The balanced equation that describes the reaction occurring is

$$NH_4^+(aq) \rightleftharpoons H^+(aq) + NH_3$$

That is, we write the formation of the aquated proton just as for any other acid in water.

The fact that a qualitative conjugate acid-base relationship exists suggests that we might be able to find a quantitative connection. Let's look in more detail at the NH$_3$-NH$_4$$^+$ conjugate acid-base pair. Each of these species reacts with water as follows:

$$NH_3 + H_2O \rightleftharpoons NH_4^+(aq) + OH^-(aq) \qquad [16.20]$$

$$NH_4^+(aq) \rightleftharpoons NH_3 + H^+(aq) \qquad [16.21]$$

Each of these equilibria is expressed by a characteristic dissociation constant:

$$K_b = \frac{[NH_4^+][OH^-]}{[NH_3]} \qquad K_a = \frac{[NH_3][H^+]}{[NH_4^+]}$$

Now we notice something very interesting and important. When Equations [16.20] and [16.21] are added together, the NH$_3$ and NH$_4$$^+$ species cancel:

$$\begin{array}{r} NH_3 + H_2O \rightleftharpoons NH_4^+(aq) + OH^-(aq) \\ \underline{NH_4^+(aq) \rightleftharpoons NH_3 + H^+(aq)} \\ H_2O \rightleftharpoons H^+(aq) + OH^-(aq) \end{array}$$

and we are left with just the autoionization of water.

To determine what we should do about the equilibrium constants for the added reactions, we make use of a rule that can be derived from the general principles governing chemical equilibria: *When two reactions are added to give a third reaction, the equilibrium constant for the third reaction is given by the product of the equilibrium constants for the two added reactions.* Thus in general,

$$\text{If reaction } 1 + \text{reaction } 2 = \text{reaction } 3$$
$$\text{then } K_1 \times K_2 = K_3$$

Applying this to our present example, if we multiply K_b and K_a, we obtain the following result:

$$K_b \times K_a = \left(\frac{[\cancel{NH_4^+}][OH^-]}{[\cancel{NH_3}]}\right)\left(\frac{[\cancel{NH_3}][H^+]}{[\cancel{NH_4^+}]}\right)$$
$$= [H^+][OH^-] = K_w$$

Thus, the result of multiplying K_b times K_a is just the ion-product constant, K_w, Equation [16.16]. This is, of course, just what we would expect, because addition of Equations [16.20] and [16.21] gave us just the autoionization equilibrium for water, for which the equilibrium constant is K_w.

The relationship we have just found is so important that it should be emphasized and called to special attention: *The product of the acid-dissociation constant for an acid and the base-dissociation constant for its conjugate base is the ion-product constant for water:*

$$K_a \times K_b = K_w \qquad\qquad [16.22]$$

SAMPLE EXERCISE 16.10

Write the equations for reaction of hydrofluoric acid and its conjugate base, fluoride ion, with water. Show that the equations add to give the autoionization equilibrium for water, and that K_a times K_b yields K_w.

Solution: The equilibria involved and their sum are

$$\begin{array}{r} HF \rightleftharpoons H^+(aq) + F^-(aq) \\ \underline{F^-(aq) + H_2O \rightleftharpoons HF + OH^-(aq)} \\ H_2O \rightleftharpoons H^+(aq) + OH^-(aq) \end{array}$$

The acid- and base-dissociation constants are

$$K_a = \frac{[H^+][F^-]}{[HF]} \qquad K_b = \frac{[HF][OH^-]}{[F^-]}$$

The product of these two is

$$K_a \times K_b = \left(\frac{[H^+][\cancel{F^-}]}{[\cancel{HF}]}\right)\left(\frac{[\cancel{HF}][OH^-]}{[\cancel{F^-}]}\right)$$
$$= [H^+][OH^-] = K_w$$

SAMPLE EXERCISE 16.11

The acid-dissociation constant for hydrofluoric acid is listed in Table 16.2 as 6.8×10^{-4}. What is the base-dissociation constant for fluoride ion?

Solution: Because we know K_a to be 6.8×10^{-4}, and K_w to be 1×10^{-14}, we have

$$K_a \times K_b = K_w$$
$$(6.8 \times 10^{-4})K_b = 1 \times 10^{-14}$$
$$K_b = \frac{1 \times 10^{-14}}{6.8 \times 10^{-4}} = 1.5 \times 10^{-11}$$

16.7
pH—a measure
of acidity

In almost every area of pure and applied chemistry the acid-base properties of water are of importance. As examples, the fate of pollutant chemicals in a water body, the rapidity with which a metal object immersed in water corrodes, and the suitability of a given water body for support of fish and plant life are all critically dependent on the acidity or basicity of the water. Because the acid-base characteristics of water are important in so many contexts, it is important to have a convenient and readily understood means of expressing the hydrogen-ion concentration. The most widely used method for doing this is called the **pH scale.** The pH is defined as the negative log in base 10, of the hydrogen-ion concentration: *

$$pH = -\log [H^+] = \log \left(\frac{1}{[H^+]} \right) \qquad [16.23]$$

As an example of the use of Equation [16.23], let us calculate the pH of a neutral solution, that is, one in which $[H^+] = [OH^-] = 1 \times 10^{-7}$ (Sample Exercise 16.6). The pH is given by

$$pH = -\log [H^+] = -\log (1 \times 10^{-7}) = -0 - (-7) = 7$$

(If you need a review of exponential notation and of the use of logs, see Appendix A). Thus, the pH of a neutral solution is 7.

Because pH is simply another means of expressing $[H^+]$, acidic and basic solutions can be distinguished on the basis of their pH values:

pH $<$ 7 in acid solutions

pH $>$ 7 in basic solutions

pH $=$ 7 in neutral solutions

The pH values characteristic of several familiar solutions are shown in Figure 16.6.

With a log table, or log function on a calculator, it is a simple matter to convert from concentration of $H^+(aq)$ to pH, and vice versa, as outlined in Sample Exercises 16.12 and 16.13.

*Usually you will see pH defined as $-\log [H^+]$, occasionally as $-\log [H_3O^+]$. As discussed in Section 16.1, the same species is involved in all cases.

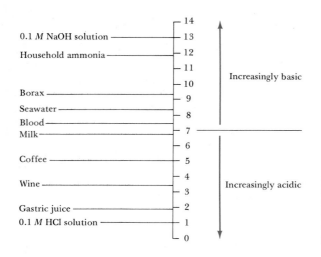

0.1 M NaOH solution — 13
Household ammonia — 12
                        11
                        10
Borax — 9
Seawater — 8
Blood — 7
Milk — 7
                        6
Coffee — 5
                        4
Wine — 3
Gastric juice — 2
0.1 M HCl solution — 1
                        0

Increasingly basic

Increasingly acidic

FIGURE 16.6 Values of pH for some more commonly encountered solutions. The pH scale in this figure is shown to extend from 0 to 14, because nearly all solutions commonly encountered have pH values in that range. In principle, however, the pH values for strongly acidic solutions can be less than 0, and for strongly basic solutions can be greater than 14.

SAMPLE EXERCISE 16.12

Calculate the pH values for the two solutions described in Sample Exercise 16.7.

Solution: In the first instance we found $[H^+]$ to be 1×10^{-12}. The pH of this solution is given by:

$$-\log (1 \times 10^{-12}) = -(-12) = 12$$

The pH of the second solution is given by

$$\begin{aligned} pH &= -\log (5 \times 10^{-6}) \\ &= -(\log 5 + \log 10^{-6}) \\ &= -(0.699 - 6.0) = 5.3 \end{aligned}$$

SAMPLE EXERCISE 16.13

A sample of freshly pressed apple juice has a pH of 3.76. Calculate $[H^+]$.

Solution: From Equation [16.23] we have

$$-\log [H^+] = 3.76$$

The simplest way to take the antilog in such a problem is to write the log as a sum of two numbers, an integer and a number less than 1. Thus, multiplying through by -1,

$$\log [H^+] = -3.76 = -4.0 + 0.24$$

The antilog of -4 is 1×10^{-4}. From a log table we find that the antilog of 0.24 is approximately 1.7. Thus,

$$[H^+] = 1.7 \times 10^{-4}$$

For most problem-solving purposes, Table 16.5, showing logs in base 10, is adequate and more convenient than looking through a larger table. A four-place table is given in Appendix B. In either table, an additional significant figure can be estimated by interpolation.

INDICATORS

Various means are available for quantitatively estimating pH. The simplest is the use of an indicator. An **indicator** is a colored substance, usually derived from plant material, that can exist in either an acid or base form. The two forms are differently colored. By adding a small amount of an indicator to a solution and noting its color, it is possible to determine whether it is in the acid or base form. If one knows the pH at which the indicator turns from one form to the other, one can then

TABLE 16.5

logs in base 10

1.0	.000	3.3	.518	5.6	.748	7.9	.898
1.1	.041	3.4	.531	5.7	.756	8.0	.903
1.2	.079	3.5	.544	5.8	.763	8.1	.908
1.3	.114	3.6	.556	5.9	.771	8.2	.914
1.4	.146	3.7	.568	6.0	.778	8.3	.919
1.5	.176	3.8	.580	6.1	.785	8.4	.924
1.6	.204	3.9	.591	6.2	.792	8.5	.929
1.7	.230	4.0	.602	6.3	.799	8.6	.934
1.8	.255	4.1	.613	6.4	.806	8.7	.939
1.9	.279	4.2	.623	6.5	.813	8.8	.944
2.0	.301	4.3	.633	6.6	.820	8.9	.949
2.1	.322	4.4	.643	6.7	.826	9.0	.954
2.2	.342	4.5	.653	6.8	.832	9.1	.959
2.3	.362	4.6	.663	6.9	.839	9.2	.964
2.4	.380	4.7	.672	7.0	.845	9.3	.968
2.5	.398	4.8	.681	7.1	.851	9.4	.973
2.6	.415	4.9	.690	7.2	.857	9.5	.978
2.7	.431	5.0	.699	7.3	.863	9.6	.982
2.8	.447	5.1	.708	7.4	.869	9.7	.987
2.9	.462	5.2	.716	7.5	.875	9.8	.991
3.0	.477	5.3	.724	7.6	.881	9.9	.996
3.1	.491	5.4	.732	7.7	.886	10.0	1.000
3.2	.505	5.5	.740	7.8	.892		

determine from the observed color whether the solution has a higher or lower pH than this value. For example, litmus, one of the most common indicators, changes color in the vicinity of pH 7. However, the color change is not very sharp. Red litmus indicates a pH of about 5 or lower, and blue litmus, a pH of about 8.2 or higher. Many other indicators change color at various pH values between 1 and 14. Some of the more commonly used are listed in Table 16.6. We see from this table that methyl orange, for example, changes color over the pH interval from 2.9 to 4.0. Below pH 2.9 it is in the acid form, which is red. In the interval from pH 2.9 to 4.0 it is gradually converted to its basic form, which has a yellow color. By pH 4.0 the conversion is complete, and the solution is

TABLE 16.6

some of the more common acid-base indicators

NAME	pH INTERVAL FOR COLOR CHANGE	ACID COLOR	BASE COLOR
Methyl Violet	0–2	Yellow	Violet
Methyl Yellow	1.2–2.3	Red	Yellow
Methyl Orange	2.9–4.0	Red	Yellow
Methyl Red	4.2–6.3	Red	Yellow
Bromthymol Blue	6.0–7.6	Yellow	Blue
Thymol Blue	8.0–9.6	Yellow	Blue
Phenolphthalein	8.3–10	Colorless	Pink
Alizarin Yellow G	10.1–12.0	Yellow	Red

yellow. Paper tape which is impregnated with various indicators, and which comes complete with a comparator color scale, is widely used for approximate determinations of pH.

SAMPLE EXERCISE 16.14

A student working in the laboratory has neglected to label two solutions and now can't remember which is which. One of the solutions contains $5 \times 10^{-4} M$ $Mg(OH)_2$, the other $0.001 M$ $HClO_4$. He has time before his instructor learns of his mistake to make one indicator test, using a sample of only one solution. Which indicators listed in Table 16.6 would be suitable?

Solution: Because $HClO_4$ is a strong acid, $[H^+]$ in a $0.001 M$ solution of $HClO_4$ is $0.001 M$. Thus, the pH of the $HClO_4$ solution is 3. A $5 \times 10^{-4} M$ solution of $Mg(OH)_2$ furnishes

$$2(5 \times 10^{-4} M) = 1 \times 10^{-3} M \, OH^-$$

to the solution. $[H^+]$ in this solution is thus

$$[H^+] = \frac{K_w}{[OH^-]}$$
$$= \frac{1 \times 10^{-14}}{1 \times 10^{-3}}$$
$$= 1 \times 10^{-11} M$$

Thus the pH of the $Mg(OH)_2$ solution is 11. Our student must choose an indicator that has a pH interval for the color change that is well within these two values. Any of the indicators in Table 16.6 from methyl orange through phenolphthalein would be suitable. Suppose, however, that he chose methyl violet. Both solutions would show the base color for this indicator, because both have a pH higher than the interval of change. Similarly, the results with alizarin yellow would at best be ambiguous. With this indicator, both solutions might give an acid response.

The pH meter is a widely used, simple instrument for rapid and accurate determination of pH. It is so common that if you go on to further study in chemistry or in an applied science you are almost certain to encounter one. A complete understanding of how a pH meter works requires a knowledge of electrochemistry, a subject we take up in Chapter 19. However, we can say at this point that a pH meter consists of a pair of electrodes that are placed in the solution to be measured and a sensitive meter for measuring small voltages, on the order of millivolts. A typical pH meter with a pair of electrodes is shown in Figure 16.7. When the electrodes are placed in the solution, they form an electrochemical cell (something like a battery) that has a voltage. The voltage of the cell is dependent on $[H^+]$; thus, by measuring the voltage we obtain a measure of $[H^+]$. Electrodes that can be used with pH meters come in all shapes and sizes, depending on their intended use, but fundamentally they are nearly all the same. One of the electrodes is a reference electrode. The one that is actually sensitive to $H^+(aq)$ is almost always a so-called glass electrode. The wire in the inner compartment of the electrode is in contact with a solution of known and fixed $H^+(aq)$ concentration.

The wall of the compartment is formed of a special thin glass that is permeable to $H^+(aq)$. As a result, the voltage that this electrode, together with a reference electrode, generates when placed in a solution depends on $[H^+]$ in the solution.

The pH meter is used by placing the clean electrodes into a reference solution of known and stable pH. The scale on the pH meter is calibrated to read pH. With a calibration control, the meter is set to read the pH corresponding to the value for the reference solution. Then the electrodes are withdrawn and carefully washed, first with distilled water, and then with a bit of the solution to be tested. The electrodes are then immersed in the test solution, and the pH reading taken.

To extend the range of possible pH measurements, much research has gone into the development of electrodes that can be used with very small quantities of solution. It is now possible to insert electrodes into single living cells in order to monitor pH of the cell medium. The pH meter is also widely used outside the laboratory. Pocket-sized models are available for use in environmental studies, monitoring of industrial effluents, and in agricultural work.

pH CHANGES DURING ACID-BASE TITRATIONS

The pH undergoes a very large change during the course of an acid-base titration. To see why this is so, let's assume we are to titrate 50.0 ml of

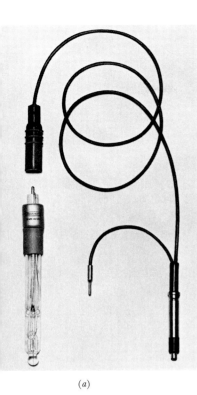

(a)

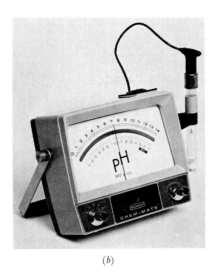

(b)

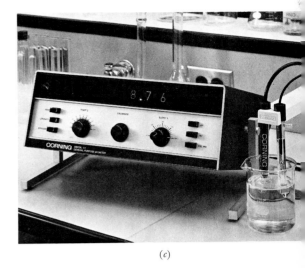

(c)

FIGURE 16.7 (a) A glass electrode used in measuring pH. (b) A pH meter of the type normally used for student work. (c) A research-type pH meter with digital display. (a and b courtesy of Beckman Instruments, Inc.; c courtesy of Corning Glass Works)

0.100 M HCl solution with 0.100 M NaOH. At the beginning of the titration the pH of the solution is 1.00, since [H$^+$] in 0.100 M HCl solution is just 0.100 M. As NaOH solution is added, the H$^+(aq)$ concentration decreases because of reaction with OH$^-$, and because of dilution. The total number of moles of H$^+(aq)$ present in the original HCl solution is

$$(0.050 \text{ l. soln}) \left(\frac{0.100 \text{ moles H}^+(aq)}{\text{liter soln}} \right)$$
$$= 5.00 \times 10^{-3} \text{ moles H}^+(aq)$$

When 45 ml of 0.100 M NaOH solution have been added, the total number of moles of H$^+(aq)$ consumed in the reaction

$$\text{H}^+(aq) + \text{OH}^-(aq) \longrightarrow \text{H}_2\text{O}$$

is

$$0.045 \text{ l. NaOH soln} \left(\frac{0.100 \text{ moles OH}^-(aq)}{\text{liter soln}} \right) \left(\frac{1 \text{ mole H}^+(aq)}{1 \text{ mole OH}^-(aq)} \right)$$
$$= 4.50 \times 10^{-3} \text{ moles H}^+(aq) \text{ consumed}$$

This leaves

$$(5.00 - 4.50) \times 10^{-3} = 5.0 \times 10^{-4} \text{ moles H}^+(aq)$$

in a volume of 95 ml. [H$^+$] is therefore

$$\frac{5.0 \times 10^{-4} \text{ moles}}{0.095 \text{ l. sol}} = 5.26 \times 10^{-3} M$$

The pH at this point is

$$-\log[5.26 \times 10^{-3}] = 2.28$$

Proceeding in this manner we could calculate the pH of the solution at various stages in the course of the titration, as in Sample Exercise 16.15.

SAMPLE EXERCISE 16.15

Calculate the pH of the titration solution when 49.9 ml NaOH solution have been added.

Solution: The number of moles of OH^- in 49.9 ml 0.100 M NaOH is

$$0.0499 \text{ l. soln} \left(\frac{0.100 \text{ moles } OH^-}{\text{liter soln}} \right)$$
$$= 4.99 \times 10^{-3} \text{ moles } OH^-$$

Thus there remains

$$(5.00 \times 10^{-3}) - (4.99 \times 10^{-3})$$
$$= 1.0 \times 10^{-5} \text{ moles } H^+(aq)$$

in 0.0999 l. solution. The concentration of $H^+(aq)$ is thus

$$\frac{1.0 \times 10^{-5} \text{ moles}}{0.0999 \text{ l.}} = 1.0 \times 10^{-4} M$$

The corresponding pH is 4.0.

We know that at the equivalence point, when the number of moles of OH^- precisely equals the number of moles of $H^+(aq)$ present originally, the pH will equal 7. This will occur when 50.0 ml of NaOH have been added. By comparing this result with the value obtained in Sample Exercise 16.15, we can see that there is a very sharp change in pH in the vicinity of the equivalence point. The change in pH as a function of added NaOH solution during the titration is shown in Figure 16.8.

Because the pH does change over such a wide interval in the close vicinity of the equivalence point, we don't need to have an indicator for the titration that changes color precisely at 7.0. For example, most

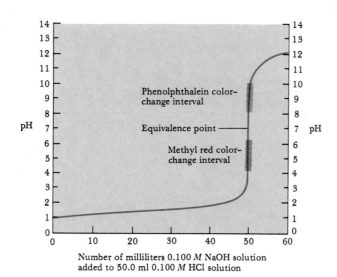

Number of milliliters 0.100 M NaOH solution
added to 50.0 ml 0.100 M HCl solution

FIGURE 16.8 The pH curve for titration of a solution of a strong acid with a solution of a strong base, in this case HCl and NaOH.

acid-base titrations are carried out using phenolphthalein. From Table 16.6 we see that this indicator changes color in the pH range 8.3 to 10. Thus a slight excess of NaOH must be present to cause the observed color change. However, it requires such a tiny excess of base to make the color change occur that no serious error is introduced. Similarly, methyl red, which changes color in the slightly acid range, could also be used. The pH intervals of color change for these two indicators are shown in Figure 16.8.

Titration of a solution of a strong base by a solution of a strong acid would yield an entirely analogous curve of pH versus added acid. In this case, however, the pH would be high at the outset of the titration, and low at its completion.

Titration of a weak acid by a strong base or of a weak base by a strong acid results in curves of pH versus added reagent that look similar to those for strong acid–strong base solutions. However, there are some interesting and important differences. These will be discussed in more detail in Chapter 17.

OTHER pX SCALES

The negative log is a convenient way of expressing the magnitudes of numbers that are generally very small. We use the convention that the negative log of a quantity is labeled p(quantity). For example, one can express the concentration of OH^- as pOH:

$$pOH = -\log [OH^-]$$

By taking the log of both sides of Equation [16.16] and multiplying through by -1 we can obtain

$$pH + pOH = -\log K_w = 14 \qquad [16.24]$$

This expression is often convenient to use.

It is also convenient to express small equilibrium constants as the negative logs. If you have occasion to look up the dissociation constant of a particular acid or base in a handbook, you will often find it expressed as pK_a or pK_b. Thus, for example, the pK_a for saccharin, used as a sugar substitute, is listed as 11.68. To obtain K_a from this value we must change the sign and find the antilog:

$$pK_a = -\log K_a = 11.68$$
$$K_a = \text{antilog} (-11.68) = \text{antilog} (-12 + 0.32)$$
$$= 2.1 \times 10^{-12}$$

Notice also that Equation [16.22] can be put in a useful form by taking the negative log:

$$pK_a + pK_b = pK_w = 14 \qquad [16.25]$$

16.8

acid-base character and chemical structure

From our discussion to this point, we have seen that when any substance is dissolved in water one of three things can happen. The $H^+(aq)$ concentration might increase, in which case the substance behaves as an acid; it might decrease (with corresponding increase in OH^-), in which case the substance is acting as a base; or, there might be no change in $[H^+]$, an indication that the substance possesses neither acid nor base character. It would be very helpful to have some general guidelines as to how acid or base characteristics relate to chemical structure, so that we might be able to predict in advance how a compound will behave on dissolving in water. However, we must expect that any simple rules we might formulate won't always work. Many different factors contribute to ionization in a polar solvent such as water. The best we can hope for are a few rules that are *almost always* obeyed.

EFFECTS OF BOND POLARITY AND BOND STRENGTH

When a substance HX transfers a proton to the solvent, an ionic rupture of the H—X bond occurs, as described by Equation [16.4]. Such a reaction will occur most readily when the H—X bond is already polarized in the following sense:

$$\overrightarrow{\text{H--X}}$$

For example, compare NH_4^+ and CH_4. These two species have the same electronic structure, that is, they are isoelectronic. Both consist of a central atom with an octet of electrons bonding four hydrogens. The difference is in the nuclear charge of the central atom. Because the nuclear charge of N is one greater than for C, the electron pairs shared with the hydrogens are more closely attracted to N in NH_4^+ than to C in CH_4. That is, the N—H bonds are more polarized than the C—H bonds. Correspondingly, ammonium ion is an acid in water, whereas methane is not:

$$NH_4^+(aq) \rightleftharpoons H^+(aq) + NH_3 \qquad K_a = 5.5 \times 10^{-10} \qquad [16.26]$$
$$CH_4 \rightleftharpoons H^+(aq) + CH_3^-(aq) \qquad \text{no reaction} \qquad [16.27]$$

Another example is provided by the comparison of carbonic acid with urea. In carbonic acid, the hydrogens that might be transferred to the water are bound to oxygen. In urea the hydrogens are bound to nitrogen, a less electronegative element:

Carbonic acid
$K_a = 4.3 \times 10^{-7}$

Urea
$K_b = 1.5 \times 10^{-14}$

The O—H bonds are more polarized than the N—H bonds; we therefore expect carbonic acid to be more acidic. It turns out that carbonic acid is a weak acid in water, whereas urea is actually a weak base. This means that the tendency of the nitrogens in urea to capture a proton from water is greater than their tendency to donate a proton to water.

One other factor of major importance in determining whether a substance acts as an acid is the strength of the H—X bond. Because this bond energy is lost when acid dissociation occurs, it follows that very strong bonds are less easily ionized than weaker ones. This factor is of importance in the case of the hydrogen halides. The H—F bond is the most polar of any H—X bond. One might therefore expect that HF would be a very strong acid, if the first rule were all that mattered. However, the energy required to dissociate HF into H and F atoms is much higher than for the other hydrogen halides, as shown in Table 8.4. As a result, HF is a weak acid, whereas all the other hydrogen halides are strong acids in water.

Because a base acts as a proton acceptor rather than as a proton donor, the factors that make for a strong acid should make for a weak base. For a substance to attract a proton it must possess an unshared pair of electrons readily available for bonding. Elements of higher electronegativity attract the unshared pairs closely to themselves, making them less available for bonding. Thus, base strength should decrease with increasing electronegativity. As an example, methanol is a weaker base in water than methyl amine:

$$
\begin{array}{c}
\text{H} \\
| \\
\text{H}-\underset{\cdot\cdot}{\text{N}}-\text{CH}_3 + \text{H}_2\text{O} \longrightarrow
\end{array}
$$

Methyl amine

$$
\left[\begin{array}{c}
\text{H} \\
| \\
\text{H}-\text{N}-\text{CH}_3 \\
| \\
\text{H}
\end{array}\right]^+ + \text{OH}^- \qquad K_b = 4.4 \times 10^{-4} \qquad [16.28]
$$

$$
\text{H}-\underset{\cdot\cdot}{\overset{\cdot\cdot}{\text{O}}}-\text{CH}_3 + \text{H}_2\text{O} \longrightarrow
$$

Methanol

$$
\left[\begin{array}{c}
\overset{\cdot\cdot}{\text{H}-\text{O}}-\text{CH}_3 \\
| \\
\text{H}
\end{array}\right]^+ + \text{OH}^- \qquad \text{no reaction} \qquad [16.29]
$$

In comparing bases, however, we must also include some consideration of bond energies. As an example of how these enter in, compare ammonia with phosphine:

$$
\begin{array}{c}
\text{H} \\
| \\
\text{H}-\underset{\cdot\cdot}{\text{N}}-\text{H} + \text{H}_2\text{O} \longrightarrow
\end{array}
$$

$$
\left[\begin{array}{c}
\text{H} \\
| \\
\text{H}-\text{N}-\text{H} \\
| \\
\text{H}
\end{array}\right]^+ + \text{OH}^- \qquad K_b = 1.8 \times 10^{-5} \qquad [16.30]
$$

$$H\overset{\overset{\displaystyle H}{|}}{\underset{..}{-}P}-H + H_2O \rightleftharpoons$$

$$\left[\begin{array}{c} H \\ | \\ H-P-H \\ | \\ H \end{array}\right]^{+} + OH^{-} \qquad \text{no significant reaction} \qquad [16.31]$$

From relative electronegativity values, Figure 8.8, we might guess that the reaction involving PH_3 would occur to a greater extent than would that involving NH_3. Phosphorus is a less electronegative element than nitrogen and should thus more readily provide a pair of electrons for bonding with the proton. But in fact, PH_3 is an extremely weak base in water. The reason for the failure of the reaction shown in Equation [16.31] to occur to a significant extent is that the P—H bond energy is not sufficiently high. As a result of these considerations, we formulate the following rule: *Almost all neutral molecules that show basic character in water involve ionizable hydroxide groups or contain a nitrogen atom as the source of base behavior.*

HYDROXIDES AND OXYACIDS

Except for the hydrogen halides and ammonium-type ions, nearly all the acids commonly encountered involve one or more O—H bonds. Let's consider, then, an OH group bound to some other atom Y, which might in turn have other groups attached to it:

$$\overset{\diagdown}{\underset{\diagup}{-}}Y-O-H$$

At one extreme, Y might be a metal such as Na, K, or Mg. The pair of electrons shared between Y and O is then completely transferred to oxygen, and an ionic compound involving OH^- is formed. Because of the charge that surrounds it, the oxygen of the OH^- ion does not strongly attract to itself the electron pair it shares with hydrogen. That is, the O—H bond in OH^- is not strongly polarized. There is therefore no tendency for the hydrogen of OH^- to be transferred to the solvent as $H^+(aq)$.

When Y is an element of intermediate electronegativity, around 2.0, the bond to O is more covalent in character, and the substance does not readily lose OH^-. Elements with electronegativities in this range include B, C, P, As, and I (Figure 8.8). Examples of acids of such elements include orthoboric acid, hypoiodous acid, and methanol:

$$\overset{\overset{\displaystyle O-H}{|}}{H-O-B-O-H} \qquad I-O-H \qquad CH_3-O-H$$

Orthoboric acid · · · · · · · Hypoiodous acid · · · · · Methanol

Such substances might behave as acids in water, depending on the ease with which the proton is lost from oxygen. As a general rule, the more strongly the group Y attracts the electron pair it shares with the oxygen, the more polar the OH bond will be, and the more acidic the substance. In the three examples just given, the central atom does not strongly attract the electron pair it shares with oxygen. The acid-dissociation constant for orthoboric acid is 6.5×10^{-10}, for hypoiodous acid, 2.3×10^{-11}; no acidic character is observed for methyl alcohol in water.

SAMPLE EXERCISE 16.16

Draw the Lewis structure for orthosilicic acid, $Si(OH)_4$. What do you predict for the acid-base properties of this substance?

Solution: The Lewis structure for $Si(OH)_4$ is as follows:

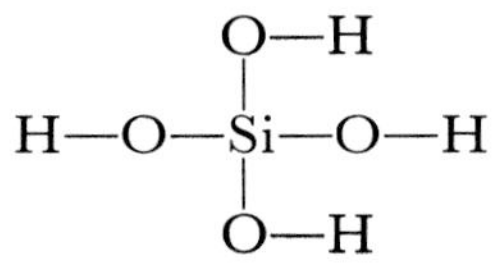

$$\begin{array}{c} O\!-\!H \\ | \\ H\!-\!O\!-\!Si\!-\!O\!-\!H \\ | \\ O\!-\!H \end{array}$$

Because silicon is an element of intermediate electronegativity (Figure 8.8), we would expect that $Si(OH)_4$ would not be strongly acidic or basic. By analogy with orthoboric acid, we might guess that it would be weakly acidic, as in fact it is:

$$K_a = 2 \times 10^{-10}$$

As the electronegativity of Y increases, or as more electron-withdrawing groups are placed on Y, the acidic properties of the substance increase. Acids in which OH groups and possibly additional oxygen atoms are bound to the central Y atom are referred to as **oxyacids.** For example, both orthoboric and hypoiodous acid shown above are oxyacids. It is possible to relate the acid strengths of oxyacids both to the electronegativity of Y and to the number of groups attached to it. *For acids that have the same structure, but differ in the electronegativity of the central atom, Y, acid strength increases with increasing electronegativity.* Examples are shown in Table 16.7. You can see that the rule is generally well obeyed. However, H_2TeO_3 is an exception. It is much more acidic than one would expect from the electronegativity value for Te.

In a series of acids that have the same central atom, Y, but differing numbers of attached groups, the acid strength increases with increasing oxidation number of the central atom. For example, in the series of oxyacids of chlorine extending from hypochlorous to perchloric acid, acid strength steadily increases:

Acid	$H\!-\!\ddot{O}\!-\!\ddot{C}l\!:$	$H\!-\!\ddot{O}\!-\!\ddot{C}l\!-\!\ddot{O}\!:$	$H\!-\!\ddot{O}\!-\!\overset{\displaystyle :\ddot{O}:}{\underset{}{C}l}\!-\!\ddot{O}\!:$	$H\!-\!\ddot{O}\!-\!\overset{\displaystyle :\ddot{O}:}{\underset{\displaystyle :\ddot{O}:}{C}l}\!-\!\ddot{O}\!:$
Chlorine oxidation numbers	Hypochlorous $+1$	Chlorous $+3$	Chloric $+5$	Perchloric $+7$

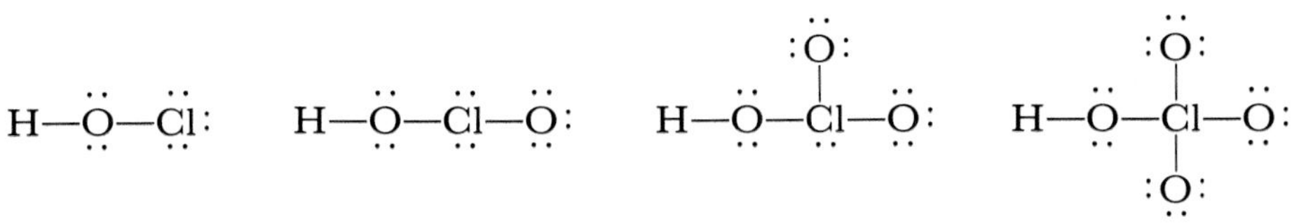

Increasing acid strength ⟶

In this series the ability of the chlorine to withdraw electrons from the OH group, and thus make the O—H bond even more polar, increases as electron-withdrawing oxygen atoms are added to the chlorine.

TABLE 16.7

acid-dissociation constants (K_a) of oxyacids in comparison with electronegativity values (EN) of central atom Y

H—O—Y	K_a	EN OF Y	$H-O-\overset{\overset{\displaystyle O}{\|}}{Y}-O-H$	K_a	EN OF Y
HOCl	3×10^{-8}	3.2	H_2SO_3	1.5×10^{-2}	2.6
HOBr	2×10^{-9}	3.0	H_2SeO_3	3.5×10^{-3}	2.6
HOI	2×10^{-11}	2.7	H_2TeO_3	3.0×10^{-3}	2.1
HOCH$_3$	~ 0	2.5[a]	H_2CO_3	4.3×10^{-7}	2.5

[a] This value is the electronegativity for carbon

CARBOXYLIC ACIDS

The effects of changes in structure on acid strength are very important in **carboxylic acids,** which have the general structure:

$$H-O-\overset{\overset{\displaystyle O}{\|}}{C}-X$$

As the electron-withdrawing properties of X increase, the acid strength increases. Some examples are listed in Table 16.8.

TABLE 16.8

dissociation constants of carboxylic acids, HCO_2X

X GROUP	NAME	K_a
$H-\overset{\overset{\displaystyle H}{\|}}{\underset{\underset{\displaystyle H}{\|}}{C}}-$	Acetic	1.8×10^{-5}
$H-\overset{\overset{\displaystyle Cl}{\|}}{\underset{\underset{\displaystyle H}{\|}}{C}}-$	Chloroacetic	1.4×10^{-3}
$H-\overset{\overset{\displaystyle Cl}{\|}}{\underset{\underset{\displaystyle Cl}{\|}}{C}}-$	Dichloroacetic	3.3×10^{-2}
$Cl-\overset{\overset{\displaystyle Cl}{\|}}{\underset{\underset{\displaystyle Cl}{\|}}{C}}-$	Trichloroacetic	2×10^{-1}
$H-$	Formic	1.8×10^{-4}
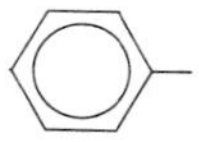 ⬡—	Benzoic	6.5×10^{-5}

SAMPLE EXERCISE 16.17

Based on the data in Table 16.8, what can you say about the electron-withdrawing properties of the methyl group, CH_3, as compared with the trichloromethyl group, CCl_3?

Solution: The value of K_a for trichloroacetic acid is much higher than for acetic acid. Replacement of hydrogens on the methyl group of acetic acid by the more electronegative chlorines causes increased electron withdrawal from the OH group, thus increasing acid strength. The trichloromethyl group, is, in effect, a more electronegative group than methyl.

POLYPROTIC ACIDS

Many substances, especially among the oxyacids, are capable of furnishing more than one proton to water. Substances of this type are called **polyprotic acids.** As an example, sulfurous acid, H_2SO_3, may react with water in two successive steps:

$$H_2SO_3(aq) \rightleftharpoons H^+(aq) + HSO_3^-(aq) \quad K_a = 1.5 \times 10^{-2} \quad [16.32]$$
$$HSO_3^-(aq) \rightleftharpoons H^+(aq) + SO_3^{2-}(aq) \quad K_a = 1.0 \times 10^{-7} \quad [16.33]$$

The values for K_a in each case show that the reactions are incomplete. Notice that loss of the second proton occurs much less readily than the first, as shown by the smaller value for K_a in the second reaction. Because the HSO_3^- ion has an overall negative charge, the central atom attracts electrons less strongly than in H_2SO_3, and the proton is less easily lost. The positively charged proton is also held by electrostatic attraction to the negatively charged ion.

As a rough rule, *the values of K_a for successive losses of protons in oxyacids decrease by about 10^{-5} for each loss.* Examples are shown in Table 16.9.

TABLE 16.9

successive acid-dissociation constants of some oxyacids

ACID	K_a
H_2CO_3	4.3×10^{-7}
HCO_3^-	5.6×10^{-11}
H_2SO_3	1.5×10^{-2}
HSO_3^-	1.0×10^{-7}
H_2SeO_3	3.5×10^{-3}
$HSeO_3^-$	5×10^{-8}
H_2CrO_4	1.8×10^{-1}
$HCrO_4^-$	3.2×10^{-7}
H_3BO_3	7.3×10^{-10}
$H_2BO_3^-$	1.8×10^{-13}
H_3PO_4	7.5×10^{-3}
$H_2PO_4^-$	6.2×10^{-8}
HPO_4^{2-}	2.2×10^{-13}

You should keep in mind that anions that have already lost one or more protons are also capable of acting as bases. Using the relationships in Section 16.6, the value for the base-dissociation constant for the anion can be calculated. Depending on which is larger, the ion may act as an acid or base when added to neutral water.

SAMPLE EXERCISE 16.18

Predict whether the salt Na_2HPO_4 will form an acidic or basic solution on dissolving in water.

Solution: The two possible reactions that HPO_4^{2-} may undergo on addition to water are

$$HPO_4^{2-}(aq) \rightleftharpoons H^+(aq) + PO_4^{3-}(aq) \quad [16.34]$$

$$HPO_4^{2-}(aq) + H_2O \rightleftharpoons$$
$$H_2PO_4^-(aq) + OH^-(aq) \quad [16.35]$$

Depending on which of these has the larger equilibrium constant, the ion will cause the solution to be acidic or basic. The value of K_a for reaction Equation [16.34], as shown in Table 16.9, is 2.2×10^{-13}. We

must calculate the value of K_b for reaction Equation [16.35] from the value of K_a for the conjugate acid formed, $H_2PO_4^-$. We make use of the relationship shown in Equation [16.22]:

$$K_a \times K_b = K_w$$

We want to know K_b for the base HPO_4^{2-}, knowing the value of K_a for the conjugate acid $H_2PO_4^-$:

$$K_b(HPO_4^{2-}) \times K_a(H_2PO_4^-) = K_w = 1 \times 10^{-14}$$

Because K_a for $H_2PO_4^-$ is 6.2×10^{-8} (Table 16.9), we calculate K_b for HPO_4^{2-} to be 1.6×10^{-7}. This is considerably larger than K_a for HPO_4^{2-}; thus, the reaction shown in Equation [16.35] would predominate over that in Equation [16.34], and the solution would be basic.

16.9
the Lewis theory of acids and bases

Throughout this chapter our emphasis has been on water as the solvent, and on the proton as the source of acidic properties. It is certainly appropriate that water should receive much of our attention; it is by far the most important solvent. Nevertheless, it is useful to think of many reactions that do not involve transfer of protons as acid-base reactions. G. N. Lewis proposed a definition of acid and base that emphasizes the shared electron pair. He proposed that *an acid be defined as an electron-pair acceptor, and a base as an electron-pair donor.*

Remember that in the Brønsted-Lowry model a base was defined as a proton acceptor. But to be a proton acceptor a substance must possess an unshared pair of electrons for binding the proton:

$$B: + H^+ \rightleftharpoons B{-}H^+$$

Thus a Brønsted-Lowry base is indeed an electron-pair donor, but only toward the proton. In the Lewis theory, a base could be a donor toward something else. For example, NH_3, which acts as a base in water, reacts with BF_3 to form a compound of a type that is often referred to as a **Lewis acid-base adduct**. This reaction occurs because BF_3 has a vacant orbital in its valence shell (Section 8.6). It therefore acts as an electron-pair acceptor (Lewis acid) toward ammonia, which donates the electron pair:

Base   Acid   Acid-base adduct

[16.36]

The Lewis theory helps us understand why a substance such as carbon dioxide reacts with water to form carbonic acid, H_2CO_3. The reaction can be pictured as an attack by a water molecule on CO_2, in which the water acts as an electron-pair donor, and the CO_2 as an electron pair acceptor:

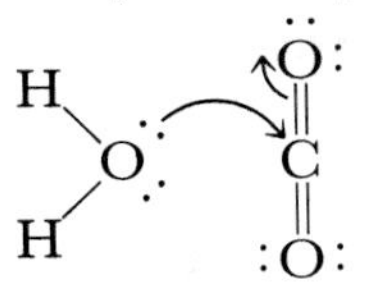

The electron pair of one of the carbon-oxygen π bonds is moved onto the oxygen to leave a vacant orbital on the carbon, which can act as electron-pair acceptor. We have shown the shift of these electrons with arrows. After forming the initial "adduct," a proton moves from one oxygen to another, thereby forming carbonic acid:

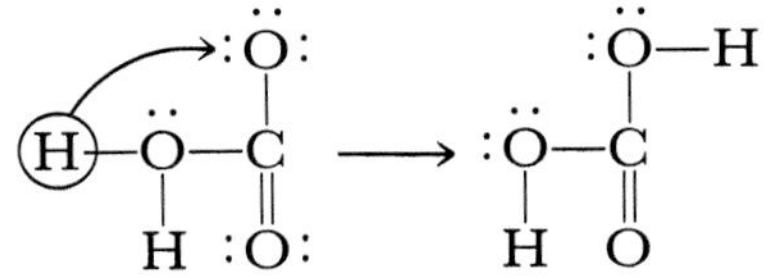

A similar kind of Lewis acid-base reaction takes place when any oxide of a nonmetal dissolves in water to form an acidic solution.

HYDROLYSIS OF METAL IONS

The Lewis theory is also helpful in explaining why solutions of many metal ions show acidic properties. For example, a solution of a salt such as $Cr(NO_3)_3$ is quite acidic. Solutions of $MgCl_2$ or $ZnCl_2$ are likewise acidic, though to a lesser extent. To understand why this is so, we must examine the interaction between a metal ion and water molecules.

Because metal ions are positively charged, they attract the unshared electron pairs of water molecules. It is primarily this interaction, referred to as **hydration**, which causes salts to dissolve in water, as explained in Section 12.2. The strength of attraction increases with the charge of the ion and is strongest for the smallest ions. The ratio of ionic charge to ionic radius provides a good measure of the extent of hydration. This ratio is listed in Table 16.10 for a selection of metal ions. The process of hydra-

TABLE 16.10

Ionic-charge/ionic-radius ratio for metal ions of various charges

METAL ION	CHARGE/IONIC RADIUS
Li^+	1.5
Na^+	1.0
Cu^{2+}	2.8
Mg^{2+}	3.1
Ca^{2+}	2.1
Zn^{2+}	2.7
Al^{3+}	6.7
Cr^{3+}	4.8
Fe^{3+}	4.7

tion is a Lewis acid-base interaction, in which the metal ion acts as a Lewis acid, and the water molecules as Lewis bases. When the water molecule interacts with the positively charged metal ion, electron density is drawn from the oxygen, as illustrated in Figure 16.9. This flow of electron density causes the O—H bond to become more polarized; as a result, water molecules bound to the metal ion are more acidic than those in the bulk solvent. The hydrated metal ion thus acts as a source of protons:

$$M(H_2O)_n{}^{z+} \rightleftharpoons M(H_2O)_{(n-1)}(OH)^{(z-1)+} + H^+(aq) \qquad [16.37]$$

In this equation z is the charge on the metal ion, and n is the number of hydrating water molecules. In the case of the $3+$ metal ions listed in Table 16.10, n is 6; for the other ions it is probably closer to 4, although the exact number is difficult to determine. The reaction shown in Equation [16.37] is often called **hydrolysis**. You should note that it represents the behavior of an acid in just the same way as Equation [16.3], which applies to HF. The "acid" in Equation [16.37] is not just a single molecule, but a collection of molecules. The main point is that this collection is able to transfer a proton to water and thus act as an acidic species. The effect is greatest for the smallest and most highly charged ions, as shown in Figure 16.9.

SAMPLE EXERCISE 16.19

It is known that corrosion of iron occurs more rapidly in acidic solution. If either $CaCl_2$ or NaCl can be used to melt ice on highways, which is likely to be more harmful in promoting corrosion of automobiles?

Solution: From Table 16.10 we see that Ca^{2+} possesses a much higher charge/radius ratio than Na^+. This means that water molecules hydrating Ca^{2+} will be more acidic than those hydrating Na^+, and that the pH of $CaCl_2$ solutions will be lower than for NaCl solutions. Thus $CaCl_2$ should be more corrosive to automobiles.

FOR REVIEW

summary

In this chapter we have considered the general properties of acidic and basic solutions, with emphasis on water as the solvent. We have seen that an acid solution is created when a substance reacts with water in such a way as to increase the concentration of solvated hydrogen ions, $H^+(aq)$. Weak acids are substances for which the reaction is incomplete, and an equilibrium is established. The extent to which the reaction proceeds is expressed by the **acid-dissociation constant**, K_a. From a knowledge of K_a it is possible to calculate $[H^+]$ in a solution of a weak acid. Bases are substances that cause an increase in the concentration of OH^- when dissolved in water. Aside from ionic hydroxides such as NaOH, bases produce an increase of OH^- by reaction with water. The extent to which a base reacts with water is measured by the **base-dissociation constant**, K_b.

Water spontaneously ionizes to a slight degree, forming $H^+(aq)$ and $OH^-(aq)$. The extent of ionization is expressed by the ion product constant for water:

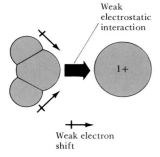

Charge/radius small

Weak electrostatic interaction

1+

Weak electron shift

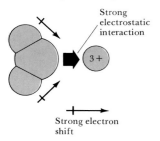

Charge/radius large

Strong electrostatic interaction

3+

Strong electron shift

FIGURE 16.9 Interaction of a water molecule with a cation of 1^+ charge or 3^+ charge. The interaction is much stronger with the smaller ion of higher charge.

$$K_w = [H^+][OH^-] = 1 \times 10^{-14}$$

In the **Brønsted-Lowry model,** an acid is a proton donor, a base a proton acceptor. Reaction of an acid with water results in formation of $H^+(aq)$ and the **conjugate base** of the acid. Reaction of a base with water results in formation of OH^- and the **conjugate acid** of the base. The conjugate bases of weak acids are relatively strong bases. The acid-dissociation constant for an acid and the base-dissociation constant for its conjugate base are related by the expression

$$K_a \times K_b = K_w$$

The acidity of an aqueous solution is conveniently expressed in terms of **pH,** defined as $-\log [H^+]$. Solutions of pH less than 7 are acid; those with pH greater than 7 are basic. The approximate pH of a solution can be determined by the use of **indicators,** or with pH paper impregnated with indicators. More accurate pH measurements are made with a pH meter. In titration of a solution of a strong base with a solution of a strong acid, or *vice versa,* the pH changes by several pH units in the immediate vicinity of the equivalence point. Use of an appropriate indicator permits accurate determination of the equivalence point.

The tendency of a substance to show acidic or basic characteristics in water can be correlated reasonably well with chemical structure. Acid character requires the presence of a highly polar H—X bond, promoting loss of hydrogen as H^+ on reaction with water. Basic character, on the other hand, requires the presence of an available pair of electrons. By considering the effects of changes in structure, it is possible to predict how a given structural change is likely to alter the acidity or basicity.

In the **Lewis theory** of acids and bases, the emphasis is on the shared electron pair rather than on the proton. An acid is defined as an electron-pair acceptor, a base as an electron-pair donor. The Lewis theory is more general than the Brønsted-Lowry model, because it applies to cases in which the proton is the acid, and to others as well.

learning goals

Having read and studied this chapter, you should be able to:

1. Explain the process that occurs when an acid dissolves in water.
2. Describe the forms in which the proton exists in water.
3. Define an acid, base, conjugate acid, and conjugate base in terms of the Brønsted-Lowry theory of acids and bases.
4. Write the acid-dissociation-constant expression for any weak acid in water.
5. Calculate $[H^+]$ for a weak acid solution in water, knowing acid concentration and K_a.
6. Explain what is meant by autoionization of water and write the ion-product-constant expression.
7. Write the base-dissociation-constant expression for a weak base in water.
8. Calculate $[H^+]$ for any weak base solution in water, knowing base concentration and K_b.
9. Explain the relationship between an acid and its conjugate base or between a base and its conjugate acid and calculate K_b from a knowledge of K_a, or *vice versa.*
10. Explain what is meant by pH and calculate either pH from a knowledge of $[H^+]$, or $[H^+]$ from a knowledge of pH.
11. Explain how indicators work and how they may be used to determine pH.

12. Explain the pH changes that occur during titration of a strong acid by a strong base, or of a strong base by a strong acid.

13. Explain how acid strength relates in a general way to the nature of the H—X bond.

14. Predict the relative acid strengths of oxyacids and oxyanions.

15. Define an acid or base in terms of the Lewis acid-base theory.

16. Predict the relative acidities of solutions of metal salts from a knowledge of metal-ion charges and ionic radii.

key terms

Among the more important terms and expressions used for the first time in this chapter are the following:

An acid-base indicator (Section 16.7) is a substance whose color changes in passing from an acidic to a basic form, or vice versa.

The acid-dissociation constant, K_a (Section 16.3), is an equilibrium constant that expresses the extent to which an acid transfers a proton to solvent water.

Autoionization of water (Section 16.4) is the process whereby water spontaneously forms low concentrations of $H^+(aq)$ and $OH^-(aq)$ ions by proton transfer from one water molecule to another.

The base-dissociation constant, K_b (Section 16.5) is an equilibrium constant that expresses the extent to which a base reacts with solvent water, accepting a proton and forming $OH^-(aq)$.

A Brønsted acid (Section 16.2) is any substance capable of acting as a source of protons.

A Brønsted base (Section 16.2) is any substance capable of acting as a proton acceptor.

A conjugate acid (Section 16.6) is a substance formed by addition of a proton to a Brønsted base.

A conjugate base (Section 16.6) is a substance formed by loss of a proton from a Brønsted acid.

Hydrolysis (Section 16.9) is a process in which a cation or anion reacts with water so as to change the pH.

The ion-product constant (Section 16.6) for water, K_w, is the product of the aquated hydrogen ion and hydroxide ion concentrations: $[H^+][OH^-] = K_w = 1 \times 10^{-14}$.

A Lewis acid (Section 16.9) is defined as an electron-pair acceptor.

A Lewis acid-base adduct (Section 16.9) is the product of reaction between a Lewis acid and a Lewis base.

A Lewis base (Section 16.9) is defined as an electron-pair donor.

An oxyacid (Section 16.8) is a compound in which one or more OH groups, and possibly additional oxygen atoms, are bonded to a central atom.

The term pH (Section 16.7) is defined as the negative log in base 10 of the aquated hydrogen-ion concentration: $pH = -\log[H^+]$.

A polyprotic acid (Section 16.8) is a substance capable of dissociating more than one proton in water; H_2SO_4 is an example.

EXERCISES

acid solutions; the hydrated proton

16.1 Hydrogen chloride gas is extremely soluble in water, but not very soluble in benzene. Explain.

16.2 Gaseous hydrogen iodide, HI, is very soluble in water. A solution containing 52.4 percent HI by weight has a density of 1.603 g cm³. What is the molarity of this solution?

16.3 How would you expect the freezing point of a 0.5 M aqueous solution of HI to compare with that of a 0.5 M solution of NaCl? With a 0.5 M solution of ethyl alcohol?

16.4   What is meant by the expression, "a strong acid in water"?

16.5   Describe the state of the proton in a dilute aqueous solution of hydrogen iodide. Is it different from the state of the proton in a dilute solution of hydrogen chloride? Explain.

16.6   The electrical conductivities of dilute solutions of hydrogen iodide increase almost in direct proportion to the concentration of HI. Furthermore, the conductivities are almost the same as those of hydrogen chloride solutions of the same concentrations. Explain these observations.

[16.7]   When very concentrated hydrogen iodide solutions are cooled to low temperatures a solid hydrate of the formula $HI \cdot 2H_2O$ is formed. Assuming that this compound is actually ionic, what do you think might be the structures of the anion and cation? Draw the Lewis structures for them.

16.8   Draw the Lewis structures for three different forms in which the proton might exist in water solution. Does it exist in any one of these forms for a long period of time? Explain.

strong and weak acids and bases

16.9   Write the equation for reaction of hydrogen iodide with water. Indicate which species in this equation are acids and which bases. Is iodide ion a strong or weak Brønsted base?

16.10   Nitrous acid, HNO_2, is a weak acid in water. Nitric acid, HNO_3, is a strong acid. Write the reaction for each with water. Which reaction, if either, proceeds to completion? Which is the stronger base, NO_2^- or NO_3^-?

16.11   What does the statement, "Hypochlorous acid is a weak acid in water" mean?

16.12   Compare a solution of acetic acid, a weak acid, with a solution of hydrogen chloride of the same concentration, with respect to electrical conductivity, freezing-point lowering, and reaction with an active metal.

16.13   On the basis of the data in Table 16.2, which of the following acids is the strongest acid in water? Which is the weakest? (a) benzoic acid; (b) nitrous acid; (c) formic acid; (d) phenol

16.14   Solutions of 0.1 M concentration of formic acid, hydrocyanic acid, and phosphoric acid are allowed to react with magnesium as indicated in Figure 16.4. From Table 16.2, predict which will react most rapidly.

16.15   Write the formula and name for the conjugate base associated with each of the following acids:

(a) nitric acid, HNO_3; (b) hydrogen sulfide, H_2S; (c) hydrobromic acid, HBr; (d) hydrocyanic acid, HCN; (e) formic acid, HCO_2H. (You may need to review nomenclature, Section 2.10.)

16.16   In each of the following pairs, indicate which is the stronger Brønsted base: (a) HS^- or S^{2-}; (b) NO_2^- or NO_3^-; (c) $C_6H_5O^-$ or $H\!-\!\overset{\displaystyle \|}{\underset{\displaystyle O}{C}}\!-\!O^-$; (d) Br^- or HCO_3^-.

16.17   In each of the following pairs, indicate which is the stronger Brønsted acid: (a) HCl or H_2CO_3; (b) HS^- or H_2SO_4; (c) HNO_3 or NO_3^-; (d) HBr or C_6H_5OH.

acid dissociation constants

16.18   Write the acid-dissociation-constant expression for the weak acid, HNO_2.

16.19   Calculate the concentration of aquated hydrogen ion in a 0.15 M solution of phenol in water (see Table 16.2 for value of K_a).

16.20   Ingestion of massive doses of vitamin C is popular in some circles as a cold preventative. From the value for K_a in Table 16.2 calculate $[H^+]$ for a solution of 1 g ascorbic acid (MW = 176 g/mole) in 250 ml solution.

16.21   The name for formic acid derives from the Latin word for "ant"; it was first isolated from the stinging red ant. Calculate the approximate value for $[H^+]$ for a 2 M solution of formic acid (K_a is given in Table 16.2).

16.22   A new compound is isolated and found to have a molecular weight of 268 g/mole. A solution of 3.5 g in 500 ml solution has a hydrogen-ion concentration of $2.6 \times 10^{-5}\ M$. Calculate the value of K_a for the new compound.

16.23   Rhubarb owes its sour taste to the presence of oxalic acid, $H_2C_2O_4$, for which K_a is 5.9×10^{-2}. If there are 1.9 mg of oxalic acid per g of rhubarb, and 200 g of rhubarb are cooked with sufficient water to make 1 l. of liquid sauce, what concentration of $H^+(aq)$ should the sauce have?

[16.24]   Extend the calculations shown in Sample Exercise 16.5 to HF solutions of 0.07, 0.03, and 0.006 M concentrations. Using these results and those in Sample Exercise 16.5, make a graph of percent ionization of the acid versus acid concentration. To what value would you expect this curve to extrapolate at zero concentration of acid?

autoionization of water

16.25   What are the relative concentrations of $H^+(aq)$ and $OH^-(aq)$ in a neutral solution? In an acid solution? In a basic solution? What is the value for the product $[H^+][OH^-]$ in a neutral solution? In an acid solution? In a basic solution?

16.26 Calculate the concentration of $H^+(aq)$ in each of the following solutions: (a) 0.01 M NaOH; (b) $3.3 \times 10^{-4} M$ KOH; (c) 0.002 M Ca(OH)$_2$.

solutions of bases

16.27 In terms of the Brønsted theory, state the difference between a weak base and a strong base in water.

16.28 The strong odor associated with fresh fish is due largely to amines. In cleaning up after a session of fish cleaning, the odor can be largely eliminated by rinsing with a dilute acid solution, for example, a solution of lemon juice. Describe the acid-base reaction that accounts for these observations.

16.29 Calculate the concentration of OH^- in a 0.10 M solution of hydroxylamine (see Table 16.3).

16.30 Calculate the percentage of pyridine that forms pyridinium ion, $C_5H_5NH^+$, in a 0.10 M solution of pyridine in water (see Table 16.3).

16.31 Aniline is an industrially important amine used in the making of dyes and many other organic chemical processes.

Aniline

Its molecular weight is 93 g/mole, and K_b is 3.8×10^{-10}. Aniline is soluble in neutral water to the extent of 3.9 g per 100 ml solution. What is $[H^+]$ in a saturated aqueous solution?

[16.32] Many drugs that have chemical structures involving organic amine groups, such as morphine, codeine, or quinine, are sold and administered as the acid salts. The anion could be Cl^- or perhaps SO_4^{2-}. As an example, quinine (let's call it Q) has a K_b of 1.1×10^{-6}. In order to isolate the quinine as the salt QH^+, the acidity of the solution must be adjusted so that the concentration of the species QH^+ is much higher than that of Q. What should $[H^+]$ be in order that $[QH^+] = 1 \times 10^3[Q]$? This may look like a much more difficult problem than it is; just use an expression like Equation [16.19], and follow it up with Equation [16.16].

conjugate acid-base relationships

16.33 Indicate whether each of the following solutions is neutral, acidic, or basic, using in some cases Table 16.2 or 16.3 for guidance as needed: (a) KBr; (b) NH$_4$Br; (c) potassium benzoate, K(CO$_2$C$_6$H$_5$); (d) KCN; (e) ammonium acetate, NH$_4$(C$_2$H$_3$O$_2$); (f) Na$_2$CO$_3$; (g) NH$_4$CN.

16.34 Using Table 16.2, calculate the value of K_b

for each of the following species: (a) formate ion, CO_2H^-; (b) dihydrogen phosphate ion, $H_2PO_4^-$; (c) acetate ion, $C_2H_3O_2^-$.

16.35 Using Table 16.3, calculate the value of K_a for each of the following acids: (a) $CH_3NH_3^+$; (b) H_2S; (c) pyridinium ion, $C_5H_5NH^+$; (d) hypochlorous acid, HOCl.

16.36 Arrange the following ions in the order of increasing base strength: (a) CN^-; (b) $C_2H_3O_2^-$; (c) Br^-; (d) NO_2^-.

[16.37] The two acid dissociations of oxalic acid are as follows:

$$H_2C_2O_4 \rightleftharpoons H^+(aq) + HC_2O_4^-(aq)$$
$$K_{a1} = 5.9 \times 10^{-2}$$

$$HC_2O_4^-(aq) \rightleftharpoons H^+(aq) + C_2O_4^{2-}(aq)$$
$$K_{a2} = 6.4 \times 10^{-5}$$

Using the rule and procedure described in the text, add these two reactions, write the net overall reaction, and the equilibrium expression for the overall reaction. What is the value for K for the overall reaction?

pH

16.38 Calculate the pH of each of the following solutions: (a) 0.01 M HBr; (b) 0.01 M KOH; (c) 0.005 M Ba(OH)$_2$; (d) $3.5 \times 10^{-5} M$ NaOH; (e) $2.2 \times 10^{-3} M$ HCl; (f) 0.35 M HNO$_3$.

16.39 Calculate $[H^+]$ for solutions with the following pH values: (a) 3.56; (b) 12.25; (c) 6.55; (d) 8.81; (e) pOH = 4.55. Indicate which of these solutions is acidic and which basic.

16.40 A sample of lake water is filtered and tested with indicators to determine its pH. It turns methyl yellow a yellow color, and turns methyl red a red color. Indicate the highest and lowest values of $[H]^+$ the solution might have.

16.41 Which of the following solutions will give a pH less than 7, which greater than 7: (a) NaCN; (b) Mg(OH)$_2$; (c) HClO$_2$; (d) KNO$_2$?

16.42 (a) Sketch the curve of pH versus added HCl for the titration of 50 ml of 0.100 M NaOH solution with 0.100 M HCl solution; indicate the equivalence point. (b) Do the same for titration of 50 ml of 0.100 M NaOH solution with 0.200 M HCl solution.

16.43 A student carrying out a titration of a solution of HCl with NaOH solution uses methyl yellow as his indicator. Which color will the solution be initially, and what color at the end of the titration? Was this a good choice of indicator? If not, is the student likely to have added too much or too little NaOH solution?

acid-base character and chemical structure

16.44 All other factors being equal, what relationship is there between the acidity of a substance H—X in water and the polarity of the H—X bond?

[16.45] From careful studies of the degree of ionization in protonic solvents other than water, it has been possible to determine that the order of acid strength among the hydrogen halides is HI > HBr > HCl > HF. How can this order be explained?

16.46 Draw the Lewis structures for nitrous acid, HNO_2, and for nitric acid, HNO_3. Which is the stronger acid in water? Why?

16.47 Draw the Lewis structures for carbonic acid, H_2CO_3, and sulfurous acid, H_2SO_3. Which is the stronger acid? Why?

16.48 Using the rule given in the text, predict the acid-dissociation constant for $H_2AsO_4^-$ and $HAsO_4^{2-}$, if K_a for H_3AsO_4 is 5.6×10^{-3}.

16.49 Borax, $Na_2B_4O_7 \cdot 10H_2O$, is often used as a conditioner in laundering. It dissolves in water to give an alkaline, or basic, solution. Write a possible reaction that could produce the basic solution.

16.50 Predict whether each of the following salts will form an acidic or basic solution in water: (a) $NaHCO_3$; (b) KH_2BO_3; (c) NH_4I; (d) $KHTeO_3$; (e) $MgHPO_4$.

[16.51] Many moderately large organic molecules containing basic nitrogen atoms are not very soluble in water as the neutral molecule, but are frequently much more soluble as the acid salt. Assuming that the pH in the stomach is 2.5, indicate whether each of the following compounds would be present in the stomach as the neutral compound or in the protonated form:

COMPOUND	K_b
Nicotine	7×10^{-7}
Caffeine	4×10^{-14}
Strychnine	1×10^{-6}
Quinine	1.1×10^{-6}

[16.52] For nicotine, shown in Table 16.3, two values of K_b are listed. From comparisons with other compounds shown in the table, indicate which value of K_b is probably associated with which nitrogen.

Lewis acids and bases

16.53 Define an acid in terms of the Lewis theory of acids and bases and compare this definition with that based on the Brønsted-Lowry theory. Which is the more general; that is, which includes the other within its scope? Explain.

16.54 Explain how the initial reaction of HCl with water might be described in terms of Lewis acid-base theory.

16.55 A metal oxide such as CaO may be considered to consist of oxide ions, O^{2-}, and Ca^{2+} cations. It reacts with water to form a basic solution. Using Lewis structures, describe the reaction of oxide ion with water in terms of the Brønsted-Lowry theory; in terms of Lewis theory.

16.56 The BF_4^- ion is a very stable ion, and salts such as $NaBF_4$ are readily soluble in water to form neutral solutions. By drawing Lewis structures, describe how the BF_4^- ion can be visualized as the Lewis acid-base adduct of BF_3 and F^-.

16.57 Draw Lewis structures to show how SO_2 might react with water to form H_2SO_3.

16.58 Salts such as $MgSO_4$, $CaSO_4$, $Mg(ClO_4)_2$, and others are often used as drying agents for gases or organic liquids. What process occurs when they remove water from these media? How can it be described in terms of Lewis acid-base theory?

16.59 Describe the process that causes a solution of $Al(NO_3)_3$ to be acidic.

16.60 In each of the following pairs of salts, indicate which you would expect to produce the more acidic solution: (a) LiI or ZnI_2; (b) $CaCl_2$ or $FeCl_3$; (c) NH_4Cl or NaCl; (d) $MgCl_2$ or $Mg(C_2H_3O_2)_2$; (e) $FeCl_2$ or $FeCl_3$.

[16.61] Write the acid-dissociation-constant expression for Al^{3+} ion in water. The value for this constant, sometimes called the hydrolysis constant, is 7×10^{-6}. Calculate the pH of a 0.05 M solution of $Al(NO_3)_3$.

general exercises

16.62 Write chemical equations that show how the following substances react with water to form either acidic or basic solutions: (a) NaH; (b) K_2O; (c) SeO_2; (d) P_2O_5.

16.63 It is said that the hydroxide ion is the strongest base possible in water. Is this a correct statement? If so, why is it so? (Hint: Imagine what would happen if a stronger base than OH^- were introduced into water.)

[16.64] Liquid ammonia, which boils at $-33°C$, has actually received quite a lot of study as a solvent medium. By analogy with water, write the equation for autoionization of this solvent. What are the formulas of the acid and base that are the counterparts of HCl and NaOH in water?

16.65 Which of the following ions would you expect to be the strongest base in water: (a) N^{3-}; (b) O^{2-}; (c) F^-? Explain your answer.

16.66 Water, liquid ammonia, and liquid HF are all polar liquids that dissolve ionic substances and support ionization of acids and bases. On the other hand, H_2S, PH_3, and HCl as liquids are relatively poor solvents. What accounts for the difference in properties of the third row hydrides as compared with the second row?

aqueous equilibria

17

Water is the solvent of preeminent importance on our planet. It is also, in a sense, the solvent of life. It is difficult to imagine how living matter in all its complexity could exist with any liquid other than water as solvent. Water occupies its position of importance not only because of its abundance, but also because of its exceptional ability to dissolve a wide variety of substances. Many equilibria exist between these substances. In the previous chapter we saw examples of weak acid and weak base equilibria. We now go on to examine important extensions of these acid-base equilibria and to consider other equilibrium processes in water.

17.1
**buffer
solutions**

Many aqueous solutions resist a change in pH upon addition of small amounts of acid or base. Such solutions are called buffer solutions, and are said to be buffered. Human blood, for example, is a complex aqueous medium with a pH buffered at about 7.4. Any significant variation of the pH from this value results in a severe pathological response and, eventually, death. As another example, the chemical behavior of seawater is determined in very important respects by its pH, buffered at about 8.1 to 8.3 near the surface. Addition of a small amount of an acid or base to either blood or seawater does not result in a large change in pH. Compare, for example, the behavior of a liter of seawater and a liter of pure water upon addition of 0.1 ml of 1 M HCl solution (1×10^{-4} moles of HCl), Figure 17.1. The pH of pure water changes by 4 units, from 7 to 3. The pH of seawater changes by only 0.6 pH units. Substances already dissolved in seawater limit the change in [H$^+$] on addition of HCl.

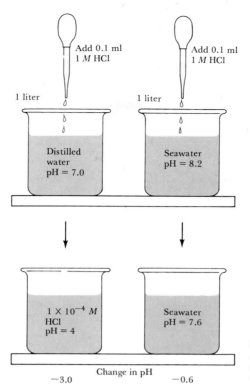

Add 0.1 ml
1 M HCl

Add 0.1 ml
1 M HCl

1 liter

1 liter

Distilled
water
pH = 7.0

Seawater
pH = 8.2

1×10^{-4} M
HCl
pH = 4

Seawater
pH = 7.6

Change in pH

−3.0

−0.6

FIGURE 17.1 A comparison of the effect of added acid on the pH of distilled water as compared with the same quantity of seawater.

To understand how buffer solutions work, let's consider a solution made by adding 0.10 mole of sodium acetate, $NaC_2H_3O_2$, to a liter of 0.10 M acetic acid solution. We shall examine this buffer solution first from a qualitative, then from a quantitative, point of view.

Sodium acetate is a strong electrolyte; that is, it is completely ionized in solution, forming Na^+ and $C_2H_3O_2^-$ ions. Acetic acid, $HC_2H_3O_2$, is a weak acid whose dissociation is governed by the equilibrium reaction given in Equation [17.1]:

$$HC_2H_3O_2(aq) \rightleftharpoons H^+(aq) + C_2H_3O_2^-(aq) \qquad [17.1]$$

The $C_2H_3O_2^-$ in this equilibrium comes from both $HC_2H_3O_2$ and $NaC_2H_3O_2$. This mixture can either react with surplus H^+ ions, or release them, according to the circumstances. For example, if a small quantity of acid is added to the solution, the equilibrium shifts to the left; acetate ion reacts with the added H^+. The solution thereby limits pH change due to added acid. On the other hand, if a small quantity of a base is added, it reacts with H^+. This reaction causes the equilibrium of Equation [17.1] to shift to the right; $HC_2H_3O_2$ dissociates to form more H^+. The solution thereby also resists change in pH due to added base.

Other buffer solutions can be prepared from other weak acids or weak bases. In each case the weak acid or base is mixed with the salt of that weak acid or base. For example, a mixture of the weak base ammonia, NH_3, and the salt ammonium chloride, NH_4Cl, serves as a buffer. We shall shortly work out a quantitative treatment of buffer solutions. In

493

general terms, the pH of a buffer solution depends on whether a weak acid or weak base is involved, the ionization constant of that weak acid or base, and the relative concentrations of the weak acid or base and its corresponding salt. The acetic acid–sodium acetate buffer that we have been considering maintains pH in the vicinity of 4.7. By comparison, a solution that is 0.10 M in NH_3 and 0.10 M in NH_4Cl is buffered at a pH of about 9.3. In addition to the pH of a buffered solution, one other aspect of interest is its capacity to absorb added acid or base without large changes in pH. This capacity depends on the amount of weak acid or base and corresponding salt present in the solution.

QUANTITATIVE CONSIDERATIONS

The pH of a buffer solution can be calculated using the methods described in Chapter 16. Calculation of the pH of a solution which is 0.10 M in $HC_2H_3O_2$ and 0.10 M in $NaC_2H_3O_2$ is shown in Sample Exercise 17.1.

SAMPLE EXERCISE 17.1

Suppose we add 8.20 g, or 0.10 moles, of sodium acetate, $NaC_2H_3O_2$, to 1 l. of a 0.10 M solution of acetic acid, $HC_2H_3O_2$. What is the pH of the resultant solution?

Solution: Because $NaC_2H_3O_2$ is a strong electrolyte, the only equilibrium that we need to consider is that for dissociation of acetic acid. This equilibrium reaction and the concentrations of the species involved are summarized below:

$$HC_2H_3O_2(aq) \rightleftharpoons H^+(aq) + C_2H_3O_2^-(aq)$$

Initial
concen-
tration:　　0.10 M　　　　0 M　　　0.10 M

Equilibrium
concen-
tration: $(0.10 - x)M$　　　$x\,M$　　$(0.10 + x)M$

Notice that the equilibrium concentration of $C_2H_3O_2^-$ is the initial amount coming from the added $NaC_2H_3O_2$, plus the amount (x) formed by dissociation of acetic acid.

The equilibrium constant expression is

$$K_a = 1.8 \times 10^{-5} = \frac{[H^+][C_2H_3O_2^-]}{[HC_2H_3O_2]}$$

(The value of the equilibrium constant is taken from Table 16.2.) Addition of the acetate salt does not change the value of the equilibrium constant. Substituting the equilibrium concentrations in this equilibrium-constant equation, we obtain:

$$\frac{(x)(0.10 + x)}{(0.10 - x)} = 1.8 \times 10^{-5}$$

We can simplify this equation by ignoring x relative to 0.10. Recall that x is the concentration of H^+ formed by dissociation of the acid. As we saw in the worked-out example in Section 16.3, x could be ignored in comparison with 0.10 M. In the present problem x will be even smaller because of the effect of added $C_2H_3O_2^-$. This simplification gives us

$$\frac{x(0.10)}{(0.10)} = 1.8 \times 10^{-5}$$

$$x = 1.8 \times 10^{-5}\,M = [H^+]$$

$$pH = -\log[1.8 \times 10^{-5}] = 4.74$$

Earlier (Section 16.3) we calculated that in a 0.10 M solution of $HC_2H_3O_2$, $[H^+]$ is $1.3 \times 10^{-3}\,M$, corresponding to pH of 2.89. The calculation of Sample Exercise 17.1 shows that the addition of $C_2H_3O_2^-$ sharply lowers $[H^+]$, with a corresponding increase in pH to 4.74. The effect of added acetate ion is consistent with qualitative predictions. By applying LeChatelier's principle to the equilibrium in Equation [17.1], we conclude that addition of $C_2H_3O_2^-$ will have the effect of shifting the equilibrium to the left, thus causing $[H^+]$ to decrease, and the pH of the solution to increase. We can also look at it from another point of view. Acetate ion is the conjugate base of the weak acid, acetic acid. Addition of a base to a solution will cause an increase in pH.

In Sample Exercise 17.1 we considered a solution in which the concentrations of the acid and conjugate base were equal. There is no need for this to be the case. It would be very useful to have a general expression that could be used to calculate the pH of a buffer solution, knowing the concentrations of acid and corresponding salt, and knowing the acid-dissociation constant for the acid. Let us consider the pH of a solution of an acid HX, and of a corresponding salt MX, where M could be Na^+, K^+, and so forth. The acid dissociation equilibrium is

$$HX(aq) \rightleftharpoons H^+(aq) + X^-(aq) \qquad [17.2]$$

and the corresponding acid dissociation constant expression is

$$K_a = \frac{[H^+][X^-]}{[HX]} \qquad [17.3]$$

Let us now take the logarithm in base 10 of both sides of the equation:

$$\log K_a = \log[H^+] + \log \frac{[X^-]}{[HX]} \qquad [17.4]$$

Multiplying through on both sides by -1, we have

$$-\log K_a = -\log[H^+] - \log \frac{[X^-]}{[HX]} \qquad [17.5]$$

We see that the first term on the right side of the equation is just pH. Using the convention that the negative log of a quantity is labeled p (quantity), we can write

$$-\log K_a = pK_a$$

Thus, we have from Equation [17.5]:

$$pK_a = pH - \log \frac{[X^-]}{[HX]} \qquad [17.6]$$

$$pH = pK_a + \log \frac{[X^-]}{[HX]} \qquad [17.7]$$

Equation [17.7] is very useful; it enables us to express the pH of a solution containing a weak acid and its conjugate base if we know K_a for the weak acid, and the relative concentrations of the acid and conjugate base.

SAMPLE EXERCISE 17.2

Calculate the pH of a solution containing 0.10 M acetic acid and 0.20 M sodium acetate.

Solution: We can use Equation [17.7] to solve this problem. Using the same arguments as in Sample Exercise 17.1, we assume that the nonionized acetic acid concentration is just equal to the initial concentration of acetic acid, and that the acetate ion concentration is due solely to sodium acetate. We have

$$K_a = 1.8 \times 10^{-5}$$
$$[X^-] = [C_2H_3O_2^-] = 0.20 \ M$$
$$[HX] = [HC_2H_3O_2] = 0.10 \ M$$

Inserting these quantities into Equation [17.7] we obtain

$$pH = -\log(1.8 \times 10^{-5}) + \log\frac{(0.20)}{(0.10)}$$
$$= 4.74 + 0.30$$
$$= 5.04$$

Note that the addition of still more acetate ion has caused the pH of the solution to increase above the value calculated in Sample Exercise 17.1.

To further illustrate the buffer action of a solution of acetic acid and acetate ion, we should compare it with the action of a solution that is not a buffer. We saw in Sample Exercise 17.1 that the pH of a solution which is 0.10 M in acetic acid and 0.10 M in sodium acetate is 4.74. A solution of this same pH is obtained by addition of 1.8×10^{-5} moles of HCl to a liter of water. Because HCl is a strong acid, 1.8×10^{-5} M HCl has an H^+ concentration of 1.8×10^{-5} M, and thus its pH is 4.74. Suppose that 1 ml

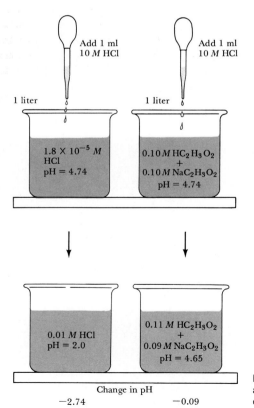

FIGURE 17.2 Comparison of the effect of added acid on a buffer solution of pH 4.74 as compared with an HCl solution of pH 4.74.

of 10 M HCl solution, that is, 0.01 mole HCl, is added to a liter of each of these two solutions of pH 4.74, as shown in Figure 17.2.

The HCl added to the solution of acetic acid and acetate ion reacts with the acetate ion. The reaction between acetate ions and H^+ is just the reverse of Equation [17.1]. If that reaction proceeds in the forward direction to only a slight extent, then the reverse reaction must proceed to a very large extent. * Thus, we can assume that all the added 0.01 mole of HCl reacts with 0.01 mole of acetate ion to form 0.01 mole of acetic acid and 0.01 mole of NaCl. The latter is, of course, inactive from an acid-base point of view. As shown in Figure 17.2, addition of the HCl simply changes the relative amounts of acetic acid and acetate ion we have to deal with in setting up the problem: Before the addition the solution was 0.10 M each in $HC_2H_3O_2$ and $C_2H_3O_2^-$. After addition, the solution is 0.11 M in $HC_2H_3O_2$ and 0.09 M in $C_2H_3O_2^-$. Using the method outlined in Sample Exercise 17.2, the pH is calculated to be 4.65, about 0.09 units lower than before addition (Figure 17.2). Clearly, this is a much smaller change than occurs on addition of acid to the dilute HCl solution.

We might have chosen to add 0.01 mole of hydroxide to the solutions illustrated in Figure 17.2. This addition would have caused the pH of the dilute HCl solution to go from 4.74 to essentially 12 (you should be able to explain why this is so), whereas the pH of the buffer solution would have increased by only about 0.09 pH units. Thus, a buffer solution responds to approximately the same degree, but in opposite direction, to acid or base addition.

Buffer solutions usually have a pH value in the range from 3 to 11. The capacity of the buffer to react with added acid or base, and thus to limit the pH change, is determined by the concentrations of acids and bases present. From Equation [17.6] we can see that the pH of an acetic acid–acetate buffer solution is constant, regardless of the concentration. Thus, solutions that are 0.2 M each in $HC_2H_3O_2$ and $C_2H_3O_2^-$ have the same pH as solutions that are 0.1 M each, or 0.02 M each in these reagents. But the ability of the solution to act as a buffer diminishes as the concentration decreases. You can verify this for yourself by calculating the pH change that would occur if the buffer solution in Figure 17.2 were initially composed of 0.02 M acetic acid and 0.02 M sodium acetate.

We have so far given examples only of acidic buffer solutions. The same idea applies to solutions of weak bases and their conjugate acids. The pH of the buffer solution in these cases is generally above 7. One problem that frequently confronts the laboratory worker is to make up a buffer of known pH. Sample Exercise 17.3 outlines a problem of this type.

---

*We can also appreciate the extent to which the reverse reaction occurs by considering the value for the equilibrium constant. The equilibrium constant for a reaction written in the reverse direction is just the reciprocal of the equilibrium constant for the forward reaction. Thus the equilibrium constant for the reaction

$$C_2H_3O_2^-(aq) + H^+(aq) \rightleftarrows HC_2H_3O_2(aq)$$

is

$$\frac{1}{1.8 \times 10^{-5}} = 5.5 \times 10^4$$

What concentration of NH_4Cl must be added to a 0.1 M solution of NH_3 to adjust the pH to 9.0?

Solution: We should first write out the equilibrium involved, and the corresponding equilibrium-constant expression.

$$NH_3(aq) + H_2O \rightleftharpoons NH_4^+(aq) + OH^-(aq)$$

$$K_b = \frac{[NH_4^+][OH^-]}{[NH_3]} = 1.8 \times 10^{-5}$$

(The value of K_b is taken from Table 16.3.)

It would be very helpful to have an expression for this system analogous to Equation [17.7] for a solution of a weak acid. Taking the log of both sides we obtain

$$\log K_b = \log[OH^-] + \log \frac{[NH_4^+]}{[NH_3]}$$

Multiplying through by -1, we obtain:

$$-\log K_b = -\log OH^- - \log \frac{[NH_4^+]}{[NH_3]}$$

$$pK_b = pOH - \log \frac{[NH_4^+]}{[NH_3]} \qquad [17.8]$$

Using the ion product constant for water we can write

$$[H^+][OH^-] = K_w = 10^{-14}$$

Taking the log of both sides and multiplying through by -1 we have (see also Section 16.7),

$$pH + pOH = 14$$

Thus, if the pH of our solution is to be 9.0, the pOH must be 5.0. We have

$$K_b = 1.8 \times 10^{-5} \qquad pK_b = 4.74 \qquad pOH = 5.0$$

Inserting these quantities into Equation [17.8] we obtain:

$$4.74 = 5.0 - \log \frac{[NH_4^+]}{[NH_3]}$$

Rearranging,

$$\log \frac{[NH_4^+]}{[NH_3]} = 0.26$$

$$\frac{[NH_4^+]}{[NH_3]} = 1.8$$

Because ammonia is a weak base, we can assume that the fraction that ionizes is very small. This is especially the case because the conjugate acid, NH_4^+ is to be added. Thus, $[NH_3]$ equals 0.10 M.

$$\frac{[NH_4^+]}{[0.10]} = 1.8$$

$$[NH_4^+] = 0.18 \ M$$

17.2 titrations of weak acids and bases

In previous material (Sections 12.2 and 16.6), we have encountered the titrations of acids and bases. In all the examples we have considered so far, a strong acid has been titrated by a strong base, or vice versa. Our discussion of buffer solutions provides the background needed to consider the more complicated case of titrations involving weak acids or bases. Titration of a weak acid by a strong base or of a weak base by a strong acid proceeds in much the same way as when only strong acids and bases are involved, but there are important differences. To see what these differences are, we shall consider the titration of a 0.10 M acetic acid solution by 0.10 M NaOH, for comparison with the titration of 0.10 M HCl by 0.10 M NaOH, discussed in Section 16.6.

A graph of pH as a function of added base in the titration of acetic acid is shown in Figure 17.3. For comparison, the pH versus added base curve in the titration of HCl is shown in color. In general, the pH in the weak acid titration is shifted upward on the acid side of the equivalence point. This is what we would expect, because the pH of a weak acid solution would be higher than that of a strong acid solution of equivalent concentration. The pH at the beginning of the titration is just the pH of a 0.10 M solution of acetic acid, which is 2.89. (Sample Exercise 16.2).

One of the most important points on the curve is the pH at the equivalence point. In the case of the HCl–NaOH titration, at the equivalence point the ionic substance in solution is NaCl. The pH is therefore 7. In the case of the titration of acetic acid by sodium hydroxide, the ionic substance present in solution at the equivalence point is $NaC_2H_3O_2$. Because acetate ion is a weak base, the pH of the solution is higher than 7, as shown in Sample Exercise 17.4.

SAMPLE EXERCISE 17.4

Calculate the pH at the equivalence point in the titration of $0.10\ M\ HC_2H_3O_2$ by $0.10\ M$ NaOH.

Solution: At the equivalence point we have $0.05\ M$ $NaC_2H_3O_2$. (Remember that the total volume of solution is doubled by addition of the NaOH solution to the $HC_2H_3O_2$ solution.) We calculate the pH of the solution of this weak base in the usual way:

$$C_2H_3O_2^-(aq) + H_2O \rightleftharpoons$$
$$HC_2H_3O_2(aq) + OH^-(aq)$$

Initial concentration:	$0.05\ M$	$0\ M$	$0\ M$
Equilibrium concentration:	$(0.05 - x)\ M$	$x\ M$	$x\ M$

The base dissociation constant for acetate ion is given by:

$$K_b = \frac{K_w}{K_a} = \frac{1 \times 10^{-14}}{1.8 \times 10^{-5}}$$
$$= 5.5 \times 10^{-10}$$

$$5.5 \times 10^{-10} = \frac{[HC_2H_3O_2][OH^-]}{[C_2H_3O_2^-]}$$
$$= \frac{(x)(x)}{(0.05 - x)} \simeq \frac{(x^2)}{0.05}$$
$$x^2 = (0.05)(5.5 \times 10^{-10})$$
$$= 2.8 \times 10^{-11}$$
$$x = 5.2 \times 10^{-6} = [OH^-]$$
$$pOH = 5.29$$

Because

$$pH + pOH = 14$$

we know that

$$pH = 8.71$$

(If you had trouble following along in this outline of how to work the problem, you should go back to Chapter 16, particularly to Sections 16.3 and 16.5).

At some intermediate point in the titration, the added NaOH has neutralized some of the $HC_2H_3O_2$, so that the solution contains a mixture of the acid and its conjugate base. To find the pH of the solution at some stage in the titration, it is easy to use Equation [17.7] as shown in Sample Exercise 17.5.

SAMPLE EXERCISE 17.5

Calculate the pH in the titration of acetic acid by sodium hydroxide after 30.0 ml of $0.100\ M$ NaOH solution have been added to 50.0 ml of $0.100\ M$ acetic acid solution.

Solution: The total number of moles of $HC_2H_3O_2$ originally in solution is

$$\left(\frac{0.100\ \text{mole}\ HC_2H_3O_2}{1\ \text{l. soln}}\right)(0.0500\ \text{l. soln}) =$$
$$5.0 \times 10^{-3}\ \text{moles}\ HC_2H_3O_2$$

Similarly, 30.0 ml of $0.100\ M$ NaOH solution contains 3.00×10^{-3} moles OH^-. During the titration, this OH^- reacts with the acetic acid, forming

3.00×10^{-3} moles of $C_2H_3O_2^-$, and leaving 2.00×10^{-3} moles of $HC_2H_3O_2$, in 80 ml solution:

$$OH^- + HC_2H_3O_2 \longrightarrow H_2O + C_2H_3O_2^-$$

The resulting molarities are thus

$$[HC_2H_3O_2] = \frac{2.00 \times 10^{-3} \text{ moles } HC_2H_3O_2}{0.0800 \text{ l.}}$$

$$= 0.0250 \, M$$

$$[C_2H_3O_2^-] = \frac{3.00 \times 10^{-3} \text{ moles } C_2H_3O_2^-}{0.0800 \text{ l.}}$$

$$= 0.0375 \, M$$

The value for pK_a of acetic acid is $-\log(1.8 \times 10^{-5}) = 4.74$. Inserting these quantities into Equation [17.7] we find

$$pH = 4.74 + \log \frac{(0.0375)}{(0.0250)}$$

$$= 4.91$$

By proceeding in such fashion the titration curve shown in Figure 17.3 could be traced out. Incidentally, you might note that at the halfway point in the titration, when $[C_2H_3O_2^-] = [HC_2H_3O_2]$, the pH equals pK_a, in this case 4.74.

The fact that the pH at the equivalence point in the titration of the weak acid is considerably higher than it is in the titration of the strong acid is important. We saw earlier that in titrating 0.10 M HCl with 0.10 M NaOH, either phenolphthalein or methyl red could be used as indicator. Although the pH of color change does not correspond precisely to the equivalence point for either indicator, both were close enough so that no significant error would be introduced by using either of them (Figure 16.8). In titration of acetic acid by NaOH, phenolphthalein is an ideal indicator, because it changes color just at the pH of the equivalence point. However, methyl red is not a good choice. The pH at the mid-range of its color change is 5.2. At this pH we are still far short of the equivalence point as shown in Figure 17.3.

If the acid to be titrated were weaker than acetic acid, the titration curve would look similar to that shown in Figure 17.3. However, the pH at the beginning of the reaction would be higher, and the pH at the

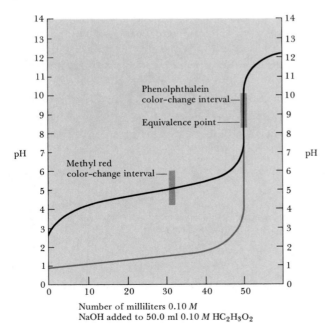

Number of milliliters 0.10 M NaOH added to 50.0 ml 0.10 M HC$_2$H$_3$O$_2$

FIGURE 17.3 The black line shows the graph of pH versus added 0.10 M NaOH solution in the titration of 0.10 M acetic acid solution. The color line segment shows the graph of pH versus added base for the titration of 0.10 M HCl.

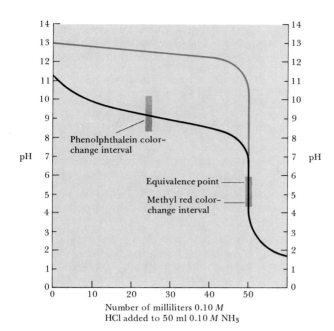

Number of milliliters 0.10 M
HCl added to 50 ml 0.10 M NH$_3$

FIGURE 17.4 The black line shows the graph of pH versus added 0.10 M HCl solution in the titration of 0.10 M ammonia. The color line segment shows the graph of pH versus added acid for the titration of 0.10 M NaOH.

equivalence point would also be higher. This means that the pH change at the equivalence point would not be as sharp, and the end point would be more difficult to detect. With very weak acids, for which K_a is less than about 1×10^{-8}, the acid-base titration is not really a good quantitative procedure when using indicators.

Titration of a weak base, for example 0.10 M NH$_3$, by a strong acid such as 0.10 M HCl solution, leads to the titration curve shown in Figure 17.4. In this particular example, the equivalence point occurs at pH 5.3. In this case methyl red would be an ideal indicator, but phenolphthalein would be a poor choice.

17.3

ionizations of polyprotic acids

Polyprotic acids are those capable of dissociating more than one ionizable hydrogen. Examples include phosphoric acid, H$_3$PO$_4$, oxalic acid, H$_2$C$_2$O$_4$, sulfuric acid, H$_2$SO$_4$, and carbonic acid, H$_2$CO$_3$. The equilibria involving aqueous solutions of these acids are generally a little more complex than when only one dissociable hydrogen is involved. Nevertheless, it is possible with the aid of a few simplifying approximations to solve most equilibrium problems of interest.

As an important example of a polyprotic acid, we'll consider carbonic acid, H$_2$CO$_3$. This acid is formed when carbon dioxide, CO$_2$, dissolves in water. The solubility of CO$_2$ in water is quite sensitive to pH. In equilibrium with pure water at 28°C and 1 atm pressure, the solubility is about 0.0034 M. There exists in water an equilibrium between dissolved CO$_2$ and what we might call the hydrated form, H$_2$CO$_3$. This latter form results from reaction of CO$_2$ with water, as described in Section 16.8. The equilibrium constant for the reaction

$$CO_2(aq) + H_2O \rightleftharpoons H_2CO_3(aq)$$ [17.9]

501

is not very large. Only about 1 percent of the dissolved CO_2 exists as H_2CO_3 rather than as CO_2. However, it makes no difference for our purposes whether we write Equation [17.10] or Equation [17.11]

$$CO_2(aq) + H_2O \rightleftharpoons H^+(aq) + HCO_3^-(aq) \qquad [17.10]$$

$$H_2CO_3(aq) \rightleftharpoons H^+(aq) + HCO_3^-(aq) \qquad [17.11]$$

for the first ionization of the acid. The form of the equilibrium constant and the value for K_{a1} would be the same, so long as we used the total concentration of CO_2 in water as the acid concentration. The common practice, therefore, is to use Equation [17.11] in describing the ionization and to assume that all the dissolved CO_2 is in the form of H_2CO_3. On this basis, the equilibrium constant for reaction Equation [17.11] has the value 4.3×10^{-7}, as listed in Table 16.2. The second ionization of a proton, from HCO_3^-, shown in Equation [17.12], occurs to a much smaller extent:

$$HCO_3^-(aq) \rightleftharpoons H^+(aq) + CO_3^{2-}(aq) \qquad [17.12]$$

The equilibrium constant for this second ionization, K_{a2}, is 5.6×10^{-11}.

Let's consider first a saturated solution of CO_2 in pure water at 28°C. What are the concentrations of all the possible species formed by ionization—that is, H^+, HCO_3^- and CO_3^{2-}? To answer this question, we must solve the equilibrium-constant expressions one at a time. For the first ionization we have

$$K_{a1} = \frac{[H^+][HCO_3^-]}{[H_2CO_3]} = 4.3 \times 10^{-7}$$

Recalling that the concentration of CO_2 in water at 28°C is 0.0034 M we have the following:

	$H_2CO_3(aq)$	$\rightleftharpoons$	$H^+(aq)$	$+$	$HCO_3^-(aq)$
Initial concentration:	0.0034 M		0 M		0 M
Equilibrium concentration:	$(0.0034 - x)M$		$x\,M$		$x\,M$

Using these equilibrium concentrations we can solve this problem in the usual way, by assuming that x is small in comparison to 0.0034 M:

$$K_{a1} = \frac{[H^+][HCO_3^-]}{[H_2CO_3]} = \frac{x^2}{0.0034} = 4.3 \times 10^{-7}$$

Solving for x gives $[H^+]$ of 3.8×10^{-5} M.

The second ionization occurs to a much smaller extent than the first. Using the values of $[HCO_3^-]$ and $[H^+]$ calculated above, and setting $[CO_3^{2-}] = y$, we have:

$$HCO_3^-(aq) \rightleftharpoons H^+(aq) + CO_3^{2-}(aq)$$

Initial concentration:	$3.8 \times 10^{-5}\ M$	$3.8 \times 10^{-5}\ M$	$0\ M$
Equilibrium concentration:	$(3.8 \times 10^{-5} - y)M$	$(3.8 \times 10^{-5} + y)M$	$y\ M$

Assuming that y is small compared to 3.8×10^{-5}, we have:

$$K_{a2} = \frac{[H^+][CO_3^{2-}]}{[HCO_3^-]} = \frac{(3.8 \times 10^{-5})(y)}{(3.8 \times 10^{-5})} = 5.6 \times 10^{-11}$$

$$y = 5.6 \times 10^{-11}\ M = [CO_3^{2-}]$$

The value calculated for y is indeed very small in comparison with 3.8×10^{-5}, showing that our assumption was justified. It also shows that the ionization of the HCO_3^- is negligible in comparison with that of H_2CO_3 as far as production of H^+ is concerned. However, it is the *only* source of CO_3^{2-}, which has a very low concentration in the solution.

Our calculations thus tell us that in a solution of carbon dioxide in water most of the CO_2 is in the form of CO_2 or H_2CO_3, a small fraction ionizes to form H^+ and HCO_3^-, and an even smaller fraction ionizes to give CO_3^{2-}. The pH of the saturated CO_2 solution is 4.42.

The situation we have just described applies when dissolved CO_2 is the only source of H^+. Often, however, it happens that the pH of the solution is controlled by the presence of other substances. For example, addition of a base to a saturated solution of CO_2 would cause formation of more HCO_3^-. With still further addition of a base, considerable CO_3^{2-} would form. The reactions that occur are:

$$H_2CO_3 + OH^- \rightleftharpoons HCO_3^- + H_2O \qquad [17.13]$$

$$HCO_3^- + OH^- \rightleftharpoons CO_3^{2-} + H_2O \qquad [17.14]$$

The fraction of total CO_2 present in solution in the form of H_2CO_3, HCO_3^-, or CO_3^{2-} varies with the pH, as shown in Figure 17.5. From this graph we can draw many useful conclusions. For example, we note that at a pH of about 5 and lower, the dissolved CO_2 is essentially all in a nonionized form. This means that the solubility of gaseous CO_2 in acidic solution is essentially independent of pH in the range of 5 and lower. On the other hand, as pH increases above 5, the solubility of CO_2 increases, because a substantial fraction of the dissolved CO_2 is converted into ionic forms. As we'll see in Chapter 18, the equilibrium between HCO_3^- and CO_3^{2-} has several important consequences for the chemistry of natural waters.

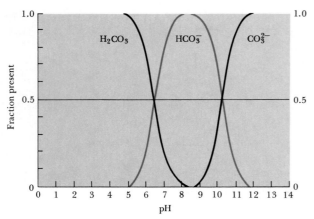

FIGURE 17.5 Relative abundances of the various species derived from dissolved CO_2 as a function of pH.

A graph similar to that shown in Figure 17.5 can be constructed for any polyprotic acid, using the calculation methods we have outlined in this chapter and in Chapter 16. You will be relieved to know that we will not go into the details of these calculations. You should, however, have a thorough qualitative understanding of why the relative concentrations change as they do with changing pH.

SAMPLE EXERCISE 17.6

In what form or forms does dissolved CO_2 exist at a pH of 7.0?

Solution: From the graph in Figure 17.5 we read the fractions present at pH = 7.0. These are H_2CO_3, 0.18; HCO_3^-, 0.82; CO_3^{2-}, 0. (The concentration of CO_3^{2-} is, of course, not exactly zero; it is merely a *very* small fraction of the total CO_2 present.)

The characteristics of human blood offer an interesting application of the aqueous CO_2 equilibria. Human blood is slightly basic, with a pH of about 7.39 to 7.45. Blood is a highly buffered medium. In a healthy person the pH never departs more than perhaps 0.2 pH units from its average value.

One of the important functions served by blood is transport of oxygen from the lungs to cells within the body, which require oxygen for their metabolic activities. There is a corresponding transport of carbon dioxide, the "exhaust gas" of cell metabolism, to the lungs where it is expelled. Normally carbon dioxide is rather slow to react with water. The reaction shown in Equation [17.9] has a half-life of several minutes at room temperature. However, there is in the red blood cells an enzyme called **carbonic anhydrase,** which catalyzes the reaction of CO_2 with water (and, of course, the reverse reaction also). Because of the action of this enzyme, the blood very rapidly absorbs CO_2 to establish an equilibrium with the carbon dioxide in the cells. Similarly, CO_2 is rapidly released in the lungs to establish an equilibrium with the air

there. Thus, through the action of carbonic anhydrase, the body is able rapidly to vent CO_2, the waste gas from the body's energy-producing chemical reactions.

We see from Figure 17.5 that at the pH of the blood, about 7.4, the dominant form of CO_2 is HCO_3^-. Thus, when CO_2 dissolves in the blood it is first converted into H_2CO_3 with the aid of carbonic anhydrase. This acid very quickly loses a proton to form HCO_3^-. Although the blood is fairly strongly buffered, there are slight changes in pH as CO_2 is absorbed and released. The circulatory system of the blood is shown schematically in Figure 17.6. Venous blood is low in oxyhemoglobin (the form in which oxygen is carried in the blood; see Figure 10.11) and high in CO_2 content. It has a lower pH than arterial blood, which has just been through the lungs. When a condition occurs in which CO_2 cannot be effectively released from the blood, as in pneumonia, an excessive concentration of CO_2 builds up, and the blood pH drops. This condition, known as **acidosis,** is harmful to tissues, which soon die. A low blood pH may also result from accumulation of other acids in the blood stream. This may occur temporarily, for example, during strenous physical exercise. Although no lasting physical harm results, the low pH developed during vigorous exercise contributes to the symptoms we call exhaustion.

17.4 solubility equilibria

All of the equilibria we have so far considered in quantitative terms have been homogeneous; that is, all of the species involved in an equilibrium have been in the same phase. The equilibrium between a solid substance and its solution is an example of heterogeneous equilibrium.

For a true equilibrium to exist between a solid and its solution, the solution must be saturated, as described in Section 12.2. As an example, consider a saturated solution of the slightly soluble salt $BaSO_4$. The relevant reaction can be written as

$$BaSO_4(s) \rightleftharpoons Ba^{2+}(aq) + SO_4^{2-}(aq) \qquad [17.15]$$

(Henceforth it will be understood that the ions written on the right are solvated, or aquated, and the (aq) notation will be dropped.) The equilibrium constant for Equation [17.15] is

$$K_c = \frac{[Ba^{2+}][SO_4^{2-}]}{[BaSO_4]_s} \qquad [17.16]$$

The terms in the numerator correspond to the molar concentrations of the ions in solution. The term in the denominator corresponds to the "concentration" of solid $BaSO_4$. Because the concentration of a pure solid is constant, $[BaSO_4]$ can be combined with K to give a new constant that we label K_{sp}:

$$K_c[BaSO_4]_s = K_{sp} = [Ba^{2+}][SO_4^{2-}] \qquad [17.17]$$

K_{sp} is called the **solubility-product constant**. For barium sulfate in equilibrium with water at a particular temperature, it is a constant regardless of whether other substances are also present in the water. The value of K_{sp} can be calculated from a knowledge of the solubility of $BaSO_4$ in water, as illustrated in Sample Exercise 17.7.

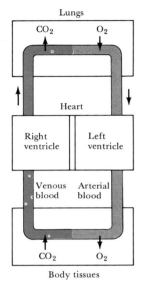

Lungs

CO_2 O_2

Heart

Right ventricle   Left ventricle

Venous blood   Arterial blood

CO_2 O_2

Body tissues

FIGURE 17.6 A schematic diagram of the blood system. Arterial blood, rich in oxyhemoglobin, is pumped through the left ventricle of the heart to the tissues. Oxygen is released, and carbon dioxide is taken up. The venous blood, rich in dissolved CO_2, is pumped by the right ventricle of the heart to the lungs, in which some of the CO_2 is released, and oxygen is taken up.

SAMPLE EXERCISE 17.7

Solid barium sulfate is shaken in contact with pure water at 25°C for several days. Each day a sample is withdrawn and analyzed for its barium and sulfate concentrations. After several days the values of $[Ba^{2+}]$ and $[SO_4^{2-}]$ are constant, indicating that equilibrium has been reached. The concentrations of Ba^{2+} and SO_4^{2-} are $1.04 \times 10^{-5}\,M$ each. What is K_{sp} for $BaSO_4$?

Solution: The analyses provide values for both Ba^{2+} and SO_4^{2-}:

$[Ba^{2+}] = 1.04 \times 10^{-5}$
$[SO_4^{2-}] = 1.04 \times 10^{-5}$

Inserting these into the expression for K_{sp}, Equation [17.17], we have:

$$\begin{aligned} K_{sp} &= [Ba^{2+}][SO_4^{2-}] \\ &= (1.04 \times 10^{-5})(1.04 \times 10^{-5}) \\ &= 1.08 \times 10^{-10} \end{aligned}$$

The rules for writing the solubility-product expression are the same as those for the writing of any equilibrium-constant expression: *The solubility product is equal to the product of the concentrations of the ions involved in the equilibrium, each raised to the power of its coefficient in the equilibrium equation.*

SAMPLE EXERCISE 17.8

Write the expression for the solubility-product constant for $Ca_3(PO_4)_2$.

Solution: We first write the equation for the solubility equilibrium:

$$Ca_3(PO_4)_2(s) \rightleftharpoons 3Ca^{2+} + 2PO_4^{3-}$$

Following the rule stated above, the power to which the Ca^{2+} concentration is raised is 3, the power to which the PO_4^{3-} concentration is raised is 2. The resulting expression for K_{sp} is

$$K_{sp} = [Ca^{2+}]^3[PO_4^{3-}]^2$$

From a knowledge of the K_{sp} value for a salt, the values for the concentrations of ions present in solution at equilibrium can be determined. Values of K_{sp} for several substances are listed in Table 17.1. Such K_{sp} values always correspond to the concentrations of ions expressed in terms of molarity. More extensive tables are given in handbooks of chemistry. Values of K_{sp} are not generally given for the more soluble salts, but only for relatively insoluble substances. For soluble compounds the solubilities are generally expressed as grams of solute per 100 g water, or per liter of solution.

TABLE 17.1

solubility-product constants in water

SALT	SOLUBILITY-PRODUCT EXPRESSION	K_{sp}
AgCl	$[Ag^+][Cl^-]$	1.56×10^{-10}
AgBr	$[Ag^+][Br^-]$	7.7×10^{-13}
AgI	$[Ag^+][I^-]$	1.5×10^{-16}
MgF_2	$[Mg^{2+}][F^-]^2$	6.4×10^{-9}
CaF_2	$[Ca^{2+}][F^-]^2$	3.4×10^{-11}
$BaSO_4$	$[Ba^{2+}][SO_4^{2-}]$	1.1×10^{-10}
$BaCrO_4$	$[Ba^{2+}][CrO_4^{2-}]$	1.6×10^{-10}
$PbSO_4$	$[Pb^{2+}][SO_4^{2-}]$	1.06×10^{-8}
MnS	$[Mn^{2+}][S^{2-}]$	1.4×10^{-15}
NiS	$[Ni^{2+}][S^{2-}]$	1.4×10^{-24}
CdS	$[Cd^{2+}][S^{2-}]$	3.6×10^{-29}
CuS	$[Cu^{2+}][S^{2-}]$	8×10^{-45}
Ag_2S	$[Ag^+]^2[S^{2-}]$	1.6×10^{-49}
$Zn(OH)_2$	$[Zn^{2+}][OH^-]^2$	1.8×10^{-14}
$Mg(OH)_2$	$[Mg^{2+}][OH^-]^2$	1.2×10^{-11}
$MgC_2O_4^a$	$[Mg^{2+}][C_2O_4^{2-}]$	8.6×10^{-5}
CaC_2O_4	$[Ca^{2+}][C_2O_4^{2-}]$	1.78×10^{-9}

^{a}$C_2O_4^{2-}$ is oxalate ion.

SAMPLE EXERCISE 17.9

From the value of K_{sp} listed in Table 17.1, determine the solubility of CaF_2 in water, in grams per liter.

Solution: The solubility equilibrium involved is

$$CaF_2(s) \rightleftharpoons Ca^{2+} + 2F^- \qquad [17.18]$$

and the corresponding solubility-product expression is

$$K_{sp} = [Ca^{2+}][F^-]^2 \qquad [17.19]$$

The dissolution of CaF_2 in water produces two F^- ions for each Ca^{2+}. Thus, if we call the concentration of Ca^{2+} in solution x, $[F^-]$ is $2x$. We have then

$$[Ca^{2+}] = x \qquad [F^-] = 2x$$
$$K_{sp} = [Ca^{2+}][F^-]^2 = 3.4 \times 10^{-11}$$
$$= (x)(2x)^2 = 3.4 \times 10^{-11}$$
$$4x^3 = 3.4 \times 10^{-11}$$
$$x^3 = 8.5 \times 10^{-12}$$

The simplest means of obtaining x is to take the log of x^3 and divide by three:

$$\log x^3 = \log (8.5 \times 10^{-12})$$
$$3 \log x = -11.07$$
$$\log x = -3.69 = -4 + 0.31$$
$$x = 2.05 \times 10^{-4}$$

Thus, the molar solubility of CaF_2 is 2.05×10^{-4} moles/liter. The mass of CaF_2 per liter that dissolves is

$$\left(\frac{2.05 \times 10^{-4} \text{ moles } CaF_2}{\text{liter soln}}\right)\left(\frac{78.1 \text{ g } CaF_2}{\text{mole } CaF_2}\right) =$$
$$1.95 \times 10^{-2} \frac{\text{g } CaF_2}{\text{liter soln}}$$

THE COMMON-ION EFFECT

The equilibrium between a solid salt and dissolved ions remains valid regardless of the relative concentrations of ions present in solution. It often happens that one of the ionic species involved in the solubility equilibrium is present in relatively much higher concentration than the other. For example, suppose that 0.01 moles of NaF, a salt soluble in water, is added to a liter of solution that is in equilibrium with solid CaF_2. The NaF solution causes the concentration of F^- to be 0.01 M. The solubility equilibrium for CaF_2, Equation [17.18], is still valid. By applying LeChatelier's principle, we see that addition of F^- should cause the equilibrium in Equation [17.18] to be shifted to the left. Some of the CaF_2 present in a saturated solution of CaF_2 in pure water is driven out of solution by addition of excess F^-. The effect of F^- on the solubility of CaF_2 is an example of the **common-ion effect**. In this case an ion that is common to a slightly soluble salt is present in excess and thus decreases the solubility of that salt. By employing the solubility-product constant we can determine quantitatively the effect of the common ion on solubility, as in Sample Exercise 17.10.

SAMPLE EXERCISE 17.10

What is the solubility of CaF_2 in a solution containing 0.01 M NaF?

Solution: To solve this problem we employ the solubility-product expression, Equation [17.19]. The value of K_{sp} is unchanged by the fact that the solution initially contains 0.01 M NaF. In Sample Exercise 17.9 the relative concentrations of Ca^{2+} and F^- were determined entirely by the solubility of CaF_2. In the present exercise we must take account of the fact that there is a second source of F^-. Let us call the concentration of Ca^{2+} in solution x. Then the concentration of F^- derived from the CaF_2 is $2x$. But there is in addition a 0.01 M contribution to the F^- concentration from the dissolved NaF. We thus have

$$[Ca^{2+}] = x \qquad [F^-] = 0.01 + 2x$$
$$K_{sp} = [Ca^{2+}][F^-]^2 = [x][0.01 + 2x]^2$$
$$= 3.4 \times 10^{-11}$$

This would be a messy problem to solve exactly, but fortunately it is possible to greatly simplify matters. Even without the common-ion effect of $0.01\ M\ F^-$, the solubility of CaF_2 in water is not very great, as illustrated by the small value of x obtained in Sample Exercise 17.9. We know from application of LeChatelier's principle that the solubility of CaF_2 will be even smaller in the presence of $0.01\ M$ NaF.

We can therefore safely assume that the $0.01\ M\ F^-$ concentration from the NaF is much greater than the small additional contribution resulting from the solubility of CaF_2. That is, we can neglect $2x$ in comparison with 0.01. We then have

$$3.4 \times 10^{-11} = (x)(0.01)^2$$
$$x = 3.4 \times 10^{-7}\ M = [Ca^{2+}]$$

This value for x represents the solubility of CaF_2 in a solution that is $0.01\ M$ in NaF. Note that it is much smaller than $2.05 \times 10^{-4}\ M$, the solubility of CaF_2 in pure water.

SOLUBILITY AND pH

When one or more of the ions involved in a solubility equilibrium is derived from a moderately strong acid or base, the solubility of the compound is dependent on the pH of the solution. For example, consider $Mg(OH)_2$, for which the solubility equilibrium is

$$Mg(OH)_2(s) \rightleftharpoons Mg^{2+} + 2OH^- \qquad [17.20]$$

The value of K_{sp} for $Mg(OH)_2$ is 1.2×10^{-11}. Suppose that solid $Mg(OH)_2$ is equilibrated with a solution buffered at a pH of 9. Then pOH is 5, that is, $[OH^-] = 1 \times 10^{-5}$. Inserting this value for $[OH^-]$ into the solubility-product expression, we have

$$K_{sp} = [Mg^{2+}][OH^-]^2 = 1.2 \times 10^{-11}$$
$$[Mg^{2+}][1 \times 10^{-5}]^2 = 1.2 \times 10^{-11}$$
$$Mg^{2+} = 0.12\ M$$

Thus, $Mg(OH)_2$ is quite soluble in a buffered, slightly basic medium. If the solution were made more acidic, the solubility of $Mg(OH)_2$ would increase, because the OH^- concentration decreases with increasing acidity. The Mg^{2+} concentration would thus increase to maintain the equilibrium condition.

The solubility of almost any salt is affected if the solution is made sufficiently acidic or basic. The effects are very noticeable, however, only when one or both ions involved is a moderately strong acid or base. The metal hydroxides we've just discussed are good examples of compounds involving a strong base, the hydroxide ion. As an additional example, the fluoride ion of CaF_2 is a moderately strong base, because it is the conjugate base of the weak acid HF. As a result, CaF_2 is more soluble in acidic solutions than in neutral or basic ones, because of the reaction of F^- with H^+ to form HF.

$$CaF_2(s) \rightleftharpoons Ca^{2+} + 2F^- \qquad [17.21]$$
$$F^- + H^+ \rightleftharpoons HF \qquad [17.22]$$

Qualitatively we can understand what occurs in terms of LeChatelier's principle: The solubility equilibrium is driven to the right because the free F^- concentration is reduced by reaction with H^+.

Which of the following substances will be more soluble in acidic solution than in basic solution? (a) $Ni(OH)_2(s)$; (b) $CaCO_3(s)$; (c) $BaSO_4(s)$

Solution: $Ni(OH)_2(s)$ will be more soluble in acidic solution, because of reaction of H^+ with the OH^- ion, forming water:

$$Ni(OH)_2(s) \rightleftharpoons Ni^{2+} + 2OH^-$$
$$2OH^- + 2H^+ \rightleftharpoons 2H_2O$$

Overall: $Ni(OH)_2(s) + 2H^+ \rightleftharpoons Ni^{2+} + 2H_2O$

Similarly, $CaCO_3(s)$ reacts with acid, liberating gaseous CO_2:

$$CaCO_3(s) \rightleftharpoons Ca^{2+} + CO_3^{2-}$$
$$CO_3^{2-} + 2H^+ \rightleftharpoons H_2CO_3 \longrightarrow CO_2(g) + H_2O$$

Overall: $CaCO_3(s) + 2H^+ \longrightarrow Ca^{2+} + CO_2(g) + H_2O$

The solubility of $BaSO_4$ is largely unaffected by changes in solution pH, because SO_4^{2-} is a rather weak base and thus has little tendency to combine with a proton. However, $BaSO_4$ is slightly more soluble in strongly acidic solutions.

Many metal hydroxides that are relatively insoluble in neutral water dissolve in *either* a strongly acidic or strongly basic medium. Such hydroxides are said to be **amphoteric.** Examples include $Al(OH)_3$, $Zn(OH)_2$, $Cr(OH)_3$, and $Sn(OH)_2$. Although these compounds are written as simple hydroxides, they generally also contain water molecules coordinated to the metal ion. The metal ion is thus surrounded by a well-defined arrangement of water molecules and hydroxide ions, forming a **complex ion.** For example, $Al(OH)_3$ is best formulated as $Al(OH)_3(H_2O)_3$; in this neutral form it is insoluble. But addition of H^+ causes reaction to form a metal-containing complex cation:

$$Al(OH)_3(H_2O)_3(s) + H^+ \rightleftharpoons Al(OH)_2(H_2O)_4^+ \qquad [17.23]$$

$$Al(OH)_2(H_2O)_4^+ + H^+ \rightleftharpoons Al(OH)(H_2O)_5^{2+} \qquad [17.24]$$

$$Al(OH)(H_2O)_5^{2+} + H^+ \rightleftharpoons Al(H_2O)_6^{3+} \qquad [17.25]$$

We can view each of these reactions as addition of a proton to one of the hydroxide ions attached to the metal ion. Each successive addition occurs less readily; as the overall positive charge on the complex increases, the tendency of the complex to accept still another proton decreases. (Remember that there is an accompanying spectator ion, the anion of the particular acid used, for example, NO_3^- or Cl^-.)

Reaction of insoluble $Al(OH)_3(H_2O)_3$ with base to form a soluble ion is caused by *removal* of H^+ from the neutral species:

$$Al(OH)_3(H_2O)_3(s) + OH^- \rightleftharpoons Al(OH)_4(H_2O)_2^- + H_2O \quad [17.26]$$

$$Al(OH)_4(H_2O)_2^- + OH^- \rightleftharpoons Al(OH)_5(H_2O)^{2-} + H_2O \quad [17.27]$$

$$Al(OH)_5(H_2O)^{2-} + OH^- \rightleftharpoons Al(OH)_6^{3-} + H_2O \quad [17.28]$$

Again, each successive reaction occurs less readily than the one before. As the charge on the complex becomes more negative, it is increasingly difficult to remove a positively charged proton.

The extent to which an insoluble metal hydroxide reacts with either acid or base varies with the particular metal ion involved. Many metal hydroxides, for example, $Ca(OH)_2$, $Fe(OH)_2$, and $Fe(OH)_3$, are capable of dissolving in acidic solution, but do not react with excess base. These hydroxides are not amphoteric.

The purification of aluminum ore in the manufacture of aluminum metal provides an interesting application of the property of amphoterism. As we have seen, $Al(OH)_3$ is amphoteric, whereas $Fe(OH)_3$ is not. Aluminum occurs in large quantities as the ore **bauxite,** which is essentially Al_2O_3 with additional water molecules. The ore is contaminated with Fe_2O_3 as an impurity. When bauxite is added to a strongly basic solution, it dissolves, because the aluminum forms complex ions of the type shown in Equations [17.26] to [17.28]. The iron oxide impurity, however, is not amphoteric and remains as a solid. The solution is filtered, getting rid of the iron impurity. Aluminum hydroxide is then precipitated by addition of acid. The purified hydroxide receives further treatment and eventually yields aluminum metal.

When writing reactions of amphoteric substances, we often show them as the oxides rather than the hydrated hydroxides. Although the various ways of showing the reactions are confusing at first, the only real differences are in the number of water molecules shown. Sample Exercise 17.12 shows how we can view the reaction of a metal oxide with base.

SAMPLE EXERCISE 17.12

Write a balanced chemical equation to show the reaction of Al_2O_3 with aqueous base to form the complex ion $Al(OH)_4(H_2O)_2^-$.

Solution: The overall, unbalanced form of the reaction is

$$Al_2O_3(s) + OH^- + H_2O \longrightarrow 2Al(OH)_4(H_2O)_2^-$$

This is *not* an oxidation-reduction equation; all species involved have the same oxidation state in reactants and products. We need merely balance for charge by providing for two hydroxides on the left, and then add sufficient H_2O on the left so that the O and H counts balance. The overall balanced equation is

$$Al_2O_3(s) + 2OH^- + 7H_2O \longrightarrow 2Al(OH)_4(H_2O)_2^-$$

17.5

formation of complex ions

We saw in Section 16.8 that the characteristics of metal ions in aqueous solution can be understood in terms of Lewis acid-base theory. The metal ions act as Lewis acids, or electron-pair acceptors, toward the water molecules, which act as Lewis bases, or electron-pair donors. Depending on pH, the coordinated water molecules may lose protons, so that the charge of the metal ion and its collection of water or hydroxide ions varies with pH.

Substances other than water that are Lewis bases might be expected also to interact with the metal ions. If such a substance is added to an aqueous solution of a metal ion, the base must compete with the large number of water molecules present for the available places around the metal ion. With many metal ions, a base such as ammonia does this very well. For example, addition of concentrated ammonia to a blue solution containing copper(II) ion results in formation of the very deep blue $Cu(NH_3)_4^{2+}$ ion:

$$Cu(H_2O)_4^{2+} + 4NH_3 \rightleftharpoons Cu(NH_3)_4^{2+} + 4H_2O \qquad [17.29]$$

Similarly, cyanide ion, CN^-, forms an even stronger complex:

$$Cu(H_2O)_4^{2+} + 4CN^- \rightleftharpoons Cu(CN)_4^{2-} + 4H_2O \qquad [17.30]$$

We have written these reactions as though they occur in a single step, removing all the water molecules at once. In fact, the replacement of the water molecules is a stepwise process. However, we generally work with solutions in which there is an excess of coordinating base. For all practical purposes, all water molecules are replaced.

We'll have more to say in Chapter 23 about the many complexes formed between metal ions and bases. Here we need only consider briefly their equilibrium aspects. The stability of a particular metal complex is measured by the tendency of the coordinated bases to be displaced by water. For example, the relative stabilities of the $Cu(NH_3)_4^{2+}$ and $Cu(CN)_4^{2-}$ complexes would be expressed by the tendencies of the reverse reactions in Equations [17.29] and [17.30] to occur. In all such equilibria it is understood that water is taking the place of the base that comes off the metal. For convenience, the water is normally left out of the equation altogether. The dissociations of the two copper complexes are then written as:

$$Cu(NH_3)_4^{2+} \rightleftharpoons Cu^{2+} + 4NH_3 \qquad [17.31]$$

$$Cu(CN)_4^{2-} \rightleftharpoons Cu^{2+} + 4CN^- \qquad [17.32]$$

The equilibrium constant expressions for the dissociations of coordinating bases from a metal ion are called **dissociation constants**, K_d. The dissociation constant expressions for the equilibria in Equations [17.31] and [17.32] are

$$K_d = \frac{[Cu^{2+}][NH_3]^4}{[Cu(NH_3)_4^{2+}]} \qquad K_d = \frac{[Cu^{2+}][CN^-]^4}{[Cu(CN)_4^{2-}]}$$

Because these two equilibria involve the same number of coordinating bases, the relative values of K_d give us a measure of the relative stabilities of the complexes. The value of K_d for the $Cu(NH_3)_4^{2+}$ complex is 2×10^{-13}; for $Cu(CN)_4^{2-}$, the value of K_d is 1×10^{-25}. We can conclude that the cyanide complex is the more stable. The dissociation constants for a number of metal complexes are given in Table 17.2. The equilibrium between a metal ion and Lewis base to form a complex ion can also be written in terms of the formation of the complex, that is, as the reverse of the reactions in Equations [17.31] or [17.32]. The equilibrium constants for the reactions written thus are called **formation constants**, K_f. They are just the reciprocals of the dissociation constants.

SAMPLE EXERCISE 17.13

From the value of K_d listed in Table 17.2, calculate the concentration of Ag^+ present in solution at equilibrium when concentrated ammonia is added to a $0.01\ M$ solution of $AgNO_3$ until the ammonia concentration is $0.20\ M$. (Neglect the small change in volume of the solution that results from adding concentrated NH_3.)

Solution: We should first write out the equilibrium reaction, and the corresponding equilibrium-constant expression:

$$Ag(NH_3)_2^+ \rightleftharpoons Ag^+ + 2NH_3$$

$$K_d = \frac{[Ag^+][NH_3]^2}{[Ag(NH_3)_2^+]} = 5.8 \times 10^{-8}$$

Because K_d is quite small, the equilibrium constant for the reverse reaction, formation of the complex, is quite large. We might begin then with the assumption that practically all the Ag^+ ion is converted into the $Ag(NH_3)_2^+$ complex. If the initial Ag^+ concentration was $0.01\ M$, then the concentration of $Ag(NH_3)_2^+$ formed is also essentially $0.01\ M$. Because 2 moles of NH_3 are released to the solution for each mole of $Ag(NH_3)_2^+$ that dissociates, we have:

	$Ag(NH_3)_2^+$	$\rightleftharpoons$ Ag^+ +	$2NH_3$
Initial concentration:	$0.01\ M$	0	$0.20\ M$
Equilibrium concentration:	$(0.01 - x)\ M$	$x\ M$	$(0.20 + 2x)\ M$

If the amount of complex that dissociates is small, x should be small relative to 0.20 or 0.01. Substituting into the equilibrium-constant expression, we have

$$\frac{[Ag^+][NH_3]^2}{[Ag(NH_3)_2^+]} = \frac{(x)(0.20)^2}{(0.01)} = 5.8 \times 10^{-8}$$

Solving for x we obtain

$$x = 1.4 \times 10^{-8} = [Ag^+]$$

The ability of metal ions to form complexes is an extremely important aspect of their chemistry. We shall see applications in later chapters to photography, metallurgy, biochemistry, chemical analysis, and other areas.

17.6
qualitative analyses for metallic elements

In this chapter we have seen several examples of equilibria involving metal ions in aqueous solution. In this final section we look very briefly at how solubility equilibria and complex formation can be used to detect the presence of particular metal ions in solution. Before the development of modern analytical instrumentation, it was necessary to analyze mixtures of metals in a sample by so-called "wet" chemical methods. For example, a metallic sample that might contain several metallic elements

TABLE 17.2

dissociation constants of some metal complex ions in water at 25°C

COMPLEX ION	K_d	EQUILIBRIUM EXPRESSION
$Ag(NH_3)_2^+$	5.9×10^{-8}	$Ag(NH_3)_2^+ \rightleftharpoons Ag^+ + 2NH_3$
$Ag(SCN)_2^-$	1×10^{-10}	$Ag(SCN)_2^- \rightleftharpoons Ag^+ + 2SCN^-$
$Ag(CN)_2^-$	1×10^{-21}	$Ag(CN)_2^- \rightleftharpoons Ag^+ + 2CN^-$
$Ag(S_2O_3)_2^{3-}$	1×10^{-13}	$Ag(S_2O_3)_2^{3-} \rightleftharpoons Ag^+ + 2S_2O_3^{2-}$
$Cu(CN)_4^{2-}$	1×10^{-25}	$Cu(CN)_4^{2-} \rightleftharpoons Cu^{2+} + 4CN^-$
$Ni(CN)_4^{2-}$	1×10^{-14}	$Ni(CN)_4^{2-} \rightleftharpoons Ni^{2+} + 4CN^-$
$Zn(CN)_4^{2-}$	1×10^{-17}	$Zn(CN)_4^{2-} \rightleftharpoons Zn^{2+} + 4CN^-$
$Fe(CN)_6^{4-}$	1×10^{-24}	$Fe(CN)_6^{4-} \rightleftharpoons Fe^{2+} + 6CN^-$
$Fe(CN)_6^{3-}$	1×10^{-31}	$Fe(CN)_6^{3-} \rightleftharpoons Fe^{3+} + 6CN^-$
$Cu(NH_3)_2^+$	1×10^{-7}	$Cu(NH_3)_2^+ \rightleftharpoons Cu^+ + 2NH_3$
$Zn(NH_3)_4^{2+}$	3×10^{-10}	$Zn(NH_3)_4^{2+} \rightleftharpoons Zn^{2+} + 4NH_3$
$Hg(NH_3)_4^{2+}$	4×10^{-20}	$Hg(NH_3)_4^{2+} \rightleftharpoons Hg^{2+} + 4NH_3$

was dissolved in a concentrated acid solution. This solution was then tested in a systematic way for the presence of various metallic ions.

Qualitative analysis involves simply determining the presence or absence of a particular metal ion. It should be distinguished from **quantitative analysis,** which involves determining how much of a given substance is present. Qualitative analysis is no longer so important as a means of analysis. It is useful, however, as a means of learning about and more clearly understanding the properties of metallic ions in solution. The analysis of a solution containing a mixture of metallic ions usually proceeds in two stages. The ions are first separated into various groups on the basis of their solubility properties. Within each group the individual ions are identified by special properties that may have to do with amphoterism, complex-ion formation, or oxidation-reduction behavior. In following through the qualitative analysis scheme below, it is important to keep in mind that the order in which the separations and tests are carried out is important. The most selective separations, that is, those which involve the smallest number of ions, are carried out first. Suppose, then, that we have an aqueous solution containing any number of metal ions, and we wish to identify which ions are present. The ions are separated into groups in the following order:

1. *Insoluble chlorides:* Addition of dilute HCl to the solution causes precipitation of $AgCl$, $PbCl_2$, and Hg_2Cl_2. All other ions remain in solution. The solution is filtered to remove any precipitate that forms, and the test for the second group is applied.

2. *Acid-insoluble sulfides:* H_2S is added to the dilute acidic solution. H_2S is a weak acid that shows two acid dissociations:

$$H_2S \rightleftharpoons HS^- + H^+ \qquad K_{a1} = 1 \times 10^{-7} \qquad [17.33]$$

$$HS^- \rightleftharpoons S^{2-} + H^+ \qquad K_{a2} = 1 \times 10^{-15} \qquad [17.34]$$

   Because the solution is acidic, both equilibria are shifted far toward the left. This means that the concentration of free S^{2-} is very low. If S^{2-} is very low, only the most insoluble metal sulfides, CuS, Bi_2S_3, CdS, HgS, As_2S_3, Sb_2S_3, and SnS_2, can precipitate. (Note the very small values of K_{sp} for some of these sulfides, Table 17.1.) Those metal ions whose sulfides are somewhat more soluble, for example NiS or MnS, remain in solution. After the solution is filtered to remove the acid-insoluble sulfides, the solution is made slightly alkaline, or basic; that is, $pH > 7$.

3. *Base-insoluble sulfides:* As the pH increases, the reactions shown in Equations [17.33] and [17.34] are shifted to the right, because $[H^+]$ decreases. This means that the concentration of free S^{2-} in equilibrium with H_2S and HS^- increases. Thus, in a basic solution containing sulfide ion, the solubility-product constants for the more soluble metal sulfides may be exceeded, and precipitation occurs. The metal ions brought down in this stage are Al^{3+}, Cr^{3+}, Fe^{3+}, Zn^{2+}, Ni^{2+}, Co^{2+}, and Mn^{2+}. (Actually the Al^{3+} and Cr^{3+} ions do not form insoluble sulfides, but instead are precipitated as the insoluble hydroxides at the same time.)

4. *Insoluble carbonates:* At this point the solution contains only metal ions from periodic table groups 1A and 2A. Addition of carbonate ion to a slightly basic solution causes precipitation of the group 2A elements Mg^{2+}, Ca^{2+}, Sr^{2+}, and Ba^{2+}, because these metals form insoluble carbonates.

5. *The alkali metal ions and* NH_4^+: The ions that remain after removal of the insoluble carbonates form a small group in which each ion can be tested for by an individual test. For example, the flame test is useful to show the presence of K^+, because the flame turns a characteristic violet color if K^+ is present.

Additional testing is necessary to determine which ions are present within each of the groups. As an example, consider the ions of the insoluble chloride group. The precipitate containing the metal chlorides is boiled in water. It happens that $PbCl_2$ is relatively soluble in hot water, whereas AgCl and Hg_2Cl_2 are not. The hot solution is filtered and a solution of Na_2CrO_4 added to the filtrate. If Pb^{2+} is present, a yellow precipitate of $PbCrO_4$ forms. The test for Ag^+ consists of treating the metal chloride precipitate with dilute ammonia. Only Ag^+ forms an ammonia complex. If AgCl is present in the precipitate it will dissolve in the ammonia solution:

$$AgCl(s) + 2NH_3 \rightleftharpoons Ag(NH_3)_2^+ + Cl^- \qquad [17.35]$$

After treatment with ammonia, the solution is filtered, and the filtrate made acidic by adding nitric acid. The nitric acid removes ammonia from solution by forming NH_4^+, thus releasing Ag^+, which should reform the AgCl precipitate:

$$Ag(NH_3)_2^+ + Cl^- + 2H^+ \rightleftharpoons AgCl(s) + 2NH_4^+ \qquad [17.36]$$

The analyses for individual ions in the acid-insoluble and base-insoluble sulfides are a bit more complex, but the same general principles are involved. The detailed procedures for carrying out such analyses are given in laboratory manuals.

FOR REVIEW

summary

In this chapter we've considered several types of important equilibria occurring in aqueous solution. One of the most important concepts in acid-base chemistry is that of the **buffer solution.** Such solutions are composed of mixtures of weak acids and bases, usually a weak acid and its conjugate base or a weak base and its conjugate acid. Addition of small amounts of additional acid or base to a buffered solution causes only small changes in pH, because the buffer reacts with the added acid or base. Buffer solutions are of great practical importance, both in the laboratory and as they occur in nature.

The titration of a solution of a weak acid by a strong base or of a weak base by a strong acid differs in important respects from a strong acid–strong base titration. The pH of the equivalence point is shifted from 7.0 by an amount that depends on the value of K_a for the weak acid or K_b for the weak base involved. For this reason it is important to choose carefully the indicator to be used in the titration. The pH at the equivalence point should correspond reasonably closely with the pH range in which the color change occurs for the indicator.

Polyprotic acids undergo successive ionizations in water. The extent to which the ionizations occur depends on the pH of the solution, which may be controlled by the presence of other substances. When no other source of H^+ or OH^- is present, each successive ionization is much less extensive than the previous one. However, if the pH is controlled by the presence of other substances the relative importances of the equilibria may be changed. The direction in which the position of equilibrium is shifted by addition of acid or base is readily predicted with the aid of LeChatelier's principle.

The equilibrium between a solid salt and its ions in solution provides an example of heterogeneous equilibrium. The **solubility-product constant** is an equilibrium constant that expresses quantitatively the extent to which the salt dissolves. Addition to the solution of an ion common to a solubility equilibrium causes the solubility of the salt to decrease. This effect is known as the **common-ion effect**. Solubility may also be affected by pH when one or more of the ions involved in the solubility equilibrium is an acid or base. For example, the solubility of MnS is increased on addition of acid, because the S^{2-} ion involved in the solubility equilibrium is a strong base and is removed by reaction with acid. **Amphoteric metal hydroxides** are those slightly soluble metal hydroxides that dissolve on addition of either acid or base. The reactions that give rise to the amphoterism are acid-base reactions involving the OH^- or H_2O groups bound to the metal ion.

Complex-ion formation involves the displacement of water molecules or OH^- ions attached to a metal ion by other Lewis bases, for example, NH_3 or CN^-. The extent to which such complex formation occurs is expressed in the **dissociation constant** for the complex, or by its reciprocal, the **formation constant**.

The fact that the ions of different metallic elements vary a great deal in the solubilities of their salts, in their acid-base behavior, and in their tendencies to form complexes can be used to separate and detect the presence of metal ions in mixtures. **Qualitative analysis** refers to determination of the presence or absence of a metal ion in a mixture of metal ions in solution. The analysis usually proceeds by separation of the ions into groups on the basis of precipitation reactions, and then by analyses for individual metal ions within each group.

learning goals

Having read and studied this chapter, you should be able to:

1. Understand the characteristics of buffer solutions.
2. Calculate the pH of a buffer solution consisting of known concentrations of a weak acid and its conjugate base or a weak base and its conjugate acid.
3. Calculate the change in pH of a simple buffer solution of known composition caused by adding a small amount of strong acid or base.
4. Prepare a buffer with a particular pH from substances with known acid- or base-dissociation constants.
5. Describe the form of the titration curve for titration of a weak acid by strong base, or weak base by strong acid, and calculate the pH at the equivalence point from a knowledge of K_a or K_b.
6. Explain the nature and extent of the successive dissociation equilibria that occur in solutions of polyprotic acids.
7. Explain the ionization equilibria involving $CO_2(H_2CO_3)$ in water, and how the relative concentrations of H_2CO_3, HCO_3^-, and CO_3^{2-} vary with pH.
8. Set up the expression for the solubility-product constant for a salt.

9. Calculate the solubility of a salt from a knowledge of K_{sp}.

10. Calculate the effect of an added common ion on the solubility of a slightly soluble salt.

11. Explain the effect of pH on a solubility equilibrium involving a moderately strong acid or base.

12. Explain the origin of amphoteric behavior and write equations describing the dissolution of an amphoteric metal hydroxide in either acidic or basic medium.

13. Explain the effect of an acid or base on the solubility of a salt when one of the ions involved in the solubility equilibrium is a moderately strong acid or base.

14. Formulate the equilibrium between a metal ion and a Lewis base to form a complex ion of a metal.

15. Explain the general principles that apply to the groupings of metal ions in the qualitative analysis of an aqueous mixture.

key terms

Among the more important terms and expressions used for the first time in this chapter are the following:

Acidosis (Section 17.3) is a pathological condition in which the pH of the blood is unacceptably low, usually due to excessive accumulation of carboxylic acids in the blood.

Amphoterism (Section 17.4) is a term used to describe the ability of certain slightly soluble metal hydroxides to dissolve in either an acidic or basic medium. The solubility results from formation of a complex ion via an acid-base reaction.

A **buffer solution** (Section 17.1) is one that undergoes a limited change in pH upon addition of a small amount of acid or base.

Carbonic anhydrase (Section 17.3) is an enzyme that catalyzes the reaction between CO_2 and water. It is present in red blood cells.

The **common-ion effect** (Section 17.4) refers to the effect of an ion common to an equilibrium in shifting the equilibrium. For example, added Na_2SO_4 decreases the solubility of the slightly soluble salt $BaSO_4$, or added $NaC_2H_3O_2$ decreases the percent ionization of $HC_2H_3O_2$.

The **formation constant** (Section 17.5) for a metal ion complex is the equilibrium constant for formation of the complex from the metal ion and base species present in solution. It is a measure of the tendency of the complex to form. The reciprocal of the formation constant is the **dissociation constant** (Section 17.5) for the complex, a measure of the tendency of the complex to dissociate into metal ion and free base species.

A **metal ion complex** (Section 17.5) consists of a metal ion and a well-defined group of ions or neutral molecules bound to the ion via a Lewis acid-base interaction.

Qualitative analysis (Section 17.6) refers to determining the presence or absence of a particular substance that might be present in a mixture.

The **solubility-product constant** (Section 17.4) is an equilibrium constant related to the equilibrium between a solid salt and its ions in solution. It provides a quantitative measure of the solubility of a slightly soluble salt.

EXERCISES

buffer solutions

17.1   Explain how a buffer solution acts to limit changes in pH on addition of acid or base.

17.2   Suppose you have two vessels, one containing a liter of buffer solution of pH 7, the other a KBr solution of pH 7. Explain how you would test a small sample from each vessel to determine which is the buffer solution

17.3   Explain with the aid of equations how a

solution of $0.1\ M\ HC_2H_3O_2$ and $0.1\ M\ NaC_2H_3O_2$ solution acts as a buffer.

17.4 Using the values of K_a and K_b listed in Tables 16.2 and 16.3, calculate the pH of each of the following solutions: (a) $0.1\ M$ benzoic acid and $0.1\ M$ sodium benzoate; (b) $0.01\ M$ pyridine and $0.01\ M$ pyridinium chloride; (c) $0.2\ M$ nitrous acid and $0.2\ M$ sodium nitrite; (d) $0.05\ M$ ammonia and $0.1\ M$ ammonium chloride.

17.5 Suppose that 1×10^{-3} moles of hydrogen iodide are added to a liter of each of the solutions listed in the preceding question. Calculate the change in pH that occurs. How much pH change should you expect to occur in each instance if 1×10^{-3} moles of NaOH were added instead of the HI? Explain.

17.6 Explain what is meant in speaking of the *capacity* of a buffer solution.

17.7 Which of the following solutions has the greatest buffering capacity and which has the least: (a) $0.01\ M\ HC_2H_3O_2$ and $0.01\ M\ NaC_2H_3O_2$, pH = 4.74; (b) $1.8 \times 10^{-5}\ M$ HBr, pH = 4.74; (c) $0.1\ M\ HC_2H_3O_2$ and $0.1\ M\ NaC_2H_3O_2$, pH = 4.74? Explain your answers.

17.8 Calculate the pH before and after addition of 1×10^{-3} moles of HCl to a liter of each of the following solutions (K_a for lactic acid, $HC_3H_5O_3$, is 8.4×10^{-4}): (a) $0.5\ M\ HC_2H_3O_3$ and $0.5\ M\ NaC_2H_3O_3$; (b) $0.05\ M\ HC_2H_3O_3$ and $0.05\ M\ NaC_2H_3O_3$; (c) $0.005\ M\ HC_2H_3O_3$ and $0.005\ M\ NaC_2H_3O_3$.

17.9 Calculate the quantity of solid NH_4Cl that should be added to a liter of a $0.15\ M$ solution of ammonia to bring the pH to exactly 9.0.

17.10 A solution that is $0.1\ M$ in $HC_2H_3O_2$ and $0.01\ M$ in $NaC_2H_3O_2$ is a good buffer toward added base, but a poor buffer toward added acid. Explain why this is so.

17.11 The optimal composition of a buffer solution made up from an acid and its conjugate base (or from a base and its conjugate acid) is one in which the acid and conjugate base (or base and conjugate acid) are present in comparable concentrations. Explain why this is so.

[17.12] Calculate the concentration of Na_2HPO_4 that should be present in a $0.15\ M$ solution of NaH_2PO_4 to form a buffer solution of pH 7.20. (Refer to Table 16.9 for needed K_a values.)

[17.13] The HPO_4^{2-} and $H_2PO_4^-$ pair form an important component of the buffer system that controls blood pH. At the pH of the blood, say 7.40, what is the $HPO_4^{2-}/H_2PO_4^-$ ratio? (Refer to Table 16.9 for needed K_a values.)

weak acid or base titration

17.14 How many milliliters of $0.025\ M$ NaOH are required to reach the equivalence point in titrating each of the following solutions: (a) 25 ml $0.035\ M$ HBr; (b) 32 ml $0.028\ M\ HC_2H_3O_2$; (c) 32 ml $0.028\ M$ chloroacetic acid?

17.15 How does titration of a weak acid by a strong base differ from titration of a strong acid by a strong base, with respect to each of the following points: (a) quantity of base required to reach equivalence point; (b) pH at the beginning of the titration; (c) pH at the equivalence point; (d) pH after addition of a slight excess of base; (e) choice of indicator for determining the equivalence point?

17.16 Which of the indicators listed in Table 16.6 might be used in titrating $0.1\ M$ ammonia solution with $0.1\ M$ HCl? Explain.

17.17 Which of the indicators listed in Table 16.6 might be used in titrating $0.05\ M$ benzoic acid with $0.1\ M$ NaOH? Explain.

17.18 Calculate the pH at the equivalence point in titration of each of the following $0.1\ M$ solutions with $0.1\ M$ KOH; (a) HNO_2; (b) benzoic acid; (c) formic acid (see Table 16.2).

17.19 Calculate the pH at the equivalence point for titration of $0.1\ M$ solutions of each of the following bases by $0.1\ M$ perchloric acid, $HClO_4$ (see Table 16.3): (a) hydroxylamine; (b) nicotine (first base dissociation only); (c) sodium phenolate, $NaOC_6H_5$ (see Table 16.2).

[17.20] It happens that acetic acid has an acid-dissociation constant of the same magnitude as the base-dissociation constant of ammonia. Based on this fact, make a sketch of the pH versus added acid curve for titration of a $0.1\ M$ solution of ammonia by a $0.1\ M$ acetic acid solution. (You don't need to do any calculations; simply make a guess based on the curves in Figures 17.3 and 17.4.) What do you expect for the pH at the equivalence point? Is this likely to be a good titration as an analytical procedure? Explain your answer.

ionizations of polyprotic acids

17.21 Suppose that a sample of pure water is saturated with CO_2 gas. Arrange the following species in the order of the relative concentrations at equilibrium, with the species in highest concentration listed first: (a) CO_3^{2-}; (b) $H(aq)^+$; (c) HCO_3^-; (d) H_2CO_3.

[17.22] When CO_2 is bubbled through a saturated brine solution (NaCl) at about $0°C$, no precipitation reaction occurs. However, when the brine solution is also saturated with respect to ammonia, $NaHCO_3$ precipitates. Write chemical equations that account for the difference in the two cases.

17.23 The carbon dioxide concentration in the blood (that is, all H_2CO_3, HCO_3^-, and CO_3^{2-}) in a healthy person at rest is in the range from 0.0025 to 0.0030 M. This is higher than the concentration of CO_2 in pure water that is in equilibrium with the gaseous CO_2 of the atmosphere. Explain how the blood is able to maintain such a relatively large total carbon dioxide content. How is this total content likely to vary with the pH of the blood?

17.24 The pH values at the crossing points for the curves giving the fractional abundances of the various products of ionization of a polyprotic acid as in Figure 17.5 can be estimated quite well by just using the pK_a values for the various acids, as listed in Table 16.8. (You might verify this for H_2CO_3 by comparing the pH values of the crossing points in Figure 17.5 with pK_a values calculated from the K_a values listed in Table 16.9.) Using this as a rough guide, sketch the curves showing fractional abundances of the various species resulting from dissociation of H_3PO_4.

[17.25] The pH of a sample of surface ocean water is 8.2. What are the relative fractions of H_2CO_3, HCO_3^-, and CO_3^{2-} in this water?

solubility equilibria

17.26 Write the expression for the solubility-product constant for the solubility equilibrium of each of the following compounds: (a) $AgBr$; (b) MgC_2O_4; (c) $Cd(IO_3)_2$; (d) MnS.

17.27 For each salt listed below, there follows the solubility in grams per liter of solution. From these data calculate the value for K_{sp} in each case. (a) $AgIO_3$—0.0283 g/l.; (b) MgF_2—0.082 g/l.; (c) $SrSO_4$—0.119 g/l.; (d) PbF_2—0.808 g/l.

17.28 The value of K_{sp} for silver thiocyanate, $AgSCN$, is 4.9×10^{-13} at $18°C$, and 4.0×10^{-12} at $40°C$. What concentration of SCN^- in solution is required to maintain a silver ion concentration of $2 \times 10^{-9}\ M$ in a solution in equilibrium with solid $AgSCN$ at each of these temperatures?

17.29 The concentration of sulfate ion in seawater is about $5.6 \times 10^{-2}\ M$. The measured concentrations of several metal ions in seawater and the solubility products for the metal sulfates measured in pure distilled water are as follows:

METAL ION	CONCENTRATION IN SEAWATER (M)	K_{sp} FOR MSO_4 IN DISTILLED WATER
Ba^{2+}	4×10^{-7}	8.7×10^{-11}
Sr^{2+}	1.5×10^{-4}	2.7×10^{-7}
Pb^{2+}	2×10^{-8}	1.1×10^{-8}

Is the solubility product for any of these salts apparently exceeded in seawater?

17.30 The value of K_{sp} for AgOH is 1.5×10^{-8}. Suppose base is added to an acidic 0.01 M solution of $AgNO_3$. At what pH will AgOH precipitate? (Neglect any volume change.)

17.31 Which of the following slightly soluble salts would you expect to dissolve to an appreciable extent upon addition of dilute acid: (a) $BaSO_4$; (b) CaF_2; (c) $Mn(OH)_2$; (d) CuI; (e) HgS; (f) $BaCO_3$. Explain your answers.

17.32 The value of K_{sp} for $Ba(IO_3)_2$ is 6.5×10^{-10}, whereas that for $AgIO_3$ is 9.2×10^{-9}. If each of these salts is separately placed in contact with pure water until equilibrium is attained, which solution has the higher metal ion concentration? If the two salts are equilibrated with a solution that has an iodate concentration of 0.01 M, which metal ion is present in higher concentration?

[17.33] The solubility product for $CaCO_3$ is 8.7×10^{-9}. The concentration of total CO_2 in deep seawater is about 0.0023 M. If the pH is 8.2, what concentration of Ca^{2+} results from the equilibrium of $CaCO_3$ with seawater? (From Figure 17.5 it is evident that most of the total CO_2 is in the form of HCO_3^-. Use Equation [17.12] to calculate the small concentration of CO_3^{2-} in equilibrium with the HCO_3^-. Then use K_{sp} for $CaCO_3$ to estimate the equilibrium Ca^{2+} concentration.)

amphoterism

17.34 Which of the following metal hydroxides are amphoteric: (a) $CsOH$; (b) $Cr(OH)_3$; (c) $Fe(OH)_3$; (d) $Sn(OH)_2$; (e) $Ba(OH)_2$.

17.35 Write balanced chemical equations for reactions of the hydrated metal hydroxide, $Zn(OH)_2(H_2O)_2$, with both acid and base to form soluble complex metal ion species.

17.36 When 6 M NaOH solution is added to a solution with equimolar concentrations of $Cd(NO_3)_2$ and $Zn(NO_3)_2$, the precipitate that first forms contains both Zn and Cd. Upon addition of still more base, however, and with stirring, part of the precipitate dissolves. The portion that remains contains Cd, but no Zn. Describe what has happened in terms of amphoteric behavior and write balanced chemical equations for the processes that occur in the first and second additions of base.

17.37 Write balanced chemical equations to represent the reaction of each of the following metal oxides with base: (a) SnO; (b) As_2O_3; (c) ZnO.

formation of complex ions; qualitative analysis

17.38 When the compound $Co(NH_3)_6Cl_3$ is dissolved in water, there is no evidence of a significant concentration of ammonia in the solution. Explain this observation.

17.39 Explain the formation of a metal ion complex such as $Cu(NH_3)_4^{2+}$ in terms of Lewis acid-base theory.

17.40 In what sense is the formation of a metal ion complex such as $Cu(NH_3)_4^{2+}$ a competition reaction?

17.41 Write the dissociation equilibrium for each of the following complex ions: (a) $Cu(NC_5H_5)_2^{2+}$; (b) $Ni(NH_3)_6^{2+}$; (c) $AgCl_2^-$; (d) $Pd(CN)_4^{2-}$; (e) AlF_4^-.

17.42 Suppose that 10^{-3} moles of $Cu(NO_3)_2$ and of $Fe(NO_3)_2$ are added to a liter of solution containing $0.1\ M$ NaCN. Which metal ion is present in higher concentration in the free form at equilibrium (see Table 17.2 for needed K_d values)?

17.43 Suppose that 1×10^{-4} moles of solid AgCl is precipitated from a solution in a qualitative analysis procedure. This is placed into 20 ml of water and concentrated ammonia is added drop by drop until all the AgCl has dissolved. What is the concentration of ammonia needed to dissolve this much solid in 25 ml of ammonia solution?

17.44 A solution containing an unknown number of metal ions is treated with dilute HCl; no precipitate forms. The pH is adjusted to about 1, and H_2S gas is bubbled through. A precipitate forms. The precipitate is filtered, and the pH of the filtrate solution adjusted to 8. Again H_2S is bubbled through and a precipitate again forms. The filtrate from this solution is treated with a dilute solution of sodium carbonate. No precipitate forms. Which of the metal ion groups discussed in Section 17.6 is present?

[17.45] A student who is in a great hurry to finish his laboratory work decides that his instructor has given him a qualitative analysis unknown that contains a metal ion from group 4 as described in Section 17.6. He therefore shortcuts all the lengthy procedures and simply adds carbonate ion to a slightly basic solution of his unknown. He observes a precipitate and concludes that a metal ion from group 4 is present. Why is this possibly an erroneous conclusion? (You might present some data from a handbook to support your answer.)

general questions

17.46 Suggest a reagent that might be added to dissolve each of the following precipitates and write a balanced chemical equation for the reaction that occurs: (a) AgBr; (b) $Mn(OH)_2$; (c) CoS; (d) CuC_2O_4, copper(II) oxalate; (e) $Sn(OH)_2$; (f) $Cu(IO_3)_2$. (Note: Iodic acid is a strong acid.)

[17.47] Calcium oxalate, CaC_2O_4, is a major component of kidney stones, small granules of insoluble salts that form in the kidney with sometimes very painful results. The two acid-dissociation constants for oxalic acid are $K_{a1} = 5.9 \times 10^{-2}$ and $K_{a2} = 6.4 \times 10^{-5}$. K_{sp} for CaC_2O_4 is 2.6×10^{-9}. If the pH of the fluid present in the kidneys is 8.3, would a slight shift toward more acidic conditions cause increased solubility of the kidney stones, assuming equilibrium conditions prevail? (Hint: To what extent does the reaction

$$C_2O_4^{2-} + H_2O \rightleftharpoons HC_2O_4^- + OH^-$$

proceed at a pH of, say, 8.0 as compared with 8.3?)

17.48 Vigorous exercise causes production of lactic acid in tissues. This lactic acid is transferred to the blood. The added lactic acid causes a reduction in the total carbon dioxide content of the blood. Explain in terms of Figure 17.5.

17.49 Sketch the curve for titration of 50 ml of $0.05\ M$ methylammonium chloride solution by $0.10\ M$ KOH. Calculate pH at the beginning of the titration, at the equivalence point, and at the point halfway to the equivalence point.

[17.50] Mixtures of Na_2HPO_4 and NaH_2PO_4 are useful for making up buffer solutions in the range of around 7. Suppose we need to have a buffer solution of pH 7.0, and this is to be made up by adding $10\ M$ HCl to a $0.10\ M$ solution of Na_2HPO_4. What volume of the HCl solution must be added to a liter of the Na_2HPO_4 solution to attain the desired pH? (Neglect the change in volume on adding the HCl.)

17.51 Explain how each of the following represents an example of the common-ion effect: (a) NaF is dissolved in a solution saturated with CaF_2 and in contact with solid CaF_2; (b) sodium acetate is added to a $0.05\ M$ solution of acetic acid; (c) the solid substance $K_2Cu(CN)_4$ is dissolved in water and NaCN added.

[17.52] Which of the following buffer solutions has the highest value of pH? Which has the lowest pH? (a) $0.1\ M$ $NaHCO_3$ and $0.1\ M$ Na_2CO_3; (b) $0.1\ M$ NaF and $0.1\ M$ HF; (c) $0.1\ M$ HCl and $0.2\ M$ $NaC_2H_3O_2$; (d) $0.1\ M$ H_3PO_4 and $0.1\ M$ NaH_2PO_4.

17.53 Which point along the curve in titration of acetic acid by NaOH, shown in Figure 17.3, corresponds to optimal buffer solution conditions? What is the pH at this point?

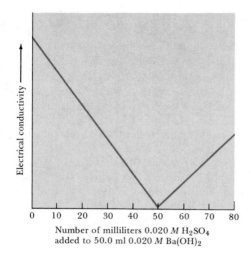

FIGURE 17.7 Electrical conductivity versus added 0.020 M H₂SO₄ in the titration of 0.020 M Ba(OH)₂.

Number of milliliters 0.020 M H_2SO_4 added to 50.0 ml 0.020 M $Ba(OH)_2$

Electrical conductivity

17.54 A sample of gaseous SO_2 is dissolved in water to form a solution which is 0.11 M. Using the values of K_a listed in Table 16.9, calculate the concentrations of HSO_3^-, SO_3^{2-}, and $H^+(aq)$ in such a solution. (Hint: You'll need to use the quadratic formula in solving the problem.)

17.55 (a) 50.0 ml of a 0.020 M solution of barium hydroxide, $Ba(OH)_2$, is titrated with a 0.020 M solution of sulfuric acid, by measuring the electrical conductivity of the solution. The graph of electrical conductivity versus volume of added H_2SO_4 is shown in Figure 17.7. Explain why the electrical conductivity varies as it does throughout the titration, and in particular why it goes to essentially zero at 50.0 ml added H_2SO_4. (Hint: You need to consider both solubility-product and acid-base equilibria.) (b) A titration of the sort described in part (a) was carried out using 0.020 M H_2SO_4 solution and a barium hydroxide solution of unknown concentration. The conductivity minimum was reached at 23.6 ml added acid. Calculate the concentration of the barium hydroxide solution.

chemistry of natural waters

In the previous two chapters we've focused on equilibria occurring in solutions, mainly in water. The emphasis on water as a solvent is highly appropriate; we live on a planet in which the aqueous environment, called the hydrosphere, dominates.

In this chapter, we shall see how the water on the surface of the planet is divided between saline, or salty, water and fresh water. We'll learn about the composition of ocean water, including something of its physical and chemical properties. The equilibria between the oceans and the gases of the atmosphere are important for the existence of life in the oceans and even on the landmasses. The oceans represent a vast, largely untapped reservoir of chemical and mineral resources.

In the course of being used, the limited fresh water at our disposal becomes contaminated. We'll see how water taken from the environment must be treated so that it is fit for human consumption and other uses. The water we return to the environment should be (but is not always) treated to remove dissolved organic matter and other substances that may be harmful to human health or to other units of the environment. Many substances released in waste waters are toxic, or may become toxic, as a result of chemical changes. It is therefore important that we know the chemical characteristics of compounds contained in waste waters and understand how they will behave in the natural environment.

Water-pollution problems often involve complex chemical considerations. The community in which you live may have to cope with problems very different from those encountered somewhere else. As a voting citizen you may have to form an opinion regarding a bond issue to build a new water-treatment plant, or you may have to decide whether a particular industrial operation in your community should be forced to add an expensive water-treatment facility. The answers to questions of this sort are usually not clear-cut, but the better your understanding of the chemical principles involved, the better your chances of forming a sound judgment.

The physical properties of water and the relationships between its solid, liquid, and gaseous phases are important in determining the roles that water plays in our natural environment. It is well to begin, then, by recalling some of the special characteristics of water described in preceding chapters. For a substance of such low molecular weight, water possesses unusually high melting and boiling points (Section 11.2). Methane, CH_4, which has about the same molecular weight as water, boils at 89 K, as compared with 373 K for water. Water also has an unusually high specific heat, 4.184 J/g-°C. The specific heat of most simple organic liquids is only about half this large. This means that a body of water can absorb a certain amount of heat with a smaller temperature increase than is the case for other liquids. The heat of vaporization of water is also exceptionally high; more heat is required to cause the evaporation of a gram of water than is needed to evaporate a gram of any other liquid. About one-third of all the energy that earth receives from the sun is used up in the evaporation of water from the surface of the oceans and other water bodies. The high heat of vaporization of water is thus obviously important in determining the conditions of life on earth. As we shall see in Section 18.4, the heat of vaporization of water is also important in determining the feasibility of recovering fresh water from salt-laden ocean water.

Ice, the solid form of water, has a very interesting and unusual structure, shown in Figure 15.3. Because this structure has so much open space, ice is less dense than liquid water. Water is one of the very few substances for which the solid phase is less dense than the liquid.

All of the unusual physical properties of water in both the liquid and solid forms are related to the formation of hydrogen bonds, as described in Section 11.3.

All of the vast layer of salty water that covers so much of the earth is connected and is of more or less constant composition. For this reason, oceanographers (scientists whose major interest is the sea) speak in terms of a world ocean, rather than of the separate oceans we learn about in elementary school geography books. The world ocean is indeed huge. Its volume is 1.35 billion cubic kilometers. It covers about 72 percent of the earth's surface. Almost all the water on earth, 97.2 percent, is in the world ocean. About 2.1 percent is in the form of ice caps and glaciers. All of the fresh water, in lakes, rivers, and ground water, amounts to only 0.6 percent. The remaining 0.1 percent is in the form of brine wells and brackish (salty) waters.

Seawater is often referred to as saline water. The **salinity** of seawater is defined as the mass in grams of dry salts present in 1 kg of seawater. In the world ocean, the salinity varies from 33 to 37, with an average of about 35. To put it another way, this means that seawater contains about

3.5 percent dissolved salts. The list of elements present in seawater is very long. However, most are present only in very low concentrations. Table 18.1 lists the eleven ionic species present in seawater at concentrations greater than 0.001 g/kg, or 1 part per million (ppm) by weight. (To convert from g/kg to ppm, multiply by 1000.) In a lower range of concentration, the elements nitrogen, lithium, rubidium, phosphorus, iodine, iron, zinc and molybdenum are present in amounts ranging from 1 ppm to 0.01 ppm. At least 50 other elements have been identified at still lower concentrations.

So far as can be determined, the composition of the oceans is constant with time. Of course, chemical data on the composition of seawater has been available for less than 100 years. If the composition were changing, it would probably be doing so on a time scale much longer than 100 years. However, there are other lines of evidence that also suggest that the chemical makeup of the oceans has not changed materially over a long period.

The constancy of composition of the oceans represents a balance of input and output processes. Water continually flows into the oceans from rivers, bringing in water with very different mineral composition from that already present. For example, the weathering of rocks leads to incorporation of aluminum, silicon, iron, or calcium in the river water. In the sea, these elements eventually become part of a biological cycle or are removed by precipitation. The average abundance of many elements in the ocean thus represents a balance between the input and output rates. The result is a more or less constant composition of the oceans as a function of time.

The **residence time** for a particular element is the average time the chemical forms of that element may be expected to remain in the sea before removal by one means or another. Some elements present in seawater exist in relatively stable chemical forms, for example, sodium as Na^+ or chlorine as Cl^-. The residence times of these elements are quite long; for example, the residence time of Na^+ is estimated to be about 300 million years. By contrast, an element such as iron is likely to be

TABLE 18.1

ionic constituents of seawater present in concentrations greater than 0.001 g/kg (1 ppm) by weight

IONIC CONSTITUENT	g/kg SEAWATER	CONCENTRATION (M)
Chloride, Cl^-	19.35	0.55
Sodium, Na^+	10.76	0.47
Sulfate, SO_4^{2-}	2.71	0.028
Magnesium, Mg^{2+}	1.29	0.054
Calcium, Ca^{2+}	0.412	0.010
Potassium, K^+	0.40	0.010
Carbon dioxide[a]	0.106	2.3×10^{-3}
Bromide, Br^-	0.067	8.3×10^{-4}
Strontium, Sr^{2+}	0.0079	9.1×10^{-5}
Boric acid, H_3BO_3	0.027	4.3×10^{-4}
Fluoride, F^-	0.001	7×10^{-5}

[a]CO_2 is present in seawater as HCO_3^- and CO_3^{2-}.

removed from solution after a relatively short time, by incorporation into a biological cycle (for example, it might become part of a clam shell or the blood of a shark) or by precipitation as a slightly soluble salt. Oceanographers estimate the residence time of iron in the seas to be only about 140 years.

CARBON DIOXIDE IN SEAWATER

The carbon dioxide present in the atmosphere (Sections 5.6 and 10.5) is in equilibrium with carbon dioxide dissolved in the ocean. The equilibrium between atmospheric CO_2 and the layer of ocean water extending to a depth of about 100 m is established in a period of about 2 years.* On the other hand, the surface waters are brought into equilibrium with deeper waters more slowly, over a period of perhaps several thousand years.

Dissolved CO_2 is an important component of the oceanic buffer system. The solubility of CO_2 in seawater is much higher than it is in pure water. Carbon dioxide is about 15 times more soluble in seawater than is O_2, and 30 times more soluble than is N_2, assuming the same pressure of gas over the solution in each case. To understand why CO_2 is so soluble in seawater, we must examine all the equilibria involved:

$$CO_2(g) \rightleftharpoons CO_2(aq) \qquad [18.1]$$

$$CO_2(aq) + H_2O \rightleftharpoons H_2CO_3(aq) \qquad [18.2]$$

$$H_2CO_3(aq) \rightleftharpoons H^+(aq) + HCO_3^-(aq) \qquad [18.3]$$

$$HCO_3^-(aq) \rightleftharpoons H^+(aq) + CO_3^{2-}(aq) \qquad [18.4]$$

Both of the ionization equilibria, Equations [18.3] and [18.4], are shifted to the right as pH increases, that is, as H^+ decreases. Because seawater is slightly alkaline, most of the CO_2 dissolved is in the form of HCO_3^- and CO_3^{2-}. As a result, the solubility equilibrium for gaseous CO_2 is thus also shifted to the right, in accordance with LeChatelier's principle.

We saw in Figure 17.5, that the form in which the dissolved CO_2 exists at equilibrium depends very much on the pH of the medium. Figure 17.5 is applicable to CO_2 dissolved in an aqueous solution that is relatively dilute with respect to ionic substances. However, when the total concentration of ionic substances in water is increased, the values for equilibrium constants are altered.† For example, the ionization equilibria of carbonic acid in seawater are shifted from the values for dilute aqueous solution, as shown in Figure 18.1. The color lines correspond to those shown in Figure 17.5. The black lines apply to seawater. We see that at the pH of seawater, about 8, there is a considerably higher concentration of CO_3^{2-} than would be present in a dilute aqueous

---

*This means that if we could somehow keep track of all the CO_2 molecules released into the atmosphere during a short time period, the fraction of them dissolved in the upper part of the ocean would be constant after a period of about two years.

†When the concentration of ions in solution is high, there is a high probability that ions of opposite charge will be found close together in the solution. Thus the assumption that each ion moves more or less independently of the others no longer holds true. As a result, the equilibrium constants are shifted in value. In accurate work, ionic equilibria in aqueous solutions of fairly high concentration must be corrected for the so-called interionic attractions. There are standard methods for doing this, but we need not concern ourselves with them.

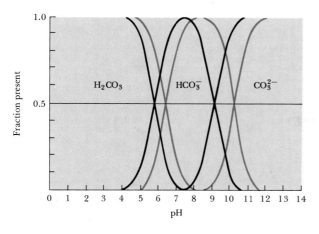

FIGURE 18.1 The black line shows the distribution of dissolved CO_2 in seawater between various ionization species, as a function of pH. The color line represents the same equilibria in water containing a very low concentration of dissolved ionic substances, as shown in Figure 17.5. Note that CO_2 is, in effect, more acidic in seawater than it is in pure water.

solution at the same pH. If oceanographers are to understand properly the equilibria involving carbonate and bicarbonate in seawater, they must employ the corrected values.

SAMPLE EXERCISE 18.1

The total concentration of CO_2 in a sample of seawater brought up from a depth of 150 m is 0.0043 M. The pH of the sample is 8.3. What are the concentrations of H_2CO_3, HCO_3^-, and CO_3^{2-} in this sample?

Solution: We employ Figure 18.1 to determine the fractions of each of the three species in seawater of pH 8.3. These are H_2CO_3, about 0; HCO_3^-, 0.86; CO_3^{2-}, 0.14. These fractions are then taken times the total CO_2 concentration to obtain the following approximate molar concentrations:

$$[H_2CO_3] = 0$$
$$[HCO_3^-] = 0.86(0.0043\ M) = 0.0037\ M$$
$$[CO_3^{2-}] = 0.14(0.0043\ M) = 0.0006\ M$$

We know, of course, that the concentration of H_2CO_3 is not precisely zero; it is merely quite small relative to the concentrations of the other two forms.

One of the most important equilibria involving dissolved CO_2 in seawater is that for formation of solid $CaCO_3$. The equilibrium between solid $CaCO_3$ and the Ca^{2+} and CO_3^{2-} ions present in the oceans is important in the development of many marine biological forms, for example, corals and shellfish. The solubility product for $CaCO_3$ in seawater at 20°C has the value 6.0×10^{-7}, whereas for $CaCO_3$ in equilibrium with pure water it is 5.0×10^{-9}. The solubility equilibrium is shifted toward increased solubility in seawater because of the effects of all the other ions present in the medium. The increase of more than 100-fold in solubility in seawater is another example of the effects of interionic attraction in aqueous media of high concentration.

The ocean appears to be supersaturated with respect to $CaCO_3$ at depths of about 1 km or less. This means that the solubility product for $CaCO_3$ is exceeded. However, the rate at which $CaCO_3$ is removed by precipitation or incorporation into animal shells or skeletal formations is quite small. At lower depths, where the concentration of Ca^{2+} is lower, the ocean is undersaturated. Carbonate shells formed near the surface sink to lower depths when the organism dies and dissolve there. At water depths greater than about 3 to 4 km, the degree of undersaturation with

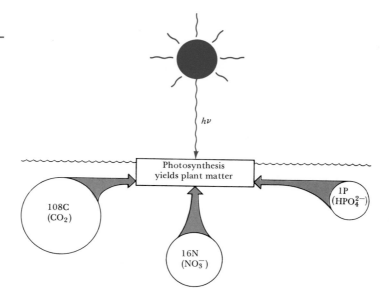

FIGURE 18.2 Schematic diagram of photosynthesis in phytoplankton at or near the surface of the sea. The relative amounts of required nutrients are indicated in the circles. The chemical forms shown are not the only ones usable, but are simply the more abundant ones.

respect to $CaCO_3$ seems to be quite large. As a result, there is very little $CaCO_3$ in sediments taken from the ocean bottom at these depths and below.

18.2
biochemical
processes
in the sea

Plants and animals present in the ocean exert an important influence on the composition of seawater. The simplest elements of the food chain are the **phytoplankton**, minute plants in which CO_2, water, and other nutrients are converted by photosynthesis into plant organic matter. Analysis of the composition of phytoplankton shows that carbon, nitrogen, and phosphorus are present in atomic ratios of $108:16:1$, as illustrated in Figure 18.2. Thus, about 108 molecules of CO_2 are required for each 16 atoms of nitrogen (usually in the form of nitrate ion) and for each single atom of phosphorus (usually present as hydrogen phosphate ion, HPO_4^{2-}). Because of its high solubility in seawater, CO_2 is always present in excess. The concentration of nitrogen or phosphorus is therefore the limiting factor in the rate of formation of organic matter via photosynthesis.

SAMPLE EXERCISE 18.2

The overall elemental composition of phytoplankton can be represented approximately by the formula $C_{108}H_{266}N_{16}O_{109}P$. Suppose that in a particular stretch of water the concentration of nitrogen is limiting for growth of phytoplankton and that the nitrogen concentration in the zone from the surface to 100 m averages 1×10^{-6} moles of nitrogen atoms per liter. Assuming that half this nitrogen is converted to phytoplankton, what is the total mass of plant matter formed in a square kilometer of ocean in the region from the surface to a depth of 100 m?

Solution: Let's first calculate the total mass of nitrogen present in the volume of water we are concerned with. This volume is

$$1 \text{ km}^2 \times 0.1 \text{ km} = 0.1 \text{ km}^3$$

Converting this to liters, we have

$$0.1 \text{ km}^3 \left(\frac{1000 \text{ m}}{1 \text{ km}}\right)^3 \left(\frac{100 \text{ cm}}{1 \text{ m}}\right)^3 \left(\frac{1 \text{ l.}}{1000 \text{ cm}^3}\right) =$$
$$1 \times 10^{11} \text{ l.}$$

The mass of nitrogen is thus
$$\left(\frac{1 \times 10^{-6} \text{ moles N}}{\text{liter}}\right)\left(\frac{14 \text{ g N}}{1 \text{ mole N}}\right)$$
$$\times (1 \times 10^{11} \text{ l.}) = 14 \times 10^5 \text{ g N}$$

The formula weight of phytoplankton is obtained by adding up all the atomic weights of the elements present, each multiplied by its coefficient in the formula. This comes to 3524 g. The formula mass of nitrogen in this formula weight is 224 g. Thus to

obtain the total mass of phytoplankton formed we have
$$\frac{1}{2}\left(\frac{14 \times 10^5 \text{ g N}}{0.1 \text{ km}^3 \text{ water}}\right)\left(\frac{3524 \text{ g C}_{108}\text{H}_{226}\text{N}_{16}\text{O}_{109}\text{P}}{224 \text{ g N}}\right)$$
$$= \frac{1.1 \times 10^7 \text{ g C}_{108}\text{H}_{226}\text{N}_{16}\text{O}_{109}\text{P}}{0.1 \text{ km}^3 \text{ water}}$$

The concentrations of nitrate and phosphate are complicated functions of ocean currents and many other factors. Vertical mixing of water is most important. This is so because the deep ocean water is higher in phosphate and nitrate than is the water near the surface. A typical profile of phosphate concentration as a function of depth is shown in Figure 18.3. The curve for nitrate concentration is very similar.

Photosynthesis occurs in the **photosynthetic zone** near the surface, where the sun's rays are strongest. Thus, in the water extending to a depth of about 150 meters, phosphate and nitrate concentrations are depleted because of the photosynthesis that has already occurred. At lower depths, dead plant and animal matter decomposes, restoring the phosphate and nitrate levels.

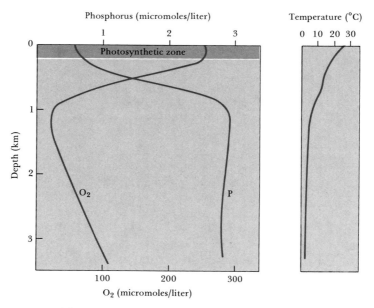

FIGURE 18.3 Concentrations of dissolved phosphorus (mostly as phosphate) and O_2 as a function of depth in the sea. The scale of phosphorus concentration is along the top of the figure; the oxygen concentration scale is along the bottom.

527

In the photosynthetic zone, the oxygen concentration is high because oxygen is released during photosynthesis. At lower depths, the oxygen level drops sharply because the oxygen is used up in oxidizing dead plant and animal matter. As shown in Figure 18.3 the oxygen concentration is at a minimum at about 1 km depth, the same region in which the phosphate level is restored to its highest values.

All the higher forms of life in the ocean are ultimately related to the phytoplankton at the base of the food chain. Thus, the abundance of fish is coupled very closely to the conditions that determine the rate of photosynthesis. In certain regions of the ocean, where seasonal upwellings of nutrient-rich lower water occur, the rate of photosynthesis may become very high. For example, the Grand Banks in the northern Atlantic, off the coast of Newfoundland, are one of the richest fishing regions in the world ocean. The abundance of fish there is due to an upwelling of nutrient-rich deep water.

18.3
raw materials
from seawater

The sea is a vast storehouse of chemicals. Each cubic mile of seawater contains 165 million tons of dissolved solids. The sea is so vast that if a substance is present in seawater to only the extent of 1 part per billion by weight, there are still 5 million tons of it in the world ocean. Nevertheless, the ocean is not used very much as a source of raw materials, because the costs of extracting the desired substances from the water are too high. We might think of the problem in the thermodynamic terms discussed in Chapter 15. Dispersal of a substance throughout the huge volume of the ocean is a spontaneous process, accompanied by large increase in entropy. To reverse that spontaneous dispersal, that is, to recover the material in a concentrated form, requires a large investment in energy.

SODIUM CHLORIDE

Because sodium chloride is the most abundant dissolved substance in seawater, it is perhaps not too surprising that a considerable amount (about 40 million tons each year) of pure sodium chloride is recovered from the sea. Seawater is filtered and then allowed to evaporate partially until the solubility of NaCl is exceeded. The solid NaCl that crystallizes is quite pure, but it may be recrystallized from fresh water for still higher purity, depending on the intended use.

BROMINE

Seawater serves as the major source of bromine. Worldwide production of this element is over 150,000 tons per year. From Table 18.1 we see that bromide ion is present in seawater at a concentration of only $8.3 \times 10^{-4}\ M$. Recovery of pure bromine thus represents a considerable concentration. The first step in removal of bromine is addition of sulfuric acid to the seawater to adjust the pH to 3.5. Then chlorine gas is added in slight excess as compared with the amount of bromine present. An oxidation-reduction reaction occurs between the dissolved chlorine gas and bromide ion:

$$\text{Cl}_2(aq) + 2\text{Br}^-(aq) \rightleftharpoons \text{Br}_2(aq) + 2\text{Cl}^-(aq) \qquad [18.5]$$

We might expect that this reaction would occur, because chlorine is a more electronegative element than bromine.

Bromine is freed by passing the seawater through a tower packed with wood strips, while air is swept through in the reverse direction. To recover the bromine from this air stream, it is treated with SO_2 and steam. A solution of hydrobromic acid and sulfuric acid is formed:

$$SO_2(g) + Br_2(g) + 2H_2O \rightleftharpoons 2HBr(aq) + H_2SO_4(aq) \qquad [18.6]$$

Bromine can be recovered from this solution by again treating with just the right amount of chlorine, as in Equation [18.5], and again sweeping the gaseous bromine out with air. The bromine is recovered from the air stream by allowing it to pass over a cold surface. The boiling point of liquid bromine is 59°C, so it can be separated from water by distillation. The dilute sulfuric acid solution left after bromine is removed, Equation [18.6], is used to acidify the incoming fresh batch of seawater.

MAGNESIUM

Magnesium is the second most abundant metallic element in sea water. The largest plant for production of magnesium from seawater in the United States is operated by the Dow Chemical Company at Freeport, Texas. In the process used, magnesium is precipitated from seawater in large settling ponds (Figure 18.4) as the hydroxide, by addition of lime,

FIGURE 18.4 An aerial view of settling ponds (*right of center*) used in precipitating $Mg(OH)_2$ from seawater. (*Courtesy of Dow Chemical U.S.A.*)

CaO. The calcium oxide employed in the process is obtained from oyster shells in nearby Galveston Bay. Oyster shells are composed of calcium carbonate. They are washed, then heated in a kiln to form lime:

$$CaCO_3(s) \longrightarrow CaO(s) + CO_2(g) \qquad [18.7]$$

Magnesium hydroxide is formed in the following reaction:

$$Mg^{2+}(aq) + CaO(s) + H_2O \longrightarrow Mg(OH)_2(s) + Ca^{2+}(aq) \quad [18.8]$$

The magnesium hydroxide that precipitates out is contaminated with calcium and sodium ions and with bicarbonate. The solid is filtered and then treated with a mixed HCl and H_2SO_4 solution. The $Mg(OH)_2$ dissolves as the solution is made acid (see Section 17.4):

$$Mg(OH)_2(s) + 2H^+(aq) \longrightarrow Mg^{2+}(aq) + 2H_2O \qquad [18.9]$$

Most of the sodium ion impurity crystallizes as NaCl, and calcium ion is precipitated as $CaSO_4$. The solution of Mg^{2+} is filtered and then concentrated in an evaporator. Solid $MgCl_2$ is eventually recovered. It is dissolved in a mixture of molten metal chlorides at 700°C in electrolysis cells (see Figure 18.5). In these cells electrical energy is supplied to cause the formation of magnesium metal and chlorine gas from the molten

FIGURE 18.5 A view of a row of cells in which molten $MgCl_2$ is electrolyzed to form Mg. The photo shows only the tops of the cells. The round vertical rods are carbon anodes. The rectangular bars are copper lines that carry up to 100,000 amps of current to the cells. (*Courtesy of Dow Chemical U.S.A.*)

FIGURE 18.6 Manganese nodules dredged from the ocean floor. (*Photo by B. J. Nixon, Deepsea Ventures, Inc.*)

metal chloride (electrolysis is discussed in more detail in Chapter 19):

$$MgCl_2(l) \xrightarrow{\text{electrical energy}} Mg(l) + Cl_2(g) \qquad [18.10]$$

The molten metal is cast into ingots of 99.9 percent purity.

DEPOSITS ON THE OCEAN FLOOR

Certain types of deposits found on the ocean floor have received much attention of late as possible sources of metals. Perhaps the most interesting are the so-called **manganese nodules.** These are found as baseball-sized lumps, scattered over large areas of the ocean floor (Figure 18.6). It has been estimated that there are more than 10^{12} tons of nodules on the floor of the Pacific Ocean. The average percent-by-weight composition of Pacific Ocean nodules is listed in Table 18.2.

It appears that the nodules form because the sea is supersaturated with respect to manganese and iron. These two elements precipitate as hydrated, colloidal hydroxides. These precipitates are very thinly dispersed in the seawater. As they settle slowly toward the bottom, other

TABLE 18.2

**average composition
of manganese nodules
from the Pacific Ocean**

ELEMENT	WEIGHT PERCENTAGE
Manganese	24
Iron	14
Silicon	9
Aluminum	3
Sodium	2.6
Magnesium	1.7
Titanium	0.7
Copper	0.5
Cobalt	0.4
Vanadium	0.05

metallic elements may become absorbed within them in slight quantities. On the bottom the precipitates gather on some object that acts as a point of nucleation. That object may be a shark's tooth, a piece of bone, or some other small object on the bottom. Somehow, over a period of many years, the nodules grow by continued precipitation on their surfaces. It has been estimated that they increase in thickness at the rate of only about 1 mm per 1000 years, or perhaps even more slowly.

The nodules are of interest not only as a possible source of manganese, but also of the less abundant elements such as cobalt, titanium, and vanadium. Means have been developed during recent years for collecting the nodules from the bottom.

The mineral **phosphorite,** which contains about 67 percent by weight of the compound $Ca_3(PO_4)_2$, is found as lumps on the ocean bottom in some regions bordering coastal waters. The powdered mineral has value as a fertilizer.

**18.4
desalination**

Although it may seem strange at first, the most valuable component of seawater may prove in the long run to be fresh water. Shortages of fresh water have developed even in countries such as the United States that are relatively well supplied with rainfall. In many regions of the United States, the demand for fresh water for home, agricultural, and industrial uses exceeds the available supply. In countries such as Israel or Kuwait, where rainfall is low, the supply is totally inadequate to meet the demands brought on by modernization and increasing population. Inevitably, humankind must look to the oceans as the source of the needed water.

Because of its high salt content, seawater is unfit for human consumption and indeed for most of the uses to which we put water. In the United States, the salt content of municipal water supplies is restricted by health codes to no more than about 0.05 percent. This is much lower than the 3.5 percent dissolved salts present in seawater, or the 0.5 percent or so present in brackish water found underground in some regions. The removal of salts from seawater or brackish water to the extent that the water becomes usable is called **desalination.**

It is an easy matter to dissolve a teaspoonful of salt in a glass of fresh water; it is not such an easy matter to take it out again, so as to recover the fresh water. Actually, there are many different ways to desalinate water, and any one of them can be made the basis of a large production plant. The challenge is to carry out the desalination with the absolute minimum energy requirement and with the least possible investment in equipment and facilities. This is important because a nation or region that must rely upon desalinated water to a large extent must compete economically with other nations that may have more abundant and cheaper sources of fresh water. A small nation such as Kuwait, situated on the Persian Gulf, which has almost no fresh-water resources, can afford to depend to a large extent on desalinated water only because it derives a very high per capita income from its oil production.

DESALINATION BY DISTILLATION

Water can be separated from dissolved salts by distillation, as described in Section 2.3. This process takes advantage of the fact that water is a volatile substance, whereas the salts are nonvolatile. The principle of distillation is simple enough, but there are many problems associated with carrying out the process on a large scale. For example, as water is distilled from a vessel containing seawater, the salts become more and more concentrated and eventually precipitate out. This causes formation of scale, which in turn causes poor heat transfer through the vessel, plugging of pipes, and so forth. The obvious solution is to discard the seawater after a certain amount of water has been distilled from it and begin again with a new batch. But unless this is done very carefully, all the heat values stored in the hot seawater will be lost, and additional heat must be then supplied to heat up the cold, incoming seawater. The heat lost represents wasted energy and higher costs. In addition, if the distillation is carried out at atmospheric pressure, the water must be heated to 100°C; at lower pressures, the boiling point of water would be lowered, and less heat would be required.

One rather successful attempt to get around some of these difficulties is called the multistage flash-distillation process, illustrated in Figure 18.7. To see how this method works, let's begin at A with warm seawater,

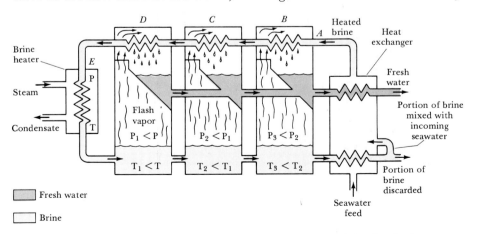

FIGURE 18.7 Schematic diagram of a multistage flash-distillation unit for desalination.

which we'll call brine. The brine passes through the coils of a condenser in B, then into C, and then into D, growing hotter in each chamber. The heat comes from condensing steam that forms on the coils in each chamber. The condensed steam, which is fresh water, is collected and pumped off. Each of the sealed tanks B, C, and D is at a lower pressure than the one following it, and so the boiling point of water is higher in each succeeding tank. In E the heated brine is heated still further by steam that passes around the coils. The steam supplied at this point represents a large fraction of the total energy input to the system. It might be supplied, for example, as the outlet steam from a large nuclear power plant. The heated brine passes from E into the chamber D. Because the pressure is lower in this chamber, some of the brine is "flashed off," or distilled to form water vapor. We know that energy is required to evaporate water. As water evaporates from our bodies the surface that remains is cooled. In the same way, the brine that remains after some water has evaporated, or flashed off, is cooler. It then passes into chamber C, in which the pressure is a little lower still. A bit more of the water evaporates, and the brine is cooled still further. In each succeeding stage, the brine becomes more concentrated in salts and lower in temperature. At the end, a portion of the brine, which is now about 7

FIGURE 18.8 A multistage flash-distillation unit. This unit is capable of producing about 9 million liters of fresh water each day. (*Courtesy of Aqua-Chem, Inc., Milwaukee, Wisconsin*)

percent salt by weight, is mixed with incoming sea water. The other portion is discarded to prevent the salt concentration from getting too high.

A large multistage flash-distillation unit is shown in Figure 18.8. This particular unit is designed to produce about 9 million liters of fresh water per day. The efficiency of a multistage flash-distillation system is limited more than anything else by the formation of scale in the hot-brine circulating system. The major culprits are calcium carbonate and magnesium hydroxide. Various additives have been employed to inhibit formation of these precipitates and thus permit use of higher temperatures. At higher temperatures, however, calcium sulfate precipitation becomes a problem. The chemist thus has an important role to play in improving this particular approach to desalination.

The large heat requirements of any distillation process are a major factor in the overall cost. In a typical multistage flash-distillation unit, the cost of the steam represents about 40 percent of the cost of the water. Many other schemes for desalinating water that bypass the need for vaporization have been advanced. In one method the water is removed from sea water by freezing. As ice freezes out of seawater, the dissolved salts are left behind. Of course the freezing process also requires energy, as anyone who has made ice cubes in a home refrigerator knows. Large-scale processes using the freezing technique are currently being tested.

DESALINATION BY REVERSE OSMOSIS

In the reverse-osmosis method of desalination, water is separated from dissolved salts by means of a membrane that is permeable to the passage of water, but not of dissolved salts. As we learned in Section 12.7, the phenomenon of osmosis depends on having a selective membrane that permits the flow of water through it, but not substances dissolved in the water. If such a membrane is placed between a brine solution and pure water, the tendency to equalize concentrations on each side would cause water to flow through the membrane into the brine solution. The flow may be countered by applying a pressure on the side of the brine solution. If the pressure is sufficiently high, the flow may be stopped altogether. The pressure required to balance the tendency of water to flow across the membrane is called the osmotic pressure. For seawater at standard conditions, the osmotic pressure is about 25 atm.

SAMPLE EXERCISE 18.3

If the osmotic pressure of seawater, with a salinity of 35, is 25 atm, what is the approximate osmotic pressure of a sample of brackish water with a salinity of 7?

Solution: From Section 12.7, we have the formula for osmotic pressure, π:

$$\pi = MRT$$

In this expression R is the molar gas constant, T is absolute temperature, and M is molarity. If we make the assumption that the molarity is proportional to salinity, then we can say that the molarity of the brackish water is in the same ratio to that of seawater as its salinity:

$$\frac{M \text{ (brackish water)}}{M \text{ (seawater)}} =$$
$$\frac{\text{salinity of brackish water}}{\text{salinity of seawater}} = \frac{7}{35}$$

Because osmotic pressure is directly proportional to molarity, the osmotic pressure of the brackish water should be 7/35 or 1/5 that of seawater, or 5 atm.

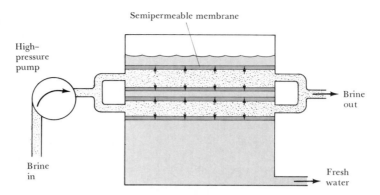

FIGURE 18.9   Schematic diagram of a reverse-osmosis process for desalination of brackish water or sea water. The pressure applied by the high pressure pump is sufficient to overcome the osmotic pressure of the brine solution compared with fresh water. As a result, fresh water flows across the semipermeable membrane. To prevent the buildup of salt near the membrane surface, the pump must continually circulate brine through the tubes. In practice the tubes may be very small in diameter, and the apparatus may consist of many thousands of tubes.

If the pressure on the brine solution is increased beyond the osmotic pressure, the direction in which water tends to flow across the membrane is reversed, and fresh water passes from the brine into the fresh water side. This process, called **reverse osmosis,** is illustrated in Figure 18.9. Seawater or brackish water is pumped at high pressure into chambers lined with semipermeable membranes. As the water passes across the membrane the local concentration of salt at the wall of the membrane increases. This causes the osmotic pressure to increase, and decreases the flow of fresh water. To prevent this, seawater must be continually pumped through. The flow of water through the membrane is proportional to the applied pressure. The maximum pressure that can be applied is limited by the characteristics of the membranes. Under excessive pressure the membranes may rupture, become fouled with impurities present in the water, or leak through an excessive amount of dissolved salts. In a typical reverse-osmosis unit, the tubes illustrated in Figure 18.9 are constructed of a porous material lined on the inside with an extremely thin film of cellulose acetate. The cellulose acetate (from which cellophane and photographic film are made) acts as the semipermeable membrane. The unit consists of many such tubes arranged in parallel array. The rate of water flow through the membrane is rather low. For example, using brackish water of about 0.5 percent dissolved salts, and a pressure of 50 atm, about 700 l. of fresh water per square meter of membrane per day can be attained. Because it requires a great many small tubes to produce a large surface area, the reverse-osmosis process has not yet been used to produce large quantities of usable water. This process does appear to have promise, with development of improved membranes, for purification of brackish waters, because brackish water requires lower pressures than does seawater.

18.5

sources of fresh water

We've seen that the total amount of fresh water on earth is not a large fraction of the total water present. What there is of it can be traced to evaporation of water from the oceans and the land, and to transpiration through the leaves of plants. The water vapor that accumulates in the atmosphere is transported via global atmospheric circulation to other latitudes, where it falls as rain or snow. The water that falls on land runs off in rivers or collects in lakes or underground caverns. Eventually it is evaporated or carried via streams and rivers back to the oceans.

The water needed for domestic uses, for agriculture, or for industrial processes is taken from naturally occurring lakes, rivers, and underground sources or from man-made reservoirs. The total water from rainfall that runs into the streams, rivers, and lakes is known as runoff. The total annual runoff of water for the United States (excluding Hawaii and Alaska), is about 1.6×10^{15} l. About one-third of this is used for industrial, agricultural, or other purposes. Because the distribution of water is quite uneven, some areas have more water than they are presently using, whereas in other areas water is in short supply. Much of the water that finds its way into municipal water systems is "used" water; it has already passed through one or more sewage systems or industrial plants.

Most of us take for granted the water that flows from the faucets in our homes. We assume that it will be clear and fit for human consumption. During the past few years, it has become increasingly evident that confidence in our water supplies may not be justified. Many municipal water systems have been shown to contain significant amounts of harmful chemicals. To more fully understand why such substances may be present in the water we drink, we must understand the kinds of treatment given water in preparing it for use. We must also examine how water is treated in industrial waste-treatment facilities and municipal sewage-disposal plants, before it is returned to nature. Let's begin with a brief look at how water is treated before it reaches the kitchen faucet.

18.6 treatment of municipal water supplies

Depending on the available supply, municipalities may draw their water from wells, rivers, lakes, or man-made reservoirs. Cities that are fortunate enough to have an abundant supply of clean, fresh water have a relatively easy task in supplying acceptable water for domestic use. On the other hand, those that must depend on a heavily contaminated source, such as a river that has been polluted upstream, face a difficult and expensive job.

The water that forms the raw material of a water-treatment system may be characterized in several different ways. The most obvious is in terms of its physical characteristics such as clarity, color, and quantity of total dissolved solids. The chemical characteristics of interest are its pH and the detailed chemical contents. In recent years, the presence of toxic substances at low concentrations has been recognized as a serious problem; we'll discuss it as a separate matter in a later section. The presence of organic matter and bacteria are measured by three tests: biological oxygen demand (BOD), the coliform count, and the algal count. The BOD measures the total volume of oxygen gas taken up by a given quantity of water in a period of 5 days at 20°C. The oxygen is consumed by microorganisms that use the oxygen to decompose complex organic molecules present in the water. The BOD test thus provides a measure of the total quantity of microorganisms in the sample, and of the nutrients available to them.

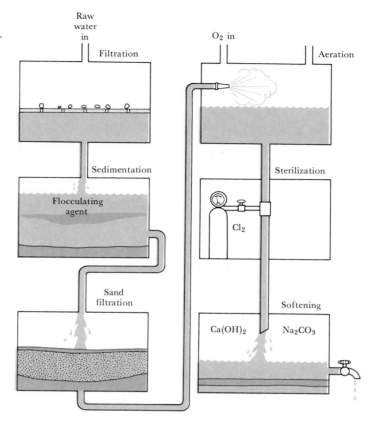

FIGURE 18.10   Stages in treatment of water for a municipal water system.

The coliform count is used to determine the presence of harmful bacteria in the water. This is done by looking for the presence of a common, harmless bacterium, *E. coli,* which is present in human excreta. The idea is that if the water is contaminated with this common bacterium, there is a possibility of contamination by pathogenic, or harmful, bacteria as well. The algal count is a biological test for the presence of microscopic plant and animal life (other than bacteria and viruses) that may be present in the water.

Treatment of water to render it fit for consumption takes place in several stages, as outlined in Figure 18.10. Not all water-treatment facilities employ all these steps, and some use less common methods different from those we shall describe. However, the scheme outlined in Figure 18.10 includes most of the techniques used.

FILTRATION AND SEDIMENTATION

The first step in water treatment is a simple screening or coarse filtration to remove solid objects, dead fish, and other debris. Following filtration, the water may be led into large basins in which finely divided sand and other minute particles are allowed to settle out. For natural sedimentation processes to be effective, the water flow must be very slow. Some particles in the water are so small, or of such low density, that they settle very slowly or not at all. To remove such particles a **flocculating agent** is added to the water. This usually is aluminum sulfate (alum), $Al_2(SO_4)_3$. When alum is added to slightly alkaline water, aluminum hydroxide forms as a spongy, gelatinous precipitate. Remember from Section 17.4 that aluminum hydroxide is best viewed as a hydrated hydroxide, that is, as $Al(OH)_3(H_2O)_x$, where x is three or larger. If the pH of the solution is properly adjusted, some ions of the formula $Al(OH)_2(H_2O)_4^+$ are incorporated in the precipitate. This causes the precipitate to have an overall positive charge. On the other hand, most of the suspended particles in the water have an excess of an overall negative charge, because of preferential adsorption of anions on their surfaces. As a result, the suspended particles are attracted to the spongy precipitate and carried down with it. The aluminum hydroxide precipitate thus acts as a kind of sweeper that removes almost all finely divided matter.

SAND FILTRATION

Water may next be filtered through a sand bed. The sand acts to screen out suspended matter and provides a support for microscopic plant life that feeds on nutrients present in the water. The minute organisms decompose dissolved organic matter and consume some of the nitrate, phosphate, and carbon dioxide present. Lower in the sand bed, various bacteria detoxify the water by consuming additional dissolved organic matter.

AERATION

Oxygen is an effective purifying agent. It attacks many forms of bacteria and hastens oxidation of dissolved organic materials. To increase the concentration of dissolved oxygen, the water may be sprayed into the air or allowed to tumble over a device that increases the surface area of the water and causes turbulence.

STERILIZATION

As a final insurance against the presence of harmful bacteria, the water may be treated with a chemical agent. The most effective commonly used agent of this type is ozone, O_3. Ozone is a powerful oxidizing agent and bactericide. One part per million of ozone in water effectively destroys all bacteria and viruses in a short while. Because ozone is too reactive to be manufactured and shipped, it must be generated at the place where it is to be used.

Chlorine is more convenient to use than ozone. It can be shipped in tanks as the liquefied gas and dispensed from the tanks through a metering device directly into the water supply. The amount used depends on the presence of other substances with which the chlorine might react and on the concentrations of bacteria and viruses to be removed. The sterilizing action of chlorine is probably due not to Cl_2 itself, but to hypochlorous acid, which forms when chlorine reacts with water:

$$Cl_2(aq) + H_2O \longrightarrow HOCl(aq) + H^+(aq) + Cl^-(aq) \qquad [18.11]$$

Although chlorine has been used to sterilize water for many years with no obvious harmful effects on those who use the water, a potential hazard has recently been discovered. Studies of the water supplies in several American cities have shown the presence of minute amounts of chloroform, $CHCl_3$, and carbon tetrachloride, CCl_4. These substances are known to be toxic. Although the levels at which they are present in the water supply are very low, the possibility exists that long-term consumption of water containing them may damage the liver and kidneys. They are believed to have been formed by reaction of organic pollutant molecules present in the water with the chlorine used for sterilization.

WATER SOFTENING

The water treatment described thus far should remove all the substances potentially harmful to health. Sometimes additional treatment is used to reduce the concentrations of Ca^{2+} and Mg^{2+}, which are responsible for **water hardness.** These ions react with soaps to form an insoluble material. Although they do not form precipitates with detergents, they adversely affect the performance of such cleaning agents. In addition, mineral deposits may form when water containing these ions is heated. When water containing Ca^{2+} and bicarbonate ions is heated, some carbon dioxide is driven off. The result is a shift in pH to higher values, and formation of insoluble calcium carbonate:

$$Ca^{2+}(aq) + HCO_3^-(aq) \xrightarrow{\text{heat}} CaCO_3(s) + CO_2(g) + H_2O \quad [18.12]$$

The solid $CaCO_3$ coats the surfaces of hot-water systems and the insides of teakettles, thereby reducing heating efficiency. Deposits of scale can be especially serious in boilers in which water is heated under pressure in pipes running through a furnace. Formation of scale reduces the efficiency of heat transfer and may result in melting of the pipes.

Not all municipal water supplies require water softening. In those that do, the water is generally taken from underground sources in which the water has had considerable contact with limestone ($CaCO_3$) and other minerals containing Ca^{2+}, Mg^{2+}, and Fe^{2+}. The **lime-soda process** is used for large-scale municipal water-softening operations. The water is treated with "lime," $Ca(OH)_2$, and "soda ash," Na_2CO_3. These two chemicals cause precipitation of calcium as $CaCO_3$ and of magnesium as $Mg(OH)_2$. The role of Na_2CO_3 is to increase the pH and to provide a source of CO_3^{2-}, if needed. If the water already contains a high concentration of bicarbonate ion, calcium can be removed as $CaCO_3$ simply by increasing the pH by addition of $Ca(OH)_2$:

$$Ca^{2+}(aq) + 2HCO_3^-(aq) + (Ca^{2+} + 2OH^-) \longrightarrow$$
$$2CaCO_3(s) + 2H_2O \quad [18.13]$$

Lime is used only to the extent that bicarbonate is present: 1 mole of $Ca(OH)_2$ for each 2 moles of HCO_3^-. When bicarbonate is not present, addition of Na_2CO_3 causes removal of Ca^{2+} as $CaCO_3$. The strongly basic carbonate ion (Sections 16.5 and 17.3) also serves to raise the pH enough to cause precipitation of $Mg(OH)_2$:

$$Mg^{2+}(aq) + 2CO_3^{2-}(aq) + 2H_2O \longrightarrow$$
$$2HCO_3^-(aq) + Mg(OH)_2(s) \quad [18.14]$$

There are two problems with the lime-soda process. First, formation of $CaCO_3$ and $Mg(OH)_2$ precipitates may take a long time, and they may not settle out very well. Second, the pH of the resulting water is too high. Usually alum is used as a flocculating agent to remove the precipitate, as described above under sedimentation. To prevent the remaining $Mg(OH)_2$ and $CaCO_3$ from later precipitating out and causing trouble, CO_2 is bubbled through the water. This has the effect of lowering the pH to around 8, thus preventing further precipitation.

18.7 waste-water treatment

Almost all the water used in municipal water systems and in industrial processes is returned eventually to natural water systems. Some of this water is changed chemically very little by its use. For example, the water used in the cooling condensers of an electrical power plant is heated several degrees and periodically has added to it a chemical to retard algal growth. Very large quantities of water cycled in this way are returned to lakes and rivers without any treatment or with only passage through a cooling tower to reduce the temperature. On the other hand, the waste waters from a pulp mill or the sewage that comes to a sewage treatment plant are very much altered from their original condition. Unless these waters are cleaned up, they will poison the lakes and rivers to which they are returned. The importance of treating waste waters and sewage becomes all the more apparent when one considers that many waters receive repeated use in the course of their journey to the sea. For example, the water taken by the city of New Orleans from the Mississippi River contains the effluents of a great many cities that lie along the banks of the Mississippi, Ohio, and Missouri rivers.

Municipal sewage treatment is generally divided into three stages, referred to as primary, secondary, and tertiary treatments. Primary treatment consists of removal of insoluble scum and debris. Then the water, which contains many dissolved substances, is mixed with the finely divided solid matter present to form a thick mess called sludge. Unfortunately, a significant fraction of all industrial and municipal waste water receives only this crude primary treatment before the sludge is treated with chlorine and dumped back into the natural water system.

Sewage sludge is a rich source of nutrients for the growth of microorganisms. In secondary treatment, microorganisms are allowed to feed upon the sludge to decompose the variety of organic molecules. In the most effective processes, air is blown through the sludge. The wastes in the sewage are converted into a thick mass of microbial material, with release of CO_2. Organic nitrogen is converted largely into nitrate or ammonium ion, and phosphorus is converted into phosphate. These ions are soluble in the water, which is eventually separated from the insoluble sludge. The sludge is drained into large beds, where it dries out. The water that is run off is relatively clear. Most importantly, the biological oxygen demand has been drastically reduced. This means that when the water is returned to the environment, it will not consume excessive quantities of the oxygen present in the natural water body.

Water that has received only primary and secondary treatment may contain relatively large quantities of phosphorus and nitrogen. These can cause damage to natural waters by promoting excessive growth of algae.

In addition, many chemicals present in sewage are not affected by secondary treatment and simply pass through and are released to the environment. The cost of removing the many metals and organic substances that might be present in waste water is high. As a result, very little waste water receives a general tertiary treatment, in which such contaminants are removed.

The substances that might contaminate waste water when it is returned to the environment are derived from all the many substances flushed down toilets, ground up in garbage disposal units, and rinsed down the drains in hospitals, stores, factories, and laboratories. In addition to these effluents, natural waters receive the water from storm drains and the runoff from cattle feedlots and from farmland dosed with fertilizers, insecticides, and weed-killing chemicals. We cannot consider here the long list of known contaminants. By way of example, we'll look at the characteristics of some of the more common metallic elements likely to be present.

**18.8
pollutant
metallic
elements**

Table 18.3 is a list of some of the more common (and in some cases more troublesome) metallic elements likely to be present in contaminated natural waters. For good measure we've added two nonmetallic elements, arsenic and selenium. The table lists the typical chemical form or oxidation state for each element. In any particular situation, the element might be present in some other form, depending on the source.

You may be surprised to note from Table 18.3 that many of the metals that can act as pollutants are actually essential in human nutrition. Copper affords a good example; the toxicity of this element is relatively low. An absence of copper(II) in the diet produces an anemic, or iron-deficient, condition, because copper is used in the body along with iron in some metabolic processes. The minimum dietary requirement seems to be on the order of 2 mg of copper per day. But a much higher intake, say about 50 mg per day or more, causes diarrhea, vomiting, and other miserable symptoms.

The metals vary a great deal in their toxicities and in the variety of toxic effects they bring about. This is so because they differ in the kinds of chemical reactions they undergo with biochemical systems. Although not all of the biochemical processes involved are understood, we have quite good evidence about what is happening in some cases. Cadmium, for instance, owes its high toxicity to the fact that it is chemically similar to zinc, a metallic element that is essential in many biochemical reactions.* Cadmium is apparently enough like zinc to take its place in biochemical systems, but once there it fails to perform precisely as zinc would.

It is worthwhile to consider mercury as a toxic substance in some detail, because several important points that could apply to any pollutant are involved. In the first place, *the toxicity of a substance may depend greatly on its chemical state.* Metallic mercury has a small but finite vapor pressure. If mercury metal is allowed to stand open in a poorly ventilated room for a long time, persons regularly occupying that room may inhale enough mercury over a period of time to show toxic symptoms of poi-

---

*For example, zinc is an essential part of carbonic anhydrase, the enzyme that catalyzes the reaction of CO_2 and water (Section 17.3).

TABLE 18.3

pollutant elements in water supplies

ELEMENT	COMMON CHEMICAL STATE	ESSENTIAL IN NUTRITION	TOXICITY	TOXIC EFFECTS	SOURCES	U.S. PUBLIC HEALTH SERVICE LIMITS PER LITER
Arsenic	AsO_2^-	No	High	Kidney failure, mental disturbance	Fossil-fuel combustion, detergents, smelting, pesticides	0.05 mg
Cadmium	Cd^{2+}	No	High	High blood pressure, kidney damage, red blood cell loss	Metal plating, mining, cigarette smoke	0.01 mg
Chromium	CrO_4^{2-}	Yes	Medium	Suspected carcinogen	Electroplating	0.05 mg
Copper	Cu^{2+}	Yes	Low	Liver damage	Mining, metal plating, copper pipes	1 mg
Iron	Fe^{2+}, Fe^{3+}	Yes	Low	Excessive intake may increase susceptibility to infection	Mineral sources, corroded metal	0.3 mg[a]
Lead	Pb^{2+}	No	High	Anemia, kidney failure, mental retardation (children), convulsions	Lead piping, lead paints, auto emissions from leaded gasoline	0.05 mg
Manganese	Mn^{2+}	Yes	Low	Not well characterized	Industrial waste, acid mine drainage	0.05 mg[a]
Mercury	Hg^{2+}, Hg_2^{2+}, CH_3Hg^+	No	High	Neurological damage, paralysis, insanity, blindness, birth defects	Chemical plant wastes, discarded mercury batteries	0.002 mg
Silver	Ag^+	No	Medium	Discoloration of skin and eyes	Electroplating, manufacturing	0.05 mg
Selenium	SeO_3^{2-}, SeO_4^{2-}	Yes	High	Liver damage, mental disturbance	Minerals, smelting operations	0.01 mg
Zinc	Zn^{2+}	Yes	Low	Not characterized	Metal-plating wastes, acid mine drainage	5 mg

[a] The limits on manganese and iron are not determined by their toxicity, but because they stain clothing and ceramic plumbing fixtures.

soning. However, ingestion of a small amount of mercury, as from a bit of silver amalgam a dentist uses to fill a cavity, is not considered a serious hazard; the metal passes through the system without undergoing chemical change. Compounds of mercury(I), such as calomel, Hg_2Cl_2, are not especially toxic, because they have very low solubility in water. The insoluble salts pass through the digestive system without any significant transfer to the bloodstream. Mercuric ion, Hg^{2+} is a very dangerous form of the element. When taken into the body as Hg^{2+}, the element affects the central nervous system, producing symptoms of insanity. Years ago mercuric nitrate, a water-soluble salt of mercury, was used to soften the fur used in the making of felt hats. The phrase, "mad as a hatter" originated from the symptoms displayed by hat workers suffering from mercury poisoning.

The most toxic of all forms of mercury are those compounds in which the element is combined with an organic group. Among these are the methyl mercury ion, CH_3Hg^+, and dimethyl mercury, $(CH_3)_2Hg$. The latter compound is a volatile, strong smelling substance that boils at 96°C. It is readily absorbed through the skin or may be inhaled as the vapor. It is regarded as one of the most poisonous substances known.

These observations show clearly that the toxicity of mercury varies a great deal with its chemical state. The second thing we must realize is that in nature *a substance may undergo chemical transformation that converts it from a relatively harmless form to a deadly one.* Figure 18.11 shows a diagram that illustrates how this happens in the case of mercury. For many years mercury metal was used in electrolysis cells for production of chlorine and sodium hydroxide. The metal escaped into the environment as the free element or as Hg^{2+}. The tiny amount of metal that escaped found its way to the bottoms of lakes and waterways. There it could react, possibly with some form of sulfur to form insoluble HgS, or possibly to form some other insoluble salt. However, the bottoms of rivers, lakes, and waterways contain abundant bacterial life, and after a time the sulfide is oxidized to sulfate, and Hg^{2+} is released to the water. In addition, if Hg_2^{2+} ion should be formed, it is capable of undergoing disproportionation* to mercury metal and Hg^{2+}:

$$Hg_2^{2+}(aq) \rightleftharpoons Hg(l) + Hg^{2+}(aq) \qquad [18.15]$$

This interconversion reaction can be catalyzed by microorganisms and thus serve as an additional source of Hg^{2+}. Other sources of mercury might also find their way into waters. For example, certain mercury-containing organic compounds have long been used to prevent fungus growth on seeds, in paper and paint products, and in other applications. These mercury compounds are toxic in their own right.† In addition, they may also be converted into other even more toxic forms.

The same waters that contain the mercury impurities also contain sewage wastes, and thus large numbers of bacteria that act upon the organic compounds may be present in the waters. These bacteria are able to react with mercury(II), adding one or two methyl groups to the metal and forming CH_3Hg^+ and $(CH_3)_2Hg$. Thus the mercury that might have entered the environment in any one of a number of forms becomes converted to highly toxic forms.

The third factor that is important in considering the effects of a pollutant is often referred to as **biological concentration.** Both CH_3Hg^+ and

*A disproportionation reaction is one in which a substance is converted into two different chemical forms, one of higher and the other of lower oxidation state than the reactant form. For example, in Equation [18.15], mercury in the +1 oxidation state is converted into the zero and +2 oxidation states.

†In February, 1976, the Environmental Protection Agency moved to ban production of nearly all mercury-containing pesticides, on the grounds that their continued use poses an unreasonable hazard to humans and to the environment in general.

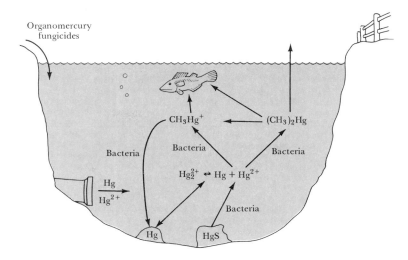

FIGURE 18.11 Chemical interconversions of mercury in natural waters polluted by mercury, sewage wastes, and other contaminants. Mercury in the water may originate from sewage-disposal outlets or in the runoff from farmlands on which organomercury fungicides have been applied. The mercury may be present in the water as the element or as Hg_2^{2+} or Hg^{2+} ion. It may also react with other substances to form an insoluble precipitate. However, bacteria act upon the mercury in these forms to convert it into dimethyl mercury, $(CH_3)_2Hg$, or methyl mercury ion, CH_3Hg^+. These species are taken up by aquatic life and are eventually concentrated in the fatty tissues of fish.

$(CH_3)_2Hg$ have a tendency to accumulate in organisms; the presence of the one or two methyl groups increases solubility in organic substances. The result is that the two species accumulate in the plants and tiny organisms on which fish feed and eventually concentrate in the fish themselves. The concentration of mercury in fish may be as much as 1000 times that in the waters from which the fish are taken. This means that, when waters contain mercury compounds at concentrations in the range of a few parts per billion, the fish that live in those waters may contain one or more parts per million of the element. U.S. and Canadian health authorities have set an upper limit of 0.5 parts per million on the mercury content of fish which may be sold or given away. In 1970 it was necessary to suspend commercial fishing in several waterways in the United States and Canada because this limit was exceeded.

Both CH_3Hg^+ and $(CH_3)_2Hg$ have a tendency to accumulate in humans who eat contaminated fish, just as they accumulated in the fish in the first place. Because they are at least partly organic in character, these compounds pass very readily through biological defenses and attack the central nervous system. They have very long retention times in the body as compared with many other poisons, and thus their effects are cumulative. Persons eating a diet high in contaminated fish acquire over a period of time ever-increasing levels of mercury.

From this example of mercury as a pollutant, you can see that the question of whether a particular substance will be a serious pollutant in the environment is really an entire set of complicated questions. Unfortunately, too many substances have been released to the environment before the proper questions were even asked, let alone answered. For example, the water in Minimata Bay in Japan was for many years polluted with mercury wastes from a chemical factory. Over a period of more than 10 years, more than 50 people in Minimata died of mercury poisoning as a result of eating fish caught in Minimata Bay. In addition, many children born in Minimata during that time were hopelessly deformed and brain-damaged.

FOR REVIEW

summary

In this chapter we've learned something of the nature and distribution of the water that covers our planet. Most of this water is in the oceans. Seawater contains about 3.5 percent by weight of dissolved salts. These salts, along with dissolved carbon dioxide, establish a buffer system that maintains the pH of seawater in the vicinity of 8.0.

The varieties of biological life which live in the sea depend on the growth of photosynthetic **phytoplankton** as the base of the food chain. The **photosynthetic zone** is near the water's surface, where the sun's rays can penetrate. The major ingredients needed for photosynthesis are CO_2 and suitable forms of nitrogen and phosphorus. Usually the availability of one or the other of these latter elements limits the rate of photosynthesis.

The sea is a vast storehouse of chemicals that remains largely untapped. Sodium chloride, the most abundant salt present in seawater, is recovered by partial evaporation. The elements bromine and magnesium are produced mainly from seawater. **Manganese nodules** and the mineral phosphorite are potentially valuable resources that could be recovered from the ocean bottom.

Because most of the world's water is in the oceans, it is perhaps inevitable that humankind must eventually look to the seas for fresh water. **Desalination** refers to the removal of dissolved salts from seawater, brine, or brackish water, so as to render it fit for human consumption. Among the many means by which desalination may be accomplished are distillation, freezing of ice from the water, and reverse osmosis. In **reverse osmosis,** pressure is applied to salt water to induce the flow of water, but not the dissolved salts, through a semipermeable membrane.

Fresh water that is available from rivers, lakes, and underground sources may require treatment to render it fit for use. The several steps which may be used in water treatment are coarse filtration, sedimentation, sand filtration, aeration, sterilization, and softening.

Waste-water treatment is applied to sewage waters or to water that has been used in an industrial operation. Municipal waste waters are given a primary treatment to remove insoluble scum, grease, and other materials. Secondary treatment consists of aeration of sewage sludge to promote the growth of microorganisms that feed on the organic compounds present in sewage. Eventually, clear water is separated from the mass of microorganisms. The water is lower in **biological oxygen demand** (BOD) than before treatment. However, it may still contain many substances that are toxic to aquatic life or to humans or that cause excessive growth of algae in natural waters. The many substances that remain in waters after secondary treatment can be removed only by extensive additional processing, referred to as tertiary treatment.

As examples of pollutant substances present in waters, we have considered some of the metallic elements. The toxicity of a particular element varies with its chemical state. For this reason, it is important to understand all the chemical transformations that an element may undergo once it has been released to the environment. An element may be released in a relatively nontoxic chemical state and then be converted in nature to a highly toxic form. In evaluating the toxicity of an element, it is important to consider also the possibility that **biological**

concentration may occur. When a toxic substance is stored cumulatively or preferentially in various species in the food chain, its concentration in the highest species in the food chain may be a thousand times what it is in the environment as a whole.

learning goals

Having read and studied this chapter, you should be able to:

1. List the more abundant ionic species present in seawater.
2. Explain the nature of the equilibrium between atmospheric CO_2 and the CO_2 dissolved in the oceans; in particular, you should be able to explain how the ionization equilibria involving H_2CO_3 are shifted by the slightly alkaline nature of seawater.
3. Describe how the growth of phytoplankton is related to the availability of nutrient nitrogen and phosphorus.
4. Describe how the concentrations of oxygen, phosphate, and nitrate as a function of ocean depth are linked to photosynthesis and decay of plant matter.
5. Write and explain the chemical reactions involved in the extraction of bromine and magnesium from seawater.
6. Explain the principles involved in the multistage flash-distillation and reverse-osmosis processes for desalination of seawater.
7. List and explain the tests applied to fresh water to determine the extent of purification needed to render it fit for human use.
8. List and explain the various stages of treatment that may be applied to a fresh-water supply.
9. Describe the chemical principles involved in the lime-soda process for reducing water hardness.
10. List and explain the stages in treatment of waste water.
11. List some of the more common metallic elements that might be present in a contaminated source of fresh water.
12. List and explain the three considerations described in the text that are important in determining the impact of a pollutant substance on the environment.

key terms

Among the more important terms and expressions used for the first time in this chapter are the following:

The **algal count** (Section 18.6) is a measure of the presence of microscopic plant and animal life.

Biological concentration (Section 18.8) refers to the tendency of a pollutant substance to accumulate in plant and animal tissues in concentrations much higher than in the immediate environment.

The **biological oxygen demand** (BOD) (Section 18.6) is a test applied to water to determine the total capacity of the water for consumption of oxygen for biological oxidations. It is given by the total volume of oxygen gas taken up by a given quantity of water during a 5-day period at $20°C$.

The **coliform count** (Section 18.6) is a test for the presence of *E. coli,* a common bacterium present in human excreta. The presence of *E. coli* in a water supply indicates contamination that may include harmful bacteria.

Desalination (Section 18.4) refers to the removal of salts from seawater, brine, or brackish water, so as to render it fit for human consumption.

A **flocculating agent** (Section 18.6) is a substance that forms a gelatinous precipitate on addition to water. The precipitate settles in the water and carries with it finely divided particulate matter that does not settle out of its own accord.

The **lime-soda process** (Section 18.6) is a method for removal of Mg^{2+} and Ca^{2+} ions from water to reduce water hardness. The substances added to the water are "lime," $Ca(OH)_2$, and "soda ash," Na_2CO_3, in amounts determined by the concentrations of the offending ions.

Manganese nodules (Section 18.3) are baseball-sized lumps of mineral material found on the ocean bottom. The nodules contain high concentrations of manganese and iron and lesser concentrations of several other metallic elements.

Multistage flash distillation (Section 18.4) is a method for desalination that involves distillation of saline water in several stages, for maximum possible conservation of heat stored in the water.

Phosphorite (Section 18.3) is a mineral composed mainly of $Ca_3(PO_4)_2$ that is found as lumps on the ocean bottom.

The **photosynthetic zone** (Section 18.2) is the region of the oceans extending from the surface to a depth of about 150 m, in which the photosynthetic growth of phytoplankton occurs.

Phytoplankton (Section 18.2) are microscopic plants that abound in the water near the ocean surface. They utilize photosynthesis to consume CO_2 and suitable forms of nitrogen and phosphorus in forming plant matter. They form the base of the food chain for biological life in the oceans.

Residence time (Section 18.1) is the average time a particular element may be expected to remain in the oceans before it is removed by one means or another.

Reverse osmosis (Section 18.4) is a method for desalination of saline water that involves application of pressure to the saline water to force passage of water, but not dissolved salts, through a semipermeable membrane.

Salinity (Section 18.1) is a measure of the salt content of seawater, brine, or brackish water. It is equal to the weight in grams of dissolved salts present in 1 kg of seawater, brine, or brackish water.

Water hardness (Section 18.6) refers to the presence in a water supply of Ca^{2+}, Mg^{2+} (and sometimes Fe^{2+}) ions. These ions form insoluble precipitates with soaps and are responsible for scale formation when the water is heated.

EXERCISES

chemical composition of sea water

18.1 Assuming a 10 percent efficiency of recovery, how many liters of seawater must be processed to obtain 1×10^8 kg of bromine in a commercial production process, assuming a bromine concentration as listed in Table 18.1?

18.2 It is estimated that the rivers of the world bring a total of 4×10^{15} g of dissolved salts to the oceans each year. What fraction of the total salts dissolved in the oceans is this annual influx?

18.3 Describe how you would carry out an experiment to determine the salinity of a sample of water taken from the ocean.

18.4 The estimated residence time of calcium in the oceans is about 2×10^6 years. Suggest a chemical reason for why this is shorter than the residence time estimated for sodium, potassium, or magnesium.

18.5 Phosphorus is present in seawater to the extent of 0.07 ppm by weight (that is, 0.07 g P/1×10^6 g H_2O). If this phosphorus is present as phosphate, calculate the corresponding molar concentration of phosphate.

18.6 Dissolved carbon dioxide is more highly ionized in seawater than in water of the same pH that contains a very low concentration of dissolved ionic substances. Why is this?

biochemical processes in the sea

18.7 The concentration of nitrate ion as a function of depth in the open sea is as shown in Figure 18.12. Account for this pattern of vertical distribution.

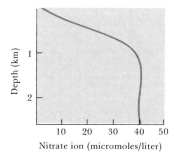

FIGURE 18.12 Concentration of nitrate ion as a function of depth.

18.8 There is some loss in total mass of biological material in proceeding through each stage of a food chain from producer to consumer. It has been estimated that the total mass of fish that can be sustained in a given volume of seawater is about 10 percent of the mass of phytoplankton produced in the volume. On this basis, how many anchovies of average mass 10 g can be sustained in the volume of water described in Sample Exercise 18.2? In the waters off Peru over 8 million tons of anchovies are taken each year. Assuming the figures above apply, what area of ocean would be required to yield this quantity of fish, if they were to be caught all at the same time?

raw materials from sea water

18.9 The solubility of sodium chloride in water is about 36 g per 100 ml of water and is nearly independent of temperature. Assuming that the other substances present in seawater do not affect the solubility of NaCl, what fraction of the water must be evaporated from seawater in settling beds before NaCl begins to crystallize out?

18.10 In the production of bromine from seawater, bromine is oxidized from bromide ion to Br_2 twice in the overall process. Suggest a reason for going through the step of reducing the bromine with SO_2 and then reoxidizing with chlorine.

18.11 In the production of magnesium from seawater, at what stage does an oxidation-reduction reaction occur? Write this reaction. What is oxidized? What is reduced?

[18.12] Oyster shells are used to form the needed CaO in the process of obtaining magnesium from seawater. Suppose that the supply of oyster shells should run out. Suggest a process by which the materials used in the process might be recycled. What additional raw materials would be needed? Write out all appropriate chemical equations.

18.13 World production of manganese is about 6 million tons per year, and of cobalt about 15,000 tons per year. Assuming that these two elements could be extracted from manganese nodules so that 50 percent of the metal present in the nodules were recovered, what mass of nodules would be required to meet world production needs for each of these two elements?

desalination

[18.14] Suppose you and a group of friends wanted to live on an idyllic little island in the Caribbean that had beautiful weather, lovely beaches, and just enough rainfall to keep a few coconut trees going and to farm with, but not enough to supply enough water to live on. The island has only a small electrical power plant and an electrically powered water pump, and no fuel sources for generating steam. Sketch a design for a simple, solar-powered desalination system using readily available structural materials such as wood, plastics, glass, and so forth as needed. The insolation, that is, the solar energy reaching the surface on this island, averages 2500 J/cm^2 per day over the entire year. The heat of vaporization of water at 40°C is about 2400 J/g. Assuming that solar energy averaged over the day can be transferred to the water with about 5 percent efficiency, estimate the area that your desalination unit must cover if it is to produce about 10,000 l. of usable water per day.

[18.15] To achieve flash evaporation of the brine in chambers D, C, and B of the flash-distillation unit shown in Figure 18.7, the pressure in chamber C must be lower than in chamber D, and lower still in chamber B. Why is this so? How do the relative temperatures vary from one chamber to the next?

[18.16] Suppose that in the large multistage flash-distillation unit shown in Figure 18.8 the efficiency of use of heat values in the steam applied to the brine heater is 20 percent, and that the steam is formed with a 20 percent efficiency from burning oil. The heat of combustion of petroleum oil is 46,000 J/g. If the unit is to produce 4 million liters of fresh water per day, and if the heat of vaporization of water in the units is 2500 J/g H_2O, how much oil must be burned per day?

water treatment

18.17 The oxygen concentration in a stream is measured at several points along the stream, and the results shown in Figure 18.13 are obtained. What can you deduce about what is occurring at point A? Why does the level of dissolved oxygen return to its original value after point B? At what point in the figure is the BOD greatest? How does the BOD at C compare with that at A?

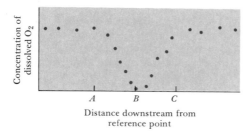

FIGURE 18.13 Oxygen concentration and BOD.

18.18 Which of the steps outlined in Figure 18.10, if any, do you think might be omitted for water taken from each of the following sources: (a) a multistage flash distillation plant; (b) an underground river running through sand and limestone formations; (c) the Mississippi River just below Memphis; (d) a reservoir that collects water from melting mountain snow.

[18.19] In many water-treatment systems in which the water is relatively contaminated, chlorine is added in concentrations as high as 5 ppm by weight. If a municipal water supply for a population of about 100,000 persons uses about 100 million liters of water per day, what weight of chlorine is required for the water sterilization? When high chlorine levels are used to ensure complete sterilization, the excess chlorine is converted to another form by adding small amounts of SO_2 to the water. Write the balanced equation for the reaction that occurs between chlorine and SO_2. (Hint: A similar reaction is described in another section of this chapter.)

[18.20] In a particular water supply, the concentration of Ca^{2+} is $2.2 \times 10^{-3} M$, and the concentration of bicarbonate ion, HCO_3^-, as determined by an acid-base titration, is $1.3 \times 10^{-3} M$. What weights of $Ca(OH)_2$ and Na_2CO_3 are needed to reduce the level of calcium ion to about one-fourth the level present originally, in 100 million liters of water?

pollutant elements

18.21 What is the maximum total mass of each of the following metals that can pass each day through a municipal water-treatment system furnishing 100 million liters of water per day, if the U.S. Public Health Service limits in Table 18.3 are not to be exceeded? (a) zinc; (b) cadmium; (c) manganese

18.22 As discussed in the text, zinc is not a toxic element under normal conditions. Yet it happens that objects made of zinc metal can under certain circumstances be the cause of a toxic metal pollution—for example, galvanized pipe, when used with water that is a little on the acidic side. By thinking of this problem in terms of periodic relationships, and with the aid of Table 18.3, suggest the source of this toxicity.

18.23 One of the most common occurrences of selenium poisoning occurs in cattle that have eaten plants that concentrate selenium from the soil. Why might this and other forms of selenium poisoning be especially evident in the area surrounding a smelting plant in which metal sulfide ores are roasted, for example:

$$2ZnS + 3O_2 \longrightarrow 2ZnO + 2SO_2$$

18.24 Describe three considerations of major importance in evaluating the possible toxic effects of a substance released to the environment.

general exercises

18.25 Complete and balance a chemical equation corresponding to each of the following verbal descriptions: (a) chlorine when added to water forms hypochlorous acid; (b) hypochlorous acid reacts with ammonia dissolved in water to form chloramine, NH_2Cl; (c) chlorine reacts with bromide ion in aqueous solution to form bromine; (d) when alum, $Al_2(SO_4)_3$, is added to slightly alkaline water a gelatinous precipitate forms; (e) mercury metal reacts in water with mercury(II) ion; (f) a sample of water containing Ca^{2+} and bicarbonate ion forms a precipitate when heated; (g) the addition of lime to seawater causes formation of a precipitate.

18.26 Using the solubility-product constant for $CaCO_3$ in seawater (6.0×10^{-7}), what is the concentration of Ca^{2+} in the sample of seawater described in Sample Exercise 18.1, assuming that the sample was in equilibrium with solid $CaCO_3$?

electrochemistry

19

Many important chemical processes utilize electricity, whereas others can be used to produce it. Because of the importance of electricity in modern society, it is useful for us to examine the subject area of electrochemistry, which deals with the relationships that exist between electricity and chemical reactions. As we shall see, our discussions of electrochemistry will provide insight into such diverse topics as the construction and operation of batteries, spontaneity of chemical reactions, electroplating, and the corrosion of metals. Because electricity involves the flow of electrons, electrochemistry focuses on reactions in which electrons are transferred from one substance to another. Such reactions are known as oxidation-reduction or "redox" reactions.

19.1

oxidation-reduction reactions: a review

Oxidation-reduction reactions were first discussed in Section 8.8. At that time we defined oxidation as an increase in oxidation number (loss of electrons) and reduction as a decrease in oxidation number (gain of electrons). If one substance gains electrons and is thereby reduced, another substance must lose electrons and be thereby oxidized. Oxidation and reduction must occur simultaneously; there cannot be one without the other. For example, consider the reaction between iron and hydrochloric acid, Equation [19.1]:

$$\overset{0}{\text{Fe}}(s) + 2\overset{+1\ -1}{\text{HCl}}(aq) \longrightarrow \overset{+2\ -1}{\text{FeCl}_2}(aq) + \overset{0}{\text{H}_2}(g) \qquad [19.1]$$

The oxidation number of each element is given above the symbol for the element. If you have forgotten the rules for finding oxidation numbers, it would be useful to refer back to Section 8.8 and review them. By looking at the oxidation states in Equation [19.1], we see that iron is oxidized while H^+ is simultaneously reduced.

In discussing oxidation-reduction reactions, it is often useful to refer to the substance causing the oxidation as the **oxidizing agent** or **oxidant**. Similarly, the substance causing the reduction is called the **reducing agent** or **reductant**. In Equation [19.1], H^+ is the oxidizing agent, because it causes the oxidation of Fe. Because Fe causes the reduction of H^+, Fe is the reducing agent. The substance reduced in a reaction is always the oxidizing agent, while the substance oxidized is always the reducing agent.

HALF-REACTIONS

Although oxidation and reduction must take place simultaneously, it is often convenient to consider them as separate processes. For example, the oxidation of Sn^{2+} by Fe^{3+}, Equation [19.2],

$$Sn^{2+}(aq) + 2Fe^{3+}(aq) \longrightarrow Sn^{4+}(aq) + 2Fe^{2+}(aq) \qquad [19.2]$$

can be considered to consist of two processes: (1) the oxidation of Sn^{2+}, Equation [19.3]; and (2) the reduction of Fe^{3+}, Equation [19.4].

Oxidation: $\qquad Sn^{2+}(aq) \longrightarrow Sn^{4+}(aq) + 2e^- \qquad [19.3]$

Reduction: $2Fe^{3+}(aq) + 2e^- \longrightarrow 2Fe^{2+}(aq) \qquad [19.4]$

Such reactions, which show either oxidation or reduction alone, are known as **half-reactions.** As shown in Equations [19.3] and [19.4], the number of electrons lost in an oxidation half-reaction must equal the number gained in a reduction half-reaction. When this condition is met and the half-reactions are balanced, they can be added to give the balanced total oxidation-reduction equation.

BALANCING EQUATIONS BY THE METHOD OF HALF-REACTIONS

In our earlier discussions, in Section 8.9, we examined how to balance oxidation-reduction equations by the method of oxidation numbers. As we have suggested above, oxidation-reduction equations can also be balanced using half-reactions. As an example, let's consider the reaction that occurs between permanganate ion, MnO_4^-, and oxalate ion, $C_2O_4^{2-}$, in acidic water solutions. When MnO_4^- is added to an acidified solution of $C_2O_4^{2-}$, the deep purple color of the MnO_4^- ion fades. Bubbles of CO_2 form, and the solution takes on the pale pink color of Mn^{2+}. We can therefore write the rough, unbalanced equation as follows:

$$MnO_4^-(aq) + C_2O_4^{2-}(aq) \longrightarrow Mn^{2+}(aq) + CO_2(g) \qquad [19.5]$$

Experiments would also show that H^+ is consumed and H_2O produced in the reaction. We shall see that this fact can be deduced in the course of balancing the equation.

To complete and balance Equation [19.3] by the method of half-reactions, we proceed through the following three steps. *Step 1* is to write two incomplete half-reactions, one involving the oxidant and the other involving the reductant.

$$MnO_4^-(aq) \longrightarrow Mn^{2+}(aq)$$
$$C_2O_4^{2-}(aq) \longrightarrow CO_2(g)$$

In *step 2*, the half-reactions are completed and balanced separately. First the atoms undergoing oxidation or reduction are balanced, then the remaining elements, and finally the charge. If the reaction occurs in acidic water solution, H^+ and H_2O can be added to either reactants or products to balance hydrogen and oxygen. Similarly, in basic solution, the equation can be completed using OH^- and H_2O. These species are in large supply in the respective solutions, and their formation as products or their utilization as reactants can easily go undetected experimentally. In the permanganate half-reaction, we already have one manganese atom on each side of the equation. However, we have four oxygens on the left and none on the right side; four H_2O molecules are needed among the products to balance the four oxygen atoms in MnO_4^-:

$$MnO_4^-(aq) \longrightarrow Mn^{2+}(aq) + 4H_2O$$

The eight hydrogen atoms that this introduces among the products can then be balanced by adding $8H^+$ to the reactants:

$$8H^+(aq) + MnO_4^-(aq) \longrightarrow Mn^{2+}(aq) + 4H_2O$$

At this stage there are equal numbers of each type of atom on both sides of the equation, but the charge still needs to be balanced. The total charge of the reactants is $+8 - 1 = +7$, while that of the products is $+2 + 4(0) = +2$. To balance the charge, 5 electrons are added to the reactant side: *

$$5e^- + 8H^+(aq) + MnO_4^-(aq) \longrightarrow Mn^{2+}(aq) + 4H_2O \qquad [19.6]$$

Proceeding similarly with the oxalate half-reaction we have

$$C_2O_4^{2-}(aq) \longrightarrow 2CO_2(g)$$

Charge is balanced by adding two electrons among the products:

$$C_2O_4^{2-}(aq) \longrightarrow 2CO_2(aq) + 2e^- \qquad [19.7]$$

In *step 3*, we multiply each equation by an appropriate factor so that the number of electrons gained in one half-reaction equals the number

---

*Although the oxidation numbers of the elements need not be used in balancing a half-reaction by this method, oxidation numbers can be used as a check. In this example MnO_4^- contains manganese in a $+7$ oxidation state. Because manganese changes from a $+7$ to a $+2$ oxidation state, it must gain five electrons just as we have already concluded.

of electrons lost in the other. The half-reactions are then added to give the overall balanced equation. In our example, the MnO_4^- half-reaction must be multiplied by 2 while the $C_2O_4^{2-}$ half-reaction must be multiplied by 5:

$$10e^- + 16H^+(aq) + 2MnO_4^-(aq) \longrightarrow 2Mn^{2+}(aq) + 8H_2O$$
$$5C_2O_4^{2-}(aq) \longrightarrow 10CO_2(g) + 10e^-$$

$$16H^+(aq) + 2MnO_4^-(aq) + 5C_2O_4^{2-}(aq) \longrightarrow$$
$$2Mn^{2+}(aq) + 8H_2O + 10CO_2(g) \qquad [19.8]$$

The balanced equation is the sum of the balanced half-reactions.

The equations for reactions that occur in basic solution can be balanced initially as if they occurred in acidic solution. The H^+ ions can then be "neutralized" by adding an equal number of OH^- ions to both sides of the equation. This procedure is illustrated in Sample Exercise 19.1.

SAMPLE EXERCISE 19.1

Complete and balance the following equation for a reaction that occurs in basic solution:

$$CN^-(aq) + MnO_4^-(aq) \longrightarrow$$
$$CNO^-(aq) + MnO_2(s)$$

Solution: The incomplete and unbalanced half-reactions are

$$CN^-(aq) \longrightarrow CNO^-(aq)$$
$$MnO_4^-(aq) \longrightarrow MnO_2(s)$$

The equations are initially balanced as if they took place in acidic solution, by appropriate addition of $H^+(aq)$ and H_2O:

$$H_2O + CN^-(aq) \longrightarrow CNO^-(aq) + 2H^+(aq)$$
$$4H^+(aq) + MnO_4^-(aq) \longrightarrow MnO_2(s) + 2H_2O$$

Because H^+ cannot exist in any appreciable concentration in basic solution, it is removed from the equations by the addition of an appropriate amount of $OH^-(aq)$. In the CN^- half-reaction, $2OH^-(aq)$ is added to both sides of the equation to "neutralize" the $2H^+(aq)$. The $2OH^-(aq)$ and $2H^+(aq)$ form $2H_2O$:

$$2OH^-(aq) + H_2O + CN^-(aq) \longrightarrow$$
$$CNO^-(aq) + 2H_2O$$

The half-reaction can be simplified because H_2O occurs on both sides of the equation. The simplified equation is:

$$2OH^-(aq) + CN^-(aq) \longrightarrow CNO^-(aq) + H_2O$$

For the MnO_4^- half-reaction, $4OH^-(aq)$ is added to both sides of the equation:

$$4H_2O + MnO_4^-(aq) \longrightarrow$$
$$MnO_2(s) + 2H_2O + 4OH^-(aq)$$

Simplifying gives

$$2H_2O + MnO_4^-(aq) \longrightarrow MnO_2(s) + 4OH^-(aq)$$

Charge is balanced by adding electrons to the side of the equation that has less negative charge so that the total charge is the same on both sides:

$$2OH^-(aq) + CN^-(aq) \longrightarrow$$
$$CNO^-(aq) + H_2O + 2e^-$$
$$3e^- + 2H_2O + MnO_4^-(aq) \longrightarrow$$
$$MnO_2(s) + 4OH^-(aq)$$

The top equation is multiplied by 3 and the bottom one by 2 to equalize electron loss and gain in the two half-reactions. The half-reactions are then added:

$$6OH^-(aq) + 3CN^-(aq) \longrightarrow$$
$$3CNO^-(aq) + 3H_2O + 6e^-$$
$$6e^- + 4H_2O + 2MnO_4^-(aq) \longrightarrow$$
$$2MnO_2(s) + 8OH^-(aq)$$

$$6OH^-(aq) + 2CN^-(aq) +$$
$$4H_2O + 2MnO_4^-(aq) \longrightarrow$$
$$3CNO^-(aq) + 3H_2O + 2MnO_2(s) + 8OH^-(aq)$$

The overall equation can be simplified, because H_2O and OH^- occur on both sides. The simplified equation is

$$3CN^-(aq) + H_2O + 2MnO_4^-(aq) \longrightarrow$$
$$3CNO^-(aq) + 2MnO_2(s) + 2OH^-(aq)$$

One of the characteristics of acids, mentioned when the concept of acids and bases was first introduced in Chapter 3, is their ability to react with certain metals to produce hydrogen gas:

$$M(s) + 2H^+(aq) \longrightarrow M^{2+}(aq) + H_2(g) \qquad [19.9]$$

Not all metals undergo such oxidation with acids. For example, copper does not. Whether or not such a reaction takes place when a metal is placed in contact with an acid depends on the ease with which the metal undergoes oxidation. In Section 19.6, we shall discuss oxidation and reduction from a quantitative standpoint. At this stage, we shall merely note that it is possible to arrange metals qualitatively in the order of their ease of oxidation. Such a list, known as an **activity series,** is shown in Table 19.1. The metals on the top are those that are most easily oxidized and whose ions are correspondingly most difficult to reduce. These are called active metals. Whether a metal will dissolve in acid depends both on temperature and on the concentration of the acid. In water, where the concentration of H^+ is very low, only the most active metals will react. Of the metals in Table 19.1, only Na, Ca, and K react at an appreciable rate with cold water. These metals, and also Zn, Al, and Mg, react with steam, whereas all the metals above hydrogen react with $1\ M\ H^+$.

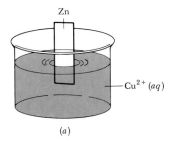

(a)

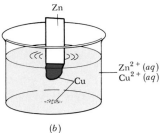

(b)

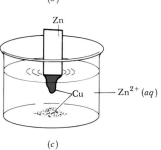

(c)

FIGURE 19.1 (a) A strip of zinc is placed in a blue solution containing Cu^{2+} ions. (b) Electrons are transferred from zinc to Cu^{2+}, and the zinc metal is eaten away and copper metal deposits. During this process, the blue color characteristic of Cu^{2+} ions in the solution fades. (c) After reaction has gone to completion, zinc is in excess, and the blue color characteristic of Cu^{2+} has disappeared.

TABLE 19.1

**activity series of hydrogen
and some common metals**

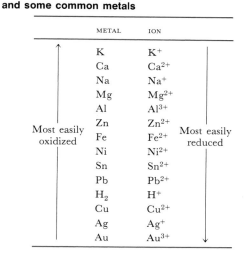

	METAL	ION	
	K	K^+	
	Ca	Ca^{2+}	
	Na	Na^+	
	Mg	Mg^{2+}	
	Al	Al^{3+}	
Most easily oxidized	Zn	Zn^{2+}	Most easily reduced
	Fe	Fe^{2+}	
	Ni	Ni^{2+}	
	Sn	Sn^{2+}	
	Pb	Pb^{2+}	
	H_2	H^+	
	Cu	Cu^{2+}	
	Ag	Ag^+	
	Au	Au^{3+}	

The activity series indicates not only the reactivities of metals with acids, but also their reactivities with each other. Consider what happens when a piece of zinc is placed in contact with a solution of Cu^{2+}. The blue color that is characteristic of Cu^{2+} ions in aqueous solutions fades, and copper metal begins to deposit on the zinc. At the same time, the zinc begins to dissolve. These transformations are shown in Figure 19.1 and are summarized by Equation [19.10]:

$$Zn(s) + Cu^{2+}(aq) \longrightarrow Zn^{2+}(aq) + Cu(s) \qquad [19.10]$$

It is instructive to consider such redox reactions as competitions for electrons. Consider what would happen if two electrons were made available to a solution containing Zn^{2+} and Cu^{2+} ions. The position of copper in the activity series indicates that it has a stronger attraction for electrons than does zinc, because it is more easily reduced. Consequently copper would end up with the electrons. In any oxidation-reduction reaction between a metal and a metal ion, the more active metal will end up in the oxidized state.

SAMPLE EXERCISE 19.2

Indicate whether a reaction will occur in the following instances: (a) $Zn(s) + Fe^{2+}(aq)$; (b) $Sn(s) + Ni^{2+}(aq)$.

Solution: (a) According to the activity series, zinc is more easily oxidized than is iron; zinc is the more active metal because it has a weaker attraction for electrons. Zinc should therefore lose electrons to iron, and a chemical reaction will be observed:

$$Zn(s) + Fe^{2+}(aq) \longrightarrow Zn^{2+}(aq) + Fe(s)$$

(b) According to the activity series, tin is more difficult to oxidize than is nickel; nickel has the weaker attraction for electrons and is the more active metal. The $Ni^{2+}(aq)$ is therefore unable to obtain electrons from $Sn(s)$, and there is no reaction.

**19.3
electrolysis
and electrolytic
cells**

Any spontaneous oxidation-reduction reaction can be used to construct a **voltaic cell,** an electrochemical device for generating electricity. On the other hand, it is possible to cause nonspontaneous reactions, such as that given by Equation [19.11], to occur by supplying electrical energy:

$$2NaCl(l) \longrightarrow 2Na(l) + Cl_2(g) \qquad [19.11]$$

In this case, the process, which is known as **electrolysis,** is performed in an apparatus known as an **electrolytic cell.** In the remainder of this chapter, we shall consider both of these types of cells. Because electrolytic cells are generally simpler in design than voltaic ones, we shall consider electrolysis first.

An electrolytic cell consists of two electrodes in a molten salt or aqueous solution of an ionic substance, as shown in Figure 19.2. The cell is driven by a battery or some other source of direct electrical current. The battery acts as an electron pump, pushing electrons into one electrode and pulling them from the other. The electrodes are in contact with

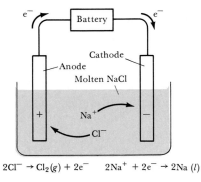

$$2Cl^- \rightarrow Cl_2(g) + 2e^- \qquad 2Na^+ + 2e^- \rightarrow 2Na\ (l)$$

FIGURE 19.2 The electrolysis of molten sodium chloride.

the solution or molten salt that is to be electrolyzed. The electrodes are the sites at which the electrochemical reactions occur. Withdrawing electrons from an electrode gives it a positive charge, whereas adding electrons to an electrode makes it negative. In the electrolysis of molten NaCl, shown in Figure 19.2, Na^+ ions pick up electrons at the negative electrode and are thereby reduced. As the Na^+ ions in the vicinity of this electrode are depleted, additional Na^+ ions migrate in. In a related fashion, there is a net movement of Cl^- ions to the positive electrode, where they give up electrons and are thereby oxidized. The electrode at which reduction occurs is called the **cathode,** whereas the electrode at which oxidation occurs is called the **anode.** * The anode and cathode reactions are just the half-reactions for oxidation and reduction, respectively. The overall reaction produced in the electrolytic cell is the sum of these half-reactions:

$$\text{Anode:} \qquad\qquad\qquad\qquad 2Cl^- \longrightarrow Cl_2(g) + 2e^-$$
$$\text{Cathode:} \qquad\qquad 2Na^+ + 2e^- \longrightarrow 2Na(l)$$
$$\overline{\text{Overall reaction: } 2Na^+ + 2Cl^- \longrightarrow 2Na(l) + Cl_2(g)} \qquad [19.12]$$

Electrolysis is used in the commercial preparation of sodium. Molten NaCl is electrolyzed in a specially designed cell called the **Downs cell.** This cell, shown in Figure 19.3, is designed to keep Na and Cl_2 from coming in contact and reforming NaCl. The Na is also prevented from coming in contact with air and forming an oxide. Calcium chloride, $CaCl_2$, is added to the NaCl to lower the melting point of the molten liquid from its normal melting point of 804°C to around 600°C.

Sodium metal cannot be prepared by electrolysis of aqueous solutions of NaCl because water is more easily reduced than are sodium cations. Consequently, electrolysis of aqueous solutions of NaCl, known as brines, produces H_2 and Cl_2 with Na^+ and OH^- ions remaining in solution. The electrode reactions are given below:

$$\text{Anode:} \qquad\qquad\qquad\qquad 2Cl^-(aq) \longrightarrow Cl_2(g) + 2e^-$$
$$\text{Cathode:} \qquad\qquad 2H_2O + 2e^- \longrightarrow H_2(g) + 2OH^-(aq)$$
$$\overline{\text{Overall reaction: } 2Cl^-(aq) + 2H_2O \longrightarrow}$$
$$Cl_2(g) + H_2(g) + 2OH^-(aq) \quad [19.13]$$

The Na^+ ion is merely a spectator ion (Section 12.6) in the electrolysis. This electrolysis is used commercially because all of the products, H_2, Cl_2, and NaOH, are commercially useful chemicals.

*Notice that the anion moves to the anode, whereas the cation moves toward the cathode. Notice also that the words "anode" and "oxidation" both begin with a vowel, whereas "cathode" and "reduction" both begin with a consonant. These observations should help you to keep the definitions straight.

SAMPLE EXERCISE 19.3

What possible electrode reactions can occur when an aqueous solution of $CuCl_2$ is electrolyzed?

Solution: At the anode we can envision oxidation of either water or Cl^-:

$$2H_2O \longrightarrow O_2(g) + 4H^+(aq) + 4e^-$$
$$2Cl^-(aq) \longrightarrow Cl_2(g) + 2e^-$$

At the cathode we can envision either reduction of Cu^{2+} or water:

$$Cu^{2+}(aq) + 2e^- \longrightarrow Cu(s)$$
$$2H_2O + 4e^- \longrightarrow 2H_2(g) + 2OH^-(aq)$$

Experimentation indicates that the products are copper and chlorine. Because the electrolysis of aqueous solutions of NaCl gave Cl_2 and not O_2 at the anode, we should have been able to predict that Cl_2 would also form in the case of $CuCl_2$. Reference to the activity series, Table 19.1, should suggest in a qualitative way that Cu^{2+} is easier to reduce than the H^+ in H_2O. Thus even at this stage we should not only be able to predict the possible electrode reactions in the electrolysis of simple substances, but also to make reasonable guesses at which reactions are most likely.

REDUCTION OF ALUMINUM

The electrolysis of aqueous solutions of the active metals such as Na, Ca, Mg, and Al, which are found near the top of Table 19.1, leads to the formation of H_2 rather than the metal. Consequently, these active metals are obtained by electrolysis of their molten salts. We have already briefly examined the electrolytic production of sodium. Let's examine one further example, the formation of aluminum. The electrolytic process used commercially to produce aluminum is known as the Hall process, named after its inventor Charles M. Hall (Figure 19.4). Aluminum oxide, Al_2O_3, is mined as the mineral bauxite. After purification, the Al_2O_3 is dissolved in molten cryolite, Na_3AlF_6, producing a solution that will conduct an electric current. The molten mixture is electrolyzed using carbon electrodes as shown in Figure 19.5. The electrode reactions are:

$$\text{Anode:} \quad C(s) + 2O^{2-} \longrightarrow CO_2(g) + 4e^- \qquad [19.14]$$
$$\text{Cathode:} \quad 3e^- + Al^{3+} \longrightarrow Al(l) \qquad [19.15]$$

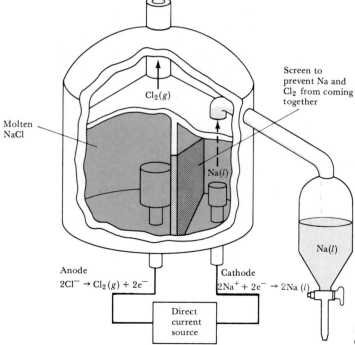

Screen to prevent Na and Cl_2 from coming together

Molten NaCl

$Cl_2(g)$

Na(l)

Na(l)

Anode
$2Cl^- \rightarrow Cl_2(g) + 2e^-$

Cathode
$2Na^+ + 2e^- \rightarrow 2Na(l)$

Direct current source

FIGURE 19.3 Downs cell used in the commercial production of sodium.

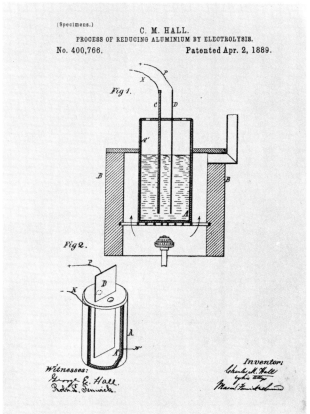

FIGURE 19.4 A photograph of Charles M. Hall (1863–1914) as a young man and the patent diagram for Hall's device for reducing aluminum. (*Courtesy of ALCOA*)

Charles M. Hall began work on the problem of reducing aluminum in about 1885, after he had learned from a professor of the difficulty of reducing ores. Prior to the development of his electrolytic process, aluminum was obtained by chemical reduction using sodium or potassium as a reducing agent. This procedure caused the cost of aluminum to be very high. As late as 1852, the cost of aluminum was $545 a pound. During the Paris Exposition in 1855, aluminum was exhibited as a rare metal in spite of the fact that it is the third most abundant element on the earth. The cost of aluminum by then had fallen to $90 a pound, which still made it more expensive than gold or silver. It is reported that the very rich of that era could flaunt their wealth by using aluminum eating utensils. Hall, who was 21 years old when he started his research, utilized handmade and borrowed equipment in his studies, and his laboratory was a woodshed near his home. In about a year's time he was able to solve the problem. Strangely, Paul Heroult, who was the same age as Hall, made the same discovery in France at approximately the same time. The basic discovery was that Al_2O_3 dissolves in cryolite, a rare mineral found in a small region in Greenland, to produce a conducting solution. As a result of the discovery of Hall and Heroult, large-scale reduction of aluminum became commercially feasible, and aluminum became a common and familiar metal. Its price subsequently fell as low as 15¢ a pound. Even today aluminum sells for less than 50¢ a pound.

Because of the rarity of natural cryolite, methods have been developed to produce Na_3AlF_6. It is also possible to convert Al_2O_3 to $AlCl_3$ by reaction with CO and Cl_2 as shown in Equation [19.16].

$$Al_2O_3(s) + 3CO(g) + 3Cl_2(g) \longrightarrow$$
$$2AlCl_3(g) + 3CO_2(g) \qquad [19.16]$$

Molten $AlCl_3$ can then be employed in the electrolysis.

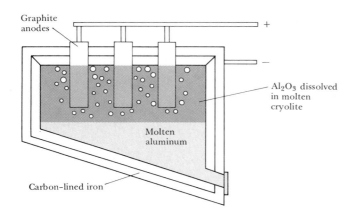

Graphite
anodes

+

−

Al$_2$O$_3$ dissolved
in molten
cryolite

Molten
aluminum

Carbon-lined iron

FIGURE 19.5   A typical Hall-process electrolysis cell used to reduce aluminum. Because molten aluminum is more dense than the molten mixture of Na$_3$AlF$_6$ and Al$_2$O$_3$, the metal collects at the bottom of the cell.

PURIFICATION OF COPPER

One of the many interesting applications of electrolysis is the refining or purification of copper metal. Commercially, copper compounds are reduced using chemical reducing agents. For example, cuprous sulfide, Cu$_2$S, can be reduced by blowing air through the molten material as shown in Equation [19.17]:

$$Cu_2S(l) + O_2(g) \longrightarrow 2Cu(l) + SO_2(g) \qquad [19.17]$$

The copper metal obtained in this way is known as blister copper; it is about 99 percent pure, containing iron, zinc, gold, and silver among the impurities. Certain impurities greatly lower the electrical conductivity of the metal. Therefore, if it is to be used to make electrical wiring, it must be further purified. This purification is done electrolytically. The blister copper is made the anode in an electrolytic cell as shown in Figure 19.6. Thin sheets of pure copper serve as the cathode, and an aqueous mixture of H$_2$SO$_4$ and CuSO$_4$ is added to the cell to permit conduction of electric current through the solution. As the current flows, copper dissolves from the anode and deposits on the cathode:

Anode: $\qquad\qquad$ $Cu(s) \longrightarrow Cu^{2+}(aq) + 2e^-$ $\qquad$ [19.18]

Cathode: $Cu^{2+}(aq) + 2e^- \longrightarrow Cu(s)$ $\qquad\qquad$ [19.19]

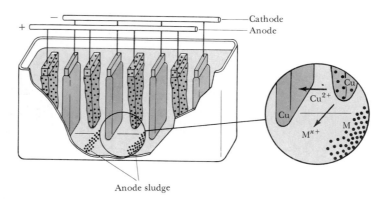

− Cathode
+ Anode

Cu^{2+}

Cu

Cu

M^{x+}

M

Anode sludge

FIGURE 19.6   Electrolysis cell for the refining of copper. Notice that as the anodes dissolve away, the cathodes grow in size.

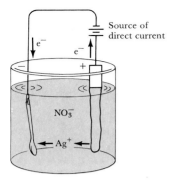

$Ag^+ (aq) + e^- \rightarrow Ag(s)$

$Ag(s) \rightarrow Ag^+(aq) + e^-$

FIGURE 19.7 Electroplating of an object with silver. The object to be plated is made the cathode, whereas the plating metal is made the anode.

Metals like zinc and iron that are more easily oxidized than copper dissolve at the anode together with the copper. Because they are not reduced as easily as copper, careful regulation of the voltage of the electrolytic cell prevents them from depositing at the cathode. Metals such as silver and gold that are not oxidized as readily as copper do not dissolve at the anode. Instead, they slough off the anode as the copper dissolves; they then accumulate as a sludge below the anode. These anode sludges are recovered periodically when the electrolysis cells are cleaned. The anode sludges are important sources of gold and silver. In fact, most gold produced in this country is obtained from them.

ELECTROPLATING

If some metal other than copper were made the cathode in the electrolytic cell that we have just described, it would become coated with copper. The "plating out" of one metal on another in an electrolytic cell is known as **electroplating**. The object to be electroplated is made the cathode of the electrolytic cell. The plating metal is made the anode. This is illustrated in Figure 19.7. Electroplating is used to protect objects against corrosion and to improve their appearance. There is amazing evidence from archaeological digs that the Parthians may have constructed batteries as early as 200 B.C. It has been speculated that these batteries were used to drive electrolytic cells for electroplating gold and silver. Figure 19.8 shows a photograph of a working replica of one of these ancient batteries that was excavated in 1936 in the Baghdad area.

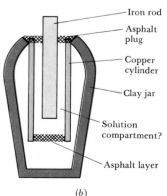

(a)

(b)

FIGURE 19.8 (a) Working replica of battery believed to date to 200 B.C. (b) A cross-section diagram of the cell. (a courtesy of The Berkshire Museum)

The quantity of chemical reaction occurring in an electrolytic cell is directly proportional to the quantity of electricity passed into the cell. For example, 1 mole of electrons will plate out 1 mole of Na metal, whereas 2 moles of electrons will plate out 2 moles of Na metal:

$$Na^+ + e^- \longrightarrow Na$$

Similarly, it requires 2 moles of electrons to produce a mole of copper from Cu^{2+} and 3 moles of electrons to produce a mole of aluminum from Al^{3+}:

$$Cu^{2+} + 2e^- \longrightarrow Cu$$

$$Al^{3+} + 3e^- \longrightarrow Al$$

The electrical charge on a mole of electrons is called a **faraday** (abbreviated $\mathcal{F}$). The faraday is named in honor of Michael Faraday, who in 1834 first stated the quantitative relationship between electrical current and the extent of electrolysis in an electrolytic cell.

The quantity of charge passing through an electrical circuit, such as that in an electrolytic cell, is generally measured in **coulombs.** There are 96,500 coulombs (coul) in a faraday:

$$1\mathcal{F} = 96,500 \text{ coul} = \text{charge of 1 mole of electrons} \qquad [19.20]$$

In terms of other, perhaps more familiar electrical units, a coulomb is the quantity of electrical charge passing a point in a circuit in 1 sec when the current is 1 ampere (amp).* Therefore, the number of coulombs passing through a cell can be obtained by multiplying the amperage and the elapsed time in seconds:

$$\text{Coulombs} = \text{amperes} \times \text{seconds} \qquad [19.21]$$

These ideas are applied in Sample Exercises 19.4 and 19.5. Although these exercises involve electrolytic cells, the same relationships can be applied to cells that produce rather than use electrical energy.

*Conversely, current is the rate of flow of electricity. An ampere is the current associated with the flow of 1 coul past a point each second.

SAMPLE EXERCISE 19.4

Calculate the amount of aluminum produced in 1.00 hr by the electrolysis of molten $AlCl_3$ if the current is 10.0 amp.

Solution: Using Equation [19.21], we can write:

$$\text{Coulombs} = (10.0 \text{ amp})(1.00 \text{ hr})$$
$$\times \left(\frac{3600 \text{ sec}}{1 \text{ hr}} \right) \left(\frac{1 \text{ coul}}{1 \text{ amp-sec}} \right)$$
$$= 3.60 \times 10^4 \text{ coul}$$

The half-reaction for the reduction of Al^{3+} is

$$Al^{3+} + 3e^- \longrightarrow Al$$

The amount of aluminum produced depends on the number of available electrons: 1 mole $Al \simeq 3\mathcal{F}$. We can therefore write:

$$\text{Grams Al} = (3.60 \times 10^4 \text{ coul}) \left(\frac{1\,\mathcal{F}}{96,500 \text{ coul}} \right)$$
$$\times \left(\frac{1 \text{ mole Al}}{3\,\mathcal{F}} \right) \left(\frac{27.0 \text{ g Al}}{1 \text{ mole Al}} \right)$$
$$= 3.36 \text{ g}$$

SAMPLE EXERCISE 19.5

A constant current was passed through a solution of Cu^{2+} for a period of 5.00 min. During this time the cathode increased in mass by 1.24 g. How many amperes of current was used?

Solution: In this case the half-reaction that we are focusing on is

$$Cu^{2+} + 2e^- \longrightarrow Cu$$

Therefore, 1 mole Cu $\simeq$ 2 $\mathfrak{F}$. Using this information, we can calculate the amperes as shown below:

$$Coulombs = (1.24 \text{ g Cu})\left(\frac{1 \text{ mole Cu}}{63.5 \text{ g Cu}}\right)$$
$$\times \left(\frac{2\mathfrak{F}}{1 \text{ mole Cu}}\right)\left(\frac{96,500 \text{ coul}}{1\mathfrak{F}}\right)$$
$$= 3.77 \times 10^3 \text{ coul}$$

$$Amperes = \frac{coul}{sec}$$
$$= \left(\frac{3.77 \times 10^3 \text{ coul}}{5.00 \text{ min}}\right)$$
$$\times \left(\frac{1 \text{ min}}{60 \text{ sec}}\right)\left(\frac{1 \text{ amp-sec}}{1 \text{ coul}}\right)$$
$$= 12.6 \text{ amps}$$

Before going on, it is useful to define some other units associated with electrical circuits. Electrons are caused to move through an electric circuit by an **electromotive force** (abbreviated emf); emf is measured in units of **volts**.* The greater the resistance of the circuit, the greater the emf required to pump the electrons through the circuit. Resistance is measured in units of **ohms.** The work performed in moving electrons through an electrical circuit is equal to the product of emf and electrical charge. The movement of 1 coul of electrical charge using an emf of 1 volt (V) corresponds to 1 J of energy. (Remember that the joule is the SI unit of energy, Section 4.1.)

$$1 \text{ J} = 1 \text{ coul-V} = 1 \text{ amp-sec-V} \qquad [19.22]$$

A watt is an amp-volt; we can therefore write:

$$1 \text{ J} = 1 \text{ amp-sec-V} = 1 \text{ watt-sec} \qquad [19.23]$$

The watt represents a certain *rate* of energy expenditure. A kilowatt-hour (kwh), a common unit of electrical energy, is equal to 3.6×10^6 J:

$$1 \text{ kwh} = (1000 \text{ watts})(1 \text{ hr})\left(\frac{3600 \text{ sec}}{1 \text{ hr}}\right)\left(\frac{1 \text{ amp-V}}{1 \text{ watt}}\right)\left(\frac{1 \text{ J}}{1 \text{ amp-sec-V}}\right)$$
$$= 3.6 \times 10^6 \text{ J}$$

*The emf is also referred to as *voltage* or *electrical potential*.

SAMPLE EXERCISE 19.6

Calculate the number of kilowatt-hours of electricity required to produce 1000 kg of aluminum by electrolysis of Al^{3+} if the required emf is 4.5 volts.

Solution:

$$\text{Coulombs} = (1000 \text{ kg Al})\left(\frac{1000 \text{ g Al}}{1 \text{ kg Al}}\right)$$

$$\times \left(\frac{1 \text{ mole Al}}{27.0 \text{ g Al}}\right)\left(\frac{3\mathcal{F}}{1 \text{ mole Al}}\right)$$

$$\times \left(\frac{96,500 \text{ coul}}{1 \mathcal{F}}\right)$$

$$= 1.07 \times 10^{10} \text{ coul}$$

A kilowatt-hour is a measure of energy. Electrical energy is obtained by multiplying coulombs and voltage, Equation [19.22]. When we do this and then apply unit conversion factors to obtain kilowatt-hours we have:

$$\text{kwh} = (1.07 \times 10^{10} \text{ coul})(4.5 \text{ volts})$$

$$\times \left(\frac{1 \text{ amp-sec}}{1 \text{ coul}}\right)\left(\frac{1 \text{ watt}}{1 \text{ amp-V}}\right)$$

$$\times \left(\frac{1 \text{ kilowatt}}{1000 \text{ watt}}\right)\left(\frac{\text{hr}}{3600 \text{ sec}}\right)$$

$$= 1.3 \times 10^4 \text{ kwh}$$

We could also use the facts that 1 J equals 1 coul-V and 1 kwh equals 3.6×10^6 J:

$$\text{kwh} = (1.07 \times 10^{10} \text{ coul})(4.5 \text{ V})$$

$$\times \left(\frac{1 \text{ J}}{1 \text{ coul-V}}\right)\left(\frac{1 \text{ kwh}}{3.6 \times 10^6 \text{ J}}\right)$$

This quantity of energy does not include the energy used to mine, transport, and process the aluminum ore, and to keep the electrolysis bath molten during electrolysis. A typical electrolytic cell used to reduce aluminum is only 40 percent efficient, 60 percent of the electrical energy being dissipated as heat. It therefore requires on the order of 15 kwh of electricity to reduce a pound of aluminum. The aluminum industry consumes about 2 percent of the electrical energy generated in the United States. Because this is used mainly for reduction of aluminum, recycling this metal would save large quantities of energy. Interestingly, the United States uses 5 percent of the world's production of aluminum to make food and beverage cans.

19.5
voltaic cells

Having discussed electrolytic cells, we now shift our attention to voltaic cells. In principle, any spontaneous redox reaction can be used to produce electricity directly. The trick is to cause the electrons to move through an external circuit as they are transferred from one substance to another. For example, consider the reduction of Cu^{2+} by zinc, which we discussed in Section 19.2:

$$Zn(s) + Cu^{2+}(aq) \longrightarrow Zn^{2+}(aq) + Cu(s)$$

If zinc and a zinc salt such as $Zn(NO_3)_2$ are placed in one compartment and copper and $Cu(NO_3)_2$ in the other as shown in Figure 19.9, no current is observed to move through the external circuit. This can be rationalized as follows. If a zinc atom loses two electrons and transfers them to the Cu^{2+} by way of the wire between the electrodes, the resultant solutions become charged. The positive charge in the zinc compartment increases as a Zn^{2+} ion goes into solution:

$$\text{Anode: } Zn(s) \longrightarrow Zn^{2+}(aq) + 2e^-$$

At the same time, the other compartment becomes negatively charged as a Cu^{2+} ion picks up the available electrons and plates out at the cathode:

$$\text{Cathode: } Cu^{2+}(aq) + 2e^- \longrightarrow Cu(s)$$

The positive charge of the anode compartment hinders entry of further Zn^{2+} ions, whereas the negative charge of the cathode compartment hinders removal of Cu^{2+} ions. Electrons instantaneously cease to flow through the external circuit unless a mechanism is provided for the

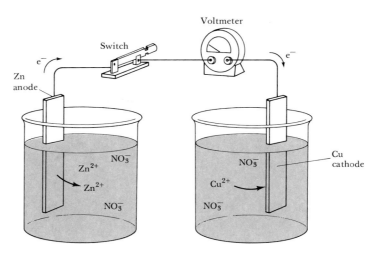

FIGURE 19.9 An incomplete voltaic cell. Movement of zinc into solution as Zn²⁺, movement of the resultant electrons through the external circuit, and their use in reducing Cu²⁺ are shown. However, such electron movement does not occur, because the half-cells would become charged. A device to permit ions to migrate from one compartment to the other is needed to complete the electrical circuit.

solutions to maintain their electrical neutrality. Electrical neutrality can be maintained if nonreactive ions are allowed to migrate into each compartment. It is possible to achieve this using a porous partition that prevents mechanical mixing of the solutions in the two compartments, but permits the passage of ions to maintain charge balance. In the laboratory, a U-tube filled with some electrolyte such as $NaNO_3$ that will not react with either solution or electrode is often used to maintain charge-balance in the compartments. The ends of the U-tube can be loosely plugged with glass-wool and the U-tube inverted so that one arm is placed in each solution as shown in Figure 19.10. The U-tube, filled in this way with electrolyte, is known as a salt bridge. The salt bridge supplies NO_3^- ions to neutralize the charge of Zn^{2+} ions that enter the solution at the anode. Similarly, as Cu^{2+} ions receive electrons at the cathode and the resultant copper atoms plate out, Na^+ ions migrate into the solution from the salt bridge to maintain the electrical neutrality of the compartment. Under these conditions electrons flow spontaneously from the anode, at which oxidation occurs, to the cathode, at which reduction occurs. Thus, an electric current is generated.*

Chemical reaction occurs in the cell shown in Figure 19.10 only so long as the electrons can move through the external circuit. If the circuit is broken using a switch, no reaction occurs. By contrast, if the $Cu(NO_3)_2$ and $Zn(NO_3)_2$ were mixed in a single container and the electrodes placed in this solution, reaction between Cu^{2+} and Zn could occur without electrons needing to move through the external circuit. Although some electrical current would be generated, there would be no way to switch off the reaction. Therefore, it would not be as useful a source of electrical energy as the cell shown in Figure 19.10.

*Because Cu metal and Zn^{2+} ions are not used in the reaction, the cell could have been constructed using some other metal as the cathode and with some electrolyte other than $Zn(NO_3)_2$ in the anode compartment.

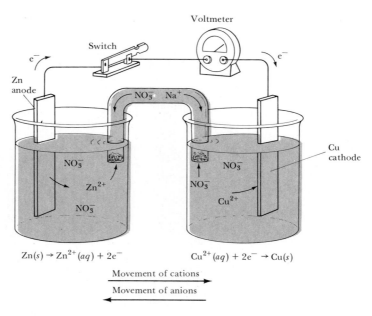

FIGURE 19.10 Complete and functioning voltaic cell using a salt bridge to complete the electrical circuit.

As a further example of a voltaic cell, consider a cell constructed to use the redox reaction between Zn and H^+, Equation [19.24].

$$Zn(s) + 2H^+(aq) \longrightarrow Zn^{2+}(aq) + H_2(g) \qquad [19.24]$$

The anode compartment can be identical to that in the previous cell. However, we cannot immerse a piece of H_2 into a solution of H^+ as we could immerse a piece of Cu into a Cu^{2+} solution. A hydrogen electrode

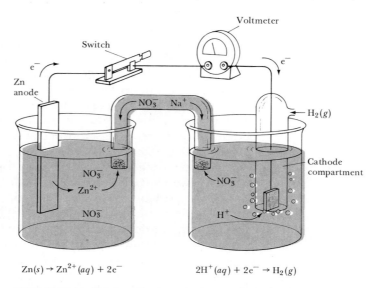

FIGURE 19.11 Voltaic cell using a hydrogen electrode.

can be constructed as shown in Figure 19.11. The electrode consists of a platinum wire and a piece of platinum foil covered with finely divided platinum that serves as a surface for the cathode reaction. This is encased in a glass tube so hydrogen gas can be bubbled over the platinum. The platinum itself does not undergo either oxidation or reduction and is therefore referred to as an inert electrode. Graphite is often used as an inert electrode in commercial voltaic cells.

SAMPLE EXERCISE 19.7

The reaction $H_2(g) + Cl_2(g) \longrightarrow 2HCl(aq)$ is used to make a voltaic cell. Diagram the cell, labeling the anode and cathode; identify the electrode reactions and show the direction of electron and ion migrations.

Solution:

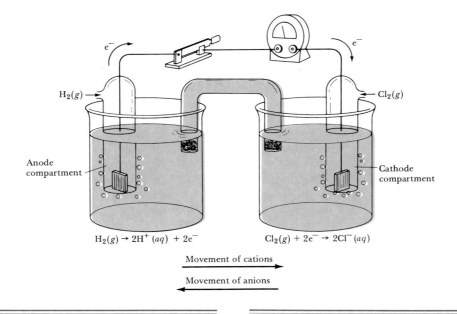

$H_2(g) \rightarrow 2H^+(aq) + 2e^-$

$Cl_2(g) + 2e^- \rightarrow 2Cl^-(aq)$

Movement of cations →

← Movement of anions

19.6 some commercial voltaic cells

Although any spontaneous redox reaction can be used as the basis of a voltaic cell, making a commercial voltaic cell that utilizes a particular redox reaction can require considerable ingenuity. Salt-bridge cells such as those that we have been discussing provide considerable insight into the operation of voltaic cells. However, they are generally unsuitable for commercial use. They have high internal resistances. As a result, if we attempt to draw a large current from them, their voltage drops sharply. Furthermore, the cells that we have pictured so far lack the compactness and ruggedness required for portability. For the most part voltaic cells are a convenient energy source whose primary virtue is portability. They cannot yet compete with other common energy sources on the basis of cost alone. The cost of electricity from a common flashlight battery is on the order of $60 per kilowatt-hour. By comparison, electrical energy from power plants normally costs the consumer less than 10¢ per kilowatt-hour.

LEAD STORAGE BATTERY

One of the most common batteries is the lead storage battery used in automobiles. The term battery is used to describe an assembly of one or more voltaic cells. A 12-V lead storage battery consists of six cells, each producing 2 V. The anode of each cell is composed of lead; the cathode is composed of lead dioxide, PbO_2, packed on a metal grid. Both electrodes are immersed in sulfuric acid. The electrode reactions that occur during discharge are as follows:

Anode: $$Pb(s) + SO_4^{2-}(aq) \longrightarrow PbSO_4(s) + 2e^-$$

Cathode: $$PbO_2(s) + SO_4^{2-}(aq) + 4H^+(aq) + 2e^- \longrightarrow$$
$$PbSO_4(s) + 2H_2O$$

$$Pb(s) + PbO_2(s) + 4H^+(aq) + 2SO_4^{2-}(aq) \longrightarrow$$
$$2PbSO_4(s) + 2H_2O \quad [19.25]$$

To increase the current output of each cell, a number of anodes and cathodes are joined together as shown in Figure 19.12. To keep the electrode plates from coming in contact with each other, they are separated by wood or glass-fiber spacers. As the above equations indicate, H_2SO_4 is used up during the discharge of a lead storage battery. Because sulfuric acid has a high density, the density of the electrolyte in the cell decreases during discharge. When fully charged the electrolyte has a density of 1.25–1.30 g/cm³. If the density falls below 1.20 g/cm³, the battery needs recharging. A service-station attendant can test the density using a hydrometer. This device, shown in Figure 19.13, has a float that sinks to a depth that is a function of the density of the liquid in which it is immersed.

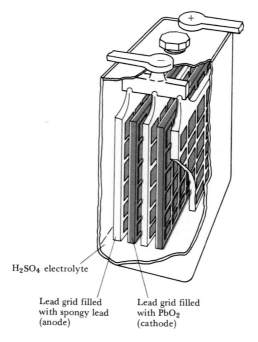

H_2SO_4 electrolyte

Lead grid filled
with spongy lead
(anode)

Lead grid filled
with PbO_2
(cathode)

FIGURE 19.12 A lead storage cell.

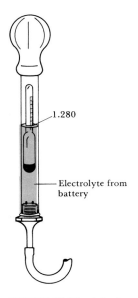

1.280

— Electrolyte from battery

FIGURE 19.13 A hydrometer of the type used to measure the density of the electrolyte in a lead storage battery.

One advantage of the lead storage cell is that it can be recharged. During recharging the cell is operated as an electrolytic cell. The voltage necessary for recharging is provided in an automobile by a generator that is driven by the engine. The energy added during electrolysis is stored for later use. The electrode reactions are the reverse of those that occur during discharge:

$$PbSO_4(s) + 2e^- \longrightarrow Pb(s) + SO_4^{2-}(aq)$$
$$\underline{PbSO_4(s) + 2H_2O \longrightarrow PbO_2(s) + SO_4^{2-}(aq) + 4H^+(aq) + 2e^-}$$
$$2PbSO_4(s) + 2H_2O \longrightarrow$$
$$Pb(s) + PbO_2(s) + 4H^+(aq) + 2SO_4^{2-}(aq) \quad [19.26]$$

Such operation is possible because the $PbSO_4$ formed during discharge is a solid and adheres to the electrodes. It is therefore in position to either receive or give up electrons during electrolysis.

DRY CELL

The **dry cell** is familiar because of its wide use in flashlights and portable radios. Because of this use, it is often referred to as a flashlight battery. It is also known as the Leclanché cell after its inventor who patented it in 1866. The anode of the dry cell consists of a zinc can that is in contact with a paste of MnO_2, NH_4Cl, and carbon. An inert cathode, consisting of a graphite rod, is immersed in the center of the paste as shown in Figure 19.14. The cell has an exterior layer of cardboard or metal to seal the cell against the atmosphere. The electrode reactions are complex, and the cathode reaction appears to vary with the rate of discharge. The reactions at the electrodes are generally represented as shown below:

Anode: $$Zn(s) \longrightarrow Zn^{2+}(aq) + 2e^- \quad [19.27]$$

Cathode: $$2NH_4^+(aq) + 2MnO_2(s) + 2e^- \longrightarrow$$
$$Mn_2O_3(s) + 2NH_3(aq) + H_2O \quad [19.28]$$

Only a fraction of the cathode material, that near the electrode, is electrochemically active because of the limited mobility of the chemicals in the cell.

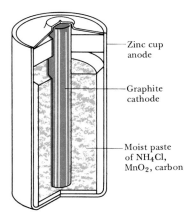

— Zinc cup anode

— Graphite cathode

— Moist paste of NH_4Cl, MnO_2, carbon

FIGURE 19.14 A cutaway view of a dry cell.

569

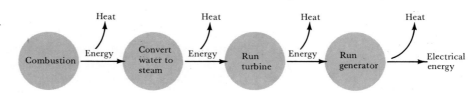

FIGURE 19.15 A schematic representation of energy flow in the course of converting fuels into electrical energy using engines and generators. The several energy conversions lead to low overall efficiency. About 40 percent of the energy of the fuel is eventually converted to electrical energy; the remainder is lost as heat.

NICKEL-CADMIUM BATTERIES

Because dry cells are not rechargeable, they have to be replaced frequently. Thus a rechargeable cell, the nickel-cadmium battery, has become increasingly popular, especially for use in battery-operated tools and calculators. The electrode reactions occurring within this cell during discharge are as follows:

Anode: $\quad Cd(s) + 2OH^-(aq) \longrightarrow Cd(OH)_2(s) + 2e^-$ [19.29]

Cathode: $\quad NiO_2(s) + 2H_2O + 2e^- \longrightarrow Ni(OH)_2(s) + 2OH^-(aq)$ [19.30]

As in the lead storage cell, the reaction products adhere to the electrodes. This permits the reactions to be readily reversed during charging. Because no gases are produced during either charging or discharging, the battery can be sealed.

FUEL CELLS

Many substances such as H_2 or CH_4 are used as fuels. The heat released in their reaction with oxygen is the source of energy. The thermal energy released by combustion is often subsequently converted to electrical energy. Because combustion reactions are oxidation-reduction reactions, it would be possible to use such reactions to generate an electrical current directly, if suitable voltaic cells could be designed. It is, in fact, advantageous to perform such a direct conversion to electrical energy. In producing electricity via combustion, heat is used to convert water to steam. The steam is used to drive a turbine that drives a generator. Energy is lost as heat in every conversion of energy from one form to another or from one material to another. Typically, a maximum of only 40 percent of the energy from combustion is converted to electricity; the remainder is lost as heat. The situation is represented schematically in Figure 19.15. Direct production of electricity from fuels via a voltaic cell should yield a higher rate of conversion of the chemical energy of the fuels into electricity. Voltaic cells that utilize conventional fuels are known as **fuel cells**.

A great deal of research has gone into attempts to develop practical fuel cells. One of the problems encountered is the high operating temperatures of most cells, which not only siphons off energy but also accelerates corrosion of cell parts. A low-temperature cell has been developed that uses H_2, but the present cost of the cell makes it too expensive for large-scale use. However, it has been used in special situations, such as in space vehicles. For example, a H_2-O_2 fuel cell was used as the primary source of electrical energy on the Apollo moon flights. The weight of the fuel cell sufficient for 11 days in space was approximately

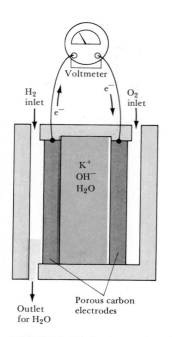

FIGURE 19.16 A cross-section of a H_2-O_2 fuel cell.

500 pounds. This may be compared to the several tons that would have been required for an engine-generator set.

The electrode reactions in the H_2-O_2 fuel cell are as follows:

Anode: $2H_2(g) + 4OH^-(aq) \longrightarrow 4H_2O + 4e^-$
Cathode: $4e^- + O_2(g) + 2H_2O \longrightarrow 4OH^-(aq)$
$$2H_2(g) + O_2(g) \longrightarrow 2H_2O \qquad [19.31]$$

The cell is illustrated in Figure 19.16. The electrodes are composed of hollow tubes of porous, compressed carbon impregnated with catalyst; the electrolyte is KOH. Because the reactants are supplied continuously, a fuel cell does not "go dead." On the other hand, fuel cells do not store electrical energy as previously discussed cells do; they are merely energy-conversion devices.

19.7
standard
potentials

The emf developed by an voltaic cell depends both on the substances that make up the cell and on their concentrations. As current is drawn from a cell, cell reactants would be consumed, and their concentrations would change. There would be a corresponding change in cell voltage, until at some point the reactants are consumed, and the cell voltage is zero.

When comparing the emf's generated by different voltaic cells, it is important that the concentrations of their reactants and products are equivalent. It is common to compare the **standard potentials** of cells. These potentials are the ones generated when the reactants and products of the redox process in the cell are at standard-state conditions—gases at 1 atm pressure, solutions at 1 M concentration, and temperature at 25°C (Section 4.6). For example, the standard potential developed by the Zn-H^+ cell (Figure 19.11) is 0.76 volts (V). This is the potential generated when $[Zn^{2+}]$ is 1 M, $[H^+]$ is 1 M, and P_{H_2} is 1 atm. The standard potential can be represented as

$$E^\circ_{cell} = 0.76 \text{ V}$$

where the superscript indicates that reactants and products are at standard-state conditions.

HALF-CELL POTENTIALS

Just as an overall cell reaction can be thought of as the sum of two half-reactions, the overall cell potential can be thought of as the sum of two half-cell potentials. For reasons that will be evident shortly, it is useful to have a set of standard potentials for half-reactions. It is impossible to measure directly the potential of an isolated half-cell. However, if one half-cell reaction is arbitrarily assigned a standard half-cell potential, the standard potentials of other half-reactions can be determined relative to the reference. The hydrogen half-reaction has been chosen as the reference half-reaction. It is assigned a standard potential of exactly 0 V:

$$H_2(1 \text{ atm}) \longrightarrow 2H^+(1 M) + 2e^- \qquad E^\circ = 0 \text{ V} \qquad [19.32]$$

The sum of the potentials of the two half-reactions of any cell equals the potential of the complete reaction generated by that cell. Since $E°$ is 0.76 V for the Zn-H$^+$ cell, the standard potential for the oxidation of Zn must be 0.76 V:

$$Zn(s) + 2H^+(aq) \longrightarrow Zn^{2+}(aq) + H_2(g)$$
$$E° = 0.76 \text{ V (measured)}$$
$$-[2H^+(aq) + 2e^- \longrightarrow H_2(g)]$$
$$-E° = 0.00 \text{ V (defined)}$$
$$\overline{Zn(s) \longrightarrow Zn^{2+}(aq) + 2e^-}$$
$$E° = 0.76 \text{ V (by difference)}$$

We shall adopt the convention of labeling the standard oxidation potential of any half-reaction as $E°_{ox}$, and the standard reduction potential as $E°_{red}$. *The standard potential for any oxidation is equal in magnitude but opposite in sign to that of the reverse reduction:*

$$Zn(s) \longrightarrow Zn^{2+}(aq) + 2e^- \qquad E°_{ox} = 0.76 \text{ V} \qquad [19.33]$$

$$Zn^{2+}(aq) + 2e^- \longrightarrow Zn(s) \qquad E°_{red} = -0.76 \text{ V} \qquad [19.34]$$

SAMPLE EXERCISE 19.8

Given that $E°_{cell}$ is 1.10 V for the Zn-Cu^{2+} cell shown in Figure 19.10, and $E°_{ox}$ is 0.76 V for the oxidation of zinc

$$Zn(s) \longrightarrow Zn^{2+}(aq) + 2e^-$$

calculate $E°_{red}$ for the reduction of copper

$$Cu^{2+}(aq) + 2e^- \longrightarrow Cu(s)$$

Solution:

$$Zn(s) + Cu^{2+}(aq) \longrightarrow Zn^{2+}(aq) + Cu(s)$$
$$E°_{cell} = 1.10 \text{ V}$$
$$-[Zn(s) \longrightarrow Zn^{2+}(aq) + 2e^-]$$
$$E°_{ox} = 0.76 \text{ V}$$
$$\overline{Cu^{2+}(aq) + 2e^- \longrightarrow Cu(s)}$$
$$E°_{red} = 0.34 \text{ V}$$

The standard potentials for other half-reactions can be established from standard cell potentials in the fashion outlined above. A number of standard reduction potentials are given in Table 19.2.

USES OF HALF-CELL POTENTIALS

Half-cell potentials indicate the ease with which a species is oxidized or reduced. The more positive the $E°$ value for a half-reaction, the greater the tendency for the reaction to occur as written. A negative reduction potential merely indicates that the species is more difficult to reduce than H$^+$, whereas a negative oxidation potential indicates that the species is more difficult to oxidize than H$_2$. Examination of the half-reactions in Table 19.2 shows that F$_2$ is the most easily reduced species and therefore the strongest oxidizing agent listed:

$$F_2(g) + 2e^- \longrightarrow 2F^-(aq) \qquad E°_{red} = 2.87 \text{ V} \qquad [19.35]$$

Lithium ion, Li$^+$, is the most difficult to reduce and therefore the poorest oxidizing agent:

$$Li^+(aq) + e^- \longrightarrow Li(s) \qquad E°_{red} = -3.05 \text{ V} \qquad [19.36]$$

TABLE 19.2

standard reduction potentials in water at 25°C

REDUCTION HALF-REACTION	STANDARD POTENTIAL (V)
$F_2(g) + 2e^- \longrightarrow 2F^-(aq)$	2.87
$Co^{3+}(aq) + e^- \longrightarrow Co^{2+}(aq)$	1.82
$H_2O_2(aq) + 2H^+(aq) + 2e^- \longrightarrow 2H_2O$	1.77
$PbO(s) + 4H^+(aq) + SO_4^{2-}(aq) + 2e^- \longrightarrow PbSO_4(s) + 2H_2O$	1.70
$MnO_4^-(aq) + 8H^+(aq) + 5e^- \longrightarrow Mn^{2+}(aq) + 4H_2O$	1.51
$Au^{3+}(aq) + 3e^- \longrightarrow Au(s)$	1.50
$Cl_2(g) + 2e^- \longrightarrow 2Cl^-(aq)$	1.36
$Cr_2O_7^{2-}(aq) + 14H^+(aq) + 6e^- \longrightarrow 2Cr^{3+}(aq) + 7H_2O$	1.33
$MnO_2(s) + 4H^+(aq) + 2e^- \longrightarrow Mn^{2+}(aq) + 2H_2O$	1.23
$O_2(g) + 4H^+(aq) + 4e^- \longrightarrow 2H_2O$	1.23
$Br_2(l) + 2e^- \longrightarrow 2Br^-(aq)$	1.09
$NO_3^-(aq) + 4H^+(aq) + 3e^- \longrightarrow NO(g) + 2H_2O$	0.96
$Ag^+(aq) + e^- \longrightarrow Ag(s)$	0.80
$Fe^{3+}(aq) + e^- \longrightarrow Fe^{2+}(aq)$	0.77
$O_2(g) + 2H^+(aq) + 2e^- \longrightarrow H_2O_2(aq)$	0.68
$MnO_4^-(aq) + 2H_2O + 3e^- \longrightarrow MnO_2(s) + 4OH^-(aq)$	0.59
$I_2(s) + 2e^- \longrightarrow 2I^-(aq)$	0.53
$O_2(g) + 2H_2O + 4e^- \longrightarrow 4OH^-(aq)$	0.40
$Cu^{2+}(aq) + 2e^- \longrightarrow Cu(s)$	0.34
$AgCl(s) + e^- \longrightarrow Ag(s) + Cl^-(aq)$	0.22
$Cu^{2+}(aq) + e^- \longrightarrow Cu^+(aq)$	0.15
$Sn^{4+}(aq) + 2e^- \longrightarrow Sn^{2+}(aq)$	0.13
$2H^+(aq) + 2e^- \longrightarrow H_2(g)$	0
$Pb^{2+}(aq) + 2e^- \longrightarrow Pb(s)$	−0.13
$Sn^{2+}(aq) + 2e^- \longrightarrow Sn(s)$	−0.14
$Ni^{2+}(aq) + 2e^- \longrightarrow Ni(s)$	−0.25
$Co^{2+}(aq) + 2e^- \longrightarrow Co(s)$	−0.28
$PbSO_4(s) + 2e^- \longrightarrow Pb(s) + SO_4^{2-}(aq)$	−0.31
$Cd^{2+}(aq) + 2e^- \longrightarrow Cd(s)$	−0.40
$Fe^{2+}(aq) + 2e^- \longrightarrow Fe(s)$	−0.44
$Cr^{3+}(aq) + 3e^- \longrightarrow Cr(s)$	−0.74
$Zn^{2+}(aq) + 2e^- \longrightarrow Zn(s)$	−0.76
$2H_2O + 2e^- \longrightarrow H_2(g) + 2OH^-(aq)$	−0.83
$Mn^{2+}(aq) + 2e^- \longrightarrow Mn(s)$	−1.18
$Al^{3+}(aq) + 3e^- \longrightarrow Al(s)$	−1.66
$H_2(g) + 2e^- \longrightarrow 2H^-(aq)$	−2.25
$Mg^{2+}(aq) + 2e^- \longrightarrow Mg(s)$	−2.37
$Na^+(aq) + e^- \longrightarrow Na(s)$	−2.71
$Ca^{2+}(aq) + 2e^- \longrightarrow Ca(s)$	−2.87
$Ba^{2+}(aq) + 2e^- \longrightarrow Ba(s)$	−2.90
$K^+(aq) + e^- \longrightarrow K(s)$	−2.93
$Li^+(aq) + e^- \longrightarrow Li(s)$	−3.05

If we consider the reverse reactions, we find that lithium is the most easily oxidized and consequently the strongest reducing agent:

$$Li(s) \longrightarrow Li^+(aq) + e^- \qquad E_{ox}^\circ = 3.05 \text{ V} \qquad [19.37]$$

Fluoride ion, F^-, is the most difficult to oxidize and therefore the poorest reducing agent:

$$2F^-(aq) \longrightarrow F_2(g) + 2e^- \qquad E^\circ_{ox} = -2.87 \text{ V} \qquad [19.38]$$

The activity series of metals, Table 19.1, corresponds to a listing of metals according to their standard oxidation potentials.

SAMPLE EXERCISE 19.9

Using Table 19.2, determine which of the following species is the strongest oxidizing agent: MnO_4^- (in acid solution), $Br_2(l)$, $Ca^{2+}(aq)$.

$$MnO_4^-(aq) + 8H^+(aq) + 5e^- \longrightarrow$$
$$Mn^{2+}(aq) + 4H_2O \qquad E^\circ_{red} = \quad 1.15 \text{ V}$$
$$Br_2(l) + 2e^- \longrightarrow 2Br^-(aq) \qquad E^\circ_{red} = \quad 1.09 \text{ V}$$
$$Ca^{2+}(aq) + 2e^- \longrightarrow Ca(s) \qquad E^\circ_{red} = -2.87 \text{ V}$$

Solution: The strongest oxidizing agent will be the species that is most readily reduced. Therefore we should compare reduction potentials. From Table 19.2 we have:

Because the reduction of MnO_4^- has the highest positive potential, MnO_4^- is the strongest oxidizing agent of the three.

Standard electrode potentials can be combined to determine whether a particular redox reaction is spontaneous under standard conditions. A positive standard potential, E°, for a reaction indicates that the reaction is spontaneous and that the oxidation-reduction process will consequently supply energy.

SAMPLE EXERCISE 19.10

Determine whether the following reactions are spontaneous under standard conditions:

(a) $Cu(s) + 2H^+(aq) \longrightarrow Cu^{2+}(aq) + H_2(g)$
(b) $Cl_2(g) + 2I^-(aq) \longrightarrow 2Cl^-(aq) + I_2(s)$

Solution: (a) We utilize Table 19.2 to obtain the necessary half-reaction potentials. Because the overall reaction converts Cu to Cu^{2+}, we use the standard oxidation potential for Cu. Because the overall reaction converts H^+ to H_2 we use the reduction potential for H^+. Adding these we obtain the standard potential for the overall reaction:

$$Cu(s) \longrightarrow Cu^{2+}(aq) + 2e^-$$
$$E^\circ_{ox} = -0.34 \text{ V}$$
$$2e^- + 2H^+(aq) \longrightarrow H_2(g)$$
$$E^\circ_{red} = \quad 0 \text{ V}$$
$$\overline{Cu(s) + 2H^+(aq) \longrightarrow Cu^{2+}(aq) + H_2(g)}$$
$$E^\circ = -0.34 \text{ V}$$

Because the standard potential is negative, the reaction is not spontaneous in the direction written. Copper does not react with acids in this fashion. However, the reverse reaction is spontaneous: Cu^{2+} can be reduced by H_2.

(b) We write equations to obtain the standard potential for the overall reaction:

$$2e^- + Cl_2(g) \longrightarrow 2Cl^-(aq)$$
$$E^\circ_{red} = \quad 1.36 \text{ V}$$
$$2I^-(aq) \longrightarrow I_2(s) + 2e^-$$
$$E^\circ_{ox} = -0.54 \text{ V}$$
$$\overline{Cl_2(g) + 2I^-(aq) \longrightarrow 2Cl^-(aq) + I_2(s)}$$
$$E^\circ = \quad 0.82 \text{ V}$$

This reaction is spontaneous and could be used to build a voltaic cell. It is often used as a qualitative test for the presence of I^- in aqueous solution. The solution is treated with a solution of Cl_2. If I^- is present, I_2 forms. If CCl_4 is added, the I_2 dissolves in the CCl_4, imparting its characteristic purple color to the CCl_4.

As the preceding discussion implies, half-cell potentials can be combined to calculate the potential produced by any voltaic cell. For example, the standard potential generated by a voltaic cell that uses the oxidation of I^- by $Cr_2O_7^{2-}$ can be determined to be 0.80 V:

$$Cr_2O_7{}^{2-}(aq) + 14H^+(aq) + 6e^- \longrightarrow 2Cr^{3+}(aq) + 7H_2O$$
$$E^\circ_{red} = \quad 1.33 \text{ V}$$
$$6I^-(aq) \longrightarrow 3I_2(s) + 6e^-$$
$$E^\circ_{ox} = \; -0.53 \text{ V}$$

$$Cr_2O_7{}^{2-}(aq) + 14H^+(aq) + 6I^-(aq) \longrightarrow$$
$$2Cr^{3+}(aq) + 7H_2O + 3I_2(s) \qquad E^\circ_{cell} = \quad 0.80 \text{ V}$$

As we see, the cell potential is positive, as it must be for the reaction to serve as the basis of a voltaic cell. Notice also that even though the iodide half-reaction must be multiplied by 6 in order to obtain the balanced equation for the reaction, the half-cell potential is *not* multiplied by 6. The standard potential does not depend on the quantity of reactants and products, but only on their concentration. Thus it doesn't matter if there are 6 moles of I$^-$ or 1 mole as long as the concentration is 1 M.

Standard potentials are also useful in dealing with electrolytic cells. In these cells, the overall cell potential is negative because the reactions involved are not spontaneous ones. The standard potential calculated for an electrolytic cell indicates the minimum voltage required for electrolysis under standard conditions.

SAMPLE EXERCISE 19.11

Calculate the minimum voltage required to cause the electrolysis of a solution of $Cu(NO_3)_2$ under standard conditions:

$$2Cu^{2+}(aq) + 2H_2O \longrightarrow$$
$$2Cu(s) + O_2(g) + 4H^+(aq)$$

Solution: The given reaction involves the reduction of Cu^{2+} and the oxidation of water. From Table 19.2, we obtain E°_{red} for Cu^{2+} (0.34 V) and for H$_2$O

(1.23 V). The standard oxidation potential for H$_2$O is therefore -1.23 V. Using these E° values we have:

$$2Cu^{2+}(aq) + 4e^- \longrightarrow 2Cu(s)$$
$$E^\circ_{red} = \quad 0.34 \text{ V}$$
$$2H_2O \longrightarrow 4e^- + O_2(g) + 4H^+(aq)$$
$$E^\circ_{ox} = \; -1.23 \text{ V}$$

$$2Cu^{2+}(aq) + 2H_2O \longrightarrow$$
$$2Cu(s) + O_2(g) + 4H^+(aq)$$
$$E^\circ_{cell} = \; -0.89 \text{ V}$$

Thus a minimum of 0.89 volts is required.

19.8
electrode potentials, free-energy changes, and equilibrium constants

We've seen that the free-energy change, ΔG, accompanying a chemical process is a measure of its spontaneity (Chapter 15). Because the cell potential indicates whether a redox reaction is spontaneous, we might expect some relationship to exist between the cell potential and the free-energy change. Indeed, this is the case; the cell potential, E, and the free-energy change, ΔG, are related by Equation [19.39]

$$\Delta G = -n\mathfrak{F}E \qquad\qquad [19.39]$$

where n is the number of moles of electrons transferred in the reaction and $\mathfrak{F}$ is Faraday's constant. For the situation in which reactants and products are in their standard state concentrations, we can write Equation [19.40]:

19.8

electrode potentials,
free-energy changes,
and equilibrium constants

electrochemistry

$$\Delta G^{\circ} = -n \mathfrak{F} E^{\circ} \qquad [19.40]$$

In using Equations [19.39] and [19.40], the most convenient unit for Faraday's constant, $\mathfrak{F}$, is the joule/volt-mole e^-:

$$\mathfrak{F} = 96{,}500 \, \frac{\text{coul}}{\text{mole } e^-} = 96{,}500 \, \frac{\text{J}}{\text{V-mole } e^-} \qquad [19.41]$$

This unit follows from the fact that a coulomb is a joule/volt, Equation [19.22].

SAMPLE EXERCISE 19.12

Use the standard electrode potentials given in Table 19.2 to calculate the standard free-energy change, ΔG°, for the following reaction:

$$2Br^-(aq) + F_2(g) \longrightarrow Br_2(l) + 2F^-(aq)$$

Solution:

$$2Br^-(aq) \longrightarrow Br_2(l) + 2e^-$$
$$E_{ox}^{\circ} = -1.09 \text{ V}$$
$$F_2(g) + 2e^- \longrightarrow 2F^-(aq)$$
$$E_{red}^{\circ} = \quad 2.87 \text{ V}$$
$$\overline{2Br^-(aq) + F_2(g) \longrightarrow Br_2(l) + F^-(aq)}$$
$$E^{\circ} = \quad 1.78 \text{ V}$$

Notice that two electrons are transferred in the reaction so $n = 2$.

$$\Delta G^{\circ} = -n \mathfrak{F} E^{\circ}$$
$$= -(2 \text{ moles } e^-)\left(96{,}500 \, \frac{\text{J}}{\text{V-mole } e^-}\right)(1.78 \text{ V})$$
$$= -3.44 \times 10^5 \text{ J} = -344 \text{ kJ}$$

Notice also that whereas a negative sign for ΔG° indicates spontaneity, a positive sign for E° indicates the same.

The free energy change, ΔG, accompanying a redox reaction is equal to the maximum amount of electrical work, w_{max}, that the reaction system can do on its surroundings. We can write this relation as follows:

$$\Delta G = -w_{max} \qquad [19.42]$$

The negative sign is used because the free energy of the system decreases as the system performs the work. The amount of electrical work performed by a redox reaction is the product of the electrical charge, Q, and the emf, E:

$$w_{max} = QE \qquad [19.43]$$

If n moles of electrons are moved, the total charge is $n\mathfrak{F}$, because $\mathfrak{F}$ is the charge on 1 mole of electrons. Therefore

$$w_{max} = n\mathfrak{F}E \qquad [19.44]$$

Equation [19.39] is obtained by substituting Equation [19.44] into equation [19.42]:

$$\Delta G = -n\mathfrak{F}E$$

It was pointed out in Section 15.7 that the standard free-energy change and the equilibrium constant for a process are related by the following equation:

$$\Delta G^{\circ} = -2.30 \, RT \log K \qquad [19.45]$$

R is the gas law constant, 8.317 J/K-mole, and T is the temperature in K. Combining this equation with Equation [19.40], we obtain Equation [19.46]:

$$\Delta G^{\circ} = -n\mathfrak{F}E^{\circ} = -2.30 \, RT \log K \qquad [19.46]$$

Rearranging this equation gives the relationship between equilibrium constant and standard electrode potential:

$$E^\circ = \frac{2.30\,RT}{n\mathfrak{F}}\log K \qquad\qquad [19.47]$$

Substituting the proper values for $\mathfrak{F}$ (96,500 J/V-mole e^-) and R (8.317 J/K-mole) and using $T = 298$ K gives Equation [19.48]:

$$E^\circ = \frac{0.0590\text{ V}}{n}\log K \qquad\qquad [19.48]$$

SAMPLE EXERCISE 19.13

Calculate the equilibrium constant at 25°C for the reaction

$$O_2(g) + 4H^+(aq) + 4Fe^{2+}(aq) \longrightarrow$$
$$4Fe^{3+}(aq) + 2H_2O$$

Solution:

$$O_2(g) + 4H^+(aq) + 4e^- \longrightarrow 2H_2O$$
$$E^\circ_{red} = \quad 1.23\text{ V}$$
$$4Fe^{2+}(aq) \longrightarrow 4Fe^{3+}(aq) + 4e^-$$
$$E^\circ_{ox} = -0.77\text{ V}$$

$$\overline{O_2(g) + 4H^+(aq) + 4Fe^{2+}(aq) \longrightarrow}$$
$$4Fe^{3+}(aq) + 2H_2O$$
$$E^\circ = 0.46\text{ V}$$

Using Equation [19.48] we have

$$\log K = \frac{nE^\circ}{0.0590\text{ V}}$$
$$= \frac{4(0.46\text{ V})}{0.0590\text{ V}}$$
$$= 31.2$$
$$K = 2 \times 10^{31}$$

Thus Fe^{2+} ions are stable in acidic solutions only in the absence of O_2 (unless a suitable reducing agent is present).

19.9
concentration and electrode potential

In practice, voltaic cells or electrolytic cells are unlikely to be operated under standard-state conditions. The relationship between electrode potentials and the concentrations of substances involved in half-reactions of the type

$$a\text{A} + b\text{B} + ne^- \longrightarrow c\text{C} + d\text{D}$$

is given by the Nernst equation, Equation [19.49],

$$E = E^\circ - \frac{2.30\,RT}{n\mathfrak{F}}\log\frac{[\text{C}]^c[\text{D}]^d}{[\text{A}]^a[\text{B}]^b} \qquad\qquad [19.49]$$

where the bracketed quantities are concentrations. Substituting T equals 298 K, together with the values of R and $\mathfrak{F}$, gives Equation [19.50]:*

$$E = E^\circ - \frac{0.0590}{n}\log\frac{[\text{C}]^c[\text{D}]^d}{[\text{A}]^a[\text{B}]^b} \qquad\qquad [19.50]$$

*Since $E = 0$ at equilibrium, Equations [19.49] and [19.50] give Equations [19.47] and [19.48].

You can see from the Nernst equation that if the concentration of a reactant is increased, the electrode becomes more positive (as in Sample Exercise 19.14). If the concentration of a product is increased, the electrode potential becomes less positive.

SAMPLE EXERCISE 19.14

Calculate the half-cell potential, measured at 25°C, for the half-reaction

$$Cu^{2+}(aq) + 2e^- \longrightarrow Cu(s)$$

if $[Cu^{2+}]$ is 5.0 M.

Solution: Using Equation [19.50] we have

$$E = 0.34 \text{ V} - \frac{0.0590 \text{ V}}{2} \log \frac{1}{[Cu^{2+}]}$$

$$= 0.34 \text{ V} - \frac{0.0590 \text{ V}}{2} \log \left(\frac{1}{5.0}\right)$$

$$= 0.34 \text{ V} - (0.0295)(-0.70)$$

$$= 0.34 \text{ V} + 0.02 \text{ V}$$

$$= 0.36 \text{ V}$$

CONCENTRATION CELLS

Because the emf of a half-cell depends on concentration, it is possible to construct a voltaic cell in which the same substances are present in both anode and cathode compartments, but at different concentrations. For example, the cathode compartment could consist of a copper electrode and a 10 M solution of $Cu(NO_3)_2$. For the reduction of a 10 M solution of Cu^{2+} at 25°C, the electrode potential is 0.37 V. The anode compartment could consist of a copper electrode and a 0.01 M solution of $Cu(NO_3)_2$. For the oxidation of a 0.01 M solution of Cu^{2+} at 25°C, the electrode potential is -0.28 V. The resultant voltaic cell, illustrated in Figure 19.17, generates an emf of 0.09 V:

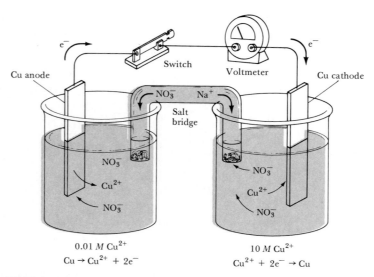

FIGURE 19.17 A concentration cell.

$$\begin{array}{ll}
\text{Cathode: } Cu^{2+}(10\ M) + 2e^- \longrightarrow Cu(s) & E_{red} = \quad 0.37\ V \\
\underline{\text{Anode: } Cu(s) \longrightarrow Cu^{2+}(0.01\ M) + 2e^-} & \underline{E_{ox} = -0.28\ V} \\
Cu^{2+}(10\ M) \longrightarrow Cu^{2+}(0.01\ M) & E_{cell} = \quad 0.09\ V
\end{array}$$

Such a cell, based on concentration differences at the two electrodes, is known as a **concentration cell.**

Certain electrodes are especially sensitive to changes in concentration. For example, the pH meter is based on a voltaic cell whose potential depends on the concentration of H^+ ion (Section 16.7).

19.10 corrosion

Before we close our discussion of electrochemistry, let's apply some of what we have learned to a very important problem, the **corrosion** of metals. Corrosion reactions are redox reactions in which a metal is attacked by some substance in its environment and converted to an unwanted compound.

One of the most familiar corrosion processes is the rusting of iron. From an economic standpoint this is a significant process. It is estimated that up to 20 percent of the iron produced annually in this country is used to replace iron objects that have been discarded because of rust damage.

The rusting of iron is known to involve oxygen; iron does not rust in water unless O_2 is present. Rusting also involves water; iron does not rust in oil, even if it contains O_2, unless H_2O is also present. Other factors such as the pH of the solution, the presence of salts, contact with metals more difficult to oxidize than iron, and stress on the iron can accelerate rusting.

The corrosion of iron is generally believed to be electrochemical in nature. A region on the surface of the iron serves as an anode at which the iron undergoes oxidation:

$$Fe(s) \longrightarrow Fe^{2+}(aq) + 2e^- \qquad E_{ox}^\circ = 0.44\ V \qquad [19.51]$$

The electrons so produced migrate through the metal to another portion of the surface that serves as the cathode. Here oxygen can be reduced:

$$O_2(g) + 4H^+(aq) + 4e^- \longrightarrow 2H_2O \qquad E_{red}^\circ = 1.23\ V \qquad [19.52]$$

Notice that H^+ is involved in the reduction of O_2. As the concentration of H^+ is lowered (that is, as pH is increased), the reduction of O_2 becomes less favorable. It is observed that iron in contact with a solution whose pH is above 9–10 does not corrode. In the course of the corrosion, the Fe^{2+} formed at the anode is further oxidized to Fe^{3+}. The Fe^{3+} forms the hydrated iron(III) oxide known as rust:*

---

*Frequently, metal compounds that are obtained from aqueous solution have water associated with them. For example, copper(II) sulfate crystallizes from water with 5 moles of water per mole of $CuSO_4$. We represent this formula as $CuSO_4 \cdot 5H_2O$. Such compounds are called hydrates. Rust is a hydrate of iron(III) oxide with a variable amount of water of hydration. We represent the variable water content by writing the formula as $Fe_2O_3 \cdot xH_2O$.

$$4Fe^{2+}(aq) + O_2(g) + 4H_2O + 2xH_2O \longrightarrow$$
$$2Fe_2O_3 \cdot xH_2O(s) + 8H^+(aq) \qquad [19.53]$$

Because the cathode is generally the area having the largest supply of O_2, the rust often deposits there. If you look closely at a shovel after it has stood outside in the moist air with wet dirt adhered to its blade, you may notice that pitting has occurred under the dirt but that rust has formed elsewhere, where O_2 is more readily available. The corrosion process is summarized by way of illustration in Figure 19.18.

The enhanced corrosion caused by the presence of salts is usually quite evident on autos in areas where there is heavy salting of roads during winter. The effect of salts is readily explained by the voltaic mechanism: The ions of a salt provide the electrolyte necessary for completion of the electrical circuit.

The presence of anodic and cathodic sites on the iron requires two different chemical environments on the surface. These can occur through the presence of impurities or lattice defects (perhaps introduced by strain on the metal). At the sites of such impurities or defects the atomic-level environment around the iron atom may permit the metal to be either more or less easily oxidized than at normal lattice sites. Thus these sites may serve as either anodes or cathodes. Ultrapure iron, prepared in such a way as to minimize lattice defects, is far less susceptible to corrosion than is ordinary iron.

Iron is often covered with a coat of paint or another metal such as tin, zinc, or chromium to protect its surface against corrosion. "Tin cans" are produced by applying a thin layer of tin over steel. The tin protects the iron only as long as the protective layer remains intact. Once it is broken and the iron exposed to air and water, tin actually promotes the corrosion of the iron. It does so by serving as the cathode in the electrochemical corrosion. As shown by the following half-cell potentials, iron is more readily oxidized than tin:

$$Fe(s) \longrightarrow Fe^{2+}(aq) + 2e^- \qquad E^\circ_{ox} = 0.44 \text{ V} \qquad [19.54]$$

$$Sn(s) \longrightarrow Sn^{2+}(aq) + 2e^- \qquad E^\circ_{ox} = 0.14 \text{ V} \qquad [19.55]$$

The iron therefore serves as the anode and is oxidized as shown in Figure 19.19.

"Galvanized iron" is produced by coating iron with a thin layer of

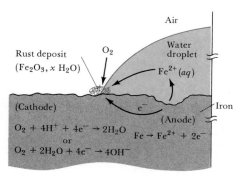

FIGURE 19.18 Corrosion of iron in contact with water.

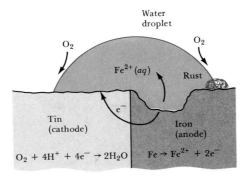

FIGURE 19.19 Corrosion of iron in contact with tin.

$$O_2 + 4H^+ + 4e^- \rightarrow 2H_2O \qquad Fe \rightarrow Fe^{2+} + 2e^-$$

zinc. The zinc protects the iron against corrosion even after the surface coat is broken. In this case the iron serves as the cathode in the electrochemical corrosion because zinc is oxidized more easily than iron:

$$\text{Zn}(s) \longrightarrow \text{Zn}^{2+}(aq) + 2e^- \qquad E^{\circ}_{\text{ox}} = 0.76 \text{ V} \qquad [19.56]$$

The zinc therefore serves as the anode and is corroded instead of the iron, as shown in Figure 19.20. Such protection of a metal by making it the cathode in an electrochemical cell is known as **cathodic protection.** Underground pipelines are often protected against corrosion by making the pipeline the cathode of a voltaic cell. Pieces of an active metal such as magnesium are buried along the pipeline and connected to it by wire as shown in Figure 19.21. In moist soil, where corrosion can occur, the active metal serves as the anode and the pipe experiences cathodic protection.

Although our discussions have centered on iron, this is not the only metal subject to corrosion. One thing that may be surprising in light of our discussions is the fact that an aluminum can, disposed of carelessly in the environment, will last so much longer than will a steel can. On the basis of the standard oxidation potentials of aluminum ($E^{\circ}_{\text{ox}} = 1.66 \text{ V}$) and iron ($E^{\circ}_{\text{ox}} = 0.44 \text{ V}$), we would expect the aluminum to be much more readily corroded. The slow corrosion of aluminum is explained by the formation of a thin, compact oxide coating that forms on its surface. This protects the underlying metal from further corrosion. Magnesium, which also has a large oxidation potential, is similarly protected. The oxide coat on iron is too porous to offer similar protection. However,

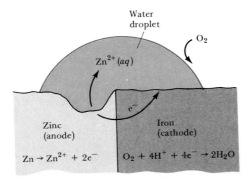

FIGURE 19.20 Cathodic protection of iron in contact with zinc.

$$\text{Zn} \rightarrow \text{Zn}^{2+} + 2e^- \qquad O_2 + 4H^+ + 4e^- \rightarrow 2H_2O$$

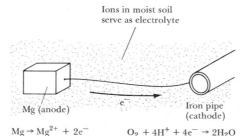

Ions in moist soil
serve as electrolyte

Mg (anode)

e^-

Iron pipe
(cathode)

$Mg \rightarrow Mg^{2+} + 2e^-$ $O_2 + 4H^+ + 4e^- \rightarrow 2H_2O$

FIGURE 19.21   Cathodic protection of a buried object such as an iron pipe.

when iron is alloyed with chromium, a protective oxide coating does form. Such alloys are called stainless steels.

SAMPLE EXERCISE 19.15

Predict the nature of the corrosion that would take place if an iron gutter were nailed to a house using aluminum nails.

Solution: A voltaic cell can be formed at the point of contact of the two metals. The metal that is most easily oxidized will serve as the anode, whereas the other metal serves as the cathode. By comparing

standard oxidation potentials of Al and Fe we see that Al will be the anode:

$$Al(s) \longrightarrow Al^{3+}(aq) + 3e^- \qquad E^\circ_{ox} = 1.66 \text{ V}$$
$$Fe(s) \longrightarrow Fe^{2+}(aq) + 2e^- \qquad E^\circ_{ox} = 0.44 \text{ V}$$

The gutter will thus be protected against corrosion in the vicinity of the nail, because the iron serves as the cathode. However, the nail would quickly corrode, leaving the gutter on the ground.

What do you think would happen if aluminum siding were nailed to a house using iron nails?

FOR REVIEW

summary

We have seen that oxidation-reduction (or redox) reactions can be considered to consist of two half-reactions, one due to **oxidation** (loss of electrons) and the other due to **reduction** (gain of electrons). The substance oxidized is called the **reducing agent,** or **reductant,** whereas the substance reduced is called the **oxidizing agent,** or **oxidant.** If the half-reactions are balanced, they can be added to obtain the balanced redox equation. The number of electrons lost by the reducing agent must equal the number gained by the oxidizing agent. Consequently, different types of atoms and total charges must each balance.

Spontaneous redox reactions can be used to generate electricity in **voltaic cells.** Conversely, electricity can be used to bring about nonspontaneous reactions in **electrolytic cells.** In the latter type of cell, electrons are "pumped" from substances having greater electron attraction to those that have less attraction. In either type of cell, the electrode at which oxidation occurs is called the **anode,** whereas the electrode where reduction occurs is called the **cathode.** The "electron pressure" moving electrons from anode to cathode is known as the **electromotive force (emf), voltage,** or **cell potential.** The amount of electrical charge possessed by a mole of electrons is known as a **faraday** (abbreviated $\mathcal{F}$), 96,500 coulombs. A **coulomb** is an **amp-sec.** The maximum work that a voltaic cell can perform is the product of the total charge it delivers, $n\mathcal{F}$, and its voltage, E: $\omega_{max} = n\mathcal{F}E$.

Electrode potentials can be assigned to half-reactions by defining the **hydrogen electrode,**

$$2H^+(1\ M) + 2e^- \longrightarrow H_2(1\ atm)$$

as having an electrode potential of exactly zero volts. The more positive the electrode potential, the greater the tendency of the reaction to occur as written. Half-reaction electrode potentials can be combined to obtain the voltage of the total redox reaction. This is useful, for example, in determining the voltages generated by voltaic cells, in determining the minimum voltages required in electrolytic cells, and in predicting whether certain redox reactions are spontaneous (positive E). Electrode potentials vary in magnitude with the concentrations of reactants and products. **Standard electrode potentials**, $E°$, are those measured under standard-state conditions (solutions at 1 M concentration and gases at 1 atm pressure). The **Nernst equation** relates electrode potentials to concentration:

$$E = E° - \frac{2.30\ RT}{n} \log \frac{[C]^c[D]^d}{[A]^a[B]^b}$$

At equilibrium

$$E = 0 \text{ and } E° = \frac{2.30\ RT}{n} \log K$$

Thus standard electrode potentials are related to equilibrium constants. Electrode potentials are also related to free-energy changes:

$$\Delta G = -n\mathcal{F}\,E$$

Our knowledge of electrochemistry allows us to design batteries and to bring about desirable redox reactions such as those involved in electroplating and in the reduction and refining of metals. Electrochemical principles also help us to understand and combat corrosion.

learning goals

Having read and studied this chapter, you should be able to:

1. Recognize redox reactions and identify the reductant and oxidant.
2. Complete and balance redox equations using the method of half-reactions.
3. Calculate the third quantity, having been given any two of the following quantities: time, current, amount of substance produced or consumed in a redox reaction.
4. Calculate the maximum work performed by an electrochemical cell, having been given emf and electrical charge (or information from which they can be determined).
5. Calculate the emf generated by a voltaic cell or the minimum emf required to cause an electrolytic cell reaction to proceed, having been given half-reaction electrode potentials.
6. Use both activity series and $E°$ values to predict whether a reaction will be spontaneous.
7. Interconvert $E°$, $\Delta G°$ and K for redox reactions.
8. Use the Nernst equation to calculate electrode potentials.
9. Diagram simple voltaic and electrolytic cells, labeling anode, cathode, and directions of ion and electron movements and giving the half-reaction occurring at each electrode.

10. Describe some widely used batteries.

11. Describe the phenomena of corrosion and cathodic protection.

key terms

Among the more important terms and expressions used for the first time in this chapter are the following:

An **anode** (Section 19.3) is an electrode at which oxidation occurs.

Cathodic protection (Section 19.10) is a means of protecting a metal against corrosion by making it the cathode in a voltaic cell. This can be done by attaching a more active metal.

A **cathode** (Section 19.3) is an electrode at which reduction occurs.

Corrosion (Section 19.10) is the process by which a metal is oxidized by substances in its environment.

A **faraday** (Section 19.4) is the total charge of a mole of electrons, 96,500 coul.

A **fuel cell** (Section 19.6) is a voltaic cell in which the substance oxidized is a combustible fuel such as H_2 or CH_4.

A **half-reaction** (Section 19.1) is an equation for either oxidation or reduction that explicitly shows the electrons involved (for example, $2H^+(aq) + 2e^- \longrightarrow H_2(g)$).

An **oxidant**, or **oxidizing agent** (Section 19.1), is a substance that is reduced and thereby causes the oxidation of some other substance.

A **reductant**, or **reducing agent** (Section 19.1), is a substance that is oxidized and thereby causes the reduction of some other substance.

A **standard electrode potential**, $E°$ (Section 19.7) is the emf of a half-reaction with all solution species at $1\,M$ concentration and all gaseous species at 1 atm, measured relative to the hydrogen electrode, for which $E°$ is exactly 0 V.

EXERCISES

redox reactions

19.1 Complete and balance the following half-reactions, indicating in each case whether oxidation or reduction is occurring:

(a) $SO_4^{2-} \longrightarrow SO_2$ (acid)

(b) $Cl^- \longrightarrow Cl_2$ (acid)

(c) $O_2 \longrightarrow OH^-$ (base)

(d) $ClO^- \longrightarrow Cl^-$ (base)

19.2 Balance the following equations, identifying in each case the oxidizing agent and the reducing agent:

(a) $H_2S + H_2O_2 \longrightarrow S + H_2O$

(b) $C + HNO_3 \longrightarrow NO_2 + H_2O + CO_2$

(c) $H_2O + PbO_2 + Cl^- + OH^- \longrightarrow ClO^- + Pb(OH)_3^-$

(d) $MnO_4^- + S^{2-} + H_2O \longrightarrow MnO_2 + S + OH^-$

19.3 Complete and balance the following equations:

(a) $Cr_2O_7^{2-} + CH_3OH \longrightarrow HCO_2H + Cr^{3+}$ (acid)

(b) $MnO_4^- + Cl^- \longrightarrow Mn^{2+} + Cl_2$ (acid)

(c) $MnO_4^- + I^- \longrightarrow MnO_2 + I_2$ (base)

(d) $H_2O_2 + ClO_2 \longrightarrow ClO_2^- + O_2$ (base)

(e) $NO_2^- + Cr_2O_7^{2-} \longrightarrow Cr^{3+} + NO_3^-$ (acid)

(f) $Cr(OH)_3 + IO_3^- \longrightarrow I^- + CrO_4^{2-}$ (base)

19.4 Hydrogen peroxide, a chemical widely used as a bleach, can act as either an oxidizing or reducing agent. Complete and balance the following equations, indicating how hydrogen peroxide is functioning in each case:

(a) $MnO_4^-(aq) + H_2O_2(aq) \longrightarrow Mn^{2+}(aq) + O_2(g)$ (acid)

(b) $I^-(aq) + H_2O_2(aq) \longrightarrow I_2(s)$ (acid)

19.5 Acid mine drainage, a type of water pollution, results from air-oxidation of pyrite, FeS_2. The oxidized products are carried into streams. Complete and balance the equation for the air-oxidation of pyrite:

$$FeS_2(s) + O_2(g) \longrightarrow$$
$$Fe^{2+}(aq) + SO_4{}^{2-}(aq)$$

electrolytic cells and the faraday

19.6 Why are different products obtained when molten $AlCl_3$ and aqueous solutions of $AlCl_3$ are electrolyzed? Predict the products in each case.

19.7 Sketch a cell for the electrolysis of a $CuSO_4$ solution using copper electrodes (the sulfate ion is not oxidized or reduced). Indicate the directions in which ions and electrons move. Give the electrode reactions and label the anode and cathode.

19.8 Sketch a cell for the electrolysis of $ZnCl_2$ solution using inert electrodes. Indicate the directions in which ions and electrons move. Give the electrode reactions and label the anode and cathode.

19.9 How many faradays of electricity are required to perform the following reductions: (a) 1 mole Fe^{2+} to Fe; (b) 1 g Sn^{4+} to Sn^{2+}; (c) 250 ml 0.10 M $Cu(NO_3)_2$ to Cu; (d) 230 ml of 0.10 N $AgNO_3$ to Ag?

19.10 Calculate Faraday's constant, $\mathfrak{F}$, given the charge on a single electron, 1.6021×10^{-19} coul.

19.11 Consider an electrolysis that results in the production of 1.00 g of copper metal from $CuSO_4$. (a) How many faradays of electricity are required? (b) How many coulombs of electricity are required? (c) If the reduction process takes 1.00 hr, how many amps are required? (d) If the reduction process uses 2.00 amp, how much time is required? (e) How much energy is required to reduce 1.00 g of copper if the cell is 100 percent efficient and requires an emf of 1.0 V? If it is only 75 percent efficient? (f) If the cell were allowed to operate for 1.00 hr at 2.00 amp, how much copper would be reduced?

19.12 Why are more faradays of electricity required to produce a gram of aluminum than a gram of copper?

19.13 How much time is required to plate a 0.010-mm-thick layer of chromium on an object whose area is 200 cm^2 if the current is 1.00 amp and the density of the chromium is 7.1 g/cm^3?

19.14 Sodium peroxyborate, $NaBO_3$, is the bleaching agent in Borateem. It is prepared by electrolytic oxidation of borax, $Na_2B_4O_7$:

$$Na_2B_4O_7 + 10NaOH \longrightarrow$$
$$4NaBO_3 + 5H_2O + 8Na^+ + 8e^-$$

How much $NaBO_3$ is prepared in 24.0 hr if the current is 20.0 amp?

19.15 How many grams of zinc are consumed in the Leclanché dry cell if it operates at a current of 0.30 amp for 1.0 hr?

19.16 The same quantity of electricity that caused 10.0 g of silver to plate out from a solution of $AgNO_3$ is passed through a solution of $CuSO_4$. How much copper will plate out?

19.17 A typical lead storage battery is rated at 80 amp-hr. How much lead is converted to $PbSO_4$ at the anode if the battery delivers 80 amp-hr of electricity? (One disadvantage of the lead storage cell, as this calculation should suggest, is its weight.)

19.18 How much energy is generated by complete discharge of a 12-V lead storage battery rated at 100 amp-hr?

[19.19] Edison's invention of the light bulb and its public demonstration in December 1879 generated considerable demand for the distribution of electricity to homes. One problem was how to measure the amount of electricity consumed by each household. Edison invented a coulometer (described in the *Journal of Chemical Education,* vol. 49, p. 627, 1972) that could be used with an AC current. Zinc plated out at the cathode of the coulometer. Every month the cathode was removed and weighed to determine the quantity of electricity used. If the cathode increased in mass by 1.8 g and the coulometer drew 0.5 percent of the current entering the home, how many coulombs of electricity were used in that month?

voltaic cells, electrode potentials, and reaction spontaneity

19.20 Sketch a voltaic cell based on the following reaction:

$$Fe(s) + Cd^{2+}(aq) \longrightarrow Fe^{2+}(aq) + Cd(s)$$

Label the anode and cathode, indicate the directions of ion and electron movements, and calculate the voltage generated by the cell under standard conditions.

19.21 Four metals, labeled A, B, C, and D, react with each other and with acids in the following way: B displaces only C from solution. Only A and D displace hydrogen from 1 M HCl. None of the metals will displace D from solution. Arrange the four metals in an activity series together with hydrogen.

19.22 Select (a) the strongest oxidizing agent, (b) the strongest reducing agent, (c) the weakest oxidizing agent, and (d) the weakest reducing agents from among the substances involved in the following half-reactions:

(a) $Br_2 + 2e^- \longrightarrow 2Br^-$

(b) $O_2 + 2H^+ + 2e^- \longrightarrow H_2O_2$

(c) $Sn^{2+} + 2e^- \longrightarrow Sn$

19.23 Arrange the following species in order of increasing strength as reducing agents: (a) F^-; (b) Al; (c) Zn; (d) Cu; (e) I^-

19.24 Why don't magnesium and aluminum metals occur naturally in the earth's crust? Why are certain other metals such as silver, gold, and platinum naturally occurring?

19.25 Predict whether the following reactions will be spontaneous under standard conditions:

(a) $Sn(s) + Fe^{2+}(aq) \longrightarrow Sn^{2+}(aq) + Fe(s)$

(b) $Sn^{2+}(aq) + 2Fe^{3+}(aq) \longrightarrow$
    $Sn^{4+}(aq) + 2Fe^{2+}(aq)$

(c) $Cu(s) + Cl_2(g) \longrightarrow Cu^{2+}(aq) + 2Cl^-(aq)$

19.26 Predict the products (if any) that will form in the following cases:

(a) $Mg^{2+} + Ag^+ \longrightarrow$

(b) $Zn + Cl_2 \longrightarrow$

(c) $Ag + Br_2 \longrightarrow$

(d) $Ni + Zn^{2+} \longrightarrow$

19.27 Consider a voltaic cell constructed with a zinc electrode in contact with $1\ M\ Zn(NO_3)_2$ and a nickel electrode in contact with $1\ M\ Ni(NO_3)_2$. Sketch the cell, labeling the anode and cathode and indicating the directions of ion and electron movements. Calculate the emf generated by the cell.

19.28 Several batteries are presently under development that use lithium as an anode material. One advantage of such cells is that they are light and yet generate a high energy. What voltage is generated by a $Li-Cl_2$ cell operating under standard conditions?

[19.29] A family owns an antique set of silverware that has a fine, dark coating of Ag_2S in the crevices of the pattern that adds to its beauty. The set is placed in a galvanized container together with soap and water in order to be cleaned. The Ag_2S disappears as the set sits in the container, leaving the silverware with the appearance of a new rather than an antique set. Explain the electrochemical processes that have occurred.

19.30 Why will a discharged lead storage battery freeze more readily in very cold weather than a fully charged one?

19.31 Bumping and shaking knocks $PbSO_4$ from the electrodes of a lead storage battery. How does this shorten the battery's life?

19.32 What would happen if HNO_3 were substi-

tuted for H_2SO_4 in the lead storage battery? What if Na_2SO_4 were substituted for H_2SO_4?

relationships between $\Delta G°$, K, and $E°$; the Nernst equation

19.33 Predict the effect of an increasing concentration of iodide ion on the emf generated by a voltaic cell utilizing the following reaction:

$$Cl_2(g) + 2I^-(aq) \longrightarrow 2Cl^-(aq) + I_2(s)$$

What is the effect of increasing the concentration of chloride ion?

19.34 Explain why the voltages of batteries are lower in cold weather than in warm weather.

19.35 What happens to the oxidizing ability of natural waters when (a) O_2 is depleted; (b) the solutions become basic.

19.36 Indicate the voltage generated by a hydrogen concentration cell utilizing the following concentrations:

$$H_2(5\ atm) + 2H^+(5\ M) \longrightarrow$$
$$H_2(0.1\ atm) + 2H^+(0.1\ M)$$

19.37 A $Zn-Cu^{2+}$ cell is operated with $[Cu^{2+}] = 5\ M$ and $[Zn^{2+}] = 0.1\ M$. Calculate the emf generated by the cell.

19.38 Calculate the equilibrium constant and free-energy change for the discharge of a lead storage cell operated under standard conditions.

[19.39] What is the emf generated by the cell shown in Figure 19.11 if $[Zn^{2+}] = 1\ M$ and $[H^+] = 0.1\ M$; if $[Zn^{2+}] = 1\ M$ and $[H^+] = 0.01\ M$? Using the Nernst equation, write a general expression for the relationship between pH and emf for this cell. (This relationship should give insight into how the pH electrodes of a pH meter, such as that discussed in Section 16.7, operate in the measurement of pH.)

19.40 Using the electrode reactions associated with the H_2-O_2 fuel cell, calculate $\Delta G°$ and the equilibrium constant for the combustion of H_2.

19.41 Calculate $\Delta G°$ and K for the reaction

$$2Na(s) + 2H_2O \longrightarrow 2NaOH(aq) + H_2(g)$$

corrosion

19.42 How does painting an iron object impede its corrosion?

19.43 The zinc on galvanized iron has been called a "sacrificial anode." What does this mean? Does chromium act similarly on iron objects that are plated with chromium?

19.44 Chrome-plating actually relies on an undercoating of nickel for the protection of iron. The chromium layer keeps the nickel from tarnishing and gives a hard, bright surface. Does the nickel protect iron by cathodic protection?

19.45 Amines are compounds related to ammonia, NH_3. One of their characteristics is their ability to function as Bronsted bases (Section 16.2). Suggest how they protect against corrosion when added to antifreeze as corrosion inhibitors.

19.46 Describe the nature of the corrosion that can occur if a steel washer is used in contact with a piece of magnesium.

19.47 Describe the nature of the corrosion that can occur if a copper pipe is fitted under the soil to a galvanized steel pipe.

19.48 Once the protective coat of paint on an iron object is scratched to expose even a microscopic area of bare metal, the metal under the paint can corrode. Explain this behavior.

general exercises

19.49 Distinguish between the following terms or symbols: (a) anode and cathode; (b) electrolytic and voltaic cells; (c) oxidation and reduction; (d) oxidation and oxidizing agent; (e) electrical charge and electrical current; (f) faraday and coulomb; (g) E and $E°$.

19.50 Which of the following reactions could be used as the basis for building a voltaic cell?

(a) $Cu(s) + Ni^{2+}(aq) \longrightarrow Ni(s) + Cu^{2+}(aq)$

(b) $MnO_4^-(aq) + 8H^+(aq) + 5Fe^{2+}(aq) \longrightarrow$
$Mn^{2+}(aq) + 4H_2O + 5Fe^{3+}(aq)$

(c) $2Cl^-(aq) + Br_2(l) \longrightarrow$
$2Br^-(aq) + Cl_2(g)$

(d) $Cl_2(g) + 2I^-(aq) \longrightarrow$
$I_2(s) + 2Cl^-(aq)$

(e) $2H^+(aq) + 2Fe^{2+}(aq) \longrightarrow$
$H_2(g) + 2Fe^{3+}(g)$

19.51 Calculate the standard free-energy change, $\Delta G°$, and the equilibrium constant, K, at 25°C for each of the reactions in question 19.50.

19.52 A Zn-Cu^{2+} cell (like that shown in Figure 19.10) is operated with $[Zn^{2+}] = 1.0\ M$. The cell generates a voltage of 0.95 V. What is the concentration of Cu^{2+}?

19.53 A Zn-H^+ cell (like that shown in Figure 19.11) has an emf of 0.65 V. What is the pH of the solution in the hydrogen compartment if $[H_2] = 1.0$ atm and $[Zn^{2+}] = 1.0\ M$?

[19.54] Several years ago, a unique proposal was made to raise the *Titanic*. The plan involved placing pontoons within the ship using a surface-controlled submarine-type vessel. The pontoons would contain cathodes and would be filled with hydrogen gas formed by the electrolysis of water. It has recently been estimated that it would require about 7×10^8 moles of H_2 to provide the bouyancy to lift the ship (*Journal of Chemical Education*, vol. 50, p. 61, 1973). (a) How many coulombs of electrical charge would be required? (b) What is the minimum voltage required to generate H_2 and O_2 if the pressure on the gases at the depth of the wreckage (2 mi) is 300 atm? (c) What is the minimum electrical energy required to raise the *Titanic* by electrolysis? (d) What is the minimum cost of the electrical energy required to generate the necessary H_2 if the electricity cost 15¢ per kilowatt-hour?

[19.55] Suppose a concentration cell is constructed using the Ag-Ag^+ half-cell reaction. In one compartment the concentration of Ag^+ is 1.0 M. In the other compartment the concentration of Ag^+ is unknown. The cell generates an emf of 0.402 V. What is the concentration of Ag^+ in this compartment? If the compartment contains a $10^{-3}\ M$ concentration of Cl^-, what is the K_{sp} of AgCl?

[19.56] A voltaic cell is constructed with one half-cell containing Ag as an electrode in contact with $AgNO_3$ solution. The other half-cell contains Cd as an electrode in contact with $Cd(NO_3)_2$. (a) Diagram the cell showing the direction of electron movement and ion migrations and labeling the anode and cathode. (b) Which electrode will gain weight and which will lose? (c) What is the standard potential generated by the cell? (d) Given the following quantities, calculate the total electrical charge that the cell can deliver (mass of silver electrode, 1.00 g; mass of cadmium electrode, 1.00 g; mass of $AgNO_3$ in solution, 1.00 g; mass of $Cd(NO_3)_2$ in solution, 1.00 g).

19.57 Cerium(IV) ion, Ce^{4+}, is an important aqueous oxidizing agent whose standard reduction potential is 1.61 V. Is this reagent stable with respect to reduction by water under standard conditions?

20 nuclear chemistry

As we have progressed through this book, our focus has been on chemical reactions, specifically reactions in which electrons play a dominate role. In this chapter, we shall consider nuclear reactions, changes in matter whose origin is the nucleus of the atom. Experts predict that we shall have to depend more and more on nuclear energy to replace our dwindling supplies of fossil fuels and to meet our rising energy demands. Thus our consideration of nuclear chemistry continues a minor theme of energy generation started in the last chapter. Even before we begin, you should have some awareness of the controversy surrounding nuclear energy—how do you feel about having a nuclear power plant in your town? Because the topic of nuclear energy evokes such emotional reaction, it is difficult to sift fact from opinion and begin to weigh pros and cons rationally. It is therefore important for any educated person of our time to have some understanding of nuclear reactions and the use of radioactive substances.

However, before we get too deeply involved in our discussions, it is useful for us to review and extend slightly some ideas introduced in Section 2.6. First, we should recall that there are two subatomic particles that reside in the nucleus, the **proton** and the **neutron**. We shall refer to these particles as **nucleons**. Recall also that all atoms of a given element have the same number of protons; this number is known as the element's **atomic number**. However, the atoms of a given element can have different numbers of neutrons and therefore different **mass numbers;** the mass number is the total number of nucleons in the nucleus. Atoms with the same atomic number but different mass numbers are known as **isotopes.**

The different isotopes of an element are distinguished by citing their mass numbers. For example, the three naturally occurring isotopes of uranium are identified as uranium-233, uranium-235, and uranium-238, where the numbers given are the mass numbers. These isotopes are also labeled, using chemical symbols, as $^{233}_{92}U$, $^{235}_{92}U$, and $^{238}_{92}U$. The superscript is the mass number, the subscript is the atomic number. Different isotopes have different natural abundances. For example, 99.3 percent of the naturally occurring uranium is uranium-238, whereas 0.7 percent is uranium-235, and only a trace is uranium-233. One reason that it now becomes important to distinguish between different isotopes is that the nuclear properties of an atom depend on the number of both protons and neutrons in its nucleus. In contrast, we have found that an atom's chemical properties are unaffected by the number of neutrons in the nucleus. Now let's begin to discuss the reactions that a nucleus can undergo.

20.1 nuclear reactions: an overview

There are several ways in which a nucleus can undergo a reaction and thereby change its identity. Some nuclei are unstable and spontaneously emit particles and electromagnetic radiation. Such spontaneous emission from the nucleus of the atom is known as **radioactivity**. The discovery of this phenomenon by Henri Becquerel in 1896 was described in Section 2.7. Those isotopes that are radioactive are known as **radioisotopes**. An example is uranium-238, which spontaneously emits **alpha rays**; these rays consist of a stream of helium-4 nuclei known as **alpha particles**. When a uranium-238 nucleus loses an alpha particle, the remaining fragment has an atomic number of 90 and a mass number of 234. It is therefore a thorium-234 nucleus. We can represent this reaction by the following nuclear equation:

$$^{238}_{92}U \longrightarrow {}^{234}_{90}Th + {}^{4}_{2}He \qquad [20.1]$$

When a nucleus spontaneously decomposes in this way, it is said to have decayed, or undergone **radioactive decay.**

Notice in Equation [20.1] that the sum of the mass numbers is the same on both sides of the equation (238 = 234 + 4). Likewise, the sum of the atomic numbers or nuclear charges on both sides of the equation is equal (92 = 90 + 2). Mass numbers and atomic numbers are similarly balanced in writing other nuclear equations. In writing nuclear equations, we are not concerned with the chemical form of the atom in which the nucleus resides. The radioactive properties of the nucleus are essentially independent of the state of chemical combination of the atom. It makes no difference whether we are dealing with the atom in the form of an element or of one of its compounds.

Another way a nucleus can change identity is to be struck by a neutron or by another nucleus. Nuclear reactions that are induced in this way are known as **nuclear transmutations.** Such a transmutation occurs when the chlorine-35 nucleus is struck by a neutron ($^{1}_{0}n$); this collision produces a sulfur-35 nucleus and a proton ($^{1}_{1}p$ or $^{1}_{1}H$). The nuclear equation for this reaction is shown in Equation [20.2]:

$$\ce{^{35}_{17}Cl + ^{1}_{0}n -> ^{35}_{16}S + ^{1}_{1}H} \qquad [20.2]$$

Notice again that the sum of mass numbers and of atomic numbers is the same on both sides of the equation. By bombarding nuclei with various particles, it is possible to prepare nuclei not found in nature. The sulfur-35 produced in Equation [20.2] is such a "man-made" isotope.

SAMPLE EXERCISE 20.1

Write a balanced nuclear equation for the nuclear transmutation in which an aluminum-27 nucleus is struck by a helium-4 nucleus producing a phosphorus-30 nucleus and a neutron.

Solution: By referring to a periodic table or a list of elements, we find that aluminum has an atomic number of 13; its chemical symbol is therefore $\ce{^{27}_{13}Al}$. The atomic number of phosphorus is 15; its chemical symbol is therefore $\ce{^{30}_{15}P}$. The balanced equation is

$$\ce{^{27}_{13}Al + ^{4}_{2}He -> ^{30}_{15}P + ^{1}_{0}n}$$

We shall take a closer look at radioactive decay in Sections 20.2 through 20.4. We shall then discuss the preparation of nuclei via nuclear transmutation in Section 20.5. In Sections 20.7 and 20.8, we shall discuss two other types of nuclear reactions. One is known as nuclear fission and the other as nuclear fusion. Fission involves the fragmentation of a large nucleus into two roughly equal-sized nuclei. Fusion involves the combination of two small nuclei to form a larger nucleus.

20.2
natural
radioactivity

As we were discussing radioactivity in the last section, two general questions might have occurred to you: Which nuclei are radioactive and what types of radiation do they emit? These are important questions, and we shall now examine them. Let us first consider the types of radiation involved in the phenomenon of radioactivity.

TYPES OF RADIOACTIVE DECAY

Emission of radiation is one of the ways by which an unstable nucleus is transformed into a stable one with less energy. The emitted radiation is the carrier of the excess energy. In Section 2.7, we discussed the three most common types of radiation emitted by radioactive substances: alpha (α), beta (β), and gamma (γ) rays.

As we noted in Section 20.1, **alpha rays** consist of streams of helium-4 nuclei known as alpha particles. Equation [20.3] gives another example of this type of radioactive decay:

$$\ce{^{222}_{86}Rn -> ^{218}_{84}Po + ^{4}_{2}He} \qquad [20.3]$$

Beta rays consist of streams of electrons. Because the beta particles are electrons, they are represented as $\ce{^{0}_{-1}e}$. The superscript zero reflects the exceedingly small mass of the electron by comparison to the mass of a nucleon. The subscript -1 indicates the negative charge of the particle, which is opposite that of the proton. Iodine-131 is an example of an isotope that undergoes decay by beta emission. This reaction is summarized by Equation [20.4]:

$$^{131}_{53}\text{I} \longrightarrow {}^{131}_{54}\text{Xe} + {}^{0}_{-1}\text{e} \qquad [20.4]$$

Emission of a beta particles can be thought of as converting a neutron into a proton, thereby increasing the atomic number of the nucleus by one:

$$^{1}_{0}\text{n} \longrightarrow {}^{1}_{1}\text{p} + {}^{0}_{-1}\text{e} \qquad [20.5]$$

However, just because electrons are ejected from the nucleus, we need not think that the nucleus is composed of these particles, any more than we consider a match to be composed of sparks simply because it gives them off when struck. The electrons come into being when the nucleus is disrupted.

Gamma rays consist of electromagnetic radiation of very short wavelength (that is, high-energy photons). The position of gamma rays in the electromagnetic spectrum is shown in Figure 6.3. Gamma rays can be represented as $^{0}_{0}\gamma$. Such radiation changes neither the atomic number nor the mass number of a nucleus. It almost always accompanies other radioactive emission, because it represents the energy lost when the remaining nucleons reorganize into more stable arrangements. Generally we shall not show the gamma rays when writing nuclear equations.

Two other types of radioactive decay that occur are positron emission and electron capture. A **positron** is a particle that has the same mass as an electron but an opposite charge.* The positron is represented as $^{0}_{1}\text{e}$. Carbon-11 is an example of an isotope that decays by positron emission:

$$^{11}_{6}\text{C} \longrightarrow {}^{11}_{5}\text{B} + {}^{0}_{1}\text{e} \qquad [20.6]$$

Emission of a positron can be thought of as converting a proton into a neutron as shown in Equation [20.7]. The atomic number of the nucleus is thereby decreased by one:

$$^{1}_{1}\text{p} \longrightarrow {}^{1}_{0}\text{n} + {}^{0}_{1}\text{e} \qquad [20.7]$$

Electron capture involves capture of an electron from the electron cloud surrounding the nucleus. Beryllium-7 undergoes decay in this fashion as shown in Equation [20.8]:

$$^{7}_{4}\text{Be} + {}^{0}_{-1}\text{e} \longrightarrow {}^{7}_{3}\text{Li} \qquad [20.8]$$

Electron capture may be thought of as converting a proton to a neutron as shown in Equation [20.9]:

$$^{1}_{1}\text{p} + {}^{0}_{-1}\text{e} \longrightarrow {}^{1}_{0}\text{n} \qquad [20.9]$$

*The positron has a very short life, because it is annihilated when it collides with an electron: $^{0}_{1}\text{e} + {}^{0}_{-1}\text{e} \longrightarrow 2^{0}_{0}\gamma$.

SAMPLE EXERCISE 20.2

Write balanced nuclear equations for the following reactions: (a) thorium-230 undergoes alpha decay; (b) thorium-231 undergoes decay to form protactinium-231.

Solution: (a) The information given in the problem can be summarized as

$$\ce{^{230}_{90}Th} \longrightarrow \ce{^{4}_{2}He} + X$$

The remaining product, X, must be deduced. Because mass numbers must have the same sum on both sides of the equation, we deduce that X has a mass number of 226. Similarly, the atomic number of X must be 88. Element number 88 is radium (refer to periodic table or list of elements). The equation is therefore as follows:

$$\ce{^{230}_{90}Th} \longrightarrow \ce{^{4}_{2}He} + \ce{^{226}_{88}Ra}$$

(b) In this case we must determine what type of particle is emitted in the course of the radioactive decay. We can write the following equation:

$$\ce{^{231}_{90}Th} \longrightarrow \ce{^{231}_{91}Pa} + X$$

The atomic numbers are obtained from a list of elements such as that given on the back inside cover. In order for the mass numbers to balance, X must have a mass number of 0. Its atomic number must be -1. The particle with these characteristics is the beta particle (electron). We therefore write the following:

$$\ce{^{231}_{90}Th} \longrightarrow \ce{^{231}_{91}Pa} + \ce{^{0}_{-1}e}$$

NUCLEAR STABILITY

There is no single rule that will allow us to predict whether a particular nucleus is radioactive and how it might decay. However, we can list some helpful empirical observations that assist us in making predictions:

1. All nuclei with 84 or more protons are unstable. For example, all isotopes of uranium, atomic number 92, are radioactive.
2. Nuclei with a total of 2, 8, 20, 50, 82, or 126 protons or neutrons are generally more stable than nuclei found near them in the periodic table. For example, there are three stable nuclei with an atomic number of 18, two with 19, five with 20, and one with 21; there are three stable nuclei with 18 neutrons, none with 19, four with 20, and none with 21. Thus there are more stable nuclei with 20 protons or neutrons than with 18, 19, or 21. The numbers 2, 8, 20, 50, 82, and 126 are called **magic numbers**. Just as enhanced chemical stability is associated with the presence of 2, 10, 18, 36, 54, or 86 electrons, the noble gas configurations, enhanced nuclear stability is associated with the magic number of nucleons.*
3. Nuclei with even numbers of both protons and neutrons are generally more stable than those with odd numbers of nucleons, as shown in Table 20.1.

*Evidence for superheavy elements with atomic numbers 116 and 126 has been found recently in some ancient rocks from the island of Madagascar. Investigators, who reported their findings at a scientific meeting in June 1976, are cautious not to claim discovery of these elements yet; however, they presently have no alternative explanations for their results. If element 126 is present, this would not be totally surprising because its nucleus has a magic number of protons.

TABLE 20.1

the number of stable isotopes with even and odd numbers of protons and neutrons

NUMBER OF STABLE ISOTOPES	PROTONS	NEUTRONS
157	Even	Even
52	Even	Odd
50	Odd	Even
5	Odd	Odd

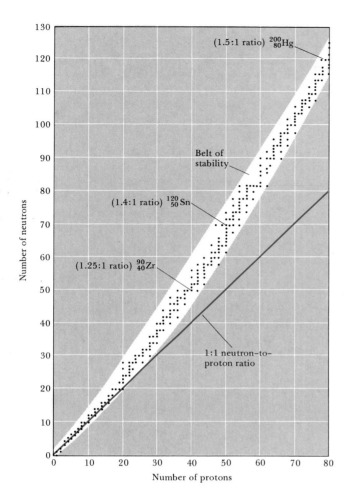

FIGURE 20.1 A plot of the number of neutrons versus the number of protons in stable nuclei. As the atomic number increases, the neutron-to-proton ratio of the stable nuclei increases. The stable nuclei are located in an area of the graph known as the belt of stability. The majority of radioactive nuclei occur outside this belt.

4. The stability of a nucleus can be correlated to a certain degree with its neutron-to-proton ratio. All nuclei with two or more protons contain neutrons. Neutrons are apparently involved in some way in holding protons together within the nucleus. As shown in Figure 20.1, the number of neutrons necessary to create a stable nucleus increases rapidly as the number of protons increases; the neutron-to-proton ratios of stable nuclei increase with increasing atomic number. The area within which all stable nuclei are found is known as the **belt of stability.**

SAMPLE EXERCISE 20.3

Would you expect the following nuclei to be radioactive: $^{4}_{2}He$, $^{39}_{20}Ca$, $^{210}_{85}At$?

Solution: Helium-4 has a magic number of both protons and neutrons (two each). We would therefore expect $^{4}_{2}He$ to be stable.

Calcium-39 has an even number of protons (20) and an odd number of neutrons (19); 20 is one of the magic numbers. Nevertheless, we should suspect that this nuclide is radioactive because the neutron-to-proton ratio is less than 1. This would place it below the belt of stability.

Astatine-210 is radioactive. Recall that there are no stable nuclei beyond atomic number 83.

The type of radioactive decay that a particular radioisotope will undergo depends to a large extent on its neutron-to-proton ratio compared to those of nearby nuclei that are within the belt of stability. Consider a nucleus whose high neutron-to-proton ratio places it above the belt of stability. This nucleus can lower its ratio and move toward the belt of stability by emitting a beta particle. Beta emission decreases the number of neutrons and increases the number of protons in a nucleus as shown in Equation [20.5].

Nuclei that have low neutron-to-proton ratios and that therefore lie below the belt of stability either emit positrons or undergo electron capture. Both modes of decay decrease the number of protons and increase the number of neutrons in the nucleus as shown in Equations [20.7] and [20.9]. Positron emission is more common than electron capture among the lighter nuclei; however, electron capture becomes increasingly common as nuclear charge increases.

Alpha emission is found primarily among nuclei whose atomic numbers are greater than 83. These nuclei would lie beyond the upper right edge of Figure 20.1, outside the belt of stability. Emission of an alpha particle moves the nucleus diagonally toward the belt of stability by decreasing the number of protons and the number of neutrons both by two. The result of each type of radioactive decay relative to a stable nucleus is shown in Figure 20.2.

SAMPLE EXERCISE 20.4

By referring to Figure 20.1, predict the mode of radioactive decay of the following nuclei: (a) $^{20}_{11}$Na; (b) $^{97}_{40}$Zr; (c) $^{235}_{92}$U.

Solution: To answer this question we use the guidelines given above in the text.

(a) This nucleus has a neutron-to-proton ratio below 1. It therefore lies below the belt of stability. It can gain stability either by positron emission or electron capture. Because the atomic number is small we might predict that the nucleus undergoes positron emission. If we refer to a standard reference like the *Handbook of Chemistry and Physics,* we find that this prediction is correct. The nuclear reaction is

$$^{20}_{11}\text{Na} \longrightarrow {}^{0}_{1}\text{e} + {}^{20}_{10}\text{Ne}$$

(b) In referring to Figure 20.1, we find that this nucleus has a neutron-to-proton ratio that is too high. We would therefore predict that it undergoes beta decay. Again the prediction is correct. The nuclear reaction is

$$^{97}_{40}\text{Zr} \longrightarrow {}^{0}_{-1}\text{e} + {}^{97}_{41}\text{Nb}$$

(c) This nucleus lies outside the belt of stability to the upper right. We might therefore predict that it would undergo alpha emission. Again the prediction is correct. The nuclear equation is

$$^{235}_{92}\text{U} \longrightarrow {}^{4}_{2}\text{He} + {}^{231}_{90}\text{Th}$$

At this point we should note that our guidelines don't always work. For example thorium-233, $^{233}_{90}$Th, which we might expect to undergo alpha decay, undergoes beta decay instead. Furthermore, a few radioactive nuclei actually lie within the belt of stability. For example, both $^{146}_{60}$Nd and $^{148}_{60}$Nd are stable and lie in the belt of stability; however $^{147}_{60}$Nd, which lies between them, is radioactive.

RADIOACTIVE SERIES

Some nuclei, like uranium-238, cannot gain stability by a single emission. Consequently, a series of successive emissions occur. As shown in Figure

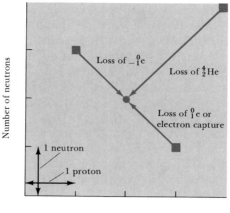

FIGURE 20.2 The result of alpha-emission ($_2^4$He), beta-emission ($_{-1}^0$e), positron emission ($_1^0$e), and electron capture on the number of protons and neutrons in a nucleus. The squares represent unstable nuclei, and the circle represents a stable one. Moving from right to left or from bottom to top, each tick mark represents an additional proton or neutron, respectively. Moving in the reverse direction indicates the loss of a proton or neutron.

20.3, the uranium-238 decays to thorium-234, which is radioactive and decays to protactinium-234. This nucleus is also unstable and subsequently decays. Such successive reactions continue until a stable nucleus, lead-206, is formed. A series of nuclear reactions that begins with an unstable nucleus and terminates with a stable one is known as a **radioactive series** or a **nuclear disintegration series**. Altogether there are three such series found in nature. In addition to the series that begins with uranium-238 and terminates with lead-206, there is one that begins with uranium-235 and ends with lead-207. The third series begins with thorium-232 and ends with lead-208.

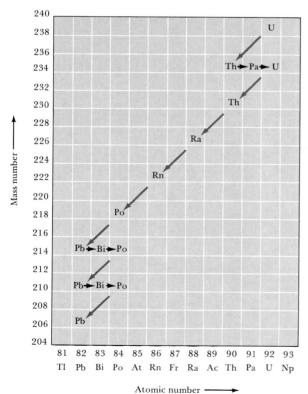

FIGURE 20.3 The nuclear disintegration series for uranium-238. The $_{92}^{238}$U nucleus decays to $_{90}^{234}$Th. Subsequent decay processes eventually form the stable $_{82}^{206}$Pb nucleus. Each of the color arrows correspond to the loss of an alpha particle. Each black arrow corresponds to the loss of a beta particle.

20.3

half-life

Again our discussions may have raised some questions in your mind. For example, why is it that some radioisotopes, like uranium-238, are found in nature, whereas others are not and must be synthesized? The key to this question is the fact that different nuclei undergo decay at different speeds. Uranium-238 undergoes decay very slowly, whereas many other nuclei such as sulfur-35 decay rapidly. To better understand the phenomenon of radioactivity, it is important to consider the rates of radioactive decay.

Radioactive decay is a first-order process. As shown in Section 13.3, first-order processes have characteristic half-lives. The **half-life** is the time required for half of any given quantity of a substance to react. The rates of decay of nuclei are commonly discussed in terms of their half-lives.

Each isotope has its own characteristic half-life. For example, the half-life of strontium-90 is 28 yr. If we started with 10.0 g of strontium-90, only 5.0 g of that isotope would remain after 28 yr. The other half of the strontium-90 would have been converted to yttrium-90 as shown in Equation (20.10):

$$\ce{^{90}_{38}Sr} \longrightarrow \ce{^{90}_{39}Y} + \ce{^{0}_{-1}e} \qquad [20.10]$$

After another 28-yr period half of the remaining 5.0 g of strontium-90 would likewise decay. The loss of strontium-90 as a function of time is shown in Figure 20.4.

Half-lives as short as millionths of a second and as long as billions of years have been observed. The half-lives of some important radioisotopes are listed in Table 20.2. One important feature of half-lives for nuclear decay is that they are unaffected by external conditions such as temperature, pressure, or state of chemical combination. Therefore, unlike chemical toxins, radioactive atoms cannot be rendered harmless by chemical reaction or by any other practical treatment. As far as we know now our only choice is merely to allow these nuclei to lose activity at their own characteristic rates. In the meantime, of course, we must take precautions to isolate the radioisotopes as much as possible because of the damage radiation can cause (see Section 20.9).

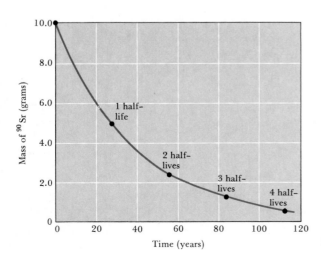

FIGURE 20.4 The decay of a 10.0 g sample of $\ce{^{90}_{38}Sr}$ ($t_{1/2}$ = 28 yr).

TABLE 20.2

some important radioactive isotopes and their half-lives and type of decay

	ISOTOPE	HALF-LIFE	TYPE OF DECAY
Natural radioisotopes	$^{238}_{92}U$	4.5×10^9 yr	Alpha
	$^{235}_{92}U$	7.1×10^8 yr	Alpha
	$^{232}_{90}Th$	1.4×10^{10} yr	Alpha
	$^{40}_{19}K$	1.3×10^9 yr	Alpha
	$^{14}_{6}C$	5,700 yr	Beta
"Man-made" radioisotopes	$^{239}_{94}Pu$	24,000 yr	Alpha
	$^{137}_{55}Cs$	30 yr	Beta
	$^{90}_{38}Sr$	28 yr	Beta
	$^{131}_{53}I$	8.1 days	Beta

SAMPLE EXERCISE 20.5

The half-life of cobalt-60 is 5.3 yr. How much of a 1.000 mg sample of cobalt-60 is left after a 15.9-yr period?

Solution: A period of 15.9 yr is three half-lives for cobalt-60. At the end of one half-life, 0.500 mg of cobalt-60 remains; 0.250 mg remains at the end of two half-lives, and 0.125 mg at the end of three half-lives.

DATING

Because the half-life of any particular nuclide is so constant, it can serve as a molecular clock and can be used to determine the ages of different objects. For example, carbon-14 has been used to determine the age of organic materials (see Figure 20.5). The procedure is based on the formation of carbon-14 by neutron capture in the upper atmosphere:

$$^{14}_{7}N + ^{1}_{0}n \longrightarrow ^{14}_{6}C + ^{1}_{1}H \qquad [20.11]$$

This reaction provides a small but reasonably constant source of carbon-14. The carbon-14 is radioactive, undergoing beta decay with a half-life of 5700 yr:

$$^{14}_{6}C \longrightarrow ^{14}_{7}N + ^{0}_{-1}e \qquad [20.12]$$

In using radiocarbon dating, it is generally assumed that the ratio of carbon-14 to carbon-12 in the atmosphere has been constant for at least 50,000 yr. The carbon-14 is incorporated into carbon dioxide, which is in turn incorporated, through photosynthesis, into more complex carbon-containing molecules within plants. The plants are eaten by animals and the carbon-14 thereby becomes incorporated within them as well. Because a living plant or animal has a constant intake of carbon compounds, it is able to maintain a ratio of carbon-14 to carbon-12 that is identical with that of the atmosphere. However, once the organism dies, it no longer ingests carbon compounds to replenish the carbon-14 that is lost through radioactive decay. The ratio of carbon-14 to carbon-12

FIGURE 20.5 Ancient manuscripts and other cultural artifacts made of organic materials are often dated by means of carbon-14 analysis.

therefore decreases. If the ratio diminishes to half that of the atmosphere, we can conclude that the object is one half-life or 5700 yr old. The method cannot be used to date objects that are over about 20,000–50,000 yr old. After this length of time, the radioactivity is too low to be measured accurately.

One of the ways that the radiocarbon-dating technique has been checked has been to compare the ages of trees both by counting their rings and by radiocarbon analysis. As a tree grows, it adds a ring each year. The old growth no longer replenishes its supply of carbon. Thus the carbon-14 decays, while the concentration of carbon-12 remains constant. Figure 20.6 shows comparisons of tree age as determined by counting rings and by radiocarbon dating. Most of the wood used was from California bristlecone pines, which reach ages up to 2000 yr. By using trees that died thousands of years ago, it is possible to make comparisons back to about 5000 B.C. As Figure 20.6 shows, the two dating methods agree within 10 percent. Furthermore, the deviations show a general pattern. The deviations suggest that the ratio of carbon-14 to carbon-12 in the atmosphere has not been strictly constant. One possible reason for changes in this ratio is a variation in cosmic ray intensity reaching earth. Cosmic rays consist of streams of particles from outer space. These rays are composed principally of protons; those that strike earth originate principally from our sun. In the upper atmosphere of our planet, these cosmic rays cause nuclear reactions that produce neutrons. These neutrons are responsible for the formation of carbon-14, as shown in Equation [20.12]. Thus the variation in the ratio of carbon-14 to carbon-12 in the atmosphere can be traced to variations in solar activity.

Other isotopes can be similarly used to date other types of objects. For example, it takes 4.5×10^9 yr for half of a sample of uranium-238 to decay to a stable product, lead-206. The age of rocks containing uranium can be determined by measuring the ratio of lead-206 to uranium-238. If the lead-206 had somehow become incorporated into the rock by normal chemical processes instead of by radioactive decay, the rock would

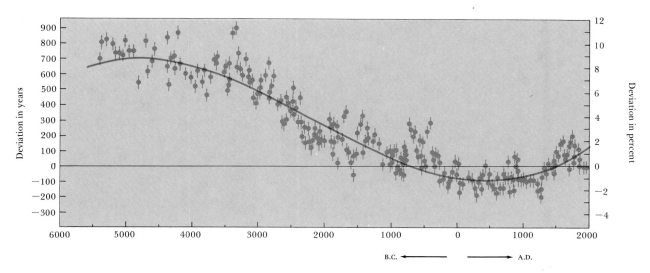

FIGURE 20.6 Comparisons of the ages of trees determined by counting rings and by carbon-14 dating. Each symbol represents a group of samples; the lines extending above and below the dots indicate the spread of deviation. Assuming that the "tree-ring age" (black line) is the true age, it is possible to determine corrections that must be applied to the "carbon-14 age" of an object. This figure shows the necessary corrections. For example, a sample whose carbon-14 content indicates an age of 7000 yr (5000 B.C.) will have an actual age of 7700 yr. The 700-yr correction is read off the color curve using the left scale. (*Redrawn with permission from an article by H. E. Suess,* Endeavour, *32, 36, 1973*)

contain large amounts of the more abundant isotope, lead-208. In the absence of large amounts of this "geonormal" isotope of lead, it is assumed that all of the lead-206 was at one time uranium-238.

The oldest rocks found on the earth are approximately 3×10^9 yr old. This age indicates that the crust of the earth has been solid for at least this length of time. Prior to the crystallization of rock, lead-206 and uranium-238 could separate. It is estimated that it required 1–1.5×10^9 yr for the earth to cool and its surface to become solid. This places the age of earth at 4.0–4.5 billion yr.

CALCULATIONS BASED ON HALF-LIFE

So far our discussion has been mainly qualitative. We now consider the topic of half-lives from a more quantitative point of view. This permits us to answer questions of the following types: How do we determine the half-life of uranium-238? We certainly don't sit around for 4.5 billion yr waiting for half of it to decay! Similarly, how do we quantitatively determine the age of an object that can be dated by radiometric means?

The rate of radioactive decay of any radioisotope is first order. It can therefore be described by Equation [20.13],

$$\text{Rate} = kN \qquad [20.13]$$

where N is the number of nuclei of a particular radioisotope, and k is the first-order rate constant. This equation can be transformed into Equation [20.14]

$$k = \frac{2.30}{t} \log \frac{N_0}{N_t}$$ [20.14]

where t is the time interval during which decay is measured, k is the rate constant, N_0 is the initial number of nuclei (at zero time), and N_t is the number remaining after the time interval. The relationship between the rate constant, k, and half-life, $t_{1/2}$, is given by Equation [20.15]:

$$k = \frac{0.693}{t_{1/2}}$$ [20.15]

If Equation [20.15] is substituted into Equation [20.14], the following relationship results:

$$\frac{0.693}{t_{1/2}} = \frac{2.30}{t} \log \frac{N_0}{N_t}$$

Rearranging this expression gives Equation [20.16]:

$$t = 3.32 t_{1/2} \log \frac{N_0}{N_t}$$ [20.16]

This equation is used in performing calculations such as those shown in Sample Exercises 20.6 and 20.7.

SAMPLE EXERCISE 20.6

A rock contains 0.257 mg of lead-206 for every milligram of uranium-238. How old is the rock?

Solution: The amount of uranium-238 in the rock when it was first formed is assumed to be the 1.000 mg present at the time of the analysis plus the quantity that decayed to lead-206.

Original $^{238}_{92}U = 1.000$ mg $+ \frac{238}{206}(0.257$ mg$)$

$= 1.297$ mg

Using Equation [20.16] and the half-life of $^{238}_{92}U$, 4.5×10^9 yr, gives the following result:

$$t = (3.32)(4.5 \times 10^9 \text{ yr}) \log \frac{1.297}{1.000}$$

$$= (3.32)(4.5 \times 10^9 \text{ yr})(0.113)$$

$$= 1.7 \times 10^9 \text{ yr}$$

SAMPLE EXERCISE 20.7

If we start with 1.00 g of strontium-90, 0.952 g will still remain after 2.00 yr. (a) What is the half-life of strontium-90? (b) How much strontium-90 would remain after 5.0 yr?

Solution: (a) Equation [20.16] can be solved for $t_{1/2}$ using $N_0 = 1.000$ g, $N_t = 0.952$ g, and $t = 2.0$ yr.

$$t_{1/2} = \frac{t}{3.32 \log \frac{N_0}{N_t}}$$

$$= \frac{2.0 \text{ yr}}{3.32 \log \frac{1.000}{0.952}} = \frac{2.0 \text{ yr}}{3.32 \log 1.05}$$

$$= \frac{2.0 \text{ yr}}{3.32(0.0212)} = 28 \text{ yr}$$

(b) Again using Equation [20.16] we have

$$\log \frac{N_0}{N_t} = \frac{t}{3.32 t_{1/2}}$$

$$= \frac{5.0 \text{ yr}}{(3.32)(28 \text{ yr})} = 0.054$$

$$\frac{N_0}{N_t} = 1.13$$

$$N_t = \frac{N_0}{1.13} = \frac{1.000 \text{ g}}{1.13} = 0.88 \text{ g}$$

**20.4
detection of
radioactivity**

A variety of methods have been devised to detect emissions from radioactive substances. Becquerel discovered radioactivity because of the effect of radiation on photographic plates. Photographic plates and film have long been used to detect radioactivity. The radiation affects photographic film in the same way as ordinary light. With care, film can be used to give a quantitative measure of activity. The greater the extent of exposure to radiation, the darker the area of the developed negative. People who work with radioactive substances carry film badges to record the extent of their exposure to radiation.

Radioactivity can also be detected and measured using a device known as a **Geiger counter.** The operation of the Geiger counter is based on the ionization of matter caused by radiation (Section 20.9). The ions and electrons produced by the ionizing radiation permit conduction of an electrical current. The basic design of a Geiger counter is shown in Figure 20.7. It consists of a metal tube filled with gas. The cylinder has a "window" made of material that can be penetrated by alpha, beta, or gamma rays. In the center of the tube is a wire. The wire is connected to one terminal of a source of direct current and the metal cylinder is attached to the other terminal. Current flows between the wire and metal cylinder whenever ions are produced by entering radiation. The current pulse created when radiation enters the tube is amplified; each pulse is counted as a measure of the amount of radiation.

Certain substances that are electronically excited by radiation can also be used as means for detecting and measuring radiation. Some substances excited by radiation give off light (fluoresce) as electrons return to their lower energy states. For example, dials of luminous watches are painted with a mixture of ZnS and a tiny quantity of $RaSO_4$. The zinc sulfide fluoresces when struck by the radioactive

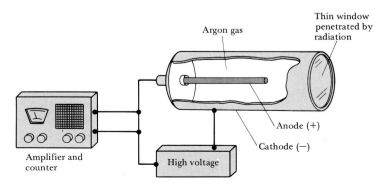

FIGURE 20.7 Schematic representation of a Geiger counter.

emissions from the radium. An instrument known as a **scintillation counter** can be used to detect and measure fluorescence and thereby the radiation that causes it.

TRACERS

Because radioisotopes can be detected so readily, they can be used to follow an element through its chemical reactions. For example, the incorporation of carbon atoms from CO_2 into glucose in photosynthesis, Equation [20.17], has been studied using CO_2 containing carbon-14:

$$6CO_2 + 6H_2O \xrightarrow[\text{chlorophyll}]{\text{sunlight}} C_6H_{12}O_6 + 6O_2 \qquad [20.17]$$

The CO_2 is said to be labeled with the carbon-14. Detection devices such as Geiger counters can then be used to follow the carbon-14 as it moves from the CO_2 through the various intermediate compounds to glucose.

Such use of radioisotopes is made possible by the fact that all isotopes of an element have essentially identical chemical properties. If a small quantity of a radioisotope is mixed with the naturally occurring stable isotopes of the same element, all of the isotopes will go through the same reactions together. Where the element goes is then revealed by the radioactivity of the radioisotope. Because the radioisotope can be used to trace the path of the element, it is called a **tracer.**

Tracers have found wide use as a diagnostic tool in medicine. For example, iodine-131 has been used to test the activity of the thyroid gland. This gland is the only important user of iodine in the body. A solution of NaI containing a small quantity of iodine-131 is drunk by the patient. The ability of the thyroid to take up the iodine is determined using a Geiger tube placed close to the thyroid, in the neck region. A normal thyroid will absorb about 12 percent of the iodine within a few hours. When radioisotopes are used in this manner, only a very small amount is used so that the patient does not receive a harmful dose of radioactivity.

20.5
preparation
of new nuclei

In 1919 Rutherford performed the first artificial conversion of one nucleus into another. He succeeded in converting nitrogen-14 into oxygen-17 using the high velocity alpha (α) particles emitted by radium. The reaction is shown in Equation [20.18]:

$$^{14}_{7}\text{N} + ^{4}_{2}\text{He} \longrightarrow ^{17}_{8}\text{O} + ^{1}_{1}\text{H} \qquad [20.18]$$

This reaction demonstrated that nuclear reactions can be induced by striking nuclei with particles such as alpha particles. Such reactions have permitted synthesis of hundreds of radioisotopes in the laboratory. As noted in Section 20.1, these conversions of one nucleus into another are called nuclear transmutations. It is common to represent such conversions by listing, in order, the target nucleus, the bombarding particle, the ejected particle, and the product nucleus. Written in this fashion, Equation [20.18] is $^{14}_{7}\text{N}(\alpha, \text{p})^{17}_{8}\text{O}$. The alpha particle, proton, and neutron are abbreviated as α, p, and n, respectively.

SAMPLE EXERCISE 20.8

Write the balanced nuclear equation for the process summarized as $_{13}^{27}\text{Al}(n, \alpha)_{11}^{24}\text{Na}$.

Solution: The n is the abbreviation for a neutron, whereas α represents an alpha particle. The neutron is the bombarding particle, and the alpha particle is a product. Therefore the nuclear equation is

$$_{13}^{27}\text{Al} + _{0}^{1}\text{n} \longrightarrow _{2}^{4}\text{He} + _{11}^{24}\text{Na}$$

Charged particles such as alpha particles must be moving very fast in order to overcome the electrostatic repulsion between them and the target nucleus. The higher the nuclear charge on either the projectile or the target, the faster the projectile must be moving to bring about a nuclear reaction. Therefore, many methods have been devised to accelerate charged particles using strong magnetic and electrostatic fields. These **particle accelerators** bear such names as the **cyclotron** and **synchrotron**. The cyclotron is illustrated in Figure 20.8. The hollow **D**-shaped electrodes are called "dees." The projectile particles are introduced into a vacuum chamber within the cyclotron. The particles are then accelerated by making the dees alternately positively and negatively charged. Magnets placed above and below the dees keep the particles moving in a spiral path until they are finally deflected out of the cyclotron and emerge to strike a target substance. Particle accelerators have been used mainly to probe the secrets of nuclear structure and to synthesize heavy elements.

Most synthetic isotopes used in quantity in medicine and scientific research are made using neutrons as projectiles. Because neutrons are neutral, they are not repelled by the nucleus; consequently, they do not need to be accelerated, as do the charged particles, in order to cause nuclear reactions (indeed they cannot be so accelerated). The necessary neutrons are produced by the reactions that occur in nuclear reactors (Section 20.7). Cobalt-60, used in radiation therapy for cancer, is produced by neutron capture. Iron-58 is placed in a nuclear reactor where it is bombarded by neutrons. The sequence of reactions shown in Equations [20.19] through [20.21] takes place:

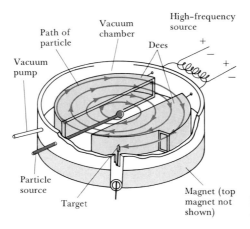

Path of particle
Vacuum chamber
High-frequency source
Dees
Vacuum pump
Vacuum
Particle source
Target
Magnet (top magnet not shown)

FIGURE 20.8 Schematic representation of a cyclotron.

$$\ce{^{58}_{26}Fe + ^1_0n -> ^{59}_{26}Fe} \qquad [20.19]$$

$$\ce{^{59}_{26}Fe ->[t_{1/2} = 45\ days] ^{59}_{27}Co + ^0_{-1}e} \qquad [20.20]$$

$$\ce{^{59}_{27}Co + ^1_0n -> ^{60}_{27}Co} \qquad [20.21]$$

The cobalt-60 so formed subsequently decays by beta emission with a half-life of 5.3 yr.

TRANSURANIUM ELEMENTS

Artificial transmutations have been used to produce the elements from atomic number 93 to 105. These are known as the **transuranium elements**, because they occur immediately following uranium in the periodic table. All of these elements have short half-lives. Consequently, if they ever existed naturally on earth they would have disappeared by radioactive decay long ago. Elements 93 (neptunium) and 94 (plutonium) were first discovered in 1940. They were produced by bombarding uranium-238 with neutrons as shown in Equations [20.22] and [20.23].

$$\ce{^{238}_{92}U + ^1_0n -> ^{239}_{92}U -> ^{239}_{93}Np + ^0_{-1}e} \qquad [20.22]$$

$$\ce{^{239}_{93}Np -> ^{239}_{94}Pu + ^0_{-1}e} \qquad [20.23]$$

Elements with larger atomic numbers are normally formed in small quantities in particle accelerators. For example, curium-242 is formed when a plutonium-239 target is struck with accelerated alpha particles:

$$\ce{^{239}_{94}Pu + ^4_2He -> ^{242}_{96}Cm + ^1_0n} \qquad [20.24]$$

20.6

mass-energy conversions

So far we have said little about the energies associated with nuclear reactions. These energies can be considered with aid of Einstein's famous equation relating mass and energy:

$$E = mc^2 \qquad [20.25]$$

In this equation E stands for energy, m for mass, and c for the speed of light, 3.00×10^8 m/sec. This equation states that the mass and energy of an object are proportional. The greater an object's mass, the greater its energy. Because the proportionality constant in the equation, c^2, is such a large number, small changes in mass are accompanied by large changes in energy.

The mass changes accompanying chemical reactions are too small to detect. For this reason, it is possible to speak of the conservation of mass in chemical reactions. The calculation of the loss of mass that accompanies the combustion of a mole of CH_4 illustrates this point. This calculation is shown in Sample Exercise 20.9.

SAMPLE EXERCISE 20.9

Calculate the mass that disappears from the system when a mole of CH_4 is combusted:

$$\ce{CH4 + 2O2 -> CO2 + 2H2O}$$

The system loses 890 kJ of energy in this process ($\Delta E = -890$ kJ).

Solution: From Equation [20.25] we can write that the change in mass, Δm, in the system is proportional to its change in energy, ΔE:

$$\Delta E = c^2 \Delta m$$

Rearranging to solve for Δm we have

$$\Delta m = \frac{\Delta E}{c^2}$$

Substituting the value given for ΔE and remembering that a negative sign is associated with an exothermic process, we have:

$$\Delta m = \frac{-890 \text{ kJ}}{(3.00 \times 10^8 \text{ m/sec})^2}\left(\frac{1000 \text{ J}}{1 \text{ kJ}}\right)$$

$$\times \left(\frac{1 \text{ kg-m}^2/\text{sec}^2}{1 \text{ J}}\right)$$

$$= -9.89 \times 10^{-12} \text{ kg}$$

The negative sign indicates mass loss. This mass change is far too small to detect.

The mass changes and the associated energy changes in nuclear reactions are much greater than in chemical reactions. The energy released through the nuclear fission of only about a pound of uranium (Section 20.7) is equivalent to that released by combustion of 1500 tons of coal.

NUCLEAR BINDING ENERGIES

It was discovered in the 1930s that the masses of nuclei are always less than the masses of the individual nucleons of which they are composed. For example, the helium-4 nucleus has a mass of 4.00150 amu. The mass of a proton is 1.00728 amu, while that of a neutron is 1.00867 amu. Consequently, two protons and two neutrons have a total mass of 4.03190 amu:

$$\text{Mass of 2 protons} = 2(1.00728 \text{ amu}) = 2.01456 \text{ amu}$$
$$\text{Mass of 2 neutrons} = 2(1.00867 \text{ amu}) = \underline{2.01734 \text{ amu}}$$
$$\text{Total mass} = 4.03190 \text{ amu}$$

The individual nucleons weigh 0.03040 amu more than the helium-4 nucleus:

$$\text{Mass of 2 protons and 2 neutrons} = 4.03190 \text{ amu}$$
$$\text{Mass of } {}^4_2\text{He nucleus} = \underline{4.00150 \text{ amu}}$$
$$\text{Mass difference} = 0.03040 \text{ amu}$$

The mass difference between the helium-4 nucleus and the two protons and two neutrons is known as the **mass defect** of the nucleus. This mass is lost in the form of energy:

$$2{}^1_1\text{p} + 2{}^1_0\text{n} \longrightarrow {}^4_2\text{He} + \text{energy} \qquad [20.26]$$

The amount of energy produced in this reaction can be calculated from the mass difference:

$$\Delta E = c^2 \, \Delta m$$

$$= (3.00 \times 10^8 \text{ m/sec})^2(-0.0304 \text{ amu})$$

$$\times \left(\frac{1.00 \text{ g}}{6.02 \times 10^{23} \text{ amu}}\right)\left(\frac{1 \text{ kg}}{1000 \text{ g}}\right)$$

$$= -4.52 \times 10^{-12}\frac{\text{kg-m}^2}{\text{sec}^2} = -4.52 \times 10^{-12}\text{ J}$$

Formation of a mole of helium-4 nuclei in this fashion would produce a tremendous quantity of energy:

$$(6.02 \times 10^{23})(4.52 \times 10^{-12}\text{ J}) = 2.74 \times 10^{12}\text{ J}$$

If we turn Equation [20.26] around, we see that 4.52×10^{-12} J would be required to break a single helium-4 nucleus into protons and neutrons. The energy calculated from the mass defect is therefore a measure of the stability of the nucleus toward decomposition into individual nucleons. The energy required to decompose a nucleus into protons and neutrons is referred to as the **binding energy** of the nucleus.

The binding energy of any nucleus can be calculated from its mass and from the masses of the nucleons of which it is composed. Insight into the relative stabilities of nuclei is gained by comparing the binding energy per nucleon for different nuclei. Three nuclei (helium-4, iron-56, and uranium-238) are compared in Table 20.3. Similar calculations for other nuclei indicate that the binding energy per nucleon increases in magnitude to about 1.4×10^{-12} J with nuclei whose mass numbers are in the vicinity of iron-56. It then decreases slowly to about 1.2×10^{-12} J for very heavy nuclei. This trend is shown in Figure 20.9. These results indicate that heavy nuclei will gain stability, and therefore give off energy, if they are fragmented. This process, known as **fission,** occurs in the atomic bomb and in nuclear power plants. Figure 20.9 also indicates that even greater amounts of energy should be available if very light nuclei are fused together. Such **fusion reactions** take place in the hydrogen bomb and are the essential energy-producing reactions in the sun. We shall look more closely at fission and fusion in the next two sections.

TABLE 20.3

mass differences and binding energies for three nuclei

NUCLEUS	MASS OF NUCLEUS (amu)	MASS OF INDIVIDUAL NUCLEONS (amu)	MASS DIFFERENCE (amu)	BINDING ENERGY (J)	BINDING ENERGY PER NUCLEON (J)
$^{4}_{2}\text{He}$	4.00150	4.03190	0.0304	4.52×10^{-12}	1.13×10^{-12}
$^{56}_{26}\text{Fe}$	55.92066	56.44938	0.52872	7.90×10^{-11}	1.41×10^{-12}
$^{238}_{92}\text{U}$	238.0003	239.9356	1.9353	2.89×10^{-10}	1.22×10^{-12}

SAMPLE EXERCISE 20.10

How much energy is lost or gained when a mole of cobalt-60 undergoes beta decay: $^{60}_{27}\text{Co} \longrightarrow {}^{0}_{-1}\text{e} +$ $^{60}_{28}\text{Ni}$? The mass of the $^{60}_{27}\text{Co}$ nucleus is 59.9381 amu, that of $^{60}_{28}\text{Ni}$ is 59.9344 amu, and that of an electron, $^{0}_{-1}\text{e}$, is 0.000549 amu.

Solution: The total mass of a mole of $_{28}^{60}\text{Ni}$ plus a mole of $_{-1}^{0}\text{e}$ is 59.9349 g:

$$\begin{array}{r} 59.9344 \text{ g} \\ 0.0005 \text{ g} \\ \hline 59.9349 \text{ g} \end{array}$$

The mass lost in the reaction is 0.0032 g:

$$\begin{array}{r} 59.9381 \text{ g} \\ -59.9349 \text{ g} \\ \hline 0.0032 \text{ g} \end{array}$$

The energy produced by the reaction can be calculated from this mass:

$$\Delta E = c^2 \, \Delta m$$

$$= (3.00 \times 10^8 \text{ m/sec})^2 (-0.0032 \text{ g}) \left(\frac{1 \text{ kg}}{1000 \text{ g}} \right)$$

$$= -2.9 \times 10^{11} \frac{\text{m}^2\text{-kg}}{\text{sec}^2} = -2.9 \times 10^{11} \text{ J}$$

By comparison, it takes only 9×10^5 J to break all the chemical bonds in a mole of water.

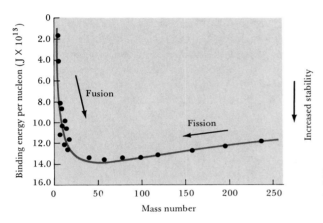

FIGURE 20.9 The average binding energy per nucleon increases to a maximum at a mass number of 50–60 and decreases slowly thereafter. As a result of these trends, fusion of light nuclei and fission of heavy nuclei are exothermic processes.

20.7
nuclear fission

The first nuclear fission to be discovered was that of uranium-235. This nucleus, as well as those of uranium-233 and plutonium-239, undergoes fission when struck by slow-moving neutrons.* The fission process is illustrated in Figure 20.10. A heavy nucleus can split in many different ways, just as can a piece of glass. Two different ways that the uranium-235 nucleus splits are shown in Equations [20.27] and [20.28]:

$$_0^1\text{n} + _{92}^{235}\text{U} \Big\langle \begin{array}{l} \to {}_{52}^{137}\text{Te} + {}_{40}^{97}\text{Zr} + 2{}_0^1\text{n} \qquad [20.27] \\ \to {}_{56}^{142}\text{Ba} + {}_{36}^{91}\text{Kr} + 3{}_0^1\text{n} \qquad [20.28] \end{array}$$

*There are other heavy nuclei that can be induced to undergo fission. However, these three are the only ones of practical importance.

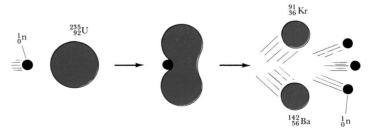

FIGURE 20.10 Schematic representation of the fission of uranium-235 showing one of its many fission patterns.

Over 200 different isotopes of 35 different elements have been found among the fission products of uranium-235. In general, these are radioactive. We shall consider the health hazards associated with radioisotopes such as these in Section 20.9.

On the average, 2.4 neutrons are produced by every fission of uranium-235. If one fission produces two neutrons, these two neutrons can cause two fissions. The four neutrons thereby released produce four fissions, and so forth as shown in Figure 20.11. The number of fissions and their associated energies quickly escalate, and, if unchecked, the result is a violent explosion. Reactions that multiply in this fashion are called **branching chain reactions.**

In order for a chain fission reaction to occur, the sample of fissionable material must have a minimum size. Otherwise, neutrons escape from the sample before they have an opportunity to strike a nucleus and cause fission. The chain stops if enough neutrons are lost. The reaction is then said to be **subcritical.** If the mass is large enough to maintain the chain reaction with a constant rate of fission, the reaction is said to be **critical.** This situation results if only one neutron from each fission is subsequently effective in producing another fission. If the mass is larger yet, few of the neutrons produced are able to escape. The chain reaction then multiplies the number of fissions and the reaction is said to be **supercritical.** The effect of mass size on whether a reaction is subcritical, critical, or supercritical is illustrated in Figure 20.12. One of the ways that an atomic bomb is triggered to produce a supercritical mass is shown in Figure 20.13. As shown, two subcritical masses are brought together by use of chemical explosives to form a supercritical mass.

What constitutes a critical mass varies for different fissionable materials. It also varies somewhat with the shape of the mass and how densely it is packed. A ball of 6 kg of plutonium-239 encased in natural uranium is a critical mass. The small size of this critical mass has raised the fear that enough material could be stolen to build an atomic bomb.

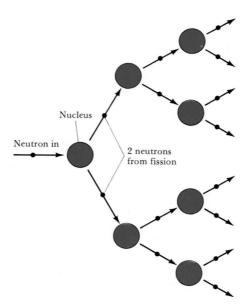

Nucleus

Neutron in

2 neutrons
from fission

FIGURE 20.11 A chain fission reaction in which each fission produces two neutrons. The process leads to an accelerating rate of fission, with the number of fissions doubling at each stage.

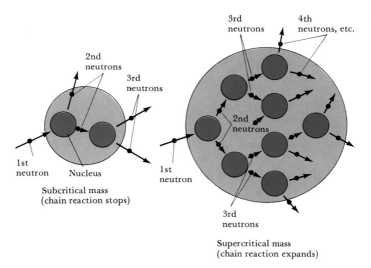

FIGURE 20.12 The chain reaction in a subcritical mass soon stops because neutrons are lost from the mass without causing fission. As the size of the mass increases, fewer neutrons are able to escape, and the chain reaction is able to expand.

We have seen that the basic design of such a bomb is amazingly simple. The fissionable materials are potentially available to any nation with a nuclear reactor. This simplicity has already resulted in the proliferation of atomic weapons. It is possible that such weapons could come into the hands of terrorist groups or groups intent on using the weapons to extort money from corporations or nations. Such events would add a frightening aspect to the nuclear age in which we live.

The fission of uranium-235 was first achieved in the late 1930s by Enrico Fermi and his colleagues in Rome and shortly thereafter by Otto Hahn and his co-workers in Berlin. Both groups were trying to produce transuranium elements. In 1938 Hahn identified barium among his reaction products. He was puzzled by this observation and questioned the identification because the presence of barium was so unexpected. He sent a detailed letter describing his experiments to Lise Meitner, a former co-worker. Meitner had been forced to leave Germany because of the antisemitism of the Third Reich, and she had settled in Sweden.

She surmised that Hahn's experiment indicated that a new nuclear process was occurring in which the uranium-235 split. She called this nuclear fission. Meitner passed word of this discovery to her nephew, Otto Frisch, a physicist working at Niels Bohr's institute in Copenhagen. He repeated the experiment, verifying Hahn's observations and finding that tremendous energies were involved. In January 1939, Meitner and Frisch published a short article describing this new reaction. In March 1939, Leo Szilard and Walter Zinn at Columbia University discovered that more neutrons are produced than were used in each

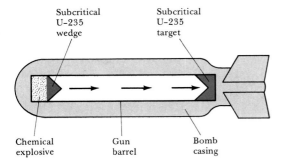

FIGURE 20.13 One design used in atomic bombs. A conventional explosive is used to bring two subcritical masses together to form a supercritical mass.

fission. As we have seen, this allows a branching chain reaction. News of these discoveries and an awareness of their potential use in explosive devices spread rapidly within the scientific community. Several scientists finally persuaded Albert Einstein, the most famous physicist of the time, to write a letter to President Roosevelt outlining the implications of these discoveries. Einstein's letter, written in August 1939, outlined the possible military applications of nuclear fission and emphasized the danger that weapons based on fission would pose if they were to be developed by the Nazis. Roosevelt judged it imperative that the United States investigate the possibility of such weapons. Late in 1941 the decision was made to build a bomb based on the fission reaction. An enormous research project, known as the "Manhattan Project," was initiated. This project led to the development of the atomic bomb and the dawning of the nuclear age.

NUCLEAR REACTORS

Nuclear fission produces the energy generated by nuclear power plants. The "fuel" of the nuclear reactor is therefore a fissionable substance such as uranium-235. Typically, uranium is enriched to about 3 percent uranium-235 and then used in the form of UO_2 pellets. These enriched uranium pellets are encased in zirconium or stainless-steel tubes. Rods composed of materials such as cadmium or boron are used to control the fission process by absorbing neutrons. These **control rods** permit a sufficient flux of neutrons to keep the reaction chain self-sustaining but yet prevent the reactor core from overheating.* The reactor is started by using a neutron source; it is stopped by inserting the control rods more deeply into the reactor core, the site of the fission (Figure 20.14). The reactor core also contains a **moderator** that acts to slow down neutrons, so that they can be captured more readily by the fuel. Finally a **cooling liquid** circulates through the reactor core to carry off the heat generated by the nuclear fission. The cooling liquid can also serve as the neutron moderator.

The design of the power plant is basically the same as that of a power plant that burns fossil fuel (except that the burner is replaced by the reactor core). In both instances, steam is used to drive a turbine that is connected to an electrical generator. Because the steam must be condensed, additional cooling water is needed. This water is generally obtained from a large source such as a river or lake. The water is returned to its source at a higher temperature than when it was removed. Power plants are therefore a significant source of thermal pollution. The nuclear power plant design shown in Figure 20.15 is the one that is currently most popular. The primary coolant, which passes through the core, is in a closed system. Subsequent coolants never pass through the reactor core at all. This lessens the chance that radioactive products could escape the core. Additionally, the reactor is surrounded by a concrete shell to shield personnel and nearby residents from radiation.

Fission products accumulate as the reactor operates. These products lessen the efficiency of the reactor by capturing neutrons. Therefore, the reactor must be stopped periodically, so that the nuclear fuel can be reprocessed. Fuel rods are removed from the reactor core and stored for several months to allow decay of short-lived nuclei. The rods are then transported in shielded containers to reprocessing plants where the fuel is

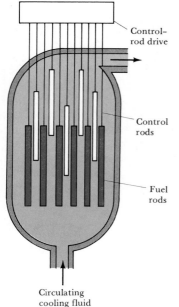

FIGURE 20.14 Reactor core showing fuel elements, control rods, and cooling fluid.

*The reactor core cannot reach supercritical levels and explode with the violence of an atomic bomb because the concentration of uranium-235 is too low. However, if the core overheats, sufficient damage might be done to release radioactive materials into the environment.

610

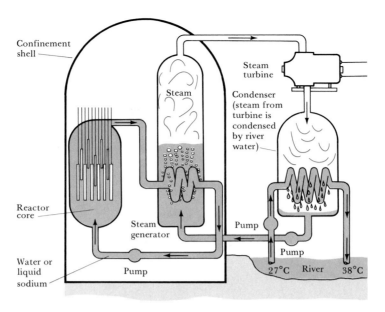

FIGURE 20.15 Basic design of a nuclear power plant. Heat produced by the reactor core is carried by a cooling fluid such as water or liquid sodium to a steam generator. The steam so produced is used to drive an electrical generator.

separated from the fission products. Because of the high radioactivity of the spent fuel, this is a hazardous operation. Storage of the leftover fission products poses a major problem because they are extremely radioactive. It is estimated that 20 half-lives are required for their radioactivity to reach levels acceptable for biological exposure. Based on the 28.8-yr half-life of strontium-90, one of the longer-lived and most dangerous of the products, the wastes must be stored for 600 yr. If plutonium-239 is not removed, storage must be for longer periods; plutonium-239 has a half-life of 24,000 yr. It is advantageous, however, to remove plutonium-239 because it can be used as a fissionable fuel. It is mainly on the problems of fuel reprocessing and waste-product storage that critics of nuclear energy focus most of their attention. They ask, for example, whether it is possible safely to store radioactive wastes for a period as long as 600 yr. The geological and political changes that can occur in 600 yr make it difficult to answer this question. The decision to rely increasingly on nuclear power will leave future generations a legacy of radioactive wastes. The basic question is whether benefits justify risks. Thus far there is no consensus among scientists concerning that question.

BREEDER REACTORS

Some scientists have estimated that it will take 40 yr or less to use all of the relatively low-cost uranium-235 available in this country. Because uranium-235 is very rare and because its supply can be exhausted so readily, methods of generating other fissionable materials are being actively investigated. Fissionable plutonium-239 and uranium-233 can be produced in nuclear reactors from nuclides that are far more abundant than uranium-235. The reactions involved are as follows:

$$^{238}_{92}\text{U} + ^{1}_{0}\text{n} \longrightarrow ^{239}_{92}\text{U}$$

$$^{239}_{92}\text{U} \longrightarrow ^{239}_{93}\text{Np} + ^{0}_{-1}\text{e} \qquad (t_{1/2} = 24 \text{ min})$$

$$^{239}_{93}\text{Np} \longrightarrow ^{239}_{94}\text{Pu} + ^{0}_{-1}\text{e} \qquad (t_{1/2} = 2.3 \text{ days})$$

$$^{232}_{90}\text{Th} + ^{1}_{0}\text{n} \longrightarrow ^{233}_{90}\text{Th}$$

$$^{233}_{90}\text{Th} \longrightarrow ^{233}_{91}\text{Pa} + ^{0}_{-1}\text{e} \qquad (t_{1/2} = 22 \text{ min})$$

$$^{233}_{91}\text{Pa} \longrightarrow ^{233}_{92}\text{U} + ^{0}_{-1}\text{e} \qquad (t_{1/2} = 27 \text{ days})$$

It is theoretically possible to build a reactor that both produces energy and converts uranium-238 or thorium-232 into fissionable fuel. We can envision a fission of a uranium-235 nucleus producing two neutrons, one causing another fission, the second initiating the change of uranium-238 into plutonium-239. Plutonium-239 is produced in ordinary reactors because of the presence of uranium-238. However, the hope is to be able to produce more fuel than the reactor uses. Reactors able to do this are still under development; they are called **breeder reactors.**

The design of breeder reactors poses many difficult technical problems. The reactor must be designed to permit very efficient use of neutrons. The necessary efficiency is difficult to attain because fission products and structural parts of the reactor absorb neutrons. The larger the number of neutrons produced per fission, the easier it is to produce energy and new fuel simultaneously. Fission of plutonium-239 using fast neutrons* provides a high neutron yield (2.9 neutrons per fission). Most attention has therefore been focused on such plutonium-239 breeders, which are known as fast-breeders. Use of plutonium introduces further design problems. Because plutonium melts at a rather low temperature (640°C), heat must be transferred very efficiently from the reactor. However, because the neutrons must be fast, the cooling agent cannot be a neutron moderator. Liquid sodium has been used. However, sodium poses problems because of its chemical reactivity. To further complicate matters, plutonium is one of the most dangerous poisons known.

20.8

nuclear fusion

As shown in Section 20.6, energy is produced when light nuclei are fused to form heavier ones. Reactions of this type are responsible for the energy produced by the sun. Spectroscopic studies indicate that the sun is composed of 73 percent H, 26 percent He, and only 1 percent of all other elements, by mass. Among the several fusion processes that are believed to occur are the following:

$$^{1}_{1}\text{H} + ^{1}_{1}\text{H} \longrightarrow ^{2}_{1}\text{H} + ^{0}_{1}\text{e} \qquad [20.29]$$

$$^{1}_{1}\text{H} + ^{2}_{1}\text{H} \longrightarrow ^{3}_{2}\text{He} \qquad [20.30]$$

$$^{3}_{2}\text{He} + ^{3}_{2}\text{He} \longrightarrow ^{4}_{2}\text{He} + 2^{1}_{1}\text{H} \qquad [20.31]$$

$$^{3}_{2}\text{H} + ^{1}_{1}\text{H} \longrightarrow ^{4}_{2}\text{He} + ^{0}_{1}\text{e} \qquad [20.32]$$

*By fast neutrons we mean those that are moving at high speed and consequently have high kinetic energy. The neutrons used in a reactor that utilizes ^{235}U are known as slow or thermal neutrons. They are the ones that have kinetic energies of about the same magnitude as those of ordinary gas molecules at room temperature.

Theories have been proposed for generation of other elements via fusion processes.

Fusion is appealing as an energy source because of the availability of light isotopes and because fusion products are generally not radioactive. It is therefore potentially a cleaner process than is fission. The problem is that high energies are required to overcome the repulsion between nuclei. The required energies are achieved by high temperatures. Fusion reactions are therefore also known as **thermonuclear reactions**. The lowest temperature required for any fusion is that needed to fuse $_1^2\text{H}$ and $_1^3\text{H}$ as shown in Equation [20.33]. This reaction requires a temperature of 40,000,000 K:

$$_1^2\text{H} + {}_1^3\text{H} \longrightarrow {}_2^4\text{He} + {}_0^1\text{n} \qquad\qquad [20.33]$$

Such high temperatures have been achieved by using an atomic bomb to initiate the fusion process. This is done in the thermonuclear, or hydrogen, bomb. Clearly, this approach is unacceptable for controlled power generation.

Numerous problems must be overcome before fusion becomes a practical energy source. Besides the high temperatures necessary to initiate the reaction, there is the problem of confining the reaction. No known structural material is able to withstand the enormous temperatures necessary for fusion. Research has centered on the use of strong magnetic fields to contain the reaction. The possibility of using lasers to generate the temperatures required for fusion has been the subject of recent studies. Although there is reason for some degree of optimism, it is impossible to tell if and when the tremendous technical difficulties involving nuclear fusion will be overcome. It is therefore not yet clear whether fusion will ever be a practical source of energy for mankind.

20.9 effects of radiation

The increased pace of synthesis and use of radioisotopes has led to increased concern over the effects of radiation on matter, particularly in biological systems. We therefore close this chapter by examining the health hazards associated with radioisotopes.

The biological damage caused by radiation is due to its ability to ionize and fragment molecules. As radiation moves through matter it gives up its energy to the molecules it encounters, thereby leaving a trail of ions and molecular debris. The chemicals produced in this fashion are very reactive, and they are consequently able to disrupt the normal operation of the body's cells. The damage caused by a radiation source outside the body depends on the penetrating ability of the radiation. Gamma rays are particularly dangerous because they penetrate human tissue very effectively, just as do X rays. Consequently, their damage is not limited to the skin. In contrast, most alpha rays are stopped by skin, and beta rays are able to penetrate only about 1 cm beyond the surface of the skin. Hence neither is as dangerous as gamma rays unless the radiation source somehow enters the body. Within the body, alpha rays are particularly dangerous because they leave a very dense trail of debris as they move through matter.

The chemical form of a radioisotope determines how easily it can enter an organism. It also determines whether the radioisotope will be retained in the body and, if so, where. Krypton-85 and strontium-90 illustrate these points. Krypton-85 is formed in nuclear fission and is released into the atmosphere during the reprocessing of nuclear fuels. Because krypton is chemically inert, no simple way of chemically immobilizing it has been devised. Once it is out in the atmosphere, the krypton-85 affects the skin and lungs. However, its lack of reactivity keeps it from getting elsewhere into the body or accumulating there. Strontium-90 is also formed by fission. Because it is an alkaline earth, strontium is able to replace calcium in its compounds. It is therefore able to enter bones where its radiation can cause bone cancer and leukemia.

Another important factor to consider is "biological concentration" (Section 18.8). As a substance passes from organism to organism through a food chain, it can become increasingly concentrated. For example, the concentration of strontium-90 may be very low in grass. However, a cow eats a great deal of it. Consequently, the concentration of strontium-90 becomes greater in the cow than in the grass. A person who then drinks milk taken from several cows could in turn end up with a concentration of strontium-90 that is higher than that found in the cows.

The effects of radiation on living systems can be classified as either **somatic** or **genetic**. Somatic damage is that which affects the organism during its own lifetime. Genetic damage is that which has a genetic effect; it harms offspring through damage to genes and chromosomes, the body's reproductive material. It is more difficult to study genetic effects than somatic ones because it may take several generations for genetic damage to become apparent. Somatic damage includes "burns," molecular disruptions similar to those produced by high temperatures. It also includes cancer. Cancer is brought about by damage to the growth-regulation mechanism of cells, which causes them to reproduce in an uncontrolled manner. In general, the tissues that show the greatest damage from radiation are those that reproduce at a rapid rate, such as bone marrow, blood-forming tissues, and lymph nodes. Leukemia is probably the major cancer problem associated with radiation.

The clinical symptoms of acute (short-term) exposure to radiation include a decrease in the number of white blood cells, fatigue, nausea, and diarrhea. Sufficient exposure can result in death from blood disorders, gastrointestinal failure, and damage to the central nervous system. In light of these effects, it is important to determine whether there are any safe levels of exposure to radiation. What are the maximum levels of radiation that we should permit from various human activities? Unfortunately, we are hampered in our attempts to set realistic standards by our lack of understanding of the effects of chronic (long-term) exposure to radiation. Two models have been proposed for such exposure. According to one of these, the linear model, the effects of radiation are proportional to exposure, even down to low exposures. According to the other, the threshold model, no effects occur below a certain threshold of exposure. The two models are compared in Figure 20.16. Much of the disagreement about the safety of various radiation sources, including nuclear power plants, boils down to arguments about which of these models describes the effects of long-term exposure.

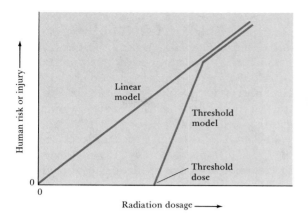

FIGURE 20.16 Comparison of two models of radiation damage. According to the linear model, any radiation dosage produces some finite risk of injury, even at low dosage. According to the threshold model, there is no risk of injury if the dosage is kept below a certain level called the threshold level.

RADIATION DOSES

A gram of radium undergoes 3.7×10^{10} nuclear disintegrations in 1 sec. This number of disintegrations is known as a **curie** (Ci). The curie is the unit used in measuring nuclear activity. For example, a 5 millicurie sample of cobalt-60 undergoes $(5 \times 10^{-3})(3.7 \times 10^{10}) = 1.8 \times 10^8$ disintegrations per second.

The damage produced by radiation depends not only on the nuclear activity but also on the energy and penetrating power of the radiation. Therefore two other units, the rad and the rem, are commonly used to measure radiation doses. (A third unit, the roentgen, is essentially the same as the rad.) A **rad** (radiation *a*bsorbed *d*ose) is the amount of radiation that deposits 1×10^{-2} J of energy per kilogram of tissue. A rad of alpha rays can produce more damage than a rad of beta rays. Consequently, the rad is often multiplied by a factor that measures the relative biological damage caused by the radiation. This factor is known as the relative biological effectiveness of the radiation, abbreviated RBE. The RBE is approximately one for beta and gamma rays and ten for alpha rays. The exact value for the RBE varies with dose rate, total dose, and type of tissue affected. The product of the number of rads and the RBE of the radiation gives the effective dosage in **rems** (*r*oentgen *e*quivalent for *m*an):

$$\text{Number of rems} = (\text{number of rads})(\text{RBE}) \qquad [20.34]$$

The effects of some short-time exposures to radiation are given in Table 20.4. The average chest X ray involves a radiation dose of 0.2 rem.

TABLE 20.4

effects of short-time exposures to radiation

DOSE (rem)	EFFECT
0–25	No detectable clinical effects
25–50	Slight, temporary decrease in white blood cell counts
100–200	Nausa; marked decrease in white blood cells
500	Death of half the exposed population within 30 days after exposure

Most radiation from human activity comes from our use of X rays; only 0.2 percent comes from nuclear power plants. Additionally, each of us receives an average of 0.1–0.2 rem each year as background radiation from natural sources such as naturally occurring radioisotopes. The present radiation-protection standards in this country permit an average exposure of 0.170 rem per year and a maximum exposure of 0.500 rem per year from human activities to individuals in the general population (exclusive of that obtained from medical sources). Critics have claimed that this standard is too high and that long-range exposure to even such small doses can cause cancer or genetic damage. Controversy over this matter persists because of our lack of knowledge of the long-range effects of radiation. However, even given a clearer picture of the effects of radiation, we would still have to weigh the risks of using X rays and nuclear power against its benefits.

FOR REVIEW

summary

Certain nuclei are **radioactive**. Most of these nuclei gain stability by emitting **alpha particles** ($^4_2$He), **beta particles** ($_{-1}^{0}$e), and/or **gamma radiation** ($^0_0\gamma$). Some nuclei undergo decay by **positron** ($^0_1$e) emission or by **electron capture**. The **neutron-proton** ratio is one factor determining nuclear stability. The presence of **magic numbers** of **nucleons** and an even number of protons and neutrons are also important. **Nuclear transmutations** can be induced by bombarding nuclei with charged particles using particle accelerators or with neutrons in a nuclear reactor.

Radioisotopes have characteristic rates of decay. These rates are generally expressed in terms of **half-lives**. The constant half-lives of nuclides permit their use in dating objects. The ease of detection of radioisotopes also permits their use as **tracers**, to follow elements through their reactions. Three methods of detection were discussed: use of **photographic film**, **scintillation counters**, and **Geiger counters**.

The energy produced in nuclear reactions is accompanied by measurable losses of mass in accordance with Einstein's relationship, $\Delta E = c^2 \Delta m$. The difference in mass between nuclei and the nucleons of which they are composed is known as the **mass defect**. The mass defect of a nuclide allows calculation of its nuclear **binding energy**, the energy required to separate the nucleus into individual nucleons. Examination of binding energies per nucleon reveals that energy is produced when heavy nuclei split (**fission**) and when light nuclei fuse (**fusion**).

Uranium-235, uranium-233, and plutonium-239 undergo fission when they capture a neutron. The resulting nuclear reaction is a **chain reaction**. If the reaction maintains a constant rate it is said to be **critical**. If it slows it is said to be **subcritical**. In the atomic bomb, subcritical masses are brought together to form a mass that is **supercritical**. In nuclear reactors the fission is controlled to generate a constant power. The reactor core consists of fissionable fuel, control rods, moderator, and cooling fluid. The nuclear power plant resembles a conventional power plant except that the core replaces the fuel burner. In **breeder reactors** more nuclear fuel is produced than is used to generate energy.

Nuclear fusion requires high temperatures because nuclei must have large kinetic energies to overcome their mutual repulsions. It has not yet been possible to generate a controlled fusion process.

The nuclear activity of a radioisotope is measured in **curies**. Radiation doses are measured in **rads** and **rems**. Radiation is able to ionize and fragment molecules. In doing so, it can cause both **somatic** and **genetic** damage. The effects of long-term exposure to radiation are not as well understood as those resulting from short-term exposure.

learning goals

Having read and studied this chapter, you should be able to:

1. Write the nuclear symbols for protons, neutrons, electrons, alpha particles, and positrons.
2. Determine the effect of different types of decay on the proton-neutron ratio and predict the type of decay that a nucleus will undergo based on its position relative to the belt of stability.
3. Complete and balance nuclear equations, having been given all but one of the particles involved.
4. Use the half-life of a substance to predict the amount of radioisotope present after a given period of time.
5. Calculate half-life, age of an object, or the remaining amount of radioisotope, having been given any two of these pieces of information.
6. Explain how radioisotopes can be used in dating objects and as tracers.
7. Explain how radioactivity is detected, including a description of the basic design of a Geiger counter.
8. Use Einstein's relation, $\Delta E = c^2 \, \Delta m$, to calculate the energy change or the mass change of a reaction, having been given one of these quantities.
9. Calculate the binding energies of nuclei, having been given their masses and the masses of protons and neutrons.
10. Explain what fission and fusion are and what types of nuclei produce energy when undergoing these processes.
11. Describe the design of a nuclear power plant, including an explanation of the role of fuel elements, control rods, moderator, and cooling fluid.
12. Explain the roles played by the chemical behavior of an isotope and its mode of radioactivity in determining its ability to damage biological systems.
13. Define the units used in discussing nuclear activity (curie) and the effects of radiation on biological systems (rem and rad).

key terms

Among the more important terms and expressions used for the first time in this chapter are the following:

Alpha particles (Section 20.1) are identical to helium-4 nuclei, consisting of two protons and two neutrons, symbol ^4_2He.

Beta particles (Section 20.2) are energetic electrons emitted from the nucleus, symbol $^0_{-1}\text{e}$.

A **breeder reactor** (Section 20.7) is a nuclear-fission reactor that produces more fissionable fuel than it consumes.

The **binding energy** (Section 20.6) of a nucleus is the energy required to decompose that nucleus into nucleons; it is usually calculated from the **mass defect** of the nucleus.

A **chain reaction** (Section 20.7) is a series of reactions in which one reaction initiates the next.

A **critical mass** (Section 20.7) is the amount of fissionable material necessary to maintain a chain reaction. Smaller masses are said to be **subcritical,** whereas larger ones are **supercritical.**

A **curie** (Section 20.9) is a measure of radioactivity: 1 curie = 3.7×10^{10} nuclear disintegrations per second.

Electron capture (Section 20.2) is a mode of radioactive decay in which an inner shell ($1s$) electron is captured by the nucleus.

Fission (Section 20.7) is the splitting of a large nucleus into two intermediate-sized ones.

Fusion (Section 20.8) is the joining of small nuclei to form larger ones.

Gamma rays (Section 20.2) are energetic electromagnetic radiation emanating from the nucleus of a radioactive atom.

Half-life (Section 20.3) is the time required for half of a sample of a particular radioisotope to decay.

The mass defect (Section 20.6) of a nucleus is the difference between the mass of that nucleus and the total masses of the individual nucleons that it contains.

A nuclear transmutation (Section 20.1) is a conversion of one kind of atom to another.

A positron (Section 20.2) is a particle with the same mass as an electron but with a positive charge, symbol $_1^0e$.

A rad (Section 20.9) is a measure of the energy absorbed from radiation by tissue or other biological material; 1 rad = transfer of 1×10^{-2} J of energy per kilogram of material.

A radioisotope (Section 20.1) is an isotope that is radioactive; that is, it is undergoing nuclear changes with emission of nuclear radiation.

A rem (Section 20.9) is a measure of the biological damage caused by radiation; rems = rads $\times$ RBE.

RBE (Section 20.9) (relative biological equivalent) is an adjustment factor used to convert rads to rems; it accounts for differences in biological effects of different particles having the same energy.

EXERCISES

nuclear reactions

20.1 Indicate the number of protons and neutrons in each of the following nuclei: (a) strontium-90; (b) $_{55}^{137}Cs$; (c) uranium-235; (d) ^{131}I.

20.2 Indicate the number of protons, neutrons, and nucleons in each of the following nuclei: (a) plutonium-239; (b) ^{14}C; (c) lead-206; (d) $_{17}^{35}Cl$.

20.3 Write balanced nuclear equations for the following transformations: (a) plutonium-241 undergoes beta decay; (b) thorium-232 decays to radium-228; (c) yttrium-84 undergoes positron emission; (d) titanium-44 undergoes electron capture.

20.4 Write balanced nuclear equations for the following transformations: (a) americium-241 undergoes alpha decay; (b) iodine-131 undergoes beta decay; (c) thorium-234 decays to protactinium-234; (d) chlorine-34 decays to sulfur-34.

20.5 Complete and balance the following nuclear equations by supplying the missing particle:

(a) $_{36}^{87}Kr \longrightarrow {}_{-1}^{0}e + ?$

(b) $_{26}^{56}Fe + {}_{-1}^{0}e \longrightarrow ?$

(c) $_{24}^{53}Cr + {}_2^4He \longrightarrow {}_0^1n + ?$

(d) $? \longrightarrow {}_{12}^{24}Mg + {}_{-1}^{0}e$

(e) $_{92}^{235}U + {}_0^1n \longrightarrow {}_{56}^{140}Ba + ? + 2{}_0^1n$

20.6 Complete and balance the following nuclear equations by supplying the missing particle:

(a) $_{92}^{235}U \longrightarrow {}_2^4He + ?$

(b) $_{29}^{64}Cu \longrightarrow {}_1^0e + ?$

(c) $_{12}^{24}Mg + {}_0^1n \longrightarrow {}_1^1H + ?$

(d) $_4^9Be + {}_1^1H \longrightarrow {}_3^6Li + ?$

(e) $_{92}^{235}U + {}_0^1n \longrightarrow {}_{40}^{99}Sr + {}_{52}^{135}Te + ?{}_0^1n$

(f) $? \longrightarrow {}_{-1}^{0}e + {}_5^7B$

20.7 Write balanced equations for the following processes: (a) $^{14}_7N(n, p)^{14}_6C$; (b) $^{15}_7N(p, \alpha)^{12}_6C$; (c) $^{35}_{17}Cl(n, p)^{35}_{16}S$.

20.8 Figure 20.3 shows the stepwise decay of uranium-238 to form the stable lead-206 nucleus. Write balanced nuclear equations for each step in this sequence.

nuclear stability

20.9 Which of the following three nuclides is

most likely to be radioactive: (a) $^{114}_{48}Cd$; (b) $^{114}_{49}In$, (c) $^{114}_{50}Sn$? Briefly justify your choice.

20.10 Which of the following nuclides would you expect to be radioactive: (a) $^{40}_{20}Ca$; (b) $^{210}_{84}Po$; (c) $^{54}_{25}Mn$? Briefly justify your choice(s).

20.11 Which member of the following pairs would you expect to be most abundant: (a) $^{14}_{6}C$ or $^{14}_{7}N$; (b) $^{16}_{8}O$ or $^{16}_{9}F$? Briefly justify your choices.

20.12 How does positron emission from $^{22}_{11}Na$ increase nuclear stability?

20.13 Which of the following will increase the proton-to-neutron ratio of iodine-131: (a) proton emission; (b) positron emission; (c) alpha particle emission; (d) beta decay; (e) gamma emission; (f) electron capture?

20.14 The neutron-to-proton ratio for maximum stability in the mass number range of 206 to 212 is 1.52. Based on this fact, would you expect bismuth-210 to emit an alpha or beta particle?

20.15 Predict whether positron or beta emission is expected for phosphorus-34 if this nucleus has a neutron-to-proton ratio that is too large to place it in the belt of stability.

20.16 Why do ^{14}C and ^{12}C have different nuclear properties, even though their chemical properties are essentially identical?

20.17 Harmful chemicals are often destroyed by chemical treatment. For example, an acid can be neutralized by a base. Why can't chemical treatment be applied to destroy the fission products produced in a nuclear reactor?

20.18 Give a reason why electron capture is more common among elements of high atomic number than among those of low atomic number.

half-life and dating

20.19 The half-life of hydrogen-3 (tritium) is 12.3 yr. How much of a 20-mg sample will remain after 24.6 yr?

20.20 If 75 percent of a radioisotope decays in 5.0 yr, what is its half-life?

20.21 The half-life of cesium-137 is 30 yr. How much cesium-137 remains in an 8.00-mg sample after 10 yr have elapsed?

20.22 The half-life of plutonium-239 is 24,000 yr. What fraction of a sample of plutonium-239 will remain after 1000 yr?

20.23 The half-life of cobalt-60 is 5.26 yr. How much of a 100-mg sample will decay in 1 yr?

20.24 What is the age of the following objects: (a) a brandy whose tritium content is 5 percent of that in living plants ($t_{1/2}$ for tritium is 12.3 yr); (b) a rock containing uranium-238 and lead-206 in a weight ratio of 1.50 to 1.00?

20.25 What is the age of the following objects: (a) a rock if its weight ratio of argon-40 to potassium-40 is 4.8 (potassium-40 decays to argon-40 with a half-life of 1.27×10^9 yr); (b) wood from an Egyptian tomb that has a carbon-14 content 0.610 times that of living plants?

20.26 A sample of iodine-131 ($t_{1/2} = 8.1$ days) had an original activity of 0.500 millicuries. What is the activity of the sample after 14 days?

20.27 A radioactive sample had an initial activity of 3112 disintegrations per second. After 42 days its activity had fallen to 980 disintegrations per second. What is the half-life of the substance?

[20.28] According to current regulations, the maximum permissible dose of strontium-90 in the body of an adult is 1 microcurie (1×10^{-6} Ci). Using the relationship activity $= kN$, calculate the number of atoms of strontium-90 this corresponds to. What mass of strontium-90 does this correspond to ($t_{1/2}$ for strontium-90 is 27.6 yr)?

mass-energy relationships

20.29 Calculate the binding energy per nucleon for the following nuclei: (a) $^{16}_{8}O$ (mass = 15.99052 amu); (b) $^{230}_{90}Th$ (mass = 229.9837 amu).

20.30 Calculate the binding energy per nucleon for the following nuclei: (a) $^{35}_{17}Cl$ (mass = 34.95952 amu); (b) $^{40}_{20}Ca$ (mass = 39.95162 amu); (c) $^{196}_{79}Au$ (mass = 195.9231 amu).

20.31 How much energy must be supplied to break a single ^7_3Li nucleus into separated protons and neutrons if this nucleus has a mass of 7.01436 amu?

20.32 Calculate the mass loss that accompanies loss of energy when a mole of graphite is combusted under standard conditions ($\Delta H° = -393.5$ kJ/mole).

20.33 It is estimated that the sun gives off 4×10^{16} J of energy each second. Calculate the mass loss that is required to produce this energy.

20.34 How much energy is produced when a single ^7_3Li and a single ^2_1H fuse?

$$^7_3Li + {}^2_1H \longrightarrow 2{}^4_2He + {}^1_0n$$

The mass of ^7_3Li is 7.01436 amu, whereas those of ^2_1H, ^4_2He, and ^1_0n are 2.01355 amu, 4.00150 amu, and 1.00867 amu, respectively.

[20.35] One of the ways uranium-235 can undergo fission is as follows:

$$^1_0n + {}^{235}_{92}U \longrightarrow {}^{142}_{56}Ba + {}^{91}_{36}Kr + 3{}^1_0n$$

The barium-142 and krypton-91 so formed undergo successive beta emissions with short half-lives to form the stable zirconium-91 and cerium-142 nuclei. The net reaction that gives these nuclei, starting with uranium-235, is

$$^{235}_{92}U \longrightarrow ^{142}_{58}Ce + ^{91}_{40}Zr + 2^1_0n + 6^0_{-1}e$$

The masses of $^{235}_{92}U$, $^{142}_{58}Ce$, $^{91}_{40}Zr$, ^1_0n, and $^0_{-1}e$ are 234.9934 amu, 141.8772 amu, 90.8833 amu, 1.00867 amu, and 0.000549 amu, respectively. Calculate the energy produced by the fission of a single $^{235}_{92}U$ nucleus in this fashion.

[20.36] Using the data in the preceding question, calculate the quantity of energy produced by a kilogram of uranium-235 if it all reacted to form cerium-142 and zirconium-91. How much coal would have to be burned to produce the same energy as a kilogram of uranium-235 decomposing as described in the preceding question. The fuel value of the coal can be taken to be 30 kJ/g.

20.37 Would fission of sulfur-32 be exothermic or endothermic? Justify your answer.

effects and uses of radioisotopes

20.38 How could radioisotopes be used to detect the location of a leak in a pipe buried under a cement floor?

20.39 How could radioisotopes be used to prove that chemical equilibrium is a dynamic and not a static process?

20.40 Explain why Rutherford observed artificial transmutation in the case of $^{14}_7N$ but not when he bombarded gold foils with alpha particles.

20.41 Why is it necessary to have very high temperatures in order to initiate a fusion reaction?

20.42 Why do alpha particles but not neutrons have to be accelerated to cause nuclear reactions?

20.43 Explain why a higher temperature is required to initiate fusion between lithium and hydrogen nuclei than is required to initiate fusion between hydrogen nuclei?

20.44 Why can't water be used as a coolant in fast-breeder reactors?

20.45 Explain why an explosive chain reaction does not occur in uranium ores.

20.46 Would you expect the following isotopes to be accumulated in the body: (a) strontium-90; (b) cesium-137; (c) iodine-131; (d) krypton-87; (e) iron-56? If so, where?

[20.47] So long as extreme precautions are taken to avoid entry of plutonium into the body, through the lungs or by ingestion, it can be handled with little concern for its radioactivity. Why is this so? Is this true of all radioactive substances—for example, cobalt-60? Explain.

general exercises

20.48 Distinguish between members of the following pairs: (a) moderator and control rod; (b) positron and proton; (c) alpha ray and beta ray; (d) critical and subcritical mass; (e) fission and fusion; (f) nucleon and neutron; (g) somatic and genetic effects; (h) rem and rad; (i) acute and chronic exposure to radiation; (j) X rays and gamma rays.

20.49 Determine the product nucleus in the following cases: (a) $^{75}_{33}As$ (α, n) —; (b) ^7_3Li (p, n) —; (c) $^{31}_{15}P$ (^2_1H, p) —.

20.50 Predict the type of emission expected from the following nuclei; (a) magnesium-23; (b) selenium-83; (c) francium-221.

[20.51] In the late nineteenth century, Marie Curie found two new elements, both radioactive, in uranium ore. These two elements were radium and polonium. Radium is in group 2A of the periodic table, whereas polonium is in group 6A. Why do these elements occur in nature together with uranium rather than with the other members of their respective periodic families?

[20.52] When a positron is annihilated by combination with an electron, two photons of equal energy result. What are the wavelengths of these photons? Are they gamma-ray photons?

20.53 How many alpha-particle emissions and how many beta-particle emissions are involved in converting thorium-232 to lead-208?

20.54 Elements with atomic numbers greater than that of uranium do not exist in nature because of their short half-lives. Yet several elements with even shorter half-lives and with atomic numbers less than that of uranium do exist in nature. How can you account for this fact?

20.55 Although potassium-40 decays to produce calcium-40 with a half-life of 1.4×10^9 yr, it is not possible to date rocks by measuring the ratio of potassium-40 to calcium-40. Why not?

20.56 Given that the masses of $^{242}_{96}Cm$ and $^{244}_{98}Cf$ are 242.0588 amu and 244.0659 amu, respectively, determine whether the following reaction is exothermic:

$$^{242}_{96}Cm + ^4_2He \longrightarrow ^{244}_{98}Cf + 2^1_0n$$

20.57 About 1 molecule out of every 10^{12} molecules of atmospheric CO_2 contains carbon-14. How many carbon-14 atoms are present in a 150-lb person if our bodies contain 18 percent carbon?

[20.58] An ancient wooden object is found to have an activity of 7.0 disintegrations per minute per gram of carbon. By contrast, the carbon in a living tree undergoes 15.3 disintegrations per minute per gram of carbon. Based on the activity of carbon-14 in the object, calculate its age. Using Figure 20.6, apply the necessary correction to this "carbon-14 age" to obtain the object's true age.

chemistry of some common nonmetals

21

For the most part, the previous chapters of this book have involved principles, such as rules for bonding, the laws of thermodynamics, the construction of electrochemical cells, and so forth. In the course of explaining these principles, we have described the chemical and physical properties of many substances. In this way, you have been exposed to a considerable amount of chemistry. However, at this point you might still find it difficult to make predictions of the chemical and physical characteristics of substances based on the chemical principles and the scattered facts you have been given. As an example, suppose that you are given a closed box labeled "Fluorine." What can you predict about the properties of the substance inside? Will it be a gas in a tank or a finely divided crystalline solid? Will it be a reactive substance or one that you can confidently expose to the air? What sorts of substances is it likely to react with? You could answer many questions on the basis of principles we have already discussed. For example, you might recall from Chapter 8 that fluorine exists as F_2 molecules; furthermore, you might deduce that F_2 is a gaseous substance because it is nonpolar and has weak intermolecular attractive forces. Remembering that fluorine is the most electronegative element, you could deduce that it is a very strong oxidizing agent; it should thus be very reactive. In short, you should now have some predictive abilities regarding the properties of chemical substances.

Just as important, you have the basis for understanding and remembering the many facts of chemistry as they come before you. In this chapter, we shall consider the chemistry of several important nonmetallic elements and their compounds—the halogens, oxygen, sulfur, nitrogen, and phosphorus. As we examine these substances, we shall encounter certain facts about them that were discussed earlier. We shall also frequently rely on principles introduced in earlier chapters to help us recognize and understand relationships among the physical and chemical properties of these substances. Thus while you are broadening your understanding of chemistry, you are also reinforcing principles learned earlier.

21.1 comparison of metals and nonmetals

If you were asked to describe some of the properties of an element, you might first find its position on the periodic table. We have seen that elements can be classified into three broad categories: metals, semimetals (or metalloids), and nonmetals (Section 2.8). In the periodic table in Chapter 7 and on the front inside cover of this book, the dividing line between metals and nonmetals is shown. In Figure 21.1, the colored boxes indicate the so-called semimetals, elements that exhibit some properties of metals and some of nonmetals. Some of the properties of metals and nonmetals are summarized in Table 21.1.

Metals have a distinguishing luster that permits us to recognize them readily. They are also characterized by good electrical and thermal conductivity and by their malleability and ductility (Section 9.6). By contrast, nonmetallic elements are not lustrous and are generally poor conductors of heat and electricity. Some nonmetals (hydrogen, nitrogen, oxygen, fluorine, and chlorine) are gases. One, bromine, is a liquid; the remaining nonmetals are solids. They exhibit a variety of colors. For example, iodine vapor is deep purple; sulfur is yellow; nitrogen and oxygen gases are colorless. The solids may be hard, like diamond, or soft and crumbly, like sulfur.

Metals and nonmetals are generally distinguished also by their structures and bonding. Metals have crystal structures in which each atom has a large number of nearest neighbors (characteristically 8 or 12). The atoms usually assume a close-packed arrangement (Section 11.4) that provides maximum filling of space. Metallic bonds were discussed in Sections 9.6 and 11.4. Although the rare gases, which are nonmetals, also crystallize in close-packed arrangements, the bonding between atoms is clearly different from that of metals. The low melting points of these rare-gas elements reflect the existence of weak van der Waals forces between atoms. In other nonmetals, the number of atoms that are bonded tightly together (nearest neighbors) is generally four or less. This

TABLE 21.1

some characteristic properties of metals and nonmetals

METALS	NONMETALS
1. Distinguishing luster.	1. Lack luster, malleability, and ductility.
2. Good conductors of heat and electricity.	2. Generally poor conductors of heat and electricity.
3. Malleable and ductile.	3. Exhibit a variety of colors.
4. Oxides are generally basic.	4. Oxides are generally acidic.
5. Tend to form cations in solution.	5. Tend to form anions in solution.

Active metals

Transition metals

Nonmetals

1A	2A	3B	4B	5B	6B	7B	8B	8B	8B	1B	2B	3A	4A	5A	6A	7A	8A
1 H 1.00797																	2 He 4.0026
3 Li 6.939	4 Be 9.0122											5 B 10.811	6 C 12.01115	7 N 14.0067	8 O 15.9994	9 F 18.9984	10 Ne 20.183
11 Na 22.9898	12 Mg 24.312											13 Al 26.9815	14 Si 28.086	15 P 30.9738	16 S 32.064	17 Cl 35.453	18 Ar 39.948
19 K 39.098	20 Ca 40.08	21 Sc 44.956	22 Ti 47.90	23 V 50.942	24 Cr 51.996	25 Mn 54.9380	26 Fe 55.847	27 Co 58.9332	28 Ni 58.71	29 Cu 63.54	30 Zn 65.37	31 Ga 69.72	32 Ge 72.59	33 As 74.9216	34 Se 78.96	35 Br 79.909	36 Kr 83.80
37 Rb 85.47	38 Sr 87.62	39 Y 88.905	40 Zr 91.22	41 Nb 92.906	42 Mo 95.94	43 Tc (99)	44 Ru 101.07	45 Rh 102.905	46 Pd 106.4	47 Ag 107.870	48 Cd 112.41	49 In 114.82	50 Sn 118.69	51 Sb 121.75	52 Te 127.60	53 I 126.9044	54 Xe 131.30
55 Cs 132.905	56 Ba 137.33	57 *La 138.91	72 Hf 178.49	73 Ta 180.948	74 W 183.85	75 Re 186.2	76 Os 190.2	77 Ir 192.2	78 Pt 195.09	79 Au 196.967	80 Hg 200.59	81 Tl 204.37	82 Pb 207.19	83 Bi 208.980	84 Po (210)	85 At (210)	86 Rn (222)
87 Fr (223)	88 Ra (226)	89 †Ac (227)	104 Rf (257)	105 Ha (260)													

Lanthanide series

58 Ce 140.12	59 Pr 140.907	60 Nd 144.24	61 Pm (147)	62 Sm 150.35	63 Eu 151.96	64 Gd 157.25	65 Tb 158.924	66 Dy 162.50	67 Ho 164.930	68 Er 167.26	69 Tm 168.934	70 Yb 173.04	71 Lu 174.97

†Actinide series

90 Th 232.038	91 Pa (231)	92 U 238.03	93 Np (237)	94 Pu (242)	95 Am (243)	96 Cm (247)	97 Bk (247)	98 Cf (249)	99 Es (254)	100 Fm (253)	101 Md (256)	102 No (253)	103 Lw (257)

FIGURE 21.1 The periodic table. Colored boxes indicate the semimetals.

FIGURE 21.2 Examples of nonmetals composed of atoms having only a single nearest neighbor.

gives rise to less densely packed arrangements than that possessed by metals. For example, nitrogen, oxygen, hydrogen, and the halogens are normally diatomic. Each atom therefore has only one nearest neighbor as seen in Figure 21.2. This relationship persists when these elements are solidified. In solid chlorine, for example, the atoms within the Cl_2 molecules are separated by 1.99 Å. The closest approach between nonbonded chlorine atoms is 2.79 Å. In the case of diamond, shown earlier in Figure 8.6, each carbon atom has four nearest neighbors. In all cases the bonds between nonmetal atoms are covalent, whereas the interactions between molecules are due to van der Waals forces. Furthermore, allotropism (Section 8.3) is common; carbon, for example, occurs in two crystalline modifications, diamond and graphite.

The oxides of metals and nonmetals also tend to differ in an important way. If we consider those oxides that dissolve in water, we find that the oxides of metals tend to form basic solutions, whereas those of nonmetals form acidic solutions. Examples are shown in Equations [21.1] and [21.2].

$$\underset{\text{Metal oxide}}{Na_2O(s)} + H_2O(l) \longrightarrow \underset{\text{Base}}{2NaOH(aq)} \qquad [21.1]$$

$$\underset{\text{Nonmetal oxide}}{CO_2(g)} + H_2O(l) \longrightarrow \underset{\text{Acid}}{H_2CO_3(aq)} \qquad [21.2]$$

Such metal oxides are called **basic anhydrides,** or basic oxides, whereas nonmetal oxides are referred to as **acidic anhydrides,** or acidic oxides. Many oxides are insoluble in water. Their acidic or basic character is then evaluated by determining whether they will dissolve in acids or bases. For example, Fe_2O_3 is insoluble in water and in base solutions such as NaOH. However, it dissolves in acid, reacting as shown in Equation [21.3]:

$$Fe_2O_3(s) + 6H^+(aq) \longrightarrow 2Fe^{3+}(aq) + 3H_2O(l) \qquad [21.3]$$

It is therefore referred to as a basic oxide.

PERIODIC TRENDS

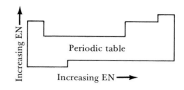

FIGURE 21.3 The general trends in electronegativity in the periodic table.

As we examine the periodic table, we should be aware of a number of general trends in chemical behavior. Recall that electronegativity increases as we move up a family and as we move from left to right across the periodic table, as shown in Figure 21.3. As a result, nonmetals tend to have higher electronegativities than do metals. Consequently, the com-

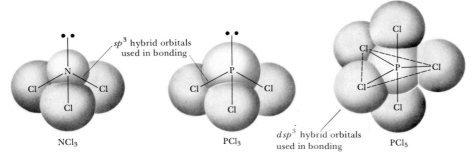

sp^3 hybrid orbitals
used in bonding

NCl₃

PCl₃

dsp^3 hybrid orbitals
used in bonding

PCl₅

FIGURE 21.4 Comparison of NCl_3, PCl_3, and PCl_5. Nitrogen is unable to form NCl_5 because it does not have *d* orbitals available for bonding.

pounds formed between metals and nonmetals tend to be ionic, with metals assuming the positive charge and nonmetals the negative charge. The solution chemistry of metals is largely that of cations, whereas the solution chemistry of nonmetals is predominately that of anions. A related trend is an increase in metallic character with an increase in atomic number within a given family.

There is one important periodic trend that we have not explicitly pointed out in our earlier discussions. The first member of each family often exhibits marked differences from subsequent members of the family. One reason for this is the smaller size and subsequently greater electronegativity of the first member. Because the first member has only the 2*s* and the three 2*p* orbitals available for bonding, it can form a maximum of four bonds. Subsequent members of the family are able to use *d* orbitals in bonding in addition to *s* and *p* orbitals; they can therefore form more than four bonds. As an example, consider the chlorides of nitrogen and phosphorus, the first two members of group 5A. Nitrogen forms a maximum of three bonds with chlorine, NCl_3. Though phosphorus can also form the trichloride compound, PCl_3, it is able in addition to form five bonds with chlorine, PCl_5. These compounds are shown in Figure 21.4.

Another difference between second-row elements and subsequent elements is the greater ability of the former to form π bonds. We can understand this, in part, in terms of atomic size. As atoms increase in size, the sideways overlap of *p* orbitals, which form the strongest type of π bond, becomes less effective. This is shown in Figure 21.5. As an illustration of this effect, consider two differences in the chemistry of carbon and silicon, the first two members of group 4A. Carbon has two crystalline allotropes, diamond and graphite. In diamond there are σ bonds between carbon atoms but no π bonds. In graphite there are π bonds resulting from sideways overlap of *p* orbitals (Section 9.3). Silicon occurs only in the diamondlike crystal form. Silicon does not exhibit a graphitelike structure because of the low stability of π bonds between silicon atoms.

We see the same type of difference in the dioxides of these elements, Figure 21.6. In CO_2 carbon forms double bonds to oxygen, thereby achieving its valence by π bonding. In contrast, SiO_2 contains no double

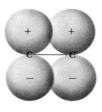

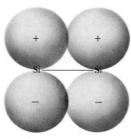

FIGURE 21.5 Comparison of π-bond formation by sideways overlap of *p* orbitals between two carbon atoms and between two silicon atoms. The distance between nuclei increases as we move from carbon to silicon because of the larger size of the silicon atom. The *p* orbitals do not overlap as effectively between two silicon atoms because of this greater separation.

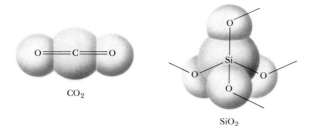

CO₂

SiO₂

FIGURE 21.6 Comparison of the structures of CO_2 and SiO_2; CO_2 has double bonds, whereas SiO_2 has only single bonds.

bonds. Instead, four oxygens are bonded to each silicon, forming an extended structure reminiscent of diamond.*

21.2
the halogens

The halogens, group 7A, consist of the elements fluorine, chlorine, bromine, iodine, and astatine. The general outer electron configuration of the family is ns^2np^5. If a halogen atom shares all of its seven outer shell electrons with a more electronegative atom, it will have an oxidation state of $+7$. It can also achieve a noble-gas configuration by gaining an electron, thereby giving it a -1 oxidation state. Correspondingly, oxidation states ranging from -1 to $+7$ are found for all but fluorine. Being the most electronegative element, fluorine exhibits only the -1 oxidation state in its compounds. The other halogens assume positive oxidation states when combined with more electronegative elements such as oxygen. The common positive oxidation states are $+1$, $+3$, $+5$, and $+7$, as seen in the series of oxyanions ClO^-, ClO_2^-, ClO_3^-, and ClO_4^-. The halogens show more pronounced similarities in their properties than do the members of any other nonmetal family except the rare gases. This is due in part to the fact that none of the halogens (with the exception of astatine) have metallic properties.

OCCURRENCE AND ISOLATION

Because of their chemical reactivities, none of the halogens occurs in nature in the elemental form. Chlorine, bromine, and iodine occur as halides in seawater and in salt deposits. The concentration of iodine in these sources is generally very small. However, iodine is concentrated by certain sea plants such as kelp. These plants can be burned and the residue used as a source of iodine. Iodine also occurs as $NaIO_3$ in deposits of Chilean saltpeter, $NaNO_3$; this compound is also used as a source of the element. Fluorine occurs in the minerals fluorspar, CaF_2, and cryolite, Na_3AlF_6. Astatine occurs in only small quantities in the decay chains of thorium and uranium. All isotopes of astatine are radioactive. The longest-lived isotope is astatine-210, which has a half-life of 8.3 hr and decays mainly by electron capture. Astatine was first synthesized by bombarding bismuth-209 with high-energy alpha particles as shown in Equation [21.4].

*The formula SiO_2 is consistent with this structure, because each oxygen is shared by two silicon atoms (not shown in Figure 21.6). For bookkeeping purposes, we may therefore count a half of each of the four oxygens that are bound to a given silicon atom as belonging to that silicon. We shall consider silicon-oxygen compounds in some detail in Chapter 22.

627

$$^{209}_{83}\text{Bi} + ^{4}_{2}\text{He} \longrightarrow ^{211}_{85}\text{At} + 2^{1}_{0}\text{n} \qquad [21.4]$$

The fact that a cyclotron must be used to synthesize the element makes it very expensive and limits its application and study.

Because of their large electrode potentials, the elements fluorine and chlorine are difficult to obtain from their halides; consequently, electrolytic oxidation is used:

$$2\text{F}^-(aq) \longrightarrow \text{F}_2(g) + 2\text{e}^- \qquad E^{\circ}_{\text{ox}} = -2.87 \text{ V}$$

$$2\text{Cl}^-(aq) \longrightarrow \text{Cl}_2(g) + 2\text{e}^- \qquad E^{\circ}_{\text{ox}} = -1.36 \text{ V}$$

Iodine and bromine are easier to obtain from their halides than are fluorine and chlorine; consequently, the necessary oxidation is normally achieved by chemical means. Chlorine can be used as the oxidizing agent, as illustrated below for the oxidation of the bromide ion:

$$\begin{aligned}
2\text{e}^- + \text{Cl}_2(g) &\longrightarrow 2\text{Cl}^-(aq) & E^{\circ}_{\text{red}} &= 1.36 \text{ V} \\
2\text{Br}^-(aq) &\longrightarrow \text{Br}_2(l) + 2\text{e}^- & E^{\circ}_{\text{ox}} &= -1.09 \text{ V} \\
\hline
\text{Cl}_2(g) + 2\text{Br}^-(aq) &\longrightarrow 2\text{Cl}^-(aq) + \text{Br}_2(l) & E^{\circ} &= 0.27 \text{ V} \quad [21.5]
\end{aligned}$$

PROPERTIES AND COMPOUNDS

Several properties of the halogens are listed in Table 21.2. Owing to its small size and high electronegativity, fluorine is very reactive. Compounds of fluorine with all elements except helium, neon, and argon have been prepared. Fluorine even reacts readily with the heavier noble gases to form KrF_2, XeF_2, XeF_4, XeF_6, and fluorides of radon.* These compounds were first prepared in the 1960s. Prior to this work, it was believed that the rare gases could form no compounds.

One of the most important compounds of fluorine is hydrogen fluoride, HF. Aqueous solutions of HF, known as hydrofluoric acid, are weakly acidic ($K_a = 6.8 \times 10^{-4}$). Hydrofluoric acid is used to etch glass, with which it reacts readily as shown in Equation [21.6]:

$$\begin{aligned}
\text{CaSiO}_3(s) + 8\text{HF}(aq) &\longrightarrow \\
&\text{H}_2\text{SiF}_6(aq) + \text{CaF}_2(aq) + 3\text{H}_2\text{O}(l) \qquad [21.6]
\end{aligned}$$

Hydrofluoric acid must therefore be stored in wax or plastic bottles.

*The high radioactivity of radon has hampered attempts to characterize the radon fluorides.

TABLE 21.2

the physical properties of the halogens

ELEMENT	MELTING POINT (°C)	BOILING POINT (°C)	PHYSICAL STATE AT 25°C	COLOR	SOLUBILITY IN H_2O (M AT 20°C)	E°_{red} (V)
Fluorine (F_2)	−223	−187	Gas	Pale yellow	—a	2.87
Chlorine (Cl_2)	−103	−35	Gas	Yellow-green	0.090	1.36
Bromine (Br_2)	−7	59	Liquid	Red-brown	0.021	1.09
Iodine (I_2)	114	183	Solid	Violet-black	0.0013	0.54

aF_2 oxidizes water.

Because of its high reactivity, F_2 is very difficult to work with. Certain metals such as copper and nickel can be used to contain F_2 because their surfaces form a protective coating of metal fluoride. Chlorine must also be handled with care. Because chlorine liquefies upon compression at room temperature, it is normally stored and handled in steel containers as a liquid.

Chlorine and the heavier halogens are reactive, though less so than fluorine. They combine directly with most elements except the rare gases. Their hydrogen halides all dissolve in water to form strong acid solutions. These halogens also form a variety of oxyacids as shown in Table 21.3.

SAMPLE EXERCISE 21.1

Determine the geometrical structure of the ClO_2^- ion.

Solution: We saw in Section 9.1 that the geometrical structures of molecules and ions can be determined by using the valence-shell electron-pair repulsion (VSEPR) model. We shall use this model to determine the structure of ClO_2^-. The Lewis-structure for the ion is $:\ddot{O}—\ddot{Cl}—\ddot{O}:^-$

Thus there are four electron pairs around chlorine; they should point toward the corners of a tetrahedron. Two of the corners are occupied by oxygen atoms, two by unshared pairs of electrons:

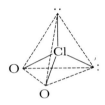

The geometrical arrangement of the atoms is thus bent.

One piece of information in Table 21.3 that bears further examination is the formula H_5IO_6 for periodic acid. It does not conform to the formulas of perchloric acid, $HClO_4$, and perbromic acid, $HBrO_4$. The size of the iodine atom is one factor that causes this difference. Iodine is large enough to be surrounded by six oxygens, and it has enough orbitals available to bond to all of them. It can therefore exist as H_5IO_6 in addition to HIO_4.

Of the oxyacids listed in Table 21.3, those of chlorine are the most common. As the number of oxygens attached to chlorine increases, the acid strength increases, a fact that is consistent with rules discussed earlier (Section 16.8). Hypochlorous acid is very weak ($K_a = 3 \times 10^{-8}$).

TABLE 21.3

the oxyacids of the halogens

OXIDATION STATE OF HALOGEN	FORMULA OF ACID				TYPE OF NAME
	F	Cl	Br	I	
+1	HOF	HOCl	HOBr	HOI	*Hypo*halous acid
+3	—	$HClO_2$	—	—	Hal*ous* acid
+5	—	$HClO_3$	$HBrO_3$	HIO_3	Hal*ic* acid
+7	—	$HClO_4$	$HBrO_4$	HIO_4, H_5IO_6	*Per*hal*ic* acid

At the other end of the scale, perchloric acid is one of the strongest acids known.

The oxidizing abilities of the oxyacids of the halogens increase as the oxidation state of the halogen is lowered. The standard reduction potentials in acid solution for the chlorine compounds can be summarized as follows:

$$ClO_4^- \xrightarrow{1.19 \text{ V}} ClO_3^- \xrightarrow{1.21 \text{ V}} HClO_2 \xrightarrow{1.64 \text{ V}} HOCl \xrightarrow{1.63 \text{ V}} Cl_2 \xrightarrow{1.36 \text{ V}} Cl^-$$

(with connecting paths labeled 1.38 V above, and 1.47 V below)

For example, the half-reaction for reduction of $HClO_2$ to $HOCl$ in acid solution has a standard potential of 1.64 V:

$$HClO_2(aq) + 2H^+(aq) + 2e^- \longrightarrow HOCl(aq) + H_2O(l)$$
$$E° = 1.64 \text{ V} \qquad [21.7]$$

By using the standard potentials just given, it is possible to calculate the standard potentials for other half-reactions. For example, we can calculate the standard potential for the reduction of ClO_4^- to Cl^- using the potentials for reduction of ClO_4^- to ClO_3^- and ClO_3^- to Cl^-:

$$2e^- + 2H^+(aq) + ClO_4^-(aq) \longrightarrow ClO_3^-(aq) + H_2O(l)$$
$$E° = 1.19 \text{ V} \qquad [21.8]$$
$$6e^- + 6H^+(aq) + ClO_3^-(aq) \longrightarrow Cl^-(aq) + 3H_2O(l)$$
$$E° = 1.38 \text{ V} \qquad [21.9]$$
$$\overline{8e^- + 8H^+(aq) + ClO_4^-(aq) \longrightarrow Cl^-(aq) + 4H_2O(l)} \qquad [21.10]$$

Adding the two half-reactions together results in the desired half-reaction. However, the correct electrode potential cannot be obtained by merely adding the corresponding $E°$ values. Instead, the standard electrode potential must first be converted to standard free-energy changes, $\Delta G° = -n\mathcal{F}E°$. The $\Delta G°$ values can then be added (equation numbers shown in brackets):

$$\Delta G° [21.10] = \Delta G° [21.8] + \Delta G° [21.9]$$
$$-n\mathcal{F}E° [21.10] = -n\mathcal{F}E° [21.8] - n\mathcal{F}E° [21.9]$$
$$-8\mathcal{F}E° [21.10] = -2\mathcal{F}(1.19 \text{ V}) - 6\mathcal{F}(1.38 \text{ V})$$
$$E° [21.10] = \frac{-2\mathcal{F}(1.19 \text{ V}) - 6\mathcal{F}(1.38 \text{ V})}{-8\mathcal{F}}$$
$$= \frac{(-10.66 \text{ V})}{-8} = 1.33 \text{ V}$$

Now you might wonder why we didn't use this procedure when we added half-reactions and their corresponding $E°$ values in Chapter 19. In Chapter 19 we were adding half-reactions to give a complete balanced equation; what we have just done above is to add two half-reactions to give a third half-reaction. When half-reactions are added to give a complete balanced equation, n, the number of electrons, is the same for all reactions; the same result is therefore obtained whether we directly add $E°$ values or use the more laborious procedures involving $\Delta G°$.

USES

Fluorine has become an important industrial chemical. It is used, for example, to prepare fluorocarbons, very stable carbon-fluorine compounds. An example is CF_2Cl_2, known as Freon-12, which is used as a refrigerant and as a propellant for aerosol cans. As we noted in Section 10.4, the effects of these substances on the ozone layer have been under recent investigation. Fluorocarbons are also used as lubricants and in plastics; Teflon is a polymeric fluorocarbon:

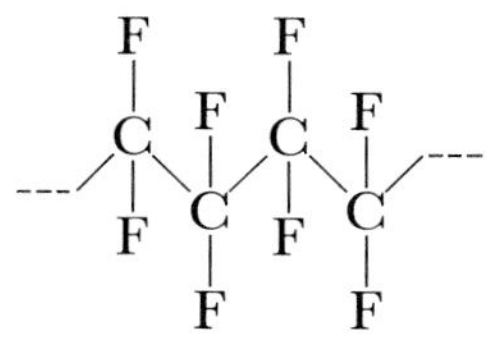

Chlorine is used primarily as a bleach in the paper and textile industries. When Cl_2 dissolves in cold dilute base it forms Cl^- and ClO^-:

$$2NaOH(aq) + Cl_2(g) \longrightarrow \\ NaCl(aq) + NaOCl(aq) + H_2O(l) \qquad [21.11]$$

Sodium hypochlorite is the active ingredient in Clorox bleach. Solid bleaching powder, which consists of $Ca(OCl)Cl$, is obtained by reaction of Cl_2 with a solution of $Ca(OH)_2$ as shown in Equation [21.12].

$$Ca(OH)_2(aq) + Cl_2(g) \longrightarrow Ca(OCl)Cl(aq) + H_2O(l) \qquad [21.12]$$

Chlorine is also used in water treatment to oxidize and thereby destroy bacteria. Recent studies have shown that such treatment of municipal sewage and drinking water produces small amounts (on the order of parts per billion) of chlorocarbon compounds. These compounds are claimed to be carcinogenic (cancer inducing) and are toxic to fish and other aquatic life. Whether they are harmful in the small amounts usually found in treated water is not known. Ozone, O_3, can be used as a substitute for Cl_2 in water treatment, but it is more costly. Chlorine is also used in the manufacture of plastics and certain insecticides such as DDT.

Bromine is used in the production of silver bromide used in photographic film. It is also used to prepare dibromoethane, $C_2H_4Br_2$, a gasoline additive used in gasoline that contains tetraethyl lead, $Pb(C_2H_5)_4$. The dibromoethane prevents formation of lead deposits in the engine by forming volatile $PbBr_2$, which is emitted in exhaust.

Iodine has not found as wide use as the other halogens. One familiar use is its addition, as NaI, to salt to form iodized salt. Iodized salt is able to provide the small amount of iodine necessary in our diets; it is essential for the formation of thyroxin, a hormone secreted by the thyroid gland.

Lack of iodine in the diet results in an enlarged thyroid gland, a condition called goiter. Iodine is also familiar as tincture of iodine, a solution of I_2 in alcohol that is used as an antiseptic.

21.3 oxygen

Oxygen, one of the most important of the elements, is vital to maintaining life as we know it, because it is essential to animal respiration. It is also essential in the combustion processes used to generate most of the energy used in our technological society. Oxygen plays an important role in the chemistries of most other elements and is found in combination with other elements in a great variety of compounds.

The electron configuration of oxygen is $[He]2s^2 2p^4$. Because it is the second most electronegative element, it exhibits principally negative oxidation states. Because the element is two electrons short of a noble-gas electron configuration, the common oxidation state is -2. The only examples of positive oxidation states occur in the oxygen-fluorine compounds OF_2 and O_2F_2.

Oxygen has two allotropes, diatomic oxygen, O_2, and triatomic oxygen, O_3 (ozone). When we speak of a molecule of oxygen, it is normally understood that we are speaking of O_2, the normal form of the element. The structure of ozone is shown in Figure 21.7.

OCCURRENCE AND ISOLATION

Oxygen is the most abundant element both in the earth's crust and in the human body. It constitutes 89 percent of water by weight. It also comprises about 50 percent of sand, clay, limestone, and igneous rocks that make up the bulk of the earth's crust. The principal source of the element is the atmosphere, which contains about 20 mole-percent O_2.

Oxygen is obtained for commercial use by the fractional distillation of liquefied air. The normal boiling point of O_2 is $-183°C$, whereas that of N_2, the other principal component of air, is $-196°C$. Where O_2 of higher purity is required, it can be obtained by electrolysis of water. Although water is inexpensive and plentiful, the cost of electricity makes this an expensive way of obtaining elemental oxygen. In choosing a commercial method of preparing a substance, cost is the most important factor. In contrast, in choosing a method for preparing small amounts of a substance in the laboratory, convenience becomes an important factor. The common laboratory preparation of O_2 involves the thermal decomposition of potassium chlorate, $KClO_3$, with MnO_2 added as a catalyst:

$$2KClO_3(s) \xrightarrow[\text{heat}]{\text{MnO}_2} 2KCl(s) + 3O_2(g) \qquad [21.13]$$

The experimental setup was shown earlier, in Figure 5.11.

Ozone can be prepared by passing electricity through dry O_2. An apparatus for accomplishing this is shown in Figure 21.8. The pungent odor of ozone gas can sometimes be detected around electrical equipment where there is a spark jump, and in the atmosphere during lightning storms.

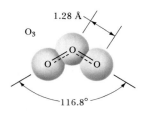

FIGURE 21.7 The structure of the ozone molecule.

O_3

1.28 Å

116.8°

632

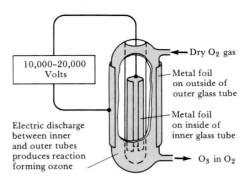

10,000–20,000 Volts

Dry O₂ gas

Metal foil on outside of outer glass tube

Metal foil on inside of inner glass tube

Electric discharge between inner and outer tubes produces reaction forming ozone

O₃ in O₂

FIGURE 21.8 An apparatus for producing ozone from O_2.

PROPERTIES AND COMPOUNDS

Oxygen in its normal elemental form, O_2, is a colorless, odorless, and tasteless gas that melts at $-218°C$ and boils at $-183°C$. It is paramagnetic (Section 9.5).

Oxygen is a chemically active substance. It combines either directly or indirectly with nearly all other elements forming oxides (O^{2-}), and occasionally peroxides (O_2^{2-}) and even superoxides (O_2^{-}). When the alkali metals are combusted in oxygen, the following oxides are the predominant products: Li_2O (oxide), Na_2O_2 (peroxide), KO_2 (superoxide), RbO_2 (superoxide), and CsO_2 (superoxide).

Potassium superoxide is used as an oxygen source in masks worn for rescue work. Moisture in the breath causes the compound to decompose to form O_2 and KOH. The KOH so formed serves to remove CO_2 from the exhaled breath:

$$2KO_2(s) + 2H_2O(l) \longrightarrow O_2(g) + H_2O_2(l) + 2KOH(aq)$$

$$2KOH(s) + CO_2(g) \longrightarrow H_2O(l) + K_2CO_3(s)$$

The most familiar peroxide is hydrogen peroxide, H_2O_2, whose structure is shown in Figure 21.9. Most hydrogen peroxide is prepared commercially by the use of a type of organic compound known as an anthraquinone. As shown in Equation [21.14], this compound is reduced by hydrogen in the presence of a palladium (Pd) catalyst:

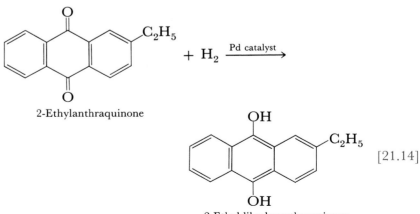

2-Ethylanthraquinone

$+ H_2$ $\xrightarrow{\text{Pd catalyst}}$

OH

C₂H₅

[21.14]

OH

2-Ethyldihydroanthraquinone

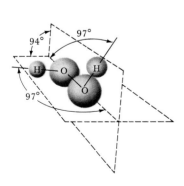

FIGURE 21.9 The structure of the hydrogen peroxide molecule.

94°   97°

97°

H — O   H

O

The product, known as a dihydroanthraquinone, is then allowed to react with oxygen, thereby regenerating the anthraquinone and producing hydrogen peroxide, Equation [21.15].

$$\text{(dihydroanthraquinone, with OH, OH and } C_2H_5 \text{ groups)} + O_2 \longrightarrow$$

$$\text{(anthraquinone, with O, O and } C_2H_5 \text{ groups)} + H_2O_2 \qquad [21.15]$$

The anthraquinone is recycled and used again, whereas the peroxide is removed and purified as necessary.

Hydrogen peroxide is capable of acting as either an oxidizing or reducing agent. Equations [21.16] and [21.17] show the half-reactions for reaction in acid solution.

$$2H^+(aq) + H_2O_2(aq) + 2e^- \longrightarrow 2H_2O(l)$$
$$E° = 1.77 \text{ V} \quad [21.16]$$

$$H_2O_2(aq) \longrightarrow O_2(g) + 2H^+(aq) + 2e^-$$
$$E° = -0.67 \text{ V} \quad [21.17]$$

In basic solution, the corresponding standard electrode potentials are 0.87 V for reduction of H_2O_2 and 0.08 V for its oxidation.

Solutions containing 3 percent by weight H_2O_2 are used in the home as a mild antiseptic; stronger solutions are used as a bleach. Hydrogen peroxide is also used as an oxidizer for rocket fuels.

USES OF OXYGEN

Oxygen is the third most widely used industrial chemical, ranking behind sulfuric acid and lime, CaO. About 32 billion pounds of the element are used annually. It is widely used as an oxidizing agent. Approximately half the oxygen produced is used in the steel industry, mainly to remove impurities from steel (Section 22.6). Oxygen is employed in medicine to quicken the life-sustaining oxidation processes. It is used together with acetylene, C_2H_2, in oxyacetylene welding. This use is based on the highly exothermic reaction between C_2H_2 and O_2, which results in temperatures in excess of 3000°C. This combustion reaction is shown in Equation [21.18].

$$2C_2H_2(g) + 5O_2(g) \longrightarrow 4CO_2(g) + 2H_2O(g)$$
$$\Delta H° = -2510 \text{ kJ} \quad [21.18]$$

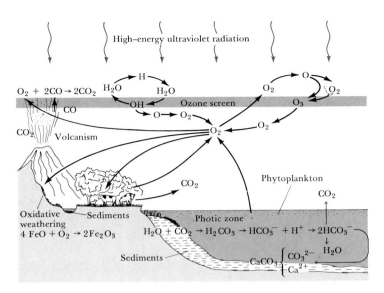

FIGURE 21.10 A simplified view of the oxygen cycle, showing some of the primary reactions involving oxygen in nature. The atmosphere, which contains O_2, is one of the primary sources of the element. Some O_2 is produced by radiation-induced dissociation of H_2O in the upper atmosphere. Some O_2 is produced by green plants from H_2O and CO_2 in the course of photosynthesis. Atmospheric CO_2, in turn, results from combustion reactions, animal respiration, and the dissociation of bicarbonate in water. The O_2 is used to produce ozone in the upper atmosphere, in oxidative weathering of rocks, in animal respiration, and in combustion reactions.

Ozone is an even more powerful oxidizing agent than O_2. However, it cannot be stored for long except at low temperatures; consequently, it must usually be used as it is generated. As we saw in Chapter 10, ozone is an important component of the upper atmosphere, in that it screens out ultraviolet radiation. In this way ozone protects the earth from the effects of these high-energy rays. However, ozone can do considerable damage in the lower atmosphere because it is such a powerful oxidizing agent. Because it causes damage to plants, animals, and structural materials, it is considered an air pollutant.

THE OXYGEN CYCLE

Oxygen accounts for about one-fourth of the atoms in living matter. Because the number of oxygen atoms is fixed, as O_2 is removed from air through respiration and other processes, it needs to be replenished. The

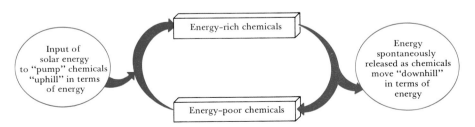

FIGURE 21.11 A schematic representation of the role of solar energy in the cycling of elements in nature.

major nonliving sources of oxygen other than O_2 are CO_2 and H_2O. A simplified picture of the movement of oxygen in our environment is shown in Figure 21.10. This figure points out both how O_2 is removed from the atmosphere and how it is replenished. Oxygen, O_2, is reformed mainly from CO_2 through the process of photosynthesis. Energy is produced when O_2 is converted to CO_2; energy must therefore be supplied to reform O_2 from CO_2. This energy is provided by the sun. Thus life on earth depends on chemical recycling made possible by solar energy. The situation is represented schematically in Figure 21.11.

21.4

sulfur

Sulfur is a member of the oxygen family. As such, both sulfur and oxygen have the same type of outer electron configuration, ns^2np^4. However, sulfur's larger size and lower electronegativity leads to several differences in chemical behavior. For example, sulfur exhibits oxidation states from -2 to $+6$, the positive oxidation states reflecting sulfur's lower electronegativity.

OCCURRENCE, ISOLATION, AND USES

Sulfur occurs in the elemental state in large underground deposits that serve as the principal source of the element. The Frasch process, illustrated in Figure 21.12, is used to obtain the element from these deposits. The method is based on the low melting point and low density of sulfur. Superheated water is forced into the deposit where it melts the sulfur. Compressed air then forces the molten sulfur up a pipe that is concentric

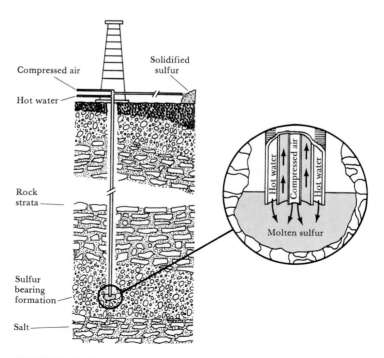

FIGURE 21.12 The mining of sulfur by the Frasch process. The process is named after Herman Frasch, who invented the process in the early 1890s. The process is particularly useful for recovering sulfur from deposits located under quicksand or water.

FIGURE 21.13 The production platform of the world's first offshore sulfur mine. The mine stands in 50 ft of water 7 mi off Grand Isle, Louisiana. (*Exxon Company, U.S.A.*)

with the ones that introduce the hot water and compressed air into the deposit. The method is particularly useful for removing sulfur from beds that lie under water or quicksand. Figure 21.13 shows an offshore sulfur mine.

Sulfur also occurs widely as sulfide and sulfate minerals and as a minor component of coal and petroleum. The presence of sulfur in coal and petroleum poses a major problem. As we saw in Section 10.6, combustion of these "unclean" fuels leads to serious sulfur oxide pollution. Likewise, operations that use sulfide minerals as sources of metals liberate sulfur oxides. Much effort has therefore been directed at removing this sulfur, and these efforts have increased the availability of sulfur. The sale of this sulfur helps partially to offset the costs of the desulfurizing processes and equipment. However, sulfur obtained from sulfur deposits by the Frasch process is about 99.5 percent pure and can be used for most commercial processes without purification. It is therefore relatively cheap. Consequently, sulfur from desulfurizing processes must be sold at prices below its cost in order to compete on the market.

Most sulfur is used in the manufacture of sulfuric acid. Sulfur is also

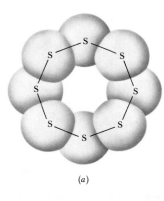

(a)

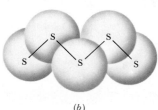

(b)

FIGURE 21.14 Top view (a) and side view (b) of the rhombic sulfur molecule.

used in vulcanizing rubber, a process that toughens rubber by introducing cross-linking between polymer chains (Section 11.5).

PROPERTIES

As we normally encounter it, sulfur is yellow, tasteless, and nearly odorless. It is insoluble in water and exists in several allotropic forms. The most thermodynamically stable form at room temperature is rhombic sulfur, which consists of puckered S_8 rings, as shown in Figure 21.14. When heated above its melting point, at 113°C, sulfur undergoes a variety of changes. The molten sulfur first contains S_8 molecules and is fluid because the rings readily slip over each other. Further heating of this straw-colored liquid causes rings to break and then join up again to form very long molecules that can become entangled. The sulfur consequently becomes highly viscous. This change is marked by a color change to dark reddish-brown. Further heating breaks the chains and the viscosity again decreases. The changes in viscosity as sulfur is heated are shown in Figure 21.15.

OXIDES AND OXYCOMPOUNDS

When sulfur is combusted in air, SO_2 forms. This gaseous substance has a choking odor and is somewhat poisonous. It is particularly toxic to lower organisms such as fungi and is consequently used for sterilizing dried fruit. Sulfur dioxide dissolves in water to form the weak diprotic sulfurous acid, H_2SO_3:

$$H_2O(l) + SO_2(g) \longrightarrow H_2SO_3(aq) \qquad [21.19]$$

In the laboratory, SO_2 is prepared by the action of hydrochloric acid on a sulfite salt as shown in Equation [21.20]

$$\begin{aligned} 2HCl(aq) + Na_2SO_3(aq) &\longrightarrow \\ SO_2(g) + H_2O(l) &+ 2NaCl(aq) \end{aligned} \qquad [21.20]$$

Combustion of sulfur in air produces small amounts of SO_3. Sulfur trioxide is of great commercial importance because it is the anhydride of sulfuric acid. In the manufacture of sulfuric acid, SO_2 is first obtained by burning sulfur. The SO_2 is then oxidized to SO_3 using a catalyst such as V_2O_5 or platinum. The SO_3 is dissolved in H_2SO_4 because it does not dissolve quickly in water. The reaction is shown in Equation [21.21]. The

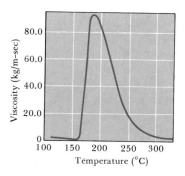

FIGURE 21.15 Graph of the change in viscosity of sulfur with temperature. By comparison, the viscosity of water at 20°C is 1.00×10^{-3} kg/m-sec (see Section 11.2).

resultant $H_2S_2O_7$ is then added to water to form H_2SO_4 as shown in Equation [21.22].

$$SO_3(g) + H_2SO_4(l) \longrightarrow H_2S_2O_7(l) \qquad [21.21]$$

$$H_2S_2O_7(l) + H_2O(l) \longrightarrow 2H_2SO_4(aq) \qquad [21.22]$$

Commercial sulfuric acid is generally 98 percent H_2SO_4 and boils at $340\,°C$. It is a colorless, oily liquid. Sulfuric acid has a number of useful properties. Foremost among these is the fact that it is a strong acid. Additionally, it is a good dehydrating agent* and a moderately good oxidizing agent.

Because of its high boiling point, H_2SO_4 can be used to prepare other acids. For example, hydrogen chloride is prepared in the laboratory by the action of concentrated sulfuric acid on sodium chloride as shown in Equation [21.23].

$$2NaCl(s) + H_2SO_4(aq) \longrightarrow 2HCl(g) + Na_2SO_4(aq) \qquad [21.23]$$

At moderately elevated temperatures, HCl escapes from the reaction mixture. It is then dissolved in water to form hydrochloric acid. An attempt to prepare HBr or HI in the same fashion yields a gaseous mixture containing Br_2 or I_2. The formation of halogen in these instances illustrates the oxidizing ability of hot H_2SO_4. The oxidation of hydrogen bromide by hot sulfuric acid is shown in Equation [21.24].

$$2HBr(aq) + H_2SO_4(aq) \longrightarrow$$
$$Br_2(l) + H_2O(l) + H_2SO_3(aq) \qquad [21.24]$$

In preparing HBr or HI, phosphoric acid, H_3PO_4, is used in place of the sulfuric acid to avoid this oxidation.

Year after year, the output of sulfuric acid has been the largest of any chemical produced in the United States. About 33 million tons are produced annually in this country. Sulfuric acid is employed in some way in almost all manufacturing. Consequently, its consumption is considered a standard measure of industrial activity. The primary uses for sulfuric acid are shown in Figure 21.16.

Only the first proton in sulfuric acid is completely ionized. The second proton ionizes only partially:

$$H_2SO_4(aq) \longrightarrow H^+(aq) + HSO_4^-(aq)$$

$$HSO_4^-(aq) \rightleftharpoons H^+(aq) + SO_4^{2-}(aq) \qquad K_a = 2 \times 10^{-2}$$

Consequently, sulfuric acid forms two series of compounds—sulfates and bisulfates (or hydrogen sulfates).

Related to the sulfate ion is the thiosulfate ion, $S_2O_3^{2-}$. The term "thio" means substitution of sulfur for oxygen. The structures of the sulfate and thiosulfate ions are compared in Figure 21.17. The thiosulfate

---

*Considerable heat is given off when sulfuric acid is diluted with water. Consequently, dilution must always be done carefully by pouring the acid into water to distribute the heat as uniformly as possible and to avoid spattering of the acid.

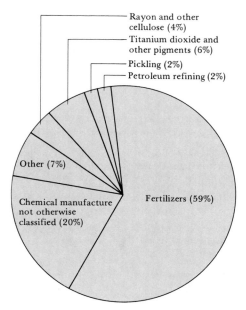

Rayon and other cellulose (4%)

Titanium dioxide and other pigments (6%)

Pickling (2%)

Petroleum refining (2%)

Other (7%)

Fertilizers (59%)

Chemical manufacture not otherwise classified (20%)

FIGURE 21.16 Sulfuric acid use in the United States.

ion is formed by boiling an alkaline solution of SO_3^{2-} with elemental sulfur as shown in Equation [21.25].

$$S(s) + SO_3^{2-}(aq) \longrightarrow S_2O_3^{2-}(aq) \qquad [21.25]$$

When acidified, the thiosulfate ion decomposes to form sulfur and H_2SO_3. The pentahydrated salt of sodium thiosulfate, $Na_2S_2O_3 \cdot 5H_2O$, is known as "hypo." It is used in photography to furnish thiosulfate ion as a complexing agent for silver ion:

$$Ag^+(aq) + 2S_2O_3^{2-}(aq) \rightleftharpoons Ag(S_2O_3)_2^{3-}(aq)$$
$$K = 1.6 \times 10^{13} \qquad [21.26]$$

Thiosulfate ion is also used in quantitative analysis as an oxidizing agent for iodine:

$$2S_2O_3^{2-}(aq) + I_2(s) \longrightarrow 2I^-(aq) + S_4O_6^{2-}(aq) \qquad [21.27]$$

SULFIDES

Sulfur forms compounds by direct combination with many elements. When the element is less electronegative than sulfur, sulfides, which contain S^{2-}, form. For example, iron(II) sulfide, FeS, forms by direct combination of iron and sulfur.

One of the most important sulfides is hydrogen sulfide, H_2S. This substance is not normally produced by direct union of the elements because it is unstable at elevated temperatures and decomposes into the elements. It is normally prepared by action of dilute sulfuric acid on iron(II) sulfide as shown in Equation [21.28].

$$FeS(s) + H_2SO_4(aq) \longrightarrow H_2S(aq) + FeSO_4(aq) \qquad [21.28]$$

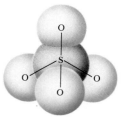

SO_4^{2-}

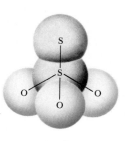

$S_2O_3^{2-}$

FIGURE 21.17 Comparison of the structures of the sulfate, SO_4^{2-}, and thiosulfate, $S_2O_3^{2-}$, ions.

A common laboratory source of H_2S is the reaction of thioacetamide with water, as shown in Equation [21.29].

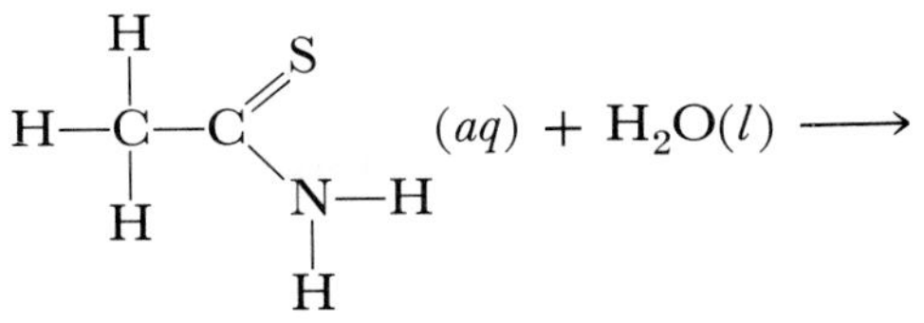

Thioacetamide

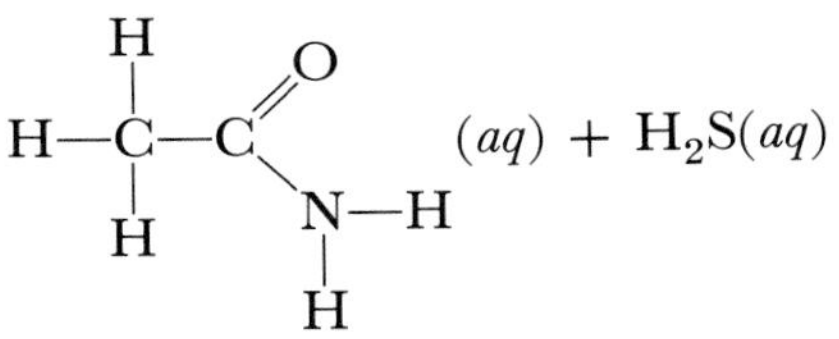

$(aq) + H_2O(l) \longrightarrow$

Acetamide

$(aq) + H_2S(aq)$ [21.29]

Hydrogen sulfide is often used in the laboratory for qualitative analysis of certain metal ions, as described in Section 17.6.

One of hydrogen sulfide's most readily recognized properties is its odor; H_2S is largely responsible for the offensive odor of rotten eggs. Hydrogen sulfide is actually quite toxic; it has about the same level of toxicity as hydrogen cyanide, the gas that was used in California's gas chambers.

The element below sulfur in the periodic table is selenium. Like sulfur, selenium exists in several allotropic forms. One form looks like a metal, supporting the idea that metallic character increases as we move down a family. This form has a very important property; it is photoconductive. This means that its electrical conductivity increases greatly when light shines on it. Because of this property selenium is used in light meters such as those in cameras. Xerox copiers also depend on the photoconductivity of selenium for their operation.

21.5 nitrogen

Nitrogen has an electron configuration of $[He]2s^2 2p^3$. A nitrogen atom is therefore able to achieve a noble-gas configuration either by loss of five electrons to a more electronegative atom or by gain of three from a less electronegative one. Consequently, the highest positive oxidation state of nitrogen is +5, whereas the highest negative state is −3. As shown in Table 21.4, all oxidation states between −3 and +5 are observed. Oxygen and fluorine have larger electronegativities than nitrogen. Consequently, it is with these elements that positive oxidation states of nitrogen are found.

OCCURRENCE, ISOLATION, PROPERTIES, AND USES

Nitrogen is a colorless, odorless, and tasteless gas. It boils at −196°C and freezes at −210°C. The most important source of nitrogen is the atmosphere, which is 78 percent N_2. Thus, N_2 can be separated from O_2 in air by fractional distillation. Nitrogen is quite nonreactive because of the strong triple bond between nitrogen atoms. Because of its lack of reac-

TABLE 21.4

oxidation states of nitrogen

OXIDATION STATE	EXAMPLE
−3	NH_3 (ammonia)
−2	N_2H_4 (hydrazine)
−1	H_2NOH (hydroxylamine)
0	N_2 (molecular nitrogen)
+1	N_2O (nitrous oxide)
+2	NO (nitric oxide)
+3	NO_2^- (nitrite ion)
+4	NO_2 (nitrogen dioxide)
+5	NO_3^- (nitrate ion)

tivity, N_2 is used in many situations where a relatively inert atmosphere is required. For example, it is used to retard filament evaporation from incandescent lamp bulbs.

The formation of nitrogen-containing compounds starting with N_2 is known as nitrogen **fixation.** As discussed in Chapter 14, the Haber process is the primary commercial means available for the fixation of nitrogen. In this process N_2 from air and H_2, which is normally obtained from the CH_4 in natural gas,* are combined to form NH_3:

$$N_2(g) + 3H_2(g) \longrightarrow 2NH_3(g) \qquad [21.30]$$

Prior to the development of the Haber process, $NaNO_3$, Chilean saltpeter, was the principal source of nitrogen used in the manufacture of nitrogen compounds. In nature, nitrogen is fixed by certain bacteria that occur in the root nodules of leguminous plants such as peas, beans, peanuts, and alfalfa. It is also fixed during lightning when N_2 and O_2 combine to form nitrogen oxides.

The demand for fixed nitrogen has increased mainly because the element is required in maintaining soil fertility. Although we are immersed in an ocean of air that principally contains N_2, our supply of food is limited more by the availability of fixed nitrogen than by that of any other plant nutrient. Thus nitrogen is used primarily for the manufacture of nitrogen-containing fertilizers. It is also used in the manufacture of explosives, plastics, and many important chemicals.

COMPOUNDS OF NITROGEN

Two of the few compounds that form directly from N_2 are Li_3N and Mg_3N_2. These compounds form when lithium or magnesium is burned in air:

*The H_2 is obtained using the following series of reactions, which occur at elevated temperatures:

$$CH_4(g) + H_2O(g) \longrightarrow CO(g) + 3H_2(g)$$
$$2CH_4 + O_2(g) \longrightarrow 2CO(g) + 4H_2(g)$$
$$CO(g) + H_2O(g) \longrightarrow CO_2(g) + H_2(g)$$

Because CH_4 is used as the source of H_2, the fixation of N_2 is closely tied to the availability of natural gas.

$$6Li(s) + N_2(g) \longrightarrow 2Li_3N(s) \qquad [21.31]$$

$$3Mg(s) + N_2(g) \longrightarrow Mg_3N_2(s) \qquad [21.32]$$

The nitride ion, N^{3-}, so formed is a strong Brønsted base, just as is O^{2-}. It therefore forms ammonia when placed in contact with water:

$$Mg_3N_2(s) + 6H_2O(l) \longrightarrow 2NH_3(aq) + 3Mg(OH)_2(s) \qquad [21.33]$$

Ammonia is one of the most important compounds of nitrogen. It is a colorless, toxic gas that has a characteristic, irritating odor. It boils at $-33°C$ and freezes at $-78°C$. In the laboratory, NH_3 is prepared by the action of NaOH on an ammonium salt as shown in Equation [21.34].

$$NH_4Cl(aq) + NaOH(aq) \longrightarrow$$
$$NH_3(g) + H_2O(l) + NaCl(aq) \qquad [21.34]$$

About 75 percent of the ammonia produced in this country is used for fertilizer.

Ammonia can be converted to nitric oxide by oxidation at elevated temperatures in the presence of platinum (Pt), which serves as a catalyst; the reaction is shown in Equation [21.35].

$$4NH_3(g) + 5O_2(g) \xrightarrow[1000°C]{\substack{Pt \\ catalyst}} 4NO(g) + 6H_2O(g) \qquad [21.35]$$

In the absence of a catalyst, NH_3 is converted to N_2 instead of NO:

$$4NH_3(g) + 3O_2(g) \xrightarrow{1000°C} 2N_2(g) + 6H_2O(g) \qquad [21.36]$$

The catalytic conversion of NH_3 to NO is the commercial route to oxygen-containing compounds of nitrogen and is part of what is known as the **Ostwald process**.* Like the Haber process, this process was developed in Germany prior to World War I. It provided a means of converting NH_3 into nitric acid, which is used in making munitions.

Nitric oxide is one of three common oxides of nitrogen. The others are N_2O (nitrous oxide) and NO_2 (nitrogen dioxide). All three of these oxides are gases. Nitrous oxide is also known as laughing gas because a person becomes somewhat giddy after inhaling only a small amount of it. This colorless gas was the first substance used as a general anesthetic. It is used as the compressed gas propellant in several aerosols and foams such as in whipped cream. It can be prepared in the laboratory by carefully heating ammonium nitrate to about 200°C:

$$NH_4NO_3(s) \longrightarrow N_2O(g) + 2H_2O(g) \qquad [21.37]$$

Nitric oxide is also a colorless gas, but unlike N_2O, it is slightly

*The Ostwald process is the process for converting ammonia to nitric acid utilizing the reactions summarized in Equations [21.35], [21.39], and [21.41].

toxic. It can be prepared in the laboratory by reduction of dilute nitric acid, using copper or iron as a reducing agent:

$$3Cu(s) + 8HNO_3(aq) \longrightarrow$$
$$3Cu(NO_3)_2(aq) + 2NO(g) + 4H_2O(l) \qquad [21.38]$$

It is also produced by direct combination of N_2 and O_2 at elevated temperatures. As we saw in Section 10.5, this reaction is a significant source of nitrogen oxide air pollutants, which form during combustion reactions in air. However, the direct combination of N_2 and O_2 is not presently used for commercial production of NO because the yield is low; as noted in Section 14.5, the equilibrium constant, K_c, at 2400 K is only 0.05.

One of the important properties of NO is its ready ability to react with O_2, forming NO_2 when exposed to air:

$$2NO(g) + O_2(g) \longrightarrow 2NO_2(g) \qquad [21.39]$$

Nitrogen dioxide is a yellow-brown gas. It is poisonous and has a choking odor. At lower temperatures it reacts with itself to form the colorless N_2O_4 as shown in Equation [21.40].

$$2NO_2(g) \rightleftharpoons N_2O_4(g) \qquad \Delta H° = -58 \text{ kJ} \qquad [21.40]$$

When dissolved in water, NO_2 forms HNO_3, nitric acid, as shown in Equation [21.41].

$$H_2O(l) + 3NO_2(g) \longrightarrow 2HNO_3(aq) + NO(g) \qquad [21.41]$$

Note that nitrogen in this reaction is both oxidized and reduced. The reduction product, NO, can be converted back into NO_2 by exposure to air and thereafter dissolved in water to prepare more HNO_3.

Nitric acid is a colorless, corrosive liquid. It is both a strong acid and a good oxidizing agent. The standard reduction potentials for conversion of nitrate into lower oxidation states in acid solution are shown below:

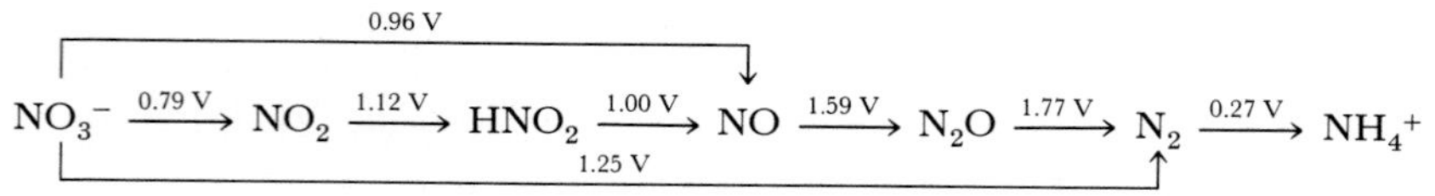

$$NO_3^- \xrightarrow{0.79 \text{ V}} NO_2 \xrightarrow{1.12 \text{ V}} HNO_2 \xrightarrow{1.00 \text{ V}} NO \xrightarrow{1.59 \text{ V}} N_2O \xrightarrow{1.77 \text{ V}} N_2 \xrightarrow{0.27 \text{ V}} NH_4^+$$

(with overall 0.96 V from NO_3^- to NO, and 1.25 V from NO_3^- to HNO$_2$)

Nitric acid boils at 86°C and freezes at −42°C. It is used in the production of plastics, drugs, nitrate fertilizers, and explosives. The development of the Haber and Ostwald processes in Germany just prior to World War I permitted Germany to make munitions even though naval blockades prevented access to traditional sources of nitrates. Among the explosives made from nitric acid are nitroglycerin, trinitrotoluene (TNT), and nitrocellulose. The reaction of nitric acid with glycerin to form nitroglycerin is shown in Equation [21.42].

$$
\begin{array}{c}
\text{H} \\
| \\
\text{H}-\text{C}-\text{OH} \\
| \\
\text{H}-\text{C}-\text{OH} + 3\,\text{HNO}_3 \\
| \\
\text{H}-\text{C}-\text{OH} \\
| \\
\text{H}
\end{array}
\longrightarrow
\begin{array}{c}
\text{H} \\
| \\
\text{H}-\text{C}\text{ONO}_2 \\
| \\
\text{H}-\text{C}\text{ONO}_2 + 3\,\text{H}_2\text{O} \\
| \\
\text{H}-\text{C}\text{ONO}_2 \\
| \\
\text{H}
\end{array}
\qquad [21.42]
$$

When nitroglycerin explodes, the reaction summarized in Equation [21.43] occurs.

$$4\text{C}_3\text{H}_5\text{N}_3\text{O}_9(l) \longrightarrow$$
$$6\text{N}_2(g) + 12\text{CO}_2(g) + 10\text{H}_2\text{O}(g) + \text{O}_2(g) \qquad [21.43]$$

A considerable amount of gaseous products form from the liquid. The sudden formation of these gases, together with their expansion resulting from the heat generated by the reaction, produces the explosion.

Plants are able to utilize several chemical forms of nitrogen, especially NH_3, NH_4^+, and NO_3^-. Common synthetic fertilizers therefore include liquid ammonia, ammonium nitrate, and urea. Ammonium nitrate, NH_4NO_3, is made by reaction of ammonia and nitric acid, as shown in Equation [21.44].

$$NH_3(aq) + HNO_3(aq) \longrightarrow NH_4NO_3(aq) \qquad [21.44]$$

Urea is made by reaction of ammonia and carbon dioxide, Equation [21.45].

$$2NH_3(aq) + CO_2(aq) \rightleftharpoons \underset{\displaystyle H_2N\overset{\displaystyle \overset{O}{\|}}{C}NH_2}{}(aq) + H_2O(l) \qquad [21.45]$$

Urea slowly releases NH_3 as it reacts with water in the soil.

Plants use nitrogen to synthesize several nitrogen-containing compounds including proteins. Proteins are made from amino acids, which are compounds of the following type:

$$
\begin{array}{c}
\quad\;\; \text{R} \;\; \text{O} \\
\quad\;\; | \quad\; \| \\
\text{H}-\text{N}-\text{C}-\text{C}-\text{O}-\text{H} \\
\quad\;\; | \quad\; | \\
\quad\;\; \text{H} \;\; \text{H}
\end{array}
$$

where R is H or any of a number of carbon-containing groups including CH_3. We shall consider these compounds in Chapter 25. Plant proteins are ingested by animals, are broken into amino acids, and are then either converted into animal protein or excreted as nitrogenous waste. Certain microorganisms are able to convert this waste back into N_2. Nitrogen is recycled in this fashion. As in the case of the cycling of oxygen, energy from the sun is required. A simplified picture of the nitrogen cycle is shown in Figure 21.18.

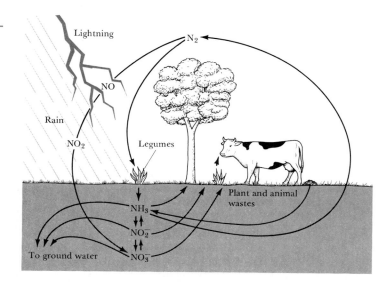

FIGURE 21.18 A simplified picture of the nitrogen cycle, showing some of the primary reactions involved in the utilization and formation of nitrogen in nature. The main reservoir of nitrogen is the atmosphere, which contains N_2. This nitrogen is fixed through the action of lightning and leguminous plants. The compounds of nitrogen reside in the soil as NH_3 (and NH_4^+), NO_2^-, and NO_3^-. All are water soluble and can be washed out of the soil by ground water. These nitrogen compounds are utilized by plants in their growth and are incorporated into animals that eat the plants. Animal waste and dead plants and animals are attacked by certain bacteria that free N_2, which escapes into the atmosphere, thereby completing the cycle.

Large-scale cultivation of nitrogen-fixing legumes and industrial fixation have increased the quantity of fixed nitrogen in the biosphere. One effect of this intrusion into the nitrogen cycle has been increased water pollution. Much fixed nitrogen ends up as nitrates in the soil. These compounds are highly water soluble. They are therefore readily washed from the soil into water bodies. Only a portion of the added fertilizer ends up being used by the plants for which it was intended. Once in a lake, nitrates stimulate plant growth, encouraging, for example, rapid growth of algae. When these plants die, their decay consumes O_2 in the water. Without O_2, fish and other oxygen-dependent organisms die. Furthermore, when the oxygen is depleted, anaerobic decomposition begins. This decomposition produces foul odors. The undecomposed plant matter falls to the bottom of the lake thereby slowly turning it into a swamp and eventually a meadow. This process of fertilization of lakes, which leads to their death, is known as **eutrophication.** It takes thousands of years for this aging process to occur naturally. Through various activities, we have added nutrients to water and thereby succeeded in greatly accelerating this aging process. Nitrogen is not the only nutrient that promotes eutrophication. We shall discuss another important plant nutrient, phosphorus, in the next section.

21.6
phosphorus

Phosphorus has an electron configuration of $[Ne]3s^23p^3$. Like nitrogen, it exhibits oxidation states ranging from -3 to $+5$. Because of its lower electronegativity, it is found more frequently in the positive oxidation states than is nitrogen. Furthermore, compounds in which phosphorus

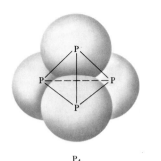

P_4

FIGURE 21.19 The structure of the P_4 molecule of white phosphorus.

has the $+5$ oxidation state are not as efficient oxidizing agents as are corresponding compounds of nitrogen. Compounds in which phosphorus has a -3 oxidation state are much stronger reducing agents than corresponding compounds of nitrogen.

OCCURRENCE, ISOLATION, AND PROPERTIES

Phosphorus occurs mainly in the form of phosphate minerals. The principal source of phosphorus is phosphate rock, which contains phosphate mainly in the form of $Ca_3(PO_4)_2$. Deposits of phosphate rock occur mainly in Florida, the western United States, North Africa, and parts of the USSR. The element is produced commercially by reduction of phosphate with coke* in the presence of SiO_2, Equation [21.46].

$$2Ca_3(PO_4)_2(s) + 6SiO_2(s) + 10C(s) \xrightarrow{1500°C}$$

$$P_4(g) + 6CaSiO_3(l) + 10CO(g) \qquad [21.46]$$

The phosphorus produced in this fashion is the allotrope known as white phosphorus. This form distills from the reaction mixture as the reaction proceeds.

White phosphorus consists of P_4 tetrahedra as shown in Figure 21.19. As we noted in Section 8.3, the 60° bond angles in P_4 are unusually small for molecules. There must consequently be much strain in the bonding, a fact that is consistent with the high reactivity of white phosphorus. This allotrope bursts spontaneously into flames if exposed to air. It is a white, waxlike solid that melts at 44.2°C and boils at 280°C. When heated in the absence of air to about 400°C, it is converted to a more stable allotrope known as red phosphorus. This form does not ignite on contact with air. It is also considerably less poisonous than the white form. Red phosphorus appears to consist of a chain structure as shown in Figure 21.20.

COMPOUNDS

Probably the most significant compounds of phosphorus are those in which the element is combined in some way with oxygen. Phosphorus(III) oxide, P_4O_6, is obtained by allowing white phosphorus to oxidize in a limited supply of oxygen. When oxidation takes place in the presence of excess oxygen, phosphorus(V) oxide, P_4O_{10}, forms. This compound is also readily formed by oxidation of P_4O_6. Phosphorus(III) oxide is often called phosphorus trioxide after its empirical formula; similarly, phosphorus(V) oxide is often called phosphorus pentoxide. These two oxides

*Coke is a form of carbon made by heating coal in the absence of air.

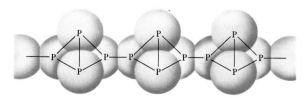

FIGURE 21.20 Red phosphorus is polymeric, possibly as shown; however, the details of its structure are not known.

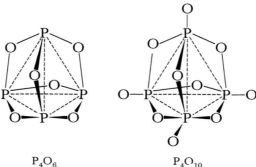

P_4O_6 P_4O_{10}

FIGURE 21.21 The structures of P_4O_6 and P_4O_{10}.

represent the two most common oxidation states for phosphorus, $+3$ and $+5$. The structural relationship between P_4O_6 and P_4O_{10} is shown in Figure 21.21. Notice the resemblance these molecules have to the P_4 molecule, Figure 21.19.

Phosphorus(V) oxide is the anhydride of phosphoric acid, H_3PO_4, a weak triprotic acid. In fact, P_4O_{10} has a very high affinity for water and is consequently used as a drying agent. Phosphorus(III) oxide is the anhydride of phosphorous acid, H_3PO_3, a weak diprotic acid. The structures of H_3PO_4 and H_3PO_3, are shown in Figure 21.22. The hydrogen atom that is attached directly to phosphorus in H_3PO_3 is not acidic.

One characteristic of phosphoric and phosphorous acids is their tendency to undergo condensation reactions when heated. A **condensation reaction** is one in which two or more molecules combine to form a larger molecule by eliminating a small molecule such as H_2O. Such a reaction in which two H_3PO_4 molecules are joined by the elimination of one H_2O molecule is shown in Equation [21.47].

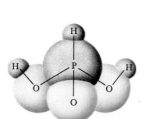

H_3PO_4

H_3PO_3

FIGURE 21.22 The structures of H_3PO_4 and H_3PO_3.

These atoms are eliminated as H_2O

[21.47]

$$2H_3PO_4 \longrightarrow H_4P_2O_7 + H_2O$$

Further condensation produces phosphates having an empirical formula of HPO_3:

$$nH_3PO_4 \longrightarrow (HPO_3)_n + nH_2O \qquad [21.48]$$

Two phosphates having this empirical formula, one cylic and the other polymeric, are shown in Figure 21.23. The three acids H_3PO_4, $H_4P_2O_7$,

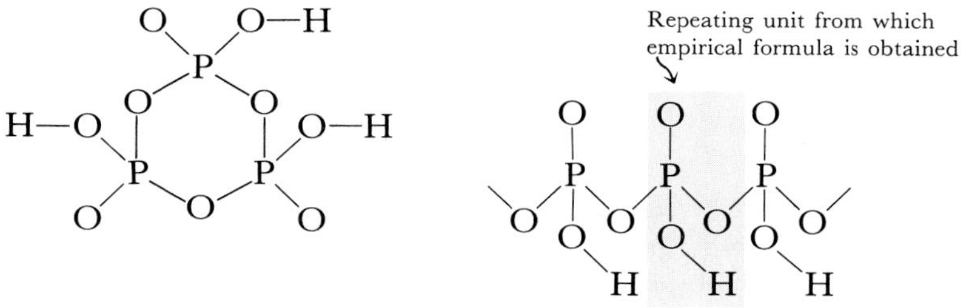

$(HPO_3)_3$
Trimetaphosphoric acid

$(HPO_3)_n$
Polymetaphosphoric acid

FIGURE 21.23 Structures of trimetaphosphoric acid and polymetaphosphoric acid.

and $(HPO_3)_n$ all contain phosphorus in its +5 oxidation state, and all are therefore called phosphoric acids. To differentiate them, the prefixes ortho, pyro, and meta are used: H_3PO_4 is orthophosphoric acid, $H_4P_2O_7$ is pyrophosphoric acid, and HPO_3 is metaphosphoric acid.

Phosphoric acid and its salts find their most important uses in detergents and fertilizers. The phosphates in detergents are often in the form of sodium tripolyphosphate, $Na_3P_3O_{10}$ (see Section 23.2). A typical detergent formulation contains 47 percent phosphate, 16 percent bleaches, perfumes, and abrasives, and 37 percent linear alkylsulfonate (LAS) surfactant such as that shown below:

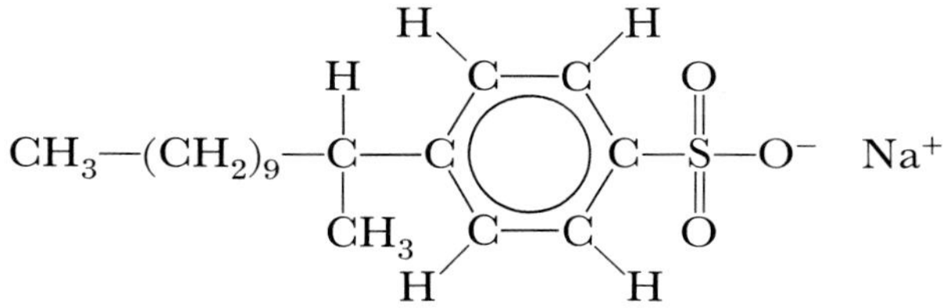

The action of such surfactants has been described in Section 12.8. The phosphate bonds with metal ions that contribute to the hardness of water. This keeps these ions from interfering with the action of the surfactant. The phosphates also keep the pH above 7 and thus prevent the surfactant molecules from protonating (gaining an H^+ ion).

Most phosphate rock mined is converted to fertilizers. The mined phosphate rock contains large amounts of sand and clay. At a treatment plant, sand, clay, and organic materials are removed from the raw ore. The resultant concentrate is shipped to plants that convert it to phosphoric acid or water-soluble phosphate fertilizers. The $Ca_3(PO_4)_2$ in phosphate rock is insoluble ($K_{sp} = 1.3 \times 10^{-32}$). It is converted to a soluble form for use in fertilizers. This can be accomplished by treating the concentrated phosphate rock with sulfuric or phosphoric acid:

$$Ca_3(PO_4)_2(s) + 2H_2SO_4(aq) \longrightarrow Ca(H_2PO_4)_2(aq) + 2CaSO_4(s) \qquad [21.49]$$

$$Ca_3(PO_4)_2(s) + 4H_3PO_4(aq) \longrightarrow 3Ca(H_2PO_4)_2(aq) \qquad [21.50]$$

The mixture formed when ground phosphate rock is treated with sulfuric acid and then dried and pulverized is known as superphosphate. The

$CaSO_4$ formed in this process is of little use in soil except when deficiencies in calcium or sulfur exist. It also dilutes the phosphorus, which is the nutrient of interest. If the phosphate rock is treated with phosphoric acid, the product contains no $CaSO_4$ and has a higher percentage of phosphorus. This product is known as triple superphosphate. Although the solubility of $Ca(H_2PO_4)_2$ allows it to be assimilated by plants, it also allows it to be washed from the soil and into water bodies, thereby contributing to eutrophication.

Phosphorus compounds are important in biological systems. The element occurs for example in RNA and DNA, the molecules responsible for control of protein biosynthesis and transmission of genetic information. It also occurs in adenosine triphosphate (ATP), which stores energy within biological cells:

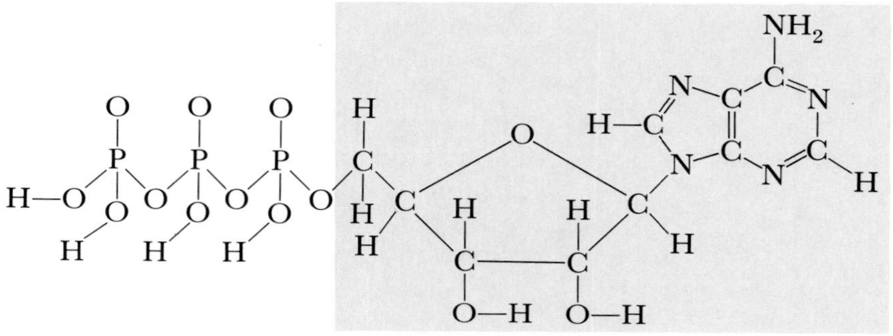

Adenosine

The P—O—P bond of the end-phosphate group is broken by hydrolysis with water, forming adenosine diphosphate (ADP). This reaction produces 33 kJ of energy:

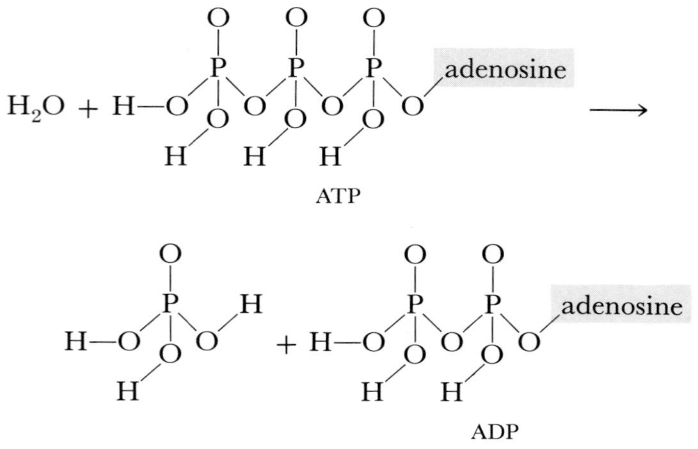

ATP

ADP

[21.51]

This energy is used to perform the mechanical work of muscle contraction and in many other biochemical reactions (see Section 15.7).

summary

There are a number of properties that distinguish metals from nonmetals. Nonmetals lack the luster, conductivity, ductility, and malleability that characterize metals. Nonmetals possess structures with few nearest neighbors. They have higher electronegativities than do metals. The soluble oxides of nonmetals are acidic anhydrides.

Electronegativity decreases with increasing atomic number in a family, and in moving from right to left through a period. Metallic character parallels electronegativity trends. Among the nonmetals, the first member of each family differs from other members of the family; the first member can form compounds that have a maximum of four bonds, and it exhibits a stronger tendency to form π bonds.

The halogens occur as diatomic molecules. All except fluorine exhibit oxidation states ranging from $+7$ to -1. Fluorine, being the most electronegative element, exhibits only the -1 oxidation state. Chlorine, bromine, and iodine occur in seawater and in salt deposits; fluorine is found in minerals such as fluorspar, CaF_2. Fluorine is a very reactive oxidizing agent; the other halogens are less reactive. The oxyhalides are also important oxidizing agents.

Oxygen is the most abundant element in the earth's crust. It occurs in the atmosphere as O_2 molecules; another elemental form is O_3, ozone. Oxygen exhibits mainly a -2 oxidation state. However, it also forms compounds in the -1 oxidation state; these are known as peroxides. The element is an important oxidizing agent.

Sulfur occurs in the elemental state in several allotropic forms; the most stable one consists of S_8 rings. The element is found in large underground deposits from which it is removed using the Frasch process. Sulfur exhibits oxidation states ranging from $+6$ to -2. Its most important compound is sulfuric acid, a strong acid; sulfuric acid is a good dehydrating agent and has a high boiling point. This acid is the most widely used industrial chemical.

Nitrogen and phosphorus exhibit oxidation states ranging from $+5$ to -3. Because it is less electronegative than nitrogen, phosphorus is found in the positive oxidation states more frequently than is nitrogen. The primary source of nitrogen is the atmosphere, where it occurs as N_2 molecules. The most important commercial process for converting N_2 into compounds is the Haber process, used to manufacture ammonia. Another important commercial process is the Ostwald process, which is used to convert NH_3 to nitric acid, HNO_3. Nitric acid is both a strong acid and a good oxidizing agent. Nitrogen compounds are important fertilizers.

Phosphorus, which is also important in fertilizers, occurs in nature in certain phosphate minerals. The element exhibits several allotropes including one known as white phosphorus, a reactive form consisting of P_4 tetrahedra. Phosphorus forms two oxides, P_4O_6 and P_4O_{10}. Their corresponding acids, phosphorous acid and phosphoric acid, show a strong tendency to undergo condensation reactions when heated.

learning goals

Having read and studied this chapter, you should be able to:

1. Identify an element as a metal, semimetal, or nonmetal based on either its position in the periodic table or its properties.
2. List the ways in which the first member of each family of the periodic table differs from subsequent members of the same family and explain the reasons for these differences.

3. Predict the relative electronegativities or metallic character of any two members of a periodic family or a period of the periodic table.

4. Predict the maximum and minimum oxidation state of each element discussed in the chapter and give an example of each.

5. Cite the most common source of each element discussed (for example, oxygen is found in air, phosphorus in phosphate rock).

6. Cite the most common molecular form of each element discussed (for example, oxygen is found as O_2, sulfur as S_8).

7. Cite at least one use for each element and its compounds (for example, sulfur is used to make sulfuric acid).

8. Describe the preparation of sulfuric acid from sulfur and nitric acid from nitrogen.

key terms

Among the more important terms and expressions used for the first time in this chapter are the following:

An **acidic anhydride** is an oxide that forms an acid when added to water; soluble nonmetal oxides are acidic anhydrides.

A **basic anhydride** is an oxide that forms a base when added to water; soluble metal oxides are basic anhydrides.

A **condensation reaction** is one in which two or more molecules combine to form larger ones by elimination of small molecules such as H_2O.

Eutrophication refers to the premature aging of lakes brought about by the presence of excess plant nutrients such as nitrates and phosphates.

The **Ostwald process** is used to make nitric acid from ammonia. The NH_3 is catalytically oxidized by O_2 to form NO; NO is air oxidized to NO_2; HNO_3 is formed by dissolving NO_2 in water.

EXERCISES

periodic trends

21.1 Summarize the ways in which the lightest member of each periodic family differs from subsequent members of the same family.

21.2 Explain why sulfur is able to form a compound with six fluorides, SF_6, whereas oxygen is able to bond to only two, OF_2.

21.3 The formulas for selenic and telluric acids are H_2SeO_4 and H_6TeO_6, respectively. Explain the difference in formulas.

21.4 Predict the trend in acidic character of the X_2O_3 oxides of group 5A. Which of these oxides should be most acidic and which one least acidic? Which member of the family should have the most metallic character?

21.5 Explain why PH_3 is a better reducing agent than NH_3. Why is HNO_3 a better oxidizing agent than H_3PO_4?

21.6 Which of the following would you expect to be the best reducing agent: (a) HF; (b) HCl; (c) HBr? Explain.

21.7 Which of the following would you expect to be the best oxidizing agent: (a) ClO_3^-; (b) BrO_3^-; (c) IO_3^-? Explain.

chemical properties

21.8 Make a table summarizing how each of the following pairs of elements differ from each other with respect to natural occurrence, preparations, properties, and uses. (a) oxygen and sulfur; (b) nitrogen and phosphorus; (c) fluorine and chlorine

21.9 Summarize the differences between metals and nonmetals.

21.10 For each of the following elements give an example of compounds containing the element in both its maximum and minimum oxidation states: (a) bromine; (b) phosphorus; (c) sulfur; (d) oxygen.

21.11 Explain why sulfuric acid never acts as a reducing agent.

21.12 Explain why elemental phosphorus is more reactive than elemental nitrogen.

21.13 Describe the geometrical structures of the

following species: (a) PO_4^{3-} ion; (b) sulfite ion; (c) H_2S molecule; (d) ClO_3^- ion; (e) SF_4 molecule; (f) PCl_5 molecule

21.14 Predict the structures of the XeF_2, XeF_4, and XeF_6 molecules.

21.15 Sulfuric acid is used in preparing other acids from their salts. What property or properties of H_2SO_4 are important in these reactions?

21.16 Why is the melting point of N_2 so much lower than that of any other member of group 5A?

nomenclature and reactions

21.17 Name the following compounds: (a) Na_3AsO_4; (b) KIO_3; (c) H_2Se; (d) SF_6; (e) $NaPO_3$; (f) Na_2O_2.

21.18 Name the following compounds: (a) HIO_3; (b) H_3PO_3; (c) BaO_2; (d) $NaNO_2$; (e) NH_3; (f) Na_2Se.

21.19 Write the chemical formulas for the following compounds: (a) sodium hypochlorite; (b) phosphorus(V) chloride; (c) potassium superoxide; (d) nitrous oxide; (e) lead(II) nitrate; (f) sulfurous acid.

21.20 Write the chemical formulas for the following compounds: (a) sodium thiosulfate; (b) hydrogen peroxide; (c) nitrogen(III) chloride; (d) nitric oxide; (e) nitrous acid; (f) potassium perchlorate.

21.21 Write the anhydride for each of the following acids: (a) H_3PO_4; (b) H_2SO_3; (c) H_2CO_3; (d) HNO_2.

21.22 Write the anhydride for each of the following compounds: (a) $NaOH$; (b) $HClO_3$; (c) $Ba(OH)_2$; (d) H_2SeO_4.

21.23 Write balanced chemical equations for each of the following processes: (a) conversion of O_2 to O_3; (b) combustion of red phosphorus in excess oxygen; (c) addition of hydrochloric acid to iron(II) sulfide; (d) addition of water to P_4O_{10}.

21.24 Write balanced chemical equations for each of the following processes: (a) conversion of Cl_2 in base to Cl^- and ClO^-; (b) hydrolysis of PCl_3 forming $P(OH)_3$; (c) addition of Li_3N to water; (d) combustion of magnesium in air (two reactions occur, each involving a principal component of air).

21.25 Write a balanced chemical equation corresponding to each of the following verbal descriptions: (a) When sodium hypochlorite is warmed, it decomposes into sodium chloride and sodium chlorate. (b) Phosphine, PH_3, a very poisonous, colorless gas, can be prepared by the reaction of white phosphorus with a concentrated aqueous solution of sodium hydroxide. Sodium hypophosphite forms together with the phosphine. (c) Nitrous acid can be prepared by adding an equimolar mixture of nitric oxide and nitrogen dioxide to water.

21.26 Write complete balanced equations for the

reactions involved in the following procedures: (a) making nitric acid starting with N_2; (b) making calcium sulfate beginning with sulfur; (c) preparing P_4O_{10} from $Ca_3(PO_4)_2$.

21.27 In which of the following cases will a reaction occur?

(a) $Br_2 + NaCl$

(b) $Cl_2 + BaI_2$

21.28 What are the products when aqueous NaF is electrolyzed?

21.29 Identify which of the reactions in question 21.24 involve oxidation and reduction and indicate the oxidation numbers that change in both reactants and products.

21.30 Identify which of the reactions in question 21.25 involve oxidation and reduction and indicate the oxidation numbers that change in both reactants and products.

general exercises

21.31 Suppose a vessel contains either N_2, O_2, or Cl_2. What types of tests would allow you to distinguish between these gases?

21.32 How many grams of phosphorus are present in the following: (a) 10.0 g of $Ca_3(PO_4)_2$; (b) 10.0 g of sodium metaphosphate.

21.33 Explain why hypophosphorous acid, $H_2PO(OH)$, is a monobasic and not a tribasic acid.

21.34 An average gallon of gasoline contains 2.2 g of lead. What is the minimum amount of $C_2H_4Br_2$ needed to remove this lead as $PbBr_2$?

21.35 How much Cl_2 is required to prepare a kilogram of I_2 from I^-?

21.36 How much current is required to prepare 1.0 kg of F_2 from molten NaF if the electrolysis time is 30.0 min?

21.37 From the electrode potentials in Table 19.2, calculate $\Delta G°$ for the reaction

$$F_2 + 2Br^- \longrightarrow Br_2 + 2F^-$$

21.38 From the electrode potentials in Table 19.2, calculate the equilibrium constant for the reaction

$$2I^- + Cl_2 \longrightarrow 2Cl^- + I_2$$

21.39 Oxygen forms not only the peroxide (O_2^{2-}) and superoxide (O_2^-) ions, but also the O_2^+ ion. The bond lengths of O_2^+, O_2, O_2^-, and O_2^{2-} are 0.112, 0.120, 0.128, and 0.149 nm, respectively. Explain this trend, using molecular-orbital theory.

21.40 Why do farmers rotate leguminous and nonleguminous crops, planting one in a field one year and the other the next?

[21.41] About 7.5×10^6 tons of nitrogen are consumed annually in the United States in the form of various synthetic nitrogen fertilizers. A single cow produces an average of 120 lb of nitrogen yearly (in the form of 0.9 tons of manure). How many cows would be needed to replace the synthetic fertilizers used? Compare this with the current population of 2×10^8 persons and 1×10^8 cattle. What difficulties does this data suggest would be encountered in totally replacing synthetic fertilizers with manure?

21.42 Using the electrode potentials given in Section 21.4, calculate the standard electrode potential for reduction of NO to N_2.

21.43 Using the electrode potentials given in Section 21.4, calculate the standard electrode potential for oxidation of NH_4^+ to N_2O.

[21.44] Using electrode potentials given in Chapter 19, calculate the minimum amount of energy required to obtain a liter of oxygen (at STP) by electrolysis of water.

[21.45] Calculate the quantity of nitric oxide present at equilibrium at 2000 K if we start with 4.00 atm of N_2 and 1.00 atm of O_2. Refer to Figure 14.3 for the equilibrium constant for this process.

[21.46] In its reaction with water, P_4O_{10} may be considered as a Lewis acid. Using structural formulas, show how the P_4O_{10} molecule might interact with an H_2O molecule in the first stage of the reaction.

21.47 Which of the following species has the most polar bonds: (a) NO_3^- or PO_4^{3-}; (b) NH_3 or PH_3?

21.48 Astatine-204 undergoes decay by electron capture. Write a nuclear equation for this process.

the lithosphere: geochemistry and metallurgy

22

In earlier chapters, we considered the properties of our atmospheric environment and of the earth's waters. In this chapter, we shall examine the characteristics of the **lithosphere,** the solid earth beneath our feet. The lithosphere provides us with most of the materials with which we feed, clothe, shelter, and amuse ourselves.

Although the bulk of the earth is solid, we have ready access only to a small area near the surface. The deepest well ever drilled is only 7.7 km deep, and the deepest mine descends only 3.4 km into the earth. In comparison, the earth has a radius of 6370 km. We shall see that many substances most useful to us are not especially plentiful in the portion of the lithosphere to which we have access. Furthermore, most of them occur in chemical forms that are not useful. Usually the compounds or elements that we desire must be separated from a large quantity of undesirable material and then chemically processed to render them useful. By consuming huge quantities of substances located close to the surface, we have literally changed the face of the earth (see Figure 22.1). Experts estimate that about 25 tons of materials are extracted from the lithosphere and processed annually to support each person in our society. The total quantity of materials mined is expected to increase because of rising populations and increasing per capita demand. In addition, because the richest sources of many substances are becoming exhausted, it will be necessary in the future to process larger volumes of lower quality raw materials. This means that the extraction of the compounds and elements we need will cost more in terms of energy and environmental impact.

FIGURE 22.1 A large open-pit mining operation.

22.1
**the structure
and composition
of the earth**

Although we have not penetrated very far into the earth, we have been able to construct a general picture of its structure and composition from indirect evidence. One line of evidence is based on comparison of the average density of the entire planet (5.5 g/cm³) with the average density of rocks on its surface (2.8 g/cm³). This comparison tells us that the innermost portions of the earth must be considerably more dense than the surface area. Further information is obtained by observing how earthquake waves are propagated through the earth. This type of study indicates that the earth has a layered arrangement consisting of four

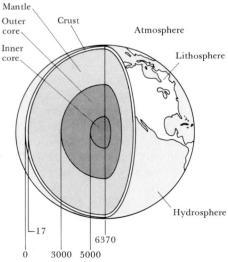

FIGURE 22.2 Regions of the earth's surface and interior.

TABLE 22.1

**the twelve most abundant
elements in the earth's crust**

ELEMENT	PERCENT BY WEIGHT
Oxygen	49.5
Silicon	25.7
Aluminum	7.5
Iron	4.7
Calcium	3.4
Sodium	2.6
Potassium	2.4
Magnesium	1.9
Hydrogen	0.9
Titanium	0.6
Chlorine	0.2
Phosphorus	0.1

distinct regions: the crust, mantle, outer core, and inner core. These regions are shown in Figure 22.2.

The outer portion, or **crust,** constitutes only about 0.4 percent of the total mass of the earth. It has an average thickness of 17 km, ranging in depth from 4 to 70 km. The term "crust" is a carry-over from the time when the entire interior of the planet was thought to be molten. The **mantle,** which lies immediately below the crust, is now believed to be almost entirely solid. It extends to a depth of about 3000 km and comprises 68.2 percent of the earth's mass. Although we have never penetrated into the mantle, it is believed that over 90 percent of it consists of four elements—magnesium, iron, silicon, and oxygen. Because the material in the mantle is at a higher pressure and temperature than that in the crust, the chemical forms in which these elements occur may differ considerably from those found in the crust. The innermost portions of the earth are known as the **inner** and **outer cores.** These portions are believed to contain about 80 percent iron. The outer core is thought to be molten, whereas the more dense inner core is thought to be solid.

Eighty-eight elements occur in the earth's crust. Twelve of these, listed in Table 22.1, make up 99.5 percent of the crust by weight.* However, if you were to analyze a shovel-full of dirt from your backyard, you would not find it to have the composition listed in Table 22.1. Elements are not distributed uniformly throughout the crust.

It is interesting to compare the order of abundance of the elements with the order of estimated world consumption shown in Table 22.2. The ranking of elements in the two lists shows little correlation; some elements that are not very abundant are widely used. Consequently, the occurrence and distribution of concentrated deposits of certain elements in the lithosphere is important. Indeed, because a modern, high-technology society such as ours has need for a broad range of substances such

*The top 1 km of the earth's crust has a total mass of approximately 10^{18} tons.

TABLE 22.2

estimated annual world consumption of elements

ELEMENT	TONS CONSUMED/YEAR
C	10^9–10^{10}
Na, Fe	10^8–10^9
N, O, S, K, Ca	10^7–10^8
H, F, Mg, Al, P, Cl, Cr, Mn, Cu, Zn, Ba, Pb	10^6–10^7
B, Ti, Ni, Zr, Sn	10^5–10^6
Ar, Co, As, Mo, Sb, W, U	10^4–10^5
Li, V, Se, Sr, Nb, Ag, Cd, I, rare earths, Au, Hg, Bi	10^3–10^4
He, Be, Te, Ta	10^2–10^3

as metals, the distribution of such deposits may become an important factor in international politics.

22.2
minerals

Most elements occur in nature in combination with other elements, that is, in compounds. Solid substances occurring in nature are referred to as **minerals.** Table 22.3 lists some common minerals. Several of these are pictured in Figure 22.3. Minerals are usually known by their common names rather than by their chemical names. They are often named by the person who first described them. Although there is no systematic method of nomenclature, the name frequently ends in *ite*. Minerals often have variable compositions owing to the substitution of one element for another within the solid (Section 22.5). Nevertheless, they have well-defined crystal structures. What we know as *rock* is merely an aggregate of different kinds of minerals.

TABLE 22.3
some common minerals

MINERAL NAME	CHEMICAL FORMULA
Bauxite	$Al_2O_3 \cdot 2H_2O$
Calcite	$CaCO_3$
Calcopyrite	$CuFeS_2$
Cinnabar	HgS
Corundum	Al_2O_3
Fluorite	CaF_2
Galena	PbS
Gypsum	$CaSO_4 \cdot 2H_2O$
Halite	$NaCl$
Hematite	Fe_2O_3
Malachite	$Cu_2(CO_3)(OH)_2$
Pyrite	FeS_2
Perovskite	$CaTiO_3$
Quartz	SiO_2
Talc	$Mg_3(Si_4O_{10})(OH)_2$
Turquoise	$CuAl_6(PO_4)_4(OH)_2 \cdot 4H_2O$
Wulfenite	$PbMoO_4$

(a)

(b)

(c)

FIGURE 22.3 Three common minerals: (a) calcite; (b) fluorite; (c) wulfenite. Note the variety of crystal shapes.

TABLE 22.4

types of minerals

GENERAL CATEGORY	EXAMPLES
Native elements	Cu, Ag, Au, Bi, Pt, Pd, S
Silicate minerals	$ZrSiO_4$, $Be_3AlSi_6O_{18}$
Nonsilicate minerals	
Oxides	Al_2O_3, Fe_2O_3, Fe_3O_4, Cu_2O, TiO_2
Hydroxides	$Mg(OH)_2$
Carbonates	$CaCO_3$, $MgCO_3$, $PbCO_3$, $ZnCO_3$
Sulfates	$BaSO_4$, $PbSO_4$, $CaSO_4$
Sulfides	Ag_2S, Cu_2S, HgS, ZnS
Halides	$NaCl$, $MgCl_2$
Phosphates	$Ca_3(PO_4)_2$

Minerals can be classified into three groups: native elements, silicate minerals, and nonsilicate minerals. Examples are shown in Table 22.4. Metals found in elemental form include silver, gold, palladium, platinum, ruthenium, rhodium, osmium, and iridium. These metals, which are in families 8 and 1B of the periodic table, are known as **noble metals** because of their lack of reactivity. All of these metals have very low standard oxidation potentials.

**22.3
silicate
minerals**

The silicate minerals are the most abundant group of minerals. It has been estimated that over 90 percent of the earth's crust consists of silicates, if quartz (SiO_2) is included. Suppose you pick up a piece of common granite rock* such as that shown in Figure 22.4 and determine its elemental composition. You would find that it contains about 50 percent oxygen and 25 percent silicon. It also contains a surprising variety of important metals. Each 100 kg of granite contains approxi-

*Granite is a mixture of small crystals of mica, quartz, and feldspars. We will discuss these minerals later in this section.

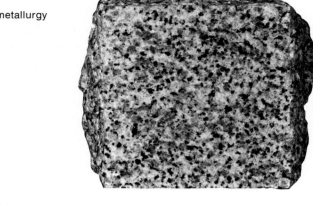

FIGURE 22.4 A piece of granite rock.

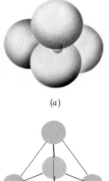

(a)

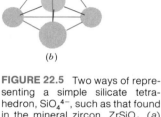

(b)

FIGURE 22.5 Two ways of representing a simple silicate tetrahedron, SiO_4^{4-}, such as that found in the mineral zircon, $ZrSiO_4$. (a) The small Si(IV) surrounded by four O^{2-} as it might appear if we could actually view the tetrahedron. (b) The geometric structure showing the orientation of the atoms.

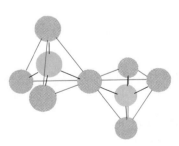

FIGURE 22.6 Geometric structure of the $Si_2O_7^{6-}$ ion, which is formed by the sharing of an oxygen atom by two silicon atoms. This ion occurs in the mineral hardystonite, $Ca_2Zn(Si_2O_7)$.

mately 8 kg of aluminum, 5 kg of iron, 90 g of manganese, 20 g of nickel, and 10 g of copper. Despite these figures, the silicate minerals are not presently economical sources of these or any other metals. The silicates are extremely stable chemical substances; a large amount of energy is required to remove metals from them. Nevertheless, they are important items of commerce and are used, for example, in the manufacture of cement and glass.

To understand why silicates are so stable, we need to examine their structures. The basic structural unit of the silicate minerals is a silicon atom bound in a tetrahedral fashion to four oxygens as shown in Figure 22.5.* This structural unit occurs as the orthosilicate ion, SiO_4^{4-}. However, this simple ion is found in very few minerals. Usually, the silicate tetrahedra are connected by sharing oxygen atoms, giving Si—O—Si bonds. When two silicate tetrahedra are so joined by sharing of an oxygen atom, the $Si_2O_7^{2-}$ ion shown in Figure 22.6 results. The situation is similar to the linking of phosphate tetrahedra, discussed in Section 21.6. By sharing oxygen atoms, silicate tetrahedra can be arranged in strands, sheets, or three-dimensional arrays. Two of the strandlike arrangements and the sheetlike arrangement are shown in Figure 22.7. Notice the empirical formula that results from each of these arrangements.

It is interesting how the cleavage properties of silicate minerals reflect the arrangement of the silicate ions. The asbestos minerals, which are fibrous, as shown in Figure 22.8, have either the double-strand chain structure or a rolled-sheet structure in which the sheets are rolled into strands. The fibrous texture of the asbestos minerals arises because the electrostatic bonds between strands are much weaker than are the covalent bonds within the strands. Talc, $Mg_3Si_4O_{10}(OH)_2$, has a structure composed of planar sheets. The relatively weak forces between sheets allows them to slip over each other much as in the case of graphite.

*Under extreme conditions a different arrangement of oxygens about silicon might be formed. For example, under high pressures and high temperatures it is possible to prepare a silicate in which six oxygens surround each silicon. Such a structure has been found among the compounds in the impact craters of meteorites. It has been postulated that such silicates may also exist in the mantle of the earth.

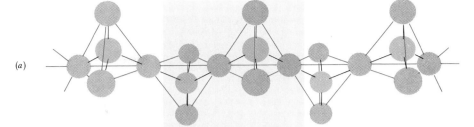

(a)

Repeating unit of chain

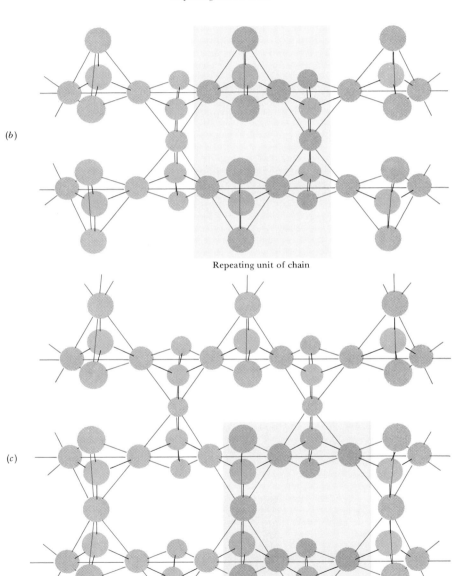

(b)

Repeating unit of chain

(c)

FIGURE 22.7 Schematic representations of chain and sheet silicates formed by linking together silicate tetrahedra: (a) single-strand silicate, which has an empirical formula of SiO_3^{2-}, as in the mineral enstatite, $Mg(SiO_3)$; (b) double-strand silicate chain, which has an empirical formula of $Si_4O_{11}^{6-}$, as in the mineral tremolite, $Ca_2Mg_5(Si_4O_{11})_2(OH)_2$; (c) sheet silicate, which has an empirical formula $Si_2O_5^{2-}$, as in the mineral talc, $Mg_3(Si_2O_5)_2(OH)_2$. Note that in (a) the repeating unit is double the empirical formula.

Repeating unit of sheet

FIGURE 22.8 Serpentine asbestos; note the fibrous character of this mineral.

Talcum powder therefore has a slippery feel. Extension of the silicate tetrahedra in three dimensions in space, so that every oxygen atom is bridging between two silicons, gives quartz, SiO_2, which is shown in Figure 22.9. Because the silicate is locked together in a three-dimensional array, much like diamond, quartz is harder than the strand- or sheet-type silicates.

Asbestos is a general term applied to a variety of silicates that are fibrous. These asbestos minerals are widely used, especially as thermal insulators. They are used, for example, to insulate furnaces and steam pipes and to make protective clothing for people exposed to high temperatures. Certain forms of asbestos pose a health hazard. Tiny fibers readily penetrate the tissues of the lungs and digestive tract where they can apparently induce cancer.

ALUMINOSILICATES

In many silicate minerals Si^{4+} ions are replaced by Al^{3+} ions within the silicate tetrahedra. This replacement produces **aluminosilicates.** In order to maintain charge balance, an extra cation such as K^+ must accompany each of these substitutions. Muscovite, $KAl_2(AlSi_3O_{10})(OH)_2$,* a mica mineral, is an aluminosilicate. Replacement of a quarter of the silicon atoms in a sheet silicate, $Si_4O_{10}^{4-}$, with aluminum produces the $AlSi_3O_{10}^{5-}$ sheets found in this mineral. In both the silicate-sheet and aluminosilicate-sheet minerals, cations are located between the sheets to balance the charge. The electrostatic attraction between these cations

*Aluminum is found in this mineral in two different environments. The first two Al^{3+} ions are located between the aluminosilicate sheets. The aluminum that is shown in parentheses is located within the sheets. It has replaced a silicon and is therefore located in an AlO_4 tetrahedron.

FIGURE 22.9 Quartz crystals. (*Courtesy of Ward's Natural Science Establishment*)

and the sheets is greater for the aluminosilicate than for the silicate because of the former's greater negative charge. Thus while the sheets in the silicate mineral talc slide readily over each other, the sheets in the aluminosilicate mica do not. Nevertheless, mica does cleave readily into sheets, as shown in Figure 22.10.

When aluminum replaces up to half of the silicon atoms in SiO_2, the **feldspar** minerals result (see Figure 22.11). The cations that compensate the extra negative charge accompanying this replacement are usually Na^+, K^+, or Ca^{2+} ions. The feldspars are the most abundant rock-forming silicates, comprising nearly 54 percent of the minerals in the earth's crust.

SAMPLE EXERCISE 22.1

The mineral anorthite is a feldspar mineral formed by replacing half of the silicon atoms in SiO_2 with aluminum and maintaining charge balance with Ca^{2+} ions. What is the simplest formula for this mineral?

Solution: If we write SiO_2 as Si_4O_8, the replacement of half of the Si^{4+} with Al^{3+} produces $Al_2Si_2O_8{}^{2-}$. One Ca^{2+} is therefore needed to maintain charge balance and the empirical formula of the mineral is $CaAl_2Si_2O_8$.

CLAY

The clay minerals are hydrated aluminum silicates having sheet-type structures. They are thought to be formed during weathering when H_2O and CO_2 slowly attack feldspars. For example, the action of H_2O and CO_2 on the feldspar mineral anorthite to form the clay mineral known as kaolinite is given in Equation [22.1].

FIGURE 22.10 A piece of mica, showing its cleavage into thin sheets.

FIGURE 22.11 Orthoclase, a feldspar mineral.

$$CaAl_2Si_2O_8(s) + 3H_2O + 2CO_2(aq) \longrightarrow$$
Anorthite

$$Al_2Si_2O_5(OH)_4(s) + Ca(HCO_3)_2(aq) \qquad [22.1]$$
Kaolinite

The ions released by such weathering reactions may be incorporated into plants and other minerals or washed into the sea. The clay minerals have small particle size and correspondingly large surface areas. They have the ability to adsorb cations on their surfaces. This provides a means of storing plant nutrients such as K^+ that can then be made available to plants through equilibrium shifts:

$$K^+\text{—clay} \rightleftharpoons K^+(aq) + clay \qquad [22.2]$$

Some metal ions may exchange for others within the crystal lattice of the clay and thereby be stored in this fashion. Often the metal ions displace hydrogen ions from the OH groups on the surface of the clay particle:

$$M^+(aq) + H\text{—O—clay} \rightleftharpoons M\text{—O—clay} + H^+(aq) \qquad [22.3]$$

This situation gives rise to pH-dependent equilibria. Notice that the higher the concentration of $H^+(aq)$, the more the equilibrium is shifted to the left. If the soil is basic, the equilibrium lies to the right, and $M^+(aq)$ is not available to plants. Thus the pH of a soil plays an important role in determining its fertility (that is, its ability to supply plants with essential nutrients). The majority of plants grow best in soil whose pH is 6–7, that is, one that is slightly acidic. Basic or alkaline soils are common in areas where there is poor rainfall or where soils drain poorly. However, it is more common for soils to be too acidic than for them to be too alkaline. The acidity of a soil is often lowered by adding lime, CaO, to adjust pH. This process is known as "liming." The CaO is a basic anhydride and therefore can react with $H^+(aq)$ as shown in Equation [22.4].

$$CaO(s) + 2H^+(aq) \longrightarrow Ca^{2+}(aq) + H_2O \qquad [22.4]$$

GLASS

As noted earlier, in Section 11.5, quartz melts at approximately 1600°C, forming a tacky liquid. In the course of melting, many silicon-oxygen bonds are broken. When the liquid is rapidly cooled, silicon-oxygen bonds are reformed before the atoms are able to arrange themselves in a regular fashion. An amorphous solid, known as quartz glass or silica glass, therefore results (see Figure 11.28). Many different substances can be added to SiO_2 to cause it to melt at a lower temperature. The common glass used in windows and bottles is known as lime glass. It contains CaO and Na_2O in addition to SiO_2 from sand. The CaO and Na_2O are produced by heating two inexpensive chemicals, limestone, $CaCO_3$, and soda ash, Na_2CO_3. These substances decompose at elevated temperatures as shown in Equations [22.5] and [22.6].

$$CaCO_3(s) \longrightarrow CaO(s) + CO_2(g) \qquad [22.5]$$

$$Na_2CO_3(s) \longrightarrow Na_2O(s) + CO_2(g) \qquad [22.6]$$

Other chemicals can be added to impart color or change the properties of the glass in other ways. Addition of CoO produces blue "cobalt glass." Partial substitution of CaO by PbO produces a high-density glass known as "flint glass" or "lead glass"; this glass is used in making lenses and cut-glass "crystal." Addition of B_2O_3 produces boro-silicate glass, which is more resistant to thermal shock and chemical attack than is lime glass; this type of glass is sold under such trade names as Pyrex and Kimax.

A recent popular development has been that of "photochromic" glasses that darken in the sun and return to their clear state in the dark. This glass contains a dispersion of AgCl or AgBr. These substances are photosensitive, in that they decompose to silver and halogen atoms in the presence of light. The finely divided silver is black. The silver and halogen atoms are kept in close proximity by the glass matrix and reform AgCl or AgBr in the dark:

$$AgCl \underset{\text{dark reaction}}{\overset{\text{light reaction, } h\nu}{\rightleftharpoons}} Ag + Cl \qquad [22.7]$$

CEMENT

Portland cement, used as a binder in making concrete, is similar in structure to glass except that it is an aluminosilicate. It is made by heating a powdered mixture of limestone, $CaCO_3$, sand, SiO_2, and clay in a kiln at about 1500°C. After the resulting mass is pulverized, a small amount of gypsum, $CaSO_4 \cdot 2H_2O$, is added. The composition of Portland cement is given in Table 22.5. The reactions involved in the setting of cement are complex, involving reaction with both water and CO_2. Concrete, like other materials containing Si—O bonds, is highly incom-

TABLE 22.5

composition of Portland cement

CONSTITUENTS[a]	PERCENTAGE RANGE
CaO	60–64
SiO_2	18–26
Al_2O_3	4–12
Fe_2O_3	2–4
MgO	1–4
Other	2

[a]The cement is analyzed as if it were a mixture of oxides. However, it is not a simple mixture of these oxides but rather a complex mixture of aluminosilicates and oxides.

pressible but lacks tensile strength.* The importance of cement and concrete in our society is indicated by the fact that the total economic value of the raw materials used annually in their manufacture is substantially higher than the value of the raw metallic ores that we process.

22.4 nonsilicate minerals

The nonsilicate minerals, though less abundant than the silicates, are important as sources of metals. The greatest quantity of metals is obtained commercially from oxides, sulfides, and carbonates; most processes used for obtaining metals utilize oxides. Both sulfide and carbonate minerals are readily converted to oxides when heated in air. Equations [22.8] and [22.9] illustrate these conversions.

$$2ZnS(s) + 3O_2(g) \longrightarrow 2ZnO(s) + 2SO_2(g) \qquad [22.8]$$

$$PbCO_3(s) \longrightarrow PbO(s) + CO_2(g) \qquad [22.9]$$

We shall discuss the reduction of metal oxides in Section 22.6.

CALCIUM CARBONATE

The most abundant nonsilicate mineral is calcite, $CaCO_3$. This mineral is pictured in Figure 22.3. It is the principal mineral in the rock known as limestone. It is also the main constituent of marble, chalk, pearls, coral reefs, and the shells of marine animals such as clams and oysters. Although $CaCO_3$ is insoluble in pure water, it does dissolve in acidic solution. The reaction with hydrochloric acid is given in Equation [22.10].

$$CaCO_3(s) + 2HCl(aq) \longrightarrow CaCl_2(aq) + H_2O + CO_2(g) \quad [22.10]$$

Because water containing CO_2 is slightly acidic (Section 18.1), $CaCO_3$ dissolves slowly in this medium:

$$CaCO_3(s) + H_2O + CO_2(aq) \rightleftharpoons Ca(HCO_3)_2(aq) \qquad [22.11]$$

This reaction occurs in nature when surface waters move underground

*Tensile strength measures the stress required to stretch a rod to the breaking point.

FIGURE 22.12 A sinkhole filled with water. (*U.S. Geological Survey*)

through limestone deposits. If the dissolving limestone underlies a comparatively thin layer of earth, sinkholes like that shown in Figure 22.12 are produced. If the limestone deposit is deep enough underground, the dissolution of the limestone produces a cave; two well-known limestone caves are the Mammoth Cave in Kentucky and the Carlsbad Cavern in New Mexico. When a solution of $Ca(HCO_3)_2$ trickles through the cracks in a cave roof, the pressure release allows escape of CO_2 and evaporation of H_2O. Loss of these substances shifts the equilibrium of Equation [22.11] to the left, forming deposits of $CaCO_3$. Those deposits that hang from the cave roof are known as stalactites, whereas those that rise from the cave floor are known as stalagmites. Figure 22.13 shows a limestone cave with prominent stalactites and stalagmites.

The chemical reaction between $CaCO_3$ and acidic solutions formed by CO_2 is responsible for the erosion of marble and limestone monuments (refer to Figure 10.7). The presence of SO_2 in polluted air accelerates this erosion as noted earlier (Section 10.5). To slow such erosion of monuments and thereby extend their lives, some have been treated with a mixture of $Ba(OH)_2$ and urea, $(NH_2)_2CO$. These two substances react with each other to form a layer of $BaCO_3$ on the monument's surface:

$$\overset{\displaystyle O}{\overset{\displaystyle \|}{H_2NCNH_2}}(aq) + H_2O \longrightarrow 2NH_3(aq) + CO_2(aq) \qquad [22.12]$$

667

$$CO_2(aq) + Ba(OH)_2(aq) \longrightarrow BaCO_3(s) + H_2O \qquad [22.13]$$

Barium carbonate ($K_{sp} = 1.6 \times 10^{-9}$) is slightly less soluble than calcium carbonate ($K_{sp} = 4.7 \times 10^{-9}$). Furthermore, when it reacts with SO_2, it forms an even more insoluble compound, $BaSO_4$, as shown in Equation [22.14].

$$2BaCO_3(s) + 2SO_2(aq) + O_2(aq) \longrightarrow 2BaSO_4(s) + 2CO_2(g)$$
$$[22.14]$$

The solubility product, K_{sp}, of $BaSO_4$ is only 1.1×10^{-10}, whereas that of $CaSO_4$ is 2.4×10^{-5}.

One of the most important reactions of $CaCO_3$ is its decomposition into CaO and CO_2 at elevated temperatures, which was given earlier in Equation [22.5]. Over 40 billion pounds of calcium oxide, known as lime or quicklime, is used in the United States each year. Because calcium oxide reacts with water to form $Ca(OH)_2$, it is an important commercial base. In this chapter, we have already discussed the use of CaO in making glass and cement and in neutralizing acidic soils (Section 22.3); in Section 22.6 we shall consider its use in high-temperature reductions of metal ores. Calcium oxide is also used in making mortar, a mixture of sand, water, and CaO used in construction to bind bricks, blocks, and rocks together. The CaO reacts with water and CO_2 to form $CaCO_3$, which binds the sand in the mortar:

$$CaO(s) + H_2O \longrightarrow Ca(OH)_2(aq) \qquad [22.15]$$

$$Ca(OH)_2(aq) + CO_2(aq) \longrightarrow CaCO_3(s) + H_2O \qquad [22.16]$$

22.5

close packing in minerals

In many minerals and other ionic compounds, the large ions in the lattice assume a close-packed arrangement identical to those discussed in Section 11.4. A close-packed array of spheres, whether hexagonal or cubic close packed, occupies 74 percent of the total volume taken up by the structure as a whole. The remaining 26 percent consists of holes between the spheres. The smaller ions in the lattice occupy these holes. For example, Fe_2O_3 consists of a cubic close-packed arrangement of oxide ions with Fe^{3+} ions in certain holes.

Examination of close-packed structures reveals that they possess two types of holes. These are shown in Figure 22.14. The first type is created by four large spheres arranged in a tetrahedral fashion. The second type is formed by six spheres arranged in an octahedral fashion. Using trigonometry, it can be shown that a **tetrahedral hole** is smaller than an **octahedral hole**. A small sphere that is 0.225 times the larger sphere in radius will fit perfectly into a tetrahedral hole. If the small sphere has a radius that is 0.414 times that of the larger spheres, it will just fit into an octahedral hole.

You may recall from our discussion of ionic radii in Section 8.2 that anions generally have larger radii than cations. It is useful then to think of the lattice as consisting of a close-packed arrangement of anions, with cations occupying one or another of the types of holes present. The size of the cation relative to that of the anion determines the type of hole that the cation occupies. The most stable arrangement is one that maximizes the number of cation-anion contacts, because these lead to electrostatic attractive forces. At the same time, however, the arrangement must also be one that prevents direct anion-anion contacts, because these contacts will generate electrostatic repulsive forces. We can examine the possibilities more closely by considering the situation in which a small cation fits just perfectly into a tetrahedral hole of a close-packed array of anions. As we have noted, this situation occurs when the ratio of the radius of the cation to that of the anion, r_c/r_a, equals 0.225. Under these conditions, the cation can just touch the four anions that surround it. Now consider what happens if the cation increases in size, so that $r_c/r_a > 0.225$. This condition forces the anions apart, thereby diminish-

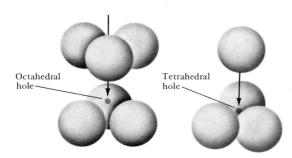

Octahedral hole

Tetrahedral hole

FIGURE 22.14 Views of the octahedral and tetrahedral holes that exist between two layers of close-packed spheres.

ing the destabilizing anion-anion contacts while maintaining the stabilizing cation-anion contacts. However, when the ratio of radii, r_c/r_a, reaches a value of 0.414, the cation is no longer most stable in a tetrahedral hole. Instead, the most stable location is the octahedral hole, which permits a greater number of stabilizing cation-anion contacts; in this location the cation can just touch six surrounding anions. Now consider what happens as the cation gets yet larger, so that $r_c/r_a > 0.414$. As before, the anions are forced apart, thereby lowering the destabilizing anion-anion contacts while maintaining the stabilizing cation-anion contacts. Lattices for which r_c/r_a falls in the range of 0.414 to 0.732 generally have cations that occupy octahedral holes. When r_c/r_a exceeds 0.732 the anions are no longer most stable in a close-packed array. Instead, they generally assume a simple cubic arrangement. This arrangement permits a cation to be in contact with eight anions, thereby increasing the number of stabilizing cation-anion contacts, as shown in Figure 22.15. These radius-ratio guidelines, which indicate the likely location of cations in a lattice composed of larger anions, are summarized in Table 22.6.

SAMPLE EXERCISE 22.2

The mineral hematite, Fe_2O_3, consists of a cubic close-packed array of oxide ions with Fe^{3+} ions occupying interstitial positions. Predict whether the iron ions are in octahedral or tetrahedral holes. (The radius of Fe^{3+} is 0.65 Å, whereas that of O^{2-} is 1.45 Å.)

Solution:

$$\frac{r_c}{r_a} = \frac{0.65 \text{ Å}}{1.45 \text{ Å}} = 0.45$$

Based on the radius ratio we would predict that Fe^{3+} ions would be located in octahedral holes. In fact, this is their location.

During the first 10 years of nuclear reactor operation, uranium metal was used as the fuel. However, fission produces two atoms for each original atom (Section 20.7). The fuel elements therefore expand, and the sheath intended to contain the fission fragments may burst under the pressure produced by the new atoms. It was then realized that use of UO_2 as the fuel greatly reduced this problem. In UO_2 the uranium atoms are arranged in a cubic close-packed array with oxygen atoms at the tetrahedral sites. The structure is relatively open, because the oxygen atoms move the uranium atoms apart slightly; in fact, the distance between uranium atoms is 42 percent greater in UO_2 than in uranium metal. Correspondingly, the octahedral holes are larger than in the metal and are available as sites for fission fragments to occupy.

TABLE 22.6

radius ratios and cation location

r_c/r_a [a]	COORDINATION NUMBER OF CATION	SPECIAL ARRANGEMENT OF ANIONS
0.225–0.414	4	Tetrahedral
0.414–0.732	6	Octahedral
0.732–1.000	8	Cubic

[a] If $r_c > r_a$, the ratio r_a/r_c gives the coordination number of the anion and the special arrangement of cations.

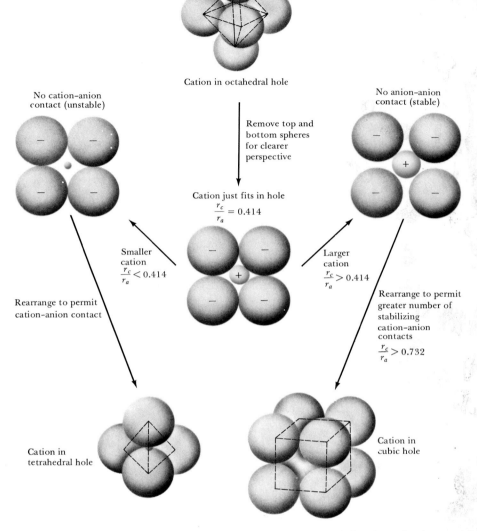

Cation in octahedral hole

No cation-anion contact (unstable)

No anion-anion contact (stable)

Remove top and bottom spheres for clearer perspective

Cation just fits in hole
$$\frac{r_c}{r_a} = 0.414$$

Smaller cation
$$\frac{r_c}{r_a} < 0.414$$

Larger cation
$$\frac{r_c}{r_a} > 0.414$$

Rearrange to permit cation-anion contact

Rearrange to permit greater number of stabilizing cation-anion contacts
$$\frac{r_c}{r_a} > 0.732$$

Cation in tetrahedral hole

Cation in cubic hole

FIGURE 22.15 A schematic representation of the effect of the relative sizes of anions and cations on the way the ions pack in the lattice. This diagram shows the effects of starting with a cation that perfectly fits into an octahedral hole in a close-packed array of anions. The changes in lattice stability and in cation location are considered as the cation gets smaller and as it gets larger.

ION SUBSTITUTION IN MINERALS

In minerals, it is not uncommon for one ion to substitute for another of similar size. For example, in the mineral olivine, Mg_2SiO_4, the Fe^{2+} ions ($r = 0.74$ Å) substitute freely for Mg^{2+} ions ($r = 0.65$ Å). Such substitution explains why this mineral and others are commonly **nonstoichiometric,** meaning that they deviate from ideal or simple chemical formulas. Substitution is most common when the radii differ by no more than 15 percent.

We have already seen that silicate minerals are not useful sources of the various metals used in modern technology. Nevertheless, the chemical sources of these metals, such as sulfide or oxide minerals, are generally found in nature mixed with silicates. The deposits that contain the metal in economically exploitable quantities are known as **ores.** The undesirable materials present in the ore are referred to as **gangue** (pronounced "gang"). As worldwide demand for metals has increased, the mining of low-grade ores has increased also. When a particularly convenient or rich ore is closed off, either because the supply is exhausted or for reasons of international politics, alternate sources must be found. Often these are less rich and consequently more costly to work. In the face of these pressures, prices have been kept fairly low mainly by increases in mining and extraction efficiency. Prices have nevertheless increased, and this has made us aware of the need to recycle materials. It is likely that urban garbage and scrap piles will become economically feasible sources of metals in the future.

The availability and cost of a metal depends on a number of factors other than abundance. These include the occurrence of concentrated ore deposits and the ease of extraction of the metal from the ore. When an element possesses useful properties, it may be in demand even though it is not readily available. Demand stimulates search for extraction processes that can improve availability. As we have noted earlier (Section 19.3), aluminum was once costly and was exhibited as a rare metal, even though its compounds were readily available. Unfortunately, much aluminum is tied up in aluminosilicates; furthermore, the Al^{3+} ion is difficult to reduce. Aluminum is uniquely useful in many applications where low density and high electrical conductivity are important characteristics. In 1886, Charles M. Hall in the United States and Paul Heroult in France independently developed a new electrolysis procedure for producing aluminum from its oxide (Section 19.3). With the advent of this procedure, the price of aluminum fell sufficiently for the metal to find widespread use.

The science of extracting metals from their ores and preparing them for use is known as **metallurgy.** Metallurgical processes are conveniently divided into three types of operations: (1) preliminary treatment, which includes concentration of the ore; (2) reduction to obtain the metal in its elemental state; and (3) refining of the metal.

PRELIMINARY TREATMENT

In the processing of many ores, the metallic element is concentrated by removing the undesirable components and often by chemically modifying the mineral. The separation may be based on differences in either the physical or chemical properties of the mineral and the gangue. For example, gold minerals can be separated from the lighter gangue by shaking the crushed ore in a stream of water on an inclined table. In its simplest form, this separation technique is used by individuals panning for gold in streams.

Flotation is a process widely used on sulfide ores, especially those of copper, zinc, and lead. In this process, the finely crushed ore is mixed with oil, water, and a detergent or flotation agent. The mineral particles are wetted by the oil, and when air is blown through the mixture they

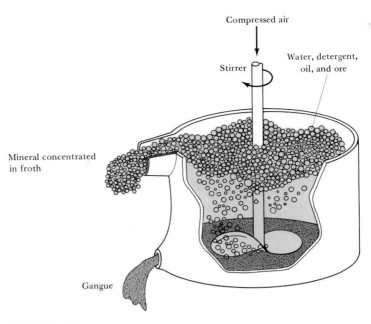

Compressed air

Stirrer

Water, detergent,
oil, and ore

Mineral concentrated
in froth

Gangue

FIGURE 22.16   A flotation tank.

float to the top of the water in the oil froth and are skimmed off as shown in Figure 22.16. The process is very economical and is widely used.

The **Bayer method** for obtaining Al_2O_3 from bauxite ore illustrates the use of chemical properties in separation. This procedure, which was discussed in Section 17.4, makes use of the amphoterism of aluminum. The extraction of magnesium from seawater, discussed in Section 18.3, provides another example of the use of chemical properties in separating substances.

Sometimes a component of an ore can be removed by **leaching,** that is, by allowing a solution to percolate through the ore and selectively dissolve certain components. For example, piles of silver and gold ore have been sprayed with dilute solutions of NaCN to leach silver and gold from them. These metals form very stable and soluble complex ions with cyanide as shown in Equations [22.17] and [22.18].

$$4Ag(s) + 8CN^-(aq) + O_2(aq) + 2H_2O \longrightarrow$$
$$4Ag(CN)_2^-(aq) + 4OH^-(aq) \qquad [22.17]$$

$$Ag_2S(s) + 4CN^-(aq) \longrightarrow 2Ag(CN)_2^-(aq) + S^{2-}(aq) \qquad [22.18]$$

The ores are placed on slabs of blacktop or concrete or on plastic sheets; the solutions that have moved through the ore can then be readily captured and later treated to remove the gold and silver.

After preliminary removal of gangue, many ores are **roasted** in air. This drives off volatile impurities, burns off organic matter, and converts carbonates and sulfides into oxides as shown earlier in Equations [22.8] and [22.9].

REDUCTION

Metals can be obtained by either electrolytic reduction or chemical reduction of their compounds. Electrolytic reduction, discussed in Section 19.3, is used commercially to obtain the more active metals, such as sodium, magnesium, and aluminum. Less active metals, such as copper, iron, and zinc, are produced commercially using chemical reducing agents. Most of these less active metals are obtained by high-temperature reduction processes in which the metals are obtained in the molten state. These processes are known as **smelting operations.**

The reduction of iron illustrates several features of a typical smelting operation. A reactor designed to operate continuously is used. Such a device, known as a blast furnace, is shown in Figure 22.17. A mixture of coke, limestone, and crushed ore, usually containing Fe_2O_3, is added to the top of the furnace. (Coke is coal that has been heated in the absence of air to drive off volatile components.) Heated air, sometimes enriched with oxygen, is forced into the furnace at the bottom. About 2 tons of ore, 1 ton of coke, and 0.3 ton of limestone are required to produce 1 ton of iron. A single furnace may produce about 2000 tons of iron a day. The air coming into the furnace reacts with carbon to form CO; this liberates a considerable amount of heat and allows the furnace to reach a temperature of about 1500°C at the bottom. Reduction to metallic iron takes place as shown in Equations [22.19] and [22.20].

$$Fe_2O_3(s) + 3C(s) \longrightarrow 2Fe(l) + 3CO(g) \qquad [22.19]$$

$$Fe_2O_3(s) + 3CO(g) \longrightarrow 2Fe(l) + 3CO_2(g) \qquad [22.20]$$

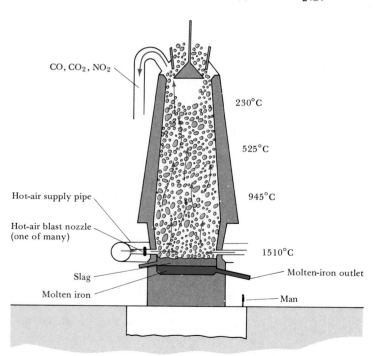

FIGURE 22.17 A blast furnace used for reduction of iron ore. Notice the approximate temperatures in the various regions of the furnace.

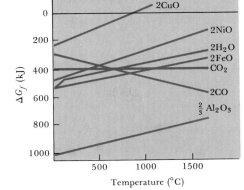

FIGURE 22.18 Stabilities of certain oxides (as measured by ΔG_f) shown as a function of temperature.

Carbon monoxide is produced not only by reaction between coke and oxygen, Equation [22.21], but also between coke and CO_2, Equation [22.22].

$$2C(s) + O_2(g) \longrightarrow 2CO(g) \qquad [22.21]$$

$$C(s) + CO_2(g) \longrightarrow 2CO(g) \qquad [22.22]$$

Carbon dioxide is produced in the reduction of the iron oxide, Equation [22.20], as well as from decomposition of limestone. The limestone serves another important role in smelting operations. An ore typically still contains much gangue when it is reduced. The CaO formed by decomposition of limestone reacts with gangue to form a liquid known as slag. One of the important reactions is given in Equation [22.23].

$$\underset{\text{Gangue}}{CaO(s) + SiO_2(s)} \longrightarrow \underset{\text{Slag}}{CaSiO_3(l)} \qquad [22.23]$$

In the smelting of iron, the slag floats on top of the molten iron, protecting it from oxidation by the incoming air. The slag and iron are removed periodically. The iron produced in the blast furnace, called **pig iron**, contains up to 5 percent carbon and as much as 2 percent other impurities such as silicon, phosphorus, and sulfur.

The thermodynamic basis for smelting operations using coke or CO as a reducing agent can be seen by reference to Figure 22.18. As this figure shows, metal oxides generally become less stable (ΔG_f more positive) as temperature increases. This follows from the fact that $\Delta G_f = \Delta H_f - T\Delta S_f$ (Section 15.5). The formation of an oxide from a metal and O_2, Equation [22.24],

$$2M(s) + O_2(g) \longrightarrow 2MO(s) \qquad [22.24]$$

generally represents a decrease in disorder, so that ΔS_f is negative. The factor $-T\Delta S_f$ therefore becomes increasingly positive with increasing temperature. In Figure 22.18 we notice that for CuO, ΔG_f becomes positive at approximately 900°C. Consequently, CuO spontaneously decomposes above this temperature. The negative slope of the CO line is noteworthy. For the formation of CO from carbon and oxygen, Equation [22.21], ΔS_f is positive. For a reduction of the type given in Equation [22.25],

$$MO(s) + C(s) \longrightarrow M(l) + CO(g) \qquad [22.25]$$

ΔG is negative, and the reaction is therefore spontaneous in all cases where the CO line is below the stability line for the metal oxide.

SAMPLE EXERCISE 22.3

Using Figure 22.18, estimate ΔG at 1500°C for the reaction

$$FeO(s) + C(s) \longrightarrow Fe(s) + CO(g)$$

Solution: For this reaction we have

$$\Delta G = \Delta G_f(CO) - \Delta G_f(FeO)$$

From Figure 22.18, we obtain

$$\Delta G_f(CO) = \frac{-550}{2} \text{ kJ/mole}$$

and

$$\Delta G_f(FeO) = \frac{-300}{2} \text{ kJ/mole}$$

Therefore

$$\Delta G = -275 \text{ kJ/mole} + 150 \text{ kJ/mole}$$
$$= -125 \text{ kJ/mole}$$

(-120 kJ/mole, to two significant figures). Notice that at 1500°C the stability line for CO in the figure is below the stability line for FeO. Correspondingly, ΔG is negative, and the reaction therefore is spontaneous.

Coke is a reasonably inexpensive reducing agent. There are occasions, however, when it is not satisfactory. It cannot be used when the reduction is not spontaneous (as for Al_2O_3) or when carbon impurities in the metal are undesirable. Under these conditions, electrolytic reduction or chemical reduction using hydrogen or active metals such as sodium, magnesium, or zinc as reducing agents is employed. These reducing agents are more expensive than carbon. Equations [22.26] through [22.28] provide examples of chemical reduction using reducing agents other than carbon.

$$WO_3(s) + 3H_2(g) \longrightarrow W(s) + 3H_2O(g) \qquad [22.26]$$

$$2Ag(CN)_2^-(aq) + Zn(s) \longrightarrow 2Ag(s) + Zn(CN)_4^{2-}(aq) \qquad [22.27]$$

$$UF_4(g) + 2Mg(l) \longrightarrow U(s) + 2MgF_2(l) \qquad [22.28]$$

REFINING

Once the metal has been obtained from its ore, it may still be necessary to further purify it. Blister copper, obtained by smelting of copper ores, is about 99 percent pure. It contains small amounts of arsenic, antimony, silver, gold, and other metals. The electrical conductivity of copper is greatly decreased by certain impurities; an arsenic content of only 0.03 percent lowers the conductivity by 14 percent. Consequently, copper used for electrical applications must be very pure. It is refined electrolytically by making the blister copper the anode in an electrolytic cell, as discussed in Section 19.3.

Similarly, pig iron contains numerous impurities that cause it to be

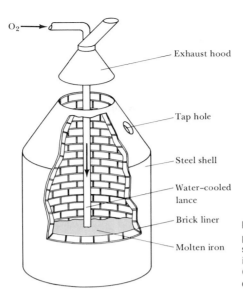

FIGURE 22.19 Diagram of an oxygen-process furnace used to convert pig iron into steel. The furnace is generally mounted so that it can be tilted to pour steel out of the tap hole. On the average such furnaces are 8–16 ft in diameter and have a capacity of 30–300 tons.

Labels for Figure 22.19:
- O₂
- Exhaust hood
- Tap hole
- Steel shell
- Water-cooled lance
- Brick liner
- Molten iron

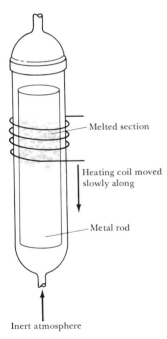

FIGURE 22.20 A zone-refining apparatus.

Labels for Figure 22.20:
- Melted section
- Heating coil moved slowly along
- Metal rod
- Inert atmosphere

brittle and of low tensile strength; it therefore has few uses. Most pig iron is therefore converted to steel; over half is converted by means of an oxygen furnace. In this process, limestone is added to the molten iron to form a slag with the phosphorus and silicon; a high pressure stream of oxygen is blown through the molten iron to burn off sulfur and excess carbon as shown in Figure 22.19. After this, small amounts of carbon and other materials may be added, depending on the type of steel desired. Small amounts of carbon increase the hardness and strength of the steel. So-called **mild steels** contain less than 0.2 percent carbon; they are malleable and ductile and are used to make cables, nails, and chains. **Medium steels** contain 0.2 to 0.6 percent carbon; they are tougher than the mild steels and are used to make girders and rails. **High-carbon steel,** used in cutlery, tools, and springs, contains 0.6 to 1.5 percent carbon. Other elements may be added to form **alloy steels.** One of the most common of these is stainless steel, which contains 0.4 percent carbon, 18 percent chromium, and 1 percent nickel.

Other purification or refining procedures include distillation, which can be used with mercury, and zone refining, which is used to purify silicon and germanium for use in semiconductors. In the zone-refining process a heating coil is passed slowly along a metal rod as shown in Figure 22.20; a narrow band of the element is thereby melted. As the molten area is swept slowly along the length of the rod the impurities concentrate in the molten region, following it to the end of the rod. The end in which the impurities are collected is cut off and the purified top portion retained.

**22.7
the physical
properties
of metals**

The physical properties of metals depend on the structure of the metal lattice, the presence of irregularities or lattice defects, and the strength of the bonds between metal atoms. Figure 22.21 shows the melting points of the metals through the transition series. Notice that the melting point

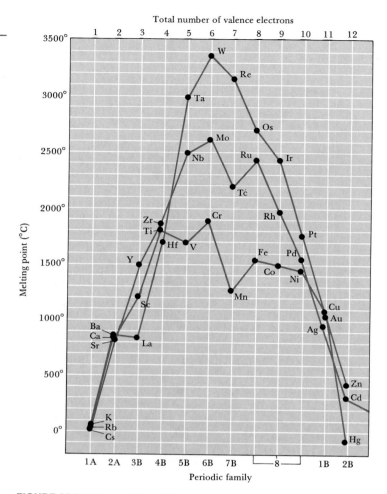

FIGURE 22.21 The melting points of metals as a function of their location in the periodic table and their number of valence electrons.

maximum in each period corresponds to outer electron configurations of $(n-1)d^3(n)s^2$ or $(n-1)d^4(n)s^2$, which are five or six electrons beyond the noble gas arrangements. Curves for the boiling points, heats of fusion, heats of vaporization, hardnesses, and densities of metals have very similar appearances. We can qualitatively understand the shape of Figure 22.21 if we assume that only the outer s and d electrons are involved in metallic bonding. Recall that large numbers of atoms interact within metallic lattices, thereby forming energy bands consisting of a series of closely spaced "molecular" orbitals (Section 9.6).

If the energy band arising from interaction of s orbitals on each atom overlaps with the band arising from the d orbitals, the situation shown in Figure 22.22 results. The overlapping bands have room for 12 electrons per metal atom. Of these, 6 will be bonding electrons, whereas the other 6 will be antibonding. On this basis the metal with 6 electrons should have the strongest bonds and consequently the highest melting point. Of course, this analysis is an oversimplification; other factors such as atomic radius, nuclear charge, and crystal form are involved.

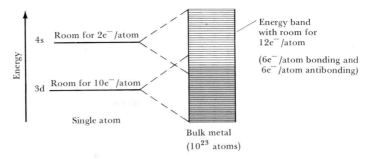

FIGURE 22.22 A band-structure model for the metals shown in Figure 22.21. This model helps explain the general trend in melting points of the metals as described in the text.

DISLOCATIONS AND GRAIN BOUNDARIES

Crystalline materials contain a variety of types of irregularities or defects that can affect their physical and chemical properties. The fewer the number of imperfections in the crystal, the greater its strength. For example, iron of ordinary purity has a tensile strength of 4000 lb/in². However, ultrapure single crystals of iron, which are largely free from impurities and lattice imperfections, have been found to have tensile strengths as high as 1,900,000 lb/in.².

One type of crystal defect that is commonly encountered is known as an **edge dislocation.** This type of defect is shown in Figure 22.23. It consists of an extra plane of particles inserted part way into the lattice.

Extra plane

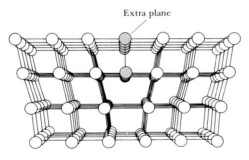

FIGURE 22.23 An edge dislocation, with the extra plane shown in color.

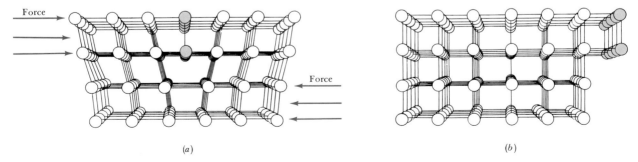

(a)

(b)

FIGURE 22.24 The presence of an edge dislocation facilitates movement of particles in the lattice. (a) Applied force causes the movement of the extra plane. (b) Additional force eventually brings the extra plane to the surface.

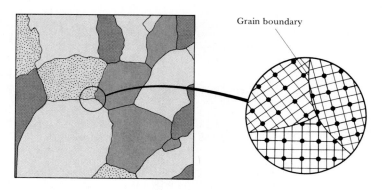

FIGURE 22.25 Pattern of typical grain structure of a metal as it might be seen under a microscope and a representation of the atomic-level arrangement of particles at the intersection of crystallites.

The presence of an edge dislocation greatly facilitates deformation of the solid as shown in Figure 22.24. The presence of these dislocations in metals therefore decreases their hardness and increases their malleability and ductility.

Real crystals generally consist of a series of crystallites or grains that are interlocked as shown in Figure 22.25. Metals, in particular, tend to be polycrystalline, consisting of many crystallites or grains. The grain structure greatly influences the properties of metals. For example, grain boundaries block movement of atoms over each other and thereby lower ductility. The finer the grain size (the larger the number of grain boundaries), the stronger the material. Fine grain size can be obtained in a number of ways, including rapid freezing of the liquid. Cold-working of a metal (as is carried out by a blacksmith hammering a piece of metal) causes stresses that remove dislocations and flatten grains; this procedure increases the hardness of the metal and renders it less ductile and malleable. Heating a cold-worked metal causes migration of atoms,

TABLE 22.7

some common alloys

PRIMARY ELEMENT	NAME OF ALLOY	COMPOSITION BY WEIGHT	PROPERTIES	USES
Bismuth	Wood's metal	50% Bi, 25% Pb, 2.5% Sn, 12.5% Cd	Low melting point (70°C)	Fuse plugs, automatic sprinklers
Copper	Yellow brass	67% Cu, 33% Zn	Ductile, takes polish	Hardware items
Iron	Stainless steel	80.6% Fe, 0.4% C, 18% Cr, 1% Ni	Resists corrosion	Tableware
Lead	Plumber's solder	67% Pb, 33% Sn	Low melting point (275°C)	Soldering joints
Nickel	Monel	69% Ni, 33% Cu, 7% Fe	Resists corrosion; bright surface	Kitchen fixtures
Silver	Sterling silver	92.5% Ag, 7.5% Cu	Bright surface	Tableware
Silver	Dental amalgam	70% Ag, 25% Pb, 3% Cu, 2% Hg	Easily worked	Dentistry
Tin	Pewter	85% Sn, 6.8% Cu 6% Bi, 1.7% Sb		Utensils

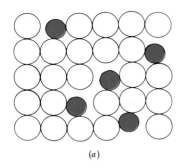

(a)

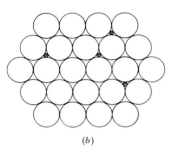

(b)

FIGURE 22.26 (a) Substitutional and (b) interstitial alloys. The open circles are the host metal, the colored circles the other component of the alloy.

which causes grains to grow in size; this procedure, known as **annealing,** decreases the hardness and increases the ductility and malleability of the metal.

ALLOYS

An **alloy** is a material that contains more than one element and has the characteristic properties of metals. The alloying of metals is of great importance, because it is one of the primary ways of modifying the properties of the pure metallic elements. For example, pure gold is too soft to be used in jewelry, whereas alloys of gold and copper are quite hard. Pure gold is 24 karat; the common alloy used in jewelry is 14 karat, meaning that it is 58 percent gold ($14/24 \times 100$). A gold alloy of this composition has suitable hardness to be used in jewelry. The alloy can be either yellow or white depending on the elements added. For the most part, pure metals are little used, and metals are consumed mainly in the form of alloys. Some further examples of alloys are given in Table 22.7.

Alloys can be classified as solution alloys, heterogeneous mixtures, and intermetallic compounds. The **solution alloys** are homogeneous mixtures with the components dispersed randomly and uniformly. Atoms of the solute can take positions normally occupied by the solvent, thereby forming a **substitutional alloy,** or they can occupy interstitial positions, thereby forming an **interstitial alloy.** These types are shown in Figure 22.26. The solution alloy can have a different crystal lattice than either of its components, and the lattice may change with a change in composition.

In **heterogeneous alloys,** the components are not dispersed uniformly. For example, in one form of steel known as pearlite, there is a heterogeneous mixture of iron and Fe_3C, known as cementite. We noted earlier that mixtures do not generally have characteristic melting points. However, in certain cases sharp and characteristic melting behavior can result. For example, pure lead melts at 237°C. Addition of tin lowers the melting point of the lead as shown in Figure 22.27. Pure tin has a melting

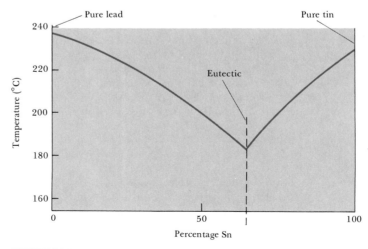

FIGURE 22.27 The melting point of lead shown as a function of added tin. The minimum melting point is at the composition 36% lead, 64% tin.

681

point of 232°C, which is lowered by the addition of lead. As Figure 22.27 shows, there is a lowest melting mixture, which consists of 64 percent tin and melts at 181°C. Such mixtures that melt at characteristic temperatures are known as **eutectic mixtures.**

Intermetallic alloys are homogeneous and have definite properties and composition. For example, copper and aluminum form a compound, $CuAl_2$, known as duralumin. Intermetallic compounds are rarely used as such, but are often found distributed through alloys just as cementite is found distributed through some steels.

FOR REVIEW

summary

In this chapter, we have examined the structure and composition of our planet. Earth can be divided into four layers, the **crust, mantle, outer core,** and **inner core.** We have noted that we have access to only a small portion of the planet, largely its surface. The solid portion to which we have access is referred to as the **lithosphere.** The solid compounds that occur naturally in the lithosphere are called **minerals. Rocks** are solid mixtures of minerals.

We can divide minerals into three general types: **native elements, silicate compounds,** and **nonsilicate compounds.** Silicate minerals are the most abundant. These minerals are based on SiO_4 tetrahedra, which, through sharing oxygen atoms, are able to link together to form chains, sheets, and three-dimensional arrays. We examined how the macroscopic properties, such as cleavage, reflect the molecular-level arrangements of several silicate minerals. In many minerals, Si^{4+} ions are replaced by Al^{3+} ions, thus forming **aluminosilicates** such as the **feldspar** minerals. Silicates are important components of glass and cement, both of which were briefly discussed. However, the silicates are not presently economically feasible sources of most metals.

Nonsilicate minerals, on the other hand, are especially important as sources of metals. The only nonsilicate mineral considered in any depth was calcium carbonate, which is the most abundant nonsilicate mineral in the lithosphere. Both $CaCO_3$ and CaO, obtained by heating $CaCO_3$, are useful in many processes discussed in this chapter.

Many compounds are conveniently viewed as consisting of a close-packed array of anions with cations located in octahedral or tetrahedral holes. The ratio of the cation radius to the anion radius is the most important factor in determining which type of hole a particular cation will occupy. Ions of similar size and the same charge can often substitute for one another in minerals.

Economically exploitable mixtures of minerals are known as **ores.** The undesirable portion of the ore is called the **gangue.** The process of extracting metals from their ores and modifying their properties is known as **metallurgy.** Three important types of processes involved in metallurgy are the **concentration** of an ore, its **reduction,** and the **refining** of the metal. Methods of concentration include **flotation, leaching,** and **roasting.** Most reductions involve carbon (coke), H_2, active metals, or electrolytic procedures. Methods of refining include electrolytic and chemical procedures, zone refining, and distillation. Much of our focus as we discussed reduction was on iron, which is obtained in a **smelting operation.** The physical properties of a metal, such as its hardness, are determined by several factors: its crystal structure, the number of electrons used in bonding, the size of the atom, the presence of defects such as **edge dislocations,** its polycrystalline character (**grain structure**), and the presence of other atoms. One of the primary ways of modifying the properties of a metal is to add other elements, thereby forming an **alloy**.

learning goals

Having read and studied this chapter, you should be able to:

1. Describe the structure of the earth.
2. List the five most abundant elements in the earth's crust.
3. Describe the structures possible for silicates and their empirical formulas (for example, silicate tetrahedra can combine through bridging oxygens to form a single-strand silicate chain whose empirical formula is SiO_3^{2-}).
4. Correlate the physical properties of certain silicate minerals, such as asbestos, with their structures.
5. Explain the changes in composition and properties that accompany substitution of Al^{3+} for Si^{4+} in a silicate.
6. Describe the role of clay and pH in soil fertility.
7. Describe the manufacture of glass and cement.
8. Describe the behavior of $CaCO_3$ in acid solutions and at high temperatures.
9. Describe the octahedral and tetrahedral holes in a close-packed array of ions and use the radius-ratio rules to predict which type of hole a particular ion will occupy.
10. Give at least two examples each of concentration, reduction, and refining processes.
11. Describe the production of steel starting with Fe_2O_3 ore.
12. Describe how the physical properties of metals depend on the presence of defects and on the strength of bonding.

key terms

Among the more important terms and expressions used for the first time in this chapter are the following:

An **alloy** (Section 22.7) is a material that contains more than one element and has the characteristic properties of metals, such as luster and electrical conductivity.

An **edge dislocation** (Section 22.7) is a type of crystal defect that occurs when an extra layer of particles is inserted part way into the crystal lattice.

Minerals (Section 22.2) are naturally occurring forms of solid elements or compounds.

An **octahedral hole** (Section 22.5) is an opening in a close-packed array of atoms or ions created by six particles arranged in an octahedral fashion.

An **ore** (Section 22.6) is an economically feasible source of a metal.

A **rock** (Section 22.2) is an aggregate of minerals of different kinds.

A **tetrahedral hole** (Section 22.5) is an opening in a close-packed array of atoms or ions created by four particles arranged in a tetrahedral fashion.

EXERCISES

silicate and nonsilicate minerals

22.1   According to the data in Table 22.2, more carbon is consumed annually than any other element. Describe two ways this element is used.

22.2   Why are so many different types of silicate minerals found in nature?

22.3   Explain why the mica minerals can be split

into thin sheets, whereas the asbestos minerals cleave in strands.

22.4 Explain why quartz and the feldspar minerals do not cleave in strands or sheets.

22.5 Explain, on the basis of molecular structure, why the melting points of CO_2 and SiO_2 are so different.

22.6 Explain the differences between CO_3^{2-} and SiO_3^{2-}.

22.7 When a person exhales through lime water (an aqueous solution of $Ca(OH)_2$), the solution becomes milky. Explain.

22.8 Write equations for the reactions that occur in each of the following cases: (a) CO_2 is bubbled through a solution of $Ba(OH)_2$; (b) $Ca(OH)_2$ is produced, starting with limestone; (c) solid $SrCO_3$ is heated.

[**22.9**] Crysotile, $Mg_3Si_2O_5(OH)_4$, is an asbestos mineral. From its formula, predict whether it consists of double strands or rolled sheets of linked silicate tetrahedra.

22.10 How is the common mica mineral $KMg_3(AlSi_3O_{10})(OH)_2$ structurally related to the mica mineral discussed in the text? How do these minerals differ from talc?

22.11 The mineral olivine is often represented by the formula $(Mg,Fe)_2SiO_4$. (a) What is meant by the (Mg,Fe) notation? (b) Predict whether the silicate portion of this mineral consists of isolated orthosilicate ions, silicate chains, or silicate sheets.

22.12 What type of silicate structure is consistent with each of the following empirical formulas: (a) SiO_4^{4-}; (b) SiO_3^{2-}; (c) $Si_2O_5^{2-}$; (d) $Si_4O_{11}^{6-}$?

22.13 What type of aluminosilicate structure is consistent with each of the following empirical formulas: (a) $AlSiO_5^{3-}$; (b) $AlSi_3O_{11}^{7-}$; (c) $AlSi_3O_8^{-}$?

22.14 What is the total charge on each of the following aluminosilicate ions: (a) $AlSiO_4$; (b) $Al_2Si_2O_{11}$?

22.15 Cyclic silicate anions, such as that shown in

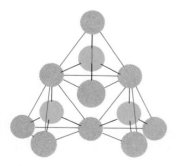

FIGURE 22.28

Figure 22.28, are found in some minerals. What is the empirical formula for the silicate ion shown?

22.16 The mineral orthoclase is a feldspar mineral formed by replacing a quarter of the Si^{4+} ions in SiO_2 with Al^{3+} and maintaining charge balance with K^+ ions. What is the simplest formula for this mineral?

22.17 If the radius of Na^+ is 0.98 Å, whereas that of Cl^- is 1.81 Å, what types of holes do the Na^+ ions occupy in a close-packed array of Cl^- ions?

22.18 If the radius of Zn^{2+} is 0.74 Å, whereas that of S^{2-} is 1.90 Å, what types of holes do the Zn^{2+} ions occupy in a close-packed array of sulfide ions?

22.19 The radius of Cs^+ is 1.67 Å, whereas that of Cl^- is 1.81 Å. What type of environment does the radius ratio of these ions suggest for Cs^+?

22.20 In the close-packed structures, there are two tetrahedral holes and one octahedral hole for each of the close-packed particles. How many of the available octahedral sites are occupied in Fe_2O_3? In NaCl?

22.21 Would you expect substitution of one of these ions for the other in minerals? (a) $Li^+(0.68$ Å$)$ for $Na^+(0.98$ Å$)$; (b) $Li^+(0.68$ Å$)$ for $Mg^{2+}(0.65$ Å$)$; (c) $Fe^{2+}(0.75$ Å$)$ for $Ni^{2+}(0.70$ Å$)$; (d) $Hf^{4+}(0.71$ Å$)$ for $Zr^{4+}(0.72$ Å$)$

metallurgy

22.22 Outline the processes and reactions that could be employed to obtain pure iron from an ore that contains FeS.

22.23 Write complete and balanced equations for each of the following reactions: (a) reduction of GeO_2 with hydrogen; (b) reduction of V_2O_5 with aluminum; (c) roasting of CuS.

22.24 Write complete and balanced equations for each of the following reactions: (a) reduction of $TiCl_4$ with magnesium; (b) reduction of Ta_2O_5 by sodium; (c) roasting of $PbCO_3$.

22.25 Rationalize the observation that the alkali metals have very low melting points and are very soft metals.

22.26 Rationalize the observation that the alkaline earth metals are harder and have higher melting points than the adjacent alkali metals.

22.27 List five metals found around your home. Are they alloys or pure metals? What properties fit them for their particular role?

22.28 How do lattice imperfections and impurities affect the strengths of solids?

22.29 How is the presence of defects in solids related to the concept of entropy?

22.30 What is the percentage of gold in 10 karat gold?

22.31 Distinguish between the following terms: (a) mantle and crust; (b) coke and coal; (c) ore and mineral; (d) smelting and roasting.

22.32 Distinguish between the following terms: (a) rock and mineral; (b) ore and gangue; (c) lime and limestone; (d) flotation and leaching; (e) feldspar and quartz.

22.33 Explain the following observations: (a) Although scandium is more abundant than silver in the earth's crust, silver is much more familiar and has been known for a longer time. (b) Carbonate rocks can be readily identified by dropping acid on them. (c) Barium is found in nature as a sulfate mineral.

22.34 Explain the following observations: (a) Limestone is more soluble in rain than in absolutely pure water. (b) Fe^{2+} exchanges readily for Mg^{2+} in many minerals. (c) Silver and gold are common components of copper ores. (d) Bauxite, $Al_2O_3 \cdot 2H_2O$, but not clay, is used as an ore for aluminum.

[22.35] What would happen in the electrolytic refining of copper if the voltage is too high? Too low?

22.36 Why is lime glass often called a supercooled liquid?

22.37 List three uses for limestone discussed in this chapter.

22.38 Using the average density and radius of earth, calculate its mass.

22.39 How many tons of O_2 would you need to burn off the carbon in a ton of pig iron that contains 4 percent carbon, leaving a steel containing 1 percent carbon. Assume that all of the carbon that is burned off forms CO_2.

22.40 Calculate the number of tons of coke that should be added to a blast furnace for each ton of Fe_2O_3, assuming that coke is 100 percent carbon.

22.41 The composition of lime glass can be approximately described by the chemical formula $Na_2O \cdot CaO \cdot 6SiO_2$. (a) How many kilograms of SiO_2 are required to make a kilogram of lime glass? (b) How many kilograms of $CaCO_3$ are required to make a kilogram of lime glass?

22.42 Calculate the amount of CaO obtained from a metric ton (1000 kg) of limestone rock if the rock is 90 percent $CaCO_3$.

22.43 Sodium carbonate is an important industrial chemical used, for example, in the manufacture of glass. About 6 million tons are used annually in this country. About half of this amount is obtained from natural deposits and half produced by the Solvay process. In the Solvay process the basic raw materials are NaCl, $CaCO_3$, and NH_3. How many tons of $CaCO_3$ are required to prepare 3 million tons of Na_2CO_3 assuming that the process allows 100 percent conversion?

chemistry of metals: coordination compounds

23

On several occasions we have discussed ions or molecules in which a metal ion is surrounded by a group of anions or molecules. An example is the $Ag(CN)_2^-$ ion discussed in connection with metallurgy in Section 22.6. Several other examples, including $Cu(NH_3)_4^{2+}$, $Cu(CN)_4^{2-}$, and $Ag(NH_3)_2^+$, were encountered in our discussions of equilibria in Section 17.5. Such ions are known as **complex ions,** or merely **complexes.** Compounds containing them are called **complex compounds,** or **coordination compounds.**

Coordination compounds play important roles in all areas of chemistry. They are utilized in medicine, agriculture, and metallurgy. They are used as catalysts and dyes and in chemical analysis and the electroplating of metals. Many complexes are essential for life; for example, hemoglobin, a complex of iron, is responsible for carrying oxygen in blood (Section 10.5). In fact, all metal ions in aqueous solution are complexes because they have water molecules bound to them. What then is the nature of these important compounds?

23.1
the structure of complexes

The molecules or ions that surround a metal ion in a complex are known as **ligands** (from the Latin word *ligare,* meaning "to bind"). Ligands are normally either anions or polar molecules. Furthermore, they normally have at least one unshared pair of valence electrons, as illustrated in the following examples:

$$:\overset{\cdot\cdot}{O}-H \qquad :\overset{\overset{\displaystyle H}{|}}{\underset{\displaystyle |}{N}}-H \qquad :\overset{\cdot\cdot}{\underset{\cdot\cdot}{Cl}}:^{-} \qquad :C\equiv N:^{-}$$

$$\underset{\displaystyle H}{}$$

In some instances, the bonding between a metal and its ligands can be understood to result from electrostatic attraction between positive ions and either negative ions or the negative end of a dipole. Correspondingly, the ability of metals to form complexes normally increases as the positive charge of the metal increases and as its size decreases. The alkali metals such as Na^+ and K^+ do not readily form complexes. On the other hand, the dipositive and tripositive transition-metal ions excel in forming complexes. In fact, transition-metal ions often form complexes more readily than their size and charge alone would suggest. For example, on the basis of size alone we might expect that Al^{3+} ($r = 0.45$ Å) would form complexes more readily than would the larger Cr^{3+} ($r = 0.62$ Å). However, Cr^{3+} forms much more stable complexes than does Al^{3+}. Thus the bonding in these complexes cannot be explained merely on the basis of electrostatic interaction between metal and ligand. It is therefore convenient to view the bonding from a covalent viewpoint. Because metal ions have empty valence orbitals, they can act as Lewis acids (electron-pair acceptors). Because ligands have unshared pairs of electrons, they are able to function as Lewis bases (electron-pair donors). The bond between metal and ligand can therefore be pictured as resulting from the sharing of a pair of electrons that was originally on the ligand:

$$Ag^+(aq) + 2 :\overset{\overset{\displaystyle H}{|}}{\underset{\displaystyle |}{N}}-H\ (aq) \longrightarrow H-\overset{\overset{\displaystyle H}{|}}{\underset{\displaystyle |}{N}}:Ag:\overset{\overset{\displaystyle H}{|}}{\underset{\displaystyle |}{N}}-H^+\ (aq) \qquad [23.1]$$

We shall examine the bonding in complexes more closely in Section 23.8.

In forming a complex, the ligands are said to coordinate to the metal or to be complexed by the metal. The central metal and the ligands bound to it constitute the **coordination sphere.** In writing the chemical formula for a coordination compound, we use square brackets to set off the groups within the coordination sphere from other parts of the compound. For example, the formula $[Cu(NH_3)_4]SO_4$ represents a coordination compound consisting of the $Cu(NH_3)_4^{2+}$ ion and the SO_4^{2-} ion. The four ammonia groups in the compound are bound directly to the copper(II) ion.

The charge of a complex is the sum of the charges on the central metal and on its surrounding ligands. In $[Cu(NH_3)_4]SO_4$ we can deduce the charge on the complex if we first recognize SO_4 as being the sulfate ion and therefore having a -2 charge. Because the compound is neutral, the complex ion must have a $+2$ charge, $Cu(NH_3)_4^{2+}$. The oxidation number of the copper must be $+2$ because the NH_3 groups are neutral:

$$+2 + 4(0) = +2$$
$$Cu(NH_3)_4^{2+}$$

SAMPLE EXERCISE 23.1

What is the oxidation number of the central metal in $[Co(NH_3)_5Cl](NO_3)_2$?

Solution: The NO_3 group is the nitrate anion and has a -1 charge, NO_3^-. The NH_3 ligands are neutral; the Cl is a coordinated chloride and therefore has a -1 charge. The sum of all the charges must be zero:

$$x + 5(0) + (-1) + 2(-1) = 0$$
$$[Co(NH_3)_5Cl](NO_3)_2$$

The charge on the cobalt, x, must therefore be $+3$.

SAMPLE EXERCISE 23.2

Given that a complex ion contains a chromium(III) bound to four water molecules and two chloride ions, write its formula.

Solution: The metal has a $+3$ oxidation number, water is neutral, and chloride has a -1 charge:

$$+3 + 4(0) + 2(-1) = +1$$
$$Cr(H_2O)_4Cl_2$$

Therefore the charge on the ion is $+1$, $Cr(H_2O)_4Cl_2^+$.

COORDINATION NUMBERS AND GEOMETRIES

The atom in a ligand that is bound directly to the metal is known as the **donor atom**. For example, nitrogen is the donor atom in the $Ag(NH_3)_2^+$ complex shown in Equation [23.1]. The number of donor atoms attached to a metal is known as its **coordination number**. In $Ag(NH_3)_2^+$, silver has a coordination number of two; in $Cr(H_2O)_4Cl_2^+$, chromium has a coordination number of six.

Some metal ions exhibit constant coordination numbers. For example, the coordination number of chromium(III) and cobalt(III) is invariably six, whereas that of platinum(II) is always four. However, the coordination numbers of most metal ions vary with the ligand. The most common coordination numbers are four and six. A very rough rule of thumb is that the common coordination number of a metal is twice its oxidation state.

The coordination number of a metal ion seems to be influenced by the relative sizes of the metal ion and the surrounding ligands. As the ligand gets larger, fewer can coordinate to the metal. This may explain why iron(II) is able to coordinate to six fluorides, FeF_6^{4-}, but to only four chlorides, $FeCl_4^-$. Ligands that place a high charge density on the metal also produce reduced coordination numbers. For example, six neutral ammonia molecules can coordinate to nickel(II) forming $Ni(NH_3)_6^{2+}$; however, only four negatively charged chlorides coordinate, $NiCl_4^{2-}$.

Four-coordinate complexes have two common geometries, tetrahedral and square planar, as shown in Figure 23.1. The tetrahedral geometry is the more common of the two and is especially common among the nontransition metals. The square-planar geometry is characteristic of transition-metal ions with eight d electrons in the valence shell, for example, platinum(II) and gold(III); it is also found in some copper(II) complexes.

Six-coordinate complexes have an octahedral geometry, as shown in

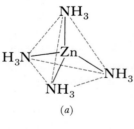

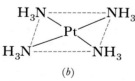

FIGURE 23.1 The structures of (a) $Zn(NH_3)_4^{2+}$ and (b) $Pt(NH_3)_4^{2+}$, illustrating the tetrahedral and square-planar geometries, respectively. These are the two common geometries for complexes in which the metal ion has a coordination number of four.

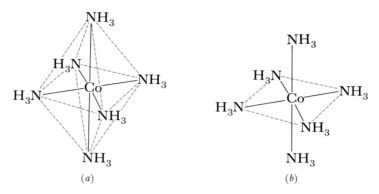

FIGURE 23.2 Two representations of an octahedral coordination sphere, the common geometric arrangement for complexes in which the metal ion has a coordination number of six.

Figure 23.2(a). Notice that the octahedron can be represented as a planar square with ligands above and below the plane, as in Figure 23.2(b).

23.2
chelates

The ligands that we have discussed so far, such as NH_3 and Cl^-, are known as **monodentate ligands** (from the Latin meaning "one-toothed"). These ligands possess a single donor atom. Some ligands have two or more donor atoms situated so that they can simultaneously coordinate to a metal ion. They are called **polydentate ligands** ("many-toothed"); because they appear to grasp the metal between two or more donor atoms, they are also known as **chelating agents** (from the Greek word *chele*, "claw"). One such ligand is ethylenediamine:

$$H_2\overset{..}{N}\diagup\overset{\displaystyle CH_2-CH_2}{}\diagdown\overset{..}{N}H_2$$

This ligand, which is abbreviated en, has two nitrogen atoms that have unshared pairs of electrons. These donor atoms are sufficiently far apart that the ligand can wrap around a metal ion with the two nitrogen atoms simultaneously complexing to the metal. The $Co(en)_3^{3+}$ ion, which contains three ethylenediamine ligands bound to the octahedral coordination sphere of cobalt(III), is shown in Figure 23.3. Notice that the ethylenediamine has been written in a shorthand notation as two nitrogen atoms connected by a line.

Ethylenediamine is an example of a bidentate ligand. Other common bidentate ligands include oxalate, $C_2O_4^{2-}$, and carbonate, CO_3^{2-}:

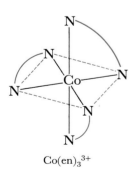

$Co(en)_3^{3+}$

FIGURE 23.3 The $Co(en)_3^{3+}$ ion showing how each bidentate ethylenediamine ligand is able to occupy two positions in the coordination-sphere.

Another common polydentate ligand is the ethylenediaminetetraacetate ion:

689

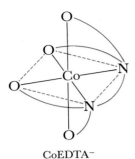

CoEDTA⁻

$$-:\overset{\displaystyle :O:}{\underset{}{\overset{\|}{O}CCH_2}}\diagdown_{NCH_2CH_2N}\diagup^{CH_2\overset{\displaystyle :O:}{\overset{\|}{C}O}:-}$$

EDTA⁴⁻

FIGURE 23.4 The CoEDTA⁻ ion showing how the ethylenediaminetetraacetate ion is able to wrap around a metal ion, occupying six positions in the coordination sphere.

This ion, abbreviated EDTA⁴⁻, has six donor atoms. It can wrap around a metal ion using all six of these donor atoms as shown in Figure 23.4.

In general, chelating agents form more stable complexes than related monodentate ligands. This is illustrated by the formation constants for $Ni(NH_3)_6^{2+}$ and $Ni(en)_3^{2+}$, shown in Equations [23.2] and [23.3]:

$$Ni(H_2O)_6^{2+}(aq) + 6NH_3(aq) \rightleftharpoons$$
$$Ni(NH_3)_6^{2+}(aq) + 6H_2O \qquad K_c = 4 \times 10^8 \qquad [23.2]$$

$$Ni(H_2O)_6^{2+}(aq) + 3en(aq) \rightleftharpoons$$
$$Ni(en)_3^{2+}(aq) + 6H_2O \qquad K_c = 2 \times 10^{18} \qquad [23.3]$$

Although the donor atom is nitrogen in both instances, $Ni(en)_3^{2+}$ has a stability constant nearly 10^{10} times larger than $Ni(NH_3)_6^{2+}$.

Complexing agents often can be added to a solution to prevent one or more of the customary reactions of a metal ion without actually removing it from the solution. For example, a metal ion that interferes with a chemical analysis can often be complexed and its interference thereby removed. In a sense the complexing agent hides the metal ion and is therefore referred to as a sequestering agent. Because chelates perform this role more effectively than do monodentate ligands, the term sequestering agent is usually reserved for chelates.

One of the most common applications of sequestering agents is in complexing cations in natural waters to keep them from interfering with the action of soap or detergent molecules. As noted in Section 18.6, Mg^{2+} and Ca^{2+} ions react with soap to form a precipitate commonly known as soap scum. Although these and other metal ions do not precipitate detergent molecules, they do complex with them, thereby interfering with their cleansing action.* Phosphates are effective and cheap sequestering agents for these ions. The most important phosphate used for this purpose is sodium tripolyphosphate:

$$Na_5 \left[O-\overset{\overset{\displaystyle O}{\|}}{\underset{\underset{\displaystyle O}{|}}{P}}-O-\overset{\overset{\displaystyle O}{\|}}{\underset{\underset{\displaystyle O}{|}}{P}}-O-\overset{\overset{\displaystyle O}{\|}}{\underset{\underset{\displaystyle O}{|}}{P}}-O \right]$$

Complexing agents can also stabilize a metal ion toward oxidation or reduction. For example, $Ni(H_2O)_6^{2+}(aq)$, which we have usually written as $Ni^{2+}(aq)$, is more easily reduced than is $Ni(NH_3)_6^{2+}(aq)$:

$$Ni(H_2O)_6^{2+} + 2e^- \longrightarrow Ni + 6H_2O$$
$$E° = -0.250 \text{ V}$$

$$Ni(NH_3)_6^{2+} + 2e^- \longrightarrow Ni + 6NH_3$$
$$E° = -0.476 \text{ V}$$

Complexing agents also enhance the solubility of metal salts. For example, AgBr, the photosensitive material in photographic film, is insoluble in water, but dissolves in the presence of thiosulfate ion, $S_2O_3^{2-}$:

$$AgBr(s) \rightleftharpoons Ag^+(aq) + Br^-(aq)$$
$$K_{sp} = 7.7 \times 10^{-13}$$

$$Ag^+(aq) + 2S_2O_3^{2-}(aq) \rightleftharpoons Ag(S_2O_3)_2^{3-}(aq)$$
$$K_c = 1.6 \times 10^{13}$$

*The action of soaps and detergents was discussed earlier, in Section 12.8; the formula of a typical detergent molecule is shown in Section 21.5.

FIGURE 23.5 — Periodic Table

1A	2A	3B	4B	5B	6B	7B	8B	8B	8B	1B	2B	3A	4A	5A	6A	7A	8A
3 Li 6.939	4 Be 9.0122											5 B 10.811	6 C 12.01115	7 N 14.0067	8 O 15.9994	1 H 1.00797	2 He 4.0026
11 Na 22.9898	12 Mg 24.312											13 Al 26.9815	14 Si 28.086	15 P 30.9738	16 S 32.064	9 F 18.9984	10 Ne 20.183
19 K 39.098	20 Ca 40.08	21 Sc 44.956	22 Ti 47.90	23 V 50.942	24 Cr 51.996	25 Mn 54.9380	26 Fe 55.847	27 Co 58.9332	28 Ni 58.71	29 Cu 63.54	30 Zn 65.37	31 Ga 69.72	32 Ge 72.59	33 As 74.9216	34 Se 78.96	17 Cl 35.453	18 Ar 39.948
37 Rb 85.47	38 Sr 87.62	39 Y 88.905	40 Zr 91.22	41 Nb 92.906	42 Mo 95.94	43 Tc (99)	44 Ru 101.07	45 Rh 102.905	46 Pd 106.4	47 Ag 107.870	48 Cd 112.41	49 In 114.82	50 Sn 118.69	51 Sb 121.75	52 Te 127.60	35 Br 79.909	36 Kr 83.80
55 Cs 132.905	56 Ba 137.33	57 *La 138.91	72 Hf 178.49	73 Ta 180.948	74 W 183.85	75 Re 186.2	76 Os 190.2	77 Ir 192.2	78 Pt 195.09	79 Au 196.967	80 Hg 200.59	81 Tl 204.37	82 Pb 207.19	83 Bi 208.980	84 Po (210)	53 I 126.9044	54 Xe 131.30
87 Fr (223)	88 Ra (226)	89 †Ac (227)	104 Rf (257)	105 Ha (260)												85 At (210)	86 Rn (222)
																71 Lu 174.97	103 Lw (257)

*Lanthanide series

58 Ce 140.12	59 Pr 140.907	60 Nd 144.24	61 Pm (147)	62 Sm 150.35	63 Eu 151.96	64 Gd 157.25	65 Tb 158.924	66 Dy 162.50	67 Ho 164.930	68 Er 167.26	69 Tm 168.934	70 Yb 173.04	71 Lu 174.97

†Actinide series

90 Th 232.038	91 Pa (231)	92 U 238.03	93 Np (237)	94 Pu (242)	95 Am (243)	96 Cm (247)	97 Bk (247)	98 Cf (249)	99 Es (254)	100 Fm (253)	101 Md (256)	102 No (253)	103 Lw (257)

Active metals — Transition metals — Nonmetals

FIGURE 23.5 The elements that are essential for life are indicated by the shaded areas. The dark color indicates the four most abundant elements in living systems (hydrogen, carbon, nitrogen and oxygen). The light color indicates the seven next most common elements. The grey shading indicates the elements needed in only trace amounts.

The thiosulfate can be visualized as shifting the first equilibrium to the right by complexing the Ag^+. Sodium thiosulfate decahydrate, $Na_2S_2O_3 \cdot 10H_2O$, known as hypo, is used in black-and-white photography to dissolve unexposed and undeveloped AgBr from the photographic film.

METALS AND CHELATES IN LIVING SYSTEMS

The elements presently known to be essential for life are shown in Figure 23.5. Over 99 percent of the atoms required by living cells are either hydrogen, oxygen, carbon, or nitrogen. Six of the essential elements are transition metals—iron, copper, zinc, manganese, cobalt, and molybdenum. These elements owe their roles in living systems mainly to their abilities to form complexes with a variety of chelating agents. For example, many enzymes, the body's catalysts, require metal ions in order to function. We shall take a close look at enzymes in Chapter 25.

Among the most important chelating agents in nature are those derived from the porphine molecule shown in Figure 23.6. This molecule can coordinate to a metal using the four nitrogen atoms as donors. Upon coordination to a metal, the two H^+ shown bonded to nitrogen are displaced. Complexes derived from porphine are called **porphyrins.** Different porphyrins contain different metals and have different substituent groups attached to the carbon atoms at the ligand's periphery. Two of the most important porphyrins are heme, which contains iron(II), and chlorophyll, which contains magnesium(II). We discussed heme earlier in Section 10.5. Hemoglobin contains four heme subunits as shown in Figure 10.10. The iron is coordinated to the four nitrogen atoms of the porphyrin and also to a nitrogen atom from the protein that composes the bulk of the hemoglobin molecule. The sixth position around the iron is occupied either by O_2 (in oxyhemoglobin, the red form) or by water (in deoxyhemoglobin, the blue form). Oxyhemoglobin is shown in Figure 23.7. As noted in Section 10.5, some groups such as CO act as poisons because of their ability to bind iron more strongly than O_2.

When we have an insufficient quantity of iron in our diet, we develop iron-deficiency anemia. The lack of iron leads to a reduction in the amount of hemoglobin; we develop what advertisements have referred to as "iron-poor blood." Without hemoglobin to transport oxygen, our body's cells are unable to produce energy. Therefore, the symptoms of anemia include weakness and drowsiness.

Plants also need iron. Iron is part of a plant enzyme that participates in the production of chlorophyll, the green pigment essential to photosynthesis. Plants suffering from a deficiency in iron develop a condition known as iron chlorosis. The effect is usually first noticed in young leaves, which have a yellow coloration. Often the plant develops chlorosis because the soil conditions interfere with the availability of the iron. For example, iron may be present in the soil, but only in insoluble forms unavailable to the plant.

Iron may occur in soil either as iron(II) or iron(III), depending on the ability of oxygen to penetrate the soil and oxidize the iron. Although both Fe^{2+} and Fe^{3+} can be used by plants, the Fe^{3+} is much less soluble than Fe^{2+} in normal soils, which usually have pH's in the vicinity of 7. For example, in typical soil, Fe^{3+} precipitates as $Fe(OH)_3$ when the pH exceeds 3, whereas Fe^{2+} precipitates as $Fe(OH)_2$ when the pH exceeds 6. Therefore plants growing in alkaline soils readily suffer from iron defi-

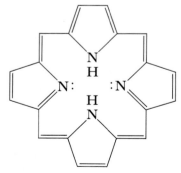

FIGURE 23.6 The structure of the porphine molecule. This molecule forms a tetradentate ligand with the loss of the two protons bound to nitrogen atoms. Porphine is the basic structure of porphyrins, compounds whose complexes play a variety of important roles in nature.

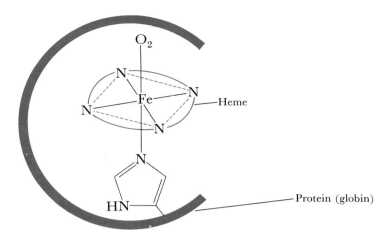

FIGURE 23.7 A schematic representation of oxyhemoglobin, showing one of the four heme units in the molecule. The iron is bound to four nitrogen atoms of the porphyrin, to a nitrogen from the surrounding protein, and to an O_2 molecule.

ciency. Simple addition of iron salts to the soil does not correct this condition, because the iron is merely precipitated. Current practice in agriculture is to add the iron in a complexed form, such as $FeEDTA^{2-}$. The chelated iron will not precipitate and is therefore available to the plant; once in the plant, the iron is removed as needed.

Some plants, such as certain strains of soybeans, secrete their own chelates to solubilize the needed iron. Similarly, mosses and lichens growing on rocks generate chelating agents to extract the metals that they need.

All living cells need iron. One mechanism that our body uses to fight invading bacteria is to withhold iron needed by the bacteria. The bacteria use powerful chelating agents to obtain their required iron. Our body, in turn, keeps its iron tightly complexed. Thus there is a confrontation between the chelating agents of our body and those of the microbial invaders. It has been found that the ability of bacteria to synthesize their chelates is suppressed at elevated temperatures. Consequently, fever is part of the body's attempt to overcome the invading bacteria.

Chelates and chelating agents have also been used as drugs. For example, they have been used to destroy bacteria by depriving them of essential metals. In this fashion the drug mimics the body's natural defenses described above. Chelating agents are also used to remove metals such as Hg^{2+}, Pb^{2+}, and Cd^{2+}, which are detrimental to health. For example, one method of treating lead-poisoning is to administer $Na_2[CaEDTA]$. The EDTA chelates the lead, allowing its removal from the body in urine.

23.3 nomenclature

Before we go too far in our discussions of complexes, it is useful to describe how these substances are named. When complexes were first discovered and few were known, they were named after the chemist who originally prepared them. A few of these names persist; for example, $NH_4[Cr(NH_3)_2(NCS)_4]$ is known as Reinecke's salt. As the number of known complexes grew, they began to be named by color. For example, $[Co(NH_3)_5Cl]Cl_2$, whose formula was then written $CoCl_3 \cdot 5NH_3$, was known as purpureocobaltic chloride after its purple color. Once the structures of complexes were more fully understood, it became possible to name them in a more systematic manner. Before we give the rules for naming complexes, let's consider two examples. We can thereby get a little better idea of how the nomenclature rules apply. The two complexes are:

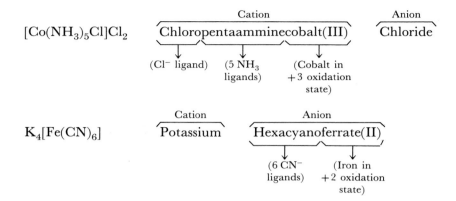

The rules of nomenclature are as follows:

1. *In naming salts, the name of the cation is given before the name of the anion.* Thus in $[Co(NH_3)_5Cl]Cl_2$ we name the $Co(NH_3)_5Cl^{2+}$ and then Cl^-.

2. *Within a complex ion, the ligands are named before the metal; anionic ligands are named before neutral ones.* Thus in the $Co(NH_3)_5Cl^{2+}$ ion, we name the chloride ligand, then the ammonia ligands, and finally the metal. The opposite order is used in writing formulas.

3. *Anionic ligands end in the letter o, whereas neutral ones ordinarily bear the name of the molecule.* Some common ligands and their names are listed in Table 23.1. Special names are given to H_2O (aquo) and NH_3 (ammine). Thus the terms chloro and ammine occur in the name for $[Co(NH_3)_5Cl]Cl_2$.

4. *A Greek prefix (for example, di-, tri-, tetra-, penta-, and hexa-) is used to indicate the number of each kind of ligand when more than one is present.* Thus in the name for $Co(NH_3)_5Cl^{2+}$ we have pentaammine, indicating five NH_3 ligands. *If the name of the ligand itself contains a Greek prefix such as* mono *or* di, *then the ligand is enclosed in parentheses and*

TABLE 23.1

some common ligands

LIGAND	LIGAND NAME
Bromide, Br^-	Bromo
Chloride, Cl^-	Chloro
Cyanide, CN^-	Cyano
Hydroxide, OH^-	Hydroxo
Carbonate, CO_3^{2-}	Carbonato
Oxalate, $C_2O_4^{2-}$	Oxalato
Ammonia, NH_3	Ammine
Ethylenediamine, en	Ethylenediamine
Water, H_2O	Aquo

alternate prefixes (bis-, tris-, tetrakis-, pentakis-, *and* hexakis-) *used.* For example, the name for $[Co(en)_3]Cl_3$ is tris(ethylenediamine)-cobalt(III) chloride.

5. *If the complex is an anion, its name ends in* -ate. For example, in $K_4[Fe(CN)_6]$ the anion is called the hexacyanoferrate(III) ion. The suffix -ate is often added to the Latin stem as in this example.

6. *The oxidation number of the metal is given in parentheses in Roman numerals following the name of the metal.* For example, the Roman numeral III is used to indicate the $+3$ oxidation state of cobalt in $Co(NH_3)_5Cl^{2+}$.

SAMPLE EXERCISE 23.3

Give the name of the following compounds: (a) $[Cr(H_2O)_4Cl_2]Cl$; (b) $Na_4[Ni(CN)_4]$.

Solution: (a) Beginning with the negative ligands, we have dichloro to indicate two Cl^-. Then we have four waters, which we indicate as tetraaquo. The oxidation state of Cr is $+3$:

$$+3 + 4(0) + 2(-1) + (-1) = 0$$
$$[Cr(H_2O)_4Cl_2]Cl$$

Thus we have chromium(III). Finally, the anion is chloride. Putting these parts together we have the compound's name: dichlorotetraaquochromium(III) chloride.

(b) The complex has four CN^-, which we indicate as tetracyano. The oxidation state of the nickel is zero:

$$4(+1) + 0 + 4(-1) = 0$$
$$K_4[Ni(CN)_4]$$

Because the complex is an anion, the metal is indicated as nickelate(0). Putting these parts together and naming the cation first we have: potassium tetracyanonickelate(0).

SAMPLE EXERCISE 23.4

Write the formula for difluorobis(ethylenediamine)-cobalt(III) perchlorate.

Solution: The complex cation contains two fluorides, two ethylenediamines, and a cobalt with a $+3$ oxidation number. Knowing this we can determine the charge on the complex:

$$+3 + 2(0) + 2(-1) = +1$$
$$Co(en)_2F_2$$

The perchlorate anion has a single negative charge, ClO_4^-. Therefore, only one is needed to balance the charge on the complex cation. The formula is thus $[Co(en)_2F_2]ClO_4$.

23.4
Isomerism

Two or more compounds that have the same formula but a different structure (that is, the same collection of atoms but arranged in different ways) are called **isomers.** Several types of isomerism are observed in coordination chemistry. A relatively rare but interesting type is known as **linkage isomerism;** this isomerism arises when ligands are capable of coordinating to a metal in two ways. For example, the nitrite ion, NO_2^-, can coordinate through either a nitrogen or an oxygen atom, as shown in Figure 23.8. Isomers differ in their physical and chemical properties. For example, the N-bound isomer shown in Figure 23.8 is yellow while the O-bound isomer is red. Other ligands that are capable of coordinating

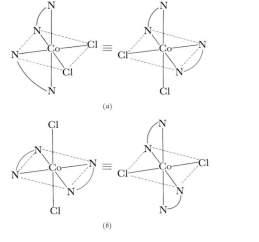

FIGURE 23.8 (a) The yellow N-bound and (b) the red O-bound isomers of $Co(NH_3)_5NO_2^{2+}$.

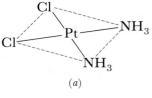

(a)

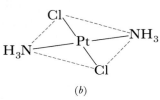

(b)

FIGURE 23.9 (a) The *cis* and (b) the *trans* geometric isomers of the square-planar $Pt(NH_3)_2Cl_2$.

through either of two donor atoms include thiocyanate, SCN^-, whose potential donor atoms are N and S.

GEOMETRIC ISOMERISM

The most important form of isomerism is known as **stereoisomerism.** Stereoisomers are isomers that have the same chemical bonds but different special arrangements. For example, in $Pt(NH_3)_2Cl_2$, the chloro ligands can be arranged either adjacent or opposite to each other, as shown in Figure 23.9. This particular form of stereoisomerism, in which the arrangements of donor atoms around the coordination sphere are different, is known as **geometric** or *cis-trans* **isomerism.** The isomer with like groups close together is called the *cis* isomer, whereas the one with like groups far apart is called the *trans* isomer.

Because all of the corners of a tetrahedron are adjacent to one another, *cis-trans* isomerism is not observed in the case of tetrahedral complexes. However, it is found in octahedral ones.

SAMPLE EXERCISE 23.5

How many geometric isomers are there for $Co(en)_2Cl_2^+$?

Solution: To answer this question, draw several octahedrons with ligands in different locations. A systematic approach would be to keep one chloride in the same location each time. Examination of these structures indicates that there are two unique geometric isomers: one with adjacent Cl^- ligands (the *cis* isomer) and one with Cl^- ligands across the metal from each other (the *trans* isomer). Two different orientations of both of these isomers are shown in Figure 23.10. It is easy to overestimate the number of geometric isomers; sometimes different orientations of a single isomer are incorrectly thought to be different isomers. Therefore, you should keep in mind that if two structures can be rotated so that they are equivalent, they are not isomers of one another. The problem of identifying isomers is com-

FIGURE 23.10 (a) The *cis* and (b) the *trans* geometric isomers of the octahedral $Co(en)_2Cl_2^+$ ion.

pounded by the difficulty we often have in visualizing three-dimensional molecules from their two-dimensional representations. It is easier to determine the number of isomers if we are working with three-dimensional models.

OPTICAL ISOMERISM

A second type of stereoisomerism is known as **optical isomerism.** Optical isomers are nonsuperimposable mirror images of one another. They bear the same resemblance to one another that our left hand bears to our right hand. If you look at your left hand in a mirror, as shown in Figure 23.11, the image is identical to your right hand. Furthermore, your two hands are not superimposable on one another. A good example of a complex that exhibits this type of isomerism is the $Co(en)_3^{3+}$ ion. The two isomers of $Co(en)_3^{3+}$ and their mirror-image relationship to one another are shown in Figure 23.12. Just as there is no way that we can twist or turn our right hand to make it look identical to our left, so also there is no way to rotate one of these isomers to make it identical to the other. If we had models of each that we could handle, perhaps we could more easily satisfy ourselves of this fact. Molecules or ions that have nonsuperimposable mirror images are said to be **chiral** (pronounced KY-rul). Enzymes, the body's catalysts, are among the most highly chiral molecules known. As noted in Section 23.2, many enzymes contain complexed metal ions.

SAMPLE EXERCISE 23.6

Does either *cis*- or *trans*-$Co(en)_2Cl_2^+$ have optical isomers?

Solution: To answer this question, we need to draw both the *cis* and *trans* isomers of $Co(en)_2Cl_2^+$ (see Figure 23.10) and then their mirror images. The mirror image of the *trans* isomer is identical to the original. Consequently *trans*-$Co(en)_2Cl_2^+$ has no optical isomer. However, the mirror image of *cis*-$Co(en)_2Cl_2^+$ is not identical to the original. Consequently, there are optical isomers for this complex.

Most of the physical and chemical properties of optical isomers are identical. The properties of two optical isomers differ only if they are in

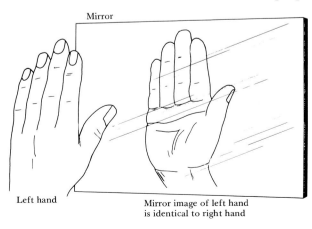

Mirror

Left hand

Mirror image of left hand is identical to right hand

FIGURE 23.11 Our hands are nonsuperimposable mirror images of each other.

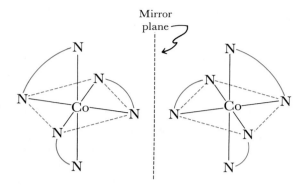

FIGURE 23.12 The two optical isomers of Co(en)$_3^{3+}$; notice that the ions are nonsuperimposable mirror images of each other.

a chiral environment—that is, one in which there is a sense of right- and left-handedness. For example, in the presence of a chiral enzyme, the reaction of one optical isomer might be catalyzed, whereas the other isomer remains totally unreacted. Consequently, one optical isomer may produce a specific physiological effect within our body, whereas its mirror image produces a different effect or none at all.

Optical isomers are usually distinguished from each other by their interaction with plane-polarized light. If light is polarized, for example by passage through a sheet of Polaroid film, the light waves are vibrating in a single plane, as shown in Figure 23.13. If the polarized light is passed through a solution containing one optical isomer, the plane of polarization is rotated either to the right (clockwise) or to the left (counterclockwise). The isomer that rotates the plane of polarization to the right is said to be **dextrorotatory**; it is labeled the dextro or *d* isomer (Latin *dexter*, "right"). Its mirror image will rotate the plane of polarization to the left; it is said to be **levorotatory** and is labeled the levo or *l* isomer (Latin *laevus*, "left"). Because of their effect on plane-polarized light, chiral molecules are said to be **optically active.**

When a substance with optical isomers is prepared in the laboratory, the chemical environment during the synthesis is not usually chiral. Consequently, equal amounts of the two isomers are obtained; the mixture is said to be **racemic.** A racemic mixture will not rotate polarized light because the rotatory effects of the two isomers cancel each other. In order to separate the isomers from the racemic mixture, they must be

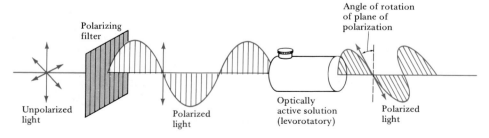

FIGURE 23.13 The effect of an optically active solution on the place of polarization of plane-polarized light. The unpolarized light is passed through a polarizing filter. The resultant polarized light thereafter passes through a solution containing a levorotatory optical isomer. As a result, the plane of polarization of the light is rotated to the left.

placed in a chiral environment. For example, one optical isomer of the chiral tartrate anion,* $C_4H_4O_6^{2-}$, can be used to separate a racemic mixture of $[Co(en)_3]Cl_3$. If d-tartrate is added to an aqueous solution of the $[Co(en)_3]Cl_3$, d-$[Co(en)_3](d$-$C_4H_4O_6)Cl$ will precipitate leaving l-$[Co(en)_3]^{3+}$ in solution.

23.5
ligand
exchange rates

If we were to examine a number of different complexes in solution, we would observe that some exchange ligands at an extremely rapid rate, whereas others do so quite slowly. For example, addition of ammonia to an aqueous solution of $CuSO_4$ produces an essentially instantaneous color change as the pale blue $Cu(H_2O)_4^{2+}$ ion is converted to the deep blue $Cu(NH_3)_4^{2+}$ ion. When this solution is acidified, the pale blue color is regenerated again at a rapid rate:

$$Cu(NH_3)_4^{2+} + 4H_3O^+ \longrightarrow Cu(H_2O)_4^{2+} + 4NH_4^+ \qquad [23.4]$$

In contrast, $Co(NH_3)_6^{3+}$ is more difficult to prepare than is $Cu(NH_3)_4^{2+}$. However, once it has been formed and is then placed in an acidic solution, reaction to form NH_4^+ takes several days. This tells us that the coordinated NH_3 groups are not readily removed from the metal. Complexes like $Cu(NH_3)_4^{2+}$ that undergo rapid ligand exchange are said to be **labile**; those like $Co(NH_3)_6^{3+}$ that undergo slow ligand exchange are said to be **inert**. The distinction between labile and inert complexes applies to how rapidly equilibrium is attained and not to the position of the equilibrium. For example, although $Co(NH_3)_6^{3+}$ is inert in acidified aqueous solutions, the equilibrium constant indicates that the complex is not thermodynamically stable under these conditions:

$$\begin{aligned} Co(NH_3)_6^{3+} + 6H_3O^+ &\rightleftharpoons \\ Co(H_2O)_6^{3+} + 6NH_4^+ \qquad K_c &\simeq 10^{20} \end{aligned} \qquad [23.5]$$

The kinetic inertness of $Co(NH_3)_6^{3+}$ can be attributed to a high activation energy for the reaction.

Cobalt(III) is among the few metal ions that consistently form inert complexes; others include chromium(III), platinum(IV), and platinum(II). Complexes of these ions maintain their identity in solution long enough to permit study of their structures and properties. They were therefore among the first complexes studied. Much of our understanding of structure and isomerism comes from studies of these complexes.

23.6
structure and
isomerism:
a historical
perspective

Many early systematic studies of coordination compounds involved complexes of cobalt(III), chromium(III), platinum(II), and platinum(IV) with ammonia as a ligand. One of the earliest reports of the preparation of an ammine complex dates from 1798 when a chemist by the name of Tassaert accidentally prepared an ammonia complex of cobalt. He had been trying to develop new methods of analyzing certain ores for cobalt

*When sodium ammonium tartrate, $NaNH_4C_4H_4O_6$, is crystallized from solution, the two isomers form separate crystals whose shapes are mirror images of each other. In 1848, Louis Pasteur achieved the first separation of a racemic mixture into optical isomers; using a microscope he picked the "right-handed" crystals of this compound from the "left-handed" ones.

and had treated a solution containing Co^{2+} with aqueous ammonia. After this solution had been allowed to stand overnight, orange-colored crystals separated from the mixture. The compound was found to have the empirical formula $CoN_6H_{18}Cl_3$. Tassaert wrote the formula as $CoCl_3 \cdot 6NH_3$, suggesting that the compound was analogous to hydrated salts like $CoCl_2 \cdot 6H_2O$.

By 1890, many ammine complexes had been prepared, and a great deal of information about them had been gathered by a number of different investigators. By this time chemists had begun to wonder about how the atoms in these complexes were connected to each other and about the possible effect of these arrangements on the properties of the complexes. Among the observations that any successful theory would have to account for were the electrical conductivity of the complexes and their behavior toward $AgNO_3$. By measuring the conductivity of solutions of the complexes, it was possible to determine the number of ions in solution; the greater the number of ions, the greater the conductivity. For example, because the molar conductivity of aqueous solutions of $CoCl_3 \cdot 5NH_3$ was about the same as that of $CaCl_2$ and other $1:2$ electrolytes, it was deduced that $CoCl_3 \cdot 5NH_3$ consisted of three ions. The number of these that were chloride ions was determined by treating solutions with $AgNO_3$. When cold, freshly prepared solutions of $CoCl_3 \cdot 5NH_3$ were treated with $AgNO_3$, two moles of $AgCl$ precipitated for each mole of complex; one chloride in the compound did not precipitate. These results are summarized in Table 23.2.

In 1893, a 26-year old Swiss chemist, Alfred Werner, proposed a theory that successfully explained these facts and became the basis for our subsequent understanding of metal complexes. However, as is often the case when new ideas are presented, especially by young men without any special reputation, Werner's ideas did not gain immediate acceptance. In fact, many chemists were convinced that Werner was wrong. Foremost among these critics was a Danish chemist, Sophus Jorgensen, who was then in his mid-fifties. Jorgensen had done a considerable amount of excellent experimental work on complexes. He continued his important contributions as he sought ways to discredit Werner's theory.

Werner's first basic postulate was that metals exhibit both primary and secondary valences. We now refer to these as the metal's oxidation state and coordination number, respectively. This postulate had no theoretical basis; it predated Lewis's theory of covalent bonding by 23 years. However, it allowed Werner to explain many experimental facts. Werner postulated a primary valence of three and a secondary valence of

TABLE 23.2

properties of some ammonia complexes of cobalt(III)

ORIGINAL FORMULATION	COLOR	IONS PER FORMULA UNIT	Cl^- IONS IN SOLUTION PER FORMULA UNIT	MODERN FORMULATION
$CoCl_3 \cdot 6NH_3$	Orange	4	3	$[Co(NH_3)_6]Cl_3$
$CoCl_3 \cdot 5NH_3$	Purple	3	2	$[Co(NH_3)_5Cl]Cl_2$
$CoCl_3 \cdot 4NH_3$	Green	2	1	$trans\text{-}[Co(NH_3)_4Cl_2]Cl$
$CoCl_3 \cdot 4NH_3$	Violet	2	1	$cis\text{-}[Co(NH_3)_4Cl_2]Cl$

six for cobalt(III). He therefore wrote the formula for $CoCl_3 \cdot 5NH_3$ as $[Co(NH_3)_5Cl]Cl_2$. The ligands within the brackets satisfied cobalt's secondary valence of six; the three chlorides satisfied the primary valence of three. Werner proposed that the chlorides within the coordination sphere of cobalt(III) are bound so tightly that they are unavailable to contribute to the compound's conductivity or to react with $AgNO_3$. Thus $CoCl_3 \cdot 5NH_3$ consists of a $Co(NH_3)_5Cl^{2+}$ ion and two Cl^- ions.

Werner also sought to deduce the arrangement of ligands around the central metal. He postulated that cobalt(III) complexes exhibit an octahedral geometry and sought to verify this postulate by comparing the number of observed isomers with the number expected for various geometries. For example, $Co(NH_3)_4Cl_2^+$ would exhibit three geometric isomers if it was planar, two if it was octahedral, and three if it was trigonal prismatic. These possibilities are shown in Figure 23.14. When

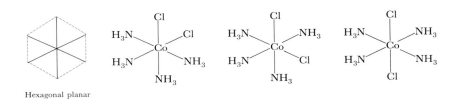

Hexagonal planar

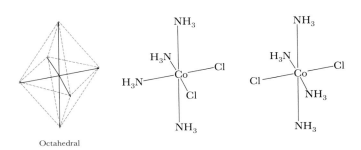

Octahedral

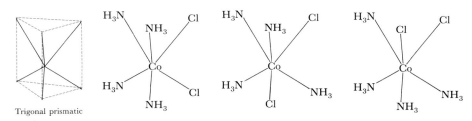

Trigonal prismatic

FIGURE 23.14 The possible isomers of $Co(NH_3)_4Cl_2^+$. A planar molecule of this composition will have three isomers; an octahedral molecule will have two; and a trigonal prismatic molecule will have three. Experimentally, two isomers are found.

he first postulated an octahedral geometry for cobalt(III), only one isomer of $Co(NH_3)_4Cl_2^+$ was known, the green *trans* isomer. In 1907, after considerable effort, Werner succeeded in isolating the violet *cis* isomer. Even before that, however, he had succeeded in isolating other *cis* and *trans* isomers of cobalt(III) complexes. The occurrence of two isomers was consistent with his postulate of octahedral geometry. Another result consistent with octahedral geometry was the demonstration that $Co(en)_3^{3+}$ and certain other complexes were optically active. As such evidence supporting Werner's ideas accumulated, Jorgensen and others finally accepted Werner's theory. In 1913 Werner was awarded the Nobel Prize in chemistry.

SAMPLE EXERCISE 23.7

Suggest the structure for $CoCl_2 \cdot 6H_2O$.

Solution: By analogy to the ammonia complexes of cobalt(III), we might write the formula of this compound as $[Co(H_2O)_6]Cl_2$. Indeed, experimental evidence indicates that the water molecules are attached to the metal as ligands. Hydrated metal salts generally have water coordinated to the metal. However, water can also be hydrogen bonded to the anion, particularly to oxyanions. For example, the familiar $CuSO_4 \cdot 5H_2O$ has four water molecules coordinated to Cu^{2+}, and one to SO_4^{2-}.

**23.7
magnetism
and color**

Werner's theory helps us to understand many properties of complexes, including isomerism and conductivity. However, Werner's theory must be extended before we can use it to explain certain other properties such as the magnetic behavior and color of transition-metal complexes. These two properties have played an important role in the further development of concepts about metal-ligand bonding. Therefore, let's briefly discuss these properties before taking a closer look at the bonding in transition-metal complexes.

MAGNETISM

In our earlier discussions of chemical bonding in Section 9.5, it was noted that substances containing unpaired electrons are **paramagnetic;** they are drawn into a magnetic field. The magnitude of the paramagnetism is related to the number of unpaired electrons.* Substances without unpaired electrons are **diamagnetic;** they are slightly repelled by a magnetic field. Therefore, one way to determine the number of unpaired electrons is to measure the effect of a magnetic field on the sample, as shown in Figure 23.15. The mass of the sample is first measured in the absence of a magnetic field and then in its presence. If the sample appears to weigh more in the magnetic field, it is being drawn into the field and is consequently paramagnetic. If the sample appears to weigh less in the magnetic field, it is being pushed upward, out of the field; it must consequently be diamagnetic. The point of interest in the complexes of the transition metals is that the number of unpaired electrons associated with a particular metal ion depends on the surrounding ligands. For example, there are no unpaired electrons in $Co(NH_3)_6^{3+}$, but there are

---

*The behavior of a sample in a magnetic field is related to its magnetic moment, μ. The magnetic moment of transition-metal complexes is proportional to the quantity $\sqrt{n(n + 2)}$, where n is the number of unpaired electrons.

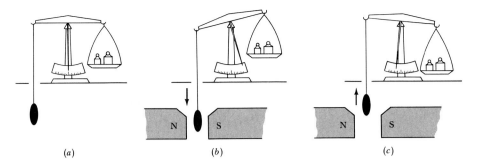

FIGURE 23.15 (*a*) A sample is weighed in the absence of a magnetic field. If it is attracted into a magnetic field (*b*) it is paramagnetic. If it is slightly repelled by a magnetic field (*c*) it is diamagnetic.

four in CoF_6^{3-}; yet both complexes contain cobalt(III). Any successful bonding theory must explain this observation.

COLOR

Another interesting characteristic of transition-metal complexes is the variety of colors that they exhibit. The colors of some complexes of cobalt(III) were shown earlier in Table 23.2. From that list, we see that color can change as the ligands surrounding the metal ion change. If we were to examine a more extensive list of substances, we would see that color also depends on the metal and its oxidation state.

Before we can attempt to explain the origin of these colors, we need to review our earlier discussion of light and introduce some new concepts. Recall first that visible light consists of electromagnetic radiation whose wavelength, λ, ranges from 400 nm to 700 nm (Figure 6.3). The energy of this radiation is inversely proportional to its wavelength, as discussed earlier, in Section 6.2:

$$E = h\nu = h(c/\lambda)$$
[23.6]

The visible spectrum is shown in Figure 23.16. The colors of the spectrum, indicated by their first letters, are shown in order of increasing energy. Notice that these letters spell out what appears to be a man's name: Roy G. Biv.

When a sample absorbs light, what we see is the sum of the re-

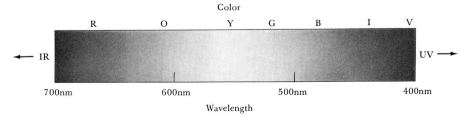

FIGURE 23.16 The visible spectrum showing the relation between color and wavelength (R = red, O = orange, Y = yellow, G = green, B = blue, I = indigo, and V = violet). Notice that the colors spell out a name: Roy G. Biv.

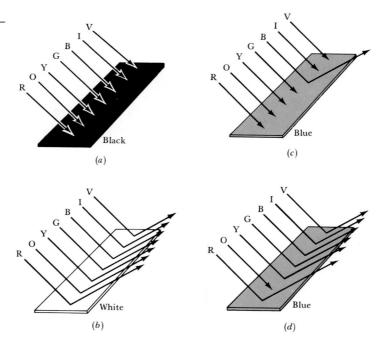

FIGURE 23.17 (a) An object is black if it absorbs all colors of light. (b) An object is white if it reflects all colors of light. (c) An object is blue if it reflects this color. (d) An object is also blue if it reflects all colors except for the complementary color of blue.

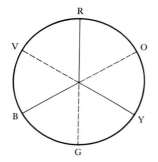

FIGURE 23.18 A crude color wheel that allows us to determine which colors are complementary to one another.

maining colors that strike our eyes. If a sample absorbs all wavelengths of visible light, none reaches our eyes from that sample. Consequently, it appears black. If the sample absorbs no light, it is white or colorless. If it absorbs all but blue, the sample appears blue. Each of these situations is shown in Figure 23.17. That figure shows one further situation; we also perceive a blue color when visible light of all colors except orange strikes our eyes. In a complementary fashion, if the sample absorbed only blue, it would appear orange; blue and orange are said to be **complementary colors.** Complementary colors can be determined with the aid of the crude color wheel shown in Figure 23.18. We can construct this color wheel by first positioning red (R), yellow (Y), and blue (B), the **primary colors.** Combinations of red and blue gives violet (V), which is shown on the dotted line between red and blue. Combination of red and yellow gives orange (O), whereas combination of blue and yellow gives green (G). These combinations are located on dotted lines between the appropriate primary colors. The colors that are complementary to one another, like orange and blue, are across the wheel from each other.

SAMPLE EXERCISE 23.8

The complex ion *trans*-$[Co(NH_3)_4Cl_2]^+$ absorbs light primarily in the red region of the visible spectrum (the most intense absorption is at 640 nm). What is the color of the complex?

Solution: Because the complex absorbs red light, its color will be the complementary color of red. From Figure 23.18, we see that this is green.

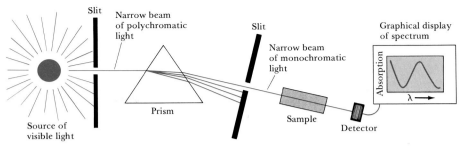

FIGURE 23.19 A schematic representation of the experimental way in which the absorption spectrum of a solution is determined. The prism is rotated so that different wavelengths of light pass through the sample. The detector measures the amount of light reaching it, and this information can be displayed as the absorption at each wavelength.

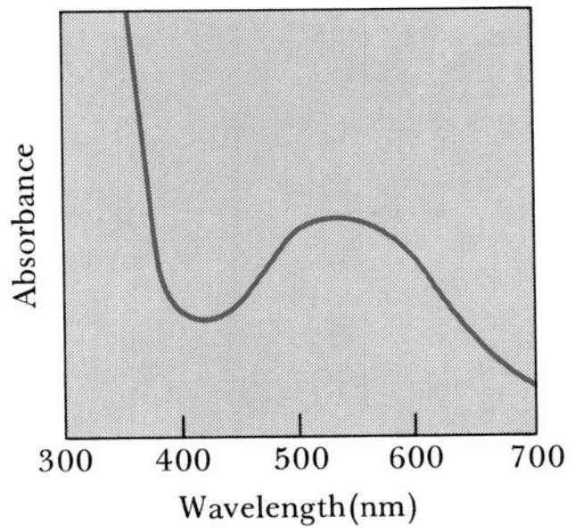

FIGURE 23.20 The visible absorption spectrum of the $Ti(H_2O)_6^{3+}$ ion.

The amount of light absorbed by a sample as a function of wavelength is known as its **absorption spectrum.** The visible absorption spectrum of a transparent sample, such as a solution of *trans*-$[Co(NH_3)_4Cl_2]^+$, can be determined as shown in Figure 23.19. The spectrum of $Ti(H_2O)_6^{3+}$, which we shall discuss in the next section, is shown in Figure 23.20. The absorption maximum of $Ti(H_2O)_6^{3+}$ is at 510 nm. Because the sample absorbs most strongly in the green and yellow regions of the visible spectrum, it appears purple.

23.8 bonding in complexes of transition metals

We have noted that the complexes of transition metals are particularly numerous and stable. We can gain further insight into these complexes by examining their bonding more closely. Several different bonding theories have been applied to transition-metal complexes. We shall consider two of these. The first, known as the valence-bond model, views the bonding as being covalent and examines the hybridization of orbitals on the metal. The second, known as the crystal-field model, examines the bonding from an ionic point of view and focuses on the effect of the surrounding ligands on the energies of the metal d orbitals.

VALENCE-BOND MODEL

As noted in Section 23.1, complexes can be visualized as resulting from the interaction of a Lewis acid with Lewis bases. The base can be considered to donate a pair of electrons into a suitable empty hybrid orbital on the metal, as shown in Figure 23.21. The resultant σ bond

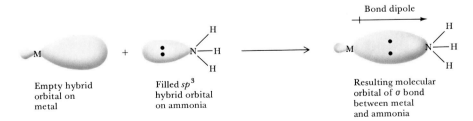

FIGURE 23.21 In the valence-bond model, the bond between a metal and donor atom is visualized as shown. The metal has an empty hybrid orbital that accepts electron density from a filled orbital on the donor.

705

23.8
bonding in complexes
of transition metals

chemistry of metals:
coordination compounds

between the metal and the donor atom of the ligand is polar covalent because the donor atom has a stronger attraction for electrons than does the metal.

The first-row transition metals have outer $3d$, $4s$, $4p$, and $4d$ orbitals that could possibly be used to form hybrid orbitals. As we noted in Section 9.2, the tetrahedral, square-planar, and octahedral geometries have associated hybridizations of sp^3, dsp^2 and d^2sp^3, respectively.

The complex $Zn(NH_3)_4^{2+}$ exemplifies tetrahedral coordination. The outer electron configuration of an isolated zinc atom and an isolated Zn^{2+} ion are given below:

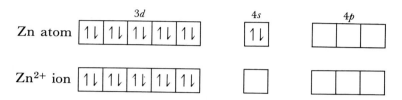

The $3d$ orbitals are full, but the $4s$ and $4p$ orbitals are empty in the Zn^{2+} ion. They are able to hybridize to form four sp^3 hybrid orbitals that can each accept a pair of electrons from a ligand:

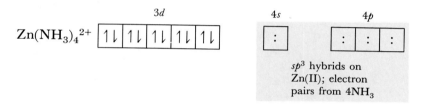

sp^3 hybrids on Zn(II); electron pairs from $4NH_3$

The $Ni(CN)_4^{2-}$ ion has a square-planar geometry and can therefore be used to illustrate dsp^2 hybridization:

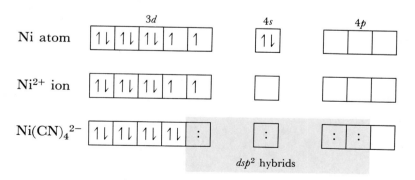

dsp^2 hybrids

In this case, the formation of an empty set of dsp^2 hybrid orbitals requires that an electron in the $3d$ orbital become paired. Whereas the isolated Ni^{2+} ion is paramagnetic with two unpaired electrons, the $Ni(CN)_4^{2-}$ ion is diamagnetic.

The $Co(NH_3)_6^{3+}$ ion has an octahedral geometry and can be used to illustrate d^2sp^3 hybridization:

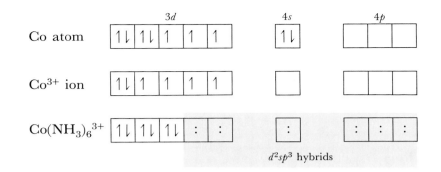

In order to form empty d^2sp^3 hybrid orbitals on the metal utilizing the $3d$ orbitals, electrons must be paired in the three unhybridized $3d$ orbitals. We are thus able to rationalize the fact that $Co(NH_3)_6^{3+}$ is diamagnetic. However, we have already pointed out in Section 23.6 that not all cobalt(III) complexes are diamagnetic; CoF_6^{3+} is paramagnetic with four unpaired electrons. This is rationalized by postulating that the $4d$ orbitals rather than the $3d$ ones are used in bonding:

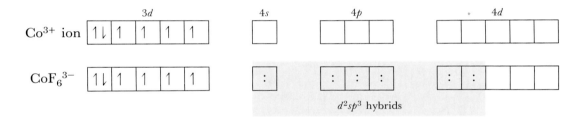

Complexes like CoF_6^{3-} in which the central metal ion has the same number of unpaired electrons as the free metal ion are said to be **high-spin complexes**. Those like $Co(NH_3)_6^{3+}$ that have fewer unpaired electrons than the free metal ion are said to be **low-spin complexes**.

SAMPLE EXERCISE 23.9

Using the valence-bond approach, show the possible electron configurations of iron in $Fe(CN)_6^{3-}$.

Solution: The electron configurations of Fe and Fe^{3+} are as follows:

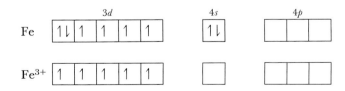

The complex could be either low spin with one unpaired electron or high spin with five unpaired electrons, as shown at the top of page 708.

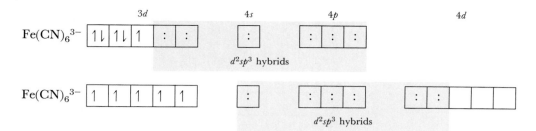

Experimentally, the complex is found to have one unpaired electron.

CRYSTAL-FIELD MODEL

One of the shortcomings of the valence-bond model is that it offers no explanation for the color of complexes. Furthermore, it is not clear on the basis of this model why some complexes are high spin and others low spin. However, the crystal-field model, which assumes ionic bonding between ligands and metal, provides an explanation for color as well as for the magnetic properties of complexes.

We can begin to construct this model by representing the ligands in an octahedral complex as small negative charges and the metal ion as a positive charge. The resultant assembly is shown in Figure 23.22, with the negative charges lying along the x-, y-, and z-axes and the metal ion at the origin. The energy of this assembly is less than that of the separated charges because of the electrostatic attraction that exists between positive and negative charges. The relative energies of the separated charges and the complex are shown in Figure 23.23. Now let's refine this model by focusing on the positive metal ion. The electrons of the metal ion are distributed in the three-dimensional space around the nucleus. All of these electrons, but especially those on the outer periphery of the metal ion, are repelled by the negatively charged ligands. If the electrons were spherically distributed, this repulsion would raise the energy of the system as shown in Figure 23.23. Now let's consider one further refinement of the model by specifically examining the interactions between the six surrounding negative charges and the electrons in the outer d orbitals. These are the outermost electrons in a transition-metal ion. We find when we consider these interactions that the five d orbitals are split into two groups of different energies as shown at the far right of Figure 23.23. This splitting is the key to the crystal-field model; we therefore need to examine both the origin of the splitting and its consequences more closely.

The shapes and orientations of the five d orbitals are shown in Figure 23.24. Notice that two orbitals, the d_{z^2} and the $d_{x^2-y^2}$, have lobes that point along the axes. The other three, the d_{xy}, the d_{xz}, and the d_{yz}, have lobes that point between the axes. Now recall that the six negative charges surrounding the metal ion lie along the axes (Figure 23.22). Thus the d_{z^2} and the $d_{x^2-y^2}$ orbitals have lobes that point directly at the charges. Consequently, there is a strong electrostatic repulsion between electrons in these orbitals and the surrounding negative charges. On the

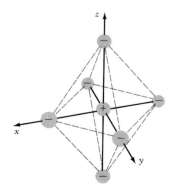

FIGURE 23.22 An octahedral array of negative charges surrounding a positive charge.

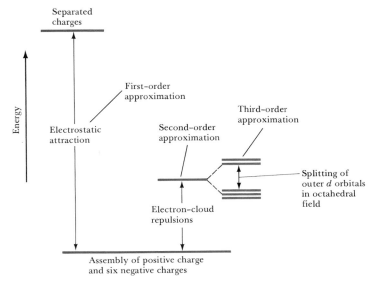

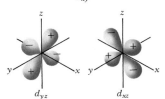

$d_z{}^2$ $d_{x^2 - y^2}$

d_{xy}

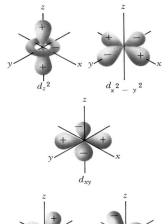

d_{yz} d_{xz}

FIGURE 23.23 In the crystal-field model, the bonding between metal ions and donor atoms is considered to be ionic. As a first approximation, the donor atoms and metal ions are considered to be point charges. The energy of the assembled ions is less than that of the separated ions because of the electrostatic attraction between positive and negative charges. As a second approximation, the metal atom is considered to have a spherically symmetric distribution of electrons at its periphery. The resultant energy is thereby increased because of repulsions between the surrounding negative charges and the electrons on the metal ion. As a third approximation, the actual distribution of electrons on the metal is considered. The outer d electrons are of two types: three orbitals are pointed between the negative charges and have lower energy; two orbitals are pointed at the negative charges and have high energy.

FIGURE 23.24 The shapes of the five d orbitals.

other hand, electrons in the d_{xy}, d_{yz}, and d_{xz} orbitals experience a smaller repulsion because they have lobes that point between the negative charges. The d_{xy}, d_{yz}, and d_{xz} orbitals are therefore of lower energy than the $d_{x^2-y^2}$ and d_{z^2} orbitals. The splitting of the d orbitals by an octahedral field of negative charges is reproduced in Figure 23.25.

The energy gap between the d orbitals, labeled Δ, is of the same order of magnitude as the energy of a photon of visible light. It is therefore

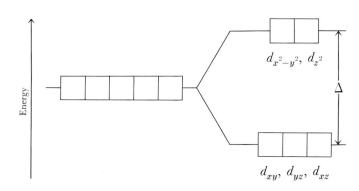

FIGURE 23.25 The energies of the d orbitals in an octahedral crystal field.

23.8
bonding in complexes
of transition metals

chemistry of metals:
coordination compounds

possible for a transition-metal complex to absorb visible light, which thereby excites an electron from the lower energy d orbitals into the higher energy ones. The $Ti(H_2O)_6^{3+}$ ion provides a simple example, because titanium(III) has only one $3d$ electron. As shown in Figure 23.20, $Ti(H_2O)_6^{3+}$ has a single absorption peak in the visible region of the spectrum. The maximum absorption is at 510 nm (3.9×10^{-19} J/molecule). Light of this wavelength causes the d electron to move from the lower energy set of d orbitals into the higher energy set, as shown in Figure 23.26.

SAMPLE EXERCISE 23.10

Al^{3+}, Zn^{2+}, and Co^{2+} ions are placed in octahedral environments. Which can absorb visible light and thereby exhibit color?

Solution: The Al^{3+} ion has an electron configuration of [Ne]. Because it has no outer d electrons, it is colorless.

The Zn^{2+} ion has an electron configuration of $[Ar]3d^{10}$. In this case all of the $3d$ orbitals are filled.

There is no room in the d_{z^2} and $d_{x^2-y^2}$ orbitals to accept an electron from a lower energy d_{xy}, d_{yz}, or d_{xz} orbital. The complex is therefore colorless.

The Co^{2+} ion has an electron configuration of $[Ar]3d^7$. In this case there is room for movement of a d electron from the lower energy d_{xy}, d_{yz}, and d_{xz} into the higher energy d_{z^2} and $d_{x^2-y^2}$ orbitals. The complex is therefore colored.

Gemstones such as ruby and emerald owe their color to the presence of trace amounts of transition-metal ions. For example, replacement of a fraction of the aluminum in the colorless mineral corundum, Al_2O_3, produces several different gems: chromium forms ruby, manganese forms amethyst, and iron forms topaz. Sapphire, which occurs in a variety of colors, but is most often blue, contains titanium and cobalt. Several other gems are produced by replacing a trace of aluminum in the colorless mineral beryl, $Be_3Al_2Si_6O_{18}$, with transition-metal ions. For example, emerald contains chromium, whereas aquamarine contains iron.

The magnitude of the energy gap, Δ, and consequently the color of a complex, depend on both the metal and the surrounding ligands. For example, $Fe(H_2O)_6^{3+}$ is yellow, $Cr(H_2O)_6^{3+}$ is violet, and $Cr(NH_3)_6^{3+}$ is yellow. Ligands can be arranged in order of their abilities to increase the energy gap, Δ. The following is an abbreviated list of common ligands arranged in order of increasing Δ:

$$Cl^- < F^- < H_2O < NH_3 < en < NO_2^- < CN^-$$

This list is known as the **spectrochemical series.**

With the spectrochemical series in mind, let's consider the CoF_6^{3-} and $Co(CN)_6^{3-}$ ions. The F^- ion, being on the low end of the spectrochemical series, is known as a weak-field ligand. The CN^-, being on the

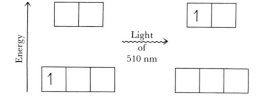

FIGURE 23.26 The $3d$ electron of $Ti(H_2O)_6^{3+}$ is excited from the lower-energy d orbitals to the higher-energy ones when irradiated with light of 510 nm wavelength.

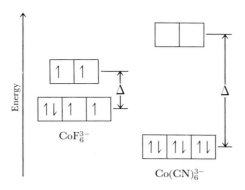

FIGURE 23.27 The population of d orbitals in the high-spin CoF_6^{3-} ion (small Δ) and low-spin $Co(CN)_6^{3-}$ ion (large Δ).

CoF_6^{3-}

$Co(CN)_6^{3-}$

high end of the series, is known as a strong-field ligand. It produces a larger energy gap than does the F^- ion. The splitting of the d orbital energies in the complexes is compared in Figure 23.27. A count of the electrons in cobalt(III) tells us that we have six electrons to place in the 3d orbitals. If no other factors were involved, these six electrons would go into the 3d orbitals one at a time, with parallel spin, until five such electrons have been added. The remaining electron would then be paired up in one of the 3d orbitals. However, because the two sets of 3d orbitals differ in energy, there is an energy penalty for adding an electron to one of the higher-energy orbitals instead of pairing it up with one of those in the lower-energy set. We thus have two conflicting energy considerations at work. If the crystal-field splitting is not too large, the electrons follow Hund's rule (Section 7.6) and remain unpaired as much as possible. This situation occurs in CoF_6^{3-}. However, in the case of $Co(CN)_6^{3-}$, the separation between the d orbital energies is so large that the electrons pair up in the lower-energy set until all three of these orbitals are filled. Thus the CoF_6^{3-} complex is high spin, the $Co(CN)_6^{3-}$ complex is low spin.

SAMPLE EXERCISE 23.11

Deoxyhemoglobin, a complex of iron(II), is blue and has four unpaired electrons. Coordination of an O_2 molecule to deoxyhemoglobin forms oxyhemoglobin, which is red and diamagnetic. Assuming that the diagram for octahedral crystal-field splitting applies, explain the relation between the color and magnetic properties in these complexes.

Solution: Deoxyhemoglobin appears blue because it absorbs orange light; this light is on the low-energy side of the visible spectrum (Figure 23.16). Consequently, the energy gap, Δ, between the d orbitals must not be large. In contrast, oxyhemoglobin is red because it absorbs green light; this light is on the high-energy end of the spectrum. Consequently, Δ must be larger in this case than in the case of deoxyhemoglobin.

The electron configuration of iron(II) is $[Ar]3d^6$. The population of the d electrons in the two cases is

shown in the following diagrams:

Deoxyhemoglobin

Oxyhemoglobin

The fact that deoxyhemoglobin is a high-spin complex, whereas oxyhemoglobin is a low-spin complex therefore correlates with their colors.

23.8
bonding in complexes
of transition metals

chemistry of metals:
coordination compounds

Thus far we have examined only the octahedral geometry. The square-planar geometry can be thought of as arising from the octahedral by removal of the negative charges from the z-axis. As the negative charges are removed, the d_{z^2}, d_{xz}, and d_{yz} orbitals, which all have a z component, become more stable, as shown in Figure 23.28. The splitting pattern for the tetrahedral case is shown in Figure 23.29; it is just the reverse of the splitting pattern for octahedral complexes. However, four tetrahedrally arranged ligands produce a much smaller splitting than do six octahedrally arranged ones. Consequently, all tetrahedral complexes are high-spin ones.

SAMPLE EXERCISE 23.12

Four-coordinate nickel(II) complexes exhibit both square-planar and tetrahedral geometries. The tetrahedral ones such as $NiCl_4^{2-}$ are paramagnetic; the square-planar ones such as $Ni(CN)_4^{2-}$ are diamagnetic. Show how the d electrons of nickel(II) populate the d orbitals in the appropriate crystal-field-splitting diagram in each of these cases.

Solution: Nickel(II) has an electron configuration of $[Ar]3d^8$. The population of the d electrons in the two geometries is given below:

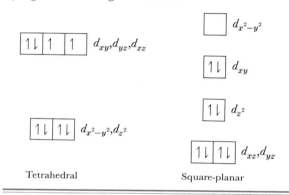

Tetrahedral   Square-planar

We have seen that the crystal-field model provides a basis for explaining many features of transition-metal complexes. In fact, it can be used to explain many observations in addition to those we have discussed.

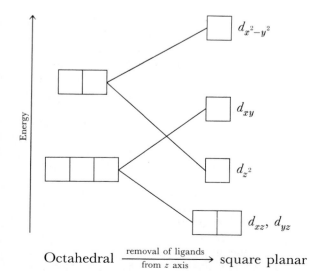

Octahedral $\xrightarrow[\text{from } z \text{ axis}]{\text{removal of ligands}}$ square planar

FIGURE 23.28 The effect on the relative energies of the d orbitals of removing the two negative charges from the z-axis of an octahedral complex. When the charges are completely removed, the square-planar geometry results.

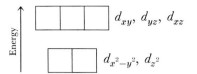

$$d_{xy}, \ d_{yz}, \ d_{xz}$$

$$d_{x^2-y^2}, \ d_{z^2}$$

FIGURE 23.29 The energies of the d orbitals in a tetrahedral crystal field.

However, there are many lines of evidence that show that the bonding between transition-metal ions and ligands must have some covalent character. Molecular-orbital theory (Sections 9.4 and 9.5) can be used to describe the bonding in complexes. However, the application of molecular-orbital theory to coordination compounds is beyond the scope of our discussion. The crystal-field model, though not entirely accurate in all details, provides a quite adequate and useful description.

FOR REVIEW

summary

Coordination compounds or **complexes** contain metal ions bonded to several surrounding anions or molecules known as **ligands**. The metal ion and its ligands comprise the **coordination sphere** of the complex. The atom of the ligand that bonds to the metal ion is known as the **donor atom**. The number of donor atoms attached to the metal ion is known as the **coordination number** of the metal ion. The common coordination numbers are four and six; the common coordination geometries are tetrahedral, square-planar, and octahedral.

If a ligand has several donor atoms that can coordinate simultaneously to the metal, it is said to be **polydentate** and is referred to as a **chelating agent.** Two common examples are ethylenediamine (en), which is potentially bidentate, and ethylenediaminetetraacetate ($EDTA^{4-}$), which is potentially hexadentate. Many biologically important molecules such as the porphyrins are complexes of chelating agents.

Isomerism is common among coordination compounds. One simple form of isomerism, known as **linkage isomerism**, occurs when a ligand is capable of coordinating to a metal through either of two donor atoms. The most common forms of isomerism are **geometric** and **optical isomerism,** two forms of **stereoisomerism.** Geometric isomers differ from one another in the relative locations of donor atoms in the coordination sphere; the most common are *cis-trans* isomers. Optical isomers differ from one another in that they are nonsuperimposable mirror images of one another. Isomers differ from one another in their chemical and physical properties; however, optical isomers differ only in the presence of a **chiral** environment. Most often optical isomers are distinguished from one another by their interactions with plane-polarized light; one isomer will rotate the plane of polarization to the right, whereas its mirror image rotates the plane to the left. The isomer that rotates the plane of polarization to the right is said to be **dextrorotatory**, whereas its isomer is **levorotatory**. Chiral molecules are said to be **optically active**. A 50–50 mixture of two optical isomers will not rotate plane-polarized light and is said to be **racemic.**

Many of the early studies that served as a basis for our current understanding of complexes involved complexes of chromium(III), cobalt(III), platinum(II), and platinum(IV). Complexes of these metal ions are **inert,** in that they undergo ligand exchange at a slow rate. Complexes undergoing rapid exchange are said to be **labile.** The historically significant experiments include studies of the molar conductivities of aqueous solutions, the behavior of these solutions toward

AgNO$_3$, and the number of isomers associated with different stoichiometries. Studies of the magnetic properties and color of transition-metal complexes have played an important role in the formulation of bonding theories. In discussing the color of a substance, it is important to remember that the color seen is complementary to the color absorbed by the substance.

The bonding in complexes can be thought of as the polar-covalent type, with the metal acting as a Lewis acid and the ligand as a Lewis base. In the **valence-bond model,** filled orbitals on the donor atoms are pictured as overlapping with empty hybrid orbitals on the metal to form bonds between the metal and the donor atoms. Although this model allows us to explain the magnetic properties of complexes, it does not provide a basis for explaining their color. Insight into both magnetic properties and color is obtained if we assume that the bonding between ligands and metal ion is ionic; we can then focus on the effect that the surrounding ligands have on the energies of the d orbitals of the metal. The resultant model is known as the **crystal-field model.**

learning goals

Having read and studied this chapter, you should be able to:

1. Determine either the charge of a complex ion, having been given the oxidation state of the metal, or the oxidation state, having been given the charge of the complex. (You will need to recognize the common ligands and their charges.)
2. Describe, with the aid of drawings, the common geometries of complexes. (You will need to recognize whether the common ligands are functioning as monodentate or polydentate ligands.)
3. Name coordination compounds, having been given their formulas, or write their formulas, having been given their names.
4. Describe the common types of isomerism.
5. Determine the possible number of stereoisomers for a complex, having been given its composition.
6. Distinguish between inert and labile complexes.
7. Describe how Werner was able to explain the conductivity, precipitation reactions, and isomerism of complexes.
8. Explain the geometry and magnetic behavior of transition-metal complexes in terms of the valence-bond model. (You will need to know the types of hybridization and their associated geometries.)
9. Explain the magnetic behavior and color of transition-metal complexes in terms of the crystal-field model. (You will need to know the way in which the d orbitals split in octahedral, square-planar, and tetrahedral fields and will also need to be familiar with the spectrochemical series.)

key terms

Among the more important terms and expressions used for the first time in this chapter are the following:

A **chelating agent** (Section 23.2) is a polydentate ligand that is capable of occupying two or more sites in the coordination sphere.

Chiral (Section 23.4) means having a nonsuperimposable mirror image; for example, we might refer to a molecule as being chiral.

A *cis* geometric arrangement (Section 23.4) refers to one with like groups adjacent to each other.

The **coordination sphere** (Section 23.1) of a complex is the volume enclosing the metal ion and its surrounding ligands.

The **coordination number** (Section 23.1) of an atom is the number of adjacent atoms to which it is directly bonded; in a complex, the coordination number of the metal ion is the number of donor atoms to which it is bonded.

The term **dextrorotatory**, or merely **dextro** or *d* (Section 23.4), is used to label a chiral molecule that rotates the plane of polarization of plane-polarized light to the right (clockwise).

A **diamagnetic** substance (Section 23.7) has no unpaired electrons and is therefore weakly repelled by a magnetic field.

The **donor atom** (Section 23.1) of a ligand is the one that bonds to the metal.

Geometric isomers (Section 23.4) have different arrangements of donor atoms around the coordination sphere.

A **high-spin** complex (Section 23.8) has the same number of unpaired electrons as does the isolated metal ion.

An **inert** complex (Section 23.5) exchanges ligands at a slow rate.

Isomers (Section 23.4) are compounds whose molecules have the same overall composition but different structures.

A **labile** complex (Section 23.3) exchanges ligands at a rapid rate.

The term **levorotatory**, or merely **levo** or *l* (Section 23.4), is used to label a chiral molecule that rotates the plane of polarization of plane-polarized light to the left (counterclockwise).

A **ligand** (Section 23.1) is an ion or molecule that coordinates to a single central metal atom or to an ion to form a complex.

A **low-spin** complex (Section 23.8) has fewer unpaired electrons than does the isolated metal ion.

Optical isomers (Section 23.4) are nonsuperimposable mirror images of one another.

The **spectrochemical series** (Section 23.8) is a list of ligands arranged in order of their abilities to split the *d* orbital energies (using the terminology of the crystal-field model).

A *trans* geometric arrangement (Section 23.4) refers to one with like groups opposite each other.

EXERCISES

structure and isomerism

23.1 Give the coordination number of the central metal ion in each of the following complex ions: (a) $AlCl_4^-$; (b) NbF_7^{2-}; (c) $Fe(CO)_5$; (d) $Co(en)_2NH_3Cl^{2+}$.

23.2 What is the charge of the complex ion formed by chromium(III) and the following ligands: (a) four NH_3 and two H_2O; (b) three CN^- and three H_2O; (c) four H_2O and two Cl^-?

23.3 What is the oxidation state of the central metal ion in the following complexes: (a) $PtCl_6^{2-}$; (b) $Cr(NH_3)_2(SCN)_4^-$; (c) $Fe(CO)_5$; (d) $[Ni(en)_2(H_2O)_2](ClO_4)_2$?

23.4 Sketch the structure of each of the following complex ions: (a) $Cu(NH_3)_4^{2+}$; (b) *cis*-$Pt(NH_3)_2Cl_2$; (c) *trans*-$Rh(en)_2Br_2^+$; (d) $Pt(NH_3)_2(C_2O_4)$; (e) $GaCl_4^-$.

23.5 Draw the structures of all distinct complexes having each of the following formulas: (a) $Co(NH_3)_3(H_2O)_3^{3+}$; (b) $ZnCl_4^{2-}$; (c) $Pt(NH_3)_2ClBr$;

(d) $Zn(NH_3)_2Cl_2$; (e) $Co(en)_2(NO_2)_2^+$; (f) $Cr(C_2O_4)_3^{3-}$.

23.6 Draw all the possible isomers for each of the following: (a) the chloroamminebis(ethylenediamine)cobalt(III) ion; (b) the trichlorotriammineplatinum(IV) ion.

23.7 Is the sulfate ion acting as a mondentate or as a bidentate ligand in the following compounds: (a) $[Co(NH_3)_4SO_4]Cl$; (b) $[Co(NH_3)_4(H_2O)SO_4]Cl$?

23.8 Arrange the following compounds in order of increasing molar conductivity: (a) $[Pt(NH_3)_4Cl_2]Cl$; (b) $[Co(NH_3)_6]Cl_3$; (c) $Co(NH_3)_3(NO_3)_3$; (d) $K[Pt(NH_3)Cl_3]$.

23.9 Which of the following compounds may exhibit optical isomerism: (a) $Pt(en)_2^{2+}$; (b) $Coen(NH_3)_4^{3+}$; (c) $Fe(EDTA)^{2-}$; (d) $Co(en)_2(NH_3)_2^{3+}$? Explain.

23.10 Two complexes have the same empirical formula: $Co(NH_3)_5SO_4Cl$. Discuss the possible structures of the two complexes and describe experiments that could be used to identify the two compounds.

23.11 Two compounds have the same empirical formula: $Co(en)_2NO_2Cl_2$. Both are $1:1$ electrolytes. When $AgNO_3$ is added to compound A, one mole of $AgCl$ precipitates. When $AgNO_3$ is added to compound B, no $AgCl$ precipitates. Draw the possible geometries for both compounds.

nomenclature

23.12 Name each of the following compounds: (a) $[Ni(NH_3)_6]Cl_2$; (b) $K_3[Co(C_2O_4)_3]$; (c) $[Co(en)_2SO_4]ClO_4$; (d) $Na_3[Co(CN)_5Br]$.

23.13 Name each of the following compounds or ions: (a) $Au(CN)_4^-$; (b) $[Co(NH_3)_6][Co(NO_2)_6]$; (c) $[Co(NH_3)_4CO_3]Cl$; (d) *trans*-$Pt(NH_3)_2Br_2$.

23.14 Write the formula for each of the following compounds or ions: (a) chloropentaaquochromium(III) chloride; (b) dinitrogenpentaammineruthenium(II) nitrate (dinitrogen is N_2); (c) chloronitrobis(ethylenediamine)cobalt(III) perchlorate; (d) tetracarbonylnickel(0) (carbonyl is CO).

23.15 Draw the possible structures for each of the following compounds or ions: (a) potassium trioxalatochromate(III); (b) *trans*-dichlorobis(ethylenediamine)rhodium(III) chloride; (c) sodium hexachlorostannate(IV); (d) trichlorotriamminecobalt(III).

23.16 Write balanced chemical equations describing the following reactions: (a) When the chloroamminebis(ethylenediamine)cobalt(III) ion is allowed to stand in water, the chloride ligand is slowly displaced by a water molecule. (b) When the trichloronitroplatinum(II) ion is treated with ammonia, *trans*-dichloronitroammineplatinum(II) is formed. (c) When the carbonatopentaamminecobalt(III) ion is treated with an acid, it is converted to the aquopentaaminecobalt(III) ion with the accompanying evolution of carbon dioxide gas. (d) When solid tris(ethylenediamine)chromium(III) thiocyanate is heated to $130°C$ in the presence of a catalytic amount of NH_4SCN, *trans*-dithiocyanatobis-(ethylenediamine)chromium(III) thiocyanate forms with the accompanying evolution of ethylenediamine as a gas.

color, magnetism, and bonding

23.17 If a sample appears blue when examined under visible light, what color will it be when examined under orange light?

23.18 Explain why a black car gets warmer inside on a sunny day than a white one.

23.19 Does a violet compound or a green one absorb higher-energy visible light?

23.20 The absorption maxima in the visible-absorption spectra of several complexes are given below. Predict the color of each complex. (a) $Co(NH_3)_6^{3+}$ (470 nm); (b) *trans*-$Co(NH_3)_4(NO_2)_2^+$ (440 nm); (c) *cis*-$Co(NH_3)_4(H_2O)_2^{3+}$ (510 nm); (d) *cis*-$Co(en)_2Cl_2^+$ (535 nm)

23.21 Give the number of outer d electrons associated with the metal ion in each of the following complexes: (a) $Pd(NH_3)_4^{2+}$; (b) $Cu(NH_3)_4^{2+}$; (c) $Cr(NH_3)_4Cl_2^+$; (d) $V(H_2O)_6^{3+}$.

23.22 Write the electron configurations of Mn and Mn^{2+} and draw the valence-bond diagram for $Mn(H_2O)_6^{2+}$ (as was done for $Fe(CN)_6^{3-}$ in Sample Exercise 23.9).

23.23 Using valence-bond theory, predict the possible number of unpaired electrons in each of the following compounds: (a) $Cr(NH_3)_6^{3+}$; (b) $Ni(NH_3)_6^{2+}$; (c) $Pt(NH_3)_4^{2+}$; (d) $GaCl_4^-$; (e) $FeCl_4^-$. What hybridization is used by the metal ion in each case?

23.24 Rationalize the observation that square-planar complexes of nickel(II) are diamagnetic, whereas the tetrahedral ones are paramagnetic.

23.25 Classify the following compounds as either high spin or low spin based on the number of unpaired electrons: (a) $Co(NH_3)_6^{2+}$ (three unpaired electrons); (b) $Mn(CN)_6^{3-}$ (two unpaired electrons).

23.26 Which of the following pairs would you expect to absorb the highest energy radiation in the visible spectrum: (a) $Ni(H_2O)_6^{2+}$ or $Ni(NH_3)_6^{2+}$; (b) $CrCl_6^{3-}$ or $Cr(NH_3)_6^{3+}$?

23.27 Using crystal-field theory, rationalize the observation that $Fe(CN)_6^{3-}$ has a single unpaired electron.

23.28 How would you experimentally determine whether a compound containing the $Co(NCS)_4^{2-}$ ion is paramagnetic or diamagnetic? How would you explain the presence of three unpaired electrons in this complex?

23.29 How would you expect the color of octahedral complexes of titanium(III) to vary as the ligands are changed from H_2O to NH_3 to CN^-?

[23.30] Determine the splitting pattern for the d orbitals of a transition metal in a linear crystal field (assume that the two negative charges approach the metal along the z-axis).

general exercises

23.31 Distinguish between the following terms or symbols: (a) *cis* and *trans;* (b) levo and dextro; (c) high spin and low spin; (d) d^2sp^3 and dsp^2; (e) mondentate and bidentate; (f) paramagnetic and diamagnetic; (g) geometric isomers and optical isomers; (h) crystal-field theory and valence-bond theory; (i) coordination number and oxidation number; (j) labile and inert.

23.32 Would you expect the presence of a strong complexing agent like CN^- to make it easier or more difficult to oxidize gold?

23.33 The silver(I) complexes $Ag(NH_3)_2^+$ and $Ag(S_2O_3)_2^{3-}$ have dissociation constants of 5.9×10^{-8} and 1.6×10^{-13}, respectively. Which would you expect to be more effective in dissolving AgBr, a $Na_2S_2O_3$ solution or an NH_3 solution of equal molarity?

23.34 Explain how oxalic acid functions in removing rust stains from cloth.

23.35 What is the ratio of complexed to uncomplexed Pb^{2+} in a solution whose equilibrium concentration of EDTA is $1 \times 10^{-4} M$ if the dissociation constant of $PbEDTA^{2-}$ is 1×10^{-18}?

23.36 A 0.250 g sample containing Ni^{2+} was titrated with $0.0120 M$ EDTA. If it took 40.0 ml to reach an endpoint (all Ni^{2+} complexed), calculate the percentage of nickel in the sample.

23.37 Explain the observation that Li^+ is more tightly hydrated than Na^+; why is Ca^{2+} more tightly hydrated than K^+?

23.38 A compound of nickel has the stoichiometry $NiCl_2 \cdot 4NH_3$. What are its possible geometries? If the compound contains two unpaired electrons, can any structures be eliminated as possibilities?

23.39 How might the colligative properties of solutions of complexes be used to give information like that obtained by measuring conductivities?

23.40 Do the p orbitals split in energy just as the d orbitals do when they are placed in an octahedral field?

23.41 Refining of nickel often involves the "Mond process." Impure nickel is treated with 15 atm of carbon monoxide gas at $80°C$. Tetracarbonylnickel-(0), which is gaseous under these conditions, forms. This complex is swept out of the reaction vessel and decomposed by increasing the temperature and lowering the pressure. The carbon monoxide is then recycled while the deposited nickel metal is collected periodically. Write equations for the reactions described above.

[23.42] Rationalize the observation that all octahedral complexes with electrons in the $d_{x^2-y^2}$ and/or d_{z^2} orbitals are labile.

[23.43] Rationalize the observation that the ionic radius of low-spin iron(II) is less than that of high-spin iron(II).

23.44 A compound is analyzed and found to contain 46.2 percent platinum, 33.6 percent chlorine, 16.6 percent nitrogen, and 3.6 percent hydrogen. The molar conductivities of freshly prepared aqueous solutions of this compound are consistent with those of 3:1 electrolytes. Addition of $AgNO_3$ to these solutions precipitates 3 moles of Cl^- as AgCl. Write the chemical formula for the compound, show its probable geometry, and name it.

24 organic compounds

Organic chemistry deals mainly with the chemical characteristics of compounds in which carbon is a principal element. There are no precise boundaries between organic chemistry and other areas of chemical science. The concept of a separate area of chemistry that could be thought of as organic developed out of the vitalist theory, which held that substances that make up living matter are fundamentally different from those that form inanimate matter. In 1828, Fredrich Wohler, a German chemist, reacted potassium cyanate, $KOCN$, with ammonium chloride, NH_4Cl; much to his surprise he obtained urea, H_2NCONH_2, a well-known substance that had been isolated from the urine of mammals. Following Wohler, many other organic substances were prepared from inorganic starting materials, and the vitalist theory gradually disappeared. Nevertheless organic chemistry continued to develop as a distinct area of chemistry. This may be explained partly by the fact that the raw materials for much of organic chemistry—oil, coal, wood, animal matter, and so forth—are of plant or animal origin. Secondly, the chemical characteristics of organic substances could be classified and systematized. The "rules" for how organic substances behave seemed to fit into a pattern that made for a classification system that was separate from that used for inorganic materials. Even though the large body of orderly chemical information and theory called organic chemistry is treated as a separate topic, we should not forget that the same laws of thermodynamics, bonding rules, and so forth govern both inorganic and organic chemical processes. It is also true that the distinctions between one area of chemistry and another are less well defined today than ever before. For example, organometallic chemistry is a very large area of modern chemistry that deals with compounds containing metallic elements bonded to organic molecules and groups.

In this chapter, we shall present a brief view of some of the elementary aspects of organic chemistry. We can do no more than hint at the magnitude of the subject. It has been estimated that there are now more than one million known organic substances. Each year several thousand new organic substances are discovered in nature or synthesized in the laboratory. These huge numbers might lead one to think that learning the subject of organic chemistry is hopelessly difficult. However, certain arrangements of atoms and groups of atoms, called **functional groups,** occur repeatedly in organic substances. These arrangements lead to particular chemical characteristics that are very similar among the compounds containing that functional group. Thus, by learning the characteristic chemical properties of these functional groups, you can understand the chemical characteristics of many organic substances.

24.1 the hydrocarbons

Hydrocarbons contain only two elements, carbon and hydrogen. With such a limited range of composition, you might suppose that there would be little variety in the chemical properties of the hydrocarbons. However, such is not the case. The key structural feature of hydrocarbons, and for that matter of most other organic substances, is the presence of stable carbon-carbon bonds. Carbon alone among the elements is able to form stable, extended chains of atoms bonded through single, double, or triple bonds. No other element is capable of forming similar structures.

The hydrocarbons can be divided into four groups, the **alkanes, alkenes, alkynes,** and **aromatic hydrocarbons.** We have already encountered at least one member from each of these series. Figure 24.1 shows the geometrical structure, empirical formula, and name of the simplest member that contains a carbon-carbon bond in each series.

The alkanes consist of carbon atoms bonded either to hydrogen or to other carbon atoms by four single bonds. These four bonds are directed toward the corners of a tetrahedron, in which the carbon is at the center. Depending on its place in the alkane structure, a carbon atom may be bonded to three hydrogens and one carbon, two hydrogens and two carbons, a hydrogen and three carbons, or four carbons. The alkenes are hydrocarbons with one or more carbon-carbon double bonds. The simplest member of the alkene series is ethylene; you might wish to review the electronic structure of ethylene, discussed in Section 9.3. In the alkynes, there is at least one carbon-carbon triple bond, as in the simplest member of the series, acetylene (Section 9.3). In the aromatic hydrocarbons the carbon atoms are connected in a planar ring structure, joined by both σ and π bonds between carbon atoms. Benzene is the best-known example of an aromatic hydrocarbon. Other examples are illustrated in Figure 9.13. The nonaromatic hydrocarbons, that is, the alkanes, alkenes, and alkynes, are referred to as **aliphatic** compounds, to distinguish them from aromatic substances.

The members of the different series of hydrocarbons exhibit different chemical behaviors, as we shall see shortly. However, the hydrocarbons are very similar in many ways. Because carbon and hydrogen are

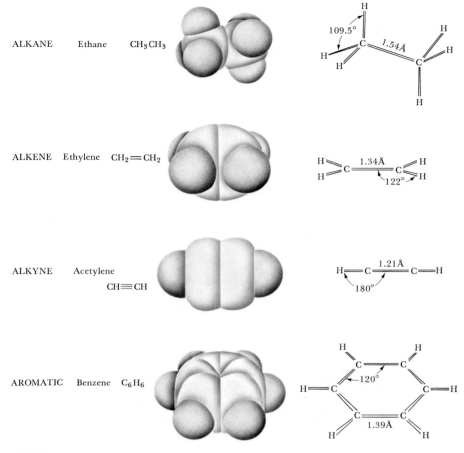

ALKANE	Ethane	CH_3CH_3
ALKENE	Ethylene	$CH_2 = CH_2$
ALKYNE	Acetylene	$CH \equiv CH$
AROMATIC	Benzene	C_6H_6

FIGURE 24.1 Names, geometrical structures, and molecular formulas of examples of each type of hydrocarbon.

not greatly different in electronegativity (2.5 for carbon, 2.2 for hydrogen), the C—H bond is not very polar. Hydrocarbons are formed entirely from C—H bonds and bonds between carbon atoms; this means that hydrocarbon molecules are relatively nonpolar. Thus, they are very much unlike water; hydrocarbons are almost completely insoluble in water. Those that are liquids are good solvents toward nonpolar molecules, but poor solvents toward ionic substances, such as sodium chloride, or polar substances, such as NH_3.

THE ALKANES

Table 24.1 lists several of the simplest alkanes. Many of these substances are familiar because of their widespread use. Methane is a major component of natural gas and is used for home heating and in gas stoves and hot-water heaters. Propane is the major component of bottled, or LP, gas used for home heating, cooking, and so forth in areas where natural gas is not available. Butane is used in the disposable lighters sold in drug stores and in the fuel cannisters for gas camping stoves and lanterns. Pentane, hexane, and heptane are found in gasolines.

The formulas for these alkanes are written in a notation called the condensed structural formula. This notation reveals the way in which atoms are bonded to one another, but does not require drawing in all the bonds. For example, the Lewis structure and abbreviated structural formula for butane, C_4H_{10}, are:

$$H-\underset{\underset{H}{|}}{\overset{\overset{H}{|}}{C}}-\underset{\underset{H}{|}}{\overset{\overset{H}{|}}{C}}-\underset{\underset{H}{|}}{\overset{\overset{H}{|}}{C}}-\underset{\underset{H}{|}}{\overset{\overset{H}{|}}{C}}-H \longrightarrow CH_3CH_2CH_2CH_3$$

Notice that each succeeding compound in the series listed in Table 24.1 is related to the one before it by the addition of a CH_2 unit. A series such as that shown in Table 24.1 is known as a **homologous series**. The general formula for all the compounds listed in the table is C_nH_{2n+2}, where n is a whole number. One of the characteristics of a homologous series is that all the compounds of the series can be described by the same general formula. We shall see several other examples of homologous series as we proceed.

The alkanes listed in Table 24.1 are called straight-chain alkanes, because all the carbon atoms are joined in a continuous chain. However, for alkanes consisting of four or more carbon atoms, other arrangements of the carbon atoms are possible. Figure 24.2 shows the Lewis structures and condensed structural formulas for all the possible structures of alkanes containing four or five carbon atoms. Notice that the two possible forms of butane have the same molecular formula, C_4H_{10}. Similarly, the three possible forms of pentane have the same molecular formula, C_5H_{12}. Compounds with the same molecular formula, but with different structures, are called **isomers**. The isomers of a given alkane differ slightly from one another in physical properties. By way of illustration, the melting and boiling points (°C) of the isomers of butane and pentane are given in Figure 24.2. The number of possible isomers increases rapidly with the number of carbon atoms in the alkane. For example, there are 18 possible isomers of octane, C_8H_{18}, and 75 possible isomers of decane, $C_{10}H_{22}$.

TABLE 24.1

first several members of the straight-chain alkane series

MOLECULAR FORMULA	NAME	BOILING POINT (°C)
CH_4	Methane	−161
CH_3CH_3	Ethane	−89
$CH_3CH_2CH_3$	Propane	−44
$CH_3CH_2CH_2CH_3$	Butane	−0.5
$CH_3CH_2CH_2CH_2CH_3$	Pentane	36
$CH_3CH_2CH_2CH_2CH_2CH_3$	Hexane	68
$CH_3CH_2CH_2CH_2CH_2CH_2CH_3$	Heptane	98
$CH_3CH_2CH_2CH_2CH_2CH_2CH_2CH_3$	Octane	125
$CH_3CH_2CH_2CH_2CH_2CH_2CH_2CH_2CH_3$	Nonane	151
$CH_3CH_2CH_2CH_2CH_2CH_2CH_2CH_2CH_2CH_3$	Decane	174

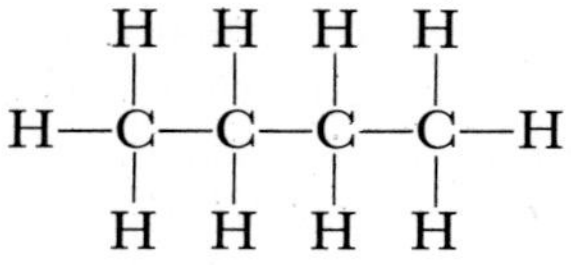

$CH_3CH_2CH_2CH_3$

n-Butane

m.p. $-135°C$
b.p. $-0.5°C$

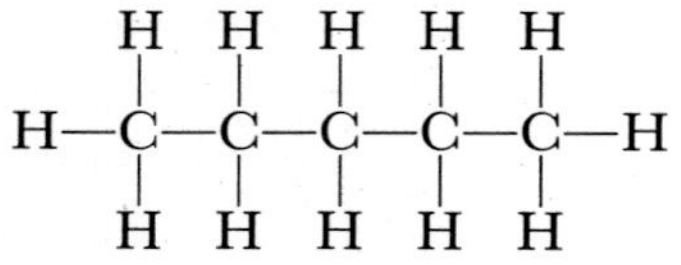

$CH_3CH_2CH_2CH_2CH_3$

n-Pentane

m.p. $-130°C$
b.p. $+36°C$

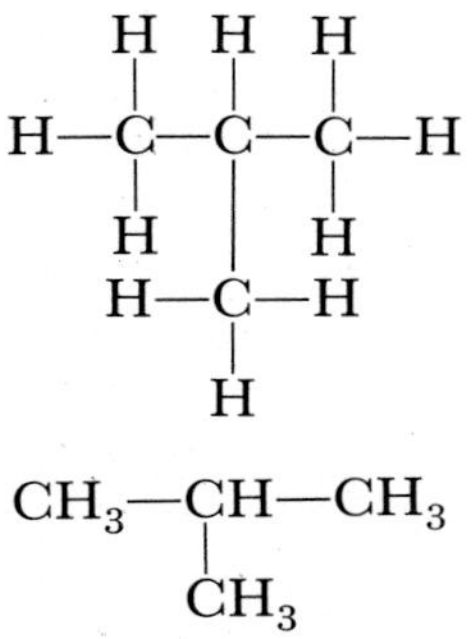

$CH_3-CH-CH_3$
$\quad\quad CH_3$

Isobutane

m.p. $-145°C$
b.p. $-10°C$

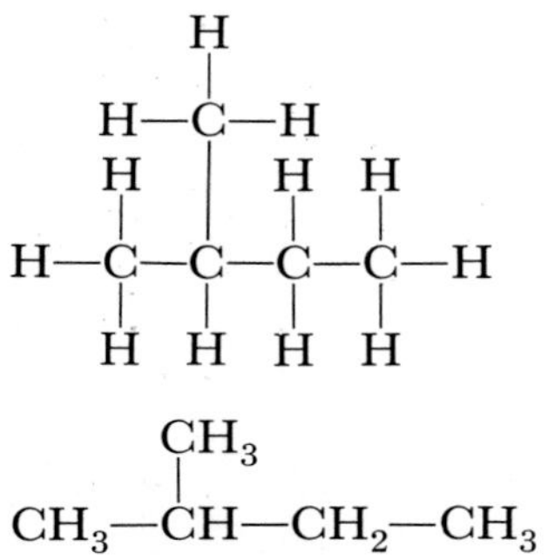

CH_3
$CH_3-CH-CH_2-CH_3$

Isopentane

m.p. $-160°C$
b.p. $+28°C$

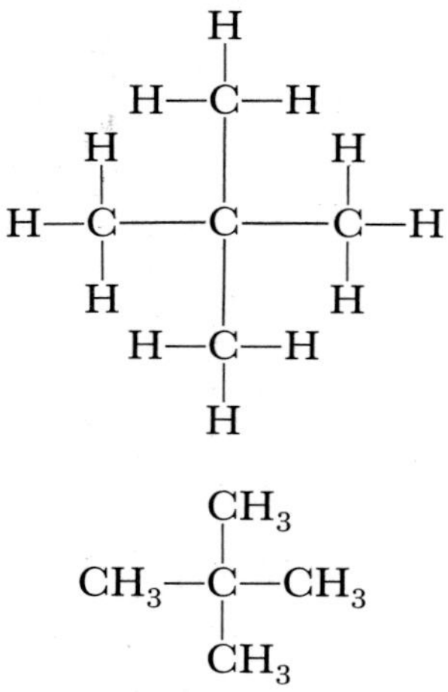

CH_3
CH_3-C-CH_3
$\quad\quad CH_3$

Neopentane

m.p. $-20°C$
b.p. $+9°C$

FIGURE 24.2 Possible structures, names, and melting and boiling points of alkanes of formula C_4H_{10} and C_5H_{12}.

NOMENCLATURE OF ALKANES

The names given to the structural isomers shown in Figure 24.2 are the so-called common names. The straight-chain isomer is referred to as the normal isomer, abbreviated by the prefix *n-*. The isomer in which one CH_3 group is branched off the major chain is labeled the iso-isomer; for example, isobutane. However, as the number of isomers grows, it becomes impossible to find a suitable prefix to denote each isomer. The need for a systematic means of naming organic compounds was recognized early in the history of organic chemistry. In 1892 an organization called the International Union of Chemistry met in Geneva, Switzerland, to formulate rules for systematic naming of organic substances. Since that time the task of keeping the rules for naming compounds up to date has fallen to the International Union of Pure and Applied Chemistry (IUPAC). It is interesting to note that through two devastating world wars and major social upheavals, the work of IUPAC has continued. Chemists everywhere, regardless of their nationality or political affiliation, subscribe to a common system for naming compounds.

The IUPAC rules for naming are most easily illustrated for the alkanes. First, the compound is named for the longest continuous chain of carbon atoms present. The name for this longest chain is that given in Table 24.1. You should memorize this list of names for the straight-chain compounds. Secondly, the locations of other groups of carbon atoms are denoted by numbering the atoms along the longest continuous chain and indicating the number of the carbon atom to which the group is attached. The numbering should be started from the end that is closer to the carbon atoms bearing the side chains. For example, the longest chain of carbon atoms in isobutane is three. This compound is thus a derivative of propane. In the IUPAC system it is called 2-methylpropane. This example raises the question of what name we give to the carbon groups attached to the longest continuous chain of carbon atoms. In the case of isobutane the group attached to the middle, or number two, carbon is CH_3, called the methyl group. But in other cases the carbon group attached along the chain may contain more than one carbon. These carbon-hydrogen groups are called radicals. We can consider them as being derived from a parent hydrocarbon by removal of a hydrogen atom. For example, the methyl group can be considered as derived from methane. Table 24.2 lists the condensed structural formulas and names of several of the more common alkyl radicals.

With the aid of Tables 24.1 and 24.2, and using the IUPAC naming rules, you should be able to name most alkanes.

SAMPLE EXERCISE 24.1

Name the following alkane:

$$CH_3—CH—CH_3$$
$$CH_3—CH—CH_2$$
$$CH_3$$

Solution: To name this compound properly, you must first find the longest continuous chain of carbon atoms. This chain, extending from the upper left CH_3 group to the lower right CH_3 group, is five carbon atoms long. The compound is thus named as a derivative of pentane. We could number the carbon atoms starting from either end. However, the IUPAC rules state that the numbering should be done so that the numbers of those carbons which bear side chains are as low as possible. This means that we should start numbering with the upper carbon. There is a methyl group on carbon number two, and one on carbon number three. The compound is thus called, 2,3-dimethylpentane.

SAMPLE EXERCISE 24.2

Write the condensed structural formula for 2-methyl-3-ethylpentane.

Solution: The longest chain of continuous carbon atoms in this compound is five. We can therefore begin by writing out a string of five C atoms:

$$C—C—C—C—C$$

We next place a methyl group on the second carbon, and an ethyl group on the middle carbon atom of the chain. Hydrogens are then added to all the other carbon atoms to make their covalencies equal to four:

$$\begin{array}{c} CH_3 \\ | \\ CH_3—CH—CH—CH_2CH_3 \\ | \\ CH_2CH_3 \end{array}$$

STRUCTURES OF ALKANES

The Lewis structures or condensed structural formulas for alkanes do not tell us anything about the three-dimensional structures of these substances. The three-dimensional structures can be shown using various kinds of models. The geometry about each carbon is tetrahedral; that is, the four groups attached to each carbon are located at the vertices of a tetrahedron about the central carbon atoms, as shown in Figure 24.3. The structure can be represented by a ball and stick model or by a model in which each bond distance from the central carbon is accurately represented by the length of the rod from the center of the tetrahedron. Models of this sort are easy to construct and show the spatial relationships between atoms quite well. Still another form of model is the so-called space-filling model, in which the atoms are represented as

TABLE 24.2

names and condensed structural formulas for several alkyl radicals

GROUP	NAME
$CH_3—$	Methyl
$CH_3CH_2—$	Ethyl
$CH_3CH_2CH_2—$	n-Propyl
$CH_3CH_2CH_2CH_2—$	n-Butyl
$\begin{array}{c} CH_3 \\ \| \\ HC— \\ \| \\ CH_3 \end{array}$	Isopropyl
$\begin{array}{c} CH_3 \\ \| \\ CH_3—C— \\ \| \\ CH_3 \end{array}$	t-Butyl

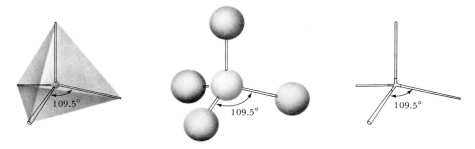

FIGURE 24.3 Representations of the three-dimensional arrangement of bonds about carbon in alkanes.

spheres. The stick and space-filling models for the alkane ethane are illustrated in Figure 24.1. Similar models for propane are shown in Figure 24.4.

One of the characteristics of the carbon-carbon single bond is that rotation about this bond is relatively free. You might imagine grasping the top left methyl group in Figure 24.4 and twisting it relative to the rest of the structure. Motion of this sort occurs very rapidly in alkanes at room temperature. Thus, a long-chain alkane is constantly undergoing motions that cause it to change its shape, something like a length of chain that is being shaken.

One possible structural form for alkanes is that in which the carbon chain forms a ring, or cycle. Alkanes with this form of structure are called **cycloalkanes.** A few examples of cycloalkanes are shown in Figure 24.5. The cycloalkane structures are sometimes drawn as simple polygons, as illustrated in Figure 24.5. In this shorthand notation, each corner of the polygon represents a CH_2 group. This method of representation is similar to that used for aromatic rings, as illustrated in Figure 9.13. In the case of the aromatic structures, each corner represents a CH group.

Carbon rings containing less than six carbon atoms are strained, because the C—C—C bond angle in the smaller rings must be less than the 109.5° tetrahedral angle. The amount of strain increases as the rings get smaller. In cyclopropane, which has the shape of an equilateral triangle, the angle is only 60°; this molecule is therefore much more reactive than either propane, its straight-chain analogue, or cyclohexane, which has no ring strain. Note that the empirical formula for the cycloalkanes is C_nH_{2n}, which is different than for the straight-chain alkanes. The cycloalkanes thus form a separate homologous series.

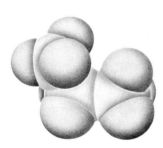

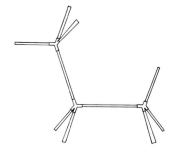

FIGURE 24.4 Three-dimensional models for propane, C_3H_8.

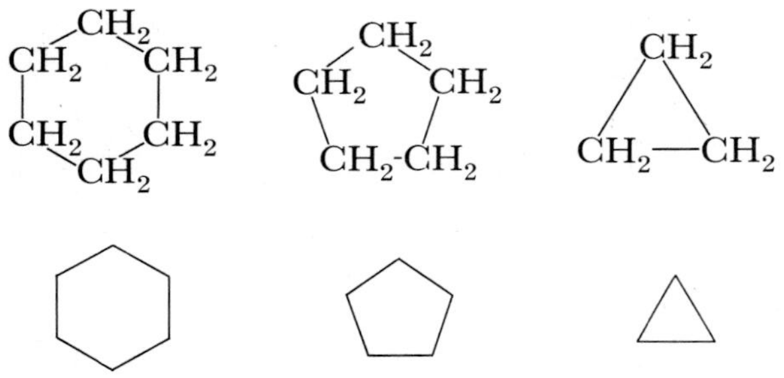

Cyclohexane   Cyclopentane   Cyclopropane   **FIGURE 24.5** Condensed structural formulas of three cycloalkanes.

ALKENES

The alkenes are close relatives of the alkanes. They differ in that there is at least one carbon-carbon double bond in the molecule. Alkenes are sometimes referred to as **olefins.** The presence of a double bond results in two fewer hydrogens than would be present in an alkane. Because the alkenes possess fewer hydrogens than are needed to form the alkane, they are said to be **unsaturated.** As we shall see a little later, the presence of the double bond confers considerably more chemical reactivity on the alkenes than is found in the alkanes. The simplest alkene is C_2H_4, called ethene, or ethylene. The next member of the series is $CH_3—CH{=}CH_2$, called propene or propylene. When there are more than three carbon atoms in the molecule, there are several possibilities for forming isomers. For example, the possible alkenes with four carbon atoms, and with molecular formula C_4H_8, are shown in Figure 24.6. The first compound shown has a branched chain; the other three all have continuous chains of four carbon atoms. In naming alkenes, the compound is named for the length of the longest continuous chain of carbon atoms, modifying the ending of the name as listed in Table 24.1 from *-ane* to *-ene.* The location of the double bond is indicated by a prefix number that designates the number of the carbon atom that is part of the double bond and is nearest an end of the chain. Thus, the compound on the left in Figure 24.6 is called a propene, because the longest continuous chain length is three carbons. The location of the double bond is obvious in this compound. The placement of the methyl group is designated by a prefix numeral; thus, the name is 2-methylpropene. The other three compounds are named as butenes, as shown.

Cis- and *trans*-2-butene are **geometrical isomers;** that is, they are compounds that have the same molecular formula and the same groups bonded to one another, but that differ in the spatial arrangement of those

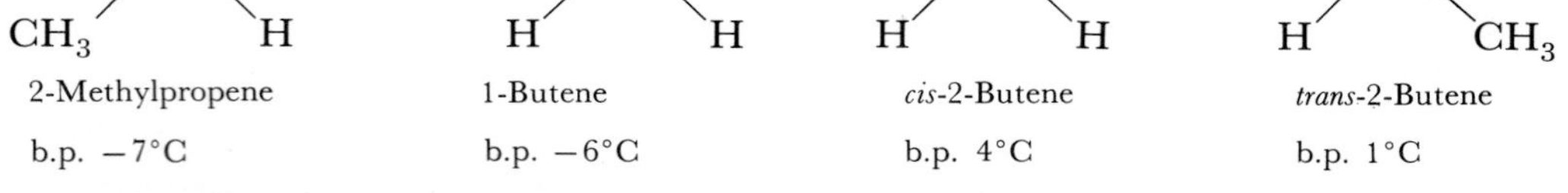

2-Methylpropene   1-Butene   *cis*-2-Butene   *trans*-2-Butene

b.p. $-7°C$ b.p. $-6°C$ b.p. $4°C$ b.p. $1°C$

FIGURE 24.6 Structures, names, and boiling points of the alkenes with molecular formula C_4H_8.

groups. In the *cis* form, the two methyl groups are on the same side of the double bond; in the *trans* form they are on opposite sides. We have already met examples of geometrical isomerism, in the chemistry of transition metal coordination compounds (Section 23.4). Geometrical isomers possess distinct physical properties and may even differ significantly in their chemical behavior in certain circumstances.

Recall from our earlier discussion of the geometry about carbon (Section 9.3) that the double bond between two carbon atoms consists of a σ and π part. Figure 24.7 shows the geometrical arrangement as in a *cis* alkene. The bonding arrangement around each carbon is planar; that is, the carbon-carbon bond axis and the bonds to the other two groups, either hydrogen or carbon, are all in a plane. It is easy to see from Figure 24.7 that rotation about the carbon-carbon double bond will not be easy. Such a rotation would cause the p orbitals which form the π bond to lose their overlap, and the π bond would be destroyed. Because rotation about the carbon-carbon bond is difficult the *cis* and *trans* isomers can be separated and studied. If the interconversion were easy, we would always have an equilibrium mixture of the two forms rather than the pure isomers.

SAMPLE EXERCISE 24.3

Name the following compound:

$$CH_3CH_2CH_2-\overset{\overset{\displaystyle CH_3}{\displaystyle |}}{CH}\overset{\displaystyle CH_3}{\underset{\displaystyle H}{\diagdown}}C=C\overset{\displaystyle CH_3}{\underset{\displaystyle H}{\diagup}}$$

Solution: The longest continuous chain of carbon atoms in this compound is seven in length. Because it possesses a double bond it is an alkene. Thus, the compound is a heptene. The double bond begins at carbon atom number two; thus the name is 2-heptene. Continuing the numbering along the chain, a methyl group is bound at carbon atom number four. Thus, the compound is 4-methyl-2-heptene. Finally, we note that the geometrical configuration at the double bond is *cis;* that is, the alkyl groups are bonded to the double bond on the same side. Thus, the full name is 4-methyl-*cis*-2-heptene.

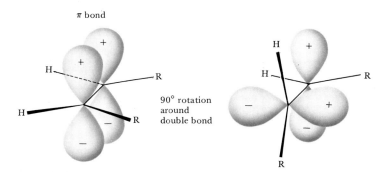

FIGURE 24.7 Schematic illustration of rotation about a carbon-carbon double bond in an alkene. The overlap of the p orbitals that form the π bond is lost in the rotation. For this reason, rotation about the carbon-carbon double bonds does not occur readily.

A comparison of alkenes with cycloalkanes affords another case of isomerism. As an example, a substance with molecular formula C_4H_8 could have one of the structures shown in Figure 24.6, or the following cyclic structure:

$$\begin{array}{ccc} CH_2 & — & CH_2 \\ | & & | \\ CH_2 & — & CH_2 \end{array}$$

This substance, cyclobutane, is a structural isomer of the butenes shown in Figure 24.6.

In substances containing two alkene functional groups, each must be located by a number. The ending of the name is altered to identify the number of functional groups: diene (two double bonds), triene (three double bonds), and so forth. For example, 1,4-pentadiene is $CH_2{=}CH{-}CH_2{-}CH{=}CH_2$.

ALKYNES

The alkynes are yet another series of unsaturated hydrocarbons, in which there is one or more carbon-carbon triple bond between carbon atoms. The empirical formula for simple alkynes is C_nH_{2n-2}. The simplest alkyne, acetylene, is a highly reactive molecule. When acetylene is burned in a stream of oxygen in the so-called oxyacetylene torch, the flame reaches a very high temperature, about 3200 K (Section 21.3). The oxyacetylene torch is widely used in welding, where high temperatures are required. Alkynes in general are highly reactive molecules. Because of their higher reactivity they are not as widely distributed in nature as the alkenes; however, they are important intermediates in many industrial processes.

The alkynes are named by identifying the longest continuous chain in the molecule containing the triple bond, and modifying the ending of the name as listed in Table 24.1 from *-ane* to *-yne*, as shown in the following sample exercise.

SAMPLE EXERCISE 24.4

Name the following compounds:

(a) $CH_3CH_2CH_2{-}C{\equiv}C{-}CH_3$

(b) $CH_3CH_2CH_2CH{-}C{\equiv}CH$
$\qquad\qquad\quad |$
$\qquad\qquad CH_2CH_2CH_3$

Solution: In (a) the longest chain of carbon atoms is six. There are no side chains. The triple bond begins at carbon atom number two (remember we always arrange the numbering so that the smallest possible number is assigned to the carbon containing the multiple bond.) Thus, the name is 2-hexyne.

In (b) the longest continuous chain of carbon atoms is seven; but because this chain does not contain the triple bond we do not count it as derived from heptane. The longest chain containing the triple bond is six, so this compound is named as a derivative of hexyne, 3-propyl-1-hexyne.

AROMATIC HYDROCARBONS

The aromatic hydrocarbons are a large and important class of hydrocarbons. The simplest member of the series is benzene (see Figure 24.1), with molecular formula C_6H_6. As we have already noted, benzene is a planar, highly symmetrical molecule. The molecular formula for benzene suggests a high degree of unsaturation. One might thus expect that benzene would be highly reactive, and that it might resemble the unsaturated hydrocarbons in reactivity. In fact, however, benzene is not at all similar to alkenes or alkynes in chemical behavior. The great stability of benzene and the other aromatic hydrocarbons as compared with alkenes and alkynes is due to stabilization of the π electrons through delocalization in the π orbitals (Section 9.3).

We can obtain an estimate of the stabilization of the π electrons in benzene by comparing the energy involved in adding hydrogen to benzene to form a saturated compound, as compared with that involved in hydrogenating simple alkenes. The hydrogenation of benzene to form cyclohexane can be represented as

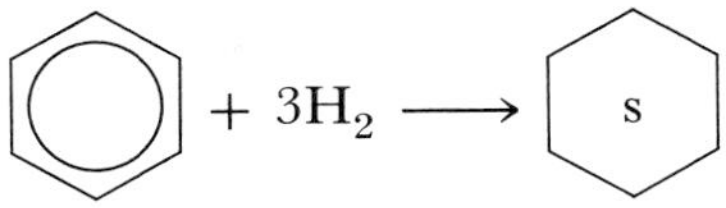

$$\text{(benzene)} + 3H_2 \longrightarrow \text{(cyclohexane, s)} \qquad \Delta H° = -208 \text{ kJ/mole} \qquad [24.1]$$

(The s in the ring on the right indicates that it is a cycloalkane, with CH_2 groups at each corner.) The enthalpy change in this reaction is -208 kJ/mole. The heat of hydrogenation of the cyclic alkene cyclohexene, is -120 kJ/mole:

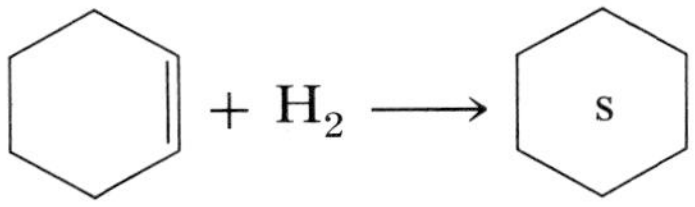

$$\text{(cyclohexene)} + H_2 \longrightarrow \text{(s)} \qquad \Delta H° = -120 \text{ kJ/mole} \qquad [24.2]$$

Cyclohexene

Similarly, the heat released on hydrogenating 1,4-cyclohexadiene is -232 kJ/mole:

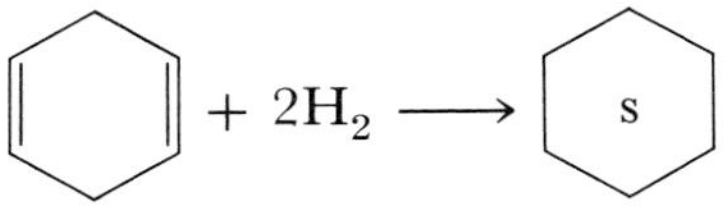

$$\text{(1,4-cyclohexadiene)} + 2H_2 \longrightarrow \text{(s)} \qquad \Delta H° = -232 \text{ kJ/mole} \qquad [24.3]$$

1,4-Cyclo-
hexadiene

From these last two reactions it would appear that the heat of hydrogenating a double bond is about 116 kJ/mole, for each bond. There is the equivalent of three double bonds in benzene. Thus we might expect that the heat of hydrogenating benzene would be about three times -116, or -348 kJ/mole, if benzene behaved as though it were "cyclohexatriene," that is, if it behaved as though it were three double bonds in a ring. Instead, the heat released is much less than this, indicating that benzene is more stable than would be expected for three double bonds. The difference of 140 kJ/mole between -348 kJ/mole

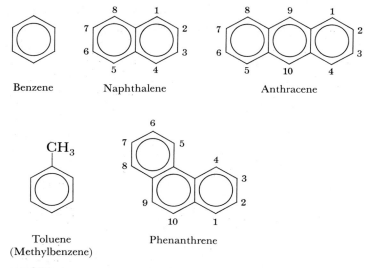

Benzene   Naphthalene   Anthracene

Toluene
(Methylbenzene)   Phenanthrene

FIGURE 24.8 Structures, names, and numbering system of several aromatic compounds.

and the observed heat of hydrogenation, -208 kJ/mole, can be ascribed to stabilization of the π electrons through delocalization in the π orbitals that extend around the ring.

There is no widely used systematic nomenclature for naming aromatic rings. Each ring system is given a common name; several aromatic compounds are shown in Figure 24.8. The aromatic rings are represented by hexagons with a circle inscribed inside to denote aromatic character. Each corner represents a carbon atom. Each carbon is bound to three other atoms—either three carbons or two carbons and a hydrogen. The hydrogen atoms are not shown. In naming derivatives of the aromatic hydrocarbons, it is often necessary to indicate the position in the aromatic ring at which some side chain or other group is located. For this purpose the numbering shown in Figure 24.8 is used.

24.2
petroleum

Petroleum is a complex mixture of organic compounds, mainly hydrocarbons, with smaller quantities of other organic compounds containing nitrogen, oxygen, or sulfur. Petroleum is formed over a period of millions of years by the decomposition of marine plants and animals.

The usual first step in the **refining,** or processing, of petroleum is separation of the crude oil into fractions on the basis of boiling points. The fractions commonly taken are shown in Table 24.3. As you might expect, the fractions that boil at higher temperatures are made up of molecules with larger numbers of carbon atoms per molecule. The fractions collected in the initial separation may require further processing to yield a usable product. For example, modifications must be made to the straight-run gasoline obtained from fractionation of petroleum to render it suitable for use as a fuel in automobile engines. Similarly, the fuel oil fraction may need additional processing to remove sulfur before it is suitable for use in an electrical power station or a home heating system.

At present, the most commercially important single product from petroleum refining is gasoline.

GASOLINE

Gasoline is a mixture of volatile hydrocarbons. Depending on the source of the crude oil, it may contain varying amounts of cyclic alkanes and aromatic hydrocarbons in addition to alkanes. Straight-run gasoline consists mainly of straight-chain hydrocarbons, which in general are not very suitable for use as fuel in an automobile engine. In an automobile engine, a mixture of air and gasoline vapor is ignited by the spark plug at the moment when the gas mixture inside the cylinder has been compressed by the piston. The burning of the gasoline should create a strong, smooth expansion of gas in the cylinder, forcing the piston outward and imparting force along the drive shaft of the engine. If the gas burns too rapidly, the piston receives a single hard slam rather than a strong, smooth push. The result is a "knocking" or pinging sound; the efficiency with which the energy of gasoline combustion is converted to power is reduced.

Gasolines are rated according to **octane number**. Gasolines with high octane numbers burn more slowly and smoothly, and thus are more effective fuels, especially in engines in which the gas-air mixture is highly compressed. It happens that the more highly branched alkanes have higher octane numbers than the straight-chain compounds; some examples are shown in Table 24.4. Because straight-run gasoline contains mostly straight-chain hydrocarbons, it has a low octane number. It is therefore subjected to a process called **cracking** to convert the straight-chain compounds into more desirable branched-chain molecules. Cracking is also used to convert some of the less volatile kerosene and fuel oil fractions into compounds with lower molecular weights that are suitable for use as automobile fuel. In the cracking process, the hydrocarbons are mixed with a catalyst and heated to 400 to 500°C. The catalysts used are naturally occurring clay minerals, or synthetic Al_2O_3–SiO_2 mixtures. In addition to forming molecules more suitable for gasoline, cracking results in the formation of hydrocarbons of lower

TABLE 24.3
hydrocarbon fractions from petroleum

FRACTION	SIZE RANGE OF MOLECULES	BOILING-POINT RANGE (°C)	USES
Gas	C_1–C_5	−160 to 30	Gaseous fuel, production of H_2
Straight-run gasoline	C_5–C_{12}	30 to 200	Motor fuel
Kerosene, fuel oil	C_{12}–C_{18}	180 to 400	Diesel fuel, furnace fuel, cracking
Lubricants	C_{16} and up	350 and up	Lubricants
Paraffins	C_{20} and up	Low melting solids	Candles, matches
Asphalt	C_{36}	Gummy residues	Surfacing roads, fuel

TABLE 24.4

octane numbers of some C$_7$ and C$_8$ hydrocarbons

NAME	CONDENSED STRUCTURAL FORMULA	OCTANE NUMBER
n-Heptane	$CH_3CH_2CH_2CH_2CH_2CH_2CH_3$	0
2-Methylhexane	$CH_3CH_2CH_2CH_2—CH—CH_3$ $\qquad\qquad\qquad\quad CH_3$	40
Methylcyclohexane	(ring structure of methylcyclohexane)	75
2,3-Dimethylpentane	$\qquad\qquad\quad CH_3$ $CH_3CH_2—CH—CH—CH_3$ $\qquad\qquad CH_3$	90
2,2,4-Trimethylpentane	$\qquad\qquad\qquad\quad CH_3$ $CH_3—CH—CH_2—C—CH_3$ $\qquad CH_3\qquad\quad CH_3$	100

molecular weight, such as ethylene and propene. These are used in a variety of processes to form plastics and other chemicals.

The octane number of a given blend of hydrocarbons can be improved by adding an **antiknock agent,** a substance that helps control the burning rate of the gasoline. The most widely used substances for this purpose are tetraethyl lead, $(CH_3CH_2)_4Pb$, and tetramethyl lead, $(CH_3)_4Pb$. The premium gasolines contain 2 or 3 ml of one of these lead compounds per gallon, with a resultant increase of 10 to 15 in octane rating. Although alkyl lead compounds are undoubtedly effective in improving gasoline performance, their use in gasolines has been drastically curtailed because of the environmental hazards associated with lead. This metal is highly toxic, and there is good evidence that the lead released from automobile exhausts is a general health hazard. Although other substances have been tried as antiknock agents in gasolines, none of these has proved to be an effective and inexpensive antiknock agent that is environmentally safe. The 1975 and later model cars are designed to operate with unleaded gasolines. The gasolines blended for these cars are made up of more highly branched components and more aromatic components, because these have relatively high octane ratings.

24.3
reactions of
hydrocarbons

The hydrocarbons are capable of undergoing a variety of reactions with other substances. Many of these reactions are of considerable importance in the chemical industry, because they lead to useful products or to substances that can in turn be converted to useful products.

OXIDATION

The most common oxidation reactions of hydrocarbons are those that result in complete oxidation to form carbon dioxide and water:

$$CH_3CH_2CH_3 + 5O_2 \longrightarrow 3CO_2 + 4H_2O \qquad [24.4]$$

$$CH_3CH{=}CH_2 + \frac{9}{2}O_2 \longrightarrow 3CO_2 + 3H_2O \qquad [24.5]$$

$$CH_3C{\equiv}CH + 4O_2 \longrightarrow 3CO_2 + 2H_2O \qquad [24.6]$$

$$C_6H_6 + \frac{15}{2}O_2 \longrightarrow 6CO_2 + 3H_2O \qquad [24.7]$$

These reactions are all highly exothermic. Combustion of the hydrocarbons results in release of energy that can be used to propel an auto or an airplane, to generate steam in the boiler of an electric power plant, or to heat a building. The tremendous demand for petroleum to meet the world's energy needs has led to the tapping of oil wells in such forbidding places as the North Sea and the North Slope of the continent in Alaska, facing the Arctic Ocean (Figure 24.9).

Controlled oxidation of hydrocarbons yields organic substances that contain oxygen in addition to carbon and hydrogen. These products of controlled oxidations will be considered later when we discuss hydrocarbon derivatives.

ADDITION REACTIONS

Under appropriate conditions, it is possible to add atoms or groups of atoms to alkenes, alkynes, or aromatic hydrocarbons by disrupting the carbon-carbon multiple bonds. For example, ethylene reacts with bromine to form 1,2-dibromoethane, as in Equation [24.8].

FIGURE 24.9 Drilling for oil on the North Slope in Alaska. (*Courtesy of Exxon Corp.*)

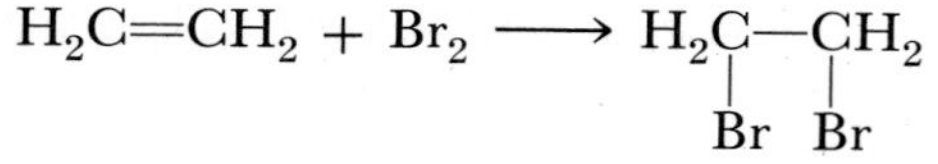

$$H_2C=CH_2 + Br_2 \longrightarrow H_2C-CH_2 \quad [24.8]$$
$$\underset{Br \ \ Br}{|\ \ \ |}$$

The pair of electrons that form the π bond in ethylene are uncoupled and are used to form two new bonds to the two bromine atoms. The σ bond between the carbon atoms remains.

Addition of halogens to alkynes also occurs readily, as in the example shown in Equation [24.9].

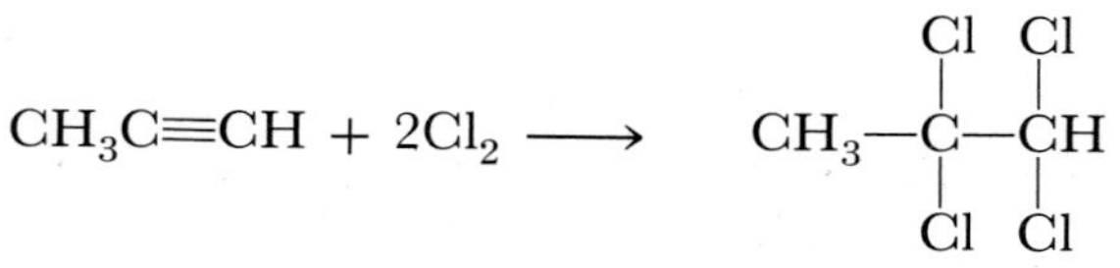

$$CH_3C\equiv CH + 2Cl_2 \longrightarrow CH_3-\overset{Cl}{\underset{Cl}{\overset{|}{C}}}-\overset{Cl}{\underset{Cl}{\overset{|}{CH}}} \quad [24.9]$$

1,1,2,2,-Tetrachloropropane

Reaction of alkenes or alkynes with hydrogen halides also results in addition:

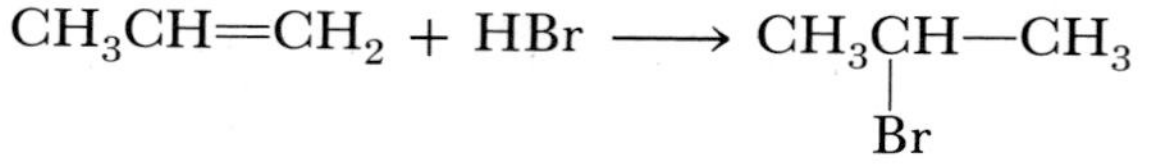

$$CH_3CH=CH_2 + HBr \longrightarrow CH_3\underset{Br}{\overset{|}{CH}}-CH_3 \quad [24.10]$$

Notice that this reaction might have proceeded to give another product, in which the bromine atom is on the end carbon. It turns out that when a hydrogen halide adds to an alkene, the more electronegative halogen atom always ends up on the carbon atom of the double bond that has the fewer hydrogen atoms. This rule, known as Markovnikoff's rule, was first formulated by the Russian chemist V. V. Markovnikoff.

SAMPLE EXERCISE 24.5

Predict the product of reaction of HCl with the following compound:

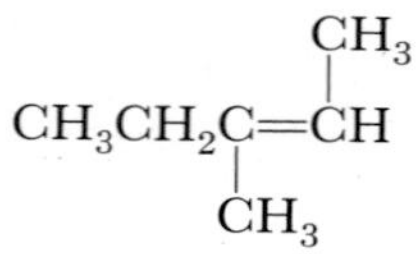

$$CH_3CH_2\underset{CH_3}{\overset{CH_3}{\overset{|}{C}}}=CH$$

Solution: The double-bonded carbon atom on the right has one hydrogen on it, the other has none. According to Markovnikoff's rule, the chloride of HCl will end up on the carbon atom on the left:

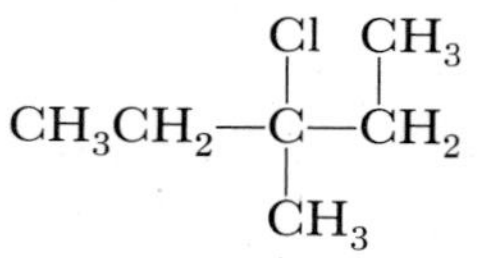

$$CH_3CH_2-\overset{Cl}{\underset{CH_3}{\overset{|}{C}}}-\overset{CH_3}{CH_2}$$

Markovnikoff's rule also applies to the addition of unsymmetrical reagents to alkynes. For example, reaction of HBr with 1-butyne yields 2,2-dibromobutane:

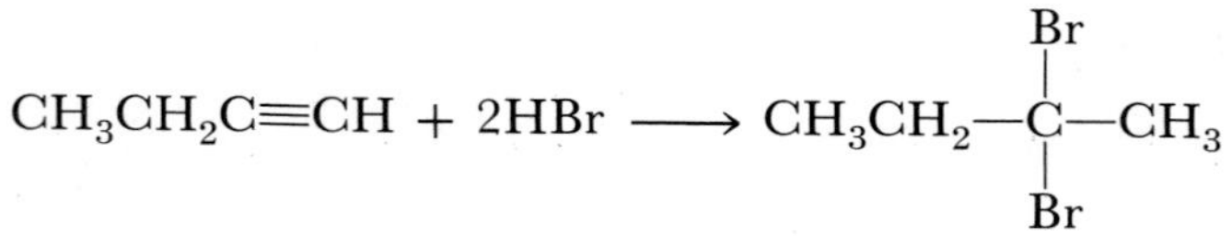

$$CH_3CH_2C\equiv CH + 2HBr \longrightarrow CH_3CH_2-\overset{Br}{\underset{Br}{\overset{|}{C}}}-CH_3 \quad [24.11]$$

Using an acid such as H_2SO_4 as catalyst, it is possible to add H_2O to a double bond. The products of such reactions are **alcohols;** that is, compounds containing an OH group bonded to carbon. Markovnikoff's rule applies here as well; we consider the water molecule as being polarized as follows: H^+—OH^-. Thus, the OH group adds to the carbon atom with fewer hydrogens:

$$CH_3CH{=}CH_2 + HOH \xrightarrow{\text{H}_2\text{SO}_4} CH_3{-}\overset{\displaystyle \overset{OH}{|}}{C}H{-}CH_3 \qquad [24.12]$$

Propene   2-Propanol

Addition of H_2 to an alkene converts it into an alkane. This reaction, referred to as hydrogenation, does not occur readily under ordinary temperature and pressure conditions. One of the reasons for the lack of reactivity of H_2 toward an alkene is the high bond energy of the H_2 bond. To promote the reaction it is necessary to use a catalyst that assists in the rupture of the H—H bond. The most widely used catalysts are heterogeneous and consist of finely divided metals on which H_2 is adsorbed. The action of these heterogeneous catalysts in reaction of H_2 with an alkene is described in detail in Section 13.7. Molecular hydrogen also reacts with alkynes in the presence of a catalyst to yield alkanes, as in the example of Equation [24.13].

$$CH_3C{\equiv}CH + 2H_2 \xrightarrow{\text{Ni}} CH_3CH_2CH_3 \qquad [24.13]$$

We have so far not mentioned addition reactions of aromatic compounds. Such reactions proceed much less readily than do reactions with alkenes or alkynes. For example, benzene does not react at all with Cl_2 or Br_2 under ordinary conditions. Under very rigorous conditions, that is, in the presence of ultraviolet light, benzene reacts with chlorine to give hexachlorocyclohexane:

1,2,3,4,5,6-Hexachlorocyclohexane

Similarly, under high hydrogen pressure and with a catalyst, it is possible to hydrogenate benzene to cyclohexane:

Benzene   Cyclohexane

ADDITION POLYMERIZATION

One of the most important addition reactions of alkenes is **addition polymerization**. In a polymerization reaction, small molecules called **monomers** react with one another to form long, chainlike molecules of high molecular weight called **polymers**. For example, 2-methylpropene (common name, isobutylene) reacts with itself in the presence of a small amount of acid catalyst to form polyisobutylene, which may contain more than a thousand isobutylene units:

$$
n\text{CH}_2{=}\overset{\displaystyle \text{CH}_3}{\underset{\displaystyle \text{CH}_3}{\text{C}}} \xrightarrow{\text{acid}} -\text{CH}_2-\overset{\displaystyle \text{CH}_3}{\underset{\displaystyle \text{CH}_3}{\text{C}}}-\text{CH}_2-\overset{\displaystyle \text{CH}_3}{\underset{\displaystyle \text{CH}_3}{\text{C}}}-\text{CH}_2-\overset{\displaystyle \text{CH}_3}{\underset{\displaystyle \text{CH}_3}{\text{C}}}-\text{CH}_2-\overset{\displaystyle \text{CH}_3}{\underset{\displaystyle \text{CH}_3}{\text{C}}}- \ldots \quad [24.14]
$$

Butylene   Polyisobutylene

The product on the right in this reaction is a rubbery, rather gummy material that can be further treated to form a wide variety of useful products. In its finished form, it is known as butyl rubber.

Other substances besides acids can catalyze the polymerization of molecules such as isobutylene. One of the most important types of catalyst is the **free radical.** A free radical is a species with an odd number of electrons. There is thus an atom in the radical that does not have its full complement of electrons. As a result, the radical is quite reactive. The simplest examples of free radicals are single atoms such as a chlorine atom, which is one electron short of completing its valence shell. Free radicals can be formed by rupture of chemical bonds. For example, peroxides, which contain O—O bonds, decompose on heating, with rupture of the O—O bond. The result is a pair of free radicals:

$$
\text{R}-\overset{..}{\underset{..}{\text{O}}}-\overset{..}{\underset{..}{\text{O}}}-\text{R} \rightleftharpoons 2\text{R}-\overset{..}{\underset{..}{\text{O}}}\cdot \qquad [24.15]
$$

In this equation R could be any of a number of different organic groups. By placing a small amount of a peroxide in contact with an alkene and then heating it, radicals that can catalyze addition reactions are generated. For example, consider the polymerization of ethylene initiated (that is, started) by a radical formed from a peroxide:

$$
\text{R}-\overset{..}{\underset{..}{\text{O}}}-\overset{..}{\underset{..}{\text{O}}}-\text{R} \rightleftharpoons 2\text{R}-\overset{..}{\underset{..}{\text{O}}}\cdot
$$

$$
\text{R}-\overset{..}{\underset{..}{\text{O}}}\cdot + \text{H}_2\text{C}{=}\text{CH}_2 \longrightarrow \text{H}-\overset{\displaystyle \text{H}}{\underset{\displaystyle \text{OR}}{\text{C}}}-\overset{\displaystyle \text{H}}{\underset{\displaystyle \text{H}}{\text{C}}}\cdot \qquad [24.16]
$$

In Equation [24.16], the R—O radical reacts with a molecule of ethylene, forming a bond to one of the carbons. For this to occur, the pair of electrons forming the π bond in ethylene must have uncoupled. One of them is involved in the bond to the O—R group, the other is on the second carbon. Thus, the product of this reaction is itself a free radical. This species goes on to react with a second molecule of ethylene:

$$H-\overset{\displaystyle H}{\underset{\displaystyle OR}{C}}-\overset{\displaystyle H}{\underset{\displaystyle H}{C}}\cdot \;+\; H_2C{=}CH_2 \longrightarrow RO-\overset{\displaystyle H}{\underset{\displaystyle H}{C}}-\overset{\displaystyle H}{\underset{\displaystyle H}{C}}-\overset{\displaystyle H}{\underset{\displaystyle H}{C}}-\overset{\displaystyle H}{\underset{\displaystyle H}{C}}\cdot \qquad [24.17]$$

By means of such successive reactions, the polymer grows in length. A reaction of this type is known as a chain reaction, because it is self-propagating. However, chain-termination reactions might occur. If two radicals come into contact, their unpaired spins might couple, and no further reaction would occur. In forming polymers by such free-radical reactions, we must control conditions carefully so that a polymer chain with the desired average length results.

In addition to acids and free radicals, other substances may serve as catalysts for addition polymerization reactions of alkenes. The choice of catalyst is important, because it influences various characteristics of the resulting polymer. For example, polyethylene (see Section 11.5), formed by polymerizing ethylene, can be a pliable material with a low melting point or a much stiffer substance with a higher melting point, depending on the nature of the catalyst. Many alkenes derived from ethylene by replacing one or more hydrogens with other groups can be polymerized to form a variety of useful materials. Some examples are listed in Table 24.5.

SUBSTITUTION REACTIONS

In a substitution reaction of a hydrocarbon, one or more hydrogen atoms are replaced by other atoms or groups. Substitution reactions are difficult to carry out with aliphatic compounds. One of the most important substitution reactions of alkanes involves replacement of hydrogen by a halogen atom. The chlorination of an alkane is a photo-initiated reaction. That is, it requires the use of light, which causes dissociation of the Cl_2 molecule, forming reactive chlorine atoms. A chlorine atom then attacks the alkane, removing a hydrogen and forming HCl and an alkyl radical.

TABLE 24.5

alkenes that undergo addition polymerization

MONOMER FORMULA	MONOMER NAME	POLYMER NAME	USES
$CH_2{=}CH_2$	Ethylene	Polyethylene	Coating for milk cartons, wire insulation, plastic bags
$CF_2{=}CF_2$	Tetrafluoroethylene	Teflon	Insulation, bearings, frying-pan surfaces
$CH_2{=}\underset{\displaystyle Cl}{C}H$	Vinyl chloride	Polyvinyl chloride (PVC)	Phonograph records, rain-wear, piping
$CH_2{=}\underset{\displaystyle CN}{C}H$	Acrylonitrile	Polyacrylonitrile (Orlon)	Rug fibers
$CH_2{=}\underset{\displaystyle C_6H_5}{C}H$	Vinyl benzene (styrene)	Polystyrene	TV lead-in wire, combs

The alkyl radical then attacks a Cl_2 molecule, forming an alkyl halide and a chlorine atom:

$$Cl_2 \xrightarrow{\text{light}} 2Cl \cdot \qquad [24.18]$$

$$Cl \cdot + CH_3CH_3 \longrightarrow HCl + CH_3CH_2 \cdot \qquad [24.19]$$

$$CH_3CH_2 \cdot + Cl_2 \longrightarrow CH_3CH_2Cl + Cl \cdot \qquad [24.20]$$

The chlorine atom formed in Equation [24.20] reacts with more ethane, continuing the cycle. Thus, for each quantum of light absorbed by a chlorine molecule, many molecules of ethyl chloride may be formed. This reaction provides an example of a **radical chain** process. One of the disadvantages of radical chain reactions such as this is that they are not very selective. As the concentration of ethyl chloride builds up in the reaction, chlorine atoms may abstract hydrogen atoms from it, so that eventually dichloroethane and even more highly chlorinated molecules may be formed. Thus, several products are formed in the reaction, and these must be carefully separated by distillation or some other separation procedure.

Substitution into alkenes or alkynes is difficult to carry out, because the presence of the reactive double or triple bond usually leads to addition reactions rather than substitution. By contrast, substitution into aromatic hydrocarbons is relatively easy. For example, when benzene is warmed in a mixture of nitric and sulfuric acid, hydrogen is replaced by the nitro group, NO_2:

$$\bigcirc + HNO_3 \xrightarrow{H_2SO_4} \bigcirc\text{--}NO_2 + H_2O \qquad [24.21]$$

More vigorous treatment results in substitution of a second nitro group into the molecule:

$$\bigcirc\text{--}NO_2 + HNO_3 \longrightarrow \bigcirc\langle {}^{NO_2}_{NO_2} + H_2O \qquad [24.22]$$

There are three possible isomers of benzene with two nitro groups attached. These three isomers are named *ortho*-, *meta*-, and *para*-dinitrobenzene:

ortho-Dinitrobenzene
m.p. 118°C

meta-Dinitrobenzene
m.p. 90°C

para-Dinitrobenzene
m.p. 174°C

Only the *meta* isomer is formed in the reaction of nitric acid with nitrobenzene.

Bromination of benzene is carried out using $FeBr_3$ as a catalyst:

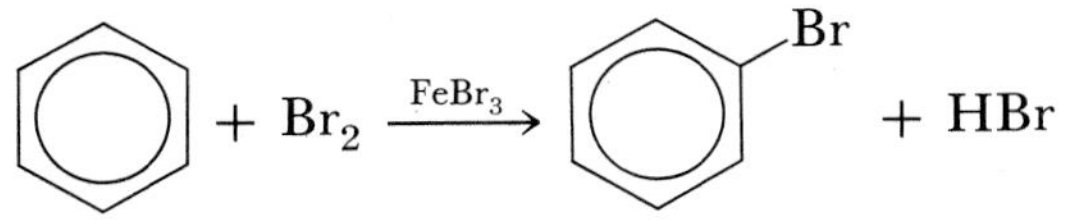

$$\text{benzene} + Br_2 \xrightarrow{\ FeBr_3\ } \text{bromobenzene(Br)} + HBr \qquad [24.23]$$

In a similar reaction, called the **Friedel-Crafts reaction,** alkyl groups can be substituted onto an aromatic ring by reaction of an alkyl halide with an aromatic compound in the presence of $AlCl_3$:

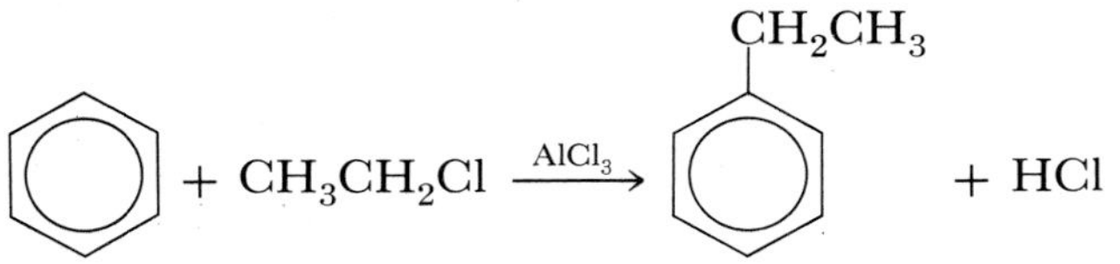

$$\text{benzene} + CH_3CH_2Cl \xrightarrow{\ AlCl_3\ } \text{(}CH_2CH_3\text{)} + HCl \qquad [24.24]$$

The substitution reactions of aromatic compounds occur via attack of a positively charged reagent on the ring. Substances such as H_2SO_4, $FeCl_3$ or $AlCl_3$ serve as catalysts by generating the positively charged species. For example, the bromination of benzene proceeds via the following steps:

$$FeBr_3 + Br_2 \rightleftharpoons FeBr_4^- + Br^+ \qquad [24.25]$$

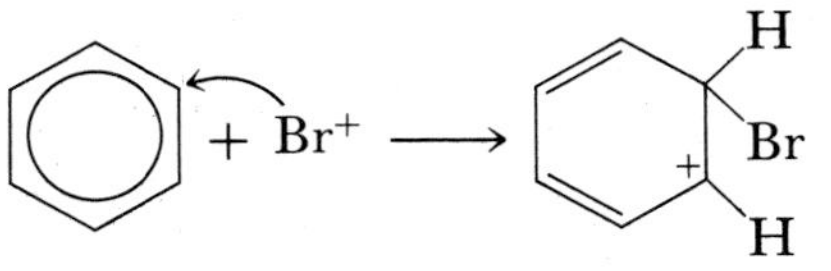

$$\text{benzene} + Br^+ \longrightarrow \text{(ring with H, Br, H; +)} \qquad [24.26]$$

$$\text{(ring with H, Br, H; +)} \xrightarrow{\ FeBr_4^-\ } \text{(Br)} + FeBr_3 + HBr \qquad [24.27]$$

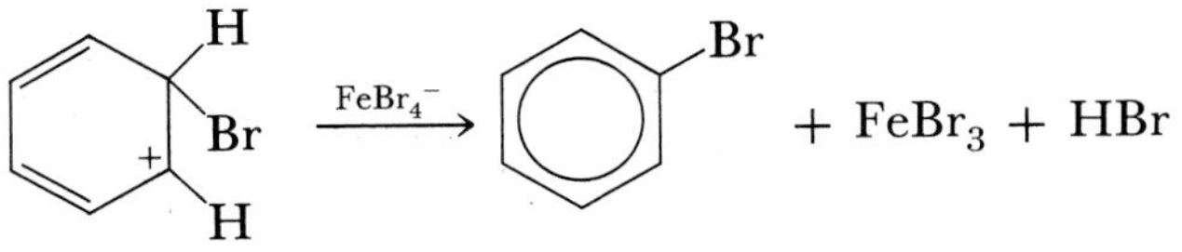

The Br^+ ion, called the **bromonium ion,** has only six electrons in its valence shell. Using a pair of π electrons from the benzene, it forms a single bond to a carbon atom. In doing so, it leaves the adjacent carbon atom with only six electrons in its valence shell. Such a species is not very stable. Loss of a proton from the carbon atom to which the bromine is attached converts the molecule to a more stable one, bromobenzene. The overall result is that bromine has replaced hydrogen on the benzene ring.

**24.4
hydrocarbon
derivatives**

In the course of our discussion of the chemical characteristics of the hydrocarbons we have introduced many substances that are hydrocarbon derivatives—that is, substances that are largely hydrocarbon in nature, but in which there is one or more functional groups, such as hydroxo

(OH), chloro (Cl), nitro (NO_2), and so forth. The chemical behavior of these hydrocarbon derivatives is often dominated by the functional group; the hydrocarbon portion of the molecule may be essentially unchanged in various chemical reactions. Let us now look briefly at the chemical characteristics of several important functional groups in organic chemistry.

ALCOHOLS

Alcohols are hydrocarbons in which one or more hydrogens have been replaced by a hydroxo group, OH. Figure 24.10 shows the structural formulas and names of several alcohols. Note that the accepted name for an alcohol ends in *-ol*. The simple alcohols are named by changing the last letter in the name of the corresponding alkane to *-ol*—for example, ethan*e* becomes ethan*ol*. Where necessary, the location of the OH group is designated by an appropriate prefix numeral that indicates the number of the carbon atom bearing the OH group, as shown in the examples in Figure 24.10.

Aliphatic alcohols are classified according to the number of carbon groups bonded to the carbon that contains the OH group. If there is only one other carbon atom, as with ethanol or 1-propanol, the alcohol is termed a **primary alcohol.** If there are two other carbon groups attached, as in 2-propanol, the alcohol is **secondary.** If there are three other carbons, as in *t*-butanol, the alcohol is **tertiary.**

Because the OH group is quite polar, the presence of the hydroxo group confers considerable polarity on the hydrocarbon molecule. Table 24.6 lists the solubilities of several alcohols in water at room temperature. In the alcohols having low molecular weights, the presence of the OH group plays an important role in determining physical properties. However, as the length of the hydrocarbon chain increases, the OH group becomes less important in determining overall behavior (see also Section 12.3).

The simplest alcohol, methanol, has many important industrial uses, and it is produced on a large scale. Carbon monoxide and hydrogen

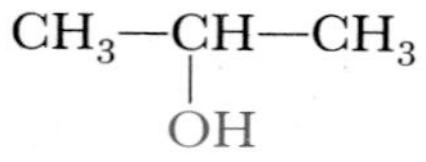

2-Propanol
(isopropyl alcohol;
rubbing alcohol)

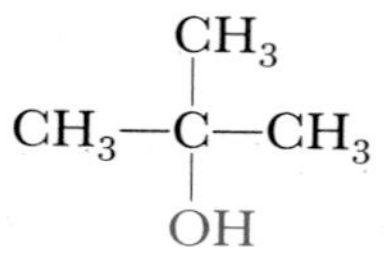

2-Methyl-2-propanol
(*t*-butanol; *t*-butyl alcohol)

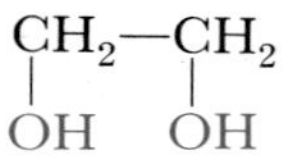

1,2-Ethanediol
(ethylene glycol)

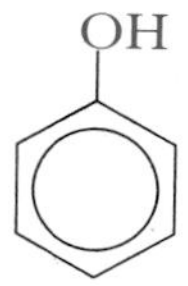

Phenol

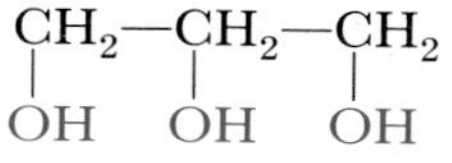

1,2,3-Propanetriol
(glycerol; glycerin)

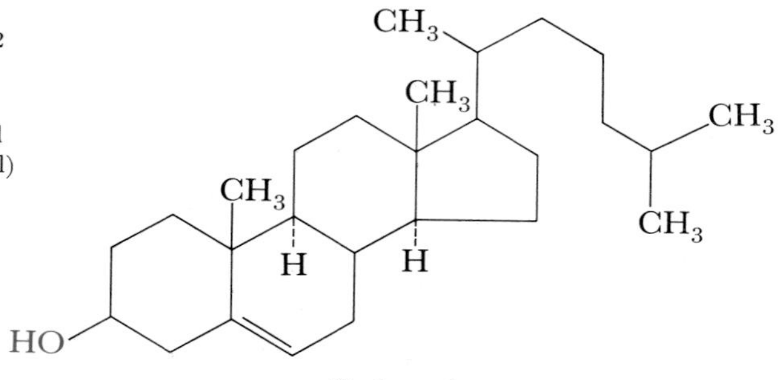

Cholesterol

FIGURE 24.10 Structural formulas of several important alcohols.

TABLE 24.6
solubilities of several straight-chain alcohols in water

ALCOHOL	BOILING POINT (°C)	SOLUBILITY IN WATER AT 25° (g/100 g H_2O)
Methanol	65	miscible
Ethanol	78	miscible
1-Propanol	97	miscible
1-Butanol	117	9
Cyclohexanol	161	5.6
1-Hexanol	158	0.6

are heated together under pressure in the presence of a metal oxide catalyst:

$$CO(g) + 2H_2(g) \xrightarrow[400°C]{200-300 \text{ atm}} CH_3OH(g) \qquad [24.28]$$

Methanol has a very high octane rating of about 110 when it is blended with gasoline. There has been considerable discussion of using it as a fuel extender by mixing it with gasolines. Methanol boils at 65°C, and freezes at −98°. It is therefore suitable in terms of volatility for use as an automotive fuel. It could probably be manufactured for about twice the present cost of gasoline (on a per mile basis), if the manufacture were carried out on a larger scale than at present. To offset its higher cost, methanol has the advantage of requiring less pollution-control equipment. A disadvantage of methanol for fuel use is that it is miscible with water in all proportions; it thus might attract water into methanol-gasoline mixtures. Secondly, methanol is quite toxic. However, gasoline is also a toxic substance, and only moderate additional precautions would need to be taken if methanol were added to gasoline.

Ethanol, C_2H_5OH, is a product of the fermentation of carbohydrates such as sugar or starch. Bacterial cultures such as yeast work in the absence of air to convert carbohydrates into a mixture of ethanol and CO_2, as shown in Equation [24.29]. In the process, they derive the energy necessary for growth:

$$C_6H_{12}O_6 \xrightarrow{\text{yeast}} 2C_2H_5OH + 2CO_2 \qquad [24.29]$$

This naturally occurring reaction is carried out under carefully controlled conditions to produce beer, wine, and other beverages in which ethanol is the active ingredient. It is often said that ethanol is the least toxic of the straight-chain alcohols. Although this is true in the strictest sense, the combination of ethanol and automobiles produces far more human fatalities each year than any other chemical agent.

Many alcohols are formed industrially by addition of the elements of water to an olefin. For example, isopropyl alcohol is formed by hydration of propylene, using sulfuric acid as a catalyst:

$$CH_3CH{=}CH_2 + H_2O \xrightarrow{H_2SO_4} CH_3{-}\underset{\underset{OH}{|}}{CH}{-}CH_3 \qquad [24.30]$$

Ethylene glycol, the major ingredient in automobile antifreeze, is formed in a two-stage reaction. First ethylene is reacted with hypochlorous acid:

$$CH_2{=}CH_2 + HOCl \longrightarrow \underset{\underset{OH \quad Cl}{\displaystyle |\quad\;\; |}}{CH_2{-}CH_2} \qquad [24.31]$$

Chlorine is removed from the product of this reaction by displacing the chloride ion with hydroxide ion:

$$\underset{\underset{OH \quad Cl}{\displaystyle |\quad\;\; |}}{CH_2{-}CH_2} + OH^- \longrightarrow \underset{\underset{OH \quad OH}{\displaystyle |\quad\;\; |}}{CH_2{-}CH_2} + Cl^- \qquad [24.32]$$

A reaction of this sort, in which hydroxide ion displaces another functional group is called **base hydrolysis.**

Phenol is the simplest example of a compound with an OH group attached to an aromatic ring. One of the most striking effects of the aromatic group is the greatly increased acidity of the proton. Phenol is about a million times more acidic in water than a typical aliphatic alcohol such as ethanol. Even so, it is not a very strong acid (K_a, 1.3×10^{-10}). Phenol is used industrially in the making of several kinds of plastics and in the preparation of dyes.

Cholesterol, shown in Figure 24.10, is an example of a biochemically important alcohol. Notice that the OH group forms only a small component of this rather large molecule. As a result, cholesterol is not very soluble in water (0.26 g per 100 ml H_2O). Cholesterol is a normal component of our bodies. However, when present in excessive amounts it may precipitate from solution. It precipitates in the gall bladder to form crystalline lumps called gall stones. It may also precipitate against the walls of veins and arteries and thus contribute to high blood pressure and other cardiovascular problems. The amount of cholesterol in our blood is determined not only by how much cholesterol we eat, but by total dietary intake. There is evidence that excessive caloric intake leads the body to synthesize excessive cholesterol.

ETHERS

Ethers can be thought of as formed from two molecules of alcohol by splitting out a molecule of water. The reaction is thus a dehydration process; it is catalyzed by sulfuric acid, which takes up water to remove it from the system:

$$CH_3CH_2{-}OH + H{-}OCH_2CH_3 \xrightarrow{H_2SO_4}$$

$$CH_3CH_2{-}O{-}CH_2CH_3 + H_2O \qquad [24.33]$$

A reaction in which water is split out from two substances is called a **condensation reaction.** An inorganic example of a condensation reaction was presented earlier (Section 21.5) in our discussion of the chemistry of phosphates. We shall encounter several additional examples in this chapter and in Chapter 25.

Ethers are used as solvents, and both dimethyl ether and tetrahydrofuran are used as solvents for organic reactions.

$$CH_3-O-CH_3$$

$$
\begin{array}{cc}
CH_2 & \!\!\!\!\!\!-CH_2 \\
| & | \\
CH_2 & CH_2 \\
& O
\end{array}
$$

Dimethyl ether   Tetrahydrofuran (THF)

OXIDATION OF ALCOHOLS; ALDEHYDES AND KETONES

It is fairly easy to oxidize alcohols. Complete oxidation results in formation of CO_2 and H_2O, as in the burning of methanol:

$$CH_3OH(g) + \tfrac{3}{2}O_2 \longrightarrow CO_2(g) + 2H_2O(g)$$

$$\Delta H° = -676 \text{ kJ/mole}$$ [24.34]

Controlled oxidation to form other organic substances that may have value is carried out by using various oxidizing agents such as air, hydrogen peroxide (H_2O_2), ozone (O_3), or potassium dichromate ($K_2Cr_2O_7$). We shall not concern ourselves with balancing most of these oxidation-reduction equations; instead we shall simply show the source of oxygen as an O in parentheses.

When a primary alcohol is oxidized, the initial product is called an **aldehyde.** For example, acetaldehyde is formed by oxidation of ethanol:

$$CH_3CH_2-OH + (O) \longrightarrow CH_3-\overset{\displaystyle O}{\overset{\displaystyle \|}{C}}-H + H_2O$$ [24.35]

Ethanol   Acetaldehyde

The key functional group present in aldehydes is the **carbonyl** group,

$$-\overset{\displaystyle O}{\overset{\displaystyle \|}{C}}-$$

When secondary alcohols are oxidized, the compound that results is called a **ketone.** For example, acetone is formed in large quantities by oxidation of isopropyl alcohol:

$$CH_3-\overset{\displaystyle OH}{\overset{\displaystyle |}{C}H}-CH_3 + (O) \longrightarrow CH_3-\overset{\displaystyle O}{\overset{\displaystyle \|}{C}}-CH_3 + H_2O$$ [24.36]

Acetone

Like aldehydes, ketones possess the carbonyl functional group. However, ketones are less reactive than are aldehydes because they do not possess the reactive C—H bond attached to the carbonyl carbon. Ketones are used extensively as solvents. Acetone, which boils at 56°C, is the most widely used. The carbonyl functional group imparts polarity to the solvent. Acetone is completely miscible with water, and yet it dissolves a wide range of organic substances. Methyl ethyl ketone, $CH_3COCH_2CH_3$, which boils at 80°C, is also used industrially as a solvent.

24.5
carboxylic
acids and
their derivatives

When a primary alcohol is oxidized with two moles of oxygen atoms per mole of alcohol, a carboxylic acid results:

$$CH_3CH_2OH + 2(O) \longrightarrow CH_3-\overset{\overset{\displaystyle O}{\|}}{C}-OH + H_2O \qquad [24.37]$$
$$\text{Acetic acid}$$

We have already discussed carboxylic acids briefly in Chapter 16. Carboxylic acids are widely distributed in nature and are important compounds in many industrial chemical processes. Figure 24.11 shows the structural formulas of several substances containing one or more carboxylic acid functional groups. Note that these substances are not named in a systematic way. The names of many acids are based on their historical origins. For example, formic acid was first prepared by extraction from ants; its name is therefore derived from the Latin word *formica,* meaning "ant." Oxalic acid was first identified as a constituent of the *oxalis* family of plants.

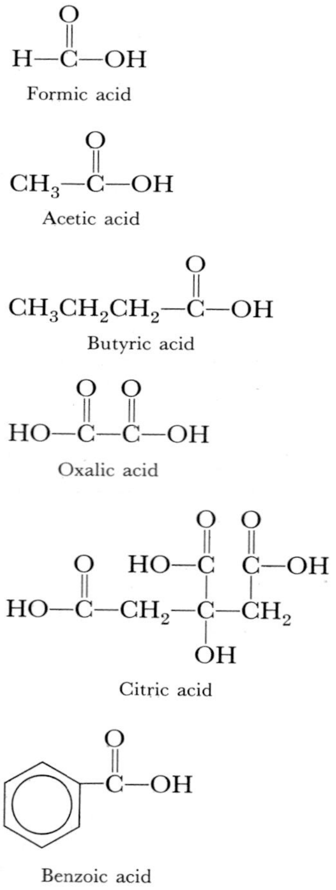

Carboxylic acids are important in the manufacture of polymers used to make fibers, films, and paints. Among the compounds having low molecular weights, acetic acid is an industrially important compound. A relatively new method for manufacture of acetic acid involves reaction of methanol with carbon monoxide, in the presence of a rhodium catalyst:

$$CH_3OH + CO \xrightarrow{\text{catalyst}} CH_3-\overset{\overset{\displaystyle O}{\|}}{C}-OH \qquad [24.38]$$

Notice that this reaction involves, in effect, the insertion of a carbon monoxide molecule between the CH_3 and OH groups. A reaction of this kind is called **carbonylation.**

Carboxylic acids may react with alcohols, with splitting out of water, to form **esters:**

$$CH_3-\overset{\overset{\displaystyle O}{\|}}{C}-OH + HO-CH_2CH_3 \longrightarrow$$
$$\text{Acetic acid} \qquad\qquad \text{Ethanol}$$

$$CH_3-\overset{\overset{\displaystyle O}{\|}}{C}-O-CH_2CH_3 + H_2O \qquad [24.39]$$
$$\text{Ethyl acetate}$$

This is another example of a condensation reaction. The reaction product, the ester, is named by using first the group from which the alcohol is derived and then the group from which the acid is derived.

$$H-\overset{\overset{\displaystyle O}{\|}}{C}-OH$$
Formic acid

$$CH_3-\overset{\overset{\displaystyle O}{\|}}{C}-OH$$
Acetic acid

$$CH_3CH_2CH_2-\overset{\overset{\displaystyle O}{\|}}{C}-OH$$
Butyric acid

$$HO-\overset{\overset{\displaystyle O}{\|}}{C}-\overset{\overset{\displaystyle O}{\|}}{C}-OH$$
Oxalic acid

$$HO-\overset{\overset{\displaystyle O}{\|}}{C}-CH_2-\underset{\underset{\displaystyle OH}{|}}{\overset{\overset{\displaystyle HO-\overset{\overset{\displaystyle O}{\|}}{C}\quad \overset{\overset{\displaystyle O}{\|}}{C}-OH}{}}{C}}-CH_2$$
Citric acid

$$\text{(benzene ring)}-\overset{\overset{\displaystyle O}{\|}}{C}-OH$$
Benzoic acid

FIGURE 24.11 Structural formulas of several familiar carboxylic acids.

Name the following esters:

(a)

$$\underset{\text{(a)}}{\text{[benzene ring]}}\text{—}\overset{\displaystyle O}{\overset{\|}{C}}\text{—OCH}_2\text{CH}_3$$

(b) $\text{CH}_3\text{CH}_3\text{CH}_2\text{—}\overset{\displaystyle O}{\overset{\|}{C}}\text{—O—}\text{[benzene ring]}$

Solution: In (a) the ester is derived from ethanol and benzoic acid. Its name is therefore ethyl benzoate. In (b) the ester is derived from phenol and butyric acid. The residue from the phenol, C_6H_5, is called the phenyl group. The ester is therefore named phenyl butyrate.

Esters generally have very pleasant odors; they are largely responsible for the pleasant aromas of fruit. Ethyl butyrate, for example, is responsible for the odor of apples. When esters are treated with acid or base in aqueous solution, they are hydrolyzed; that is, the molecule is split into its alcohol and acid components:

$$\underset{\text{Methyl propionate}}{\text{CH}_3\text{CH}_2\text{—}\overset{\displaystyle O}{\overset{\|}{C}}\text{—O—CH}_3} + \text{Na}^+ + \text{OH}^- \longrightarrow$$

$$\underset{\text{Sodium propionate}}{\text{CH}_3\text{CH}_2\text{—}\overset{\displaystyle O}{\overset{\|}{C}}\text{—O}^-} + \text{Na}^+ + \underset{\text{Methanol}}{\text{CH}_3\text{OH}} \qquad [24.40]$$

In this example, the hydrolysis was carried out in basic medium. The products of the reaction are the sodium salt of the carboxylic acid and the alcohol.

Amides are rather closely related to carboxylic acids and esters. In a simple amide, the OH group of the carboxylic acid is replaced by an NH_2 group, as in these examples:

$$\underset{\text{Acetamide}}{\text{CH}_3\overset{\displaystyle O}{\overset{\|}{C}}\text{—NH}_2} \qquad \underset{\text{Benzamide}}{\text{[benzene ring]}\text{—}\overset{\displaystyle O}{\overset{\|}{C}}\text{—NH}_2}$$

The nitrogen may have one or more organic groups attached in place of the hydrogens. The amide linkage,

$$\text{R—}\overset{\displaystyle O}{\overset{\|}{C}}\text{—}\underset{\displaystyle H}{\overset{\displaystyle |}{N}}\text{—R}'$$

where R and R′ are organic groups, is the key functional group in the structures of proteins, about which we shall have more to say in the next chapter. This same linkage is also very important in many synthetic polymeric materials formed by condensation polymerization.

CONDENSATION POLYMERS

A **condensation polymer** is formed by condensation reactions between molecules in which the functional groups are arranged so as to give long-chain, polymeric products. The most important condensation polymers are the nylons. A substance with two carboxylic acid functional groups is reacted with a substance with two amine functional groups, to yield a polymer. For example, nylon-66 is formed by heating a six-carbon diacid with a six-carbon diamine:

$$\text{---COOH} \quad H_2N\text{---}(CH_2)_6\text{---}NH_2 \quad HOOC\text{---}(CH_2)_4\text{---}COOH \quad H_2N\text{---}(CH_2)_6\text{---}NH_2 \quad HOOC\text{---}(CH_2)_4\text{---}COOH \quad H_2N\text{---}$$

$$\downarrow \text{heat}$$

$$\text{---}\overset{O}{\overset{\|}{C}}\text{---}NH\text{---}(CH_2)_6\text{---}NH\text{---}\overset{O}{\overset{\|}{C}}\text{---}(CH_2)_4\text{---}\overset{O}{\overset{\|}{C}}\text{---}NH\text{---}(CH_2)_6\text{---}NH\text{---}\overset{O}{\overset{\|}{C}}\text{---}(CH_2)_4\text{---}\overset{O}{\overset{\|}{C}}\text{---}NH\text{---} \qquad [24.41]$$

The polymer is formed by loss of water from between each pair of reacting functional groups. The resulting polymer is labeled 66 because there are six carbon atoms in the sections between each NH group along the chain.

Condensation polymers can also result from formation of ester linkages; that is, by condensation of carboxylic acids with alcohols. In forming a polymer, it is important to use difunctional acids and alcohols. By varying the acid and alcohol functions, various kinds of polyester materials can be formed. The familiar polyester Dacron, from which so much of our clothing is produced, is a condensation polymer of ethylene glycol and terephthalic acid:

$$\text{---HOCH}_2\text{CH}_2\text{OH} \quad HO\overset{O}{\overset{\|}{C}}\text{---}\bigcirc\text{---}\overset{O}{\overset{\|}{C}}OH \quad HOCH_2CH_2OH \quad HO\overset{O}{\overset{\|}{C}}\text{---}\bigcirc\text{---}\overset{O}{\overset{\|}{C}}OH\text{---} \xrightarrow{\text{heat}}$$

Ethylene glycol   Terephthalic acid

$$\text{---OCH}_2\text{CH}_2\text{O}\overset{O}{\overset{\|}{C}}\text{---}\bigcirc\text{---}\overset{O}{\overset{\|}{C}}OCH_2CH_2O\overset{O}{\overset{\|}{C}}\text{---}\bigcirc\text{---}\overset{O}{\overset{\|}{C}}OCH_2CH_2O\overset{O}{\overset{\|}{C}}\text{---}\bigcirc\text{---}\overset{O}{\overset{\|}{C}}O\text{---} \qquad [24.42]$$

Dacron

The polymer formed under typical reaction conditions may have 80 to 100 units per molecule. This material can be melted without decomposition. The molten polymer is forced through tiny holes, and then it cools and solidifies in the form of fibers. As the tiny fibers are stretched, the long molecules are forced to lie in a more or less parallel arrangement. The drawn or stretched fibers are then spun into a yarn.

summary

In this chapter we have studied the chemical and physical characteristics of simple organic substances. **Hydrocarbons** are composed of only carbon and hydrogen. There are four major classes of hydrocarbons. The **alkanes** are composed of only carbon-hydrogen and carbon-carbon single bonds. The **alkenes** contain one or more carbon-carbon double bonds. **Alkynes** contain one or more carbon-carbon triple bonds. Aromatic hydrocarbons contain cyclic arrangements of carbon atoms bonded through both σ and π bonds. **Isomers** are substances that possess the same molecular formula but that differ in some other respect. Isomers may differ in the bonding arrangements of atoms within the molecule or in the geometrical arrangements of groups. **Geometrical isomerism** is possible in the alkenes because of restricted rotation about the C=C double bond.

The major sources of hydrocarbons are fossil fuels such as coal and oil. Crude oil is separated by distillation into several fractions according to variations in volatility. The most important fraction is gasoline, used as automotive fuel. Less volatile fractions are **cracked**; that is, they are heated with catalysts to convert them to more volatile substances that have lower molecular weights and are suitable for gasoline. In the same process, straight-chain hydrocarbons are caused to rearrange to the more desirable branched-chain isomers.

Combustion of hydrocarbons is a highly exothermic process. The chief use of hydrocarbons is as a source of heat energy via combustion. The unsaturated hydrocarbons, the alkenes and alkynes, readily undergo **addition reactions** to the carbon-carbon multiple bonds. By contrast, addition reactions to aromatic hydrocarbons are difficult to carry out. Addition polymerization of alkenes is a source of many useful synthetic materials.

Substitution reactions are those in which one or more hydrogens of a hydrocarbon are replaced by some other atom or group. The substitution reactions of aliphatic (nonaromatic) hydrocarbons are not easily carried out. However, substitution reactions of aromatic hydrocarbons are easily carried out in the presence of acid catalysts.

The chemistry of hydrocarbon derivatives is often dominated by the nature of their **functional groups**. The functional groups we have considered are summarized here:

$$\underset{\text{Alcohol}}{R-O-H} \qquad \underset{\text{Aldehyde}}{R-\overset{\displaystyle O}{\overset{\|}{C}}-H} \qquad \underset{\text{Alkene}}{\overset{\diagdown}{\underset{\diagup}{C}}=\overset{\diagup}{\underset{\diagdown}{C}}}$$

$$\underset{\text{Alkyne}}{-C\equiv C-} \qquad \underset{\text{Amide}}{R-\overset{\displaystyle O}{\overset{\|}{C}}-N\diagup^{\diagdown}} \qquad \underset{\substack{\text{Carboxylic} \\ \text{Acid}}}{R-\overset{\displaystyle O}{\overset{\|}{C}}-O-H}$$

$$\underset{\text{Ester}}{R-\overset{\displaystyle O}{\overset{\|}{C}}-O-R'} \qquad \underset{\text{Ether}}{R-O-R'} \qquad \underset{\text{Ketone}}{R-\overset{\displaystyle O}{\overset{\|}{C}}-R'}$$

(Remember that R and R′ represent some hydrocarbon group—for example, methyl, CH_3, or phenyl, C_6H_5.)

747

Alcohols are hydrocarbon derivatives containing one or more OH groups. **Ethers** are related compounds that are formed by a **condensation reaction** of two molecules of alcohol, with splitting out of water. Oxidation of primary alcohols can lead to formation of **aldehydes**; oxidation of secondary alcohols leads to formation of **ketones**.

Carboxylic acids are organic acids containing the —COOH functional grouping. They can form esters by a condensation reaction with alcohols, or they can form **amides** by condensation reaction with amines. **Condensation polymers** are formed by condensation reactions involving molecules with functional groups on both ends.

learning goals

Having read and studied this chapter, you should be able to:

1. List the four groups of hydrocarbons and draw the structural formula of an example from each group.
2. List the names of the first ten members of the alkane series of hydrocarbons.
3. Write the structural formula of an alkane, alkene, or alkyne, given its systematic (IUPAC) name.
4. Name an alkane, alkene, or alkyne, given its structural formula.
5. Give an example of geometrical isomerism in alkenes.
6. Explain why aromatic hydrocarbons do not readily undergo addition reactions as do the alkenes and alkynes.
7. List and describe the fractions obtained in petroleum refining.
8. Explain the general relationship between octane number and structure in alkanes and describe the methods used to increase the octane numbers of straight-run gasolines.
9. Give examples of addition reactions of alkenes and alkynes, showing the structural formulas of reactants and products.
10. Describe Markovnikoff's rule and apply it to the addition reactions of alkenes and alkynes.
11. Give an example of addition polymerization of an alkene.
12. Write the steps in the photo-initiated chlorination of an alkane.
13. Give two or three examples of substitution of an aromatic compound.
14. Write structural formulas of molecules that are examples of an alcohol, an ether, an aldehyde, a ketone, a carboxylic acid, an ester, and an amide.
15. Describe the general character of condensation reactions and give an example of condensation polymerization.

key terms

Among the more important terms and expressions used for the first time in this chapter are the following:

Addition polymerization (Section 24.3) is a reaction in which alkenes add together end-to-end by opening of the carbon-carbon double bond, to form long polymeric chain molecules.

An **addition reaction** (Section 24.3) is one in which a reagent adds to the two carbon atoms of a carbon-carbon multiple bond.

Alcohols (Section 24.4) are hydrocarbons in which one or more hydrogens have been replaced by a hydroxo group, OH.

Aliphatic hydrocarbons (Section 24.1) are those that are not aromatic; that is, the alkanes, cycloalkanes, alkenes, and alkynes.

Alkanes (Section 24.1) are compounds of carbon and hydrogen containing only single carbon-carbon bonds.

Alkenes (Section 24.1) are hydrocarbons containing one or more carbon-carbon double bonds.

Alkynes (Section 24.1) are hydrocarbons containing one or more carbon-carbon triple bonds.

An **antiknock agent** (Section 24.2) is a substance added to automotive fuel to increase its octane number. This agent slows the rate of burning of fuel in the cylinder.

Aromatic hydrocarbons (Section 24.1) are hydrocarbon compounds that contain a planar, cyclic arrangement of carbon atoms linked by both σ and π bonds.

Base hydrolysis (Section 24.4) is a chemical reaction involving the replacement of a negatively charged functional group by aqueous hydroxide ion.

The **carbonyl** (Section 24.5) functional group, $C=O$, is a characteristic feature of several organic functional groups such as ketones or carboxylic acids.

Carbonylation (Section 24.5) is a reaction in which the carbonyl functional group is introduced into a molecule.

A **condensation polymer** (Section 24.5) is a substance having a high molecular weight that is formed by elimination of water from between molecules.

In a **condensation reaction** (Section 24.4), water is eliminated from between two molecules, with the joining together of the two molecules, usually through an —O— or —NH— bond.

Cracking (Section 24.2) is a catalytic process in which straight-chain alkanes are converted into more highly branched hydrocarbons with lower molecular weights.

Cycloalkanes (Section 24.1) are saturated hydrocarbons of general formula C_nH_{2n}, in which the carbon atoms form a closed ring.

A **functional group** (introduction) is an atom or group of atoms that has characteristic chemical properties.

Geometrical isomers (Section 24.1) have the same molecular formulas and functional groups, but differ in the geometrical arrangements of atoms. For example, *cis*- and *trans*-2-butenes are geometrical isomers.

In **homologous series** (Section 24.1), compounds contain common structural elements, but differ in the number of atoms making up the molecule. Members of a homologous series have the same general formula. For example, the aliphatic saturated alcohols have the general formula $C_nH_{2n+2}O$.

A **hydrocarbon** (Section 24.1) is a compound composed of carbon and hydrogen. Substituted hydrocarbons may contain functional groups composed of atoms other than carbon and hydrogen.

Markovnikoff's rule (Section 24.3) states that when a reagent of general formula XY adds to an alkene, the more negatively charged species, X or Y, ends up on the carbon with the fewer attached hydrogen atoms. For example, on addition of HCl to propene, Cl ends up on the middle carbon atom.

Octane number (Section 24.2) is a measure of the burning qualities of an automotive fuel. The higher the octane number, the more desirable the substance as a component of gasoline.

An **olefin** (Section 24.1) is a compound containing one or more carbon-carbon double bonds.

Petroleum (Section 24.2) is a complex mixture of hydrocarbons formed over a period of millions of years by decomposition of plant matter.

Hydrocarbon **radicals** (Section 24.1) are hydrocarbon molecules from which a hydrogen atom has been removed.

A **radical chain reaction** (Section 24.3) proceeds by reaction of a free radical with a stable molecule, forming a new stable molecule and a new radical. The newly formed radical continues the chain by reacting itself.

Refining (Section 24.2) is the process in which crude petroleum oil is separated into various component fractions on the basis of varying volatility or ease of distillation.

In a substitution reaction (Section 24.3), one atom or functional group bonded to a larger molecule is replaced by another.

Unsaturated hydrocarbons (Section 24.1) are nonaromatic compounds of carbon and hydrogen that contain one or more carbon-carbon multiple bonds.

EXERCISES

hydrocarbons

24.1   Write the molecular formulas of the compounds in the alkane, alkene, alkyne, and aromatic hydrocarbon homologous series that contain ten carbon atoms.

24.2   Draw the condensed structural formulas of all the straight-chain alkenes of molecular formula C_7H_{14}. Name each compound.

24.3   Write the molecular structures of all the alkanes of molecular formula C_7H_{16}. Name each compound.

24.4   Draw the condensed structural formulas of as many aliphatic hydrocarbons as you can think of that could have molecular formula C_5H_8. Name each compound.

24.5   What are the characteristic bond angles about carbon in an alkane; about the carbon-carbon double bond in an alkene; about the carbon-carbon triple bond in an alkyne?

24.6   Would you expect cyclohexyne to be a stable compound? Explain.

24.7   Draw a structural formula for, and name, one isomer of each of the following compounds: (a) cis-2-hexene; (b) 1,2-dibromobenzene; (c) n-pentane; (d) methylcyclopentane; (e) 2-chlorobutane; (f) 1,1-dichloroethylene.

[24.8]   You should recall from Chapter 9 that an aromatic compound such as benzene can be described in terms of more than one equivalent Lewis structure. Naphthalene can be described in terms of three Lewis structures, each involving five carbon-carbon double bonds in the aromatic rings. Draw these three Lewis structures. From these structures, what do you predict regarding the relative lengths of the various carbon-carbon bonds in the molecule (see Section 9.3)?

24.9   Why are aromatic compounds so much less reactive than alkenes or alkynes toward reagents such as Br_2?

24.10   Indicate whether each of the following molecules is capable of cis-trans geometrical isomerism. For those that are, draw the structures of the isomers: (a) 2,3-dichlorobutane; (b) 2,3-dichloro-2-butene; (c) 1,2-dichlorobenzene; (d) 1,4-dichloro-2-butyne.

24.11   Write condensed structural formulas for each of the following compounds: (a) 5-methyl-cis-3-heptene; (b) trans-2-pentene; (c) ethylcyclopentane; (d) 3-chloropropyne; (e) 1,5-dichloroanthracene; (f) 2-iodo-1-phenylbutane.

24.12   In each of the following pairs, indicate which molecule is the more reactive and give a reason for the greater reactivity: (a) butane, cyclobutane; (b) cyclohexane, cyclohexene; (c) benzene, 1-hexene; (d) 2-hexyne, 2-hexene.

[24.13]   The heat of combustion of decahydro-naphthalene, $C_{10}H_{18}$, is 6286 kJ/mole. The heat of combustion of naphthalene, $C_{10}H_8$, is 5157 kJ/mole, and $CO_2(g)$ and $H_2O(l)$ are products. From these data calculate the heat of hydrogenation of naphthalene. (Hint: First calculate the heat of formation of each compound.) By comparing this value with the data presented on page 729, indicate whether there is evidence for aromatic character in naphthalene from the value for the heat of hydrogenation.

petroleum

24.14   Indicate briefly the meaning of each of the following terms: (a) cracking; (b) isomerization; (c) antiknock agent; (d) octane number; (e) branched-chain alkane; (f) refining.

24.15   Describe two ways in which the octane number of a gasoline consisting of alkanes can be increased.

24.16   In places where the average temperature varies widely with the season, gasoline used during the winter contains a greater proportion of hydrocarbons with lower molecular weights than does gasoline used during the summer. Suggest a reason for this.

reactions of hydrocarbons

24.17 Give an example, in the form of a balanced equation, of each of the following chemical reactions: (a) substitution reaction of an alkane; (b) oxidation of an alkene; (c) addition reaction of an alkyne; (d) substitution reaction of an aromatic hydrocarbon.

24.18 Why do addition reactions occur more readily with alkenes and alkynes than with aromatic hydrocarbons?

24.19 Using the standard heats of formation listed in Appendix E, calculate the heat of combustion per mole of ethane, of ethylene, and of acetylene, assuming gaseous products.

24.20 Predict the product or products of each of the following reactions. Name the product in each case.

(a) $CH_3CH_2-C{\equiv}C-H + HBr \longrightarrow$

(b) $CH_3C{=}CHCH_3 + H_2O \xrightarrow{H_2SO_4}$
         |
        CH_3

(c) $cis\text{-}CH_3CH{=}CHCH_3 + H_2 \xrightarrow{Ni}$

(d) $CH_3-C{=\!=\!=}C-C_2H_5 + HI \longrightarrow$
        | |
       CH_3 CH_3

24.21 Draw a short section of the polymer chain formed by addition polymerization of vinyl chloride.

24.22 Bromination of butane requires irradiation. Write a series of reaction steps that can account for the bromination of butane under irradiation conditions.

24.23 When cyclopropane is treated with HI, *n*-propyl iodide is formed. A similar type of reaction does not occur with cyclopentane or cyclohexane. How do you account for the reactivity of cyclopropane?

[**24.24**] The chain mechanism described on page 738 for chlorination of ethane is incomplete, because it omits chain-terminating reactions—that is, reactions that result in loss of radicals and thus prevent the chain reaction from continuing. Write two possible chain-terminating reactions that might occur in the chlorination of ethane.

24.25 Draw the structures for all the possible isomeric forms of chloronitromethyl benzene.

oxygen derivatives of hydrocarbons

24.26 Alcohols of low molecular weight are moderately soluble in water, whereas ethers of about the same molecular weight are not. Explain.

24.27 Reaction of an alcohol with H_2SO_4 under dehydrating conditions can result in loss of water to form an ether, or, by reversal of the reaction in Equation [24.30], to form an alkene. Draw the structural formulas for the possible products of reaction of 2-butanol with H_2SO_4.

24.28 Draw structural formulas for the possible products of controlled oxidation of the following: (a) *n*-propanol; (b) cyclopentanol; (c) 2-butanol; (d) ethylene glycol.

24.29 Write structural formulas for each of the following compounds: (a) ethyl butyrate; (b) propionamide; (c) methylphenylketone; (d) benzaldehyde.

24.30 Draw the condensed structural formula for a section of the polymer formed by the condensation reaction between terephthalic acid and 1,4-diaminobutane.

[**24.31**] What type of chemical reaction would essentially reverse the formation of a condensation polymer? Under what conditions would you expect such a reaction to occur?

general exercises

[**24.32**] An unknown organic compound is found on analysis to have the empirical formula $C_5H_{12}O$. It is slightly soluble in water. Upon careful oxidation it is converted into a compound of empirical formula $C_5H_{10}O$, which behaves chemically like a ketone. Indicate two or more reasonable structures for the unknown.

24.33 Indicate briefly what is meant by each of the following terms: (a) aromatic hydrocarbon; (b) amide; (c) condensation reaction; (d) cycloalkane; (e) unsaturated hydrocarbon; (f) addition polymer.

24.34 Indicate briefly what is meant by each of the following terms: (a) aliphatic hydrocarbon; (b) *cis* isomer; (c) functional group; (d) carboxylic acid; (e) condensation polymer; (f) addition reaction.

24.35 Draw as many compounds as you can that have the molecular formula C_4H_8O. Indicate which of the compounds are functional-group isomers; which are geometrical isomers.

24.36 Draw a sketch of the geometrical structure of propene. Indicate the hybridization of the orbitals employed by the carbon atoms in forming the bonds in the molecule. Which are σ bonds? Which are π bonds?

24.37 Write a balanced equation and show the condensed structural formulas for reactants and products in each of the following reactions. Where not given, indicate reagents used to carry out the reaction. (a) Methyl benzoate is hydrolyzed in aqueous base.

(b) 1,5-Pentanediol is treated to form a cyclic ether. (c) 1-Butene is converted in two steps to methylethylketone. (d) Naphthalene is converted to 1-ethylnaphthalene. (e) *meta*-Dibromobenzene is formed from benzene. (f) Propene is converted into a polymeric substance.

24.38 Beginning with 3-methyl-1-butyne, show how you could prepare each of the following compounds:

(a) 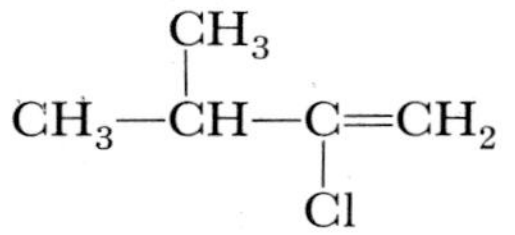
$$CH_3-CH-C=CH_2$$
with CH_3 above the CH and Cl below the C

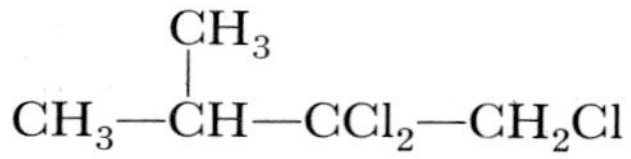

(b)
$$CH_3-CH-CCl_2-CH_2Cl$$
with CH_3 above the CH

(c)
$$CH_3-CH-CH-CH_3$$
with CH_3 above the first CH and Cl below the second CH

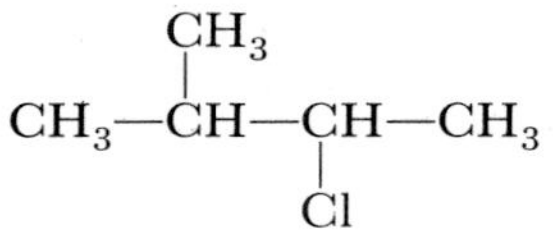

[24.39] An unknown substance is found to contain only carbon and hydrogen. It is a liquid that boils at 49°C at 1 atm pressure. Upon analysis it is found to contain 85.7 percent carbon, 14.3 percent hydrogen. At 100°C and 735 torr pressure, the vapor of the unknown has a density of 2.21 g/l. When it is dissolved in hexane solution and bromine water added, no reaction occurs. Suggest the identity of the unknown compound.

24.40   We have seen in this chapter a few examples of reactions that might serve as a starting point for synthesis of organic compounds when the world's supply of fossil fuels has been exhausted. Starting with CO and H_2 and using reaction types described in this chapter, show how one can form methanol, formaldehyde, and acetic acid.

24.41   Give the products of the reaction of 2-butene with each of the following reagents: (a) H_2SO_4, water; (b) Cl_2; (c) HOCl; (d) H_2/catalyst; (e) O_2, flame; (f) itself (polymerization catalyst).

24.42   For each of the following compounds, draw a structural isomer; for those that can have a geometrical isomer, draw a geometrical isomer as well: (a) *cis*-2-pentene; (b) 1,1-dichloroethene; (c) 3-methylcyclopentene; (d) acetone, C_2H_6O; (e) *para*-xylene (1,4-dimethylbenzene); (f) 1-pentene-3-ol; (g) 1,2,3-trifluorocyclopropane; (h) ethyl acetate; (i) 1,4-pentadiene; (j) 1,4-cyclohexadiene.

the biosphere

25

In earlier chapters of this text, we have considered various aspects of the physical world in which we live. We have discussed chemistry of the atmosphere (Chapter 10), of natural waters (Chapter 18), and of the solid earth (Chapter 22). In this final chapter we consider the biosphere, which is defined as that part of the earth in which living organisms are formed and live out their life cycles. The biosphere is not distinct from the atmosphere, natural waters, or the solid earth; rather, it is an integral part of it in those places where conditions permit life to exist. For living organisms to be sustained, there must be a supply of available energy, because the growth and maintenance of organisms requires energy. Secondly, there must be an adequate supply of water, because organisms are composed largely of water and use it for exchange of materials with their environment.

25.1

energy requirements of organisms

Living organisms require energy for their maintenance, growth, and reproduction. The ultimate source of this energy is the sun. However, as living matter proliferated on earth, and as organisms became more and more specialized, many of them developed the capacity for obtaining energy indirectly, by utilizing the energy stored in other organisms. Thus, for example, our bodies have essentially no capacity for directly utilizing solar energy. We consume animal and plant materials to obtain substances that our bodies can utilize as energy sources.

There are two distinct reasons why living systems need energy. First, organisms rely on substances readily available in their surroundings for synthesis of compounds needed for their existence. Most of these reactions are endothermic. To make such reactions proceed, energy must be supplied from an outside source. Secondly, living organisms are highly organized. The complexity of all the substances that make up even the simplest of single-cell organisms and the relationships between all the many chemical processes occurring are truly amazing. In thermodynamic terms, this means that living systems are very low in entropy as compared with the raw materials from which they are formed. The orderliness that is characteristic of living systems is attained by expenditure of energy.

Recall from the discussion in Section 15.5 that the free energy change, ΔG, is related to both the enthalpy and the entropy changes that occur in a process:

$$\Delta G = \Delta H - T\Delta S$$

If the entropy change in a process that builds up a living organism is negative (in other words, if a more highly ordered state results), the contribution to ΔG is positive. This means that the process becomes less spontaneous. Thus, both the enthalpy and the entropy changes that result in the formation, maintenance, and reproduction of living systems are such that the overall process is nonspontaneous. To overcome the positive values for ΔG associated with the essential processes, living systems must be coupled to some outside source of energy that can be converted into a form useful for driving the biochemical processes. The ultimate source of this needed energy is the sun.

The major means for conversion of solar energy into forms that can be utilized by living organisms is **photosynthesis**. The photosynthetic reaction that occurs in the leaves of plants is conversion of carbon dioxide and water to carbohydrate, with release of oxygen:

$$6CO_2 + 6H_2O \xrightarrow{48\ h\nu} C_6H_{12}O_6 + 6O_2 \qquad [25.1]$$

Note that formation of a mole of sugar, $C_6H_{12}O_6$, requires the absorption and utilization of 48 photons. This needed radiant energy comes from the visible region of the spectrum (Figure 6.3). The photons are absorbed by photosynthetic pigments in the leaves of plants. The key pigments are the chlorophylls; the structure of the most abundant chlorophyll, called chlorophyll-a is shown in Figure 25.1. Chlorophyll is a coordination compound; it contains a Mg^{2+} ion bound to four nitrogen atoms arranged around the metal in a planar array. The nitrogen atoms are part of a porphyrin ring (Section 23.2). Notice that there is a series of alternating double bonds in the ring that surrounds the metal ion. This system of alternating, or conjugated, double bonds gives rise to the strong absorptions of chlorophyll in the visible region of the spectrum. Figure 25.2 shows the absorption spectrum of chlorophyll as compared with the distribution of solar energy at the earth's surface. Chlorophyll is green in color because it absorbs red light (maximum absorption at 665 nm) and blue light (maximum absorption at 430 nm) and transmits green light.

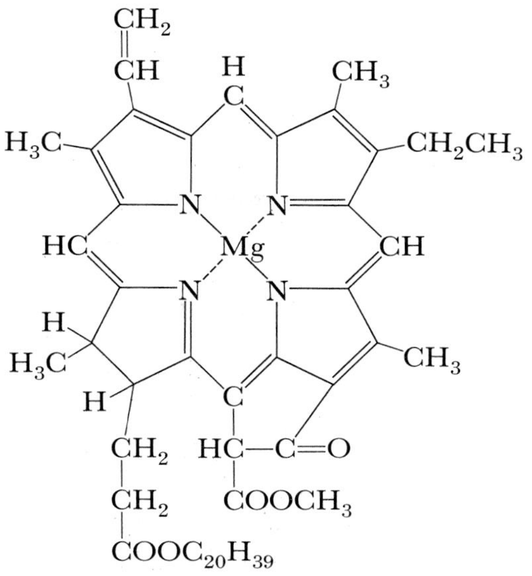

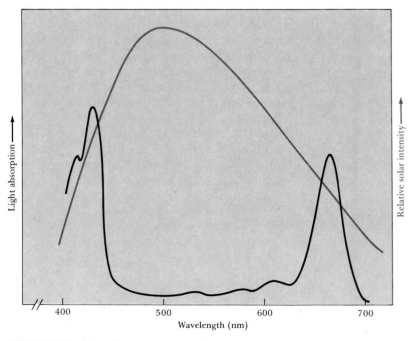

FIGURE 25.1 The structure of chlorophyll-a. All chlorophyll molecules are essentially alike; they differ only in details of the side chains.

The solar energy absorbed by chlorophyll is converted in a complex series of steps into chemical energy. This stored energy is then used to drive the reaction in Equation [25.1] to the right, a direction in which it is highly endothermic. Thus, plant photosynthesis is nature's solar-energy-conversion machine; all living systems on earth are dependent on

FIGURE 25.2 Absorption spectrum of chlorophyll (black curve), in comparison with the solar radiation at ground level (blue curve).

it for continued existence. A field of corn in the Midwest at the height of its growing season converts several percent of all the incident solar radiation into plant matter. It has been suggested that by cultivating about 6 percent of the land surface in the United States and using optimal growing conditions, we could obtain sufficient energy to satisfy all the energy needs of modern society.

Let us now turn our attention to the materials from which living organisms are formed. At some level of concentration, in some organism somewhere, almost every element seems to have a role to play in the biosphere. However, the elements of major importance in terms of their abundances in living systems are carbon, hydrogen, oxygen, nitrogen, phosphorus, and sulfur. Many other elements, including many metallic elements, are present in lesser quantities (refer back to Figure 23.5).

Much of the material present in living systems is in the form of **macromolecules,** polymers of high molecular weight. These **biopolymers** can be classified into three broad groups: proteins, carbohydrates, and nucleic acids. The proteins and carbohydrates, along with a class of molecules called fats and oils, are the major sources of energy in the food supply of animals. In addition, polymeric carbohydrates are the major construction material that gives form to plant systems, whereas proteins perform a similar role in animal systems. The nucleic acids are the storehouse of information that determines the form of a living system and that regulates its reproduction and development.

25.2
proteins

Proteins are macromolecular substances present in all living cells. They serve as the major structural component in animal tissues; they are a major component of skin, cartilage, nails, and the skeletal muscles. Enzymes, the catalysts for the biochemical reactions occurring in all living systems, are proteins. Proteins transport vital substances within the body. For example, hemoglobin, which carries O_2 from the lungs to cells, is a protein. The antibodies that serve as a defense mechanism against undesirable substances within the organism are also composed of proteins.

The molecular weights of proteins vary from about 10,000 to over 50 million. Simple proteins are composed entirely of amino acids; other proteins, called **conjugated proteins,** are made up of simple proteins bonded to other kinds of biochemical structures. The most abundant elements present in proteins are carbon (50 to 55 percent), hydrogen (7 percent), oxygen (23 percent), and nitrogen (16 percent). Sulfur is present in most proteins to the extent of about 1 or 2 percent; phosphorus either is absent or is present to only a very slight extent.

Simple proteins are linear polymers of amino acids. The characteristic polymeric linkage in proteins is the amide linkage, introduced in Section 24.5:

$$
\begin{array}{c}
\quad\quad O \\
\quad\quad \| \\
R-C-N-R' \\
\quad\quad | \\
\quad\quad H
\end{array}
$$

This particular functional group is called the **peptide** linkage when it is formed from amino acids. Proteins are sometimes referred to as **polypeptides.** Usually, however, the term polypeptide refers to a polymer of amino acids that has a molecular weight of less than 10,000. To see how the peptide linkage might serve to form a polymer we must investigate the structures of the amino acids, substances from which proteins are formed.

AMINO ACIDS

Twenty amino acids are found to occur commonly in various proteins. All of these are α-amino acids; that is, the amino group is located on the carbon atom immediately adjacent to the carboxylic acid functional group. The general formula for an amino acid is as follows:

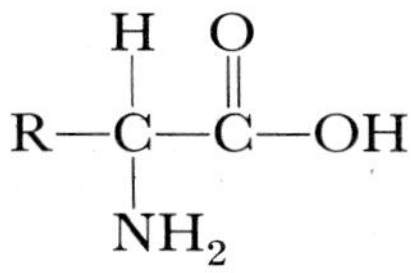

$$
\begin{array}{c}
\quad\ \ \text{H}\quad\ \ \text{O} \\
\quad\ \ | \qquad\ || \\
\text{R}-\text{C}-\text{C}-\text{OH} \\
\quad\ \ | \\
\quad\ \ \text{NH}_2
\end{array}
$$

The amino acids differ in the nature of the R group. Figure 25.3 shows the structural formulas of the 20 amino acids found in most proteins. Although certain of the amino acids are more common than others, most large proteins contain most of the amino acids.

You can see from the condensed structural formula shown above that the α-carbon atom, which bears both the amino and carboxylic acid groups, has attached to it four different groups.* As a result, amino acid molecules are **chiral.** As we pointed out in Section 23.4, a chiral molecule is one that is not superimposable with its mirror image. Let us consider a particular amino acid, say alanine, and suppose that we have two such molecules that are mirror images of one another, as in Figure 25.4. This figure also shows two familiar objects that are mirror images of one another, a pair of hands. Your right and left hands are not superimposable. That is, if there were some way that you could hollow out your left hand and place it face down on a surface, there is no way that you could slide your right hand, also face down, into the left. (This is, of course, why you can't wear your right-hand glove on your left hand.) In the same way, one of the mirror-image forms of alanine cannot be superimposed on the other.

The two molecules of a chiral substance that are mirror images of one another are called **enantiomers.** Because two enantiomers are not exactly the same, they are isomers of one another. This type of isomerism is called **configurational isomerism,** or **optical isomerism.** The two enantiomers of a pair are sometimes labeled R- (from the Latin *rectus,* "right") or S- (from the Latin *sinister,* "left") to distinguish them. Another widely used notation for distinguishing the enantiomers uses the prefix labels D- (from the Latin *dexter,* "right") or L- (from the Latin *laevus,* "left"). The enantiomers of a chiral substance possess the same physical properties, such as solubility, melting point, and so forth. Their chemical behavior

*The sole exception is glycine, for which R = H. For this amino acid, there are two H atoms on the α-carbon atom.

Amino acids with hydrocarbon side chains

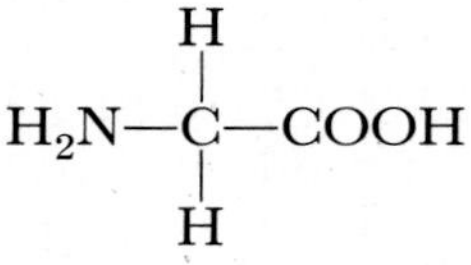

$$H_2N-\underset{\underset{H}{|}}{\overset{\overset{H}{|}}{C}}-COOH$$

Glycine (gly)

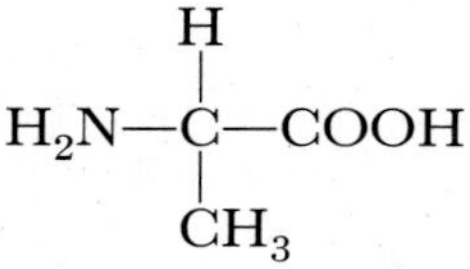

$$H_2N-\underset{\underset{CH_3}{|}}{\overset{\overset{H}{|}}{C}}-COOH$$

Alanine (ala)

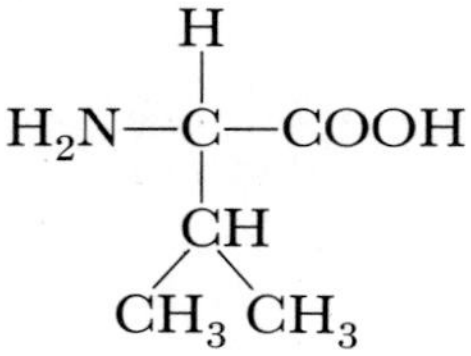

$$H_2N-\underset{\underset{\underset{CH_3 \quad CH_3}{}}{CH}}{\overset{\overset{H}{|}}{C}}-COOH$$

Valine (val)

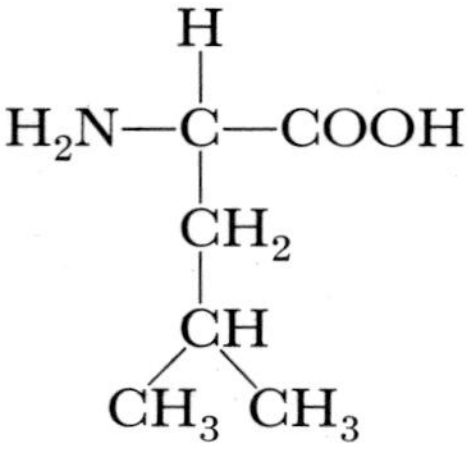

$$H_2N-\underset{\underset{\underset{\underset{CH_3 \quad CH_3}{}}{CH}}{CH_2}}{\overset{\overset{H}{|}}{C}}-COOH$$

Leucine (leu)

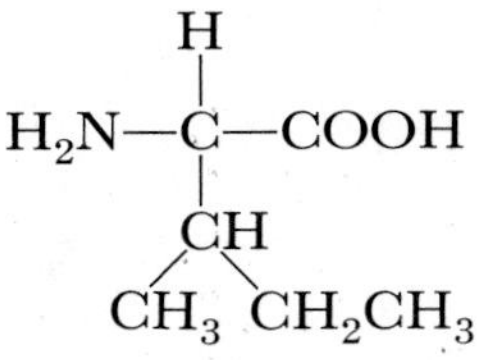

$$H_2N-\underset{\underset{\underset{CH_3 \quad CH_2CH_3}{}}{CH}}{\overset{\overset{H}{|}}{C}}-COOH$$

Isoleucine (ile)

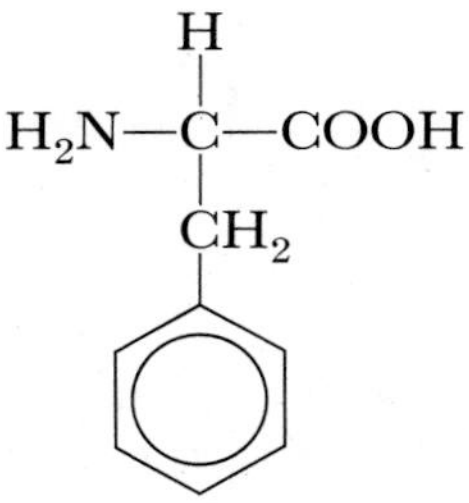

$$H_2N-\underset{\underset{}{CH_2}}{\overset{\overset{H}{|}}{C}}-COOH$$

Phenylalanine (phe)

$$HN-\overset{\overset{H}{|}}{C}-COOH$$

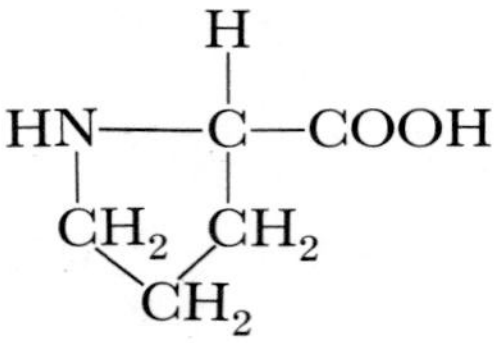

Proline (pro)

Amino acids with polar, neutral side chains

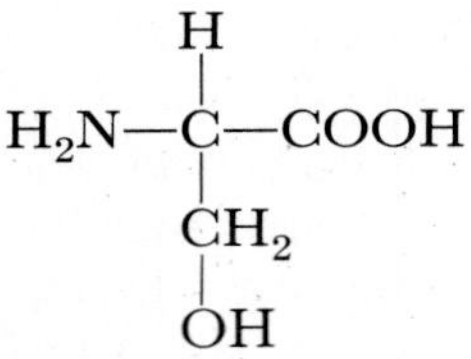

$$H_2N-\underset{\underset{OH}{CH_2}}{\overset{\overset{H}{|}}{C}}-COOH$$

Serine (ser)

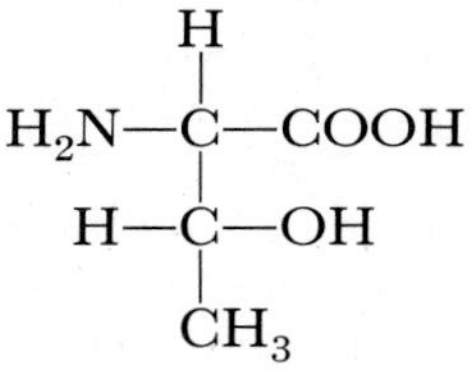

$$H_2N-\underset{\underset{CH_3}{H-C-OH}}{\overset{\overset{H}{|}}{C}}-COOH$$

Threonine (thr)

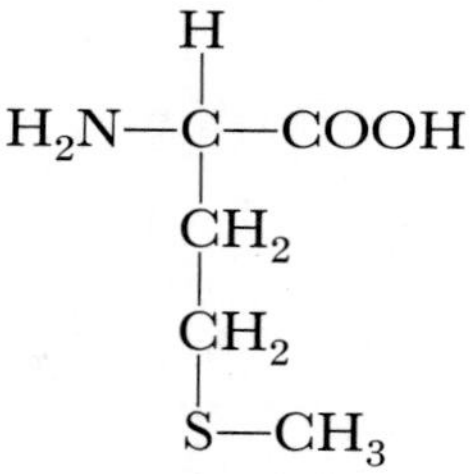

$$H_2N-\underset{\underset{\underset{S-CH_3}{CH_2}}{CH_2}}{\overset{\overset{H}{|}}{C}}-COOH$$

Methionine (met)

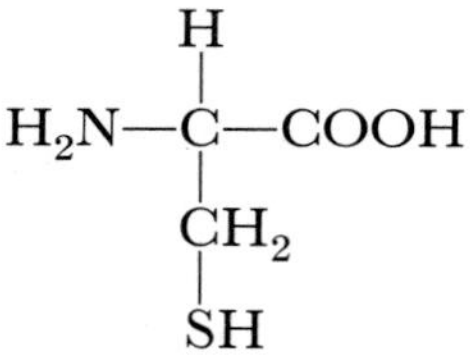

$$H_2N-\underset{\underset{SH}{CH_2}}{\overset{\overset{H}{|}}{C}}-COOH$$

Cysteine (cys)

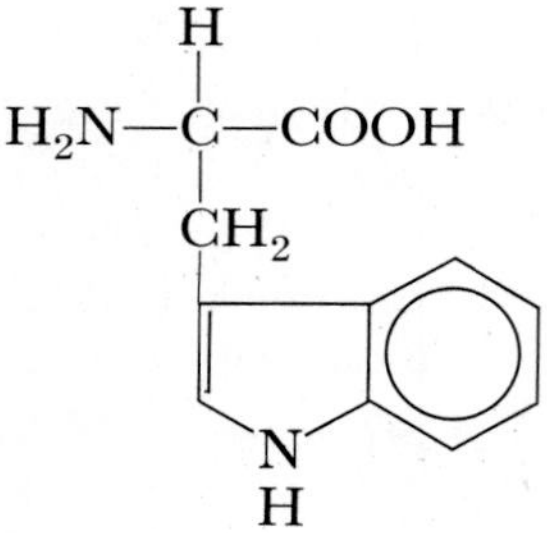

$$H_2N-\underset{\underset{}{CH_2}}{\overset{\overset{H}{|}}{C}}-COOH$$

Tryptophan (trp)

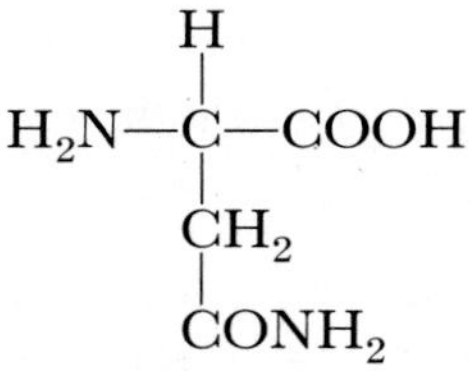

$$H_2N-\underset{\underset{CONH_2}{CH_2}}{\overset{\overset{H}{|}}{C}}-COOH$$

Asparagine (asn)

$$H_2N-\underset{\underset{\underset{CONH_2}{CH_2}}{CH_2}}{\overset{\overset{H}{|}}{C}}-COOH$$

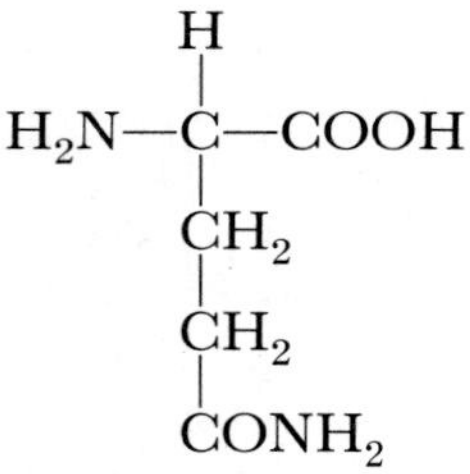

Glutamine (gln)

Amino acids with acidic or basic side chain functional groups

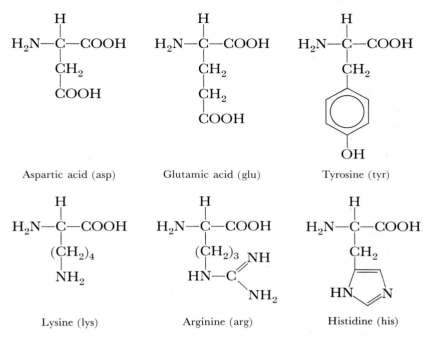

Aspartic acid (asp)   Glutamic acid (glu)   Tyrosine (tyr)

Lysine (lys)   Arginine (arg)   Histidine (his)

FIGURE 25.3 Condensed structural formulas of the amino acids, with the three-letter abbreviation for each acid.

toward ordinary chemical reagents is also the same. However, they differ in chemical reactivity toward other chiral molecules. It is a striking fact that all the amino acids in nature are of the *S*-, or L-, configuration at the carbon center (except glycine, which is not chiral). Only amino acids of

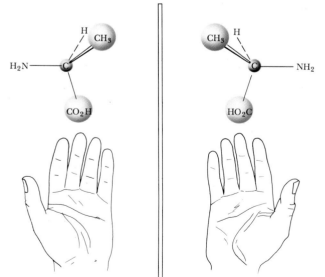

FIGURE 25.4 An illustration of chiral character. The pair of hands shown are mirror images of one another and are not superimposable. In the same way, the two alanine molecules that are mirror images are also not superimposable.

this specific configuration at the chiral carbon center are biologically effective in forming polypeptides and proteins in most organisms; peptide linkages are formed in cells under such specific conditions that enantiomeric molecules can be distinguished.

Chiral substances differ from one another in their effect on the plane of polarized light. We saw an example of this in Section 23.4, which dealt with the isomerism of coordination compounds. Suppose a beam of polarized light is passed through a solution containing a chiral substance such as alanine, using the arrangement shown in Figure 23.13. A solution of one of the enantiomers causes the plane of polarization to be rotated in one direction; a solution of the other enantiomer has the opposite effect. In each case the amount of rotation is proportional to the concentration of the solution and the length of the cell containing it. A solution containing equal concentrations of the two enantiomers, called a **racemic mixture,** causes no net rotation. Enantiomers of a chiral substance are often called **optical isomers** because of their effect on polarized light.

The relative amounts of the various amino acids in a protein vary with the nature and function of the protein material. Amino acids that have hydrocarbon side chains predominate in the insoluble, fiberlike proteins such as silk, wool, and collagen (a tissue-supporting protein). The amino acids with polar side chains are relatively more abundant in the water-soluble proteins. In addition, the polar groups often play an important role in determining the overall structure of the protein molecule, about which we shall have more to say later. Amino acids with acid or base side chains also help to increase water solubility. In addition, the functional groups on the side chains can act as sources of acid or base character in reactions catalyzed by acids or bases.

PROTEIN STRUCTURE

We have noted that the characteristic functional group in proteins is the amide, or peptide, linkage. This linkage is formed by a condensation reaction (Sections 24.4 and 24.5) between two amino acid molecules. As an example, alanine and glycine might form the dipeptide glycylalanine:

$$H_2N-\underset{\underset{\text{Glycine}}{H}}{\overset{H}{\underset{|}{\overset{|}{C}}}}-\overset{O}{\overset{\|}{C}}-OH \ + \ HN-\underset{\underset{\text{Alanine}}{CH_3}}{\overset{H}{\underset{|}{\overset{|}{C}}}}-\overset{O}{\overset{\|}{C}}-OH \longrightarrow$$

$$H_2N-\underset{\underset{\text{Glycylalanine}}{H}}{\overset{H}{\underset{|}{\overset{|}{C}}}}-\overset{O}{\overset{\|}{C}}-\underset{H}{\overset{}{\underset{|}{N}}}-\underset{CH_3}{\overset{H}{\underset{|}{\overset{|}{C}}}}-\overset{O}{\overset{\|}{C}}-OH \ + \ H_2O \qquad [25.2]$$

Notice that the acid that furnishes the carboxyl group is named first, with a *-yl* ending; then the amino acid furnishing the amino group.

Draw the structural formula for histidylglycine.

Solution: The name for this dipeptide tells us that the amino group of glycine and the carboxylic acid function of the histidine are involved in formation of the peptide bond. The structure is therefore:

$$H_2N-\underset{\underset{HN\diagdown\underset{}{\diagup}N}{\overset{|}{CH_2}}}{\overset{\overset{H}{|}}{C}}-\overset{\overset{O}{\parallel}}{C}-\underset{\underset{H}{|}}{\overset{\overset{H}{|}}{N}}-\underset{\underset{H}{|}}{\overset{\overset{H}{|}}{C}}-\overset{\overset{O}{\parallel}}{C}-OH$$

Because 20 amino acids are commonly found in proteins and because the protein chain may consist of hundreds of amino acids, the number of possible arrangements of amino acids is huge beyond imagination. Nature makes use of only certain of these combinations, but even so the number of different proteins is very large.

The arrangement, or sequence, of amino acids along the chain determines the **primary structure** of a protein. This primary structure gives the protein its unique identity. A change of even one amino acid can alter the biochemical characteristics of the protein. For example, sickle-cell anemia is a genetic disorder resulting from a single misplacement in a protein chain in hemoglobin. The chain that is affected contains 146 amino acids. The first seven in the chain are valine, histidine, leucine, threonine, proline, glutamic acid, and glutamic acid. In a person suffering from sickle-cell anemia, the sixth amino acid is valine instead of glutamic acid. This one substitution of an amino acid with a hydrocarbon side chain for one that has an acidic functional group in the side chain alters the solubility properties of the hemoglobin, and normal blood flow is impeded (see also Section 12.8).

Proteins in living organisms are not simply long, flexible chains with more or less random shape. Rather, the chain coils or stretches in particular ways that are essential to the proper functioning of the protein. This aspect of the protein structure is called the **secondary structure**. One of the most important and common secondary structure arrangements is the **α-helix**, first proposed by Linus Pauling and R. B. Corey. The helix arrangement is shown in schematic form in Figure 25.5. Imagine winding a long protein chain in a helical fashion around a long cylinder. The helix is held in position by hydrogen-bond interactions between an N—H bond and the oxygen of a carbonyl function located directly above or below. The pitch of the helix and the diameter of the cylinder must be such that (1) no bond angles are strained and (2) the N—H and C=O functional groups on adjacent turns are in proper position for hydrogen bonding. An arrangement of this kind is possible for some amino acids along the chain, but not for others. Large protein molecules may contain segments of chain that have the α-helical arrangement interspersed with

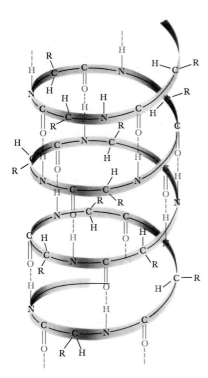

FIGURE 25.5 The α-helix structure for a protein. The symbol R represents any one of the several side chains shown in Figure 25.3.

sections in which the chain is in a random coil. The overall shape of the protein, determined by all the bends, kinks, and sections of rodlike α-helical structure, is called the **tertiary structure.** You might suppose that a long, complex protein molecule could adopt almost any shape under a given set of conditions, but this is not the case. Certain foldings of the protein chain lead to lower-energy (more stable) arrangements than do other folding patterns. For example, a protein dissolved in aqueous solution folds in such a way that the nonpolar hydrocarbon portions are tucked within the molecule, whereas the more polar, acidic and basic side-chain functional groups are projected into the solution.

When proteins are heated above the temperatures characteristic of living organisms, or when they are subjected to unusual acid or base conditions, they begin to lose their particular tertiary and secondary structure. As this happens the protein loses its biological activity; it is said to be **denatured.** When denaturation occurs under very mild conditions, it is often reversible. That is, a return to normal conditions results in a return of biological activity. On the other hand, denaturation may also be irreversible; chemical changes that permanently alter the character of the protein may occur. As an example, the protein material in the white of an egg is irreversibly denatured when the egg is placed for a time in boiling water.

With this brief introduction to proteins, let us now consider the characteristics of a very important group of proteins, the enzymes.

25.3

enzymes

The human body is characterized by an extremely complex system of interrelated chemical reactions. All of these reactions must occur at carefully controlled rates, so that thousands of individual chemical components are maintained at proper concentration levels and the system is able to respond as needed to demands made upon it. Every one of the many thousands of chemical reactions occurring in a biochemical system can be represented by an ordinary chemical equation. Indeed, many of the same reactions can be carried out under ordinary laboratory conditions. What is extraordinary about biochemical systems is that so many reactions occur with great rapidity at moderate temperatures. We cited in Chapter 13 the example of sugar, which is quite rapidly oxidized in the body at 37°C to carbon dioxide and water. In the laboratory, sugar does not react with oxygen at room temperature at a measurable rate. It must be heated to rather high temperature, for example in a burner flame, before it burns.

The oxidation of sugar in the body occurs by means of a series of more than two dozen biochemically catalyzed steps. The catalyst in each step is called an **enzyme**. Enzymes are formed in living cells and are protein in nature. The molecular weights of enzymes range from a low of perhaps 10,000 to a high of about 1 million. The enzyme may consist of just a single protein chain or may involve several chains that are loosely held together.

The reaction that the enzyme catalyzes occurs at a specific site on the protein; this is called the **active site**. The substances that undergo reaction at this site are called **substrates**. Besides the substrate, other substances, called **cofactors** or **coenzymes**, may be required in the enzyme-catalyzed reaction. For example, the enzyme may require the presence of Mg^{2+} or some other metal ion, or it may require the presence and participation of a small organic molecule.

In many enzyme-catalyzed reactions, the coenzyme is one of the vitamins. An enzyme without its required coenzyme is called an **apoenzyme**. The combination of apoenzyme and coenzyme is called the **holoenzyme**:

$$\text{Apoenzyme} + \text{coenzyme} \longrightarrow \text{holoenzyme}$$

Enzymes possess certain characteristics not common to other types of catalysts. In the first place, they have a rather special sensitivity to temperature. Experimental studies have shown that the activity of a particular enzyme maximizes at around the normal temperature of the organism in which the enzyme is found. Figure 25.6 illustrates a typical activity-versus-temperature curve. Often it happens that when the temperature is raised above the usual operating temperature of the enzyme, the activity temporarily increases, but then subsequently decreases. The secondary and tertiary structures of the protein chain, on which the activity of the active site depends, are the result of many weak forces that induce the chain to take up that particular arrangement. Heating causes the protein chain to come undone, as it were; the enzyme is denatured and loses its activity completely.

A second respect in which enzymes are special is that they often show a sharp change in activity with change in the acidity or basicity of the solution. This suggests that acid-base reactions are important in the catalyzed reaction. Thirdly, enzymes may be very specific in catalyzing

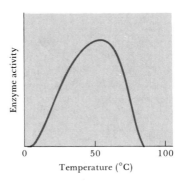

FIGURE 25.6 Enzyme activity as a function of temperature. At the higher temperature at which the activity of the enzyme drops to zero, the protein is denatured.

exclusively a particular kind of reaction and no other, or even in catalyzing a particular reaction for only one compound and no other. The
degree of specificity of enzymes varies widely. For example enzymes
called carboxypeptidases are rather general in their action.* They catalyze the hydrolysis of polypeptides into amino acids:

$$\underset{R'}{\overset{H}{\underset{|}{C}}}-\underset{}{\overset{O}{\overset{||}{C}}}-\underset{H}{\overset{H}{\underset{|}{N}}}-\underset{R}{\overset{H}{\underset{|}{C}}}-\overset{O}{\overset{||}{C}}-OH + H_2O \xrightarrow{\text{Carboxypeptidase}}$$

$$\underset{R'}{\overset{H}{\underset{|}{C}}}-\overset{O}{\overset{||}{C}}-OH + H_2N-\underset{R}{\overset{H}{\underset{|}{C}}}-\overset{O}{\overset{||}{C}}-OH \qquad [25.3]$$

This equation shows only the end member of a peptide chain; the
carboxypeptidase attacks only the amide group at the end of the chain.
However, it is active regardless of the nature of the particular side chains
R and R'. Although carboxypeptidases catalyze the hydrolysis of peptides, they are not at all active in the hydrolysis of fats; an entirely
separate set of enzymes is responsible for the latter reaction. The high
degree of specificity that enzymes possess is necessary to maintain some
degree of independence of all the reactions occurring in a complex
organism.

Finally, many enzymes differ from ordinary nonbiochemical catalysts in that they are enormously more efficient. The number of individual, catalyzed-reaction events occurring per second at a particular active
site, called the **turnover number,** is generally in the range 10^3/sec to
10^7/sec. Such large turnover numbers are indicative of reactions with
very low activation energies.

SAMPLE EXERCISE 25.2

Solutions of dilute (about 3 percent) hydrogen peroxide are stable for long periods of time, especially if
a small amount of a stabilizer is added. However,
when such a solution is poured onto an open cut or
wound, the hydrogen peroxide decomposes very
rapidly with evolution of oxygen and formation of
water:

$$2H_2O_2 \longrightarrow O_2 + 2H_2O$$

How can you account for this result?

Solution: Because the reaction occurs so rapidly on
contact with the open wound, some chemical species
present at the wound must catalyze the decomposition of hydrogen peroxide. That substance is an
enzyme called peroxidase. Peroxidase is present to
ensure that hydrogen peroxide does not accumulate in
cells, because it could interfere with many other cell
reactions. Peroxidase is an efficient enzyme, with a
very high turnover number.

Biochemists have been trying for a long time to discover how and
why enzymes are so fantastically efficient and selective. During the past

---

*The names of most enzymes end in *-ase*. The *-ase* ending is attached to the name of the
substrate on which the enzyme acts, or to the name of the type of reaction that the enzyme catalyzes.
Thus, *peptidases* are enzymes that act on peptides; *lipase* is an enzyme that acts on lipids; a
transmethylase is an enzyme that catalyzes transfer of a methyl group.

20 years, the development of new experimental techniques for studying the structures of molecules and for following the rates of very rapid reactions has led to a much better understanding of how enzymes work.

The high turnover numbers observed for enzymes suggest that the substrate molecules cannot be very tightly bound to the enzyme; if they were, they might block the active site. Reaction would then be slow, because the active site is not quickly cleared out. Most enzyme systems that have been studied behave as though there were an equilibrium between the substrate (S) and the active site (E); we can write this equilibrium as an equation:

$$E + S \underset{k_{-1}}{\overset{k_1}{\rightleftharpoons}} ES \qquad \text{[25.4]}$$

The symbol ES represents a species in which the substrate is attached in some way to the enzyme. This **enzyme-substrate complex,** as it is called, then reacts to give product (P) and free active site:

$$ES \xrightarrow{k_2} E + P \qquad \text{[25.5]}$$

In this picture, it is assumed that the substrate molecules may come off and on the active site very rapidly compared with the rate at which they undergo reaction to form products P. This means that the equilibrium described by Equation [25.4] is rapidly established between enzyme and substrate. It is also supposed that the equilibrium is such that most of the enzyme sites are not occupied by S when the substrate is present at normal concentrations. Now suppose that we study and graph (Figure 25.7) the rate of the enzyme-catalyzed reaction as a function of increasing concentration of substrate S. When S is present in low concentration, most of the enzyme active sites are not in use. Increasing the concentration of S thus shifts the equilibrium in Equation [25.4] to the right, so that a larger number of enzyme-substrate complexes are formed. This in turn increases the overall rate of the reaction, because that rate depends on the concentration of ES, $[ES]$:

$$\text{Rate} = \frac{\Delta[P]}{\Delta t} = k_2[ES] \qquad \text{[25.6]}$$

However, with still further increases in the concentration of S, a sizable fraction of the active sites are occupied. Adding still more S thus does not result in the same degree of increase in $[ES]$, and the rate does not increase as much. Eventually, with still higher concentration of S, *all* the active sites are effectively occupied. Further increases in S cannot result in faster reaction, because all available active sites are in use.

Many enzyme systems have been found to obey the sort of behavior illustrated in Figure 25.7. This strongly indicates that an enzyme-substrate complex is involved in the action of enzymes.

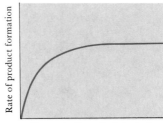

Rate of product formation

Concentration of substrate S

FIGURE 25.7 The rate of product formation as a function of concentration in a typical enzyme-catalyzed reaction. The rate of product formation is proportional to substrate concentration in the region of low concentration. At high substrate concentrations, the active sites on the enzyme are all complexed; further addition of substrate does not affect reaction rate.

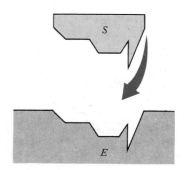

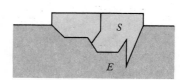

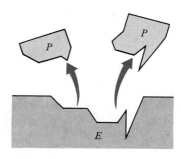

FIGURE 25.8 Lock-and-key model for formation of the enzyme-substrate complex.

THE LOCK-AND-KEY MODEL

One of the earliest models for the enzyme-substrate interaction was the **lock-and-key model** illustrated in Figure 25.8. The substrate is pictured as fitting neatly into a special place on the protein (the active site) tailored more or less specifically for that particular substrate. The catalyzed reaction occurs while the substrate is bound to the enzyme, and the reaction products then depart. Clearly, this picture of enzyme action has much in common with the models for heterogeneous catalysis that were discussed in Section 13.8. The difference is that an element of greater specificity is brought into the picture.

The lock-and-key model for enzyme action goes a long way toward explaining many aspects of enzyme action, but it cannot be the whole story. As the substrate molecules are drawn into the active site, they are somehow activated so that they are capable of extremely rapid reaction. This activation may result from the withdrawal or donation of electron density at a particular bond by the enzyme. In addition, in the process of fitting into the active site, the molecule may be distorted by the portions of the enzyme protein chain near the active site and thus made more reactive. If coenzymes are involved, the enzyme may promote reaction by binding both the coenzyme and substrate to the same site or to adjacent sites, thus keeping the reactants in close proximity. Enzymes, like all proteins, contain many acidic and basic sites along the chain, and these may be involved in acid-base reactions with the substrate. It is clear that the enzyme-coenzyme-substrate interactions can be very complex and are often difficult to unravel.

One of the more interesting corollaries of the lock-and-key model for enzyme action is that certain molecules should be capable of inhibiting the enzyme. Suppose that a certain molecule is capable of fitting the active site on the enzyme but is not reactive for one reason or another. If it is present in the solution along with substrate, this molecule competes with substrate for binding at the active sites. The substrate is thus prevented from forming the necessary enzyme-substrate complex, and the rate of product formation decreases. Metals that are highly toxic—for example, lead and mercury—probably operate as enzyme inhibitors. The heavy metal ions are especially strongly bound to sulfur-containing groups on protein side chains. By complexing very strongly to these sites on proteins, they interfere with the normal reactions of enzymes.

SAMPLE EXERCISE 25.3

Many bacteria employ *para*-aminobenzoic acid (PABA) in a vital enzyme-catalyzed metabolic reaction. Sulfa drugs such as sulfanilamide (*para*-amino-benzenesulfonamide) kill bacteria. Explain this in terms of the concept of the enzyme-substrate complex.

$$H_2N-\bigcirc-\overset{\overset{\displaystyle O}{\|}}{\underset{\underset{\displaystyle O}{\|}}{S}}-NH_2$$

Sulfanilamide
(*para*-aminobenzenesulfonamide)

$$H_2N-\bigcirc-\overset{\overset{\displaystyle O}{\|}}{C}-OH$$

para-Aminobenzoic acid (PABA)

Solution: The structural formulas reveal that PABA and sulfanilamide are very similar in shape and in the locations of polar functional groups. The sulfa drug acts as an inhibitor by binding at the active site in the enzyme and blocking access by PABA. Without this needed ingredient, the metabolic processes of the bacteria cease.

25.4 carbohydrates

The carbohydrates are a very important class of naturally occurring substances found in both plant and animal matter. The name **carbohydrate** (hydrate of carbon) comes from the empirical formulas for most substances in this class; they can be written as $C_x(H_2O)_y$. For example, **glucose,** the most abundant carbohydrate, has molecular formula $C_6H_{12}O_6$, or $C_6(H_2O)_6$. The carbohydrates are not really hydrates of carbon; rather, they are polyhydroxy aldehydes and ketones. The structural formula for glucose can be written as shown in Figure 25.9.

Notice that four of the carbon atoms in this molecule, numbered 2, 3, 4, and 5, are chiral. Thus, there are many configurational isomers of glucose. Several naturally occurring sugars differ from glucose only in the configuration at one of these four chiral carbon atoms. The fact that these sugars have different biological properties is another example of the extreme specificity of biochemical systems. Many sugars are optically active substances; their solutions cause the plane of polarization of polarized light to be rotated, as illustrated in Figure 23.13.

When glucose is placed in solution, the aldehyde functional group at one end of the molecule reacts with the hydroxy group at the other end to form a cyclic structure, called a **hemiacetal:**

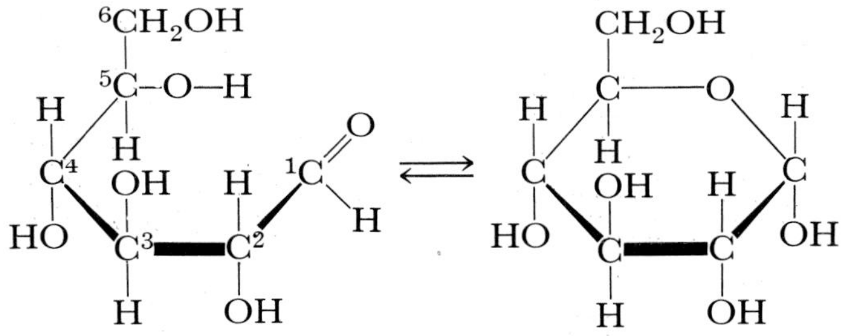

Glucose   Glucopyranose   [25.7]

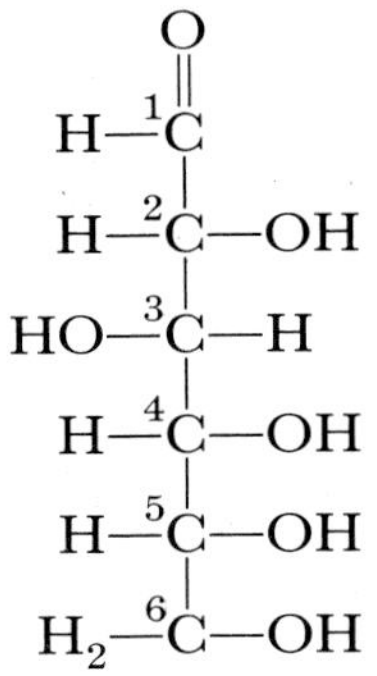

Note that in forming the ring structure in Equation [25.7], the OH group on carbon 1 can be on the same side of the ring as the OH group of carbon 2 (the α form) or on the opposite side (the β form). This seemingly minor difference is very important in distinguishing starch from cellulose, as we shall see shortly.

Six-membered cyclic hemiacetal structures are called **pyranoses.** The name glucopyranose indicates that the sugar glucose forms a six-membered ring. Sugars may also form five-membered rings called **furanoses.**

Some sugars are ketones rather than aldehydes. One important example is **fructose,** shown here in its straight-chain (linear) form as well as the five-membered cyclic form called fructofuranose:

FIGURE 25.9 Structure of glucose.

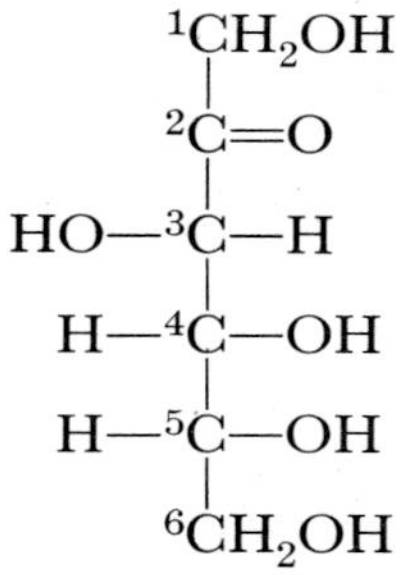

$^1CH_2OH$

$^2C{=}O$

$HO{-}^3C{-}H$

$H{-}^4C{-}OH$

$H{-}^5C{-}OH$

$^6CH_2OH$

Fructose

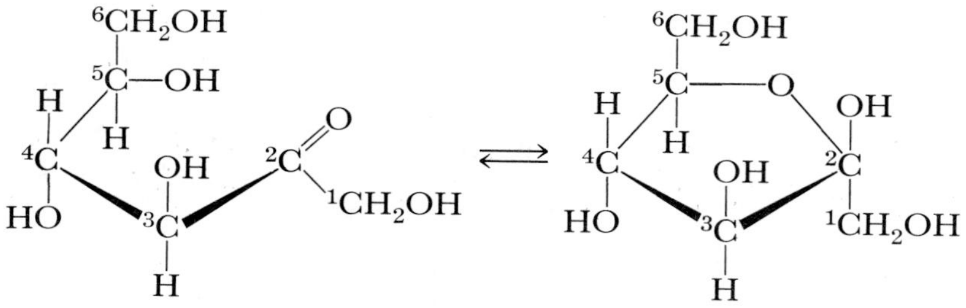

Fructose $\rightleftharpoons$ Fructofuranose   [25.8]

Both glucose and fructose are examples of simple sugars, also called **monosaccharides**. Two or more such units can be joined together to form **polysaccharides**.

SAMPLE EXERCISE 25.4

Draw the structural formula for fructopyranose.

Solution: The name of the substance for which we are to draw the structure tells us that it involves a six-membered ring. Thus, beginning with fructose, for which the linear formula was given above, we complete a six-membered ring by closure with the OH group on the terminal carbon atom:

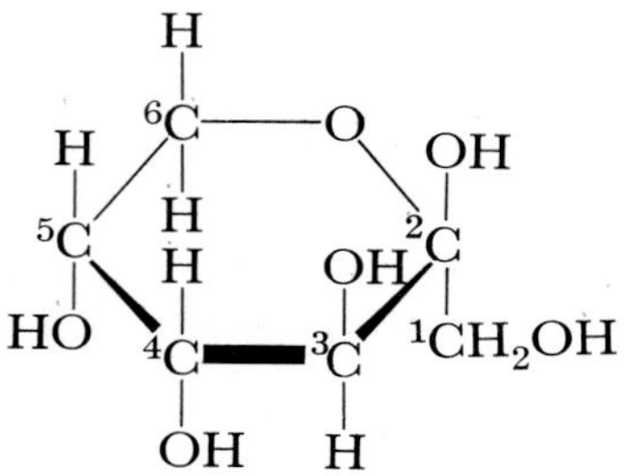

Notice that we might have drawn this structure so that the OH group of carbon atom 2 was on the opposite side of the ring as compared with the OH group of carbon 3. Thus there are two forms of fructopyranose.

DISACCHARIDES

Disaccharides are formed by the joining together of two monosaccharide units. The monosaccharides are linked by a condensation reaction, in which a molecule of water is split out from between two hydroxo groups, one on each sugar. Because there are several hydroxo groups, there are several ways in which disaccharides can be formed. The structures of three common disaccharides, **sucrose** (table sugar), **maltose** (malt sugar), and **lactose** (milk sugar) are shown in Figure 25.10. The word "sugar" makes us think of sweetness; all sugars are sweet, but they differ in the degree of sweetness we perceive when we taste them. Sucrose is about six times sweeter than lactose, about three times sweeter than maltose, slightly sweeter than glucose, but only about half as sweet as fructose. Disaccharides can be hydrolyzed; that is, they can be reacted with water in the presence of an acid catalyst to form the monosaccharides. When sucrose is hydrolyzed, the mixture of glucose and fructose that forms, called *invert sugar,* * is sweeter to the taste than the original sucrose. The sweet syrup present in canned fruits and candies is largely invert sugar formed from hydrolysis of added sucrose.

*The term "invert sugar" comes from the fact that the rotation of the plane of polarized light by the glucose-fructose mixture is in the opposite direction, or inverted, from that of the sucrose solution.

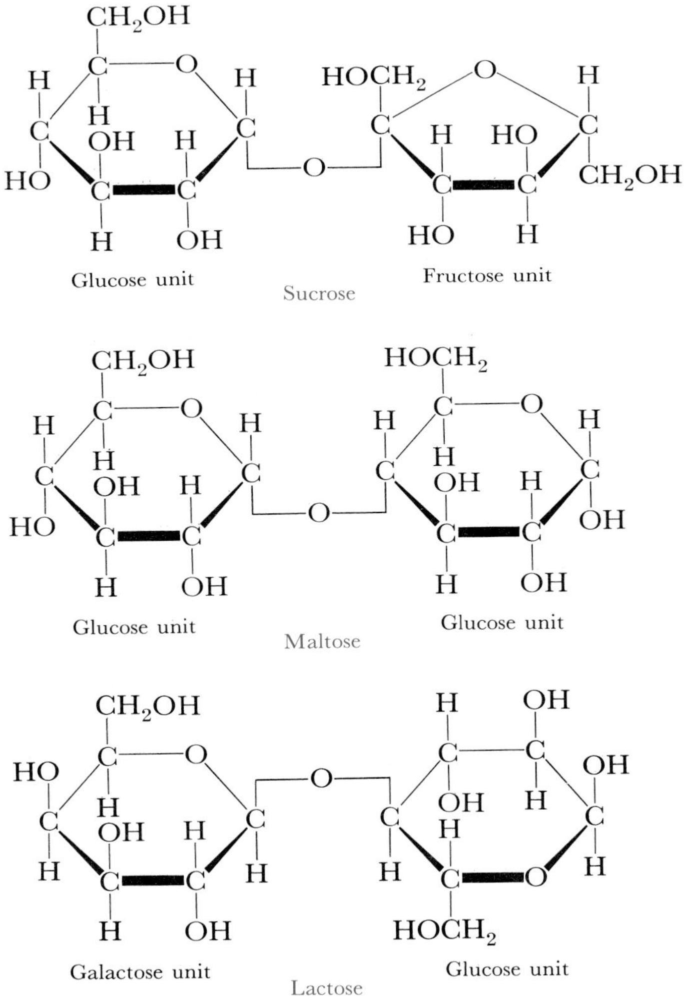

FIGURE 25.10 Structures of three disaccharides.

POLYSACCHARIDES

Polysaccharides are made up of several monosaccharide units joined together by a bonding arrangement similar to those shown for disaccharides in Figure 25.10. The most important polysaccharides are starch, glycogen, and cellulose, which are formed from repeating glucose units.

Starch is not a pure substance; the term refers to a group of polysaccharides found in plants. Starches serve as a major method of food storage in plant seeds and tubers. Corn, potatoes, wheat, and rice all contain substantial amounts of starch. These plant products serve as major sources of needed food energy for humans. Enzymes within the digestive system catalyze the hydrolysis of starch to glucose.

Some starch molecules are linear chains, whereas others are branched. All contain the type of chain structure illustrated in Figure 25.11. It is particularly important to note the geometrical arrangement in the joining of one ring to another; the glucose units are in the α form.

Glycogen is a starchlike substance synthesized in the body. Glycogen molecules vary in molecular weight from about 5000 to more than 5 million. Glycogen acts as a kind of energy bank in the body. It is concentrated in the muscles and liver. In muscles, it serves as an immediate source of energy; in the liver, it serves as a storage place for glucose and helps to maintain a constant glucose level in the blood.

Cellulose forms the major structural unit of plants. Wood is about 50 percent cellulose; cotton fibers are almost entirely cellulose. Cellulose consists of a straight chain of glucose units, with molecular weights averaging more than 500,000. The structure of cellulose is shown in Figure 25.12. At first glance, this structure looks very similar to that of starch. However, the differences in the arrangements of bonds that join the glucose units are important. Notice that glucose in cellulose is in the β form. Enzymes that readily hydrolyze starches do not hydrolyze cellulose. Thus, you might chew up and swallow a pound of cellulose and receive no caloric value from it whatsoever, even though the heat of combustion per unit weight is essentially the same for both cellulose and starch. A pound of starch, on the other hand, would represent a substantial caloric intake. The difference is that the starch is hydrolyzed to glucose, which is eventually oxidized with release of energy. Cellulose, on the other hand, is not hydrolyzed by any enzymes present in the body, and so it passes through the body unchanged. Many bacteria contain enzymes, called cellulases, that hydrolyze cellulose. These bacteria are present in the digestive systems of grazing animals, such as cattle, that utilize cellulose for food.

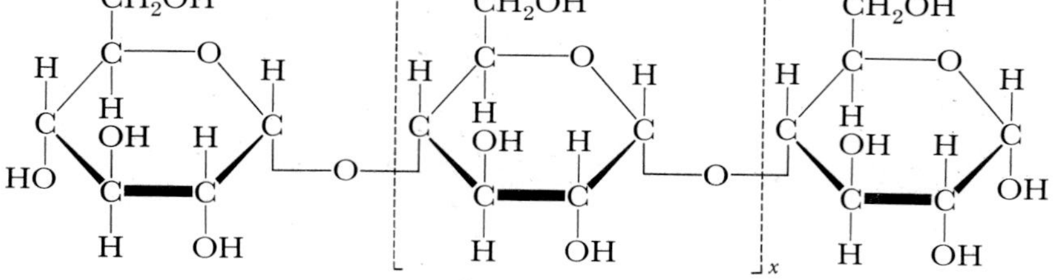

FIGURE 25.11 Structure of starch molecules.

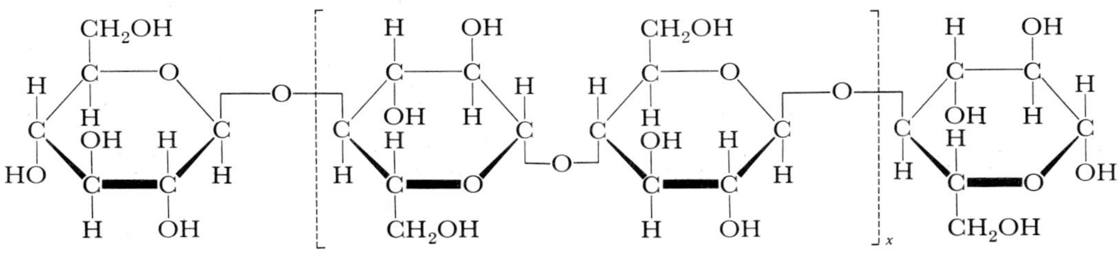

FIGURE 25.12 Structure of cellulose.

25.5
fats and oils

Both plant and animal systems need means of storing energy in various chemical forms so that this energy supply can be tapped as needed later on. In plant seeds, the stored energy is used to promote rapid growth after germination. In animals that hibernate during cold periods, the stored energy is needed when other sources of food are absent or scarce. One of the most important classes of compounds used for energy storage are the **fats** and **oils**. These substances are esters of long-chain carboxylic acids with 1,2,3-trihydroxypropane (glycerol) and are called **triglycerides**. Their general structural formula is shown in Figure 25.13. The groups R_1, R_2, and R_3 can be alike or different. The triglycerides are found as fat deposits in animals and as oils in nuts and seeds. Those that are liquid at room temperature are generally referred to as oils; those that are solid are called fats.

Hydrolysis of a fat or oil yields glycerol and the carboxylic acids of the R_1, R_2, and R_3 chains. The acids are commonly referred to as fatty acids. Most commonly the fatty acids contain from 12 to 22 carbons. It is an interesting fact that nearly all fatty acids contain an even number of carbon atoms, including the carboxylic acid carbon. Several of the more common fatty acids are listed in Table 25.1. The oils, which are obtained

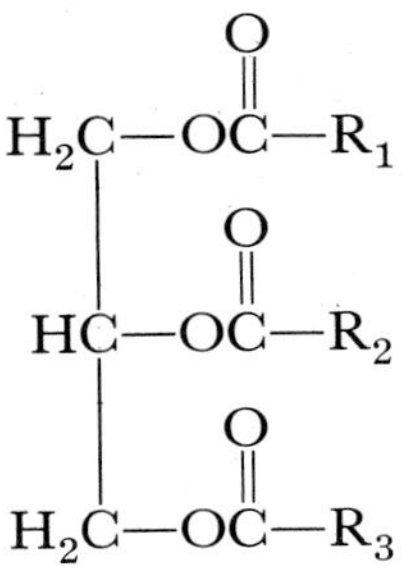

TABLE 25.1

name, source, and structure of several fatty acids

FIGURE 25.13 Structure of tri-glycerides. The groups R_1, R_2, and R_3 are hydrocarbon chains of from 3 to 21 carbons. The chains may be entirely saturated or may contain one or more *cis* olefin groups.

FORMULA OR STRUCTURE[a]	NAME	SOURCE
$CH_3(CH_2)_{10}COOH$	Lauric acid	Coconuts
$CH_3(CH_2)_{12}COOH$	Myristic acid	Butter
$CH_3(CH_2)_{16}COOH$	Stearic acid	Animal fats
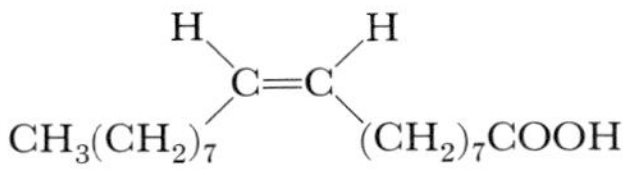 Oleic acid structure	Oleic acid	Corn oil
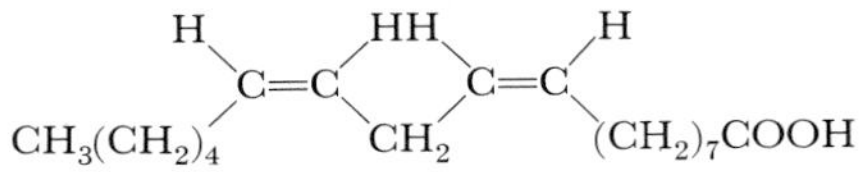 Linoleic acid structure	Linoleic acid	Vegetable oils
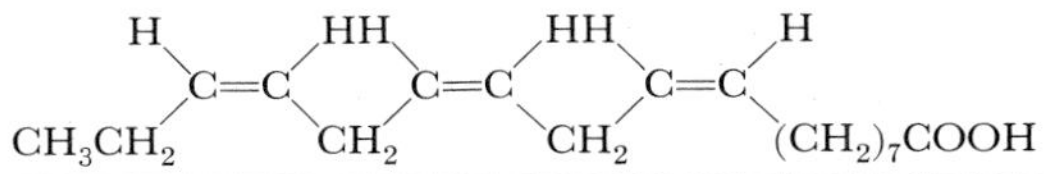 Linolenic acid structure	Linolenic acid	Linseed oil

[a]Notice that the *cis* configuration is adopted at the double bonds in the unsaturated fatty acids.

mainly from plant products (corn, peanuts, and soybeans), are formed primarily from unsaturated fatty acids. On the other hand, animal fats (beef, butter, and pork) contain mainly saturated fatty acids.

Some of the fat extracted from cottonseed, corn, or soybean oil is utilized as liquid cooking oil. Other plant-derived fats are hydrogenated to convert the carbon-carbon double bonds in the acid chains to carbon-carbon single bonds. In the process, the liquid fats are converted to solids; they are used to produce oleomargarine, peanut butter, vegetable shortening, and other similar food products. As an example, trilinolein, a major constituent of cottonseed oil, is hydrogenated to form tristearin, which is solid at room temperature:

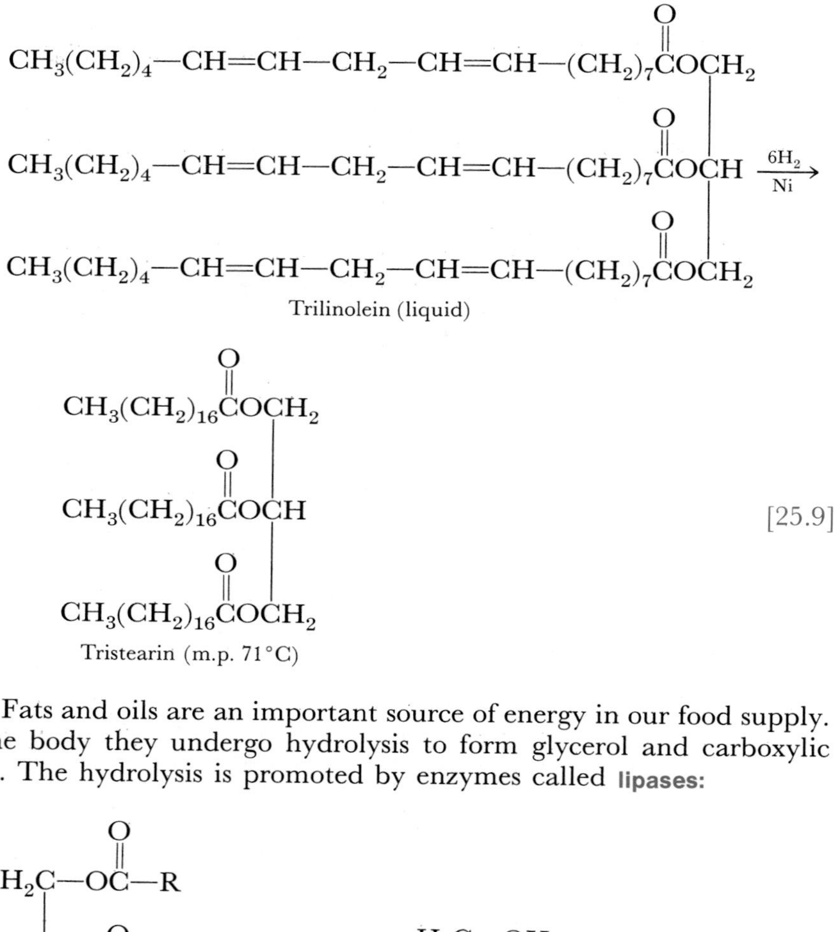

$$CH_3(CH_2)_4-CH{=}CH-CH_2-CH{=}CH-(CH_2)_7\overset{\overset{\displaystyle O}{\|}}{C}OCH_2$$

$$CH_3(CH_2)_4-CH{=}CH-CH_2-CH{=}CH-(CH_2)_7\overset{\overset{\displaystyle O}{\|}}{C}OCH \xrightarrow[\text{Ni}]{6H_2}$$

$$CH_3(CH_2)_4-CH{=}CH-CH_2-CH{=}CH-(CH_2)_7\overset{\overset{\displaystyle O}{\|}}{C}OCH_2$$

Trilinolein (liquid)

$$CH_3(CH_2)_{16}\overset{\overset{\displaystyle O}{\|}}{C}OCH_2$$

$$CH_3(CH_2)_{16}\overset{\overset{\displaystyle O}{\|}}{C}OCH$$

$$CH_3(CH_2)_{16}\overset{\overset{\displaystyle O}{\|}}{C}OCH_2$$

Tristearin (m.p. 71°C)

[25.9]

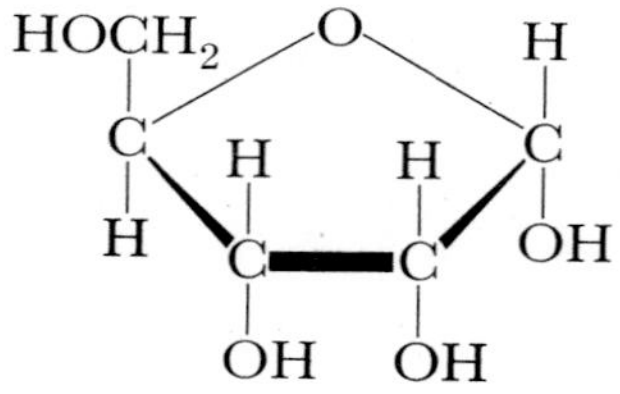

Ribose

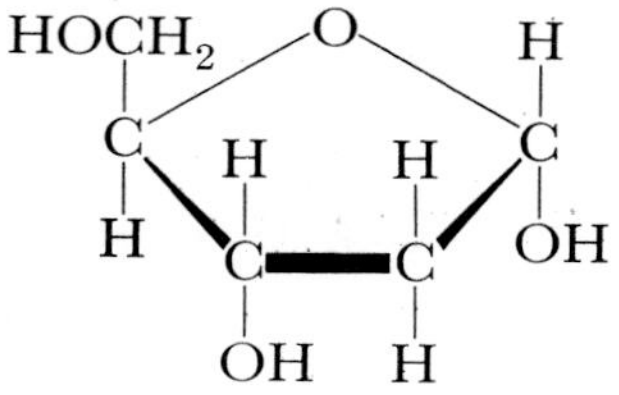

Deoxyribose

FIGURE 25.14 Ribose and deoxyribose, the two sugars found in RNA and DNA, respectively.

Fats and oils are an important source of energy in our food supply. In the body they undergo hydrolysis to form glycerol and carboxylic acids. The hydrolysis is promoted by enzymes called **lipases**:

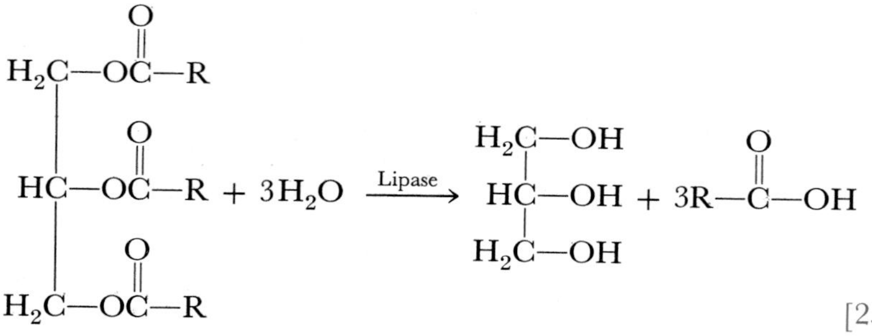

$$
\begin{array}{l}
H_2C-O\overset{\overset{\displaystyle O}{\|}}{C}-R \\[4pt]
HC-O\overset{\overset{\displaystyle O}{\|}}{C}-R + 3H_2O \xrightarrow{\text{Lipase}} \\[4pt]
H_2C-O\overset{\overset{\displaystyle O}{\|}}{C}-R
\end{array}
\quad
\begin{array}{l}
H_2C-OH \\
HC-OH + 3R-\overset{\overset{\displaystyle O}{\|}}{C}-OH \\
H_2C-OH
\end{array}
$$

[25.10]

This hydrolysis reaction, which takes place in aqueous solution, is hampered by the fact that fats and oils are essentially insoluble in water.

As a result, not much hydrolysis occurs in the stomach. The gall bladder excretes compounds called bile salts into the small intestine to assist in the hydrolysis. The bile salts break up the larger droplets of fats into an emulsion (a suspension of very small droplets), so that hydrolysis can proceed more rapidly.

25.6 nucleic acids

The nucleic acids are a class of biopolymers present in nearly all cells. They are classified into two groups—**deoxyribonucleic acids (DNA)** and **ribonucleic acids (RNA)**. DNA molecules are very large; their molecular weights may range from 6 to 16 million. RNA molecules are much smaller, with molecular weights in the range from 20,000 to 40,000. DNA is found primarily in the nucleus of the cell; RNA is found outside the nucleus, in the surrounding fluid called the cytoplasm.

Both DNA and RNA are polymers of a basic repeating unit called a **nucleotide.** To understand the structure of the polymer, we must first examine the structure of the nucleotide units. The nucleotide consists of three parts: (1) a phosphoric acid molecule, (2) a sugar in the furanose (five-membered ring) form, and (3) a nitrogen-containing organic base with a ring structure analogous to the ring structures of aromatic molecules. The sugar molecule found in RNA is ribose, shown in Figure 25.14. In DNA the sugar is deoxyribose, which differs from ribose only in the substitution of hydrogen for an —OH group on one carbon, as illustrated in Figure 25.14. In DNA the organic base may be any one of the four shown in Figure 25.15. An example of a complete nucleotide (deoxyadenosine) formed from these three components is illustrated in Figure 25.16. Substitution of one of the other bases for adenine in this structure produces any one of the other three nucleotides that make up DNA.

Just as a polypeptide is formed by a condensation reaction between amino acids, creating a peptide bond, a polynucleotide is formed by a

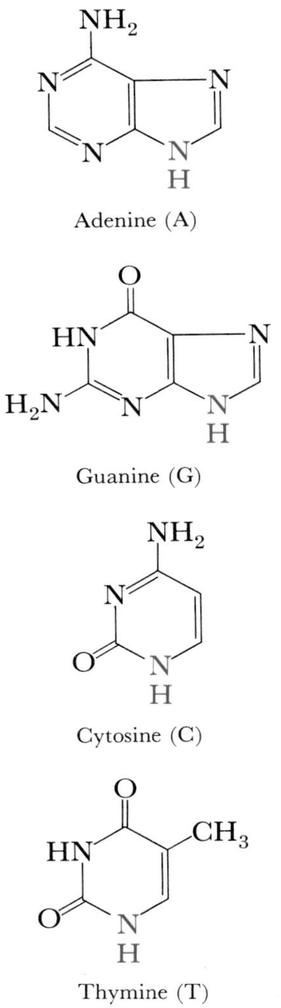

Adenine (A)

Guanine (G)

Cytosine (C)

Thymine (T)

FIGURE 25.15 The four organic bases present in DNA. In DNA each base is attached to a ribose molecule through a bond from the nitrogen shown in color.

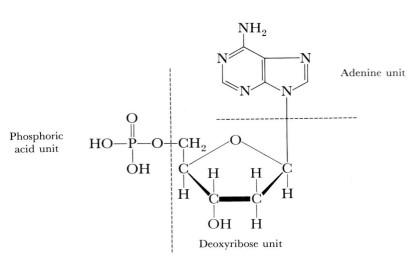

Adenine unit

Phosphoric acid unit

Deoxyribose unit

FIGURE 25.16 Structure of deoxyadenosine, a nucleotide formed from phosphoric acid, deoxyribose, and an organic base, adenine.

773

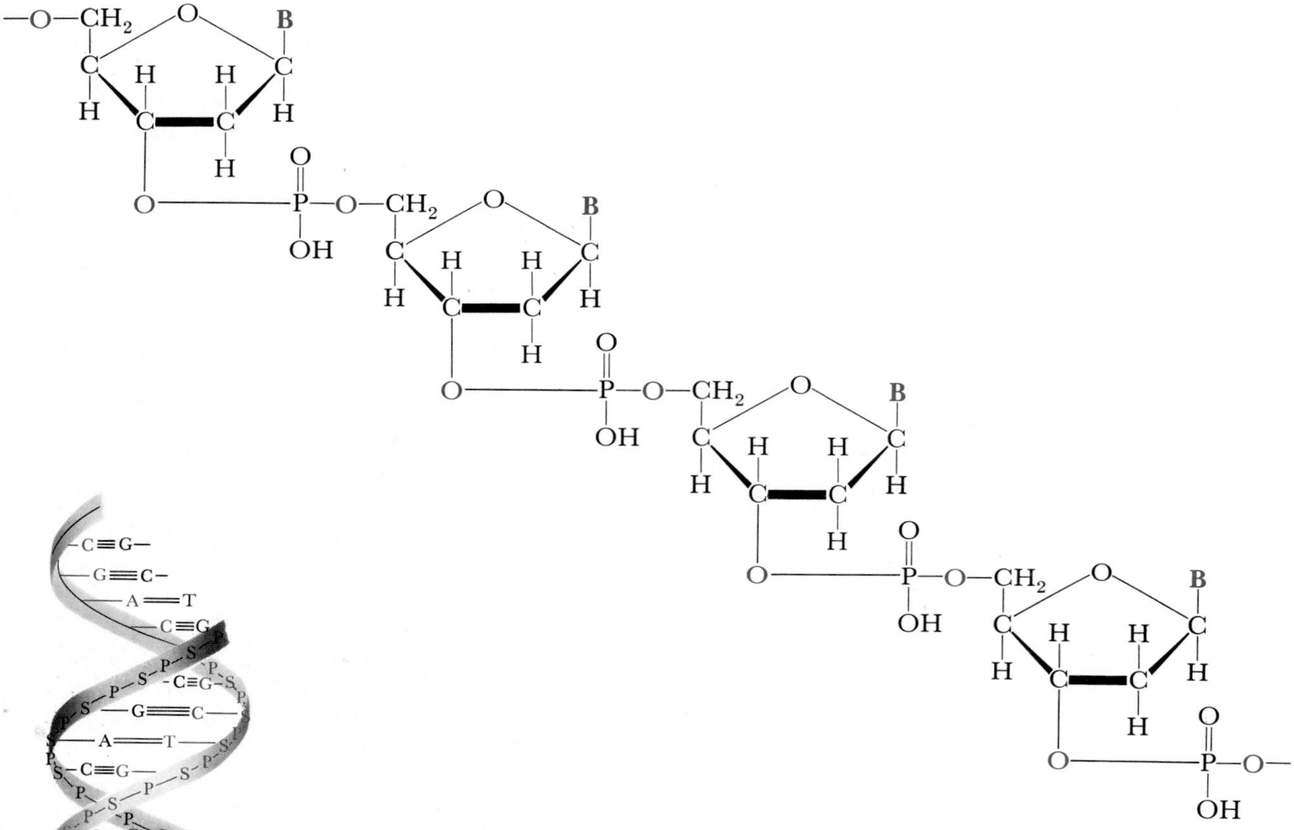

FIGURE 25.17 Structure of a polynucleotide. The symbol B represents any one of the four bases shown in Figure 25.16. Because the sugar in each nucleotide is deoxyribose, this polynucleotide is of the form found in DNA.

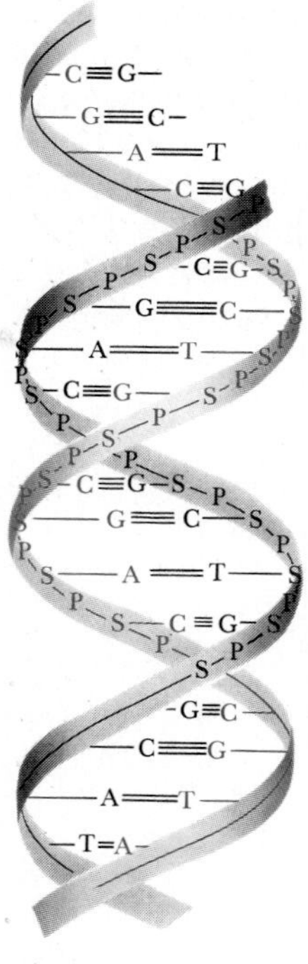

FIGURE 25.18 Schematic illustration of the double-stranded helical structure for DNA. The hydrogen-bond interactions between complementary base pairs is represented as illustrated in Figure 25.19.

condensation reaction creating a phosphate ester bond. The condensation reaction results in elimination of water from between an OH group of the phosphoric acid and an OH group of the sugar. The formation of the polymeric chain is shown schematically in Figure 25.17. The organic bases in this illustration are simply represented as B; they might be the same or different bases.

DNA molecules consist of two linear strands of the sort shown in Figure 25.17, which are wound together in the form of a **double helix**. It is very confusing to look at a model of double-stranded DNA in which all the atoms are represented. However, we can see from a more schematic

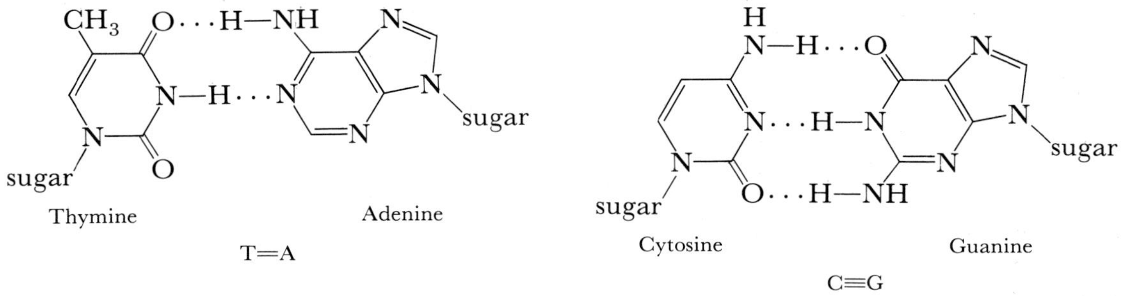

FIGURE 25.19 Hydrogen bonding between complementary base pairs. The hydrogen bonds shown here are responsible for formation of the double-stranded helical structure for DNA, as shown in Figure 25.18.

illustration, Figure 25.18, how the DNA molecule is constructed. Remember that the polymeric strand itself consists of alternating sugar and phosphate groups, represented as —S— and —P—, respectively. The various bases are attached to the polymeric strand at each sugar unit. The key to the double-helix structure for DNA is the formation of hydrogen bonds between bases on the two chains. When adenine and thymine are located opposite one another along the chain, they form a strong hydrogen bond as shown in Figure 25.19. Similarly, guanine and cytosine form especially strong hydrogen bonds with one another. This means that the two strands of DNA are complementary; opposite each adenine on one chain there is always a thymine; opposite each guanine there is a cytosine. The double-helix structure for DNA, first proposed by James Watson and Francis Crick in 1953, has proved to be the key to understanding protein synthesis in cells, the means by which viral particles infect cells, the means by which genetic information is transmitted in reproduction of cells, and many other problems of central importance in molecular biology. Those themes are beyond the scope of this book; however, if you are interested in study in the area of life sciences, you will learn a good deal about such matters in other courses.

FOR REVIEW

summary

In this chapter, we have taken a brief look at some of the major constituents of the biosphere—that part of the physical world in which organisms live out their life cycles. The maintenance of life requires a source of energy as well as appropriate environmental conditions. The ultimate source of the needed energy is the sun. Plants convert solar energy to chemical energy in the process called **photosynthesis**. Solar energy is absorbed by a plant pigment called chlorophyll and then utilized to form carbohydrate and O_2 from CO_2 and H_2O.

Proteins are polymers of **amino acids**. They form a major structural element in animal systems. **Enzymes**, the catalysts of biochemical reactions, are protein in nature. All naturally occurring proteins are formed from some 20 amino acids. The amino acids are **chiral** substances; that is, they are capable of existing as nonsuperimposable mirror-image isomers called **enantiomers**. Usually, only one of the enantiomers is found to be biologically active. Protein structure is determined by the sequence of amino acids in the chain, the coiling or stretching of the chain, and the overall shape of the complete molecule. All these aspects of protein structure are important in determining its biological activity. Heating or other forms of treatment may inactivate, or **denature,** the protein.

In this chapter, we also have considered the action of enzymes—their specificity, a model for how they may operate, and the ways in which they are sometimes inhibited.

Carbohydrates, which are formed from polyhydroxy aldehydes and ketones, are the major structural constituent of plants and a source of energy in both plants and animals. The three most important groups of carbohydrates are **starch,** which is found in plants, **glycogen,** which is found in mammals, and **cellulose,** which is also found in plants. All are **polysaccharides;** that is, they are polymers of a simple sugar, glucose.

Fats and **oils,** along with proteins and carbohydrates, form the important sources of energy in our food supply. The fats and oils are esters of long chain acids with glycerol, 1,2,3-trihydroxypropane; they are often called **triglycerides.** The major sources of these substances are animal fats and the oils of plant seeds, such as corn, peanuts, cottonseeds, and soybeans.

The **nucleic acids** are biopolymers of high molecular weight that carry the genetic information necessary for cell reproduction. In addition, the nucleic acids control cell development through control of protein synthesis. The nucleic acids consist of a polymeric backbone of alternating phosphate and ribose sugar groups, with organic bases attached to the sugar molecules. The DNA polymer is a double-stranded helix that is held together by hydrogen bonding between matching organic bases situated across from one another on the two strands.

learning goals

Having read and studied this chapter, you should be able to:

1. Describe two distinct reasons why energy is required by living organisms.
2. List the several functions of proteins in living systems.
3. Define the terms chiral and enantiomer and draw the enantiomer of a given chiral molecule.
4. Write the reaction for formation of a peptide bond between two amino acids.
5. Explain the structures of proteins in terms of primary, secondary, and tertiary structure.
6. Explain the following characteristics of enzymes in terms of what is known about proteins: specificity; temperature dependence of enzyme activity, as shown in Figure 25.6; inhibition.
7. Explain the lock-and-key mechanism for enzyme action.
8. Distinguish the cyclic hemiacetal and linear forms of a sugar and describe the difference between the pyranose and furanose forms of a sugar.
9. Describe the manner in which monosaccharides are joined together to form polysaccharides.
10. Enumerate the major groups of polysaccharides and indicate their sources and general functions.
11. Describe the structures of fats and oils and list the sources of these substances.
12. Draw the structures of any of the nucleotides that make up the polynucleotide DNA.
13. Describe the nature of the polymeric unit of polynucleotides.
14. Describe the double-stranded structure of DNA and explain the principle that determines the relationship between bases in the two strands.

key terms

Among the more important terms and expressions used for the first time, or in a new context, in this chapter are the following:

The **active site** (Section 25.3) is the specific site in an enzyme at which the enzyme-catalyzed reaction occurs.

An **amino acid** (Section 25.2) is a carboxylic acid that contains an amino (—NH$_2$) group attached to the carbon atom adjacent to the carboxylic acid (CO$_2$H) functional group. Twenty different amino acids are important in living systems.

The **biosphere** (introduction) is that part of the earth in which living organisms can exist and live out their life cycles.

Carbohydrates (Section 25.4) are a class of substances formed from polyhydroxy aldehydes or ketones.

Cellulose (Section 25.4) is a polysaccharide of glucose; it is the major structural element in plant matter.

A **chiral** molecule (Section 25.2) is one that is not superimposable on its mirror image.

Chlorophyll (Section 25.1), found in plant leaves, is a plant pigment that plays a major role in conversion of solar energy to chemical energy in photosynthesis.

A **coenzyme**, or **cofactor** (Section 25.3), is a substance that is needed along with the enzyme if an enzyme-catalyzed reaction is to occur.

Configurational isomers (Section 25.2) are molecules that differ only by being nonsuperimposable mirror images of one another. Such isomers are often called **optical isomers** because their solutions affect the plane of polarized light differently.

Denaturation (Section 25.2), is the loss of biological activity in a protein because of disruption of its tertiary structure by heating, by the action of acids or bases, or by other influence.

Deoxyribonucleic acid, DNA (Section 25.6), is a polynucleotide in which the sugar component is deoxyribose (in the furanose-ring form).

The **double-helix** structure for DNA (Section 25.6) involves the winding of two DNA polynucleotide chains together in a helical arrangement. The two strands of the double helix are complementary, in that the organic bases on the two strands are paired for optimal hydrogen-bond interaction.

Enantiomers (Section 25.2) are two mirror-image molecules of a chiral substance. The enantiomers are nonsuperimposable.

Enzymes (Section 25.3) are proteins that act as catalysts in biochemical reactions.

Fats and oils (Section 25.5) are esters of long-chain carboxylic acids and the alcohol 1,2,3-trihydroxypropane (glycerol). These esters are also called **triglycerides**.

A **furanose** (Section 25.4) is a cyclic sugar molecule in the form of a five-membered ring.

Glucose (Section 25.4), a polyhydroxy aldehyde of formula CH$_2$OH(CHOH)$_4$CHO, is the most important of the monosaccharides.

Glycogen (Section 25.4) is the general name given to a group of polysaccharides of glucose that are synthesized in mammals and used as a means of carbohydrate energy storage.

In the **α-helix** structure (Section 25.2) for a protein, the protein is coiled in the form of a helix, with hydrogen bonds between C=O and N—H groups on adjacent turns.

A **hemiacetal** (Section 25.4) is a bonding arrangement formed by reaction of an aldehyde functional group with an alcohol. The hemiacetal linkage is formed when aldehyde sugars form the cyclic structure.

Lipases (Section 25.5) are enzymes that catalyze the hydrolysis of fats.

In the **lock-and-key model** for enzyme action (Section 25.3), the substrate molecule is pictured as fitting rather specifically into the active site on the enzyme. It is assumed that in being bound to the active site the substrate is somehow activated for reaction.

Macromolecules (Section 25.1) are polymeric molecules of high molecular weight. In referring to substances of biochemical origin, the term biopolymers is often used.

A **monosaccharide** (Section 25.4) is a simple sugar, most commonly containing six carbon atoms. The joining together of monosaccharide units by a condensation reaction results in formation of polysaccharides.

Nucleic acids (Section 25.6) are polymers of nucleotides that have high molecular weights.

A **nucleotide** (Section 25.6) is formed from a molecule of phosphoric acid, a sugar molecule, and an organic base. Nucleotides form linear polymers called DNA and RNA that are involved in protein synthesis and cell reproduction.

Photosynthesis (Section 25.1) is the process occurring in plant leaves by which light energy is used to convert CO_2 and water to carbohydrates and oxygen.

Polarized light (Section 25.2) is electromagnetic radiation in which the waves that form the light move in a single plane.

The **primary structure** of a protein (Section 25.2) refers to the sequence of amino acids along the protein chain.

Proteins (Section 25.2) are biopolymers formed from amino acids.

A **pryranose** (Section 25.4) is a cyclic sugar molecule in the form of a six-membered ring.

Ribonucleic acid, RNA (Section 25.6), is a polynucleotide in which ribose (in the furanose-ring form) is the sugar component.

The **secondary structure** of a protein (Section 25.2) refers to the manner in which the protein is coiled or stretched.

Starch (Section 25.4) is the general name given to a group of polysaccharides that act as energy-storage substances in plants.

The **tertiary structure** of a protein (Section 25.2) refers to the overall shape of the molecule—specifically, the manner in which sections of the chain fold back upon themselves or intertwine.

Turnover number (Section 25.3) refers to the number of individual reaction events per unit time at a particular active site in an enzyme.

EXERCISES

energy requirements

25.1 Explain the relationship between entropy and the energy needs of organisms.

25.2 Name at least four processes occurring in an organism (such as yourself) that require energy. These may occur at the molecular level or may involve the complete organism.

25.3 What is the role of pigments in photosynthesis? What characteristics would you expect their absorption spectra to have?

25.4 The amount of solar radiation at the surface of a Midwest cornfield in July averages 1885 J/cm² per day. Assuming that 3 percent of this radiation is converted into dry weight of plant matter, what is the total energy value stored in plant matter over a 30-day period on an acre of land? (An acre is 4047 m².)

[25.5] Write a balanced equation for the overall photosynthesis reaction. Using the standard heat of combustion for crystalline glucose (2814 kJ/mole) and assuming that this same value would apply to oxidation of aqueous glucose to CO_2 and H_2O, calculate the heat of formation of glucose. From this, calculate the standard enthalpy change in your balanced photosynthesis reaction.

proteins

25.6   Name at least four distinct roles for proteins in animal organisms.

25.7   Explain what is meant by the following terms: (a) chiral molecule; (b) optical isomer; (c) enantiomer; (d) polarized light.

25.8   Draw the two enantiomeric forms of aspartic acid.

25.9   Draw the two enantiomeric forms of cysteine.

25.10   A 0.1 M solution of D-alanine causes rotation of 18° of the plane of polarized light in a polarimeter. What rotation would you expect for a 0.05 M solution of L-alanine; for a solution that is 0.1 M each in D-alanine and L-alanine?

25.11   Write reactions showing formation of aspartylcysteine and cysteinylasparagine from the constituent amino acids.

25.12   Draw the structure of the tripeptide tyr-ile-ser.

25.13   Draw the structure of the tripeptide pro-leu-gly.

25.14   Describe what is meant by the terms primary, secondary, and tertiary structures of proteins.

25.15   Describe the role of hydrogen bonding in determining the α-helix structure of a protein.

25.16   What is meant by the term denaturation? Describe how increased temperature could lead to denaturation.

enzymes

25.17   Provide a definition or description of each of the following terms: (a) enzyme; (b) apoenzyme; (c) denaturation; (d) active site; (e) holoenzyme; (f) specificity; (g) peptidase; (h) turnover number.

25.18   Normally the rates of enzyme-catalyzed reactions increase linearly with increase in substrate concentration. What does this tell us about the position of equilibrium in Equation [25.4]? Explain.

25.19   In terms of the lock-and-key model, what characteristics must an enzyme inhibitor possess?

25.20   In the enzyme-substrate model, what must the values for k_1 and k_{-1} be in relationship to k_2?

[25.21]   Suppose that a particular enzyme-substrate combination gives the sort of behavior shown in Figure 25.8. The concentration of enzyme in solution is $2 \times 10^{-7}\,M$, and the limiting rate of product formation is $4.7 \times 10^{-4}\,M/\text{sec}$. Assuming one active site per enzyme molecule, what is the turnover number? How does turnover number relate to k_2 under these conditions?

carbohydrates

25.22   The structural formula for the linear form of galactose is shown here. Draw the structure of the pyranose form of this sugar.

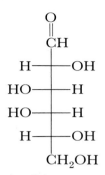

Galactose

25.23   What is the product or products of hydrolysis of starch? Of cellulose? Of sucrose?

25.24   What aspect of the structures of starch and cellulose is responsible for the fact that starch is digestible in our bodies and cellulose is not?

25.25   An aqueous solution of table sugar, when placed in a beam of plane-polarized light, causes rotation of the plane of polarized light. Explain.

fats and oils

25.26   Draw the structural formula for the triglyceride of oleic acid.

25.27   The traditional diet of Eskimos is very high in fats and oils, but it contains a much higher percentage of unsaturated fats and oils than does a typical American diet. What aspect of the traditional Eskimo environment would tend to promote the presence of the unsaturated compounds? Explain.

25.28   Write a balanced chemical equation for complete hydrogenation of trilinolein. If partial hydrogenation were to occur, with uptake of only 4 moles of H_2 per mole of fat, draw the structural formulas of two possible products.

25.29   Fats and oils belong to a general class of compounds called lipids, which are distinguished by the fact that they are soluble in organic solvents. Discuss, in terms of their comparative structures, why fats belong to this class, whereas carbohydrates such as the disaccharides do not.

nucleic acids

25.30   Describe a nucleotide. Draw the structural formula for deoxycytidine monophosphate in which cytosine is the organic base.

25.31 When samples of double-stranded DNA are analyzed, the quantity of adenine present equals that of thymine. Similarly, the quantity of guanine equals that of cytosine. Explain the significance of these observations.

general exercises

25.32 It is found that mild heating of an enzyme first increases its activity, but that additional temperature increase results in loss of activity. Cooling the solution results in return of much of the activity. Explain these observations in terms of the general properties of proteins.

25.33 Write equations to show the digestion of the tripeptide gly-ala-leu to its constituent amino acids in two steps under the influence of carboxypeptidase.

[25.34] Explain why the energy requirements of biochemical systems lead to the need for coupled reactions (see also Section 15.7).

25.35 Protein enzymes are sold commercially for use as meat tenderizers and for stain removal (for example, bloodstains) from cloth. What kind of enzymes do you expect that these are? Explain how they work.

25.36 What meaning would you attach to the term denaturation as applied to a DNA molecule?

25.37 Describe a simple test that could be used to determine whether a sample of a fat contained any unsaturation. (Hint: the answer is in Chapter 24.)

25.38 Describe the biochemical reaction catalyzed by each of the following enzymes: (a) carboxypeptidase; (b) cellulase; (c) ribonuclease; (d) lipase.

25.39 One of the many remarkable enzymes in the human body is *carbonic anhydrase,* which catalyzes the release of dissolved carbon dioxide from the blood into the air of our lungs. If it were not for this enzyme, the body could not rid itself rapidly enough of the CO_2 accumulated by cell metabolism. The enzyme has a molecular weight of 30,000, contains one atom of zinc per protein molecule, and catalyzes the dehydration (release to the air) of up to 10^7 CO_2 molecules per second. Which components of this description correspond to the terms holoenzyme, apoenzyme, cofactor, and turnover number?

[25.40] The enzyme *invertase* catalyzes the conversion of sucrose, a disaccharide, to invert sugar. When the concentration of invertase is 3×10^{-7} M, and the concentration of sucrose is 0.01 M, invert sugar is formed at the rate of 2×10^{-4} M/sec. When the sucrose concentration is doubled, the rate of formation of invert sugar is doubled also. Assuming that the enzyme-substrate model is operative, is the fraction of enzyme tied up as complex large or small? Explain. Addition of innositol, another sugar, causes a decrease in rate of formation of invert sugar. Suggest a mechanism by which this occurs.

appendices

appendix a

mathematical operations

A.1 EXPONENTIAL NOTATION

The numbers used in chemistry are often either extremely large or extremely small. Such numbers are conveniently expressed in the form

$$N \times 10^n$$

where N is a number between 1 and 10, and n is the exponent. Some examples of this exponential notation are as follows:

1,200,000 is 1.2×10^6 (read "one point two times ten to the sixth power")

0.000604 is 6.04×10^{-4} (read "six point oh four times ten to the negative fourth power")

A positive exponent, as in our first example, tells us how many times a number must be multiplied by 10 to give the long form of the number:

$$1.2 \times 10^6 = 1.2 \times 10 \times 10 \times 10 \times 10 \times 10 \times 10 \quad \text{(six tens)}$$
$$= 1,200,000$$

It is also convenient to think of the positive exponent as the number of places the decimal must be moved to the *left* to give a number greater than 1 and less than 10: if we begin with 3450 and move decimal three places to left, we end up with 3.45×10^3.

In a related fashion, a negative exponent tells us how many times we must divide a number by 10 to give the long form of the number:

$$6.04 \times 10^{-4} = \frac{6.04}{10 \times 10 \times 10 \times 10}$$
$$= 0.000604$$

It is convenient to think of the negative exponent as the number of places the decimal place must be moved to the *right* to give a number greater than 1 but less than 10: if we begin with 0.0048 and move decimal three places to right, we end up with 4.8×10^{-3}.

In the system of exponential notation, with each shift of the decimal point one place to the right, the exponent *decreases* by 1:

$$4.8 \times 10^{-3} = 48 \times 10^{-4}$$

Similarly, with each shift of the decimal point one place to the left, the exponent *increases* by 1:

$$4.8 \times 10^{-3} = 0.48 \times 10^{-2}$$

In working with exponents, it is important to know that $10^0 = 1$. The following rules are useful for carrying exponents through calculations:

1. *Addition and Subtraction.* In order to add or subtract numbers expressed in exponential notation, the powers of 10 must be the same:

$$(5.22 \times 10^4) + (3.21 \times 10^2) = (522 \times 10^2) + (3.21 \times 10^2)$$
$$= 525 \times 10^2$$
$$= 5.25 \times 10^4$$
$$(6.25 \times 10^{-2}) - (5.77 \times 10^{-3}) = (6.25 \times 10^{-2}) - (0.577 \times 10^{-2})$$
$$= 5.67 \times 10^{-2}$$

2. *Multiplication and Division.* When numbers expressed in exponential notation are multiplied, the exponents are added; when numbers expressed in exponential notation are divided, the exponent of the divisor is subtracted from the exponent of the dividend:

$$(5.4 \times 10^2)(2.1 \times 10^3) = (5.4)(2.1) \times 10^{2+3}$$
$$= 11 \times 10^5$$
$$= 1.1 \times 10^6$$
$$(1.2 \times 10^5)(3.22 \times 10^{-3}) = (1.2)(3.22) \times 10^{5-3}$$
$$= 3.8 \times 10^2$$

$$\frac{3.2 \times 10^5}{6.5 \times 10^2} = \frac{3.2}{6.5} \times 10^{5-2} = 0.49 \times 10^3$$
$$= 4.9 \times 10^2$$

$$\frac{5.7 \times 10^7}{8.5 \times 10^{-2}} = \frac{5.7}{8.5} \times 10^{7-(-2)} = 0.67 \times 10^9$$
$$= 6.7 \times 10^8$$

3. *Powers and Roots.* When numbers expressed in exponential notation are raised to a power, the exponents are multiplied by the power; when the roots of numbers expressed in exponential notation are taken, the exponents are divided by the root:

$$(1.2 \times 10^5)^3 = 1.2^3 \times 10^{5 \times 3}$$
$$= 1.7 \times 10^{15}$$
$$3\sqrt{2.5 \times 10^6} = \sqrt[3]{2.5} \times 10^{6/3}$$
$$= 1.3 \times 10^2$$

A.2 LOGARITHMS

The common, or base-10, logarithm (abbreviated log) of any number is the power to which 10 must be raised to equal the number. For example, the common logarithm of 1000 (written log 1000) is 3, because raising 10 to the third power gives $1000: 10^3 = 1000$. Further examples are:

$$\text{Log } 10^5 = 5$$
$$\text{Log } 1 = 0$$
$$\text{Log } 10^{-2} = -2$$

In these examples, the logarithm can be obtained by simple inspection. However, it is not possible to obtain the logarithm of a number like 3.2 by inspection. The logarithms of such numbers can be obtained from a log table such as that given in Table 1. This table can be used to find the logs of numbers from 1 to 10. To do so, we locate the first digit of the number in the first vertical column. We then move horizontally to the column headed by the second digit of the number. In this way we find that the log of 3.2 is 505. Because decimals are omitted from the table, the log of 3.2 is actually 0.505. Further examples are:

$$\text{Log } 2.50 = 0.398$$
$$\text{Log } 2.55 = 0.406$$

In using Table 1 to find the log of 2.55, we must estimate, or interpolate, the value from the logs given for 2.5 and 2.6. Appendix B is a table of four-place logarithms from which the logarithm of 2.55 can be read

TABLE 1

three-place common logarithms

	0	1	2	3	4	5	6	7	8	9
1	000	041	079	114	146	176	204	230	255	279
2	301	322	342	362	380	398	415	431	447	462
3	477	491	505	519	532	544	556	568	580	591
4	602	613	623	634	644	653	663	672	681	690
5	699	708	716	724	732	740	748	756	763	771
6	778	785	792	799	806	813	820	826	833	839
7	845	851	857	863	869	875	881	887	892	898
8	903	909	914	919	924	929	935	940	945	949
9	954	959	964	969	973	978	982	987	991	996

directly as 0.4065. (In using Appendix B, you should mentally insert a decimal point between the two-digit numbers that form the first column.)

The logarithm of a number that is less than 1 or greater than 10 can be determined by writing the number first in standard exponential notation as the following examples show:

$$
\begin{aligned}
\text{Log } 450 &= \log (4.50 \times 10^2) \\
&= \log 4.50 + \log 10^2 \\
&= 0.653 + 2 = 2.653
\end{aligned}
$$

$$
\begin{aligned}
\text{Log } 0.0673 &= \log (6.73 \times 10^{-2}) \\
&= \log 6.73 + \log 10^{-2} \\
&= 0.828 - 2 \\
&= -1.172
\end{aligned}
$$

Check these examples yourself, using Appendix B.

Because logarithms are exponents, mathematical operations involving logarithms follow the rules for the use of exponents:

1. Multiplication and division:

$$\text{Log } ab = \log a + \log b$$

$$\text{Log } \frac{a}{b} = \log a - \log b$$

2. Powers and roots:

$$\text{Log } a^n = n(\log a)$$

$$\text{Log } a^{1/n} = \left(\frac{1}{n}\right)(\log a)$$

Obtaining Antilogarithms. The process of finding a number given its logarithm is known as obtaining an antilogarithm. It is the reverse of taking a logarithm, as the following examples show:

1. Find the number whose logarithm is 5.322.

$$
\begin{aligned}
\text{Antilog } 5.322 &= \text{antilog } 0.322 \times \text{antilog } 5 \\
&= 2.10 \times 10^5
\end{aligned}
$$

2. Find the number whose logarithm is -2.133.

$$
\begin{aligned}
\text{Antilog } (-2.133) &= \text{antilog } 0.867 \times \text{antilog } (-3) \\
&= 7.37 \times 10^{-3}
\end{aligned}
$$

pH Problems. In general chemistry, logarithms are used most frequently in working pH problems. The pH is defined as $-\log [H^+]$, as discussed in Section 16.7. The following sample exercise illustrates this application.

SAMPLE EXERCISE 1

(a) What is the pH of a solution whose hydrogen-ion concentration is 0.015 M? (b) If the pH of a solution is 3.80, what is its hydrogen-ion concentration?

Solution:

(a) $pH = -\log [H^+]$

$ = -\log (0.015)$

$ = -\log (1.5 \times 10^{-2})$

$ = -\log 1.5 - \log (10^{-2})$

$ = -0.18 + 2 = 1.82$

(b) $pH = -\log [H^+] = 3.80$

$\log [H^+] = -3.80$

$[H^+] = \text{antilog} (-3.80)$

$ = \text{antilog } 0.20$

$ \times \text{antilog} (-4)$

$ = 1.6 \times 10^{-4} \, M$

Natural Logarithms. Natural, or base e, logarithms (abbreviated ln) are the power to which e, which has the value $2.71828 \ldots$, must be raised to equal a number. The relation between common and natural logarithms is as follows:

$$\ln a = 2.303 \log a$$

A.3 QUADRATIC EQUATIONS

An algebraic equation of the form $ax^2 + bx + c = 0$ is called a quadratic equation. The two solutions to such an equation are given by the quadratic formula:

$$x = \frac{-b \pm \sqrt{b^2 - 4ac}}{2a}$$

SAMPLE EXERCISE 2

Find x if $2x^2 + 4x = 1$.

Solution: To solve the given equation for x, we must first put it in the form

$ax^2 + bx + c = 0$

and then use the quadratic formula. If

$2x^2 + 4x = 1$

then

$2x^2 + 4x - 1 = 0$

Using the quadratic formula, where $a = 2$, $b = 4$, and $c = -1$, we have

$x = \dfrac{-4 \pm \sqrt{(4)(4) - 4(2)(-1)}}{2(2)}$

$ = \dfrac{-4 \pm \sqrt{16 + 8}}{4} = \dfrac{-4 \pm \sqrt{24}}{4}$

$ = \dfrac{-4 \pm 4.899}{4}$

The two solutions are

$x = \dfrac{0.899}{4} = 0.225$

$x = \dfrac{-8.899}{4} = -2.225$

Often in chemical problems the negative solution has no physical meaning, and only the positive answer is used.

A.4 GRAPHS

Often the clearest way to represent the interrelationship between two variables is to graph them. Usually the variable that is being experi-

TABLE 2

interrelation between pressure and temperature

TEMPERATURE (°C)	PRESSURE (atm)
20.0	0.120
30.0	0.124
40.0	0.128
50.0	0.132

mentally varied, called the independent variable, is shown along the horizontal axis (x-axis). The variable that responds to the change in the independent variable, called the dependent variable, is then shown along the vertical axis (y-axis). For example, consider an experiment in which we vary the temperature of an enclosed gas and measure its pressure. The independent variable is temperature, whereas the dependent variable is pressure. The data shown in Table 2 could be obtained by means of this experiment. These data are shown graphically in Figure 1. The relationship between temperature and pressure is linear. The equation for any straight-line graph has the form

$$y = ax + b$$

where a is the slope of the line and b is the intercept with the y-axis. In the case of Figure 1, we could say that the relationship between temperature and pressure takes the form

$$P = aT + b$$

where P is pressure in atm, and T is temperature in °C. As shown on

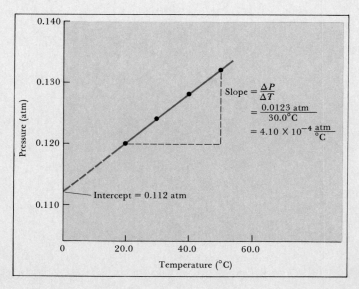

$$\text{Slope} = \frac{\Delta P}{\Delta T}$$
$$= \frac{0.0123 \text{ atm}}{30.0 \text{°C}}$$
$$= 4.10 \times 10^{-4} \frac{\text{atm}}{\text{°C}}$$

Intercept = 0.112 atm

FIGURE 1

Figure 1, the slope is $4.10 \times 10^{-4} \frac{\text{atm}}{°\text{C}}$, and the intercept, the point where the line crosses the y-axis, is 0.112 atm. Therefore the equation for the line is:

$$P = \left(4.10 \times 10^{-4} \frac{\text{atm}}{°\text{C}}\right) T + 0.112 \text{ atm}$$

	0	1	2	3	4	5	6	7	8	9
10	0000	0043	0086	0128	0170	0212	0253	0294	0334	0374
11	0414	0453	0492	0531	0569	0607	0645	0682	0719	0755
12	0792	0828	0864	0899	0934	0969	1004	1038	1072	1106
13	1139	1173	1206	1239	1271	1303	1335	1367	1399	1430
14	1461	1492	1523	1553	1584	1614	1644	1673	1703	1732
15	1761	1790	1818	1847	1875	1903	1931	1959	1987	2014
16	2041	2068	2095	2122	2148	2175	2201	2227	2253	2279
17	2304	2330	2355	2380	2405	2430	2455	2480	2504	2529
18	2553	2577	2601	2625	2648	2672	2695	2718	2742	2765
19	2788	2810	2833	2856	2878	2900	2923	2945	2967	2989
20	3010	3032	3054	3075	3096	3118	3139	3160	3181	3201
21	3222	3243	3263	3284	3304	3324	3345	3365	3385	3404
22	3424	3444	3464	3483	3502	3522	3541	3560	3579	3598
23	3617	3636	3655	3674	3692	3711	3729	3747	3766	3784
24	3802	3820	3838	3856	3874	3892	3909	3927	3945	3962
25	3979	3997	4014	4031	4048	4065	4082	4099	4116	4133
26	4150	4166	4183	4200	4216	4232	4249	4265	4281	4298
27	4314	4330	4346	4362	4378	4393	4409	4425	4440	4456
28	4472	4487	4502	4518	4533	4548	4564	4579	4594	4609
29	4624	4639	4654	4669	4683	4698	4713	4728	4742	4757
30	4771	4786	4800	4814	4829	4843	4857	4871	4886	4900
31	4914	4928	4942	4955	4969	4983	4997	5011	5024	5038
32	5051	5065	5079	5092	5105	5119	5132	5145	5159	5172
33	5185	5198	5211	5224	5237	5250	5263	5276	5289	5302
34	5315	5328	5340	5353	5366	5378	5391	5403	5416	5428
35	5441	5453	5465	5478	5490	5502	5514	5527	5539	5551
36	5563	5575	5587	5599	5611	5623	5635	5647	5658	5670
37	5682	5694	5705	5717	5729	5740	5752	5763	5775	5786
38	5798	5809	5821	5832	5843	5855	5866	5877	5888	5899
39	5911	5922	5933	5944	5955	5966	5977	5988	5999	6010
40	6021	6031	6042	6053	6064	6075	6085	6096	6107	6117
41	6128	6138	6149	6160	6170	6180	6191	6201	6212	6222
42	6232	6243	6253	6263	6274	6284	6294	6304	6314	6325
43	6335	6345	6355	6365	6375	6385	6395	6405	6415	6425
44	6435	6444	6454	6464	6474	6484	6493	6503	6513	6522
45	6532	6542	6551	6561	6571	6580	6590	6599	6609	6618
46	6628	6637	6646	6656	6665	6675	6684	6693	6702	6712
47	6721	6730	6739	6749	6758	6767	6776	6785	6794	6803
48	6812	6821	6830	6839	6848	6857	6866	6875	6884	6893
49	6902	6911	6920	6928	6937	6946	6955	6964	6972	6981

	0	1	2	3	4	5	6	7	8	9
50	6990	6998	7007	7016	7024	7033	7042	7050	7059	7067
51	7076	7084	7093	7101	7110	7118	7126	7135	7143	7152
52	7160	7168	7177	7185	7193	7202	7210	7218	7226	7235
53	7243	7251	7259	7267	7275	7284	7292	7300	7308	7316
54	7324	7332	7340	7348	7356	7364	7372	7380	7388	7396
55	7404	7412	7419	7427	7435	7443	7451	7459	7466	7474
56	7482	7490	7497	7505	7513	7520	7528	7536	7543	7551
57	7559	7566	7574	7582	7589	7597	7604	7612	7619	7627
58	7634	7642	7649	7657	7664	7672	7679	7686	7694	7701
59	7709	7716	7723	7731	7738	7745	7752	7760	7767	7774
60	7782	7789	7796	7803	7810	7818	7825	7832	7839	7846
61	7853	7860	7868	7875	7882	7889	7896	7903	7910	7917
62	7924	7931	7938	7945	7952	7959	7966	7973	7980	7987
63	7993	8000	8007	8014	8021	8028	8035	8041	8048	8055
64	8062	8069	8075	8082	8089	8096	8102	8109	8116	8122
65	8129	8136	8142	8149	8156	8162	8169	8176	8182	8189
66	8195	8202	8209	8215	8222	8228	8235	8241	8248	8254
67	8261	8267	8274	8280	8287	8293	8299	8306	8312	8319
68	8325	8331	8338	8344	8351	8357	8363	8370	8376	8382
69	8388	8395	8401	8407	8414	8420	8426	8432	8439	8445
70	8451	8457	8463	8470	8476	8482	8488	8494	8500	8506
71	8513	8519	8525	8531	8537	8543	8549	8555	8561	8567
72	8573	8579	8585	8591	8597	8603	8609	8615	8621	8627
73	8633	8639	8645	8651	8657	8663	8669	8675	8681	8686
74	8692	8698	8704	8710	8716	8722	8727	8733	8739	8745
75	8751	8756	8762	8768	8774	8779	8785	8791	8797	8802
76	8808	8814	8820	8825	8831	8837	8842	8848	8854	8859
77	8865	8871	8876	8882	8887	8893	8899	8904	8910	8915
78	8921	8927	8932	8938	8943	8949	8954	8960	8965	8971
79	8976	8982	8987	8993	8998	9004	9009	9015	9020	9025
80	9031	9036	9042	9047	9053	9058	9063	9069	9074	9079
81	9085	9090	9096	9101	9106	9112	9117	9122	9128	9133
82	9138	9143	9149	9154	9159	9165	9170	9175	9180	9186
83	9191	9196	9201	9206	9212	9217	9222	9227	9232	9238
84	9243	9248	9253	9258	9263	9269	9274	9279	9284	9289
85	9294	9299	9304	9309	9315	9320	9325	9330	9335	9340
86	9345	9350	9355	9360	9365	9370	9375	9380	9385	9390
87	9395	9400	9405	9410	9415	9420	9425	9430	9435	9440
88	9445	9450	9455	9460	9465	9469	9474	9479	9484	9489
89	9494	9499	9504	9509	9513	9518	9523	9628	9533	9538
90	9542	9547	9552	9557	9562	9566	9571	9576	9581	9586
91	9590	9595	9600	9605	9609	9614	9619	9624	9628	9633
92	9638	9643	9647	9652	9657	9661	9666	9671	9675	9680
93	9685	9689	9694	9699	9703	9708	9713	9717	9722	9727
94	9731	9736	9741	9745	9750	9754	9759	9763	9768	9773
95	9777	9782	9786	9791	9795	9800	9805	9809	9814	9818
96	9823	9827	9832	9836	9841	9845	9850	9854	9859	9863
97	9868	9872	9877	9881	9886	9890	9894	9899	9903	9908
98	9912	9917	9921	9926	9930	9934	9939	9943	9948	9952
99	9956	9961	9965	9969	9974	9978	9983	9987	9991	9996

appendix c
conversion factors

QUANTITY	SI UNIT	OTHER UNITS	CONVERSION FACTORS[a]
Length	Meter (m)	Angstrom (Å)	$1\,\text{Å} = 10^{-10}\,\text{m} = 10^{-8}\,\text{cm} = 10^{-1}\,\text{nm}$
		Micron (μ)	$1\,\mu = 10^{-6}\,\text{m} = 1\mu\text{m}$
		Inch (in.)	$1\,\text{in.} = 2.54 \times 10^{-3}\,\text{m}$
Mass	Kilogram (kg)	Atomic mass unit (amu)	$1\,\text{amu} = 1.66053 \times 10^{-27}\,\text{kg}$
		Pound (lb)	$1\,\text{lb} = 0.45359237\,\text{kg}$
Time	Second (sec)	Minute (min)	$1\,\text{min} = 60\,\text{sec}$
Temperature[b]	Kelvin (K)	°Celsius (°C)	$1\,°\text{C} = 1\,\text{K}$
		°Fahrenheit (°F)	$1\,°\text{F} = \frac{5}{9}\,\text{K}$
Volume	Cubic meter (m^3)	Liter (l.)	$1\,\text{l.} = 10^{-3}\,\text{m}^3 = 1\,\text{dm}^3 = 10^3\,\text{cm}^3$
		Quart (qt)	$1\,\text{qt} = 0.463 \times 10^{-4}\,\text{m}^3$
		Cubic inch (in.^{3})	$1\,\text{in.}^3 = 1.6387 \times 10^{-6}\,\text{m}^3$
Force	Newton (N)—kg-m/sec^2	Dyne (dyn)	$1\,\text{dyn} = 10^{-5}\,\text{N}$
Pressure	Pascal (Pa)—kg/m-sec^2	Bar	$1\,\text{bar} = 10^5\,\text{Pa}$
		Atmosphere (atm)	$1\,\text{atm} = 1.01325 \times 10^5\,\text{Pa}$
		Torr	$1\,\text{torr} = 133.322\,\text{Pa}$
		mm Hg	$1\,\text{mm Hg} = 133.322\,\text{Pa}$
		lb/in.^{2}	$1\,\text{lb/in.}^2 = 6893\,\text{Pa}$
Energy	Joule (J) $\text{kg-m}^2/\text{sec}^2$	Erg	$1\,\text{erg} = 10^{-7}\,\text{J}$
		Calorie (cal)	$1\,\text{cal} = 4.184\,\text{J}$
		British thermal unit (Btu)	$1\,\text{Btu} = 1055.056\,\text{J}$
		Electron volt (ev)	$1\,\text{ev} = 1.6022 \times 10^{-19}\,\text{J}$

[a]See also Tables 1.2 and 1.3.

[b]Temperatures in degrees Celsius (t_c) or degrees Fahrenheit (t_f) can be converted to temperatures in degrees Kelvin (K) by using the equations:

$$t_c = \text{K} - 273.15$$
$$t_f = \tfrac{9}{5}\text{K} - 459.69$$

Density: 0.99987 g/cm^3 at 0°C
1.00000 g/cm^3 at 4°C
0.99707 g/cm^3 at 25°C
0.95838 g/cm^3 at 100°C

Heat of fusion: 6.02 kJ/mole at 0°C

Heat of vaporization: 44.94 kJ/mole at 0°C
44.02 kJ/mole at 25°C
40.67 kJ/mole at 100°C

Specific heat: Ice (−3°C)—2.092 J/g°C
Water at 14.5°C—4.184 J/g°C
Steam (100°C)—1.841 J/g°C

Vapor pressure (mm Hg):

T(°C)	VP	T(°C)	VP	T(°C)	VP	T(°C)	VP
0	4.58	21	18.65	35	42.2	92	567.0
5	6.54	22	19.83	40	55.3	94	610.9
10	9.21	23	21.07	45	71.9	96	657.6
12	10.52	24	22.38	50	92.5	98	707.3
14	11.99	25	23.76	55	118.0	100	760.0
16	13.63	26	25.21	60	149.4	102	815.9
17	14.53	27	26.74	65	187.5	104	875.1
18	15.48	28	28.35	70	233.7	106	937.9
19	16.48	29	30.04	80	355.1	108	1004.4
20	17.54	30	31.82	90	525.8	110	1074.6

SUBSTANCE	ΔH_f° (kJ/mole)	ΔG_f° (kJ/mole)	S° (J/mole-K)
Al(s)	0.00	0.00	28.32
Al$_2$O$_3$(s)	−1669.8	−1576.5	51.00
Ag$^+$(aq)	105.90	77.11	73.93
AgCl(s)	−127.0	−109.70	96.11
Ba(s)	0.00	0.00	63.2
Br(g)	111.8	82.38	174.9
Br$^-$(aq)	−120.9	−102.8	80.71
Br$_2$(g)	30.71	3.14	245.3
Br$_2$(l)	0.00	0.00	152.3
C(g)	718.4	672.9	158.0
C (diamond)	1.88	2.84	2.43
C (graphite)	0.00	0.00	5.69
CCl$_4$(g)	−106.7	−64.0	309.4
CCl$_4$(l)	−139.3	−68.6	214.4
CO(g)	−110.5	−137.3	197.9
CO$_2$(g)	−393.5	−394.4	213.6
C$_2$H$_2$(g)	226.7	209.2	200.8

SUBSTANCE	ΔH_f° (kJ/mole)	ΔG_f° (kJ/mole)	S° (J/mole-K)
$C_2H_4(g)$	52.30	68.11	219.4
$C_2H_6(g)$	−84.68	−32.89	229.5
$C_3H_8(g)$	−103.85	−23.47	269.9
$CH_3OH(g)$	−201.2	−161.9	237.6
$CH_3OH(l)$	−238.6	−166.23	126.8
$C_2H_5OH(l)$	−277.7	−174.76	160.7
$CH_3COOH(l)$	−487.0	−392.4	159.8
$C_6H_6(l)$	49.0	124.5	172.8
$C_6H_6(g)$	82.9	129.7	269.2
$CaCO_3$ (calcite)	−1207.1	−1128.76	92.88
$CaO(s)$	−635.5	−604.17	39.75
$Cl_2(g)$	0.00	0.00	222.96
$Cu(s)$	0.00	0.00	33.30
$Fe(s)$	0.00	0.00	27.15
$Fe^{2+}(aq)$	−87.86	−84.93	113.4
$Fe^{3+}(aq)$	−47.69	−10.54	293.3
$Fe_2O_3(s)$	−822.16	−740.98	89.96
$H(g)$	217.94	203.26	114.60
$HBr(g)$	−36.23	−53.22	198.49
$HCl(g)$	−92.30	−95.27	186.69
$HF(g)$	−268.61	−270.70	173.51
$HI(g)$	25.94	1.30	206.3
$H_2(g)$	0.00	0.00	130.58
$H_2O(g)$	−241.8	−288.61	188.7
$H_2O(l)$	−285.85	−236.81	69.96
$H_2S(g)$	−20.17	−33.01	205.6
$Hg(g)$	60.83	31.76	174.89
$Hg(l)$	0.00	0.00	77.40
$I(g)$	106.60	70.16	180.66
$I_2(g)$	62.25	19.37	260.57
$I_2(s)$	0.00	0.00	116.73
$K(g)$	89.99	61.17	160.2
$KNO_3(s)$	−492.70	−393.13	288.1
$Mg(s)$	0.00	0.00	32.51
$MnO_2(s)$	−519.6	−464.8	53.14
$NH_3(g)$	−46.19	−16.66	192.5
$NO(g)$	90.37	86.71	210.62
$NO_2(g)$	33.84	51.84	240.45
$N_2(g)$	0.00	0.00	191.50
$N_2O(g)$	81.55	103.59	220.0
$N_2O_4(g)$	9.66	98.28	304.3
$NaCl(s)$	−410.9	−384.0	72.33
$NaHCO_3(s)$	−947.7	−851.8	102.1
$Na_2CO_3(s)$	−1130.9	−1047.7	−136.0
$O(g)$	247.5	230.1	161.0
$O_2(g)$	0.00	0.00	205.0
$O_3(g)$	142.3	163.4	237.6
$S(s,$ rhombic$)$	0.00	0.00	31.88
$SO_2(g)$	−296.9	−300.4	248.5
$SO_3(g)$	−395.2	−370.4	256.2
$Zn(s)$	0.00	0.00	41.6
$ZnO(s)$	−348.0	−318.2	43.9

CHAPTER 1

1.1 (a); (b); (c); (g); (h).   1.7 (a) extensive; (b) extensive; (c) intensive;
(d) intensive; (e) intensive; (f) extensive; (g) intensive; (h) intensive.   1.9 meter;
kilogram; cubic meter; second.   1.11 (a) 10^{-3}; (b) 10^{-6}; (c) 10^{-2}; (d) 10^3.
1.13 8.77×10^{-3} km; 8.77×10^3 mm.   1.15 (a) 2; (b) 4; (c) 2: (d) 3; (e) 2.
1.17 (a) 3.5×10^4; (b) 3.500×10^4.   1.19 201 m.   1.21 \$4.83/kg.   1.23 5.00 l.
1.25 88.5 km/hr.   1.27 0.33 g.   1.29 1.4×10^5 cm^3; 1.4×10^2 l.; 0.14 m^3.
1.31 (a) 22°C; (b) −18°C; (c) 14°F; (d) 298 K; (e) 287 K.   1.33 19.3 g/ml.
1.35 54.4 kg.   1.37 3.9 l.   1.39 0.7 g/cm^3.

CHAPTER 2

2.1 (a) O; (b) H; (c) N; (d) Au; (e) S; (f) He; (g) C; (h) Na.   2.3 (a) element;
(b) mixture; (c) mixture; (d) compound; (e) compound; (f) compound; (g) mixture;
(h) mixture; (i) mixture.   2.6 (a) liquid; (b) solid; (c) solid; (d) gas; (e) liquid;
(f) liquid; (g) gas; (h) gas; (i) gas.   2.9 (a) physical; (b) chemical; (c) chemical;
(d) physical; (e) physical; (f) chemical.   2.11 (a) physical; (b) physical;
(c) chemical; (d) chemical; (e) chemical; (f) physical.   2.13 (a) lithium;
(b) aluminum; (c) neon.

2.15

^{19}F$^-$	^{40}Ar	^{27}Al
9	18	13
10	22	14
10	18	13
−1	0	0

2.17   (17p$^+$ 18n)   17 $\big)$ e$^-$: $^{35}_{17}$Cl

(17p$^+$ 20n)   17 $\big)$ e$^-$: $^{37}_{17}$Cl

2.19 (a) They have the same number of electrons and the same nuclear charge.
(b) The alpha particle is a helium-4 nucleus and as such has a mass number of 4
and an atomic number of 2; a beta particle is an electron and as such has a mass
number of 0 and a charge of −1.   2.21 (a) deflected more; (b) deflected less;
(c) deflected more.   2.25 Na—alkali metal; F—halogen; K—alkali metal; Ne—noble
gas; Ca—alkaline earth; Ar—noble gas.   2.27 K$^+$; Mg^{2+}; Al^{3+}; F$^-$; O^{2-}.
2.29 (a) H$_2$O; (b) O$_2$; (c) CaI$_2$.   2.31 (a) molecular; (b) both; (c) both;
(d) simplest; (e) simplest; (f) molecular.   2.33 (a) CaCO$_3$; (b) NaOH; (c) K$_2$SO$_4$;
(d) FeCl$_3$; (e) HNO$_2$; (f) HBr; (g) NH$_4$Br; (h) H$_3$PO$_3$; (i) Na$_2$SO$_3$.
2.36 (a) strontium nitrate; (b) tin(IV) bromide or stannic bromide; (c) silver nitrate;
(d) potassium cyanide; (e) aluminum hydroxide; (f) sodium hydrogen sulfate or
sodium bisulfate; (g) ammonium sulfate; (h) iodic acid; (i) copper(I) bromide or
cuprous bromide.

CHAPTER 3

3.1 O$_2$ represents a single molecule consisting to two oxygen atoms; 2O represents
two individual oxygen atoms.   3.2 (a) Ba(OH)$_2$ + 2HCl $\longrightarrow$ BaCl$_2$ + 2H$_2$O;
(b) 3CaCl$_2$ + 2Na$_3$PO$_4$ $\longrightarrow$ Ca$_3$(PO$_4$)$_2$ + 6NaCl; (c) 2CH$_3$OH + 3O$_2$ $\longrightarrow$
2CO$_2$ + 4H$_2$O; (d) NH$_4$NO$_3$ + NaOH $\longrightarrow$ NH$_3$ + H$_2$O + NaNO$_3$;
(e) PCl$_5$ + 4H$_2$O $\longrightarrow$ H$_3$PO$_4$ + 5HCl.   3.4 (a) 4NH$_3$ + 3O$_2$ $\longrightarrow$ 2N$_2$ + 6H$_2$O;
(b) 2CH$_3$OH + 3O$_2$ $\longrightarrow$ 2CO$_2$ + 4H$_2$O; (c) 4C$_3$H$_5$(NO$_3$)$_3$ $\longrightarrow$
6N$_2$ + O$_2$ + 12CO$_2$ + 10H$_2$O.   3.6 (a) 2C$_2$H$_6$ + 7O$_2$ $\longrightarrow$ 4CO$_2$ + 6H$_2$O;
(b) 2C$_3$H$_7$OH + 9O$_2$ $\longrightarrow$ 6CO$_2$ + 8H$_2$O; (c) 2HNO$_3$ + Ca(OH)$_2$ $\longrightarrow$
2H$_2$O + Ca(NO$_3$)$_2$; (d) 2Li + 2H$_2$O $\longrightarrow$ 2LiOH + H$_2$;
(e) H$_2$SO$_4$ + 2KOH $\longrightarrow$ 2H$_2$O + K$_2$SO$_4$.   (a) 98.0 amu; (b) 46.0 amu;
(c) 32.0 amu; (d) 34.0 amu.   3.10 A molecular weight is the mass of a molecule. A
formula weight is the mass of a formula unit, which can represent a molecule if the
formula is a molecular formula.   3.12 20.2.   3.14 (a) 34.0 g; (b) 98.0 g; (c) 342 g;
171 g.   3.15 2.04% H; 32.7% S; 65.3% O.   3.17 (a) 0.102 mole H$_2$SO$_4$; (b) 0.156
mole O$_2$; (c) 0.149 mole Ne; (d) 2.19×10^{-4} mole Ne; (e) 10.0 moles H$_2$SO$_4$;
(f) 0.500 mole O$_2$.   3.19 (a) 1.88×10^{23} atoms; (b) 7.37×10^{22} atoms.

3.22 (a) 183 amu; (b) 183 g; (c) 26.2%; (d) 5.46×10^{-3} moles; (e) 3.29×10^{21} molecules; (f) 2.96×10^{22} C atoms; (g) 3.04 g. 3.25 $C_{10}H_{14}N_2$. 3.27 (a) CS_2; (b) CH_2O; (c) Na_2HPO_3. 3.29 CH_2. 3.31 83.5 g. 3.33 36.7 lb. 3.35 (a) 7.20 g H_2O; no O_2; 3.14 g H_2. 3.37 714 g vinegar. 3.40 (a) $1.5\,M$; (b) $0.060\,M$; (c) $0.67\,M$; (d) $0.13\,M$. 3.43 (a) Dissolve 6.80 g $AgNO_3$ in sufficient water to form 200 ml soln. (b) Dilute 100 ml of $1.00\,M$ HCl to a total volume of 500 ml. 3.45 (a) 0.0125 mole; (b) 0.471 mole; (c) 0.0236 mole. 3.47 (a) 0.20 mole; (b) 5.0×10^{-3} mole; (c) 1.5×10^{-2} mole. 3.48 1.61. 3.50 $3.68\,M$. 3.56 45.5 kg CO_2 produced; 51.7 kg O_2 consumed.

CHAPTER 4

4.1 The 1000 kg car traveling at 80 km/hr has the most energy (2.5×10^5 J; 6.0×10^4 cal). The second car has an energy of 1.2×10^5 J (2.9×10^4 cal). 4.2 1.5×10^4 J (3.5×10^3 cal). 4.4 (a) both (because of composition and motion); (b) potential energy; (c) kinetic energy. 4.5 -556.5 kJ; -327.9 kJ; 1429 kJ. 4.7 (a) -92.38 kJ (exothermic); (b) -822.16 kJ (exothermic); (c) 236.4 kJ (endothermic); (d) -50.3 kJ (exothermic). 4.11 -246.2 kJ. 4.13 -830 kJ. 4.16 1.3×10^3 J. 4.18 44.2 kJ. 4.20 -3270 kJ/mole. 4.22 540 g/day. 4.25 Propane—50.3 kJ/g; methane—55.6 kJ/g. 4.30 680 J/g.

CHAPTER 5

5.2 169.7 cm. 5.5 gaseous N_2: 3.72×10^{-20} cm³/molecule; liquid N_2: 5.75×10^{-23} cm³/molecule. 5.6 slightly more than 2 tanks. 5.8 0.059 l. 5.9 9.7 g/l. 5.12 yes. 5.14 368 l. air. 5.17 28.97 g/mole. 5.19 hydrogen—2.014; argon—39.980. 5.20 17.8 l. 5.22 3.60 l. 5.24 C_6H_{12}. 5.27 in proportion to $T^{1/2}$. 5.29 12 g/mole. 5.32 for CO, 1.018; for CO_2, 1.011. 5.35 (a) I_2; (b) SnH_4; (c) CO_2. 5.40 57.4 Å³. 5.42 3.08 F_2 holes for each H_2 hole. 5.43 lifting capacity = 86 g. 5.45 2610 g H_2O. 5.48 1.35×10^9 molecules/cm³. 5.50 10.5.

CHAPTER 6

6.1 1.28 sec. 6.3 267 m; 3.04 m. 6.5 6.67×10^{14} Hz. 6.7 2.23×10^{-18} J. 6.10 297 nm; copper would not be suitable. 6.15 -6.05×10^{-20} J. 6.16 1.88 μ; in the infrared. 6.17 Ionization potential will be larger; larger nuclear charge. 6.21 0.123 nm. 6.25 (a) and (d) are proper descriptions of $3p$ and $11s$ orbitals. 6.29 for $3s$ orbital, 4 nodes; for $2p$, one node. 6.33 (a) 16; (c) none. 6.35 (a) 7; (d) 3. 6.39 146 kJ/mole. 6.42 (b) equal; (c) equal.

CHAPTER 7

7.2 five "violations": Ar, K; Co, Ni; Te, I; Th, Pa; U, Np. Larger number of neutrons in elements of lower atomic number in each pair. 7.5 11,808 kJ/mole. 7.7 in hydrogen, value of n only; in other elements, by n and l. 7.9 larger average distance from nucleus for $2p$ than for $2s$ electron. 7.10 H and Li would separate; He and Be would not. 7.13 (a) six (b) six (c) nine. 7.15 (b) $(n-1)d^{10}ns^1$. 7.16 (b) Ca must lose two electrons; (d) H must add one electron. 7.18 (a) chlorine; (b) elements Co, Rh, and Ir; (c) Nb. 7.20 (a) two; (b) three. 7.22 (a) excited state; (c) ground state. 7.24 H_2SeO_4 and H_2CrO_4. The prediction is more reliable for Se, but both compounds are known. 7.26 Atomic size decreases from left to right because of increasing effective nuclear charge. 7.28 Effective nuclear charge for highest-energy electron is lower in Y than in Sc. 7.30 Ionization potential increases in series from K to Kr, as a result of increasing effective nuclear charge. 7.33 Mg^{2+} has smaller radius than Na^+; same number of electrons, larger nuclear charge. 7.35 The $5s$ electrons of Cd experience larger effective nuclear charge than those of Sr because of incomplete shielding of $4d$ electrons in Cd. 7.37 $4p^35s^2$, $4d^{10}5p^3$ 7.40 n must be five or greater. There are 9 g orbitals for each value of $n \geqslant 5$. There are no known elements with g electrons.

CHAPTER 8

8.3 There is a smaller coulomb attraction in KCl because of larger distance between ionic centers. Thus, there is lower lattice energy. 8.4 (b) $2Be + O_2 \longrightarrow 2BeO$ (d) $Zn + F_2 \longrightarrow ZnF_2$. 8.5 (a) CaO stable; (b) $AlCl_2$ unstable; (e) Rb_2O stable;

(f) MnF_2 stable (see Table 8.1). **8.7** (b) TiF_4; (d) $Cd_3(PO_4)_2$; (f) $KMnO_4$.
8.9 (a) Ca^{2+}, yes; (b) In^+, no; (c) Ga^{3+}, no; (d) Sc^{3+}, yes; (e) Se^{2-}, yes.
8.11 Ionic radius decreases in the order $Cl^- > K^+ > Ca^{2+}$; increasing nuclear charge in an isoelectronic sequence. **8.13** Increasing effective nuclear charge due to incomplete shielding by 3d electron. Thus ionic radius decreases in series.
8.20 $:N\equiv N:$; $H—\ddot{N}=\ddot{N}—H$; $H—\ddot{N}—\ddot{N}—H$; $H—\ddot{N}—H$
$\qquad\qquad\qquad\qquad\qquad\qquad\quad |\quad |\qquad\qquad |$
$\qquad\qquad\qquad\qquad\qquad\qquad\; H\; H\qquad\quad H$

8.24 Resonance structures for O_3 suggest O—O bond between single and double bond. Bond distance decreases with increasing bond order (that is, the number of shared electron pairs).

8.25 (b)

8.28 Sb is most likely to exhibit expanded valence shell. **8.30** (a) BF_3; (c) $GeCl_4$; (e) MgO. **8.33** Sr in SrO; greatest electronegativity difference. **8.35** CF_4, CH_4 (nonpolar) $< CF_3Cl < CF_2Cl_2 < CF_2H_2$. **8.37** (a) -1, HCl; (c) -2, CdS (e) $+1$, NaCl; (f) $+5$, H_3PO_4. **8.39** (a), (c), (d). **8.41** (a) S is oxidized, -2 to 0. N is reduced, $+5$ to $+2$. (c) Cr is oxidized, $+3$ to $+6$. Cl is reduced, $+1$ to -1. (g) Mn is oxidized, $+2$ to $+7$. Pb is reduced, $+4$ to $+2$. **8.43** (a) titanium (II) ion; (e) sulfite ion; (h) iodite ion. **8.45** (a) calcium phosphate; (b) hypophosphorous acid; (c) aluminum phosphite. **8.48** (a) dichromium hexaoxide, chromium (VI) oxide; (b) niobium pentafluoride, niobium (V) fluoride; (c) selenium dioxide, selenium (IV) oxide; (d) titanium trifluoride, titanium (III) fluoride; (e) silver difluoride, silver (II) fluoride; (f) diphosphorus tetrachloride, phosphorous (II) chloride. **8.50** $[:N=C=N:]^-$. **8.52** MgO has greater lattice energy due to higher ionic charges and comparatively small interionic distances. **8.55** $2KMnO_4 + 5H_2C_2O_4 + 3H_2SO_4 \longrightarrow K_2SO_4 + 2MnSO_4 + 10CO_2 + 8H_2O$. $KMnO_4$ solution is $5.50 \times 10^{-3}\, M$.

CHAPTER 9

9.1 Factor of primary importance is electrostatic repulsions between electron pairs. **9.3** Repulsive effect of unshared pairs is greater than that of shared pair; thus HNH angle is squeezed down by effect of unshared pair on N. The effect is amplified in H_2O because there are two unshared pairs. **9.4** (a) pyramidal; (b) tetrahedral; (c) approximately tetrahedral; (d) approximately tetrahedral; (e) tetrahedral. **9.6** (a) NO_2^- should be bent, ONO angle about $120°$. (c) SO_4^{2-} tetrahedral. **9.9** In a many-electron atom, orbitals of same n and different l differ in energy. Thus $2s$ orbital is lower in energy than $2p$. **9.12** (a) sp, two orbitals along $+$ and $-\ x$ axes; (b) sp^3, tetrahedral; (c) sp^3d^2, octahedral. **9.14** (a) sp^3; (b) sp^3; (c) sp^3d; (d) sp^3d. **9.17** Resonance stabilization in benzene prevents opening of C—C double bonds to form C—Br bonds. **9.20** The C—C distance in acetylene is smaller than in ethylene or ethane. Bond distance decreases with increase in number of shared electron pairs. **9.22** (a) 2; (b) 4; (c) 4; (d) 10. **9.24** (a) Li_2; (b) N_2; (c) Be_2^+. **9.26** (a) 3, $1\sigma + 2\pi$; (b) 1; (c) 2.5, $0.5\sigma + 2\pi$. **9.30** O_2^+ and SCl are paramagnetic; odd-electron species. **9.32** No. Absence of many neighbors, extended delocalization. **9.35** Au and Ag are closer in radii, more similar in chemical properties than Sn and Pb. **9.39** (a) pyramidal, hybridization approximately sp^3; (b) tetrahedral, sp^3 (c) octahedral, sp^3d^2; (d) octahedral, sp^3d^2; (e) trigonal plane, $sp^2 + p$; (f) pyramidal, sp^3; (g) T-shaped; sp^3d; (h) tetrahedral, sp^3. **9.41** P has available a $3d$ orbital to mix with $3s$ and $3p$; N has no $2d$ orbital to mix with the $2s$ and $2p$.

CHAPTER 10

10.2 the troposphere; the atmosphere is much denser at lower altitudes (see Figure 5.10). **10.4** higher at surface. **10.6** $28.8\, g/mole$; $24.0\, g/mole$. **10.8** 368 nm. **10.10** (a) $O_2 + h\nu \longrightarrow O + O$; (d) $CCl_2F_2 + h\nu \longrightarrow CClF_2 + Cl$. **10.12** (b) $N_2 + O^+ \longrightarrow NO^+ + N$; (c) $NO^+ + e^- \longrightarrow N + O$. **10.14** (a) $-35\, kJ$;

(b) -555 kJ; (c) -502 kJ; (d) -814 kJ.   **10.16** Shorter wavelength radiations are absorbed at higher elevations. Longer wavelengths pass through less dense upper atmosphere, but are absorbed in denser stratosphere   **10.19** $CCl_2F_2 + h\nu \longrightarrow CClF_2 + Cl$; $Cl + O_3 \longrightarrow ClO + O_2$; $ClO + O \longrightarrow Cl + O_2$.   **10.23** 4.1×10^7 g SO_2.   **10.24** 3.9×10^5 g sulfate.   **10.26** Oxidation in rain (fog droplets), oxidation on particle surfaces, photochemical oxidation   **10.27** (b) $NO + O_3 \longrightarrow NO_2 + O_2$; (c) $SO_2 + H_2O \rightleftharpoons H_2SO_3$, $2H_2SO_3 + O_2 \longrightarrow 2H_2SO_4$, $H_2SO_4 + 2NH_3 \longrightarrow (NH_4)_2SO_4$; (e) $2NO + O_2 \longrightarrow 2NO_2$, $NO_2 + h\nu \longrightarrow NO + O$, $O + O_2 + M \longrightarrow O_3 + M^*$.   **10.30** 9.5 percent   **10.35** (a) $N_2^+ + e^- \longrightarrow N + N$; (b) $H_2O + h\nu \longrightarrow H + OH$; (c) $O_2 + h\nu \longrightarrow O + O$, $O + O_2 + M \longrightarrow O_3 + M^*$; (d) $O^+ + N_2 \longrightarrow NO^+ + N$, $NO + O_2^+ \longrightarrow NO^+ + O_2$; (e) $2SO_2 + O_2 \longrightarrow 2SO_3$; (f) $NO + O_3 \longrightarrow NO_2 + O_2$.   **10.37** Metals end up as M^+ via charge transfer reactions.   **10.38** At 175 km, $\mu = 860$ m/sec; 0.35 sec per collision. At 75 km, $\mu = 360$ m/sec; 5.6×10^{-6} sec per collision.

CHAPTER 11

11.1 The evaporation of alcohol is endothermic; as the alcohol evaporates it absorbs heat from your arm.   **11.3** The wind hastened the endothermic evaporation of water from the bodies of the climbers, thereby lowering their body temperatures.   **11.5** The attractive forces between water molecules are stronger than those between alcohol molecules.   **11.8** (a) approximately $57°C$; (b) approximately $67°C$.   **11.12** 7.8×10^{-2} J/m^2.   **11.15** (a) London dispersion; (b) hydrogen bonding; (c) London dispersion; (d) hydrogen bonding; (e) hydrogen bonding; (f) London dispersion; (g) London dispersion; (h) dipole-dipole; (i) metallic.   **11.17** $CH_4 < CF_4 < CCl_4 < CBr_4 < CI_4$.   **11.19** H_2O and HF.   **11.21** Propane is more volatile than butane (lower molecular weight).   **11.23** (a) ionic; (b) London dispersion; (c) metallic; (d) London dispersion; (e) covalent; (f) London dispersion; (g) hydrogen bonding.   **11.25** Face-centered cube has four particles; body-centered cube has two particles.   **11.26** 0.584 Å.   **11.28** 3.18 g/cm^3.   **11.32** It would increase the mechanical strength, hardness, melting point, and density of the polyethylene.   **11.35** The water in the cup never gets above $100°C$ as it boils; the water thereby keeps the paper cup below the temperature necessary for it to ignite.   **11.38** 133.4 kJ.   **11.41** (a) solid; (b) equilibrium mixture of liquid and vapor; (c) equilibrium mixture of vapor, liquid, and solid; (d) vapor.   **11.45** Less energy is needed to permit particles to move past each other (melting; heat of fusion) than to separate all of the particles from each other (vaporization).   **11.49** In diamond, the bonds between particles (carbon atoms) are covalent. In water the bonds are much weaker hydrogen bonds.

CHAPTER 12

12.1 (a) 0.025 mole; (b) 0.025 mole; (c) 0.0086 mole.   **12.3** (a) 1.1 g; (b) 68 g; (c) 0.10 g.   **12.5** (a) $0.40\,N$; (b) $0.10\,M$.   **12.7** $X_{H_2SO_4} = 0.80$; wt % = 96%.   **12.10** Mg^{2+} because it is smaller and has a higher charge.   **12.12** (a) CO_2; (b) CH_3OH; (c) $AlCl_3$.   **12.15** KCl, $Fe(NO_3)_2$, NaOH, and $Ca(NO_3)_2$.   **12.17** 1.10 mole/l.   **12.18** decrease.   **12.20** (a) strong electrolyte (acid); (b) strong electrolyte (base); (c) weak electrolyte (base); (d) nonelectrolyte; (e) weak electrolyte (acid); (f) strong electrolyte (salt); (g) nonelectrolyte; (h) weak electrolyte (acid); (i) strong electrolyte (salt).   **12.22** (a) $HCl + NaHCO_3 \longrightarrow NaCl + CO_2 + H_2O$; (b) *N.R.*; (c) $3HCl + Al(OH)_3 \longrightarrow 3H_2O + AlCl_3$; (d) *N.R.*; (e) $H_2S + Cu(NO_3)_2 \longrightarrow CuS + 2HNO_3$.   **12.25** (a) $Ca^{2+}(aq) + CO_3^{2-}(aq) \longrightarrow CaCO_3(s)$; (b) $Pb^{2+}(aq) + SO_4^{2-}(aq) \longrightarrow PbSO_4(s)$; (c) $Ni^{2+}(aq) + 2OH^-(aq) \longrightarrow Ni(OH)_2(s)$.   **12.28** 8.0 torr.   **12.31** freezing point = $-2.58°C$; boiling point = $100.72°C$.   **12.38** 1.4 atm.   **12.39** 3600 g/mole.   **12.41** In the presence of salt the colloidal silt particles coagulate.   **12.44** The suspended particles in a colloid are larger than the solute particles of a true solution; the suspended particles are large enough effectively to scatter light.   **12.48** If H_2O were linear, it would not have a dipole moment; if it was not polar, it would not be a good solvent for polar substances.   **12.51** Molality = $0.91\,m$; mole fraction = 0.016; molarity = $0.88\,M$.   **12.55** 320 g/mole; $C_{21}H_{30}O_2$.

CHAPTER 13

13.2 rate $= -\Delta[I_2]/\Delta t = +\frac{1}{2}\,\Delta[HI]/\Delta t$. Moles/l. per sec, that is M/sec, or if pressure is used as the concentration unit, atm/sec.   13.4 (a) rate doubles; (b) rate increases by 1.41; no effect on rate constant.   13.5 $t_{1/2} = 1120$ sec.   13.6 $k = 2.00 \times 10^{-3}$/sec; $t_{1/2} = 346$ sec.   13.8 $k = 2.07 \times 10^4$/sec.   13.10 No effect on k or on $t_{1/2}$. 13.12 initial rate $= 5.2 \times 10^{-14}$ moles/l.-sec.   13.15 reaction A has higher E_a. 13.17 $E_a = 24.5$ kJ/mole.   13.20 Mechanism (c).   13.26 (a) E_a too high for uncatalyzed reaction; (b) E_a is lower on Pt surface. Heat of reaction taken up by Pt maintains high temperature.   13.28 (a) $[A]^2$ and rate increase by 4; (b) rate and k; (c) $[A]^2$ increases by 4, $[B]$ decreases by 4.   No change in rate; (d) Rate and k increase.   13.30 (a) overall order $\frac{3}{2}$—order in Cl_2 is $\frac{1}{2}$; (b) Overall order is 1—order in Cl_2 is -1.   13.33 $k[H_2S]$ is pseudo–first-order, rate constant $= 1.5 \times 10^{-6}$/sec. $t_{1/2} = 4.6 \times 10^5$/sec.   13.36 rate $= 6 \times 10^9$ moles/l.-sec.

CHAPTER 14

14.2 (a) $K_c = \dfrac{[H_2O]^2[SO_2]^2}{[H_2S]^2[O_2]^3}$, units are M^{-1}; (c) $K_c = \dfrac{[ICl]}{[I_2][Cl_2]}$, no units.

14.5 $K_c = \dfrac{[PCl_3][Cl_2]}{[PCl_5]} = 0.0415\,M$.   14.6 0.012 atm.   14.8 2.24×10^4 atm^2.

14.10 no effect on K_c—equilibrium would shift to left.   14.12 Melting point decreases with increase in pressure.   14.15 At 850°C, $K_c = 14.1$ atm. K_c increases with increase in temperature. Reaction is endothermic in forward direction.   14.17 (a) K_c decreases with increase in temperature; (b) K_c increases with increase in temperature; (c) K_c decreases with increase in temperature   14.20 Major factor is high bond energy of N_2 bond.   14.21 Patient claim is false; catalyst does not affect position of equilibrium.   14.23 (a) true; (b) false; (c) false.   14.26 (a) shift to left; (b) no shift; (c) shift to left.   14.27 Reaction is far from true equilibrium; reduction does occur. 14.29 (a) fractional coverage ~ 1; (b) competition between N_2 and H_2S for active sites.   14.31 $[H_2] = 0.00476\,M$; $[I_2] = 0.00214\,M$; $[HI] = 0.0240\,M$.   14.32 K_c decreases with increasing temperature; more methanol with increased pressure. 14.34 $\Theta = 0.73$.

CHAPTER 15

15.1 No—expansion of ideal gas ($\Delta H = 0$); melting of ice at $+5$°C (endothermic). 15.3 A and B equally distributed in both flasks—$\Delta H = 0$; ΔS positive; ΔG negative.   15.5 (a) spontaneous; (b) spontaneous; (c) nonspontaneous; (d) spontaneous; (e) nonspontaneous.   15.8 (a) about zero; (b) negative; (c) positive; (d) negative; (e) about zero.   15.10 (a) Ar(g) at 25°C; (b) $H_2O(g)$; (c) mixture; (d) $CCl_4(g)$.   15.12 Free-energy change is independent of pathway. 15.14 (a) -28.9 kJ; (b) $+84.2$ kJ; (c) -140 kJ; (d) -207.2 kJ.   15.16 (a) false; (b) true; (c) true; (d) false.   15.18 K_c should decrease at higher temperatures. 15.20 $\Delta G° = 659.1$ kJ; $K_c = 2.7 \times 10^{-116}$.   15.22 $\Delta S° = 254.3$ J/K; $\Delta H° = -59.8$ kJ; $\Delta G° = -135.5$ kJ; $K_c = 5.75 \times 10^{23} = [O_2]^{3/2}$; units are atm. 15.25 (a) nonspontaneous; (b) spontaneous; (c) spontaneous.   15.27 (a) $Br_2(g)$ higher because gas is more random than liquid; (c) O_3 higher because there are more ways of storing energy in vibrations and rotations.   15.29 $\Delta G° = -24.6$ kJ; $\Delta S° = -221.5$ J/K. Spontaneous at 298 K, less so at higher temp. Increased pressure shifts equilibrium to right.

CHAPTER 16

16.2 $6.55\,M$ 16.4 Reaction with water to forming $H^+(aq)$ proceeds essentially to completion   16.6 Both are strong acids; conductivities are due mainly to $H^+(aq)$, essentially the same in each solution.   16.9 $HI(aq) \rightleftharpoons H^+(aq) + I^-(aq)$. HI and $H^+(aq)$ are acids; $I^-(aq)$ is a very weak base.   16.12 Freezing-point lowering is twice as large for the HCl solution; electrical conductivity and rate of reaction with active metal are much greater.   16.13 strongest acid is HNO_2; weakest is phenol. 16.15 (a) NO_3^-, nitrate ion; (c) Br^-, bromide ion; (e) HCO_2^-, formate ion. 16.17 (a) HCl; (b) H_2SO_4; (c) HNO_3; (d) HBr.   16.19 $[H^+] = 4.4 \times 10^{-6}$. 16.21 $[H^+] = 1.8 \times 10^{-2}\,M$.   16.23 You must use full quadratic equation.

$[H^+] = 3.9 \times 10^{-3} M$ 16.26 (a) $1 \times 10^{-12} M$; (b) $3 \times 10^{-11} M$; (c) $2.5 \times 10^{-12} M$.
16.29 $[OH^-] = 3.3 \times 10^{-5} M$. 16.31 $[H^+] = 7.9 \times 10^{-10} M$ 16.33 (a) neutral;
(b) acidic; (c) basic; (d) basic; (e) neutral; (f) basic; (g) basic.
16.34 (a) $K_b = 5.9 \times 10^{-11}$; (b) $K_b = 1.3 \times 10^{-12}$; (c) $K_b = 5.5 \times 10^{-10}$.
16.36 $CN^- > C_2H_3O_2^- > NO_2^- > Br^-$. 16.38 (a) 2; (c) 12; (e) 2.66. 16.40 between
2.3 and 4.2. 16.43 not a good indicator; too little NaOH will be used.

16.46 Lewis structures are $H—O—\overset{..}{N}=O$ and $H—O—N—O$; electron withdrawal

by added oxygen causes HNO_3 to be more acidic. 16.48 predicted: 5×10^{-8},
5×10^{-13}; observed: 1.7×10^{-7}, 4×10^{-12}. 16.50 (a) basic; (b) basic; (c) acidic;
(d) probably acidic; (e) basic. 16.51 All except caffeine are in the protonated
form. 16.58 Water molecules form hydrates with metal ions. Lewis acid-base
interaction between water molecules (bases) and metal ions (acids). 16.60 (a) ZnI_2;
(b) $FeCl_3$; (c) NH_4Cl; (d) $MgCl_2$; (e) $FeCl_3$. 16.61 pH = 3.22. 16.64 $NH_3 \rightleftharpoons$
$H^+(solv) + NH_2^-(solv)$. The symbol (solv) indicates solvated species. Acid is HCl or
NH_4Cl; base is $NaNH_2$. 16.65 N^{3-}.

CHAPTER 17

17.2 Add a drop of 6 M HCl to samples of each and then measure pH.
17.4 (a) 4.19; (b) 5.23; (c) 3.35; (d) 8.95. 17.7 Solution c has greatest buffer
capacity; solution b has the least. 17.9 14.4 g. 17.12 0.15 M 17.14 (a) 35.0 ml;
(b) 35.8 ml; (c) 35.8 ml. 17.16 methyl red. 17.18 (a) 8.0; (b) 8.44; (c) 8.23.
17.21 $[H_2CO_3] > [H^+] > [HCO_3^-] > [CO_3^{2-}]$ 17.23 Because blood is buffered
at a pH of about 7.4, most dissolved CO_2 is in the form HCO_3^-, whereas in pure
water most dissolved CO_2 is in the form H_2CO_3. 17.26 (a) $K_{sp} = [Ag^+][Br^-]$;
(c) $K_{sp} = [Cd^{2+}][IO_3^-]^2$. 17.27 (a) $K_{sp} = 1 \times 10^{-8}$; (c) $K_{sp} = 4.2 \times 10^{-7}$.
17.28 pH = 8.18. 17.32 In pure water, $[Ba^{2+}] = 5.4 \times 10^{-4} M$, $[Ag^+] = 9.6 \times$
$10^{-5} M$. In 0.01 M IO_3^- solution, $[Ba^{2+}] = 6.5 \times 10^{-6} M$, $[Ag^+] = 9.2 \times 10^{-7} M$.
17.34 $Cr(OH)_3$ and $Sn(OH)_2$. 17.35 $Zn(OH)_2(H_2O)_2(s) + H^+(aq) \rightleftharpoons$
$Zn(OH)(H_2O)_3^+(aq)$; $Zn(OH)_2(H_2O)_2(s) + OH^-(aq) \rightleftharpoons Zn(OH)_3(H_2O)^-(aq) + H_2O$.
$Zn(OH)_2(H_2O)_2(s) + OH^-(aq) \rightleftharpoons Zn(OH)_3(H_2O)^-(aq) + H_2O$.
17.37 (a) $2SnO(s) + 2OH^-(aq) + 4H_2O \rightleftharpoons 2Sn(OH)_3(H_2O)^-$.
17.41 (a) $Cu(C_5H_5N)_2^{2+} \rightleftharpoons Cu^{2+} + 2C_5H_5N$; (c) $AgCl_2^- \rightleftharpoons Ag^+ + 2Cl^-$.
17.42 $[Cu^{2+}] = 1.5 \times 10^{-24} M$, $[Fe^{2+}] = 1.9 \times 10^{-21} M$.
17.43 $[NH_3] = 7.8 \times 10^{-2} M$. 17.46 (a) ammonia; $AgBr(s) + 2NH_3 \rightleftharpoons$
$Ag(NH_3)_2^+ + Br^-$; (c) 6 M HCl; $CoS(s) + 2H^+ \rightleftharpoons Co^{2+} + H_2S(aq)$;
(e) 6 M NaOH; $Sn(OH)_2(s) + OH^- \longrightarrow Sn(OH)_3^-(aq)$. 17.49 At beginning,
pH = 5.97; at half-equivalence point, pH = 10.6; at equivalence point,
pH = 11.58. 17.50 3.83 ml 10 M HCl. 17.51 $[H^+] = [HSO_3^-] = 0.0338 M$;
$[SO_3^{2-}] = 1 \times 10^{-7} M$.

CHAPTER 18

18.1 1.5×10^{13} l. 18.4 Ca^{2+} is incorporated into marine organisms or precipitates
as $CaCO_3(s)$. 18.7 Photosynthesis near surface consumes NO_3^-; decomposition of
plant and animal matter at lower depths restores NO_3^-. 18.9 About 92 percent of
water must be removed. 18.11 Redox reaction is electrolysis; species oxidized is
Cl^-; species reduced is Mg^{2+}. 18.13 to produce needed Mn, 4.5×10^{13} g nodules;
to produce needed Co, 6.8×10^{12} g nodules. 18.16 5.4×10^9 g oil.
18.18 (a) omit all but sterilization; (c) omit only water softening. 18.20 4.8×10^6 g
$Ca(OH)_2$; 4.2×10^6 g Na_2CO_3. 18.22 Culprit is Cd, a member of group 2B along
with Zn. 18.25 (a) $Cl_2(g) + H_2O \longrightarrow HOCl(aq) + H^+(aq) + Cl^-(aq)$;
(b) $HOCl(aq) + NH_3(aq) \longrightarrow NH_2Cl(aq) + H_2O$; (e) $Hg(aq) + Hg^{2+} \longrightarrow$
Hg_2^{2+}—or you could write Hg(l) instead of Hg(aq).

CHAPTER 19

19.1 (a) $2e^- + 4H^+ + SO_4^{2-} \longrightarrow SO_2 + 2H_2O$ (reduction); (b) $2Cl^- \longrightarrow$
$Cl_2 + 2e^-$ (oxidation); (c) $4e^- + 2H_2O + O_2 \longrightarrow 4OH^-$ (reduction);
(d) $2e^- + H_2O + ClO^- \longrightarrow Cl^- + 2OH^-$ (reduction).
19.2 (a) $H_2S + H_2O_2 \longrightarrow S + 2H_2O$ (H_2S is reducing agent, H_2O_2 is oxidizing

agent); (b) $C + 4HNO_3 \longrightarrow CO_2 + 4NO_2 + 2H_2O$ (C is reducing agent, HNO_3 is oxidizing agent); (c) $H_2O + PbO_2 + OH^- + Cl^- \longrightarrow Pb(OH)_3^- + ClO^-$ (Cl^- is reducing agent, PbO_2 is oxidizing agent); (d) $4H_2O + 2MnO_4^- + 3S^{2-} \longrightarrow 2MnO_2 + 3S + 8OH^-$ (S^{2-} is reducing agent, MnO_4^- is oxidizing agent).
19.6 When molten $AlCl_3$ is electrolyzed using inert electrodes, the possible reactants are Al^{3+} and Cl^-. The products are Al metal and Cl_2 gas. When aqueous $AlCl_3$ is electrolyzed there is an additional reactant, H_2O. It is easier to reduce H_2O than Al^{3+}. Therefore the products are H_2 gas and Cl_2 gas. **19.9** (a) 2 $\mathcal{F}$; (b) 0.0168 $\mathcal{F}$; (c) 0.050 $\mathcal{F}$; (d) 0.023 $\mathcal{F}$; **19.11** (a) 3.15×10^{-2} $\mathcal{F}$; (b) 3.04×10^3 coul; (c) 8.44×10^{-1} amp; (d) 1.52×10^3 sec; (e) 3.0×10^3 J, 4.0×10^3 J; (f) 1.18 g. **19.16** 2.94 g. **19.21** Starting with the least active, the list is C B H A D. **19.22** (a) Br_2; (b) Sn; (c) Sn^{2+}; (d) Br^-. **19.25** (a) $E° = -0.30$ V (nonspontaneous); (b) $E° = 0.64$ V (spontaneous); (c) $E° = 1.02$ V (spontaneous). **19.27** Anode is Zn; cathode is Ni; electrons move from Zn to Ni; cations move toward Ni and anions move toward Zn. **19.33** Voltage increases as $[I^-]$ increases and decreases as $[Cl^-]$ increases. **19.36** 0.15 V. **19.38** $K = 1.4 \times 10^{68}$; $\Delta G = -388$ kJ. **19.43** (a) The Zn is more easily oxidized than Fe; it therefore acts as the anode and is corroded; Cr would serve the same role. **19.46** Because Mg has a larger oxidation potential, it would serve as the anode. The Mg will corrode but will protect the Fe from corrosion. **19.52** 8.3×10^{-6} M. **19.55** 1.55×10^{-7} M; 1.55×10^{-10}.

CHAPTER 20

20.1 (a) 38 protons, 52 neutrons; (b) 55 protons, 82 neutrons; (c) 92 protons, 143 neutrons; (d) 53 protons, 78 neutrons. **20.3** (a) $^{241}_{94}Pu \longrightarrow {}^{0}_{-1}e + {}^{241}_{95}Am$; (b) $^{232}_{90}Th \longrightarrow {}^{228}_{88}Ra + {}^{4}_{2}He$; (c) $^{84}_{39}Y \longrightarrow {}^{0}_{1}e + {}^{84}_{38}Sr$; (d) $^{44}_{22}Ti + {}^{0}_{-1}e \longrightarrow {}^{44}_{21}Sc$. **20.5** (a) $^{56}_{37}Rb$: (b) $^{56}_{25}Mn$; (c) $^{56}_{26}Fe$; (d) $^{24}_{11}Na$; (e) $^{94}_{36}Kr$. **20.7** (a) $^{14}_{7}N + {}^{1}_{0}n \longrightarrow {}^{1}_{1}H + {}^{7}_{5}B$; (b) $^{15}_{7}N + {}^{1}_{1}H \longrightarrow {}^{4}_{2}He + {}^{12}_{6}C$; (c) $^{35}_{17}Cl + {}^{1}_{0}n \longrightarrow {}^{1}_{1}H + {}^{35}_{16}S$. **20.9** $^{114}_{49}Cd$, because it has both an odd number of protons and an odd number of neutrons. **20.11** (a) From what has been said in the text, we might expect $^{14}_{6}C$ (even number of both protons and neutrons) to be more abundant than $^{14}_{7}N$ (odd number of both protons and neutrons); however, $^{14}_{6}C$ has a high neutron-to-proton ratio and is therefore radioactive; $^{14}_{7}N$ is the more abundant of the two; (b) $^{16}_{8}O$, because it has a magic number of both protons and neutrons (8). **20.13** Only beta decay increases the proton-to-neutron ratio. **20.19** 5 mg. **20.21** 6.35 mg. **20.24** 3.7×10^9 yr. **20.26** 0.151 millicuries. **20.29** (a) 1.15×10^{-11} J/nucleon; (b) 1.23×10^{-12} J/nucleon. **20.31** 6.30×10^{-12} J. **20.34** 2.43×10^{-12} J. **20.38** A Geiger counter could be used to determine where a radioactive substance escapes from the pipe. **20.41** High temperatures are necessary to impart the nuclei with sufficient kinetic energy to overcome their mutual repulsions. **20.46** (a) accumulate in bone, replacing Ca; (b) not accumulate to any great degree because of water solubility; (c) accumulate in thyroid; (d) not accumulate; (e) accumulate in iron-containing compounds like hemoglobin. **20.53** 6 alpha particles and 4 beta particles. **20.55** Calcium-40 is the geonormal isotope of calcium and can enter rocks from the environment.

CHAPTER 21

21.1 The lightest member has a maximum coordination number of 4 in covalent compounds; it can π-bond more easily than heavier members; it has a smaller size and higher electronegativity; it has less metallic character. **21.3** Both compounds have a central atom in the +6 oxidation state; the difference in formula can be rationalized in terms of the larger size of Te compared to Se. **21.5** Phosphorus($-$III) is larger in size and lower in EN than nitrogen($-$III); PH_3 consequently loses electrons more easily than does NH_3 and is a better reducing agent; nitrogen(V) is smaller and more electronegative than phosphorus(V); consequently HNO_3 gains electrons more readily than does H_3PO_4 and is thus a better oxidizing agent. **21.9** See Table 21.1. **21.10** (a) minimum is -1 as in NaBr, maximum is $+7$ as in $NaBrO_4$; (b) minimum is -3 as in PH_3, maximum is $+5$ as in H_3PO_4; (c) minimum is -2 as in Na_2S, maximum is $+6$ as in H_2SO_4; (d) minimum is -2 as in Na_2O, maximum is $+2$ in OF_2. **21.11** Sulfuric acid contains sulfur in its maximum oxidation state, $+6$.   **21.13** (a) tetrahedral; (b) pyramidal (tetrahedral with one lone pair of electrons); (c) bent (tetrahedral with two lone pairs of electrons); (d) pyramidal (tetrahedral with one lone pair); (e) trigonal bipyramidal with equatorial lone pair; (f) trigonal bipyramidal.   **21.17** (a) sodium arsenate; (b) potassium iodate; (c) hydrogen selenide or hydroselenic acid; (d) sulfur(VI) fluoride or sulfur hexafluoride; (e) sodium metaphosphate; (f) sodium peroxide.   **21.19** (a) $NaClO$; (b) PCl_5; (c) KO_2; (e) $Pb(NO_3)_2$; (f) $KClO_4$.   **21.21** (a) P_4O_{10}; (b) SO_2; (c) CO_2; (d) N_2O_3.   **21.23** (a) $3O_2 \longrightarrow 2O_3$; (b) $4P + 5O_2 \longrightarrow P_4O_{10}$; (c) $FeS + 2HCl \longrightarrow H_2S + FeCl_2$; (d) $6H_2O + P_4O_{10} \longrightarrow 4H_3PO_4$.   **21.25** (a) $3NaClO \longrightarrow 2NaCl + NaClO_3$; (b) $P_4 + 3NaOH + 3H_2O \longrightarrow 3NaH_2PO_2 + PH_3$; (c) $NO + NO_2 + H_2O \longrightarrow 2HNO_2$.   **21.27** (a) no reaction; (b) $2Cl_2 + 2BaI_2 \longrightarrow 2BaCl_2 + 2I_2$.   **21.30** (a) redox. $Cl(+1) \longrightarrow Cl(+5)$ and $Cl(-1)$; (b) redox, $P_4(0) \longrightarrow P(+1)$ and $P(-3)$; (c) redox. $N(+2)$ and $N(+4) \longrightarrow N(+3)$.   **21.31** Color, boiling points, density, ability to support combustion.   **21.32** (a) 2.02 g; (b) 3.04 g.   **21.47** (a) PO_4^{3-}; (b) NH_3.

CHAPTER 22

22.1 as a fuel, reducing agent, and raw material in synthesizing carbon compounds.   **22.3** See Section 22.3.   **22.5** CO_2 consists of discrete CO_2 molecules that interact with each other through van der Waals interactions; in constrast, SiO_2 consists of an extended array of silicon and oxygen atoms bound by covalent bonds; thus weak van der Waals forces must be overcome to melt CO_2, whereas covalent bonds must be broken in the case of SiO_2.   **22.8** (a) $CO_2 + Ba(OH)_2 \longrightarrow BaCO_3 + H_2O$; (b) $CaCO_3 \longrightarrow CaO + H_2O$; $CaO + H_2O \longrightarrow Ca(OH)_2$; (c) $SrCO_3 \longrightarrow SrO + CO_2$.   **22.9** rolled sheets.   **22.12** (a) single tetrahedron; (b) single strand; (c) sheet; (d) double strand.   **22.14** (a) -1; (b) -8.   **22.17** octahedral.   **22.21** (a) no (different sizes); (b) no (different charges); (c) yes; (d) yes.   **22.23** (a) $GeO_2 + 2H_2 \longrightarrow Ge + 2H_2O$; (b) $3V_2O_5 + 10Al \longrightarrow 5Al_2O_3 + 6V$; (c) $CuS + O_2 \longrightarrow Cu + SO_2$.   **22.25** Each atom has only a single valence electron to use in bonding to its neighbors.   **22.30** 42%.   **22.34** (a) Rain is slightly acidic; (b) they have same charge and about same size; (c) they are in same periodic family (similar properties); (d) oxides are easier to reduce than are silicates.   **22.37** Manufacture of glass, mortar, cement; as building material (marble, limestone); in smelting, for neutralizing acidic soils.

CHAPTER 23

23.1 (a) 4; (b) 7; (c) 5; (d) 6.   **23.2** (a) $+3$; (b) 0; (c) $+1$.   **23.3** (a) $+4$; (b) $+3$; (c) 0; (d) $+2$.   **23.7** (a) bidentate; (b) monodentate.   **23.9** (a) no; (b) no; (c) yes; (d) yes.   **23.12** (a) hexaamminenickel(II) chloride; (b) potassium trioxalatocobaltate(III); (c) sulfatobis(ethylenediamine)cobalt(III) perchlorate; (d) sodium bromopentacyanocobaltate(III).   **23.14** (a) $[Cr(H_2O)_5Cl]Cl_2$; (b) $[Ru(NH_3)_5N_2](NO_3)_2$; (c) $[Co(en)_2Cl(NO_2)]ClO_4$; (d) $Ni(CO)_4$.   **23.17** black.   **23.19** violet.   **23.21** (a) 8; (b) 9; (c) 3; (d) 2.   **23.23** (a) 3; (b) 2; (c) 0; (d) 0.   **23.25** (a) high spin; (b) low spin.   **23.29** change toward red end of the visible spectrum.   **23.44** $[Pt(NH_3)_5Cl]Cl_3$; octahedral; chloropentaammineplatinum(IV) chloride.

CHAPTER 24

24.1 Alkane, $C_{10}H_{22}$; alkene, $C_{10}H_{20}$; alkyne, $C_{10}H_{18}$; aromatic, $C_{10}H_8$ (naphthalene).   **24.3** There are nine isomers.   **24.5** in alkanes, 109°; in alkenes, 120°; in alkynes, 180°.   **24.7** Examples: (a) *trans*-2-hexene or *cis*- or *trans*-3-hexene; (c) 2-methylbutane or 2,2-dimethylpropane; (e) 1-chlorobutane (*n*-butyl chloride).   **24.9** Aromatic compounds are especially stable because of energy from delocalization of π electrons in aromatic ring (resonance energy).   **24.10** Only 2,3-dichloro-2-butene is capable of *cis-trans* isomerism.   **24.12** (a) cyclobutane more reactive, because of ring strain; (b) cyclohexene, because of presence of double bond; (c) 1-hexene, because benzene is stabilized by resonance energy of aromatic ring; (d) 2-hexyne, because alkynes are more reactive than alkenes.   **24.15** Add an antiknock agent, such as tetraethyl lead, or increase content of highly branched hydrocarbons.   **24.17** (a) $CH_3CH_3 + Cl_2 \longrightarrow CH_3CH_2Cl + HCl$;

(c) $CH_3C \equiv CH + Br_2 \longrightarrow CH_3CBr = CHBr$.   **24.19** For ethane, heat of combustion is $-1428\,kJ/mole$; for ethylene, $-1323\,kJ/mole$; for acetylene, $-1256\,kJ/mole$.   **24.23** Cyclopropane is highly strained and therefore has extra reactivity.   **24.24** $CH_3CH_2 \cdot + Cl \cdot \longrightarrow CH_2CH_2Cl$; $Cl \cdot + Cl \cdot \longrightarrow Cl_2$.
24.27 $CH_3CH_2CH{-}O{-}CHCH_2CH_3$ and $CH_2{=}CHCH_2CH_3$ **24.31** The reverse of
$$\qquad\qquad\quad \underset{CH_3}{|}\qquad\quad \underset{CH_3}{|}$$
a condensation reaction is a hydrolysis. Commercial polymers are not very susceptible to hydrolysis, but in general strongly acidic or basic aqueous conditions favor the reaction.   **24.35** You should find about 12 substances, including aldehydes, ketones, cyclic ethers, and alcohols.   **24.38** (a) addition of 1 mole of HCl; (b) addition of 1 mole of HCl, then 1 mole Cl_2; (c) addition of 1 mole of HCl, then hydrogenation.   **24.39** cyclopentane, C_5H_{10}.   **24.41** (a) 2-butanol; (b) 1,2-dichlorobutane; (c) 3-chloro-2-butanaol; (d) butane; (e) carbon dioxide, water;
(f) $-CH{-}CH{-}CH{-}CH{-}CH-$
$$\quad \underset{CH_3}{|}\ \ \underset{CH_3}{|}\ \ \underset{CH_3}{|}\ \ \underset{CH_3}{|}\ \ \underset{CH_3}{|}$$

CHAPTER 25

25.2 Heart action, movement, maintenance of body temperature, transport of oxygen, synthesis of new cells, and so forth.   **25.4** $6.86 \times 10^{10}\,J$.   **25.6** Structural elements in animal tissues; catalysts for biochemical reactions; transport functions (for example, O_2 transport); antibodies.   **25.10** $0.05\,M$ solution rotates plane of light half as much, in opposite sense. Equimolar solution shows no optical activity.
25.12

$$H_2N{-}CH{-}\overset{O}{\overset{\|}{C}}{-}\overset{H}{\overset{|}{N}}{-}CH{-}\overset{O}{\overset{\|}{C}}{-}\overset{H}{\overset{|}{N}}{-}CH{-}\overset{O}{\overset{\|}{C}}{-}OH$$

25.14 Primary structure refers to sequence of amino acids; secondary structure refers to distribution of sections of helical versus other configurations along the chain. Tertiary structure refers to the overall shape of the protein molecule.
25.17 (a) apoenzyme, an enzyme without some essential component, usually a small molecule or ion; (e) holoenzyme, entire enzyme system, consisting of protein and any essential lower-molecular-weight components, called cofactors, or coenzymes; (g) peptidase, an enzyme that catalyzes hydrolysis of a protein chain.
25.20 Equilibrium must be established rapidly relative to rate of reaction itself.
25.21 $2.3 \times 10^3/sec$.   **25.23** Both starch and cellulose are formed from glucose; sucrose is formed from a molecule of glucose and molecule of fructose.
25.25 Sucrose contains chiral centers, carbon atoms containing four different bound groups. These molecules are formed in nature with only one configuration at these carbon centers, so overall the molecule is optically active.
25.28 $[CH_3(CH_2)_4CH{=}CHCH_2CH{=}CH(CH_2)_7COOCH_2]_3 + 6H_2 \longrightarrow$
Trilinolein
$$[CH_3(CH_2)_{16}COOCH_2]_3$$
Tristearin

Hydrogenation with only 4 moles of H_2 will result in fatty acid side chains with a total of two double bonds remaining. These double bonds might be in the same side chain, or one each in two different side chains. Furthermore, it turns out that the location of the single double bond that remains in a side chain might be in any of several places along the chain. Thus, many isomers are possible.   **25.31** Adenine forms hydrogen-bonding interactions with thymine on the other strand; similarly, guanine and cytosine. Thus, for each adenine there must be a thymine; for each guanine a cytosine.

25.33

$$H_2N-CH_2-\overset{\overset{\displaystyle O}{\|}}{C}-\overset{\overset{\displaystyle H}{|}}{N}-\overset{\overset{\displaystyle H}{|}}{\underset{\underset{\displaystyle CH_3}{|}}{C}}-\overset{\overset{\displaystyle H}{|}}{\underset{}{C}}-\overset{\overset{\displaystyle O}{\|}}{C}-\overset{\overset{\displaystyle H}{|}}{N}-\overset{\overset{\displaystyle H}{|}}{\underset{\underset{\underset{\displaystyle HC(CH_3)_2}{|}}{CH_2}}{C}}-\overset{\overset{\displaystyle H}{|}}{\underset{}{C}}-\overset{\overset{\displaystyle O}{\|}}{C}-OH \longrightarrow$$

$$H_2N-CH_2-\overset{\overset{\displaystyle O}{\|}}{C}-\overset{\overset{\displaystyle H}{|}}{N}-\overset{\overset{\displaystyle H}{|}}{\underset{\underset{\displaystyle CH_3}{|}}{C}}-\overset{\overset{\displaystyle H}{|}}{\underset{}{C}}-\overset{\overset{\displaystyle O}{\|}}{C}-OH \;+\; H_2N-\overset{\overset{\displaystyle H}{|}}{\underset{\underset{\underset{\displaystyle HC(CH_3)_2}{|}}{CH_2}}{C}}-\overset{\overset{\displaystyle O}{\|}}{C}-OH$$

$$\downarrow$$

$$H_2N-CH_2-\overset{\overset{\displaystyle O}{\|}}{C}-OH \;+\; H_2N-\overset{\overset{\displaystyle H}{|}}{\underset{\underset{\displaystyle CH_3}{|}}{C}}-\overset{\overset{\displaystyle O}{\|}}{C}-OH$$

25.35 These enzymes are peptidases; their function is to cleave polypeptide chains. **25.37** Shake sample of fat with Br_2 in organic solvent. Loss of color shows addition of Br_2 to double bonds. **25.38** (a) carboxypeptidases (there are several different ones) catalyze the hydrolysis of polypeptide (or protein) chains, in such a manner that the amino acid with a free carboxyl group is hydrolyzed off first (see solution 25.33 for an example); (b) cellulose catalyzes the hydrolysis of the polysaccharide cellulose; (c) ribonuclease catalyzes the hydrolysis of ribonucleic acids (RNA)—the products of the hydrolysis are the individual nucleotides; (d) a lipase catalyzes hydrolysis of fats into glycerol and fatty acids or fatty acid salts.

index

Italic page numbers indicate illustrations or structures.

a

Body temperature, 310
Bohr theory, 157–161
Boiling point, 312
Boiling-point elevation, 361
Boiling-point-elevation constant, 361
Bomb calorimeter, 104
Bond:
    covalent, 205, 212–226
    ionic, 204, 205–209
    metallic, 205, 271–273
Bond angle, 214, 239
Bond distance, 239
    atomic radii from, 198
    carbon-carbon bonds, 253
    nitrogen-nitrogen bonds, 215
Bond-dissociation energy, 268
Bonding orbital, 260
    charge distribution in, 262
Bond order, diatomic molecules, 267
Bond polarity: 222
    acid strength and, 478
Bond strength, acid character and, 479
Boron trifluoride, Lewis structure, *221*
Boyle, R., 7, 116
Boyle's law, 119–121
Brackish water, 532
    purification, 536
Bragg's law, 324
Brass, 680
Breeder reactor, 611
Bromine:
    catalyst for H_2O_2 decomposition, 394
    occurrence, 627
    from seawater, 528
    uses, 631
Bromonium ion (Br^+), 739
Brønsted, J., 456
Brønsted acid, 456
Brønsted base, 466
Brønsted-Lowry theory, 456
Brownian motion, 116
B subgroup elements, 191
    ionic radii, 211
    ionization potentials, 196
Buffer solutions, 492–498
    capacity, 497
    pH calculations, 495
    seawater, 524
Butane:
    isomers, 722
    uses, 720
Butenes, 726
Butyric acid, 744

C

Cadmium batteries, 570
Caffeine, structure, 276
Calcite, *659*, 666
Calcium carbonate:
    decomposition equilibrium, 419
    occurrence, 666
    in seawater, 525
    in SO_2 removal, 295
Calcium sulfite, in SO_2 removal, 295
Calorie, definition, 96

Calorimeter:
    bomb, *104*
    "coffee-cup," *105*
Calorimetry, 103–106
Cannizzaro, S., 70, 116
Capillary action, 315
Carbohydrates, 767–771
    fuel values, 106
Carbon:
    allotropy, 216
    bond distance, 253
    comparison with silicon, 626
    hybridization, 253
    hybrid orbitals, 248
    in steels, 677
    structure, 216
Carbon-12, atomic weight standard, 68
Carbon-14 dating, 597–599
Carbon dioxide:
    in atmosphere, 128, 301, *303*
    in blood, 504
    comparison with SiO_2, 626
    dissociation, 414
    phase diagram, 337
    pressure-volume behavior, 139
    reaction with water, 485
    in seawater, 524
Carbonic acid:
    aqueous equilibria, 501–504
    in blood, 504
    pH calculations, 501
    in seawater, 524
Carbonic anhydrase, 504, 542
Carbon monoxide, 298–301
    in atmosphere, 298
    catalytic converter, 398
    in cigarette smoke, 299
    health effects, 301
    ore reduction, 675
    in urban atmospheres, 300
Carbonylation, 744
Carbonyl group, 743
Carboxyhemoglobin, 299
Carboxylic acids, 744–746
    dissociation constants, 482
    in fats, 772
Carboxypeptidase, 764
Carcinogens, 256
Carlsbad Cavern, 667
Catalysis, 393–400
    and air pollution, 397–400
    enzymes, 762
    heterogeneous, 396
    homogeneous, 394
Catalyst, 289, 393
    cracking, 731
    effect on activation energy, 395
    effect on equilibrium, 417
    enzymes, 762
    heterogeneous, 292
    homogeneous, 394
Catalytic converter, 398
Cathode, 557
Cathode ray, 37
Cathode-ray tube, 38, *39*
Cathodic protection, 581
Cation, 50, 209
    hydration, 348

Cation (*cont.*)
    naming, 50, 694
Cell potential, 571
    equilibrium constant, 576
    free-energy change, 575
Cellulose, 770
Celsius temperature scale, 15
Cement, composition of, 665
Cesium chloride, structure, *206*
Chain reaction, nuclear, 608
Characteristic wavelength, 162
Charge-transfer reactions, 285
Charles's law, 121
Chelating agents, 689–693
    antibacterial action, 693
    in living systems, 692
    uses, 690
Chemical change, 26
Chemical equations, 61–64
Chemical formula, 46
Chemical properties, 26
Chemical reactions, 64–67, 95
    energy changes, 95
    entropy changes, 438
    free-energy changes, 440–448
Chemistry:
    definition, 1
    historical roots, 2–7
Chemists, activities of, 2
Chilean saltpeter, 627, 642
Chiral, 697
Chlorine:
    addition reactions, 734
    in bromine production, 529
    by electrolysis, 557
    occurrence, 627
    photodissociation, 737
    preparation, 628
    reaction with alkanes, 737
    reaction with benzene, 735
    in stratosphere, 290
    uses, 631
    water treatment, 539, 631
Chlorine dioxide, Lewis structures, *220*
Chlorine oxide, reaction with ozone, 290
Chlorofluoromethanes, 290, 631
Chloroform, reaction with Cl_2, 392
Chlorophyll, 754, *755*
Chlorosis, iron, 691
Chlorox, 631
Cholesteric liquid crystals, 333
Cholesterol, *740,* 742
Chromatography:
    column, 30
    definition, 29
    gas, 31
    paper, 30
Chromium:
    electron configuration, 188
    in water supplies, 543
Cis-trans isomerism, 696
Citric acid, 744
Clausius, R., 134
Clays, 663–655
    cracking catalysts, 731
Climate, relation to CO_2, 302
Close packing, 325
    cubic, *328*

physical and chemical constants

Atomic mass unit	$1 \text{ amu} = 1.66053 \times 10^{-24} \text{ g}$
	$6.022169 \times 10^{23} \text{ amu} = 1 \text{ g}$
Avogadro's number	$N = 6.022169 \times 10^{23}/\text{mole}$
Boltzmann's constant	$k = 1.38062 \times 10^{-23} \text{ J/K}$
Electron rest mass	$m_e = 0.00054859 \text{ amu}$
	$= 9.109558 \times 10^{-28} \text{ g}$
Electronic charge	$e = 1.6022 \times 10^{-19} \text{ coul}$
Faraday's constant	$\mathscr{F} = Ne = 96{,}487 \text{ coul/equivalent}$
Gas constant	$R = Nk = 8.3144 \text{ J/K-mole}$
	$= 0.082053 \text{ l.-atm/K-mole}$
Neutron rest mass	$m_\text{n} = 1.008665 \text{ amu}$
	$= 1.67492 \times 10^{-24} \text{ g}$
Pi	$\pi = 3.1415926536$
Planck's constant	$h = 6.6262 \times 10^{-34} \text{ J-sec}$
Proton rest mass	$m_\text{p} = 1.007277 \text{ amu}$
	$= 1.672614 \times 10^{-24} \text{ g}$
Velocity of light (in vacuum)	$c = 2.997925 \times 10^8 \text{ m/sec}$